Constants of Inorganic Substances
A Handbook
Revised and Augmented Edition

R.A. Lidin

L.L. Andreeva

V.A. Molochko

Edited by Rostislav A. Lidin

Professor of Inorganic Chemistry
M.V. Lomonosov Academy of Fine
Chemical Technology
Moscow, Russia

begell house, inc.
New York • Wallingford (U.K.)

0724265
CHEM

Constants of Inorganic Substances: Handbook, Revised and Augmented Edition

Library of Congress Cataloging-in-Publication Data

Lidin,R.A. (Rostislav Aleksandrovich)
   [Spravochnik po neorganicheskoĭ khimil.
   English]
      Constants of inorganic substances : handbook
/R.A. Lidin, L.L. Andreeva, V.A. Molochko. --
Rev. and augm. ed./edited by Rostislav A. Lidin.
         p.   cm.
      Rev. ed. of: Handbook of inorganic chemistry,
2nd rev. ed. 1993.
      Includes bibliographic references (p. -).
      ISBN 1-56700-041-X (hardcover)
      1. Chemistry, Inorganic--Handbooks, manuals,
 etc. I. Andreyeva, Larisa L. II. Molochko, Vadim A. III. Lidin,
R.A.(Rostislav Aleksandrovich). Handbook of inorganic chemistry. IV.
Title.
QD155.5.L5313 1995
546--dc20   95-42941
                                        CIP

*This book is dedicated to the memory
of outstanding pedagogues and scientists,
and our dear teachers, Professors
K. V. Astakhov, M. Kh. Karapet'yants,
and E. S. Sarkisov.*

Names and Symbols

English name	Latin name	Symbol	English name	Latin name	Symbol
Actinium	Actinium	$_{89}$Ac	Europium	Europium	$_{63}$Eu
Aluminum	Aluminium	$_{13}$Al	Fermium	Fermium	$_{100}$Fm
Americium	Americium	$_{95}$Am	Fluorine	Fluorum	$_{9}$F
Antimony	Stibium	$_{51}$Sb	Francium	Francium	$_{87}$Fr
Argon	Argon	$_{18}$Ar	Gadolinium	Gadolinium	$_{64}$Gd
Arsenic	Arsenicum	$_{33}$As	Gallium	Gallium	$_{31}$Ga
Astatine	Astatium	$_{85}$At	Germanium	Germanium	$_{32}$Ge
Barium	Barium	$_{56}$Ba	Gold	Aurum	$_{79}$Au
Berkelium	Berkelium	$_{97}$Bk	Hafnium	Hafnium	$_{72}$Hf
Beryllium	Beryllium	$_{4}$Be	Helium	Helium	$_{2}$He
Bismuth	Bismuthum	$_{83}$Bi	Holmium	Holmium	$_{67}$Ho
Boron	Borum	$_{5}$B	Hydrogen	Hydrogenium	$_{1}$H
Bromine	Bromum	$_{35}$Br	Indium	Indium	$_{49}$In
Cadmium	Cadmium	$_{48}$Cd	Iodine	Iodum	$_{53}$I
Calcium	Calcium	$_{20}$Ca	Iridium	Iridium	$_{77}$Ir
Californium	Californium	$_{98}$Cf	Iron	Ferrum	$_{26}$Fe
Carbon	Carboneum	$_{6}$C	Krypton	Krypton	$_{36}$Kr
Cerium	Cerium	$_{58}$Ce	Kurchatovium	Kurtchatovium	$_{104}$Ku
Cesium	Caesium	$_{55}$Cs	Lanthanum	Lanthanum	$_{57}$La
Chlorine	Chlorum	$_{17}$Cl	Lawrencium	Lawrencium	$_{103}$Lr
Chromium	Chromium	$_{24}$Cr	Lead	Plumbum	$_{82}$Pb
Cobalt	Cobaltum	$_{27}$Co	Lithium	Lithium	$_{3}$Li
Copper	Cuprum	$_{29}$Cu	Lutetium	Lutetium	$_{71}$Lu
Curium	Curium	$_{96}$Cm	Magnesium	Magnesium	$_{12}$Mg
Dysprosium	Dysprosium	$_{66}$Dy	Manganese	Manganum	$_{25}$Mn
Einsteinium	Einsteinium	$_{99}$Es	Mendelevium	Mendelevium	$_{101}$Md
Erbium	Erbium	$_{68}$Er	Mercury	Mercurius	$_{80}$Hg

of the Elements

English name	Latin name	Symbol	English name	Latin name	Symbol
Molybdenum	Molybdaenum	$_{42}$Mo	Samarium	Samarium	$_{62}$Sm
Neodymium	Neodymium	$_{60}$Nd	Scandium	Scandium	$_{21}$Sc
Neon	Neon	$_{10}$Ne	Selenium	Selenium	$_{34}$Se
Neptunium	Neptunium	$_{93}$Np	Silicon	Silicium	$_{14}$Si
Nickel	Niccolum	$_{28}$Ni	Silver	Argentum	$_{47}$Ag
Nielsbohrium	Nielsbohrium	$_{105}$Ns	Sodium	Natrium	$_{11}$Na
Niobium	Niobium	$_{41}$Nb	Strontium	Strontium	$_{38}$Sr
Nitrogen	Nitrogenium	$_{7}$N	Sulfur	Sulfur	$_{16}$S
Nobelium	Nobelium	$_{102}$No	Tantalum	Tantalum	$_{73}$Ta
Osmium	Osmium	$_{76}$Os	Technetium	Technetium	$_{43}$Tc
Oxygen	Oxygenium	$_{8}$O	Tellurium	Tellurium	$_{52}$Te
Palladium	Palladium	$_{46}$Pd	Terbium	Terbium	$_{65}$Tb
Phosphorus	Phosphorus	$_{15}$P	Thallium	Thallium	$_{81}$Tl
Platinum	Platinum	$_{78}$Pt	Thorium	Thorium	$_{90}$Th
Plutonium	Plutonium	$_{94}$Pu	Thulium	Thulium	$_{69}$Tm
Polonium	Polonium	$_{84}$Po	Tin	Stannum	$_{50}$Sn
Potassium	Kalium	$_{19}$K	Titanium	Titanium	$_{22}$Ti
Praseodymium	Praseodymium	$_{59}$Pr	Tungsten	Wolframium	$_{74}$W
Promethium	Promethium	$_{61}$Pm	Uranium	Uranium	$_{92}$U
Protactinium	Protactinium	$_{91}$Pa	Vanadium	Vanadium	$_{23}$V
Radium	Radium	$_{88}$Ra	Xenon	Xenon	$_{54}$Xe
Radon	Radon	$_{86}$Rn	Ytterbium	Ytterbium	$_{70}$Yb
Rhenium	Rhenium	$_{75}$Re	Yttrium	Yttrium	$_{39}$Y
Rhodium	Rhodium	$_{45}$Rh	Zinc	Zincum	$_{30}$Zn
Rubidium	Rubidium	$_{37}$Rb	Zirconium	Zirconium	$_{40}$Zr
Ruthenium	Ruthenium	$_{44}$Ru			

Periodic Table of the Elements

	I A / I B	II A / II B	III A / III B	IV A / IV B	V A / V B
1	**H** 1 1.008 Hydrogen				
2	**Li** 3 6.941 Lithium	**Be** 4 9.012 Beryllium	**B** 5 10.811 Boron	**C** 6 12.011 Carbon	**N** 7 14.007 Nitrogen
3	**Na** 11 22.990 Sodium	**Mg** 12 24.305 Magnesium	**Al** 13 26.982 Aluminum	**Si** 14 28.086 Silicon	**P** 15 30.974 Phosphorus
4	**K** 19 39.098 Potassium / 29 **Cu** 63.546 Copper	**Ca** 20 40.078 Calcium / 30 **Zn** 65.39 Zinc	21 **Sc** 44.956 Scandium / **Ga** 31 69.723 Gallium	22 **Ti** 47.88 Titanium / **Ge** 32 72.61 Germanium	23 **V** 50.942 Vanadium / **As** 33 74.922 Arsenic
5	**Rb** 37 85.468 Rubidium / 47 **Ag** 107.868 Silver	**Sr** 38 87.62 Strontium / 48 **Cd** 112.411 Cadmium	39 **Y** 88.906 Yttrium / **In** 49 114.82 Indium	40 **Zr** 91.224 Zirconium / **Sn** 50 118.710 Tin	41 **Nb** 92.906 Niobium / **Sb** 51 121.75 Antimony
6	**Cs** 55 132.905 Cesium / 79 **Au** 196.967 Gold	**Ba** 56 137.327 Barium / 80 **Hg** 200.59 Mercury	^{57}La – ^{71}Lu * / **Tl** 81 204.383 Thallium	72 **Hf** 178.49 Hafnium / **Pb** 82 207.2 Lead	73 **Ta** 180.948 Tantalum / **Bi** 83 208.980 Bismuth
7	**Fr** 87 223.020 Francium	**Ra** 88 226.025 Radium	^{89}Ac – ^{103}Lr **	104 **Ku** [261] Kurchatovium	105 **Ns** [262] Nielsbohrium

* Lanthanoids

57 **La** 138.906 Lanthanum	58 **Ce** 140.115 Cerium	59 **Pr** 140.908 Praseodymium	60 **Nd** 144.24 Neodymium	61 **Pm** 144.913 Promethium	62 **Sm** 150.36 Samarium	63 **Eu** 151.965 Europium	64 **Gd** 157.25 Gadolinium

** Actinoids

89 **Ac** 227.028 Actinium	90 **Th** 232.038 Thorium	91 **Pa** 231.036 Protactinium	92 **U** 238.029 Uranium	93 **Np** 237.048 Neptunium	94 **Pu** 244.064 Plutonium	95 **Am** 243.061 Americium	96 **Cm** 247.070 Curium

According to IUPAC Commission (1995 year):

element 104	Dubnium	Db	element 107	Bohrium	Bo
element 105	Joliotium	Jl	element 108	Hahnium	Hn
element 106	Rutherfordium	Rf	element 109	Meitnerium	Mt

A VI B	A VII B	A VIII B		
	(H)	**He** 2	Symbol Atomic number	
		4.003		
		Helium	**Na** 11	
O 8	**F** 9	**Ne** 10	22.990	
15.999	18.998	20.180	Sodium	
Oxygen	Fluorine	Neon		
S 16	**Cl** 17	**Ar** 18	Name Relative atomic mass	
32.066	35.453	39.948		
Sulfur	Chlorine	Argon		

24 **Cr**	25 **Mn**	26 **Fe**	27 **Co**	28 **Ni**
51.996	54.938	55.847	58.933	58.69
Chromium	Manganese	Iron	Cobalt	Nickel
Se 34	**Br** 35	**Kr** 36		
78.96	79.904	83.80		
Selenium	Bromine	Krypton		
42 **Mo**	43 **Tc**	44 **Ru**	45 **Rh**	46 **Pd**
95.94	97.907	101.07	102.906	106.42
Molybdenum	Technetium	Ruthenium	Rhodium	Palladium
Te 52	**I** 53	**Xe** 54		
127.60	126.904	131.29		
Tellurium	Iodine	Xenon		
74 **W**	75 **Re**	76 **Os**	77 **Ir**	78 **Pt**
183.85	186.207	190.2	192.22	195.08
Tungsten	Rhenium	Osmium	Iridium	Platinum
Po 84	**At** 85	**Rn** 86		
208.982	209.987	222.018		
Polonium	Astatine	Radon		
106 —	107 —	108 —	109 —	
[263]	[262]	[265]	[266]	
—	—	—	—	

65 **Tb**	66 **Dy**	67 **Ho**	68 **Er**	69 **Tm**	70 **Yb**	71 **Lu**
158.925	162.50	164.930	167.26	168.934	173.04	174.967
Terbium	Dysprosium	Holmium	Erbium	Thulium	Ytterbium	Lutetium

97 **Bk**	98 **Cf**	99 **Es**	100 **Fm**	101 **Md**	102 **No**	103 **Lr**
247.070	251.080	252.083	257.095	258.099	259.101	260.105
Berkelium	Californium	Einsteinium	Fermium	Mendelevium	Nobelium	Lawrencium

Relative atomic masses are based on $^{12}C = 12$ and conform to the 1987 IUPAC report values rounded to the third decimal figure. Number in [ ] indicate the most stable isotope.

Contents

Abbreviations and Notations

amor	amorphous	r	readly soluble
bl	blue	rhomb	(ortho)rhombic
blk	black	sk-bl	sky-blue
bp	boiling point	sld	solid
brn	brown	soft	softening
cl	colorless	soln	aqueous solution
cub	cubic	subl	sublimation
d	difficultly soluble	t	with heating
dec	decomposes,	tetr	tetragonal
	with decomposition	tricl	triclinic
dk	dark	trig	trigonal
grn	green	viol	violet
hex	hexagonal	wh	white
hydr	(cristal) hydrate	yel	yellow
i	insoluble	$\rightarrow$	converts into
lq	liquid	+/−	react/does not react
lt	light	<	less than
mon	monoclinic	>	greater than
mp	melting point	∞	miscible
p	under excess pressure		in all proportions
psv	passivates	...	no data available

Preface

Inorganic chemistry comprises several dozens of thousands of compounds. This handbook includes the data on more than 4000 substances, ions, and radicals chosen with due regard for their industrial and scientific importance.

The Handbook consists of six sections arranged in the conventional tabulated form. The first Section gives the formulas and the names of substances, their relative molecular masses, and important physical properties such as phase-transition temperature, color, aggregate state, and density along with the data on substance reactivity (chemical properties) with respect to most commonly used solutions and reagents (water, ethanol, hydrochloric, sulfuric, and nitric acids, sodium hydroxide, and ammonia hydrate). The entry that describe naturally occurring substances (minerals) also include their mineralogical names, symmetry, and hardness.

The following sections of the Handbook characterize the atomic, molecular, and thermodynamic properties of atoms, molecules (formula units), radicals, and cations and anions of those inorganic substances that can exist either as individual substances or in aqueous solution. There are also data on relative atomic masses of elements, properties of natural and radioactive isotopes, electronic configurations of atoms, energies of ionization, and affinities to electrons for atoms and molecules, binding energies and bond lengths, structure (geometric form) of constituent molecules and ions of various substances, including coordination compounds. The Handbook also lists thermodynamic constants of the substances in all their aggregate states (gas, liquid, solid state, aqueous solution), redox potentials, acidity and basicity constants, stability constants of complexes in aqueous solution and solubility in water.

The last Section deals with the nomenclature of inorganic substances. The rules to construct systematic chemical formulas of inorganic substances and their names in accordance with the IUPAC recommendations are formulated and exemplified. There are also detailed lists of nonsystematic names of substances and classification names of groups of substances still widely used in chemical literature.

All the tables in all the Sections are composed in accordance with the following principle. The chemical formulas (the first column) are arranged in the alphabetical order of the symbols of constituent elements that constitute these formulas. Each table is preceded by a concise introduction containing modern definitions of chemical terms and notions and all the necessary comments to the data included into the Section.

All the constants cited in the Handbook can be classified as the informative reference data. The values of the constants are taken from the major reference editions. In the cases where such highly reliable data were absent, the preference was given to most consistent of the known data. The numerical values are given without indicating the errors of their determination because they were rounded off within the accuracy necessary for practical calculations and estimations.

All the materials included into the American edition were revised and complemented with modern data; the relative molecular masses were brought into correspondence with the International Table published in 1987. Two new Sections were written specially for the American edition: "Index of Minerals" and "Enthalpy and Entropy of Phase Transitions".

When writing this book, the authors used many years of experience of scientific and pedagogical work at the Lomonosov Academy of Fine Chemical Technology under the guidance of Professors K. V. Astakhov, M. Kh. Karapet'yants, and E. S. Sarkisov. The authors hope that this book will be a worthy tribute to their memory and will pass the test of time.

The authors will be grateful for all the critical remarks and suggestions that can improve this Handbook.

R. A. Lidin
L. L. Andreeva
V. A. Molochko

1

Inorganic Substances.
Physical Properties and Reactivity

Formulas and Names [31, 32, 37]

The *formulas* of substances are written in conformity with the nomenclature rules (cf. Sections 6.2.1., 6.3.1., 6.4., and 6.5.). For compounds having a complex anion or a heteropolyanion, the formulas are written in an inverted form (anion, cation). e.g., $[Ag(CN)_2,K$ instead of $K[Ag(CN)_2]$, so that all complex compounds with the central silver atom are to be found among other silver-containing compounds, etc.

The formulas of crystallohydrates are placed either on a separate line or combined with the formulas of anhydrous substances depending on the completeness of information for a given substance. That the same formula is not infrequently employed for crystallohydrates and their respective anhydrous compounds can be explained by the fact that the principal (anhydrous) substance can not be obtained by subjecting the crystallohydrate to thermal dehydration which causes crystallohydrate decomposition without transition to the anhydrous state. The number of water molecules generally reflects the result of crystallization from an aqueous solution at 20°C.

The composition of aquacomplexes (complex compounds that contain aqualigands only in the inner sphere) is likewise rendered by the crystallohydrate formulas, in so far as the number of crystal water molecules not invariably corresponds to the coordination number of the central atom, while no information on the division of hydration zones is available. If, however, a particular compound is known to be a complex compound containing in the inner sphere both water and other ligands, its composition should be represented by a coordination formula. Thus, the compound having the composition $CrBr_3 \cdot 6H_2O$ should be rendered by the formula $[Cr(H_2O)_4Br_2]Br \, (\cdot 2H_2O)$.

The berthollide state is indicated when the degree of departure from stoichiometry is known (cf. Section 6.5), and in this case no general berthollide notation ($\approx$) is used.

The *names* of substances placed after formulas are recommended [cf. Sections 6.2.2., 6.2.3., and 6.3.2.]. The names of crystallohydrates are not listed, since they can be readily formed in accordance with the examples in Sec. 6.4. The names of berthollides, in which the content of elements is indicated using the letter x, e.g., $Cu_{2-x}O$, correspond to the ideal composition of the compound (Sec. 6.5), so the above berthollide shall be named copper(I) oxide (Cu_2O).

Detailed information as regards the types of crystal systems (syngonies) and on possible polymorphism phenomena is presented in "Index of minerals" in this Part and also in Sections 4.1., 4.2. "Index of trivial names" at the end of this Part lists non-nomenclature names widely used in the technical literature to the present day (cf. also Sec. 6.6.).

Relative Molecular Masses M_r [1, 37]

The values of relative molecular masses (in amu) are rounded off to the second decimal figure (or to the third decimal figure for elementary substances, provided the value of the relative atomic mass of the respective element is equal to or greater than this accuracy, (cf. Sec. 2.1).

If the formula of a crystallohydrate is placed together with the formula of an anhydrous substance on the same line, the listed value M_r pertains to the anhydrous substance. Thus, for $Al(BrO_3)_3 \, (\cdot 9H_2O)$ the listed value of $M_r = 410.69$ corresponds to the anhydrous aluminum bromate $Al(BrO_3)_3$. In the case of berthollides whose composition is described using the letter x, the listed values of M_r pertain to the ideal composition; for example, for $Cu_{2-x}O$ the cited value of $M_r = 143.09$ corresponds to the formula Cu_2O.

Phase Transition Temperature [2–5, 28, 30, 36]

The following phase transitions are considered: melting (reverse transition – crystallization from melt), boiling (equilibrium condensation) and sublimation (desublimation), and also the formation of chemically new phases in the thermal decomposition of substances without melting or boiling. In the case of amorphous substances, their softening point is listed.

The values of phase transition temperatures are listed in the increasing order and expressed in Celsius degrees (°C is omitted). The listed temperatures refer, as a rule, to the standard atmospheric pressure.

The notation "$-H_2O$ 200" shows that the crystallohydrate in question loses at the indicated temperature all of the crystal water and converts into an anhydrous substance.

The notation "486 $\rightarrow$ AgCl" in the case of $AgClO_4$ ($\cdot H_2O$) means that, on decomposition, said substance yields a solid residue of the composition AgCl (gaseous phases – by-products of decomposition – are generally not indicated).

More detailed information on phase transition temperatures is given in Sec. 4.2.

Color and Density [3–10, 27–29, 35]

The specified *color* generally relates to the state of aggregation of a substance at room temperature, but in some instances it pertains to a range of temperatures, in which the given state of aggregation exists. If the substance is a liquid or gas, the state of aggregation is listed next to the color notation (for solids, the state of aggregation is not recorded, but is implied).

The *density* of solid substances and liquids is expressed in g/cm^3 (the listed density of solids is that at room temperature, while the temperature, at which the density of a liquid was determined, is given in a superscript), and the density of gases is given in g/l at normal physical conditions.

Solvents and Reagents [3–4, 6–10, 28–30, 33, 35, 36]

The reactivity of substances towards common liquid solvents and reagents is characterized as follows.

A substance is regarded to be *soluble* in a given solvent if its interaction with the solvent involves only crystal lattice destruction (for solids), solvation (hydration), electrolytic dissociation, and reversible protolysis. All of the substances that are dissolved in a given solvent are divided into three groups: *readily soluble* (yield at least 0.1M solutions), *difficultly soluble* (yield 0.1–0.001M solutions), and *practically insoluble* substances (yield maximum 0.001M solutions).

A substance is considered as *reacting* with a given reagent if their interaction is an irreversible chemical reaction. As a result of this reaction, the substance passes into a solution not in the molecular or ionic form (as in the case of soluble substances), but as reaction products characterized by their own solubility in the given solvent. Depending on the conditions (temperature, reactant and solvent amounts), the reaction can yield not only a single-phase liquid solution of reaction products, but also a multiphase system (solution + precipitate + gas). If there occurs no chemical reaction between a substance and a reagent, the substances is assumed to be non-reactive.

This line, therefore, shows whether a substance retains its chemical individuality on contact with a specific solvent or reagent.

The solubility of substances is characterized for a *cold* (at room temperature, viz., 18–25°C) and a *warm solvent* (80–100°C) solvent, the solubilities characteristics being separated by a slash (first comes solubility in a cold solvent, and then that in a hot solvent). If solubilities in the cold and the hot solvent are qualitatively the same, only one common solubility characteristic is listed. Solubility of substances in HCl, H_2SO_4, HNO_3, NaOH, and $NH_3 \cdot H_2O$ solutions is listed for dilute solutions only.

The presence or absence of a reaction between a substance and a reagent is indicated by "+" or "−", respectively. For water and alcohol the reaction of a substance with a cold and a hot reagent is indicated, while for other reagents the reaction of a substance with a dilute (10%) and a concentrated solution is stated, the reactivity characteristics being separated by a slash. When the reactivities are identical, one common sign is used in either case.

More detailed (quantitative) data on the solubility of solid substances in water (readily soluble and difficultly soluble substances), as well as on the composition of crystallohydrates formed from an aqueous solution are presented in Section 5.2. The solubility of practically water-insoluble substances is evaluated in Section 4.6. Information on the solubility of liquid and gaseous substances in water is given in Section 5.3.

The examples of employing the data included in the column "water" are presented below. Silver, silver(I) acetate, aluminum and tetraaluminum tricarbide display the following properties towards water: Ag "−"; $AgCH_3COO$ "d/r"; Al "psv"; Al_4C_3 "+". The sign "−" for Ag indicates the absence of silver reaction with water (for the Ag_2O/Ag pair $E° > 0$, cf. Section 4.3). The salt $AgCH_3COO$ is difficultly soluble ("d" ahead of the slash), but readily soluble in hot water ("r" behind the slash), these characteristics being corroborated by quantitative data (Sections 4.6, 5.2). For Al, the notation "psv" (passivation) means that, on contact with water, the surface of aluminum becomes coated with a stable film, thereby preventing the aluminum-water reaction, although this interaction is thermodynamically feasible (for the $Al(OH)_3/Al$ pair $E° < 0$; cf. Section 4.3). Al_4C_3 undergoes with water an irreversible reaction (the sign "+"), viz., irreversible hydrolysis. This reaction yields $Al(OH)_3$ and CH_4; the first product is a solid substance which is practically insoluble in water, while the second product comprises a difficultly soluble gas (cf. Nos. 72, 419 in this table and Sections 4.6., 5.3.).

The same substances display the following properties in an aqueous solution of sulfuric acid: Ag "−/+"; $AgCH_3COO$ "+"; Al "+"; Al_4C_3 "+". Silver fails to react with H_2SO_4 in a dilute solution (the sign "−" ahead of the slash), since for the Ag_2SO_4/Ag pair $E° > 0$ (cf. Section 4.3.), but reacts with H_2SO_4 in a concentrated solution (the sign "+" behind the slash) and yields the slightly soluble solid salt, Ag_2SO_4, and SO_2 gas (for properties of these products cf. Nos. 55, 2467 in table and also Sections 4.6., 5.2.). Silver(I) acetate $AgCH_3COO$ irreversibly reacts with both dilute and concentrated solutions of H_2SO_4 (the sign "+"), the reaction products being Ag_2SO_4 and CH_3COOH (cf No. 425 in this table and Section 4.4.). Aluminum reacts with a solution of H_2SO_4 (the sign "+") and yields, apart from the soluble salt $Al_2(SO_4)_3$ (cf. No. 121 in this table and also Section 5.2), also gases, namely H_2 in a dilute solution and SO_2 in a concentrated solution of H_2SO_4. Tetraaluminum tricarbide, Al_4C_3, reacts with the $H_2SO_4 + H_2O$ mixture of any composition (the sign "+"), since it undergoes complete hydrolysis even in pure water (see above); the reaction products are $Al_2(SO_4)_3$ and CH_4.

It should be noted that the data listed in this column provide an answer only as regards the presence or absence of dissolution or/and a chemical reaction when substances contact solvents/reagents. To evaluate more fully how substances behave in the presence of solutions or reagents, recourse should be made to information on the chemistry of elements and compounds thereof (selected monographs providing this type of information are listed in References at the end of this handbook) and also to the data contained in other Sections of the present reference handbook.

Formula Index

Formula index at the end of this table is useful for finding compounds whose formulas are either unknown or inaccurate. For example, it is desired to find the compound named sodium chlorate (if the name is obsolete, its nomenclature equivalent should be established from the data of Section 6.8.). One route comprises seeking this compound among Na-containing compounds having Nos. 1692–1828 (137 lines in the table), but is excessively laborious. Under the heading "Cl", none of the above numbers are listed because in the compound of interest chlorine is contained in the anion (chlorate). If chlorate is known to contain oxygen and chlorine, the compounds in the rubric "Cl" can be excluded from consideration, and in other rubrics (for ClO^-, ClO_2^-, ClO_3^-, and ClO_4^- ions) there are compounds having Nos. 1723, 1724, 1725 and 1726–1727 which are embraced by the Nos. 1692–1828 interval for Na, so the sought-for compound can be found among five compounds, the compound under No. 1725 being the chlorate of interest. Another example comprises finding in the table a compound having the formula Fe_3O_4. Among compounds of the formula Fe_xO_y (Nos. 965–968), no such formula is present, but the interval of the rubric "Fe" (Nos. 900–987) includes No. 947 from the rubric "O^{2-}". The compound No. 947 in the table is the sought-for compound, the correct formula being $(Fe^{II}Fe_2^{III})O_4$.

Index of Minerals

Index of minerals at the end of this Part lists various minerals, inclusive those corresponding to the naturally occurring substances cited in the table.

The crystal system (syngony) and name are indicated for each mineral. Also listed are hardness values in compliance with the modified Mohs scale, wherein use is made of the following minerals as hardness standards:

1 – Talc $Mg_3(Si_2O_5)_2(OH)_2$
2 – Gypsum $CaSO_4 \cdot 2H_2O$
    Halite $NaCl$
3 – Calcite $CaCO_3$ (trig)
    Galena PbS_{1+x}
4 – Fluorite CaF_2

5 – Apatite $Ca_5(PO_4)_3(Cl,OH,F)$
    Scheelite $CaWO_4$
6 – Orthoclase $K(AlSi_3O_8)$
    Magnetite $(Fe^{II}Fe_2^{III})O_4$
7 – α-Quartz α-SiO_2 (trig)

8 – Topaz $Al_2(SiO_4)(OH,F)_2$
9 – Corundum Al_2O_3
10 – Diamond C (cub)

No.	Formula and name	M_r	Phase transition temperature
1	**Ac Actinium**	227.028	mp 1050, bp 3300
2	$AcBr_3$ Actinium(III) bromide	466.74	subl 800
3	$AcCl_3$ Actinium(III) chloride	333.39	subl 960
4	AcF_3 Actinium(III) fluoride	284.02	...
5	$Ac(NO_3)_3$ Actinium(III) nitrate	413.04	$600 \rightarrow Ac_2O_3$
6	Ac_2O_3 Actinium(III) oxide	502.05	mp ca. 2500
7	$Ac(OH)_3$ Actinium(III) hydroxide	278.05	$1100 \rightarrow Ac_2O_3$
8	**Ag Silver**	107.868	mp 961.93; bp 2170
9	$(Ag^IAg^{III})O_2$ Silver(I)-silver(III) dioxide	247.73	$>100 \rightarrow Ag$
10	Ag_3AsO_3 Silver(I) orthoarsenite	446.52	mp 150 dec
11	Ag_3AsO_4 Silver(I) arsenate	462.52	mp ca. 830 dec
12	$AgAsS_2$ Silver(I) dithiometaarsenite	246.92	mp 417, dec t
13	Ag_3AsS_3 Silver(I) trithioorthoarsenite	494.72	mp 488, dec t
14	$AgBr$ Silver(I) bromide	187.77	mp 432, bp 1505 dec
15	$AgBrO_3$ Silver(I) bromate	235.77	dec ca. 360
16	Ag_2C_2 Silver(I) acetylide	239.76	$>20 \rightarrow Ag,C$
17	$AgCH_3COO$ Siver(I) acetate	166.91	$>300 \rightarrow Ag$
18	$AgCN$ Silver(I) cyanide	133.89	mp ca. 350, dec t
19	Ag_2CN_2 Silver(I) cyanamide	255.76	dec t
20	$[Ag(CN)_2],K$ Potassium dicyanoargentate(I)	199.00	dec 250–420
21	$AgCNO$ Silver(I) fulminate	149.89	dec >120
22	Ag_2CO_3 Silver(I) carbonate	275.74	$218 \rightarrow Ag$
23	$Ag_2C_2O_4$ Silver(I) oxalate	303.75	$110 \rightarrow Ag_2CO_3$
24	$AgCl$ Silver(I) chloride	143.32	mp 455, bp 1550
25	$AgClO_2$ Silver(I) chlorite	175.32	$>250 \rightarrow AgCl,AgClO_4$
26	$AgClO_3$ Silver(I) chlorate	191.32	mp 230, dec 270
27	$AgClO_4$ ($\cdot H_2O$) Silver(I) perclorate	207.32	$486 \rightarrow AgCl$
28	Ag_2CrO_4 Silver(I) chromate	331.73	$t \rightarrow Ag,Cr_2O_3$
29	$Ag_2Cr_2O_7$ Silver(I) dichromate	431.72	$t \rightarrow Ag,Cr_2O_3$
30	AgF ($\cdot 2H_2O$) Silver(I) fluoride	126.87	mp 435, bp ca. 1000
31	AgF_2 Silver(II) fluoride	145.86	mp 690, dec >700
32	Ag_2F Disilver fluoride	234.73	$>115 \rightarrow Ag,AgF$
33	$Ag(HCOO)$ Silver(I) formate	152.89	$>300 \rightarrow Ag$
34	$Ag_2H_3IO_6$ Silver(I) trihydroorthoperiodate	441.66	mp 60 dec
35	$Ag_2H_4TeO_6$ Silver(I) tetrahydroorthotellurate	443.36	dec >200
36	AgI Silver(I) iodide	234.77	mp 554, bp 1506 (p)
37	$AgIO_3$ Silver(I) iodate	282.77	mp >200, dec t
38	$AgIO_4$ Silver(I) metaperiodate	298.77	dec 180
39	Ag_5IO_6 Silver(I) orthoperiodate	762.24	dec >320
40	$AgMnO_4$ Silver(I) permanganate	226.80	$>160 \rightarrow Ag,MnO_2$
41	AgN_3 Silver(I) azide	149.89	mp 252, $300 \rightarrow Ag$
42	Ag_3N Trisilver nitride	337.61	ca. $165 \rightarrow Ag$
43	$AgNCS$ Silver(I) thiocyanate	165.95	dec 170
44	$AgNO_2$ Silver(I) nitrite	153.87	$140–160 \rightarrow Ag$
45	$AgNO_3$ Silver(I) nitrate	169.87	mp 209.7; ca. $300 \rightarrow Ag$
46	$Ag_2N_2O_2$ Silver(I) hyponitrite	275.75	$110 \rightarrow Ag$
47	Ag_2O Silver(I) oxide	231.74	$>160 \rightarrow Ag$
48	$AgOCN$ Silver(I) cyanate	149.89	dec 147

No.	Color and density	Solubility and reactivity						
		water	alcohol	HCl	H_2SO_4	HNO_3	NaOH	$NH_3 \cdot H_2O$
1	wh, 10.07	+	+	+	+	+	+	+
2	wh, 5.85	r	r	−	−/+	−/+	+	+
3	wh, 4.81	d	i	d/+	+	−	+	+
4	wh, 7.88	i	i	−	−/+	−	−	−
5	wh	r	r	−	−	−	+	+
6	wh, 9.19	−	−	+	+	+	−/+	−
7	wh	i	i	+	+	+	i	i
8	wh, 10.494	−	−	−	−/+	+	−	−
9	dk-grey, 7.48	−	−	−/+	−/+	+	−/+	−/+
10	lt-yel	i	i	+	+	r/+	i/+	i/+
11	dk-brn, 6.66	i	i	+	+	i/+	i/+	i/+
12	grey-blk	i	...	−	−/+	−/+	+	+
13	dk-red, 5.53	i	...	−	−/+	+	−/+	−/+
14	lt-yel, 6.47	i	i	i	i/+	i	−/+	−/+
15	wh, 5.21	d	i	+	+	−	+	+
16	wh	+	+	+	+	+	+	+
17	wh, 3.26	d/r	i/d	+	+	−	+	+
18	wh, 3.95	i	i	i	i/+	i/+	−/+	−/+
19	yel	i	i	+	+	+	−/+	−/+
20	wh, 2.36	r	r	−	−/+	−/+	−	−
21	grey	d/r	i/d	+	+	−/+	+	+
22	lt-yel, 6.08	i/d	i	+	+	+	−/+	+
23	wh, 5.03	i	i	−/+	−/+	−/+	−/+	+
24	wh, 5.56	i	i	i	i	i	i/+	+
25	yel	r	d	+	+	+	+	+
26	wh, 4.43	r	i	+	+	−	+	+
27	wh, 2.81	r	r	+	+	−	+	+
28	red, 5.63	i	i	+	+	+	i/+	i/+
29	dk-red, 4.77	d/+	...	+	+	−/+	−/+	+
30	wh, 5.85	r	r	+	+	−/+	+	+
31	lt-bl, 4.57	+	+	+	+	+	+	−/+
32	yel-grn, 8.64	+	i	+	+	+	+	+
33	wh	i	...	−/+	−/+	+	−/+	+
34	yel, 5.68	d	i	+	d/+	−	+	+
35	yel	i	i	−/+	−/+	+	−/+	+
36	yel, 5.67	i	r	−	−/+	−/+	i	i
37	wh, 5.53	i	...	−/+	−	−	−	−/+
38	yel, 5.57	+	...	+	+	+	+	+
39	blk	i	i	−/+	−/+	+	−/+	−/+
40	blk, 4.4	d	+	−/+	−/+	−/+	−/+	+
41	wh	i	...	−/+	−/+	−/+	−/+	+
42	dk-brn, 9.0	−	i	+	+	+	−	−/+
43	wh	i	...	i/+	−/+	−/+	−/+	+
44	lt-yel, 4.45	d/+	i	+	+	−/+	+	+
45	wh, 4.35	r	r	+	+	r	+	+
46	yel, 5.75	i	...	+	+	−/+	−/+	−/+
47	brn-blk, 7.14	−	−	+	+	+	−	+
48	wh, 4.0	i/+	...	+	+	+	i/+	+

No.	Formula and name	M_r	Phase transition temperature
49	$AgPO_3$ Silver(I) metaphosphate	186.84	mp 482 dec
50	Ag_3PO_4 Silver(I) orthophosphate	418.57	mp 849
51	$Ag_4P_2O_7$ Silver(I) diphosphate	605.41	mp 585
52	$AgReO_4$ Silver(I) perrhenate	358.07	mp 455
53	Ag_2S Silver(I) sulfide	247.80	dec >350; mp 845 (*p*)
54	Ag_2SO_3 Silver(I) sulfite	295.80	100 → Ag_2S,Ag_2SO_4
55	Ag_2SO_4 Silver(I) sulfate	311.80	mp 660, dec >1000
56	$Ag_2(SO_3S)$ Silver(I) thiosulfate	327.87	dec >250
57	$[Ag(SO_3S)_2],Na_3$ Sodium bis(thiosulfato)argentate(I)	401.10	dec *t*
58	$(AgSb)S_2$ Silver-antimony disulfide	293.75	mp 509, dec *t*
59	Ag_2Se Silver(I) selenide	294.70	mp 897, dec *t*
60	Ag_2SeO_4 Silver(I) selenate	358.69	dec *t*
61	Ag_2Te Silver(I) telluride	343.34	mp 960
62	Ag_2TeO_3 Silver(I) tellurite	391.33	mp 450
63	**Al Aluminum**	26.982	mp 660.37; bp 2500
64	AlAs Aluminum monoarsenide	101.90	mp 1740
65	$AlAsO_4$ ($\cdot 2H_2O$) Aluminum arsenate	165.90	dec *t*
66	$AlAsO_4\cdot 8H_2O$	310.02	$-H_2O$ 335
67	AlB_2 Aluminum diboride	48.60	dec 1400
68	AlB_{12} Aluminum dodecaboride	156.71	mp ca. 2200
69	$AlBr_3$ Aluminum bromide	266.69	mp 97.5; bp 265
70	$AlBr_3\cdot 6H_2O$	374.78	mp 93, dec >100
71	$Al(BrO_3)_3$ ($\cdot 9H_2O$) Aluminum bromate	410.69	mp hydr 62.3; dec 100
72	Al_4C_3 Tetraluminum tricarbide	143.96	dec ca. 2200
73	$Al(CH_3COO)_3$ Aluminum acetate	204.11	dec >370
74	$Al(CN)_3$ Aluminum cyanide	105.04	dec 100
75	$Al_2(C_2O_4)_3$ ($\cdot 4H_2O$) Aluminum oxalate	318.02	hydr 290 → Al_2O_3
76	$[Al(C_2O_4)_3],K_3$ ($\cdot 3H_2O$) Potassium trioxalatoaluminate(III)	408.33	$-H_2O$ 120
77	$AlCl_3$ Aluminum chloride	133.34	subl 183, mp 192.6
78	$AlCl_3\cdot 6H_2O$	241.43	100 → $AlCl(OH)_2$
79	$[AlCl_4],In$ Indium(I) tetrachloroaluminate(III)	283.61	mp 268
80	$[AlCl_4],NH_4$ Ammonium tetrachloroaluminate(III)	186.83	mp 303
81	$[AlCl_4],Na$ Sodium tetrachloroaluminate(III)	191.78	mp 156
82	$Al(Cl)O$ Aluminum chloride-oxide	78.43	>500 → $AlCl_3,Al_2O_3$
83	$Al(ClO_3)_3$ ($\cdot 6H_2O$) Aluminum chlorate	277.33	dec hydr *t*
84	$Al(ClO_4)_3$ ($\cdot 6H_2O$) Aluminum perchlorate	325.33	mp hydr 121, $-H_2O$ 178
85	$AlCl(OH)_2$ ($\cdot 4H_2O$) Aluminum chloride-dihydroxide	96.45	>350 → $Al(Cl)O$
86	$[AlD_4],Li$ Lithium tetradeuteridoaluminate(III)	41.98	mp 175 dec
87	AlF_3 ($\cdot H_2O$) Aluminum fluoride	83.98	subl 1279
88	$[AlF_6],K_3$ Potassium hexafluoroaluminate(III)	258.26	subl ca. 1300
89	$[AlF_6],Li_3$ Lithium hexafluoroaluminate(III)	161.79	mp ca. 900
90	$[AlF_6],(NH_4)_3$ Ammonium hexafluoroaluminate(III)	195.09	dec 305
91	$[AlF_6],Na_3$ Sodium hexafluoroaluminate(III)	209.94	mp ca. 1009
92	AlH_3 Aluminum hydride	30.01	dec >105
93	$Al(H)Br_2$ Aluminum hydride-dibromide	187.80	dec >55
94	$Al(H)_2Br$ Aluminum dihydride-bromide	108.90	dec >35

(Continued)

No.	Color and density	Solubility and reactivity						
		water	alcohol	HCl	H_2SO_4	HNO_3	NaOH	$NH_3 \cdot H_2O$
49	wh, 6.37	i	i	+	+	+	+	+
50	yel, 6.38	i	i	+	−/+	i/r	−	+
51	wh, 5.31	i	i	i/+	i/+	i/+	−	+
52	wh, 6.96	d	i	+	+	−/+	+	+
53	blk, 7.23	i	i	+	+	+	i	i
54	wh	i/+	...	+	+	+	i/+	+
55	wh, 5.45	d	i	+	d	d	+	+
56	wh	i/+	...	+	+	+	−/+	−
57	wh	r	i	−/+	−/+	−/+	−	−
58	grey-blk, 5.2	i	...	−/+	−/+	+	−/+	−/+
59	blk, 8.19	i	i	−/+	−/+	−/+	i	−/+
60	wh, 5.72	d	i	+	−	−	−/+	+
61	grey-bl, 8.35	i	i	−/+	−/+	−/+	i	−/+
62	lt-yel	i/+	i	−/+	−/+	+	−/+	+
63	wh, 2.70	psv	−	+	+	+/psv	+	+
64	orange, 3.81	+	+	+	+	+	+	+
65	wh, 3.25	i	i	+	+	+	−/+	−
66	wh, 3.01	i	i	+ .	+	+	−/+	−
67	dk-red, 3.19	−	i	−/+	−	−/+	−/+	−
68	blk, 2.55	−	i	−	−	−/+	−	−
69	wh, 3.21	r/+	r	−	−/+	−	+	+
70	wh, 2.54	r/+	r	−	−/+	−	+	+
71	wh hydr	r	d	−/+	−	−	+	+
72	yel, 2.36	+	...	+	+	+	+	+
73	wh	r/+	r	r	r/+	r/+	+	+
74	wh	+	+	+	+	+	+	+
75	wh hydr	r	...	−	−/+	−/+	+	+
76	wh hydr	r	...	−	−/+	−/+	−/+	−/+
77	wh, 2.47	r/+	r	r/d	−/+	−	+	+
78	wh, 2.40	r	r	r	−	−	+	+
79	lt-yel	+	+	+	+	+	+	+
80	wh	+	...	+	+	+	+	+
81	wh	+	...	+	+	+	+	+
82	wh	+	+	+	+	+	+	+
83	wh hydr	r	...	+	r/+	r/+	+	+
84	wh hydr, 2.02	r	d	−	−	−	+	+
85	wh hydr	r	d	+	+	+	+	+
86	wh	+	+	+	+	+	+	+
87	wh, 2.88(2.17)	d/r	i	d/r	d/+	d/r	+	+
88	wh	i	i	−/+	−/+	−/+	+	−/+
89	wh	i	i	−/+	−/+	−/+	+	−/+
90	wh, 1.78	r	i	−/+	−/+	−/+	+	−/+
91	wh, 2.98	i	i	−/+	−/+	−/+	+	−/+
92	wh	+	+	+	+	+	+	+
93	cl lq	+	+	+	+	+	+	+
94	cl lq	+	+	+	+	+	+	+

No.	Formula and name	M_r	Phase transition temperature
95	Al(H)Cl$_2$ Aluminum hydride-dichloride	98.90	dec >60
96	Al(H)$_2$Cl Aluminum dihydride-chloride	64.45	dec >40
97	Al(H)I$_2$ Aluminum hydride-diiodide	281.80	mp 80
98	Al(H)$_2$I Aluminum dihydride-iodide	155.90	mp 35
99	[AlH$_4$],Li Lithium tetrahydridoaluminate(III)	37.96	dec 125
100	[AlH$_4$],Na Sodium tetrahydridoaluminate(III)	54.00	mp 183
101	AlI$_3$ Aluminum iodide	407.69	mp 188.3; bp 382.5
102	AlI$_3$·6H$_2$O	515.78	dec 185
103	Al(IO$_3$)$_2$NO$_3$ (·6H$_2$O) Aluminum diiodate-nitrate	438.79	dec hydr >100
104	AlN Aluminum mononitride	40.99	mp 2427
105	Al(N$_3$)$_3$ Aluminum azide	153.05	dec >200
106	AlNH$_4$(SO$_4$)$_2$ Aluminum-ammonium sulfate	237.15	dec 500
107	AlNH$_4$(SO$_4$)$_2$·12H$_2$O	453.33	mp hydr 95, −H$_2$O >200
108	Al(NO$_3$)$_3$ (·6H$_2$O) Aluminum nitrate	212.99	>150 → Al$_2$O$_3$
109	Al(NO$_3$)$_3$·9H$_2$O	375.13	mp 73.5; dec 134
110	Al$_2$O$_3$ (·0.25H$_2$O) Aluminum oxide	101.96	mp 2053, bp >3000
111	Al$_{2.67}$O$_4$ 2.67-Aluminum tetraoxide	136.04	1000 → Al$_2$O$_3$
112	Al(OH)$_3$ Aluminum hydroxide	78.00	200 → γ-AlO(OH)
113	α-AlO(OH) Aluminum metahydroxide	59.99	575 → Al$_2$O$_3$
114	γ-AlO(OH)	59.99	300 → Al$_{2.67}$O$_4$
115	AlP Aluminum monophosphide	57.96	mp 1700, dec 2000
116	AlPO$_4$ Aluminum orthophosphate	121.95	mp ca. 1650
117	AlPO$_4$·2H$_2$O	157.98	dec >1500
118	Al(PO$_3$)$_3$ Aluminum metaphosphate	263.90	mp 1240
119	Al$_3$(PO$_4$)$_2$(OH)$_3$ (·5H$_2$O) Trialuminum diorthophosphate-trihydroxide	321.91	dec >1000
120	Al$_2$S$_3$ Aluminum sulfide	150.16	subl t, mp 1120 (p)
121	Al$_2$(SO$_4$)$_3$ Aluminum sulfate	342.15	mp 770 dec
122	Al$_2$(SO$_4$)$_3$·13.5H$_2$O	585.35	dec >290
123	Al$_2$(SO$_4$)$_3$·16H$_2$O	630.39	mp 86 dec
124	Al$_2$(SO$_4$)$_3$·18H$_2$O	666.42	−H$_2$O >400
125	Al(SO$_4$)OH Aluminum sulfate-hydroxide	140.05	dec t
126	Al$_2$SO$_4$(OH)$_4$ (·7H$_2$O) Dialuminum sulfate-tetrahydroxide	218.05	dec t
127	Al$_2$Se$_3$ Aluminum selenide	290.84	dec t
128	α-Al$_2$(SiO$_4$)O Dialuminum orthosilicate-oxide	162.05	dec t
129	β-Al$_2$(SiO$_4$)O	162.05	dec t
130	γ-Al$_2$(SiO$_4$)O	162.05	mp 1545 dec
131	Al$_2$Si$_2$O$_7$ (·2H$_2$O) Aluminum disilicate	222.13	−H$_2$O ca. 400
132	Al$_2$Si$_2$O$_7$·4H$_2$O	330.22	330 → Al$_2$Si$_2$O$_7$·2H$_2$O
133	Al$_2$Te$_3$ Aluminum telluride	436.76	dec t
134	**Am Americium**	243.061	mp 1292, bp 2880
135	Am$_2$(C$_2$O$_4$)$_3$ (·10H$_2$O) Americium(III) oxalate	750.18	hydr >300 → Am$_2$O$_3$
136	AmCl$_3$ Americium(III) chloride	349.42	mp 850, bp 1750
137	AmF$_3$ Americium(III) fluoride	300.06	mp 1393, bp 2070
138	Am(NO$_3$)$_3$ Americium(III) nitrate	429.07	t → Am$_2$O$_3$
139	AmO$_2$ Americium(IV) oxide	275.06	dec >1000

(Continued)

No.	Color and density	Solubility and reactivity						
		water	alcohol	HCl	H_2SO_4	HNO_3	NaOH	$NH_3 \cdot H_2O$
95	cl lq	+	+	+	+	+	+	+
96	cl lq	+	+	+	+	+	+	+
97	wh	+	+	+	+	+	+	+
98	wh	+	+	+	+	+	+	+
99	wh, 0.72	+	−	+	+	+	+	+
100	wh	+	+	+	+	+	+	+
101	wh, 3.98	r/+	r	r	−/+	+	+	+
102	lt-yel, 2.63	r/+	r	−	−/+	+	+	+
103	wh hydr	d	...	+	−	−	+	+
104	wh, 3.12	−/+	i	−/+	−/+	−/+	−/+	−/+
105	wh	+	...	+	+	+	+	+
106	wh, 2.04	r	i	−	r	−	+	+
107	wh, 1.64	r	i	−	r	−	+	+
108	wh, 1.89	r	r	r	r	r	+	+
109	wh	r	r	r	r	r	+	+
110	wh, 3.97	−	i	−/+	−/+	−/+	−/+	−
111	wh	+	...	+	+	+	−/+	−
112	wh, 2.42	i	i	+	+	+	+	−/+
113	wh, 3.35	−	−	−/+	−/+	−/+	−/+	−
114	wh, 3.01	−	−	−/+	−/+	−/+	−/+	−
115	lt-grey, 2.41	−/+	...	+	+	+	+	−/+
116	wh, 2.56	i	i	−	−/+	−	−/+	−
117	wh, 2.54	i	i	i/+	i/+	i/+	+	−
118	wh, 2.78	i	i	−	−/+	−	−/+	−
119	wh, 2.3	i	i	+	+	+	+	−/+
120	wh, 2.02	+	...	+	+	+	+	+
121	wh, 2.71	r	d	r	r/+	r	+	+
122	wh	r	d	r	r	r	+	+
123	wh, 1.69	r	d	r	r	r	+	+
124	wh, 1.65	r	d	r	r	r	+	+
125	wh	r	...	+	+	+	+	+
126	wh	r	...	+	+	+	+	+
127	yel	+	...	+	+	+	+	+
128	wh, 3.61	−	...	−/+	−/+	−/+	−/+	−
129	wh, 3.25	−	...	−/+	−/+	−/+	−/+	−
130	wh, 3.18	−	...	−/+	−/+	−/+	−/+	−
131	wh, 2.61	i	...	−/+	−/+	−/+	−/+	−
132	wh, 2.1	i	...	−/+	−/+	−/+	−/+	−
133	dk-brn	+	...	+	+	+	+	+
134	wh, 11.7	−/+	...	+	+/psv	+/psv	−/+	−
135	pink hydr	i	...	−/+	i/+	−/+	i/+	i
136	pink	r	...	−	−/+	−	+	+
137	pink	d/r	d	−/+	−/+	−/+	+	+
138	pink	r	...	−	−	r	+	+
139	dk-brn	−	...	+	+	+	−	−

No.	Formula and name	M_r	Phase transition temperature
140	Am$_2$O$_3$ Americium(III) oxide	534.12	mp 2200
141	Am(OH)$_3$ Americium(III) hydroxide	294.08	$t \to$ Am$_2$O$_3$
142	Am(OH)$_4$ Americium(IV) hydroxide	311.09	$t \to$ AmO$_2$
143	**Ar Argon**	39.948	mp -189.4; bp -185.8
144	**α-As Arsenic**, grey	74.922	subl 615, mp 817 (p)
145	**β-As Arsenic**, black	74.922	270 $\to$ α-As
146	**As$_4$ Arsenic**, yellow	299.69	358 $\to$ α-As
147	AsBr$_3$ Arsenic tribromide	314.63	mp 31.2; bp 221
148	AsCl$_3$ Arsenic trichloride	181.28	mp -16.2; bp 131.4
149	AsF$_3$ Arsenic trifluoride	131.92	mp -5.94; bp 57.8
150	AsF$_5$ Arsenic pentafluoride	169.91	mp -79.8; bp -52.8
151	[AsF$_6$],H ($\cdot$H$_2$O) Hydrogen hexafluoroarsenate(V)	189.92	dec hydr 193
152	[AsF$_6$],Li Lithium hexafluoroarsenate(V)	195.85	dec 350
153	AsH$_3$ Arsine	77.95	mp -116.92; bp -62.47
154	AsI$_3$ Arsenic triiodide	455.63	mp 141, bp 371
155	AsN Arsenic mononitride	88.93	dec ca. 300
156	As$_2$O$_3$(amor) Arsenic trioxide	197.84	soft t; subl 195
157	α-As$_2$O$_3$	197.84	mp 314, bp 461
158	β-As$_2$O$_3$	197.84	mp 278, bp 461
159	As$_2$O$_5$ Diarsenic pentaoxide	229.84	dec >300
160	As(O)F$_3$ Arsenic oxide-trifluoride	147.92	mp -68, bp -26
161	AsP Arsenic monophosphide	105.90	dec 750
162	As$_2$S$_3$ Diarsenic trisulfide	246.04	mp 310, bp 723
163	As$_2$S$_5$ Diarsenic pentasulfide	310.17	subl ca. 190, dec 500
164	α-As$_4$S$_4$ Tetraarsenic tetrasulfide	427.95	267 $\to$ β-As$_4$S$_4$
165	β-As$_4$S$_4$	427.95	mp 321, bp 534
166	As$_2$Se$_3$ Diarsenic triselenide	386.72	mp 360
167	**At$_2$ Diastatine**	419.974	mp 244, bp 309
168	**Au Gold**	196.967	mp 1064.43; bp 2947
169	(AuAg)Te$_4$ Gold-silver tetratelluride	815.24	dec >800
170	(AuAg$_3$)Te$_2$ Gold-trisilver ditelluride	775.77	dec >500
171	AuBr Gold(I) bromide	276.87	dec ca. 200
172	AuBr$_3$ Gold(III) bromide	436.68	>150 $\to$ AuBr
173	[AuBr$_4$],K ($\cdot$2H$_2$O) Potassium tetrabromoaurate(III)	555.68	dec hydr 120
174	Au$_2$C$_2$ Gold(I) acetylide	417.96	dec >25
175	AuCN Gold(I) cyanide	222.99	dec t
176	Au(CN)$_3$ ($\cdot$3H$_2$O) Gold(III) cyanide	275.02	dec hydr 50
177	[Au(CN)$_2$],K Potassium dicyanoaurate(I)	288.10	dec >250
178	[Au(CN)$_4$],K ($\cdot$1.5H$_2$O) Potassium tetracyanoaurate(III)	340.14	$-$H$_2$O 200, dec 250
179	AuCl Gold(I) chloride	232.42	289 $\to$ Au
180	AuCl$_3$ ($\cdot$2H$_2$O) Gold(III) chloride	303.33	dec 150, mp 288 (p)
181	[AuCl$_4$],Cs Cesium tetrachloroaurate(III)	471.68	dec ca. 460
182	[AuCl$_4$],H ($\cdot$4H$_2$O) Hydrogen tetrachloroaurate(III)	339.79	dec hydr 120
183	[AuCl$_4$],K ($\cdot$0.5H$_2$O) Potassium tetrachloroaurate(III)	377.88	$-$H$_2$O 100, dec ca. 350
184	[AuCl$_4$],NH$_4$ ($\cdot n$H$_2$O) Ammonium tetrachloroaurate(III)	356.82	$-$H$_2$O t, mp 520

(Continued)

No.	Color and density	Solubility and reactivity						
		water	alcohol	HCl	H_2SO_4	HNO_3	NaOH	$NH_3 \cdot H_2O$
140	orange	−	...	+	+	+	−	−
141	pink	i	...	+	+	+	−	−
142	brn-blk	i	...	+	+	+	−	−
143	cl gas, 1.7837	d/i	d/i	−	−	−	−	−
144	grey, 5.73	−	...	−	−/+	−/+	−/+	−
145	blk, 4.9	−	...	−	−	−/+	−/+	−
146	yel, 2.03	−	...	−/+	−	−/+	−/+	−
147	wh, 3.54	+	r	+	+	+	+	+
148	cl lq, 2.16[20]	+	r	+	+	+	+	+
149	cl lq, 2.73[15]	+	r	+	+	+	+	+
150	cl gas, 7.71	+	+	+	+	+	+	+
151	wh	r	...	−	−/+	−/+	+	+
152	wh	r	r	−	−/+	−/+	+	+
153	cl gas, 3.5023	d	...	−/+	−/+	−/+	−	−
154	red, 4.39	−/+	r	+	−/+	+	+	+
155	orange-red	−	...	−	−	−/+	−/+	−
156	wh	+	d	−/+	−/+	+	+	−
157	wh, 4.15	+	d	−/+	−/+	+	+	−
158	wh, 3.74	+	d	−/+	−/+	+	+	−
159	wh, 4.32	+	r	+	+	+	+	+
160	cl gas	+	d/+	+	+	+	+	+
161	red	+	i	+	+	+	+	+
162	dk-yel, 3.43	−/+	r	−	−/+	+	−/+	−/+
163	yel	+	i	−/+	−/+	+	+	+·
164	red-brn, 3.56	−	...	−	−/+	−/+	−/+	−
165	red-brn, 3.25	−	...	−	−/+	−/+	−/+	−
166	brn, 4.75	−/+	...	−/+	−/+	−/+	+	+
167	...	−	...	−/+	−/+	+	+	−
168	yel, 19.29	−	−	−	−	−	−	−
169	dk-grey, 8.1	−	...	−	−/+	−/+	−	−
170	dk-grey, 8.85	−	...	−	−/+	−/+	−	−
171	yel, 7.9	i/+	+	+	+	+	+	+
172	brn	+	r	+	+	+	+	+
173	red hydr, 4.08	r	r	−	−/+	−/+	+	+
174	yel	−/+	...	+	+	+	+	+
175	yel, 7.14	i	i	i/r	i/r	r	i/+	i/+
176	wh hydr	+	r	−/+	−	+	+	+
177	wh, 3.45	r	d	−	−/+	+	−	−
178	wh	r	d	+	+	+	−	−
179	lt-yel, 7.4	i/+	+	+	+	+	+	+
180	dk-red, 4.67	+	r	+	+	+	+	+
181	yel	d	r	r	−	−	+	−/+
182	lt-yel hydr, 3.9	r	r	r	−	−	+	+·
183	yel, 3.75	r	r	r	−	−	+	−/+
184	yel hydr	r	r	r	−	−	+	+

No.	Formula and name	M_r	Phase transition temperature
185	[AuCl$_4$],Na (·2H$_2$O) Sodium tetrachloroaurate(III)	361.77	dec hydr 100
186	Au(Cl)O Gold chloride-oxide	248.42	290 → Au
187	AuF$_3$ Gold(III) fluoride	253.96	dec 500
188	AuF$_5$ Golg(V) fluoride	291.96	mp 75
189	AuI Gold(I) iodide	323.87	dec 120
190	AuI$_3$ Gold(III) iodide	577.68	>20 → AuI
191	AuNCS Gold(I) thiocyanate	255.06	dec 140
192	[Au(NO$_3$)$_4$],H (·3H$_2$O) Hydrogen tetranitratoaurate(III)	445.99	mp hydr 72.5 dec
193	Au$_2$O$_3$ Gold(III) oxide	441.93	dec >160
194	Au$_2$O$_3$·nH$_2$O	–	100 → AuO(OH)
195	[Au(OH)$_4$],K (·H$_2$O) Potassium tetrahydroxoaurate(III)	304.09	dec hydr >300
196	AuO(OH) Gold metahydroxide	229.97	ca. 150 → Au$_2$O$_3$
197	Au$_2$S Gold(I) sulfide	426.00	240 → Au
198	Au$_2$(S$_2$) Gold(I) disulfide(2−)	458.07	>140 → Au$_2$S
199	Au$_2$S$_3$ Gold(III) sulfide	490.13	200 → Au$_2$S
200	[Au(SO$_3$S)$_2$],Na$_3$ (·2H$_2$O) Sodium bis(thiosulfato)aurate(I)	490.20	−H$_2$O 150, dec 300
201	Au$_2$(SeO$_4$)$_3$ Gold(III) selenate	822.80	dec t
202	AuTe$_2$ Gold ditelluride	452.17	mp 464, dec >500
203	**B Boron**	10.811	mp 2075, bp 3700
204	(BAs)O$_4$ Boron-arsenic tetraoxide	149.73	subl ca. 700
205	BBr$_3$ Boron tribromide	250.52	mp −45.84; bp 89.8
206	BBr$_2$I Boron dibromide-iodide	297.52	bp 125
207	B$_4$C Tetraboron carbide	55.26	mp 2350, bp >3500
208	B(CH$_3$COO)$_3$ Boron triacetate	187.94	mp 149
209	B(CH$_3$O)$_3$ Trimethoxyborane	103.91	mp −29, bp 68.7
210	B(C$_2$H$_5$O)$_3$ Triethoxyborane	145.99	bp 118.6
211	BCl$_3$ Boron trichloride	117.17	mp −107, bp 12.5
212	B$_4$Cl$_4$ Tetraboron tetrachloride	185.06	mp 95, dec 200
213	[BD$_4$],Na Sodium tetradeuteridoborate(III)	41.86	mp >300
214	BF$_3$ Boron trifluoride	67.81	mp −128.36; bp −100.3
215	[BF$_4$],(ClF$_2$) Difluorochlorine(III) tetrafluoroborate(III)	160.25	mp 30
216	[BF$_4$],Cs Cesium tetrafluoroborate(III)	219.71	mp 555
217	[BF$_4$],K Potassium tetrafluoroborate(III)	125.90	mp 570, dec 930
218	[BF$_4$],NH$_4$ Ammonium tetrafluoroborate(III)	104.84	subl 350, mp 487 (p)
219	[BF$_4$],Na Sodium tetrafluoroborate(III)	109.79	mp 384, dec >400
220	[BF$_4$],(XeF$_5$) Pentafluoroxenon(VI) tetrafluoroborate(III)	313.08	mp 90
221	B$_2$H$_6$ Diborane(6)	27.67	mp −165.5; bp −92.5
222	B$_4$H$_{10}$ Tetraborane(10)	53.32	mp −120, bp 18 dec
223	B$_5$H$_9$ Pentaborane(9)	63.13	mp −45.6; bp 60
224	B$_5$H$_{11}$ Pentaborane(11)	65.14	mp −128.6; dec 25
225	B$_6$H$_{10}$ Hexaborane(10)	74.95	mp −65.1; dec 25
226	B$_6$H$_{12}$ Hexaborane(12)	76.96	mp −82, dec 25
227	B$_9$H$_{15}$ Nonaborane(15)	112.42	mp 3

(*Continued*)

No.	Color and density	Solubility and reactivity						
		water	alcohol	HCl	H_2SO_4	HNO_3	NaOH	$NH_3 \cdot H_2O$
185	yel hydr	r	r	r	−	−	+	+
186	red	−/+	−	−/+	−/+	−/+	+	+
187	orange	+	+	+	+	+	+	+
188	red-brn	+	+	+	+	+	+	+
189	yel-grn, 8.25	i/+	+	+	+	+	+	+
190	dk-grn	i/+	...	−/+	−/+	+	+	−/+
191	dk-yel	+	...	+	+	+	+	+
192	yel hydr	+	+	+	+	+	+	+
193	brn-blk, 10.61	−	−	+	−/+	−/+	−/+	−
194	lt-brn	i	−	+	−/+	−/+	i/+	i
195	lt-yel hydr	r/+	r	+	+	+	r	r
196	lt-brn	i	−	+	−/+	−/+	+	i
197	brn-blk, 9.0	i	...	−	−/+	−/+	−	−
198	blk-brn	i	...	−	−/+	−/+	−	−
199	blk, 8.75	+	i	−/+	−/+	+	+	+
200	wh hydr, 3.09	r	i	−/+	−/+	−/+	−	−
201	yel	+	...	+	+	+	+	+
202	yel, 8.6–9.35	−	−	−	−/+	−/+	−/+	−
203	grey-blk, 2.340	−	−	−	−	−/+	−	−
204	wh, 3.64	−/+	...	−	−/+	−/+	−/+	−
205	cl lq, 2.65^0	+	+	+	+	+	+	+
206	cl lq	+	+	+	+	+	+	+
207	blk, 2.55	−	−	−	−	−	−/+	−
208	wh	−/+	d	+	+	+	+	+
209	cl lq	+	∞	+	+	+	+	+
210	cl lq	+	∞	+	+	+	+	+
211	cl lq, 1.43^0	+	+	+	+	+	+	+
212	yel	+	...	+	+	+	+	+
213	wh	r/+	r	+	+	+	−	−
214	cl gas, 3.209	+	+	+	+	+·	+	+
215	wh	+	...	+	+	+	+	+
216	wh, 3.20	d/i	i	−	−/+	−	−/+	−
217	wh, 2.51	d/+	i	−	−/+	−	−/+	−
218	wh, 1.871	r/+	...	−	−/+	−	+	−
219	wh, 2.47	r/+	d	−	−/+	−	+	−
220	wh	+	...	+	+	+	+	+
221	cl gas, 1.234	+	+	+	+	+	+	+
222	cl gas, 2.379	+	+	+	+	+	+	+
223	cl lq, 0.61^0	+	+	+	+	+	+	+
224	cl lq	+	+	+	+	+	+	+
225	cl lq, 0.69^0	+	+	+	+	+	+	+
226	cl lq	+	+	+	+	+	+	+
227	cl lq	+	+	+	+	+	+	+

No.	Formula and name	M_r	Phase transition temperature
228	$B_{10}H_{14}$ Decaborane(14)	122.22	mp 99.7; bp 213 dec
229	$B_{16}H_{20}$ 16-Borane(20)	193.14	mp 99
230	$B_{18}H_{22}$ 18-Borane(22)	216.77	mp 180
231	$B_{20}H_{16}$ 20-Borane(16)	232.35	mp 197
232	$[BH_4]_3$,Al Aluminum tetrahydridoborate(III)	71.51	mp −64.5; bp 44.5
233	$[BH_4]_2$,Be Beryllium tetrahydridoborate(III)	38.70	subl 91, dec >120
234	$[BH_4]$,Cs Cesium tetrahydridoborate(III)	147.75	dec >570
235	$[B_{10}H_{10}]$,Cs_2 Dicesium decaboranate(10)	384.00	dec 600
236	$[B_{12}H_{12}]$,Cs_2 Dicesium dodecaboranate(12)	407.64	dec >800
237	$[BH_4]_4$,Hf Hafnium tetrahydridoborate(III)	237.86	mp 29, bp 118
238	$[BH_4]$,K Potassium tetrahydridoborate(III)	53.94	dec 500
239	$[BH_4]$,Li Lithium tetrahydridoborate(III)	21.78	dec 278
240	$B_3H_6N_3$ Borazine	80.50	mp −56, bp 55
241	$[BH]_4$,Na Sodium tetrahydridoborate(III)	37.83	mp 400 dec
242	$[BH_4]_4$,Ti Titanium(IV) tetrahydridoborate(III)	107.25	dec 25
243	$[BH_4]_4$,U Uranium(IV) tetrahydridoborate(III)	297.40	dec 70
244	$[BH_4]_2$,Zn Zinc(II) tetrahydridoborate(III)	95.08	dec 85
245	$[BH_4]_4$,Zr Zirconium(IV) tetrahydridoborate(III)	150.60	mp 29, bp 128
246	BI_3 Boron triiodide	391.52	mp 49.7; bp 209.5
247	α-BN Boron mononitride	24.82	dec >1000, mp 2727 (p)
248	β-BN	24.82	mp >3200 (p)
249	$[B(NH_3)F_3]$ Amminetrifluoroboron	84.84	dec 125, mp 162 (p)
250	$[BN_2]$,Li_3 Lithium dinitridoborate(III)	59.65	mp 870
251	B_2O_3 Diboron trioxide	69.62	mp 450, bp >2000
252	B_2O_3(amor)	69.62	soft >200
253	$B(OH)_3$ Boron trihydroxide	61.83	>70 → HBO_2; mp 170 (p)
254	$[B(OH)F_3]$,H ($\cdot H_2O$) Hydrogen hydroxotrifluoroborate(III)	85.82	mp hydr 6, dec hydr >20
255	$[B(OH)_4]$,Li ($\cdot 6H_2O$) Lithium tetrahydroxoborate(III)	85.78	mp hydr 47
256	$[B(OH)_4]$,Na ($\cdot 2H_2O$) Sodium tetrahydroxoborate(III)	101.83	mp hydr 57, dec 306
257	$[B_2(O_2)_2(OH)_4]$,Na_2 ($\cdot 6H_2O$) Sodium diperoxotetrahydroxodiborate(III)	199.63	dec hydr >60
258	BP Boron monophosphide	41.79	mp >2000 dec
259	$(BP)O_4$ Boron-phosphorus tetraoxide	105.78	mp 1600
260	B_2S_3 Diboron trisulfide	117.82	mp >310
261	$[B(SO_3F)_4]$,K Potassium tetrakis(fluorosulfonato)borate(III)	446.15	mp 65 dec
262	B_3Si Triboron silicide	60.52	mp ca. 2500
263	$[BW_{12}O_{40}]$,H_5 ($\cdot 30H_2O$) Hydrogen 40-oxododecawolframoborate(III)	2862.01	mp hydr 48.5; dec 60
264	**Ba Barium**	137.327	mp 727, bp ca. 1860
265	Ba_3As_2 Tribarium diarsenide	561.83	mp >1000
266	$(BaAs_2)O_6$ Barium-diarsenic hexaoxide	383.17	dec 500
267	$Ba_2As_2O_7$ Barium heptaoxodiarsenate(V)	536.49	dec 800
268	$Ba_3(AsO_4)_2$ Barium arsenate	689.82	mp 1605
269	BaB_6 Barium hexaboride	202.19	mp 2270
270	$BaBr_2$ Barium bromide	297.14	mp 857, bp 1980

(Continued)

No.	Color and density	Solubility and reactivity						
		water	alcohol	HCl	H_2SO_4	HNO_3	NaOH	$NH_3 \cdot H_2O$
228	wh, 0.94	+	+	+	+	+	+	+
229	wh	+	r	+	+	+	+	+
230	wh	+	r	+	+	+	+	+
231	wh	+	r	+	+	+	+	+
232	cl lq, 0.55^{20}	+	r	+	+	+	+	+
233	wh	+	r	+	+	+	+	+
234	wh, 2.40	d/+	d	+	+	+	−	−
235	wh	r	d	−	−	−	−	−
236	wh	r	d	−	−	−	−/+	−
237	wh	+	r	+	+	+	+	+
238	wh, 1.18	d/+	...	+	+	+	−	−
239	wh, 0.67	r/+	...	+	+	+	−	−
240	cl lq, 0.824^0	+	...	+	+	+	+	+
241	wh, 1.07	r/+	r	+	+	+	−	−
242	grn	+	...	+	+	+	+	+
243	grn	+	...	+	+	+	+	+
244	wh	+	i	+	+	+	+	+
245	wh	+	...	+	+	+	+	+
246	wh, 3.35	+	r	−	−/+	−/+	+	...
247	wh, 2.34	−	−	−	−	−	−/+	−
248	wh, 2.29	−	−	−	−	−	−	−
249	wh, 1.85	+	+	+	+	+	+	+
250	wh	+	...	+	+	+	+	+
251	wh, 2.46	−/+	...	−	−	−	+	−
252	wh, 1.84	+	+	+/−	+/−	+/−	+	+
253	wh, 1.44	r	r	−	−	−	+	+
254	cl lq hydr, 1.63^{20}	r	...	+	+	+	+	+
255	wh hydr	r/+	...	+	+	+	r	r
256	wh, 1.91(1.74)	r/+	...	+	+	+	r	r
257	wh hydr	d/+	...	+	+	+	−	−
258	lt-brn	−	−	−	−/+	−/+	−/+	−
259	wh, 2.52	−/+	i	−	−/+	−/+	−/+	−
260	wh, 1.55	+	+	+	+	+	+	+
261	wh	r	...	+	+	+	+	...
262	blk, 2.52	−	...	−	−/+	−	−/+	−
263	wh hydr, 3.0	r/+	r	−	−	−	+	+
264	wh, 3.60	+	+	+	+	+	+	+
265	brn, 4.1	+	...	+	+	+	+	+
266	wh	i/+	...	+	+	+	−	−
267	wh	i/+	...	i/+	i/+	i/+	i/+	i/+
268	wh, 5.10	i	...	i/+	i/+	i/+	i	i
269	dk-grey, 4.36	−	−	−	−	+	−	−
270	wh, 4.78	r	d	r	+	r	r	r

No.	Formula and name	M_r	Phase transition temperature
271	BaBr$_2$·2H$_2$O	333.17	−H$_2$O 120
272	Ba(BrO$_3$)$_2$ (·H$_2$O) Barium bromate	393.13	−H$_2$O 180, dec 270
273	BaC$_2$ Barium acetylide	161.35	mp >1780, dec 2000
274	Ba(CH$_3$COO)$_2$ (·3H$_2$O) Barium acetate	255.42	−H$_2$O >144, mp 450 dec
275	BaCO$_3$ Barium carbonate	197.34	dec >1200; mp 1555 (p)
276	BaC$_2$O$_4$ (·H$_2$O) Barium oxalate	225.35	dec hydr 400
277	BaCa(CO$_3$)$_2$ Barium-calcium dicarbonate	297.42	dec ca. 1000
278	BaCl$_2$ Barium chloride	208.23	mp 961, bp ca. 2050
279	BaCl$_2$·2H$_2$O	244.26	−H$_2$O 113
280	Ba$_2$(Cl)N Dibarium chloride-nitride	324.11	mp 965
281	Ba(ClO)$_2$ (·2H$_2$O) Barium hypochlorite	240.23	dec hydr ca. 185
282	Ba(ClO$_2$)$_2$ Barium chlorite	272.23	dec 235
283	Ba(ClO$_3$)$_2$ (·H$_2$O) Barium chlorate	304.23	−H$_2$O 120, mp 414 dec
284	Ba(ClO$_4$)$_2$ (·3H$_2$O) Barium perchlorate	336.23	−H$_2$O 260, mp 505
285	BaCrO$_4$ Barium chromate	253.32	mp 1380
286	BaCr$_2$O$_7$ (·2H$_2$O) Barium dichromate	353.31	−H$_2$O 120
287	BaF$_2$ Barium fluoride	175.32	mp 1368, bp 2260
288	BaFeO$_4$ (·H$_2$O) Barium ferrate	257.17	dec hydr 120
289	BaH$_2$ Barium hydride	139.34	dec >600
290	BaHAsO$_4$ (·H$_2$O) Barium hydroarsenate	277.25	−H$_2$O 150
291	BaHPO$_4$ Barium hydroorthophosphate	233.31	dec 400
292	Ba(H$_2$PO$_4$)$_2$ Barium dihydroorthophosphate	331.30	dec >340
293	Ba(HS)$_2$ (·4H$_2$O) Barium hydrosulfide	203.48	−H$_2$O 50, dec t
294	BaI$_2$ Barium iodide	391.14	mp 711, bp 1900
295	BaI$_2$·7.5H$_2$O	526.25	dec >540
296	Ba(IO$_3$)$_2$ (·H$_2$O) Barium iodate	487.13	−H$_2$O 130, dec t
297	BaMnO$_4$ Barium manganate	256.26	dec >700
298	Ba(MnO$_4$)$_2$ Barium permanganate	375.20	220 → BaMnO$_4$
299	BaMoO$_4$ Barium molybdate	297.26	mp 1460, bp 1730
300	Ba(N$_3$)$_2$ (·6H$_2$O) Barium azide	221.37	dec hydr >120
301	Ba$_3$N$_2$ Tribarium dinitride	440.00	dec >500
302	Ba(NCS)$_2$ (·3H$_2$O) Barium thiocyanate	253.50	−H$_2$O 130, dec t
303	Ba(NH$_2$)$_2$ Barium amide	169.37	mp 290, dec t
304	[Ba(NH$_3$)$_6$]I$_2$ Hexaamminebarium(II) iodide	493.32	dec >20
305	BaN$_2$O$_2$ (·4H$_2$O) Barium hyponitrite	197.34	−H$_2$O 100; 500 → BaO
306	Ba(NO$_2$)$_2$ (·H$_2$O) Barium nitrite	229.34	−H$_2$O 115, dec >200
307	Ba(NO$_3$)$_2$ Barium nitrate	261.34	mp 594, dec 800
308	Ba(NO$_3$)$_2$·H$_2$O	279.35	dec hydr ca. 100
309	BaO Barium oxide	153.33	subl 1377, mp ca. 2020
310	BaO$_2$ (·8H$_2$O) Barium peroxide	169.33	−H$_2$O 100, dec >790
311	Ba(O$_2$)$_2$ Barium superoxide	201.32	dec 50
312	Ba(OH)$_2$ Barium hydroxide	171.34	mp 408, dec 780
313	Ba(OH)$_2$·8H$_2$O	315.46	mp 78, −H$_2$O 125
314	Ba(PH$_2$O$_2$)$_2$ (·H$_2$O) Barium phosphinate	267.30	dec hydr ca. 100
315	Ba$_2$P$_2$O$_7$ Barium diphosphate	448.60	mp 1430, t → Ba$_3$(PO$_4$)$_2$
316	Ba$_3$(PO$_4$)$_2$ Barium orthophosphate	601.92	mp 1605
317	Ba(ReO$_4$)$_2$ Barium perrhenate	637.73	mp 799

(*Continued*)

No.	Color and density	Solubility and reactivity						
		water	alcohol	HCl	H_2SO_4	HNO_3	NaOH	$NH_3 \cdot H_2O$
271	wh, 3.58	r	r	r	+	r	r	r
272	wh, 3.99	d	i	−	+	−	−	−
273	grey-blk, 3.75	+	+	+	+	+	+	+
274	wh, 2.47(2.02)	r	d	r	+	r	r	−
275	wh, 4.43	i	i	+	+	+	−	−
276	wh hydr, 2.66	i	i	i/+	+	i/+	i	−
277	wh, 3.65	i	i	+	+	+	−	−
278	wh, 3.86	r	i	i	+	r	r	r
279	wh, 3.10	r	i	r	+	r	r	r
280	yel-grey	+	...	+	+	+	+	+
281	wh hydr	+	+	+	+	+	+	+
282	wh	+	+	+	+	+	+	+
283	wh hydr, 3.18	r	d	−/+	+	r	−	−
284	wh, 3.2(2.74)	r	r	r	+	−	−	−
285	yel, 4.50	i	i	+	+	+	i	i
286	yel-red hydr	r	...	−	+	−	+	+
287	wh, 4.89	d	i	−	+	d	−	−
288	red hydr	i	+	+	+	−	−	+
289	wh, 4.15	+	+	+	+	+	+	+
290	wh, 3.93(1.93)	d/+	...	+	+	+	−/+	−/+
291	wh, 4.17	d	...	+	+	+	−/+	−/+
292	wh, 2.9	d/+	...	+	+	+	−/+	−/+
293	wh	r/+	i	+	+	+	+	+
294	wh, 5.15	r	r	−	+	−/+	r	r
295	wh, 2.6	r	r	−	+	−/+	r	r
296	wh, 5.0(4.66)	i	i	+	−/+	−	−	−
297	blk-grn, 4.85	i/+	...	+	+	+	−	−
298	blk-viol, 3.77	d	+	r/+	+	r/+	r/+	+
299	wh, 4.65	i	...	−	−/+	−/+	−	−
300	wh hydr, 2.94	r	d	+	+	+	r	r
301	grey, 4.78	+	...	+	+	+	+	+
302	wh hydr, 2.27	r	r	r	+	−/+	r	−
303	wh	+	...	+	+	+	+	+
304	wh	r/+	...	+	+	+	−	r
305	wh, 3.89(2.74)	r/+	...	−/+	−/+	−/+	−	−
306	wh, 3.23(3.17)	r	d	−/+	+	−/+	−	−
307	wh, 3.24	r	i	r	+	i	r	r
308	wh	r	i	−	+	r	−	−
309	wh, 5.72	+	+	+	+	+	+	+
310	wh, 4.96(2.29)	i/+	i	+	+	+	−	−
311	yel-grn	+	i	+	+	+	+	+
312	wh, 4.5	r	i	+	+	+	r	r
313	wh, 2.18	r	d	+	+	+	r	r
314	wh hydr, 2.90	r	i	−	+	−/+	−	−
315	wh, 3.9	d/+	...	+	+	+	−	−
316	wh, 4.1	i	...	i/r	i/+	i/r	i	i
317	wh, 5.91	r	r	−	+	−	−	−

No.	Formula and name	M_r	Phase transition temperature
318	BaS Barium sulfide	169.39	mp ca. 2000 dec
319	$Ba(S_2)$ Barium disulfide(2−)	201.46	mp 925 dec
320	$Ba(S_3)$ Barium trisulfide(2−)	233.53	mp 554 dec
321	$Ba(S_4)$ ($\cdot 2H_2O$) Barium tetrasulfide(2−)	265.59	dec 300
322	$BaSO_3$ Barium sulfite	217.39	dec t
323	$BaSO_4$ Barium sulfate	233.39	mp 1580 dec
324	BaS_2O_6 ($\cdot 2H_2O$) Barium dithionate	297.45	$-H_2O$ 120, dec >140
325	$BaS_2O_6(O_2)$ ($\cdot 4H_2O$) Barium peroxodisulfate	329.45	dec hydr 100
326	$Ba(SO_3S)$ ($\cdot H_2O$) Barium thiosulfate	249.46	dec hydr 100
327	BaSe Barium selenide	216.29	mp 1780
328	$BaSeO_4$ Barium selenate	280.28	dec ca. 1500
329	$BaSiO_3$ ($\cdot 6H_2O$) Barium metasilicate	213.41	$-H_2O$ t, mp >1600
330	BaTe Barium telluride	264.93	mp 1510
331	$BaTeO_4$ ($\cdot 4H_2O$) Barium metatellurate	328.92	$-H_2O$ 400
332	$(BaTi)O_3$ Barium-titanium trioxide	233.20	mp 1616
333	$Ba_2V_2O_7$ Barium heptaoxodivanadate(V)	488.53	mp 863
334	$BaWO_4$ Barium wolframate	385.17	mp 1475, bp 1730
335	Ba_3XeO_6 Barium hexaoxoxenonate(VI)	639.27	dec 250
336	$(BaZr)O_3$ Barium-zirconium trioxide	276.55	mp ca. 2690
337	**Be Beryllium**	9.012	mp 1287, bp 2507
338	$(BeAl_2)O_4$ Beryllium-dialuminum tetraoxide	126.97	mp 1870
339	$BeBr_2$ ($\cdot 4H_2O$) Beryllium bromide	168.82	mp 509, bp 540
340	Be_2C Diberyllium carbide	30.04	mp 2150 dec
341	$Be(CH_3COO)_2$ Beryllium acetate	127.10	mp 295 dec
342	$[Be(C_5H_5)_2]$ Bis(cyclopentadienyl)beryllium	139.20	subl >45
343	$Be_4(CH_3COO)_6O$ Tetraberyllium hexaacetate-oxide	406.31	mp 285, bp 331
344	$BeCO_3$ ($\cdot 4H_2O$) Beryllium carbonate	69.02	$-H_2O$ 100, dec 180
345	$[Be(C_2O_4)_2],Be$ ($\cdot 3H_2O$) Beryllium dioxalatoberyllate(II)	194.06	$-H_2O$ 220; 350 → BeO
346	$BeCl_2$ Beryllium chloride	79.92	mp 415, bp 550
347	$BeCl_2 \cdot 4H_2O$	151.98	mp 96, dec 176
348	$Be(ClO_4)_2$ ($\cdot 4H_2O$) Beryllium perchlorate	279.97	dec hydr 250
349	$Be_4(ClO_4)_6O$ Tetraberyllium hexaperchlorate-oxide	648.74	mp 165
350	BeF_2 Beryllium fluoride	47.01	mp 803, bp 1175
351	$[BeF_4],K_2$ Potassium tetrafluoroberyllate(II)	163.20	mp 791
352	$[BeF_4],(NH_4)_2$ Ammonium tetrafluoroberyllate(II)	121.08	dec 230
353	$[BeF_4],Na_2$ Sodium tetrafluoroberyllate(II)	130.98	mp 578
354	BeH_2 Beryllium hydride	11.03	dec >100
355	BeI_2 ($\cdot 4H_2O$) Beryllium iodide	262.82	mp 480, bp 530
356	Be_3N_2 Triberyllium dinitride	55.05	mp ca. 2200 dec
357	$BeNH_4AsO_4$ ($\cdot 1.5H_2O$) Beryllium-ammonium arsenate	165.97	dec 400
358	$[Be(NH_3)_4]Cl_2$ Tetraamminneberyllium(II) chloride	148.04	dec >200
359	$BeNH_4PO_4$ ($\cdot H_2O$) Beryllium-ammonium ortho-phosphate	122.02	700 → $Be_2P_2O_7$
360	$Be(NO_3)_2$ ($\cdot 4H_2O$) Beryllium nitrate	133.02	mp hydr 60, dec hydr >160
361	$Be_4(NO_3)_6O$ Tetraberyllium hexanitrate-oxide	424.07	subl >125
362	BeO Beryllium oxide	25.01	mp 2580, bp 4260

(*Continued*)

No.	Color and density	Solubility and reactivity						
		water	alcohol	HCl	H_2SO_4	HNO_3	NaOH	$NH_3 \cdot H_2O$
318	wh, 4.36	r/+	i	+	+	+	+	+
319	yel	r/+	...	+	+	+	−	−
320	dk-yel	r/+	...	+	+	+	−	−
321	yel-red, 2.99	r/+	i	+	+	+	−	−
322	wh	i	...	+	+	+	i	i
323	wh, 4.50	i	...	i	i	i	i	i
324	wh hydr, 4.54	r	i	+	+	+	−	−
325	wh hydr	r/+	...	+	+	+	+	+
326	wh hydr, 3.45	i	i	+	+	−/+	−	−
327	wh, 5.02	+	...	+	+	+	+	+
328	wh, 4.61	i	...	+	+	−	−	−
329	wh, 4.40(2.58)	i/+	...	−/+	−/+	−/+	−	−
330	lt-yel, 5.13	i/+	...	+	+	+	i	i
331	wh, 4.48	i	...	−/+	−/+	−/+	−	−
332	wh, 6.08	−	i	−/+	−/+	−/+	−	−
333	wh	i	...	−/+	−/+	−/+	−/+	...
334	wh, 5.04	i	...	−/+	−/+	−/+	−/+	...
335	wh	i	...	+	+	+	−	−
336	wh	−	i	+	+	+	−	−
337	grey, 1.85	psv/+	−	+	+/psv	+/psv	−/+	−
338	yel-grn, 3.76	−	...	−	−	−	−/+	−/+
339	wh, 3.47	r/+	r	r	−/+	−	+	+
340	yel-red, 1.90	+	...	+	+	+	+	+
341	wh	i/+	i	−	−	−	−	...
342	wh	−/+	r	−	−/+	−/+	+	+
343	wh	−/+	−	+	+	+	+	+
344	wh	d/+	i	+	+	+	+	+
345	wh	r	...	−	−/+	−/+	+	+
346	wh, 1.90	r/+	r	r	−/+	−	+	+
347	wh	r/+	r	r	−/+	−	+	+
348	wh hydr	r	...	−	−	r	+	+
349	wh	+	...	+	+	+	+	+
350	wh, 1.99	r	d	r	r/+	r	+	+
351	wh	r	i	−/+	−/+	−/+	+	−
352	wh	r	i	−/+	−/+	−/+	+	−
353	wh	d	i	−/+	−/+	−/+	+	−
354	wh	+	+	+	+	+	+	+
355	wh, 4.33	r/+	r	r	−/+	−/+	+	+
356	wh	−/+	i	+	+	+	−/+	−
357	wh	i	i	−/+	−/+	−/+	−/+	−
358	wh	r/+	+	+	+	+	+	r
359	wh	i	i	−/+	−/+	−/+	−/+	−
360	wh hydr	r	r	r	r	r	+	+
361	wh	+	i	+	+	+	+	+
362	wh, 3.02	−	...	−/+	−/+	−/+	−/+	−

No.	Formula and name	M_r	Phase transition temperature
363	Be(OH)$_2$ Beryllium hydroxide	43.03	dec ca. 200
364	BeS Beryllium sulfide	41.08	dec 1800
365	BeSO$_4$ ($\cdot$2H$_2$O) Beryllium sulfate	105.07	mp 540, dec ca. 600
366	BeSO$_4\cdot$4H$_2$O	177.13	$-$H$_2$O ca. 400
367	BeSe Beryllium selenide	87.97	dec t
368	BeSeO$_4$ ($\cdot$4H$_2$O) Beryllium selenate	151.97	$-$H$_2$O 213, dec 730
369	BeSiO$_4$ Beryllium orthosilicate	110.11	mp 1560 dec
370	BeTe Beryllium telluride	136.61	dec t
371	**Bi Bismuth**	208.980	mp 271.44; bp 1564
372	BiAsO$_4$ Bismuth(III) arsenate	347.90	...
373	(Bi^{III}Bi^V)O$_4$ ($\cdot$2H$_2$O) Bismuth(III)-bismuth(V) tetraoxide	481.96	$-$H$_2$O 180, dec 305
374	BiBr$_3$ Bismuth(III) bromide	448.69	mp 218, bp 461
375	Bi(Br)O Bismuth bromide-oxide	304.88	dec ca. 500
376	Bi$_2$(C$_2$O$_4$)$_3$ ($\cdot$7H$_2$O) Busmuth(III) oxalate	682.01	$-$H$_2$O 150, dec 800
377	Bi$_2$CO$_3$(OH)$_4$ Dibismuth carbonate-tetrahydroxide	546.00	dec >670
378	BiCl$_3$ ($\cdot$H$_2$O) Bismuth(III) chloride	315.34	mp 232, bp 441
379	Bi$_{24}$Cl$_{28}$ 24-Bismuth 28-chloride	6008.20	mp 163, dec 300
380	Bi(Cl)O Bismuth chloride-oxide	260.43	dec ca. 600
381	BiClO$_4$(OH)$_2$ Bismuth perchlorate-dihydroxide	342.44	$-$H$_2$O 100
382	BiF$_3$ Bismuth(III) fluoride	265.97	mp 727, bp 900
383	BiF$_5$ Bismuth(V) fluoride	303.97	mp 151; bp 230
384	BiI$_3$ Bismuth(III) iodide	589.69	mp 407.7; bp 542 dec
385	Bi(I)O Bismith iodide-oxide	351.88	dec ca. 500
386	Bi$_2$(MoO$_4$)$_3$ Bismuth(III) molybdate	897.77	mp 643
387	Bi(NO$_3$)$_3$ ($\cdot$5H$_2$O) Bismuth(III) nitrate	394.99	mp hydr 75.5 dec
388	BiNO$_3$(OH)$_2$ Bismuth nitrate-dihydroxide	305.00	260 $\rightarrow$ Bi$_2$O$_3$
389	Bi$_2$O$_3$ Bismuth(III) oxide	465.96	mp 825, bp 1890
390	Bi$_2$O$_5$ ($\cdot n$H$_2$O) Bismuth(V) oxide	497.96	$-$H$_2$O 120, dec >350
391	Bi(O)F Bismuth oxide-fluoride	243.98	dec ca. 500
392	Bi(OH)$_3$ Bismuth(III) hydroxide	260.00	100 $\rightarrow$ BiO(OH)
393	BiO(OH) Bismuth metahydroxide	241.99	>150 $\rightarrow$ Bi$_2$O$_3$
394	BiPO$_4$ Bismuth(III) orthophosphate	303.95	dec t
395	Bi$_2$S$_3$ Bismuth(III) sulfide	514.16	mp 685 dec
396	Bi$_2$(SO$_4$)$_3$ Bismuth(III) sulfate	706.15	dec >400, mp 710 (p)
397	Bi$_2$Se$_3$ Bismuth(III) selenide	654.84	mp 706, bp 1007
398	Bi$_2$Te$_3$ Bismuth(III) telluride	800.76	mp 585, bp 1172
399	**Bk Berkelium**	247.070	mp ca. 1050, bp >2630
400	BkO$_2$ Berkelium(IV) oxide	279.07	dec 500
401	**Br$_2$ Dibromine**	159.808	mp $-$7.25; bp 59.82
402	Br$_2\cdot$7.67H$_2$O	297.92	mp 6 dec
403	[BrAg$_2$]NO$_3$ Di{silver(I)}bromine(-I) nitrate	357.64	mp 182
404	[Br(Br)Cl],Cs Cesium bromochlorobromate(I)	328.17	mp 191 (p)
405	[Br(Br)$_2$],Cs Cesium dibromobromate(I)	372.62	mp 180
406	[Br(Cl)$_2$],Cs Cesium dichlorobromate(I)	283.72	dec 150, mp 205 (p)
407	BrF Bromine monofluoride	98.90	mp $-$33, bp 20 dec
408	BrF$_3$ Bromine trifluoride	136.90	mp 8.8; bp 125.75
409	BrF$_5$ Bromine pentafluoride	174.89	mp $-$60.5; bp 40.76

(Continued)

No.	Color and density	Solubility and reactivity						
		water	alcohol	HCl	H_2SO_4	HNO_3	NaOH	$NH_3 \cdot H_2O$
363	wh, 1.92	i	...	+	+	+	+	i
364	lt-grey, 2.36	+	...	+	+	+	+	+
365	wh, 2.44	r	i	−	r/+	−	+	+
366	wh, 1.71	r	i	−	r	−	+	+
367	lt-grey, 4.32	−/+	...	+	+	+	−/+	−
368	wh hydr, 2.03	r	...	−	−	−	+	+
369	wh, 2.98	i	i	−	−/+	−	−/+	−
370	lt-grey, 5.09	−/+	...	+	+	+	−/+	−
371	wh, 9.790	−	−	−	−/psv	+/psv	−	−
372	wh, 7.14	i	i	−	−	−/+	−	−
373	brn	−	...	+	+	+	−	−
374	yel-orange, 5.6	+	r	−	−/+	−	+	+
375	wh, 8.1	−	i	+	+	+	−	−
376	wh	i	i	+	+	+	−	−
377	wh, 6.86	i	i	+	+	+	−	−
378	wh, 4.75	+	r	+	−	−	+	+
379	brn	+	+	+	+	+	+	+
380	wh, 7.72	−	...	+	+	+	−	−
381	wh	r	...	+	+	+	+	+
382	wh, 5.32	i/+	r	i/r	i/r	i/r	−	−
383	wh	+	...	+	+	+	+	+
384	dk-brn, 5.78	i/+	d	−/+	−/+	+	i	i
385	dk-red, 7.92	−	i	+	+	+	−	−
386	yel, 6.07	i	...	+	+	+	−	−
387	wh hydr, 2.83	+	...	−	−	−	+	+
388	wh, 4.93	i	...	+	+	+	−	−
389	yel, 8.90	−	...	+	+	+	−	−
390	red, 5.10	−	...	+	+	+	+	−
391	wh, 7.5	−	i	+	+	+	−	−
392	wh, 4.36	i	...	+	+	+	i	i
393	wh	−	...	+	+	+	i	i
394	wh, 6.32	i	i	+	−/+	−/+	−/+	−
395	brn-blk, 6.38	i	...	−	−	−/+	−	−
396	wh, 5.08	+	...	−	−	−	+	+
397	dk-grey, 6.82	i	i	−/+	−/+	+	−	−
398	blk, 7.7	i	i	−/+	−/+	+	−	−
399	wh, 14.8	−/+	...	+	+/psv	+/psv	−	−
400	yel	−	...	+	+	+	−	−
401	dk-red lq, 3.12^{20}	r/+	+	−	−	−	+	+
402	dk-red, 1.49	+	+	+	+	+	+	+
403	wh	+	...	+	+	+	+	+
404	yel-red	+	...	+	+	+	+	+
405	dk-yel	+	...	+	+	+	+	+
406	yel	+	...	+	+	+	+	+
407	red lq	+	+	+	+	+	+	+
408	lt-yel lq, 2.843^9	+	+	+	+	+	+	+
409	cl lq, 2.57^0	+	+	+	+	+	+	+

No.	Formula and name	M_r	Phase transition temperature
410	[BrF$_4$],K Potassium tetrafluorobromate(III)	194.99	dec >350
411	BrO$_2$F Bromine dioxide-fluoride	130.90	mp −9, dec >50
412	[Br(O)$_2$(F)$_2$],K Potassium dioxodifluorobromate(V)	189.00	dec 87
413	**α-C Graphite**	12.011	mp ca. 3800; bp ca. 4000
414	**β-C Diamond**	12.011	1800 → α-C
415	CBr$_4$ Carbon tetrabromide	331.63	mp 92.5; bp 189.5 dec
416	CCl$_4$ Carbon tetrachloride	153.82	mp −22.96; bp 76.75
417	CCl$_2$O Carbon dichloride-oxide	98.92	mp −118.8; bp 7.56
418	CF$_4$ Carbon tetrafluoride	88.00	mp −183.6; bp −128
419	CH$_4$ Methane	16.04	mp −182.5; bp −161.6
420	C$_2$H$_2$ Acetylene	26.04	mp −81 (p); bp −83.8
421	C$_2$H$_4$ Ethylene	28.05	mp −169.15; bp −103.7
422	C$_6$H$_6$ Benzene	78.11	mp 5.51; bp 80.10
423	CH$_3$C(H)O Acetaldehyde	44.05	mp −123.5; bp 20.2
424	CH$_3$CN Acetonitrile	41.05	mp −45.72; bp 81.6
425	CH$_3$COOH Acetic acid	60.05	mp 16.75; bp 118.1
426	CHCl$_3$ Chloroform	119.38	mp −63.5; bp 61.1
427	C$_2$H$_8$N$_2$ Ethylenediamine	60.10	mp 8.5; bp 117
428	C$_5$H$_5$N Pyridine	79.10	mp −41.7; bp 115.2
429	CH$_3$OH Methanol	32.04	mp −93.9; bp 64.96
430	C$_2$H$_5$OH Ethanol	46.07	mp −114.15; bp 78.39
431	CI$_4$ Carbon tetraiodide	519.63	mp 171 dec
432	C$_2$N$_2$ Dicyan	52.04	mp −27.83; bp −21.15
433	C$_6$N$_6$ Hexacyan	156.11	mp 119, bp 262
434	(CN)Br Cyanogen bromide	105.92	mp 51.3; bp 61.3
435	(CN)Cl Cyanogen chloride	61.47	mp −6.9; bp 12.66
436	(CN)F Cyanogen fluoride	45.02	mp −82, bp −46
437	C(NH$_2$)$_2$O Carbamide	60.06	mp 132.7; dec t
438	(CN)I Cyanogen iodide	152.92	subl >45, mp 146 (p)
439	CO Carbon monooxide	28.01	mp −205.02; bp −191.5
440	CO$_2$ Carbon dioxide	44.01	subl −78.476
441	COF$_2$ Carbon oxide-difluoride	66.01	mp −114, bp −83.1
442	CS$_2$ Carbon disulfide	76.14	mp −111,9; bp 46.24
443	C$_3$S$_2$ Tricarbon disulfide	100.17	mp −1, dec 90
444	CSCl$_2$ Carbon sulfide-dichloride	114.98	bp 73.5
445	CS(NH$_2$)$_2$ Thiocarbamide	76.12	mp 176, dec t
446	CSO Carbon sulfide-oxide	60.08	mp −138.82; bp −50.24
447	CSe$_2$ Carbon diselenide	169.93	mp −45.5; bp 127
448	CSe(NH$_2$)$_2$ Selenocarbamide	123.02	mp 202 dec
449	C(Se)O Carbon selenide-oxide	106.97	mp −124.4; bp −21.7
450	C(Se)S Carbon selenide-sulfide	123.04	mp −85; bp 84.5
451	C(Te)S Carbon telluride-sulfide	171.68	mp −54; dec >20
452	**Ca Calcium**	40.078	mp 842, bp 1495
453	(CaAl$_2$)O$_4$ Calcium-dialuminum tetraoxide	158.04	mp 1602 dec
454	Ca$_3$As$_2$ Tricalcium diarsenide	270.08	dec >1300
455	Ca(AsO$_2$)$_2$ Calcium metaarsenite	253.92	dec >1200
456	Ca$_3$(AsO$_4$)$_2$ (·9H$_2$O) Calcium arsenate	398.07	mp 1500 dec

(Continued)

No.	Color and density	Solubility and reactivity						
		water	alcohol	HCl	H_2SO_4	HNO_3	NaOH	$NH_3 \cdot H_2O$
410	wh	+	...	+	+	+	+	+
411	cl lq	+	...	+	+	+	+	+
412	wh	+	...	+	+	+	+	+
413	blk, 2.27	−	−	−	−	−/+	−	−
414	wh, 3.52	−	−	−	−	−	−	−
415	lt-yel, 3.42	i/+	r	−	−/+	−/+	+	+
416	cl lq, 1.594[20]	d/+	∞	−	−	−	−/+	−
417	cl gas, 4.523	+	+	−	−	−	+	+
418	cl gas	i	...	−	−	−	−	−
419	cl gas, 0.7168	d	d	−	−	−	−	−
420	cl gas, 1.1716	d	r	−	−	−	−	−
421	cl gas, 1.245	d	r	−	−	−	−	−
422	cl lq, 0.879[20]	d	∞	−	−/+	−/+	−	−
423	cl lq, 0.788[20]	∞	∞	−/+	−/+	−/+	−	−
424	cl lq, 0.784[20]	∞	∞	−/+	−/+	−/+	−	−
425	cl lq, 1.049[20]	∞	∞	−	−	−/+	+	+
426	cl lq, 1.483[20]	d	∞	−	−/+	−/+	−	−
427	cl lq, 0.901[20]	r	r	+	+	+	−	−
428	cl lq, 0.984[20]	∞	∞	+	+	+	−	−
429	cl lq, 0.791[20]	∞	∞	−	−	−	−/+	−
430	cl lq, 0.7893[20]	∞	∞	−	−	−	−/+	−
431	red, 4.34	i/+	r/+	−	−/+	+	+	+
432	cl gas, 2.335	d/+	r/+	−/+	−/+	−/+	+	...
433	wh	+	r/+	−/+	−/+	−/+	+	−
434	wh, 2.015	+	r	+	+	+	+	+
435	cl gas, 1.222	+	r	+	+	+	+	+
436	cl gas	+	...	+	+	+	+	+
437	wh, 1.335	r	r	+	+	+	−/+	−
438	wh, 2.84	+	r	+	+	+	+	+
439	cl gas, 1.250	d	r	−	−	−	−/+	−
440	cl gas, 1.977	d	d	−	−	−	+	+
441	cl gas, ca. 2.95	+	+	+	+	+	+	+
442	cl lq, 1.261[22]	d/+	∞	−	−/+	+	+	+
443	red lq, 1.27[20]	+	∞	−	−/+	+	+	+
444	red lq, 1.509[15]	+	+	−	−/+	−/+	+	+
445	wh	r	r	−	−/+	−/+	−/+	−
446	cl gas, 2.72	d/+	r	−	−/+	−/+	+	...
447	yel lq, 2.66[25]	i/+	∞	−	−/+	+	+	+
448	wh	r	r	−	−/+	−/+	−/+	−
449	cl gas	+	r	−	−/+	+	+	+
450	yel lq, 1.99[20]	i	d	−	−/+	+	+	+
451	red lq	i	i	−	+	+	+	+
452	wh, 1.54	+	+	+	+	+	+	+
453	wh, 3.67	−/+	...	−/+	−/+	−/+	−/+	−
454	red, 3.03	−/+	...	−/+	−/+	−/+	−/+	−
455	wh	d	...	−/+	−/+	+	+	...
456	wh	i	...	i/r	i/+	i/r	−/+	i

No.	Formula and name	M_r	Phase transition temperature
457	CaB_6 Calcium hexaboride	104.94	mp 2235
458	$Ca(BO_2)_2$ ($\cdot 6H_2O$) Calcium metaborate	125.70	$-H_2O$ 350, mp 1162
459	$CaBr_2$ Calcium bromide	199.89	mp 742, bp 1830
460	$CaBr_2 \cdot 6H_2O$	307.98	mp 38.2; dec 180–200
461	$Ca(BrO_3)_2$ ($\cdot H_2O$) Calcium bromate	295.88	$-H_2O$ 180, dec >200
462	CaC_2 Calcium acetylide	64.10	mp 2160, dec >2200
463	$Ca(CH_3COO)_2$ ($\cdot 2H_2O$) Calcium acetate	158.17	$-H_2O$ 100, dec t
464	$CaCN_2$ Calcium cyanamide	80.10	dec 1150 ; mp 1300 (p)
465	$Ca(CN)_2$ Calcium cyanide	92.11	mp 640
466	$CaCO_3$ ($\cdot H_2O$) Calcium carbonate	100.09	ca. 900 $\rightarrow$ CaO
467	CaC_2O_4 ($\cdot H_2O$) Calcium oxalate	128.10	$-H_2O$ 200, dec t
468	$CaCl_2$ Calcium chloride	110.98	mp 782, bp ca. 1960
469	$CaCl_2 \cdot 2H_2O$	147.01	$-H_2O$ >200
470	$CaCl_2 \cdot 6H_2O$	219.07	mp 29.92 dec
471	$Ca(ClO)_2$ ($\cdot 3H_2O$) Calcium hypochlorite	142.98	mp hydr 86, dec hydr 180
472	$Ca(ClO_2)_2$ Calcium chlorite	174.98	dec 450
473	$Ca(ClO_3)_2$ ($\cdot 2H_2O$) Calcium chlorate	206.98	$-H_2O$ 110, dec 340
474	$Ca(ClO_4)_2$ ($\cdot 4H_2O$) Calcium perchlorate	238.98	$-H_2O$ 120; dec >300
475	$CaCrO_4$ ($\cdot 2H_2O$) Calcium chromate	156.07	$-H_2O$ 200, mp ca. 1000
476	$(CaCr_2)O_4$ Calcium-dichromium tetraoxide	208.07	mp 2170
477	CaF_2 Calcium fluoride	78.07	mp 1419, bp ca. 2530
478	$(CaFe_2)O_4$ Calcium-diiron tetraoxide	215.77	mp 1220
479	CaH_2 Calcium hydride	42.09	mp ca. 1000 dec
480	$Ca(HCOO)_2$ ($\cdot 2H_2O$) Calcium formate	130.11	dec 400–495
481	$CaHPO_4$ Calcium hydroorthophosphate	136.06	>900 $\rightarrow$ $Ca_2P_2O_7$
482	$CaHPO_4 \cdot 2H_2O$	172.09	>360 $\rightarrow$ $Ca_2P_2O_7$
483	$Ca(H_2PO_4)_2$ ($\cdot H_2O$) Calcium dihydroorthophosphate	234.05	$-H_2O$ 109, dec 150–200
484	$Ca(HS)_2$ ($\cdot 6H_2O$) Calcium hydrosulfide	106.23	dec hydr >20
485	CaI_2 Calcium iodide	293.89	mp 783, bp 1760
486	$CaI_2 \cdot 6H_2O$	401.98	dec ca. 40
487	$Ca(IO_3)_2$ ($\cdot H_2O$) Calcium iodate	389.88	dec 540
488	$CaMg(CO_3)_2$ Calcium-magnesium carbonate	184.40	dec ca. 750
489	$CaMgSiO_4$ Calcium-magnesium orthosilicate	156.47	mp ca. 1500
490	$CaMg(SiO_3)_2$ Calcium-magnesium metasilicate	216.55	mp 1392
491	$Ca(MnO_4)_2$ ($\cdot 4H_2O$) Calcium permanganate	277.95	dec ca. 200
492	$CaMoO_4$ Calcium molybdate	200.01	mp 1520
493	$Ca(N_3)_2$ Calcium azide	124.12	dec ca. 140
494	Ca_3N_2 Tricalcium dinitride	148.25	mp 1195
495	$CaNH_4AsO_4$ ($\cdot 6H_2O$) Calcium-ammonium arsenate	197.04	dec hydr 140
496	$[Ca(NH_3)_6]I_2$ Hexaamminecalcium(II) iodide	396.07	dec 96
497	$Ca(NO_2)_2$ ($\cdot 4H_2O$) Calcium nitrite	132.09	$-H_2O$ 100
498	$Ca(NO_3)_2$ Calcium nitrate	164.09	mp 561 dec
499	$Ca(NO_3)_2 \cdot 4H_2O$	236.15	mp 42.5; $-H_2O$ 208
500	CaO Calcium oxide	56.08	mp ca. 2614, bp 2850
501	CaO_2 ($\cdot 8H_2O$) Calcium peroxide	72.08	$-H_2O$ >130, dec >250
502	$Ca(OH)_2$ Calcium hydroxide	74.09	580 $\rightarrow$ CaO
503	Ca_3P_2 Tricalcium disphosphide	182.18	dec >1250

No.	Color and density	Solubility and reactivity						
		water	alcohol	HCl	H_2SO_4	HNO_3	NaOH	$NH_3 \cdot H_2O$
457	dk-grey, 2.33	−	i	−	−	+	−	−
458	wh hydr, 2.70(1.88)	+	...	+	+	+	+	−
459	wh, 3.35	r	r	r	+	r	−/+	−
460	wh, 2.30	r	d	r	+	r	+	−
461	wh hydr, 3.33	r	...	−/+	+	−	+	−
462	wh, 2.22	+	i	+	+	+	+	+
463	wh	r	d	r	+	r	+	+
464	wh, 2.29	r/+	i	+	+	+	+	−
465	wh	r/+	...	r/+	+	r	r/+	...
466	wh, 2.93(2.42)	i	...	+	+	+	i	i
467	wh hydr, 2.2	i	...	i/+	−/+	−/+	−/+	i
468	wh, 2.51	r	r	r	+	r	+	+
469	wh, 2.15	r	r	r	+	r	+	+
470	wh, 1.68	r	r	r	+	r	+	+
471	wh, 2.35(2.1)	r/+	+	−/+	−/+	+	+	+
472	wh, 2.71	+	i	+	+	+	+	+
473	wh hydr, 2.71	r	r	−/+	+	r	+	−
474	wh, 2.65	r	r	r	+	−	+	−
475	yel	r	r	+	+	+	+	−
476	dk-grn, 4.8	−	...	−	−/+	−	−/+	−
477	wh, 3.18	i	...	i	i/+	i/r	i	i
478	dk-red, 5.08	−	...	−	−/+	−	−/+	−
479	wh, 1.90	+	+	+	+	+	+	+
480	wh, 2.02	r	i	r	+	−/+	+	+
481	wh, 2.89	d/+	i	+	+	r	+	...
482	wh, 2.31	d/+	i	+	+	r	+	...
483	wh hydr, 2.22	d/+	...	+	+	r	+	...
484	wh	r/+	i	+	+	+	+	−
485	wh, 3.96	r	r	r	+	+	+	+
486	lt-yel, 2.55	r	r	r	+	+	+	+
487	wh, 4.52	d	i	−/+	−/+	d	−/+	...
488	wh, 2.86	i	...	+	+	+	−/+	−
489	wh, 3.2	i	...	−/+	−/+	+	+	−
490	wh, 3.28	i	...	+	+	+	−/+	−
491	dk-red, 2.4	r	+	r/+	+	r/+	r/+	+
492	wh, 4.46	i/+	i	+	+	+	−	−
493	wh	r	d	+	+	+	+	−/+
494	dk-brn, 2.62	+	d	+	+	+	+	+
495	wh hydr, 1.91	d	...	+	+	+	+	−
496	wh	r/+	r	+	+	+	−	r
497	wh, 2.23(1.67)	r	r	−/+	+	−/+	+	−
498	wh, 2.36	r	r	r	+	r	+	−
499	wh, 1.86	r	r	r	+	r	+	+
500	wh, 3.35	+	...	+	+	+	+	+
501	wh, 2.92(1.70)	−/+	i	+	+	+	−/+	...
502	wh, 2.08	d	i	+	+	+	d	d
503	red-brn, 2.51	+	i	+	+/−	+/−	+	+

No.	Formula and name	M_r	Phase transition temperature
504	$Ca(PH_2O_2)_2$ Calcium phosphinate	170.05	dec >200
505	$Ca(PO_3)_2$ Calcium metaphosphate	198.02	mp 984
506	$Ca_2P_2O_7$ ($\cdot 5H_2O$) Calcium diphosphate	254.10	mp 1358
507	$Ca_3(PO_4)_2$ Calcium orthophosphate	310.17	mp 1670
508	$Ca(PO_3F)$ ($\cdot 2H_2O$) Calcium fluoroorthophosphate	138.05	$-H_2O$ 170, dec 400
509	$Ca_5(PO_4)_3F$ Pentacalcium triorthophosphate-fluoride	504.30	dec ca. 1800
510	CaS Calcium sulfide	72.14	mp ca. 2450 dec
511	$CaSO_3$ ($\cdot 2H_2O$) Calcium sulfite	120.14	$-H_2O$ 100, dec 550
512	$CaSO_4$ Calcium sulfate	136.14	mp 1450 dec
513	$CaSO_4 \cdot 2H_2O$	172.17	128 → $CaSO_4 \cdot 0.5H_2O$
514	$CaSO_4 \cdot 0.5H_2O$	145.15	$-H_2O$ 163
515	$Ca(SO_3S)$ ($\cdot 6H_2O$) Calcium thiosulfate	152.21	dec hydr ca. 250
516	CaSe Calcium selenide	119.04	mp ca. 1400 dec
517	$CaSeO_4$ ($\cdot 2H_2O$) Calcium selenate	183.03	dec ca. 1350
518	$CaSi_2$ Calcium disilicide	96.25	dec ca. 1000
519	$CaSiO_3$ ($\cdot H_2O$) Calcium metasilicate	116.16	mp 1544
520	Ca_2SiO_4 ($\cdot H_2O$) Calcium orthosilicate	172.24	mp 2130
521	Ca_3SiO_5 Calcium pentaoxosilicate(IV)	228.32	mp 2070 dec
522	$Ca_3Si_2O_7$ ($\cdot 3H_2O$) Calcium disilicate	288.40	mp 1464
523	CaTe Calcium telluride	167.68	dec ca. 1600
524	$CaTeO_3$ ($\cdot 2H_2O$) Calcium tellurite	215.68	dec >960
525	$CaTeO_4$ Calcium metatellurate	231.67	mp ca. 1000
526	$(CaTi)O_3$ Calcium-titanium trioxide	135.96	mp 1980
527	$CaWO_4$ Calcium wolframate	287.92	mp 1580
528	$(CaZr)O_3$ Calcium-zirconium trioxide	179.30	mp ca. 2350
529	**Cadmium**	112.411	mp 321.108; bp 766.5
530	Cd_3As_2 Tricadmium diarsenide	487.08	mp 721
531	$CdBr_2$ ($\cdot 2H_2O$) Cadmium(II) bromide	272.22	mp 565, bp 863
532	$Cd(CH_3COO)_2$ ($\cdot 2H_2O$) Cadmium(II) acetate	230.50	$-H_2O$ 130, mp 256 dec
533	$Cd(CN)_2$ Cadmium(II) cyanide	164.45	dec >200
534	$[Cd(CN)_4],K_2$ Potassium tetracyanocadmate(II)	294.68	mp ca. 450
535	$CdCO_3$ Cadmium(II) carbonate	172.42	>300 → CdO
536	CdC_2O_4 ($\cdot 3H_2O$) Cadmium(II) oxalate	200.43	dec 340
537	$CdCl_2$ ($\cdot 2.5H_2O$) Cadmium(II) chloride	183.32	mp 568.5; bp 964
538	$Cd(ClO_3)_2$ ($\cdot 2H_2O$) Cadmium(II) chlorate	279.31	mp hydr 80
539	CdF_2 Cadmium(II) fluoride	150.41	mp 1072, bp 1753
540	$Cd(H_2PO_4)_2$ ($\cdot 2H_2O$) Cadmium(II) dihydroorthophosphate	306.38	dec hydr 100
541	CdI_2 Cadmium(II) iodide	366.22	mp 388, bp 744
542	$Cd(IO_3)_2$ ($\cdot 6H_2O$) Cadmium(II) iodate	462.21	dec 800
543	$Cd(MnO_4)_2$ ($\cdot 6H_2O$) Cadmium(II) permanganate	350.28	dec hydr 95
544	$CdMoO_4$ Cadmium(II) molybdate	272.35	mp ca. 1600
545	$Cd(NH_2)_2$ Cadmium(II) amide	144.46	dec 120
546	$Cd(NH_4)_2(SO_4)_2$ ($\cdot 6H_2O$) Cadmium(II)-diammonium sulfate	340.61	$-H_2O$ 100
547	$Cd(NO_3)_2$ Cadmium(II) nitrate	236.42	mp 353; 700 → CdO
548	$Cd(NO_3)_2 \cdot 4H_2O$	308.48	mp 59.3; $-H_2O$ 132

(Continued)

No.	Color and density	Solubility and reactivity						
		water	alcohol	HCl	H_2SO_4	HNO_3	NaOH	$NH_3 \cdot H_2O$
504	lt-grey	r	i	−	+	−/+	+	+
505	wh, 2.82	i	...	+	−/+	+	−	−
506	wh, 3.09(2.25)	i	i	−	+	+	−	−
507	wh, 3.14	i	i	i/r	+	i/r	−	−
508	wh	d	...	−	+	+	+	+
509	wh, 3.2	i	i	i/+	+	i/r	−	−
510	wh, 2.59	d/+	...	+	+	+	+	+
511	wh	i	...	+	+	+	−	−
512	wh, 2.96	d	...	−	d/r	−	−	−
513	wh, 2.32	d	...	−	d/r	−	−	−
514	wh, 2.70	d	...	−	d/r	−	−	−
515	wh hydr, 1.87	r/+	r	+	+	+	+	+
516	wh, 3.82	i/+	...	+	+	+	+	−/+
517	wh, 2.93(2.68)	r	...	−	+	−	−/+	−
518	dk-grey, 2.5	−/+	−	+	+	+	+	...
519	wh, 2.92	i	i	+	+	+	−/+	−
520	wh	i/+	i	+	+	+	−/+	−
521	wh	i/+	i	+	+	+	−/+	−
522	wh	i	i	+	+	+	−	−
523	wh, 7.59	i/+	...	+	+	+	+	−/+
524	wh	d	...	+	+	+	+	...
525	wh	i	i	−/+	−/+	−/+	−	−
526	wh, 4.10	−	...	−	−/+	−/+	−	−
527	wh, 6.01	d	i	−/+	−/+	−/+	−	−
528	wh, 4.78	−	...	−/+	+	−/+	−	−
529	wh, 8.642	−	−	+	+	+	−	−
530	dk-grey, 6.21	−	i	−/+	−/+	−/+	−	−
531	wh, 5.20	r	r	−	−/+	−	+	+
532	wh, 2.34(2.01)	r	r	r	−	−	+	+
533	wh, 2.23	d	...	−	−	−	+	+
534	wh, 1.85	r	r	−	−/+	−/+	−/+	−
535	wh, 4.96	i	...	+	+	+	−	−/+
536	wh, 3.32	i	i	+	+	+	−	+
537	wh, 4.05(3.33)	r	d	r	−	−	+	+
538	wh hydr, 2.28	r	r	−/+	−/+	−/+	+	+
539	wh, 6.33	r	r	r	−	−	+	+
540	wh hydr, 2.74	r	i	−	−	−	+	+
541	wh, 5.67	r	r	r	−/+	+	+	+
542	wh, 6.43	d	i	+	−/+	−	+	+
543	viol hydr, 2.81	r	+	−/+	−	−	+	+
544	wh, 5.35	i	i	−/+	−/+	−/+	−	+
545	wh, 3.05	+	+	+	+	+	+	+/r
546	wh hydr, 2.06	r	i	−	r	−	+	+
547	wh	r	r	r	r	r	+	+
548	wh, 2.46	r	r	r	r	r	+	+

No.	Formula and name	M_r	Phase transition temperature
549	CdO Cadmium(II) oxide	128.41	subl ca. 900 dec
550	Cd(OH)$_2$ Cadmium(II) hydroxide	146.43	>300 → CdO
551	Cd$_3$P$_2$ Tricadmium diphosphide	399.18	mp 700
552	Cd$_2$P$_2$O$_7$ (·2H$_2$O) Cadmium(II) diphosphate	398.76	mp 900
553	CdS Cadmium(II) sulfide	144.48	mp ca. 1480
554	CdSO$_4$ Cadmium(II) sulfate	208.47	mp 1135 dec
555	CdSO$_4$·2.67H$_2$O	256.57	dec >475
556	CdSb Cadmium monostibide	234.16	mp 456
557	CdSe Cadmium(II) selenide	191.37	mp 1250
558	CdSeO$_4$ (·2H$_2$O) Cadmium(II) selenate	255.37	−H$_2$O 170
559	CdSiO$_3$ Cadmium(II) metasilicate	188.49	mp 1242
560	Cd$_2$SiO$_4$ Cadmium(II) orthosilicate	316.90	mp 1252
561	CdTe Cadmium(II) telluride	240.01	mp 1090
562	**Ce Cerium**	140.115	mp 804, bp ca. 3450
563	CeB$_6$ Cerium hexaboride	204.98	mp 2190 dec
564	CeBr$_3$ (·7H$_2$O) Cerium(III) bromide	379.83	mp 735, bp 1560
565	CeC$_2$ Cerium dicarbide	164.14	mp 2250
566	Ce$_2$(C$_2$O$_4$)$_3$ (·10H$_2$O) Cerium(III) oxalate	544.28	dec hydr >500
567	CeCl$_3$ (·7H$_2$O) Cerium(III) chloride	246.47	mp 822, bp 1650
568	CeF$_3$ Cerium(III) fluoride	197.11	mp 1430, bp 2180
569	CeF$_4$ Cerium(IV) fluoride	216.11	mp >650
570	CeF$_4$·H$_2$O	234.12	dec 270
571	CeI$_3$ (·9H$_2$O) Cerium(III) iodide	520.83	mp 755, bp 1400
572	Ce$_2$(MoO$_4$)$_3$ Cerium(III) molybdate	760.04	mp 973
573	CeNH$_4$(SO$_4$)$_2$ (·4H$_2$O) Cerium(III)-ammonium sulfate	350.28	−H$_2$O 150
574	Ce(NO$_3$)$_3$ (·6H$_2$O) Cerium(III) nitrate	326.13	mp hydr 39, dec hydr >200
575	Ce(NO$_3$)$_3$OH (·3H$_2$O) Cerium trinitrate-hydroxide	343.13	dec hydr 200–550
576	CeO$_2$ (·nH$_2$O) Cerium(IV) oxide	172.11	−H$_2$O t, mp 2700 (p)
577	Ce$_2$O$_3$ Cerium(III) oxide	328.23	mp 2180
578	Ce(OH)$_3$ Cerium(III) hydroxide	191.14	dec 400–500
579	CePO$_4$ Cerium(III) orthophosphate	235.09	...
580	Ce$_2$S$_3$ Cerium(III) sulfide	376.43	mp ca. 2000
581	Ce(SO$_4$)$_2$ (·8H$_2$O) Cerium(IV) sulfate	332.24	−H$_2$O 195, dec 550
582	Ce$_2$(S)O$_2$ Dicerium sulfide-dioxide	344.29	mp ca. 1950
583	Ce$_2$(SO$_4$)$_3$ (·8H$_2$O) Cerium(III) sulfate	568.42	−H$_2$O 450, dec 850
584	Ce$_2$(SeO$_4$)$_3$ (·8H$_2$O) Cerium(III) selenate	709.10	...
585	Ce$_2$(WO$_4$)$_3$ Cerium(III) wolframate	1023.77	mp 1079
586	**Cf Californium**	251.080	mp 900; bp 1227
587	CfBr$_3$ Californium(III) bromide	490.79	mp 700
588	CfCl$_3$ Californium(III) chloride	357.44	mp 545
589	**Cl$_2$ Dichlorine**	70.906	mp −101.03; bp −34.1
590	Cl$_2$·5.75H$_2$O	174.49	mp 9.6 dec
591	[ClAg$_2$]NO$_3$ Di{silver(I)}chlorine(−I) nitrate	313.19	mp 160
592	ClF Chlorine monofluoride	54.45	mp −155.6; bp −100.1
593	ClF$_5$ Chlorine pentafluoride	130.44	mp −103, bp −14
594	Cl$_2$F$_6$ Dichlorine hexafluoride	184.89	mp −76.31; bp 11.76
595	[ClF$_4$],NO Nitrosyl tetrafluorochlorate(III)	141.45	mp >−25 dec

(Continued)

No.	Color and density	Solubility and reactivity						
		water	alcohol	HCl	H_2SO_4	HNO_3	NaOH	$NH_3 \cdot H_2O$
549	brn, 8.15	−	...	+	+	+	−/+	−
550	wh, 4.79	i	...	+	+	+	i/+	+
551	grey-grn, 5.60	−	...	+	+	+	−/+	−/+
552	wh hydr, 4.97	i	i	+	+	+	−	−/+
553	yel, 4.82	i	...	−/+	−/+	+	−	−
554	wh, 4.69	r	i	r	r	r	+	+
555	wh, 3.09	r	i	r	r	r	+	+
556	grey	−	i	−	−/+	−/+	−	−
557	red-brn, 5.81	i	...	+	+	+	−	−
558	wh hydr, 3.63	r	i	r	r	r	+	+
559	wh, 4.93	i	i	−/+	−/+	−/+	−/+	−
560	wh	i	i	+	−/+	−/+	+	+
561	dk-brn, 6.20	i	...	−	−/+	+	−	−
562	grey, 6.668	−/+	−	+	+	+	−	−
563	lt-bl	−	...	−/+	−/+	−/+	+	−/+
564	wh	r/+	r	−/+	−/+	−	+	+
565	red, 5.23	+	...	+	+	+	+	+
566	wh hydr	i	i	−/+	i/+	−/+	i	−
567	wh, 3.97(3.92)	r/+	r	r	−	−	+	+
568	wh, 6.16	i	i	i	−/+	−	−	−
569	wh, 4.77	i	...	+	−/+	−/+	+	−
570	wh, 4.76	i	...	+	−/+	−/+	+	−
571	yel	r/+	r	r	−/+	−/+	+	+
572	yel, 4.83	i	i	−/+	−/+	−/+	−/+	−
573	wh hydr, 2.52	r	...	r	r	r	+	+
574	wh hydr	r	r	−	−	r	+	+
575	red hydr	r/+	...	+	+	+	+	+
576	lt-yel, 7.13	−	...	−/+	−/+	−	−	−
577	yel, 6.86	−/+	...	+	+	+	−	−
578	wh	i	...	+	+	+	i	i
579	red-yel, 5.22	i	−	+	+	+	−	−
580	dk-viol, 5.02	i/+	...	+	+	+	−	−
581	dk-yel, 3.91	r/+	...	+	+	r	+	+
582	brn-blk, 5.99	−	...	+	+	+	−	−
583	wh, 3.91(2.89)	r/d	...	r	r	r	+	+
584	wh, 4.46	r	i	r	r/+	r/+	+	+
585	yel, 6.77	i	...	−/+	−/+	−/+	−/+	−
586	wh	+	...	+	+	+	+	+
587	...	r/+	...	−	−/+	−	+	+
588	...	r/+	...	−	−	−	+	+
589	yel-grn gas, 3.214	+	+	+	+	+	+	+
590	yel, 1.29	+	+	+	+	+	+	+
591	wh	+	...	+	+	+	+	+
592	cl gas	+	+	+	+	+	+	+
593	cl gas	+	+	+	+	+	+	+
594	grn lq, 1.866[10]	+	+	+	+	+	+	+
595	wh	+	+	+	+	+	+	+

No.	Formula and name	M_r	Phase transition temperature
596	Cl_3N Trichlorine nitride	120.37	mp −27, dec 70
597	ClO_2 Chlorine dioxide	67.45	mp −59.5; bp 11,0
598	Cl_2O Dichlorine oxide	86.91	mp −116, bp 2.2
599	Cl_2O_6 Dichlorine hexaoxide	166.90	mp 3.5; dec >20
600	Cl_2O_7 Dichlorine heptaoxide	182.90	mp −90, bp 83
601	$Cl(O)F_3$ Chlorine oxide-trifluoride	108.45	mp −42, bp 29
602	ClO_2F Chlorine dioxide-fluoride	86.45	mp −115, bp −6
603	ClO_2F_3 Chlorine dioxide-trifluoride	124.45	mp −81, bp −22
604	ClO_3F Chlorine trioxide-fluoride	102.45	mp −146; bp 46.7
605	**Cm Curium**	247.070	mp 1345, bp ca. 3200
606	$Cm_2(C_2O_4)_3$ (·$10H_2O$) Curium(III) oxalate	758.19	dec hydr ca. 360
607	CmF_3 Curium(III) fluoride	304.06	mp 1406
608	$Cm(NO_3)_3$ Curium(III) nitrate	433.08	dec t
609	CmO_2 Curium(IV) oxide	279.07	>380 → Cm_2O_3
610	Cm_2O_3 Curium(III) oxide	542.14	mp ca. 2270
611	$Cm(OH)_3$ Curium(III) hydroxide	298.09	dec t
612	**Co Cobalt**	58.933	mp 1494, bp 2960
613	$(CoAl_2)O_4$ Cobalt-dialuminum tetraoxide	176.89	mp 1960
614	$CoAs_2$ Cobalt diarsenide	208.78	850 → Co_2As
615	$CoAs_{3-x}$ Cobalt triarsenide	283.70	800 → $CoAs_2$
616	Co_2S Dicobalt arsenide	192.79	mp 958
617	$Co_3(AsO_4)_2$ (·$8H_2O$) Cobalt(II) arsenate	454.64	−H_2O 120, dec 1000
618	$Co(As)S$ Cobalt arsenide-sulfide	165.92	1000 → Co_2As,CoS
619	$CoBr_2$ Cobalt(II) bromide	218.74	mp 678, bp 927
620	$CoBr_2$·$6H_2O$	326.83	mp 48, −H_2O 130–140
621	$[Co(C_5H_5)_2]$ Bis(cyclopentadienyl)cobalt	189.12	mp ca. 174
622	$Co(CH_3COO)_2$ (·$4H_2O$) Cobalt(II) acetate	177.02	−H_2O 140, dec t
623	$Co(CH_3COO)_3$ Cobalt(III) acetate	236.07	dec 100
624	$[Co(C_2H_8N_2)_3]Br_3$ (·$3H_2O$) Tris(ethylenediamine)cobalt(III) bromide	478.95	mp hydr 271, dec >400
625	$[Co(C_5H_5)_2]OH$ Bis(cyclopentadienyl)cobalt(III) hydroxide	206.13	dec t
626	$[Co^{II}(CN)_6],Co_2$ (·$7H_2O$) Cobalt(II) hexacyanocobaltate(II)	332.91	−H_2O >280, dec 450
627	$[Co(CN)_6],K_3$ Potassium hexacyanocobaltate(III)	332.34	dec >250
628	$[Co(CN)_6],K_4$ Potassium hexacyanocobaltate(II)	371.43	dec t
629	$CoCO_3$ Cobalt(II) carbonate	118.94	350 → CoO
630	CoC_2O_4 (·$2H_2O$) Cobalt(II) oxalate	146.95	dec hydr t
631	$[Co_2(CO)_8]$ Octacarbonyldicobalt	341.95	mp 51, >60 → $[Co_4(CO)_{12}]$
632	$[Co_4(CO)_{12}]$ Dodecacarbonyltetracobalt	571.85	dec 300
633	$[Co(CO)_2C_5H_5]$ Dicarbonyl(cyclopentadienyl)cobalt	180.05	mp −22; bp 139.5
634	$[Co(CO)_3NO]$ Tricarbonylnitrosylcobalt	172.97	mp −1; bp 78.5 dec
635	$CoCl_2$ Cobalt(II) chloride	129.84	mp 740, bp 1049
636	$CoCl_2$·$6H_2O$	237.93	mp 87, −H_2O 140
637	$Co(ClO_3)_2$ (·$6H_2O$) Cobalt(II) chlorate	225.83	mp hydr 50, dec 100
638	$Co(ClO_4)_2$ (·$6H_2O$) Cobalt(II) perchlorate	257.83	mp hydr 170
639	$(Co^{II}Co_2^{III})O_4$ Cobalt(II)-dicobalt(III) oxide	240.80	>900 → CoO

(Continued)

No.	Color and density	Solubility and reactivity						
		water	alcohol	HCl	H_2SO_4	HNO_3	NaOH	$NH_3 \cdot H_2O$
596	yel lq, 1.65^{20}	+	+	+	+	+	+	+
597	yel-grn gas, 3.214	i/+	+	+	+	+	+	+
598	dk-yel gas, 3.89	+	+	+	+	+	+	+
599	dk-red lq, 2.02^4	+	+	+	+	+	+	+
600	cl lq, 1.86^0	+	i	+	+	+	+	+
601	cl lq	+	+	+	+	+	+	+
602	cl gas	+	+	+	+	+	+	+
603	cl gas	+	+	+	+	+	+	+
604	cl gas	d	...	−	−	−	+	+
605	wh, 13.51	−/+	−	+	+	+	−	...
606	wh hydr	i	...	i	−/+	i/+	+	+
607	wh, 9.7	i	i	−	i/r	i/r	i/+	...
608	wh	r	...	r	r	r	+	+
609	blk	−	...	+	+	+	−	−
610	dk-grn	−	...	+	+	+	−	−
611	wh	i	...	+	+	+	i	i
612	grey, 8.84	−	−	+	+/psv	+/psv	−	−
613	bl	−	i	−	−/+	−	−/+	−
614	wh, 7.4	−	...	−/+	−/+	+	−	−
615	wh, 6.5	−	...	−/+	−/+	+	−	−
616	grey, 8.28	−	...	−	−	+	−	−
617	dk-red hydr, 2.95	i	i	i/r	i/r	i/r	i	−
618	lt-red, 6.33	−	i	−/+	−/+	−/+	−	−
619	grn, 4.91	r	r	r	−/+	−	+	+
620	red-viol, 2.46	r	r	r	−/+	−	+	+
621	viol	−	r	−	−/+	−/+	−	−
622	dk-red hydr, 1.71	r	d	r	−	−	+	+
623	blk-grn	+	+	+	+	+	+	+
624	yel, 1.85	r	...	−/+	−/+	−/+	−/+	r
625	yel	d	r	−	−	−	−	−
626	dk-pink hydr, 1.87	i	i	−	−/+	−/+	+	+
627	lt-yel, 1.88	r	i	−/+	−/+	−/+	−	−
628	red-brn, 2.04	r	i	−/+	−/+	−/+	−/+	−
629	red, 4.13	i	i	+	+	+	i	i
630	pink-red hydr, 3.02	i	...	i	−/+	−/+	i	−/+
631	orange-red, 1.73	−/+	r	−	−/+	+	−	−
632	blk	−	d	−	−/+	+	−	−
633	dk-red lq	−/+	r	−/+	−/+	−/+	−/+	−/+
634	dk-red lq	+	∞	+	+	+	+	+
635	sk-bl, 3.37	r	r	r	−/+	r	+	+
636	pink, 1.92	r	r	r	−/+	r	+	+
637	red hydr, 1.92	r/+	r	−/+	−/+	+	+	+
638	red, 3.33	r	r	−	−/+	−/+	+	+
639	blk-grey, 6.07	−	...	−	+	−	−	−

No.	Formula and name	M_r	Phase transition temperature
640	$(Co^{II}Co_2^{III})S_4$ Cobalt(II)-dicobalt(III) sulfide	305.06	mp 625; >680 → CoS
641	CoF_2 Cobalt(II) fluoride	96.93	mp 1127, bp ca. 1740
642	$CoF_2 \cdot 4H_2O$	168.99	$-H_2O$ 100, dec 200
643	CoF_3 ($\cdot 3.5H_2O$) Cobalt(III) fluoride	115.93	dec 350
644	$[CoH(CO)_4]$ Hydrotetracarbonylcobalt	171.98	mp -20, bp 10 dec
645	$Co(HCOO)_2$ ($\cdot 2H_2O$) Cobalt(II) formate	148.97	$-H_2O$ 140, dec 175
646	CoI_2 Cobalt(II) iodide	312.74	mp 516, bp 760
647	$CoI_2 \cdot 6H_2O$	420.83	mp 27, $-H_2O$ 130
648	$Co(IO_3)_2$ Cobalt(II) iodate	408.74	dec 200
649	$Co(IO_3)_2 \cdot 6H_2O$	516.83	mp 61 dec
650	Co_2N Dicobalt nitride	131.87	>250 → Co
651	$Co(NCS)_2$ ($\cdot 3H_2O$) Cobalt(II) thiocyanate	175.10	$-H_2O$ 105
652	$[Co(NCS)_4],(NH_4)_2$ ($\cdot 4H_2O$) Ammonium tetrakis(thiocyanato)cobaltate(II)	327.35	dec hydr t
653	$[Co(NH_3)_6]Br_2$ Hexaamminecobalt(II) bromide	320.93	dec 250
654	$[Co(NH_3)_6]Cl_2$ Hexaamminecobalt(II) chloride	232.03	>150 → $CoCl_2$
655	$[Co(NH_3)_6]Cl_3$ Hexaamminecobalt(III) chloride	267.48	dec >420
656	$[Co(NH_3)_4Cl_2]Cl$ ($\cdot H_2O$) Tetraamminedichlorocobalt(III) chloride	233.42	$-H_2O$ 100, dec >360
657	$[Co(NH_3)_5Cl]Cl_2$ Pentaamminechlorocobalt(III) chloride	250.45	dec >450
658	$[Co(NH_3)_6]I_2$ Hexaamminecobalt(II) iodide	414.93	dec >120
659	$[Co(NH_3)_6]I_3$ Hexaamminecobalt(III) iodide	541.83	dec < 100
660	$[Co(NH_3)_3(NO_2)_3]$ Triamminetrinitrocobalt	248.04	dec 150
661	$[Co(NH_3)_6](NO_3)_3$ Hexaamminecobalt(III) nitrate	347.13	dec >100
662	$[Co(NH_3)_5NO_2]Cl_2$ Hexaamminenitrocobalt(III) chloride	261.00	dec >210
663	$Co(NH_4)_2(SO_4)_2$ Cobalt(II)-diammonium sulfate	228.20	dec ca. 200
664	$Co(NH_4)_2(SO_4)_2 \cdot 6H_2O$	395.23	$-H_2O$ 80
665	$[Co(NH_3)_6]_2(SO_4)_3$ ($\cdot 5H_2O$) Hexaamminecobalt(III) sulfate	610.42	$-H_2O$ 150
666	$Co(NO_3)_2$ ($\cdot 6H_2O$) Cobalt(II) nitrate	182.94	mp hydr 55, >180 → $(Co^{II}Co_2^{III})O_4$
667	$[Co(NO)_3]$ Trinitrosylcobalt	148.95	dec 100
668	$[Co(NO_2)_6],K_3$ ($\cdot 1.5H_2O$) Potassiumhexanitro cobaltate(III)	452.26	dec >200
669	$[Co(NO_2)_6],K_2Na$ ($\cdot H_2O$) Dipotassium-sodium hexanitrocobaltate(II)	436.15	mp hydr 135
670	$[Co(NO_2)_6],Na_3$ Sodium hexanitrocobaltate(III)	403.93	dec >250
671	$Co_{1-x}O$ Cobalt(II) oxide	74.93	mp 1810; dec 2800
672	Co_2O_3 ($\cdot nH_2O$) Cobalt(III) oxide	165.86	hydr 250 → CoO(OH)
673	$Co(OH)_2$ Cobalt(II) hydroxide	92.95	>170 → CoO
674	CoO(OH) Cobalt metahydroxide	91.94	600 → $(Co^{II}Co_2^{III})O_4$
675	CoP Cobalt monophosphide	89.91	dec >600
676	Co_2P Dicobalt phosphide	148.84	dec >750
677	$Co_3(PO_4)_2$ ($\cdot 8H_2O$) Cobalt(II) orthophosphate	366.74	$-H_2O$ 200
678	CoS_{1+x} ($x = 0.04–0.13$) Cobalt(II) sulfide	91.00	mp 1118

(Continued)

No.	Color and density	Solubility and reactivity						
		water	alcohol	HCl	H_2SO_4	HNO_3	NaOH	$NH_3 \cdot H_2O$
640	dk-grey, 4.86	i	i	−/+	−/+	−/+	−	−
641	pink-red, 4.43	r/+	i	−	−/+	−	+	+
642	pink, 2.19	r/+	i	−	−/+	−/+	+	+
643	lt-brn, 3.88(2.31)	+	i	+	+	+	+	+
644	yel lq	+	...	+	+	+	+	+
645	red hydr, 2.13	r	d	r	−	−	+	+
646	grey-blk, 5.58	r	r	r	−/+	−/+	+	+
647	dk-red, 2.90	r	r	r	−/+	−/+	+	+
648	viol-sk-bl, 5.01	d	...	+	−/+	+	+	+
649	red, 3.69	d	...	+	−/+	+	+	+
650	grey-blk, 6.4	−/+	...	+	+	+	+	+
651	red-viol	r	r	−/+	−/+	−/+	+	+
652	bl hydr	r	r	−/+	−/+	−/+	+	+
653	dk-pink, 1.87	+	...	+	+	+	+	r
654	lt-red, 1.48	+	i	+	+	+	+	r/i
655	dk-brn, 1.70	r/+	i	+	+	+	+	r
656	grn, 1.86(1.85)	d	i	+	+	+	+	−/+
657	dk-red, 1.78	d/+	i	+	+	+	+	−/+
658	pink, 2.10	+	...	+	+	+	+	r
659	dk-red, 2.75	r	i	+	+	+	+	r/+
660	yel, 1.99	+	...	+	+	+	+	....
661	yel, 1.80	d	i	+	+	+	+	r
662	yel-brn, 1.80	r	i	+	+	+	+	−/+
663	bl-viol	r	i	−	−/+	−	+	+
664	red, 1.90	r	i	−	−/+	−	+	+
665	dk-yel, 1.80	d	i	+	+	+	+	r
666	red hydr, 1.87	r	r	−	−	r	+	+
667	blk	+	...	+	+	+	+	+
668	yel	d/+	i	−	−/+	−/+	...	...
669	yel hydr, 1.63	d	i	−	−/+	−/+	...	...
670	yel	r/+	d	+	+	+	−/+	−
671	dk-grn, 6.47	−	−	+	+	+	−/+	−
672	dk-brn hydr, 5.18	i	i	+	−/+	−/+	i	+
673	pink-red, 3.60	i	i	+	+	+	i/+	−/+
674	dk-brn, 4.60	i	i	+	−/+	−/+	i	−/+
675	grey-blk, 7.4	−/+	...	−/+	−/+	−/+	−/+	−
676	grey-blk, 6.4	−	...	−	−/+	−/+	−	−
677	pink, 2.59(2.77)	d	i	+	+	+	−/+	+
678	blk-grey, 5.45	i	i	−/+	−/+	−/+	−	−

No.	Formula and name	M_r	Phase transition temperature
679	Co(S$_2$) Cobalt(II) disulfide(2−)	123.07	mp 953 dec (→ CoS)
680	CoSO$_4$ (·7H$_2$O) Cobalt(II) sulfate	155.00	−H$_2$O 420, dec >600
681	CoSe Cobalt(II) selenide	137.89	mp 1055
682	CoSeO$_4$ (·7H$_2$O) Cobalt(II) selenate	201.89	dec hydr >500
683	CoSi Cobalt monosilicide	87.02	mp 1395
684	CoSi$_2$ Cobalt disilicide	115.11	mp 1327
685	Co$_2$Si Dicobalt silicide	145.95	mp 1332
686	Co$_2$SiO$_4$ Cobalt(II) orthosilicate	209.95	mp 1415
687	**Cr Chromium**	51.996	mp 1890, bp 2680
688	CrB Chromium monoboride	62.81	mp 2070
689	CrBr$_2$ Chromium(II) bromide	211.80	mp 842
690	CrBr$_3$ Chromium(III) bromide	291.71	subl 600–698
691	CrBr$_3$·6H$_2$O	399.80	dec t
692	[Cr(C$_5$H$_5$)$_2$] Bis(cyclopentadienyl)chromium	182.19	mp 173.5
693	[Cr(C$_6$H$_6$)$_2$] Dibenzenechromium	208.22	mp 284; >300 → Cr
694	[Cr(CN)$_6$],K$_3$ Potassium hexacyanochromate(III)	325.40	dec >150
695	[Cr(CO)$_6$] Hexacarbonylchromium	220.06	mp 155 (p), dec 120–200
696	CrCl$_2$ Chromium(II) chloride	122.90	mp 824, bp 1330
697	CrCl$_3$ Chromium(III) chloride	158.36	mp 1150, dec 1300
698	CrCl$_3$·6H$_2$O	266.45	mp 95, dec 300
699	Cr(Cl)O Chromium chloride-oxide	103.45	800 → Cr$_2$O$_3$,CrCl$_3$
700	CrCl$_2$O$_2$ Chromium dichloride-dioxide	154.90	mp −97, bp 117
701	CrCl$_3$O Chromium trichloride-oxide	174.35	>60 → CrCl$_2$O$_2$,CrCl$_3$
702	CrF$_2$ Chromium(II) fluoride	89.99	mp 894; bp 1820
703	CrF$_3$ Chromium(III) fluoride	108.99	mp ca. 1400
704	CrF$_4$ Chromium(IV) fluoride	127.99	mp ca. 200
705	CrF$_5$ Chromium(V) fluoride	146.99	mp 30
706	(Cr$_2$Fe)O$_4$ Dichromium-iron tetraoxide	223.84	mp 2200
707	[Cr(H$_2$O)$_4$Br$_2$]Br (·2H$_2$O) Tetraaquadibromochromium(III) bromide	399.80	dec hydr t
708	[Cr$_2$(H$_2$O)$_2$(CH$_3$COO)$_4$] Diaquatetraacetatodichromium	376.20	−H$_2$O 100
709	[Cr(H$_2$O)$_3$Cl$_3$] Triaquatrichlorochromium	212.40	dec >200
710	[Cr(H$_2$O)$_4$Cl$_2$] Tetraaquadichlorochromium	194.96	dec >113
711	[Cr(H$_2$O)$_4$Cl$_2$]Cl (·2H$_2$O) Tetraaquadichlorochromium(III) chloride	230.42	mp hydr 83; dec 650
712	[Cr(H$_2$O)$_5$Cl]Cl$_2$ (·H$_2$O) Pentaaquachlorochromium(III) chloride	248.43	dec hydr >50
713	[Cr(H$_2$O)$_4$(OH)$_2$] Tetraaquadihydroxochromium	158.07	dec >150
714	CrI$_2$ Chromium(II) iodide	305.81	mp 795, dec 870
715	CrI$_3$ Chromium(III) iodide	432.71	mp 857; >670 → CrI$_2$
716	CrI$_3$·9H$_2$O	594.84	dec 350
717	CrN Chromium mononitride	66.00	>1500 → Cr$_2$N
718	Cr$_2$N Dichromium nitride	118.00	mp 1650
719	[Cr(NCS)$_6$],K$_3$ (·4H$_2$O) Potassium hexakis(thiocyanato)chromate(III)	517.79	−H$_2$O 110; dec t
720	[Cr(NH$_3$)$_6$]Cl$_3$ Hexaamminechromium(III) chloride	260.54	dec >600

(Continued)

No.	Color and density	Solubility and reactivity						
		water	alcohol	HCl	H_2SO_4	HNO_3	NaOH	$NH_3 \cdot H_2O$
679	grey-blk, 4.27	i	i	−/+	−/+	−/+	−	−
680	red-pink, 3.71(1.95)	r	d	−	r	r	+	+
681	red-viol, 7.65	i	i	−/+	−/+	−/+	i	−
682	red hydr, 2.14	r	d	−	−	−	+	+
683	red, 6.30	−	...	+	−/+	−/+	−/+	−
684	dk-bl, 5.3	−	...	+	−/+	−/+	−/+	−
685	grey, 7.28	−	...	+	−/+	−/+	−/+	−
686	viol, 4.63	i	...	+	+	+	−/+	−/+
687	grey, 7.140	−	...	+	+/psv	+/psv	−	−
688	grey, 6.17	−	...	−	−/+	−/+	−/+	−
689	wh, 4.36	r	r	r	r/+	+	+	+
690	grn-blk, 4.25	r/+	r	−/+	−/+	−	+	+
691	viol, 5.4	r/+	r	−/+	−/+	−	+	+
692	red	−/+	r	−	−/+	+	−/+	...
693	brn-blk	+	r	+	+	+	+	+
694	lt-yel, 1.71	r	i	−	−/+	−/+	−	−
695	wh, 1.77	−	i	−	−	−/+	−	−
696	wh, 2.90	r	d	r	r/+	+	+	+
697	viol-red, 2.76	r/+	d	−/+	−	r	+	+
698	grey-sk-bl, 1.76	r/+	r	−/+	−	r	+	+
699	viol-grn, 3.50	−	−	−	−	−	−/+	−
700	dk-red lq, 1.911[20]	+	+	+	+	+	+	+
701	dk-red	+	...	+	+	+	+	+
702	grn, 4.11	d	i	−	−/+	−	+	+
703	grn, 3.78	d	i	−/+	−/+	−	+	+
704	brn, 2.89	+	i	+	+	+	+	+
705	red	+	+	+	+	+	+	+
706	brn-blk, 4.97	−	...	−/+	−/+	−/+	−/+	−
707	grn hydr	r	r	+/−	−/+	−	+	+
708	red ($t \rightarrow$ brn)	d/+	d	+	+	+	−	−
709	brn	+	...	+	+/−	+/−	+	+
710	dk-sk-bl	+	d	+	+	+	+	+
711	dk-grn hydr, 1.585	r	r	+/−	+/−	+/−	+	+
712	lt-grn hydr, 1.76	r	r	+/−	+/−	+/−	+	+
713	yel	i	...	+	+	+	i	i
714	red-brn, 5.02	r	...	r	r/+	+	+	+
715	blk, 4.92	r/+	...	i/+	i/+	i/+	+	+
716	viol	i	...	i/+	i/+	i/+	+	+
717	blk, 5.8	−	−	−	−/+	−/+	−	−
718	blk	−	−	−	−/+	−/+	−	−
719	red-viol hydr, 1.71	r	r	−	−/+	−/+	−	−
720	orange-yel, 1.59	r/+	d	+	+	+	+	−

No.	Formula and name	M_r	Phase transition temperature
721	[Cr(NH$_3$)$_5$Cl]Cl$_2$ Pentaamminechlorochromium(III) chloride	243.51	dec >550
722	[Cr(NH$_3$)$_2$(NCS)$_4$],NH$_4$ (·H$_2$O) Ammonium diamminetetrakis(thiocyanato)chromate(III)	336.43	−H$_2$O 100
723	CrNH$_4$(SO$_4$)$_2$ (·12H$_2$O) Chromium-ammonium sulfate	262.16	mp hydr 94, dec >100
724	[Cr(NO)$_4$] Tetranitrosylchromium	172.02	dec >20; mp 38.5 (p)
725	Cr(NO$_3$)$_3$ (·9H$_2$O) Chromium(III) nitrate	238.01	mp hydr 38.5; dec hydr >125
726	CrO Chromium(II) oxide	68.00	>700 → Cr$_2$O$_3$
727	CrO$_2$ Chromium(IV) oxide	83.99	dec 450
728	CrO$_3$ Chromium(VI) oxide	99.99	mp 195, dec 220–500
729	Cr$_2$O$_3$ Chromium(III) oxide	151.99	mp 2340, bp 3000
730	CrOF$_4$ Chromium oxide-tetrafluoride	143.99	mp 55
731	CrO$_2$F$_2$ Chromium dioxide-difluoride	121.99	mp 30
732	Cr(OH)$_3$ Chromium(III) hydroxide	103.02	100 → CrO(OH)
733	[Cr(OH)$_6$],Na$_3$ Sodium hexahydroxochromate(III)	223.01	dec >200
734	CrO(OH) Chromium metahydroxide	85.00	1000 → Cr$_2$O$_3$
735	CrP Chromium monophosphide	82.97	dec >700
736	[Cr(PF$_3$)$_6$] Hexakis(trifluorophosphorus)chromium	579.80	subl 20, mp 193
737	CrPO$_4$ Chromium(III) orthophosphate	146.97	dec >1000
738	CrPO$_4$·6H$_2$O	255.06	mp 100, −H$_2$O 300
739	CrS Chromium(II) sulfide	84.06	mp 1550
740	Cr$_2$S$_3$ Chromium(III) sulfide	200.19	mp ca. 1000, dec 1350
741	CrSO$_4$ (·7H$_2$O) Chromium(II) sulfate	148.06	dec hydr >100
742	Cr$_2$(SO$_4$)$_3$ Chromium(III) sulfate	392.18	dec >700
743	Cr$_2$(SO$_4$)$_3$·18H$_2$O	716.45	mp 80, −H$_2$O 115
744	CrTl(SO$_4$)$_2$ (·12H$_2$O) Chromium(III)-thallium(I) sulfate	448.50	mp hydr 92
745	**Cs Cesium**	132.905	mp 28.7; bp 667.6
746	CsAl(SO$_4$)$_2$ (·12H$_2$O) Cesium-aluminum sulfate	352.01	mp hydr 117
747	CsBr Cesium bromide	212.81	mp 636, bp 1297
748	CsBrO$_3$ Cesium bromate	260.81	mp 420 dec
749	Cs(CH$_3$COO) Cesium acetate	191.95	mp 194
750	CsCN Cesium cyanide	158.92	dec t
751	Cs$_2$CO$_3$ Cesium carbonate	325.82	mp 793 (p), dec >620
752	CsCl Cesium chloride	168.36	mp 645, bp 1302
753	CsClO$_3$ Cesium chlorate	216.36	mp 388, dec t
754	CsClO$_4$ Cesium perchlorate	232.35	dec 700
755	Cs$_2$CrO$_4$ Cesium chromate	381.80	mp 982
756	Cs$_2$Cr$_2$O$_7$ Cesium dichromate	481.80	dec ca. 600
757	CsCr(SO$_4$)$_2$ (·12H$_2$O) Cesium-chromium(III) sulfate	377.03	mp hydr 116
758	CsF (·1.5H$_2$O) Cesium fluoride	151.90	mp 703, bp 1251
759	CsFe(SO$_4$)$_2$ (·12H$_2$O) Cesium-iron(III) sulfate	380.88	dec hydr 175
760	Cs$_2$Fe(SO$_4$)$_2$ (·6H$_2$O) Dicesium-iron(II) sulfate	513.78	dec hydr t
761	CsGa(SO$_4$)$_2$ (·12H$_2$O) Cesium-gallium(III) sulfate	394.75	dec hydr 250
762	CsH Cesium hydride	133.91	mp ca. 400 (p)

(Continued)

No.	Color and density	Solubility and reactivity						
		water	alcohol	HCl	H_2SO_4	HNO_3	NaOH	$NH_3 \cdot H_2O$
721	dk-red, 1.69	r/+	d	i/+	+	+	+	+
722	dk-red	r/+	r	+	+	+	+	+
723	viol hydr, 1.72	d	r	−/+	d/r	−	+	+
724	blk-brn	+	r	+	+	+	+	+
725	dk-red hydr	r	r	−/+	r	r	+	+
726	blk	−	...	+	+	+	−	...
727	blk	−	...	−	−	−/+	−	−
728	red, 2.70	+	r/+	+	+	+	+	+
729	dk-grn, 5.21	−	−	−	−	−	−	−
730	red	−/+	+	+	+	+	+	+
731	red-viol	+	+	+	+	+	+	+
732	grey-sk-bl, 2.9	i	i	+	+	+	+	i
733	grn	+	i	+	+	+	+/r	+
734	grn	i	i	+	+	+	+	i
735	dk-grey, 5.7	−	...	−/+	−	−/+	−/+	−
736	wh	i/+	r	+	+	−/+	+	...
737	grey-brn, 3.05	i	i	−/+	−/+	−/+	−/+	−
738	grn-viol, 2.12	i	i	−/+	−/+	−/+	−/+	−
739	blk, 4.85	i	...	+	+	+	+	+
740	blk, 3.77	i	+	i/+	i/+	+	+	+
741	sk-bl hydr	r	d	r	−/...	+	+	+
742	lt-pink, 3.01	r	r	−/+	r	r	+	+
743	viol, 1.7	r	r	−/+	r	r	+	+
744	viol hydr, 2.39	r	i	−	r	−	+	+
745	yel, 1.873	+	+	+	+	+	+	+
746	wh hydr, 1.97	d	i	−/+	d/r	−	+	+
747	wh, 4.44	r	r	r	−/+	−	−	−
748	wh, 4.11	r	...	−	−	−	−	−
749	wh	r	r	r	r	r	r	r
750	wh, 2.93	r	i	r	r	r	r	r
751	wh	r	r	+	+	+	−	−
752	wh, 3.99	r	r	r	r/+	r	−	r
753	wh, 3.57	r	...	−/+	−	−	−	−
754	wh, 3.33	d	d	−	−	−	−	−
755	yel, 4.24	r	...	+	+	+	r	r
756	orange-red	d/r	i	r/+	r	r	+	+
757	viol hydr, 2.06	r	...	−/+	r	−	+	+
758	wh, 4.12	r	i	−	−	−	−	−
759	lt-viol hydr, 2.06	r	i	−	r	−	+	+
760	lt-grn hydr, 2.79	r	i	−	r	−/+	+	+
761	wh hydr, 2.11	d	i	−	r	−	+	+
762	wh, 3.41	+	+	+	+	+	+	+

No.	Formula and name	M_r	Phase transition temperature
763	$CsHCO_3$ Cesium hydrocarbonate	193.92	dec 175
764	$Cs(HCOO)$ Cesium formate	177.92	mp 265
765	$CsHSO_4$ Cesium hydrosulfate	229.98	ca. 220 $\rightarrow Cs_2S_2O_7$
766	CsI Cesium iodide	259.81	mp 632, bp 1280
767	$CsIO_3$ Cesium iodate	307.81	mp 565
768	$CsIO_4$ Cesium periodate	323.81	dec >900
769	$CsMnO_4$ Cesium permanganate	251.84	dec 320
770	$CsMn(SO_4)_2$ ($\cdot 12H_2O$) Cesium-manganese(III) sulfate	379.97	mp hydr 40 dec
771	CsN_3 Cesium azide	174.93	mp 326, dec 390
772	Cs_3N Tricesium nitride	412.72	dec t
773	$CsNH_2$ Cesium amide	148.93	mp 261, dec ca. 500
774	$CsNO_2$ Cesium nitrite	178.91	mp 398
775	$CsNO_3$ Cesium nitrate	194.91	mp 414, 585 $\rightarrow CsNO_2$
776	$CsNO_3 \cdot HNO_3$	257.92	mp 104; 650 $\rightarrow CsNO_2$
777	$CsNO_3 \cdot 2HNO_3$	320.93	mp 39; 670 $\rightarrow CsNO_2$
778	CsO_2 Cesium superoxide	164.90	mp 515, dec >400
779	CsO_3 Cesium ozonide	180.90	dec 82
780	Cs_2O Cesium oxide	281.81	>300 $\rightarrow Cs_2O_2$
781	Cs_2O_2 Cesium peroxide	297.81	mp 594; dec 640
782	$CsOH$ ($\cdot H_2O$) Cesium hydroxide	149.91	mp hydr 180, mp 346
783	Cs_2S ($\cdot 4H_2O$) Cesium sulfide	297.88	dec hydr 150
784	$Cs_2(S_2)$ Cesium disulfide(2−)	329.94	mp 460
785	$Cs_2(S_3)$ Cesium trisulfide(2−)	362.01	mp 217
786	$Cs_2(S_5)$ Cesium pentasulfide(2−)	426.14	mp 210
787	$Cs_2(S_6)$ Cesium hexasulfide(2−)	458.21	mp 185
788	Cs_2SO_4 Cesium sulfate	361.87	subl 900, mp 1019
789	$Cs_2S_2O_7$ Cesium disulfate	441.94	mp 280
790	Cs_3Sb Tricesium stibide	520.47	mp 725
791	Cs_2SeO_4 Cesium selenate	408.77	mp 985
792	$CsV(SO_4)_2$ ($\cdot 12H_2O$) Cesium-vanadium(III) sulfate	375.97	mp hydr 82, $-H_2O$ 230
793	**Cu Copper**	63.546	mp 1084.5; bp 2540
794	Cu_3As Tricopper arsenide	265.56	mp 830
795	$Cu_3(AsO_4)_2$ ($\cdot 4H_2O$) Copper(II) arsenate	468.40	$-H_2O$ 410, mp 1100
796	Cu_3AsS_4 Copper(I) tetrathioarsenate	393.82	dec t
797	$CuBr$ Copper(I) bromide	143.45	mp 504, bp 1350
798	$CuBr_2$ ($\cdot 4H_2O$) Copper(II) bromide	223.35	mp 498 dec
799	$Cu(BrO_3)_2$ ($\cdot 6H_2O$) Copper(II) bromate	319.35	$-H_2O$ 200
800	Cu_2C_2 Copper(I) acetylide	151.11	>20 $\rightarrow$ Cu,C
801	$CuCN$ Copper(I) cyanide	89.56	mp 473, dec 550
802	$Cu(CN)_2$ Copper(II) cyanide	115.58	dec >20
803	CuC_2O_4 ($\cdot H_2O$) Copper(II) oxalate	151.56	420 $\rightarrow$ CuO
804	$Cu_2CO_3(OH)_2$ Copper(II) carbonate-dihydroxide	221.11	200 $\rightarrow$ CuO
805	$CuCl$ Copper(I) chloride	99.00	mp 430, bp 1212
806	$CuCl_2$ Copper(II) chloride	134.45	mp 596, bp 993 dec
807	$CuCl_2 \cdot 2H_2O$	170.48	$-H_2O$ 110
808	$Cu(ClO_3)_2$ ($\cdot 6H_2O$) Copper(II) chlorate	230.45	mp 65, dec 100

(*Continued*)

No.	Color and density	Solubility and reactivity						
		water	alcohol	HCl	H_2SO_4	HNO_3	NaOH	$NH_3 \cdot H_2O$
763	wh	r	r	+	+	+	+	+
764	wh	r	r	r	−/+	−/+	−	−
765	wh, 3.35	r	i	−	r	−	+	+
766	wh, 4.51	r	r	r	−/+	−/+	r	r
767	wh, 4.85	d	...	−/+	d	d	−	−
768	wh, 4.26	d	d	d	d	d	d	d
769	viol, 3.60	d	+	−/+	−	−	−	−/+
770	dk-red hydr	+	i	+	+	+	+	+
771	wh, 2.93	r	d	r	−/+	−/+	r	−
772	dk-red	+	...	+	+	+	+	+
773	wh, 3.44	+	...	+	+	+	+	+
774	yel	r	r	r/+	−/+	r/+	r	r
775	wh, 3.69	r	...	r	r	r	r	r
776	wh	+	+	+	+	+	+	+
777	wh	+	+	+	+	+	+	+
778	yel-orange, 3.80	+	+	+	+	+	+	+
779	red, 3.19	+	+	+	+	+	+	+
780	orange, 4.68	+	+	+	+	+	+	+
781	wh, 4.74	+	+	+	+	+	+	+
782	wh, 3.68	r	r	+	+	+	r	r
783	wh hydr	r/+	r	+	+	+	r	−
784	yel	r/+	r	+	+	+	−	−
785	orange	r/+	r	+	+	+	−	−
786	red, 2.81	r/+	r	+	+	+	−	−
787	red-brn	r/+	r	+	+	+	−	−
788	wh, 4.24	r	i	r	r	r	r	r
789	wh	r	...	−	−	−	−	−
790	blk	+	...	+	+	+	+	+
791	wh, 4.45	r	...	−	−	−	−	−
792	red hydr, 2.03	d	i	−	d	−	−/+	−
793	red, 8.920	−	−	−	−/+	+	−	+
794	dk-grey	−	...	−	−/+	−/+	−/+	−
795	bl-grn hydr	i/+	i	r/+	r	r	i	+
796	grey-blk, 4.44	i	...	−/+	−/+	−/+	−/+	−/+
797	wh, 4.72	i/+	...	−/+	−/+	−	−	+
798	grn-blk, 4.77	r	r	r/+	−/+	−	+	+
799	grn-sk-bl hydr, 2.58	r	...	−/+	−	−	+	+
800	red-brn	−/+	+	+	+	+	+	+
801	wh, 2.92	i	...	i/+	i/+	i/+	+	+
802	brn-yel	i	...	r/+	r	r	+	+
803	sk-bl hydr	i	...	−/+	−/+	−/+	−/+	+
804	dk-grn, 4.05	i/+	i	+	+	+	−	−/+
805	wh, 4.14	i/+	...	i/+	−/+	−/+	−	+
806	dk-brn, 3.39	r	r	r/+	r	r	+	+
807	sk-bl, 2.51	r	r	r/+	r	r	+	+
808	grn hydr	r	r	−/+	r	−/+	+	+

No.	Formula and name	M_r	Phase transition temperature
809	$Cu(ClO_4)_2$ Copper(II) perchlorate	262.44	dec 230
810	$Cu(ClO_4)_2 \cdot 6H_2O$	370.53	mp 82, dec 120
811	$CuCrO_2$ Copper(I) dioxochromate(III)	147.54	dec ca. 1700
812	CuF_2 Copper(II) fluoride	101.54	mp 950 dec
813	$CuF_2 \cdot 2H_2O$	137.57	dec 280
814	CuH Copper(I) hydride	64.55	dec >20
815	$Cu(HCOO)_2$ ($\cdot 4H_2O$) Copper(II) formate	153.58	dec t
816	$[Cu_2(H_2O)_2(CH_3COO)_4]$ Diaquatetraacetatodicopper	399.30	dec 240
817	CuI Copper(I) iodide	190.45	mp 605, bp 1320
818	$Cu(IO_3)_2$ ($\cdot 0.67H_2O$) Copper(II) iodate	413.35	$-H_2O$ 240, dec 290
819	Cu_3N Tricopper nitride	204.65	dec 450
820	CuNCS Copper(I) thiocyanate	121.63	mp 1084
821	$[Cu(NH_2CH_2COO)_2]$ Diglycinatocopper	211.66	mp 130, dec t
822	$[Cu(NH_3)_4](NO_3)_2$ Tetraamminecopper(II) nitrate	255.68	dec 200
823	$[Cu(NH_3)_4]SO_4$ ($\cdot H_2O$) Tetraamminecopper(II) sulfate	227.73	dec hydr 250
824	$Cu(NO_3)_2$ Copper(II) nitrate	187.55	mp 226 (p), dec >170
825	$Cu(NO_3)_2 \cdot 3H_2O$	241.60	mp 114.5; dec 170
826	$Cu(NO_3)_2 \cdot 6H_2O$	295.64	mp 26, dec < 100
827	CuO Copper(II) oxide	79.55	dec 1026, mp 1447 (p)
828	$Cu_{2-x}O$ Copper(I) oxide	143.09	mp 1240, dec 1800
829	Cu_2O_3 Copper(III) oxide	175.09	dec 400
830	$Cu(OH)_2$ Copper(II) hydroxide	97.56	$80 \rightarrow CuO$
831	$[Cu(OH)_4],K$ Potassium tetrahydroxocuprate(III)	170.67	dec >200
832	$[Cu(OH)_4],Na_2$ Sodium tetrahydroxocuprate(II)	177.55	dec >200
833	Cu_3P Tricopper phosphide	221.61	mp 1022
834	CuS Copper(II) sulfide	95.61	dec 220
835	$Cu_{2-x}S$ ($0.06 \le x \le 0.2$) Copper(I) sulfide	159.16	mp 1130
836	Cu_2S Copper(I) sulfide	159.16	dec ca. 1000
837	$CuSO_4$ Copper(II) sulfate	159.61	dec 650
838	$CuSO_4 \cdot 5H_2O$	249.68	$-H_2O$ 200–250
839	Cu_3Sb Tricopper stibide	312.39	mp 687
840	CuSe Copper(II) selenide	142.51	mp 382 dec
841	Cu_2Se Copper(I) selenide	206.05	mp 1112
842	$CuSeO_4$ ($\cdot 5H_2O$) Copper(II) selenate	206.50	$-H_2O$ 150
843	Cu_2Te Copper(I) telluride	254.69	mp 855
844	**D_2 Dideuterium**	4.028	mp −254.5; bp −249.49
845	DBr Deuterium bromide	81.92	mp −87.63; bp −66.85
846	DCl Deuterium chloride	37.47	mp −114.7; bp −84.75
847	DF Deuterium fluoride	21.01	mp −83.6; bp 18.36
848	DI Deuterium iodide	128.92	mp −51.82; bp −36.2
849	D_2O Deuterium oxide	20.03	mp 3.813; bp 101.43
850	D_3PO_4 Deuterium orthophosphate	101.01	mp 49.45
851	D_2S Deuterium sulfide	36.09	mp −86.01
852	D_2SO_4 Deuterium sulfate	100.09	mp 14.35
853	**Dy Dysprosium**	162.50	mp 1409, bp 2587
854	$DyBr_3$ ($\cdot 6H_2O$) Dysprosium(III) bromide	402.21	mp 881, bp 1485
855	$Dy(BrO_3)_3$ ($\cdot 9H_2O$) Dysprosium(III) bromate	546.20	mp hydr 75, $-H_2O$ 140

(Continued)

No.	Color and density	Solubility and reactivity						
		water	alcohol	HCl	H_2SO_4	HNO_3	NaOH	$NH_3 \cdot H_2O$
809	yel-grn	r	r	–	–	–	+	+
810	lt-sk-bl, 2.23	r	r	–	–	–	+	+
811	grey-blk, 5.24	i	i	–/+	–	+	–/+	–
812	wh, 4.23	r/+	r	r/+	r	r	+	+
813	lt-bl, 2.93	r/+	r	r/+	r	r	+	+
814	red-brn	+	+	+	+	+	+	+
815	sk-bl, 1.83	r/+	r	r	r	r	+	+
816	dk-grn, 1.88	r	r	–/+	–/+	–/+	–	–
817	wh, 5.62	i	...	–/+	–/+	+	–	+
818	grn, 5.24	d	i	–/+	–/+	r	+	+
819	dk-grn, 5.84	+	...	+	+	+	+	+
820	wh, 2.84	i	i	i/+	i/+	i/+	i	+
821	sk-bl	i	i	+	+	+	–	+
822	dk-bl, 1.91	r	r	+	+	+	–/+	r
823	bl hydr, 1.81	r/+	i	+	+	+	–/+	r
824	wh	r	r	r/+	–	r	+	+
825	bl, 2.32	r	r	r/+	–	r	+	+
826	bl, 2.07	r	r	r/+	–	r	+	+
827	blk, 6.32	–	i	+	+	+	–	–/+
828	red, 6.1	–	i	+	+	+	+	+
829	dk-red	–	...	+	–	–	+	...
830	sk-bl, 3.37	i	...	+	+	+	i/+	+
831	red	+	+	+	+	+	d	+
832	bl	+	+	+	+	+	r	–/+
833	grey	–	...	–	–/+	+	–	–
834	blk, 4.68	i	i	i	–/+	+	i	–
835	blk, 5.5–5.7	i	i	–	–	+	–	+
836	blk-grey, 5.8	i	i	–	–	+	–	+
837	wh, 3.60	r	i	r/+	r	r	+	+
838	bl, 2.28	r	i	r/+	r	r	+	+
839	grey, 8.51	–	...	–/+	–/+	–/+	–	–
840	blk-grn, 5.99	i	i	–/+	–/+	+	–/+	–/+
841	blk, 7.49	i	...	+	+	+	+	+
842	sk-bl hydr, 2.56	r	i	+	+	+	+	+
843	bl, 7.27	i	...	+	+	+	+	+
844	cl gas	i	i	–	–	–	–	–
845	cl gas	r	r	r	r/+	r	+	+
846	cl gas	r	r	r	r	r	+	+
847	cl gas	r	r	r	r	r	+	+
848	cl gas	r	r	r	r/+	r/+	+	+
849	cl lq, 1.107[20]	∞	∞	r	r	r	r	+
850	wh, 1.91	r	r	r	r	r	+	+
851	cl gas	d	r	d	d/+	d/+	+	+
852	cl lq, 1.8572[25]	∞	+	r	r	r	+	+
853	wh, 8.559	psv/+	...	+	+	+	–	–
854	wh	r	r	–	–/+	–	+	+
855	yel hydr	r	d	–/+	–	–/+	+	+

No.	Formula and name	M_r	Phase transition temperature
856	$Dy_2(C_2O_4)_3$ $(\cdot10H_2O)$ Dysprosium(III) oxalate	589.05	hydr >550 $\rightarrow$ Dy_2O_3
857	$DyCl_3$ $(\cdot6H_2O)$ Dysprosium(III) chloride	268.86	mp 653, bp 1539
858	DyF_3 Dysprosium(III) fluoride	219.49	mp 1157, bp >2200
859	DyI_3 $(\cdot9H_2O)$ Dysprosium(III) iodide	543.22	mp 983, bp 1320
860	$Dy(NO_3)_3$ $(\cdot5H_2O)$ Dysprosium(III) nitrate	348.51	mp hydr 88.6; dec t
861	Dy_2O_3 Dysprosium(III) oxide	373.00	mp 2400, bp ca. 4300
862	$Dy(OH)_3$ Dysprosium(III) hydroxide	213.52	800 $\rightarrow$ Dy_2O_3
863	Dy_2S_3 Dysprosium(III) sulfide	421.20	mp ca. 1490
864	$Dy_2(SO_4)_3$ $(\cdot8H_2O)$ Dysprosium(III) sulfate	613.19	$-H_2O$ 360, dec ca. 900
865	**Er Erbium**	167.26	mp 1522, bp 2857
866	$ErBr_3$ $(\cdot9H_2O)$ Erbium(III) bromide	406.97	mp 950, bp 1460
867	$Er(CH_3COO)_3$ $(\cdot4H_2O)$ Erbium(III) acetate	344.39	dec hydr 400
868	$Er_2(C_2O_4)_3$ $(\cdot6H_2O)$ Erbium(III) oxalate	598.57	hydr 575 $\rightarrow$ Er_2O_3
869	$ErCl_3$ Erbium(III) chloride	273.62	mp 774, bp 1500
870	$ErCl_3\cdot6H_2O$	381.71	mp 153
871	ErF_3 Erbium(III) fluoride	224.25	mp 1146, bp 2230
872	ErI_3 $(\cdot6H_2O)$ Erbium(III) iodide	547.98	mp 1015, bp 1280
873	$Er(NO_3)_3$ $(\cdot5H_2O)$ Erbium(III) nitrate	353.27	$-H_2O$ 130, dec t
874	Er_2O_3 Erbium(III) oxide	382.52	mp >2200
875	$Er(OH)_3$ Erbium(III) hydroxide	218.28	850 $\rightarrow$ Er_2O_3
876	Er_2S_3 Erbium(III) sulfide	430.72	mp 1730
877	$Er_2(SO_4)_3$ Erbium(III) sulfate	622.71	dec ca. 900
878	$Er_2(SO_4)_3$ $(\cdot8H_2O)$	766.83	$-H_2O$ 400
879	**Es Einsteinium**	252.083	mp 860
880	**Eu Europium**	151.965	mp 826, bp 1440
881	$EuBr_2$ Europium(II) bromide	311.77	mp 683, bp 1880
882	$EuBr_3$ $(\cdot6H_2O)$ Europium(III) bromide	391.68	mp 705, dec t
883	$Eu_2(C_2O_4)_3$ $(\cdot10H_2O)$ Europium(III) oxalate	567.98	hydr >550 $\rightarrow$ Eu_2O_3
884	$EuCl_2$ $(\cdot2H_2O)$ Europium(II) chloride	222.87	mp 854, bp 2060
885	$EuCl_3$ $(\cdot6H_2O)$ Europium(III) chloride	258.32	mp 626 dec
886	EuF_2 Europium(II) fluoride	189.96	mp 1416, bp 2400
887	EuF_3 Europium(III) fluoride	208.96	mp 1276, bp 2280
888	EuI_2 Europium(II) iodide	405.77	mp 580, bp 1775
889	EuI_3 $(\cdot9H_2O)$ Europium(III) iodide	532.68	mp 880, dec t
890	$Eu(NO_3)_3$ $(\cdot6H_2O)$ Europium(III) nitrate	337.98	mp hydr 85, dec hydr t
891	EuO Europium(II) oxide	167.96	mp 1700
892	Eu_2O_3 Europium(III) oxide	351.93	mp >2200
893	$Eu(OH)_2$ $(\cdot H_2O)$ Europium(II) hydroxide	185.98	570 $\rightarrow$ EuO
894	$Eu(OH)_3$ Europium(III) hydroxide	202.99	820 $\rightarrow$ Eu_2O_3
895	EuS Europium(II) sulfide	184.03	mp ca. 2400
896	$EuSO_4$ Europium(II) sulfate	248.03	dec t
897	$Eu_2(SO_4)_3$ Europium(III) sulfate	592.12	dec 1600
898	$Eu_2(SO_4)_3\cdot8H_2O$	736.24	$-H_2O$ 375
899	**F_2 Difluorine**	37.996	mp -219.699; bp -188.2
900	**Fe Iron**	55.847	mp 1539, bp ca. 3200
901	$(FeAl_2)O_4$ Iron-dialuminum tetraoxide	173.81	mp 1440
902	$FeAs$ Iron monoarsenide	130.77	mp 1030

(Continued)

No.	Color and density	Solubility and reactivity						
		water	alcohol	HCl	H_2SO_4	HNO_3	NaOH	$NH_3 \cdot H_2O$
856	yel hydr	i	i	−/+	−/+	−/+	i	−
857	wh, 3.617	r	r	r	−	−	+	+
858	wh	i	i	i	−	−	−	−
859	yel-grn	r	r	r	−/+	−/+	+	+
860	yel hydr	r	r	−	−	r	+	+
861	yel-grn, 7.81	−	−	+	+	+	−	−
862	yel	i	i	+	+	+	i	i
863	yel	i/+	...	+	+	+	−	−
864	yel	d	...	−	d	−	+	+
865	wh, 9.062	psv/+	...	+	+	+	−	−
866	viol	r	r	−	−/+	−	+	+
867	wh hydr, 2.11	r	r	−	r	−	+	+
868	red hydr, 2.64	i	i	−/+	i/+	−/+	−	i
869	pink, 4.1	r	r	r	−	−	+	+
870	pink	r	r	r	−	−	+	+
871	pink	i	i	i	−	−	−	−
872	viol	r	r	r	−/+	−/+	+	+
873	red hydr	r	r	−	−	r	+	+
874	pink, 8.64	−/+	−	+	+	+	−	−
875	pink	i	i	+	+	+	i	i
876	yel-brn, 6.05	i/+	...	+	+	+	i	i
877	wh, 3.68	r/d	...	−	r	−	+	+
878	pink, 3.22	r/d	...	−	r	−	+	+
879	...	+	...	+	+	+	+	+
880	wh, 5.244	psv/+	...	+	+	+	−	−
881	wh	r/+	...	−	−/+	−	+	+
882	wh	r	r	r	−/+	−	+	+
883	wh, 4.89	i	i	−/+	−/+	i/+	−	i
884	wh, 4.88	r/+	...	r	−	−	+	+
885	yel, 4.89 (4.47)	r	r	r	−	−	+	+
886	lt-yel, 6.50	i	...	i	−	−	i	−
887	wh	i	i	i	−	−	i	−
888	wh, 5.50	r/+	...	−	−/+	−/+	+	+
889	wh	r	r	r	−/+	−/+	+	+
890	wh hydr	r	r	−	−	r	+	+
891	red, 8.2	+	...	+	+	+	+	+
892	lt-pink, 6.55	−	...	+	+	+	−	−
893	wh hydr, 4.67	r/+	...	+	+	+	r	r
894	wh	i	i	+	+	+	i	i
895	brn-viol	i/+	...	+	+	+	i	−
896	wh, 4.98	i/+	...	−	i	−	i	i
897	wh, 4.95	d	...	−	d	−	+	+
898	lt-pink	d	...	−	d	−	+	+
899	lt-grn gas, 1.696	+	+	+	+	+	+	+
900	grey, 7.864	−	−	+	+/psv	+/psv	−/+	−
901	dk-grn, 4.40	−	−	−/+	−	−/+	−/+	−
902	lt-grey, 7.83	−	i	+	+	+	−	−

No.	Formula and name	M_r	Phase transition temperature
903	FeAs$_2$ Iron diarsenide	205.69	mp 990
904	FeAsO$_4$ ($\cdot$2H$_2$O) Iron(III) arsenate	194.77	$-$H$_2$O 200, dec t
905	Fe(As)S Iron arsenide-sulfide	162.84	dec 770
906	FeB Iron monoboride	66.66	mp 1650
907	FeBr$_2$ Iron(II) bromide	215.66	mp 684 (p), bp 927
908	FeBr$_2\cdot$6H$_2$O	323.75	dec >100
909	FeBr$_3$ Iron(III) bromide	295.56	139 $\rightarrow$ Fe
910	FeBr$_3\cdot$6H$_2$O	403.65	mp hydr 27, dec 100
911	Fe$_3$C Triiron carbide	179.55	mp 1650 dec ($\rightarrow$ Fe,C)
912	[Fe$_5$C(CO)$_{15}$] Carbon(15-carbonyl)pentairon	711.40	dec 110
913	[Fe(C$_5$H$_5$)$_2$] Bis(cyclopentadienyl)iron	186.04	mp 174, bp 249, dec 400
914	[Fe(C$_5$H$_5$)$_2$]NO$_3$ Bis(cyclopentadienyl)iron(III) nitrate	248.04	dec t
915	Fe(CH$_3$COO)$_2$ ($\cdot$4H$_2$O) Iron(II) acetate	173.94	dec hydr t
916	[Fe(CN)$_6$],Ba$_2$ ($\cdot$6H$_2$O) Barium hexacyanoferrate(II)	486.61	$-$H$_2$O >120
917	[Fe^{III}(CN)$_6$],Fe Iron(III) hexacyanoferrate(III)	267.80	dec >250
918	[Fe^{II}(CN)$_6$],Fe$_2$ Iron(II) hexacyanoferrate(II)	323.65	dec 500
919	[Fe(CN)$_6$],H$_3$ Hydrogen hexacyanoferrate(III)	214.98	dec 100
920	[Fe(CN)$_6$],H$_4$ Hydrogen hexacyanoferrate(II)	215.99	dec 190
921	[Fe(CN)$_6$],K$_3$ Potassium hexacyanoferrate(III)	329.25	dec 350–400
922	[Fe(CN)$_6$],K$_4$ Potassium hexacyanoferrate(II)	368.35	mp hydr 70, dec >87
923	[Fe^{II}(CN)$_6$],KFe Potassium-iron(III) hexacyanofer-rate(II)	306.90	dec >200
924	[Fe(CN)$_6$],Na$_4$ ($\cdot$10H$_2$O) Sodium hexacyanoferrate(II)	303.92	dec hydr >100
925	[Fe(CN)$_6$],Tl$_4$ ($\cdot$2H$_2$O) Thallium(I) hexacyanofer-rate(II)	1029.49	dec hydr 150
926	FeCO$_3$ Iron(II) carbonate	115.86	dec >280–490
927	FeC$_2$O$_4$ ($\cdot$2H$_2$O) Iron(II) oxalate	143.87	dec hydr >160
928	[Fe(CO)$_5$] Pentacarbonyliron	195.90	mp $-$20, bp 103
929	Fe$_2$(C$_2$O$_4$)$_3$ ($\cdot$5H$_2$O) Iron(III) oxalate	375.75	dec hydr >100
930	[Fe$_2$(CO)$_9$] Nonacarbonyldiiron	363.78	mp 100; 120 $\rightarrow$ [Fe(CO)$_5$]
931	[Fe$_3$(CO)$_{12}$] Dodecacarbonyltriiron	503.66	mp 140 dec ($\rightarrow$ Fe)
932	[Fe(C$_2$O$_4$)$_3$],K$_3$ ($\cdot$3H$_2$O) Potassium trioxalatofer-rate(III)	437.20	$-$H$_2$O 100, dec 230
933	[Fe(C$_2$O$_4$)$_3$],(NH$_4$)$_3$ ($\cdot$3H$_2$O) Ammonium trioxalato-ferrate(III)	374.02	dec hydr 165
934	[Fe(C$_2$O$_4$)$_3$],Na$_3$ ($\cdot$5.5H$_2$O) Sodium trioxalatofer-rate(III)	388.87	dec hydr 300
935	FeCl$_2$ ($\cdot$2H$_2$O) Iron(II) chloride	126.75	mp 674, bp 1023
936	FeCl$_2\cdot$4H$_2$O	198.81	mp 90, dec >120
937	FeCl$_3$ Iron(III) chloride	162.21	mp 307.5; bp 316 dec
938	FeCl$_3\cdot$6H$_2$O	270.30	mp 37, bp 219
939	Fe(Cl)O Iron chloride-oxide	107.30	>300 $\rightarrow$ Fe$_2$O$_3$,FeCl$_3$
940	Fe(ClO$_4$)$_2$ ($\cdot$6H$_2$O) Iron(II) perchlorate	254.75	dec hydr >100
941	Fe(ClO$_4$)$_3$ ($\cdot$10H$_2$O) Iron(III) perchlorate	354.19	dec hydr >120
942	(FeCu)S$_2$ Iron-copper disulfide	183.53	mp 1000, dec >1300
943	FeF$_2$ ($\cdot$8H$_2$O) Iron(II) fluoride	93.84	$-$H$_2$O 100, mp 1102, bp 1837

(*Continued*)

No.	Color and density	Solubility and reactivity						
		water	alcohol	HCl	H_2SO_4	HNO_3	NaOH	$NH_3 \cdot H_2O$
903	lt-grey, 7.45	−	...	−	−/+	−/+	−	−
904	grn hydr, 3.18	i	i	r	i/r	i/r	i	i
905	wh, 6.07	−	i	+	+	+	−/+	−
906	grey, 7.15	−	...	−	−/+	+	−	−
907	yel-brn, 4.62	r	r	r	−/+	−/+	+	+
908	lt-grn hydr	r	r	r	−/+	−/+	+	+
909	red-blk	r/+	r	r	−/+	−	+	+
910	red	r/+	r	r	−/+	−	+	+
911	grey, 7.66	−	...	+	+	+	−	−
912	blk	−/+	i	+	+	+	−/+	−
913	orange-yel	−/+	r	−	−/+	+	−	−
914	dk-sk-bl	r	r	−	−	−	−	−
915	lt-grn hydr	r	r	r	−	−	+	+
916	yel hydr, 2.67	d	i	−	−	−/+	−	−
917	brn	d	...	−/+	−/+	−/+	−	−
918	wh	i/+	...	−	−/+	−/+	−	−
919	lt-brn	r	r	−/+	−/+	−/+	+	+
920	wh, 1.54	r/+	r	−	−	−/+	+	+
921	dk-red, 1.89	r	i	−/+	−/+	−/+	−	−
922	lt-yel hydr, 1.85	r	i	−	−/+	−/+	−	−
923	bl	i	...	−/+	−	−/+	+	+
924	yel hydr, 1.46	r	i	−	−	−/+	−	−
925	yel hydr, 4.64	d	i	−	−	−/+	−	−
926	wh, 3.8	i	...	+	+	+	−	−
927	lt-yel hydr, 2.28	d	...	−	−/+	−/+	+	+
928	lt-yel lq, 1.457[20]	i/+	r	−	−/+	−/+	+	−
929	lt-yel hydr, 2.28	r	i	r/+	−/+	r/+	+	+
930	yel, 2.09	−/+	d	−	−/+	−/+	+	−
931	dk-grn-blk	−/+	d/+	−	−/+	−/+	−	−
932	grn hydr, 2.13	r	i	−/+	−/+	−/+	+	+
933	grn hydr, 1.78	r	i	−/+	−/+	−/+	+	+
934	grn hydr, 1.97	r	r	−/+	−/+	−/+	+	+
935	wh, 3.16	r	r	r	−	−	+	+
936	grn-sk-bl, 1.93	r	r	r	−	−	+	+
937	lt-grn, 2.90	r/+	r	r	−	−	+	+
938	dk-yel	r/+	r	r	−	−	+	+
939	orange-red, 3.1	−/+	−	+	+	+	−/+	−
940	lt-grn hydr	r	r	−	−	−/+	+	+
941	pink hydr	r/+	r	−	−	r	+	+
942	yel, 4.2	i	i	−/+	−/+	−/+	−/+	−
943	wh, 4.09(2.09)	d	i	d/r	d/+	d/r	−	−

No.	Formula and name	M_r	Phase transition temperature
944	FeF_3 Iron(III) fluoride	112.84	mp 1027, bp 1327
945	$FeF_3 \cdot 4.5H_2O$	193.91	dec >100
946	$(Fe^{II}Fe^{III})F_5$ Iron(II)-iron(III) fluoride	206.68	dec 1200
947	$(Fe^{II}Fe_2^{III})O_4$ Iron(II)-diiron(III) oxide	231.54	mp 1538 dec
948	$[Fe(H_2O)_4Cl_2]Cl$ Tetraaquadichloroiron(III) chloride	234.27	mp 37, bp >280
949	$[Fe(H_2O)_3(NCS)_3]$ Triaquatris(thiocyanato)iron	284.14	dec t
950	FeI_2 Iron(II) iodide	309.66	mp 587, bp 827
951	$FeI_2 \cdot 4H_2O$	381.72	mp 90, dec >100
952	$Fe_{2+x}N$ ($0 < x \leq 0.5$) Diiron nitride	125.70	$350 \rightarrow Fe_3N$
953	Fe_3N Triiron nitride	181.55	$420 \rightarrow Fe_4N$
954	Fe_4N Tetrairon nitride	237.40	$640 \rightarrow Fe$
955	$Fe(NCS)_2$ ($\cdot 3H_2O$) Iron(II) thiocyanate	172.02	dec hydr t
956	$FeNH_4(SO_4)_2$ Iron(III)-ammonium sulfate	266.01	$480 \rightarrow Fe_2(SO_4)_3$
957	$FeNH_4(SO_4)_2 \cdot 12H_2O$	482.19	mp 40, $-H_2O$ 230
958	$Fe(NH_4)_2(SO_4)_2$ ($\cdot 6H_2O$) Iron(II)-diammonium sulfate	284.05	dec hydr 100–110
959	$Fe(NO_3)_2$ ($\cdot 6H_2O$) Iron(II) nitrate	179.86	mp hydr 60.5 dec
960	$Fe(NO_3)_3$ ($\cdot 6H_2O$) Iron(III) nitrate	241.86	mp hydr 47.2; bp hydr 125.1
961	$Fe(NO_3)_3 \cdot 9H_2O$	403.99	mp 50.1; $-H_2O$ >150
962	$[Fe(NO)_4]$ Tetranitrosyliron	175.87	dec >100
963	$[Fe(NO^+)(CN)_5],Na_2$ ($\cdot 2H_2O$) Sodium nitrosylium-pentacyanoferrate(II)	261.92	mp 687
964	$(Fe_2Ni)O_4$ Diiron-nickel tetraoxide	234.38	mp 1660 dec
965	$Fe_{1-x}O$ Iron(II) oxide	71.85	mp 1368
966	Fe_2O_3 Iron(III) oxide	159.69	$1390 \rightarrow (Fe^{II}Fe_2^{III})O_4$
967	$Fe_2O_3 \cdot nH_2O$	–	>200 $\rightarrow FeO(OH)$
968	$Fe_{2.67}O_4$ 2.67-Iron tetraoxide	213.11	$1200 \rightarrow Fe_2O_3$
969	$Fe(OH)_2$ Iron(II) hydroxide	89.86	dec >150
970	$FeO(OH)$ Iron metahydroxide	88.85	$500 \rightarrow Fe_2O_3$
971	FeP Iron monophosphide	86.82	dec >1150
972	Fe_2P Diiron phosphide	142.67	mp 1365
973	Fe_3P Triiron phosphide	198.52	mp 1166 dec
974	$[Fe(PF_3)_5]$ Pentakis(trifluorophosphorus)iron	495.69	mp 45 (p), subl >20
975	$FePO_4$ ($\cdot 2H_2O$) Iron(III) orthophosphate	150.82	dec hydr >250
976	$Fe_3(PO_4)_2$ ($\cdot 8H_2O$) Iron(II) orthophosphate	357.48	dec hydr 100–180
977	$Fe_{1-x}S$ ($0.1 \leq x \leq 0.2$) Iron(II) sulfide	87.91	mp 1195
978	$\alpha\text{-}Fe(S_2)$ Iron(II) disulfide(2−)	119.98	$450 \rightarrow \beta\text{-}Fe(S_2)$
979	$\beta\text{-}Fe(S_2)$	119.98	dec >1170
980	$FeSO_4$ ($\cdot 5H_2O$) Iron(II) sulfate	151.91	dec ca. 350
981	$FeSO_4 \cdot 7H_2O$	278.01	mp 64, $-H_2O$ 300
982	$Fe_2(SO_4)_3$ ($\cdot 9H_2O$) Iron(III) sulfate	399.88	dec >600
983	$FeSi$ Iron monosilicide	83.93	mp 1405
984	$FeSiO_3$ Iron(II) metasilicate	131.93	mp 1146
985	Fe_2SiO_4 Iron(II) orthosilicate	203.78	mp 1217
986	$FeWO_4$ Iron(II) wolframate	303.69	mp >1700
987	$(Fe_2Zn)O_4$ Diiron-zinc tetraoxide	241.08	mp 1590
988	**Fm Fermium**	257.095	...

(Continued)

No.	Color and density	Solubility and reactivity						
		water	alcohol	HCl	H_2SO_4	HNO_3	NaOH	$NH_3 \cdot H_2O$
944	lt-grn, 3.87	i/d	i	–	–/+	–	i/d	–
945	lt-pink	i/d	i	–	–	–	i/d	–
946	sk-bl-grey, 3.34	d	i	–	–/+	–/+	–/+	–
947	blk, 5.11	–	–	+	–/+	–/+	–	–
948	yel-brn	r/+	r	–/+	–	–	+	+
949	dk-red	r/+	r	–	–/+	–/+	+	–
950	red-brn, 5.32	r	r	–	–/+	–/+	+	+
951	grn, 2.87	r	r	–	–/+	–/+	+	+
952	grey-blk, 6.57	–	...	–/+	–/+	–/+	–	–
953	grey-blk	–	...	–/+	–/+	–/+	–	–
954	grey-blk, 5.02	–	...	–/+	–/+	–/+	–	–
955	grn hydr	r	r	r	–/+	–/+	+	+
956	wh, 2.49	r	i	–	r	–	+	+
957	wh, 1.71	r	i	r	r	–	+	+
958	lt-grn hydr, 1.86	r	i	–	r	–/+	+	+
959	lt-grn hydr	r	...	–	–	r/+	+	+
960	wh hydr, 1.68	r/+	r	–/+	–	r	+	+
961	wh, 1.68	r/+	+	–/+	–	r	+	+
962	blk	+	...	+	+	+	+	+
963	dk-red hydr, 1.72	r	i	–/+	–/+	–/+	–/+	–/+
964	brn-blk, 5.2	–	i	–/+	–/+	–/+	–	–
965	blk, 5.75	–/+	–	+	+	+	–	–
966	red-brn, 5.26	–	...	+	+	+	–	–
967	brn, 3.65	i	...	+	+	+	i/+	i
968	dk-brn, 4.88	–	i	+	+	+	–	–
969	wh, 3.4	i	...	+	+	+	i/+	i
970	lt-brn, 4.37	i	i	+	+	+	i/+	i
971	grey-blk, 6.07	–/+	...	–/+	–/+	–/+	–/+	–
972	grey-blk, 6.56	–	...	–	–/+	–/+	–	–
973	grey-blk, 6.74	–	...	–	–	–/+	–	–
974	lt-yel	–/+	r	+	+	+	+	...
975	lt-red hydr, 2.76	i	...	i/r	r	r	i	–
976	wh hydr, 2.68	i	...	r	r	r	i	–
977	blk, 4.62	i	...	+	+	+	i	i
978	dk-yel, 4.9	i	i	–	+	+	–	–
979	yel, 5.02	i	i	–	+	+	–	–
980	wh, 3.14	r	i	–	r	–/+	+	+
981	lt-grn, 1.90	r	i	–	r	–/+	+	+
982	wh, 3.10(2.1)	r/+	r/+	r	r	–	+	+
983	yel-grey, 6.1	–	...	–	–	–	–/+	–
984	grey-grn, 3.5	i	...	–/+	–/+	–/+	–/+	–
985	grn-yel, 4.34	i	...	–/+	–	–	–/+	–
986	blk, 7.5	d	i	–/+	–/+	–/+	–	–
987	blk, 5.34	–	–	–/+	–/+	–/+	–/+	–
988	...		...	...	...	...	...	...

No.	Formula and name	M_r	Phase transition temperature
989	**Fr Francium**	223.020	mp 21, bp 660
990	FrClO$_4$ Francium perchlorate	322.47	dec >800
991	**Ga Gallium**	69.723	mp 29.78; bp 2403
992	GaAs Gallium monoarsenide	144.65	mp 1238
993	GaBr$_3$ Gallium(III) bromide	309.44	mp 121.5; bp 279
994	[Ga^{III}Br$_4$],Ga Gallium(I) tetrabromogallate(III)	459.06	mp 165
995	GaCl$_3$ Gallium(III) chloride	176.08	mp 77.8; bp 201.3
996	[Ga^{III}Cl$_4$],Ga Gallium(I) tetrachlorogallate(III)	281.26	mp 176, bp 535
997	Ga(ClO$_4$)$_3$ (·6H$_2$O) Gallium(III) perchlorate	368.07	dec 175
998	GaF$_3$ Gallium(III) fluoride	126.72	subl >800, bp 950
999	GaF$_3$·3H$_2$O	180.76	dec >140
1000	GaI$_3$ (·H$_2$O) Gallium(III) iodide	450.44	mp 212, bp 346
1001	[Ga^{III}I$_4$],Ga Gallium(I) tetraiodogallate(III)	647.06	mp 221, bp 580
1002	GaN Gallium mononitride	83.73	subl >800, mp 1500
1003	GaNH$_4$(SO$_4$)$_2$ (·12H$_2$O) Gallium(III)-ammonium sulfate	279.89	dec hydr >250
1004	[GaN$_2$],Li$_3$ Lithium dinitridogallate(III)	118.56	dec 800
1005	Ga(NO$_3$)$_3$ (·8H$_2$O) Gallium(III) nitrate	255.74	−H$_2$O >40, dec 110–200
1006	Ga$_2$O$_3$ Gallium(III) oxide	187.44	mp ca. 1725
1007	Ga(OH)$_3$ Gallium(III) hydroxide	120.74	400 → GaO(OH)
1008	GaO(OH) Gallium metahydroxide	102.73	>540 → Ga$_2$O$_3$
1009	GaP Gallium monophosphide	100.70	mp 1465
1010	GaPO$_4$ (·2H$_2$O) Gallium(III) orthophosphate	164.69	mp 1670
1011	Ga$_2$S$_3$ Gallium(III) sulfide	235.64	mp ca. 1110
1012	Ga$_2$(SO$_4$)$_3$ (·18H$_2$O) Gallium(III) sulfate	427.63	dec hydr >520
1013	GaSb Galliumantimony	191.47	mp 712
1014	Ga$_2$Se$_3$ Gallium(III) selenide	376.33	mp 1020
1015	Ga$_2$Te$_3$ Gallium(III) telluride	522.25	mp 790
1016	**Gd Gadolinium**	157.25	mp 1312, bp 3272
1017	GdBr$_3$ (·6H$_2$O) Gadolinium(III) bromide	396.96	mp 778, bp 1490
1018	Gd$_2$(C$_2$O$_4$)$_3$ (·10H$_2$O) Gadolinium(III) oxalate	578.55	hydr >500 → Gd$_2$O$_3$
1019	GdCl$_3$ (·6H$_2$O) Gadolinium(III) chloride	263.61	mp 609, bp 1580
1020	GdF$_3$ Gadolinium(III) fluoride	214.24	mp 1232, bp 2280
1021	GdI$_3$ Gadolinium(III) iodide	537.96	mp 929, bp 1340
1022	Gd(NO$_3$)$_3$ (·5H$_2$O) Gadolinium(III) nitrate	343.26	mp hydr 92, t → Gd$_2$O$_3$
1023	Gd$_2$O$_3$ Gadolinium((III) oxide	362.50	mp 2350
1024	Gd(OH)$_3$ Gadolinium(III) hydroxide	208.27	920 → Gd$_2$O$_3$
1025	Gd$_2$S$_3$ Gadolinium(III) sulfide	410.70	mp 1885
1026	Gd$_2$(SO$_4$)$_3$ (·8H$_2$O) Gadolinium(III) sulfate	602.69	−H$_2$O 400, dec >650
1027	Gd$_2$(SeO$_4$)$_3$ (·8H$_2$O) Gadolinium(III) selenate	743.37	−H$_2$O 130
1028	**Ge Germanium**	72.61	mp 937, bp ca. 2850
1029	GeBr$_2$ Germanium(II) bromide	232.42	subl 385, mp 143 (*p*)
1030	GeBr$_4$ Germanium(IV) bromide	392.23	mp 26.1; bp 186.8
1031	Ge(CH$_3$COO)$_4$ Germaniun(IV) acetate	308.79	mp 156
1032	Ge(CN)$_4$ Germanium(IV) cyanide	176.68	dec 80
1033	GeCl$_2$ Germanium(II) chloride	143.52	dec 75–460
1034	GeCl$_4$ Germaniuum(IV) chloride	214.42	mp −49.5; bp 83.1

(*Continued*)

No.	Color and density	Solubility and reactivity						
		water	alcohol	HCl	H$_2$SO$_4$	HNO$_3$	NaOH	NH$_3$·H$_2$O
989	wh, 2.3–2.5	+	+	+	+	+	+	+
990	wh	i	...	–	–	–	i	–
991	wh, 5.904	psv/+	...	+	+	+	+	+
992	dk-grey, 5.35	–/+	i	+	+	+	+	...
993	wh, 3.69	+	...	–	–/+	–	+	+
994	wh, 3.47	+	...	+	+	+	+	+
995	wh, 2.47	r/+	r	+	–/+	–	+	+
996	wh, 2.42	+	...	+	+	+	+	+
997	wh	r	r	–	–	–	+	+
998	wh, 4.47	i/d	...	–	–/+	–	+	+
999	wh	i/d	...	d/r	d/r	d/r	+	+
1000	yel, 4.15	+	...	+	+/–	+/–	+	+
1001	yel	+	...	+	+	+	+	+
1002	yel-brn, 6.10	–	...	–	–	–	–/+	–
1003	wh hydr, 1.78	r	...	–	–	–	+	+
1004	lt grey	+	...	+	+	+	+	+
1005	wh	r	r	–	–	–	+	+
1006	wh, 6.16	–	...	–/+	–/+	–/+	+	–
1007	wh	i	...	+	+	+	+	i/+
1008	wh	i	...	+	+	+	+	i/+
1009	yel-orange, 2.48	–	...	–	–/+	–/+	–/+	...
1010	wh, 3.26	i	...	i	r	r/–	i	–
1011	yel, 3.75	+	...	+	+	+	+	+
1012	wh hydr	r	r	–	r	–	+	+
1013	lt-grey, 5.6	–	i	–/+	+	–/+	–/+	–
1014	brn, 4.92	i/+	...	+	–/+	–/+	–/+	–
1015	blk, 5.57	i/+	...	+	–/+	–/+	–/+	–
1016	wh, 7.886	psv/+	...	+	+	+	–	–
1017	wh hydr, 2.84	r	r	–	–/+	–	+	+
1018	wh hydr	i	i	–/+	i/+	–/+	i	i
1019	wh, 4.52(2.42)	r	r	r	–	–	+	+
1020	wh	i	i	d/r	d/r	d/r	i	i
1021	lt-yel	r	r	–	–/+	–/+	+	+
1022	wh hydr, 2.41	r	r	–	–	r	+	+
1023	wh, 7.41	–	...	+	+	+	–	–
1024	wh	i	...	+	+	+	i	i
1025	yel, 3.8	i/+	...	+	+	+	–/+	–
1026	wh, 4.14(3.01)	d	...	–	d/r	–	+	+
1027	wh hydr, 3.31	r	...	–	–	–	+	+
1028	lt-grey, 5.350	–	...	–	–/+	–/+	–	–
1029	wh	+	r	+	+	+	+	+
1030	wh, 3.13	+	r	+	+/–	+	+	+
1031	wh	+	d	+	+	+	+	+
1032	wh	+	r	+	+	+	+	+
1033	wh	+	i	+	+	+	+	+
1034	cl lq, 1.880[20]	–/+	r	+	+/–	+	+	+

No.	Formula and name	M_r	Phase transition temperature
1035	GeF_2 Germanium(II) fluoride	110.61	mp 110, dec >160
1036	GeF_4 Germanium(IV) fluoride	148.60	mp −15 (p), subl −36.5
1037	$GeF_4 \cdot 3H_2O$	202.65	dec 150
1038	$[GeF_6],K_2$ Potassium hexafluorogermanate(IV)	264.79	mp 730, bp ca. 835
1039	GeH_4 Monogermane	76.64	mp −165.8; bp −88.5
1040	Ge_2H_6 Digermane	151.27	mp −109, bp 30
1041	Ge_3H_8 Trigermane	225.89	mp −105.6; bp 110.7
1042	Ge_4H_{10} Tetragermane	300.52	bp 177
1043	Ge_5H_{12} Pentagermane	375.15	bp 235
1044	GeI_2 Germanium(II) iodide	326.42	mp 460 (p), dec 440
1045	GeI_4 Germanium(IV) iodide	580.23	mp 146, bp 348
1046	Ge_3N_4 Trigermanium tetranitride	273.86	dec 450–1400
1047	α-GeO_2 Germanium(IV) oxide	104.61	1033 → β-GeO_2
1048	β-GeO_2 $(\cdot nH_2O)$	104.61	−H_2O 380, mp 1116
1049	GeS Germanium(II) sulfide	104.68	mp 655
1050	GeS_2 Germanium(IV) sulfide	136.74	mp 825
1051	GeSe Germanium(II) selenide	151.57	mp 667
1052	$GeSe_2$ Germanium(IV) selenide	230.53	mp 707 dec
1053	GeTe Germanium(II) telluride	200.21	mp 725
1054	**H_2 Dihydrogen**	2.016	mp −259.19; bp −252.87
1055	H_3AsO_4 $(\cdot 0.5H_2O)$ Arsenic acid	141.94	mp hydr 35.5; −H_2O 120
1056	HBO_2 Metaboric acid	43.82	mp 236
1057	HBr Hydrogen bromide	80.91	mp −86.91; bp −66.77
1058	HCN Hydrogen cyanide	27.03	mp −13.3; bp 25.65
1059	H_2CN_2 Hydrogen cyanamide	42.04	mp 43
1060	$H_2C_2O_4$ $(\cdot 2H_2O)$ Oxalic acid	90.03	mp hydr 101.5; mp 189.5
1061	HCOOH Formic acid	46.03	mp 8.3; bp 100.8
1062	HCl Hydrogen chloride	36.46	mp −114.0; bp −85.08
1063	$HClO_4$ $(\cdot H_2O)$ Perchloric acid	100.46	mp −101, bp 120.5
1064	HD Deuteriohydrogen	3.02	mp −256.5; bp −251.02
1065	HF Hydrogen fluoride	20.01	mp −83.36; bp 19.52
1066	Hydrogen iodide	127.91	mp −50.9; bp −35.4
1067	HIO_3 Iodic acid	175.91	mp 110, 220 → I_2O_5
1068	HIO_4 Metaperiodic acid	191.91	subl 110
1069	$H_4I_2O_9$ Hydrogen nonaoxodiiodate(VII)	401.83	>100 → HIO_4
1070	H_5IO_6 Orthoperiodic acid	227.94	mp 122 dec
1071	$HMnO_4$ $(\cdot 2H_2O)$ Permanganic acid	119.94	dec 3, dec hydr 20
1072	HN_3 Hydrogen azide	43.03	mp −80, bp 35.7
1073	HNCO Hydrogen cyanate-N	43.03	mp −80, bp 23.6
1074	HNCS Hydrogen thiocyanate	59.09	mp 5 dec
1075	HNO_3 Nitric acid	63.01	mp −41.6; bp 82.6 dec
1076	$H_2N_2O_2$ Hyponitrous acid	62.03	dec >−6
1077	H_2O Water	18.02	mp 0.00; bp 100.00
1078	H_2O_2 Hydrogen peroxide	34.01	mp −0.43; dec 150
1079	HOF Hydrogen fluorooxygenate(0)	36.01	mp −117, dec >20
1080	$H(PH_2O_2)$ Phosphinic acid	66.00	mp 26.5; dec 140
1081	$H_2(PHO_3)$ Phosphonic acid	82.00	mp 74, dec 200

(Continued)

No.	Color and density	Solubility and reactivity						
		water	alcohol	HCl	H_2SO_4	HNO_3	NaOH	$NH_3 \cdot H_2O$
1035	wh	r/+	...	−/+	−/+	−/+	+	...
1036	cl gas, 6.65	+	...	+	+/−	+	+	...
1037	wh	i/+	...	+	+	+	+	−
1038	wh, 3.32	d/r	i	+	+	+	+	...
1039	cl gas, 3.420	+	r	+	+	+	+	+
1040	cl lq, 1.98^{-109}	+	...	+	+	+	+	+
1041	cl lq, 2.2^{-106}	−	...	+	−/+	−/+	−/+	...
1042	cl lq	−	...	+	−/+	−/+	−/+	...
1043	cl lq	−	...	+	−/+	−/+	−/+	−
1044	dk-yel, 5.37	r/+	...	+	+	+	+	+
1045	dk-orange, 4.32	+	...	+/−	+/−	+	+	+
1046	lt-brn, 5.25	−	...	−	−	−	−	−
1047	wh, 6.24	−	...	−	−	−	−	−
1048	wh, 4.70	−/+	...	+	+	+	+	+
1049	grey-blk, 4.01	i	...	+	+	+	+	−
1050	wh, 2.94	+	i	+	+/−	+/−	+	+
1051	brn-yel, 5.31	i	i	+	−/+	−/+	+	−
1052	orange, 4.56	−	...	−/+	−/+	−/+	−/+	−/+
1053	grey-brn	i/+	...	−/+	−/+	−/+	−/+	−
1054	cl gas, 0.08988	i	d	i	i	i	i	i
1055	wh hydr, 2.25	r/+	r	r	r	r	+	−/+
1056	wh, 1.79	+	...	+	+	+	+	+
1057	cl gas, 3.6445	r	r	r	r/+	r	+	+
1058	cl lq, 0.699^{22}	∞/+	∞	r/+	r/+	r/+	+	+
1059	wh	r/+	r	−	−/+	−/+	+	+
1060	wh	r	r	r/+	r/+	r/+	+	+
1061	cl lq, 1.220^{20}	∞	∞	r	r/+	r/+	+	+
1062	cl gas, 1.6391	r	r	r	r	r/+	+	+
1063	cl lq, 1.664^{20}	∞	r	r	r	r	+	+
1064	cl gas	i	d	i	i	i	i	i
1065	cl lq, 0.99^{13}	r	r	r	r	r/+	+	+
1066	cl gas, 5.7891	r	r	r	r/+	r/+	+	+
1067	wh, 4.63	r	r	r/+	r	r	+	+
1068	wh	r	...	r	r	r	+	+
1069	wh	r/+	r	r/+	r/+	r/+	+	+
1070	wh	r	r	r	r	r	+	+
1071	viol	r/+	r	+	−/+	−/+	+	+
1072	cl lq, 1.13^{20}	∞	∞	r	r	r/+	+	+
1073	cl lq, 1.140^{20}	i/+	r	...	−/+	−/+	+	+
1074	wh	r/+	d	−/+	−/+	−/+	+	+
1075	cl lq, 1.503^{25}	∞	+	r/+	r/+	r	+	+
1076	wh	r/+	r	−	−/+	−/+	+	+
1077	cl lq, 1.00^4	∞	∞	r	r	r	r	r
1078	cl lq, 1.448^{20}	∞	∞	r	r/+	r/+	+	+
1079	cl lq	+	+	+	+	+	+	+
1080	wh, 1.49	r	r	r	r/+	r/+	+	+
1081	wh, 1.65	r	r	r	r/+	r/+	+	+

No.	Formula and name	M_r	Phase transition temperature
1082	$H_2(P_2H_2O_5)$ Hydrogen dihydropentaoxodiphosphate(III)	145.98	mp 38, dec 130
1083	$H(PHO_2F)$ Fluorophosphonic acid	83.99	mp −94, bp 112.4
1084	HPO_3 Metaphosphoric acid	79.98	mp ca. 40 (polymer)
1085	H_3PO_4 Orthophosphoric acid	97.99	mp 42.35; dec 150
1086	$H_4P_2^{IV}O_6$ Hydrogen hexaoxodiphosphate(IV)	161.97	mp 73 dec
1087	$H_4P_2O_7$ Diphosphoric acid	177.97	mp 61, dec 300
1088	HPO_2F_2 Difluorometaphosphoric acid	101.98	mp −96.5; dec >100
1089	H_2PO_3F Fluoroorthophosphoric acid	99.99	mp −30, bp 185 dec
1090	$HPO_2(NH_2)_2$ Diaminometaphosphoric acid	96.03	mp 100 dec
1091	$H_2PO_3(NH_2)$ Aminoorthophosphoric acid	97.01	$100 \rightarrow NH_4PO_3$
1092	$H_4Re_2O_9$ Hydrogen nonaoxodirhenate(VII)	520.44	$160 \rightarrow Re_2O_7$
1093	H_2S Hydrogen sulfide	34.08	mp −85.54; bp −60.35
1094	$H_2(S_2)$ Disulfane	66.15	mp −89.6; bp 70.7
1095	$H_2(S_3)$ Trisulfane	98.21	mp −54, dec 69
1096	$H_2(S_4)$ Tetrasulfane	130.28	soft >−85
1097	$H_2(S_5)$ Pentasulfane	162.35	mp −50
1098	H_2SO_4 Sulfuric acid	98.08	mp 10.4; bp 296 dec
1099	$H_2SO_4 \cdot H_2O$	116.09	mp 8.48; bp 290
1100	$H_2SO_4 \cdot 2H_2O$	134.11	mp −39.5; bp 167
1101	$H_2S_2O_7$ Disulfuric acid	178.14	mp 35.22; dec t
1102	HSO_3Cl Chlorosulfonic acid	116.52	mp −80.5; bp 152 dec
1103	$HSO_3(NH_2)$ Aminosulfonic acid	97.09	mp 207 dec
1104	$H_2SO_3(O_2)$ Peroxosulfuric acid	114.08	mp 45
1105	$H_2S_2O_6(O_2)$ Peroxodisulfuric acid	194.14	mp 65 dec
1106	H_2SO_3S Thiosulfuric acid	114.15	dec −78
1107	H_2Se Hydrogen selenide	80.98	mp −65.72; bp −41.5
1108	H_2SeO_3 Selenious acid	128.97	mp 66.5 dec 70
1109	H_2SeO_4 ($\cdot H_2O$) Selenic acid	144.97	mp hydr 26, mp 62.4
1110	$HTcO_4$ Technetic acid	162.91	$>160 \rightarrow Tc_2O_7$
1111	H_2Te Hydrogen telluride	129.62	mp −51, bp −1.8 dec
1112	H_2TeO_3 Tellurous acid	177.61	$>40 \rightarrow TeO_2$
1113	H_6TeO_6 Orthotelluric acid	229.64	mp 136, $220 \rightarrow TeO_3$
1114	**He Helium**	4.003	mp −271 (p), bp −269
1115	**Hf Hafnium**	178.49	mp 2230, bp ca. 4620
1116	HfB_2 Hafnium diboride	200.11	mp 3250
1117	$HfBr_4$ Hafnium(IV) bromide	498.11	subl 332, mp 420 (p)
1118	$HfC_{0.94}$ Hafnium 0.94-carbide	189.78	mp 3890
1119	$[Hf(C_5H_5)_2Cl_2]$ Bis(cyclopentadienyl)dichlorohafnium	379.59	mp 231.5
1120	$HfCl_4$ Hafnium(IV) chloride	320.30	subl 317, mp 435 (p)
1121	$HfCl_2O$ ($\cdot 8H_2O$) Hafnium dichloride-oxide	265.40	−H_2O 65, dec >300
1122	HfF_4 ($\cdot 3H_2O$) Hafnium(IV) fluoride	254.48	subl 974
1123	$[HfF_6],K_2$ Potassium hexafluorohafnate(IV)	370.67	mp 608 dec
1124	$[HfF_7],(NH_4)_3$ Ammonium heptafluorohafnate(IV)	365.59	dec >240
1125	$[Hf(H_2O)_4(NO_3)_2(OH)_2]$ Tetraaquadinitratodihydroxohafnium	408.57	$640 \rightarrow HfO_2$
1126	$[Hf(H_2O)_4(SO_4)_2]$ Tetraaquadisulfatohafnium	442.67	dec 800

No.	Color and density	Solubility and reactivity						
		water	alcohol	HCl	H_2SO_4	HNO_3	NaOH	$NH_3 \cdot H_2O$
1082	wh	+	...	+	+	+	+	+
1083	cl lq, 1.58[25]	r/+	r	r	r/+	r/+	+	+
1084	wh, 2.2–2.5	r/+	r	r/+	r/+	r/+	+	+
1085	wh, 1.83	r	r	r	r	r	+	+
1086	wh	r/+	...	+	+	+	+	+
1087	wh	r/+	r	r	r	r	+	+
1088	cl lq, 1.583[25]	+	...	+	+	+	+	+
1089	cl lq, 1.818[25]	+	...	+	+	+	+	+
1090	wh	r/+	...	+	+	+	+	+
1091	wh	+	...	+	+	+	+	+
1092	lt-yel, 4.87	+	...	−	−	−	+	+
1093	cl gas, 1.539	d	r	d	d/+	d/+	+	+
1094	lt-yel lq, 1.334[20]	+	r	−	−/+	−/+	+	+
1095	lt-yel lq, 1.491[20]	+	r	−	−/+	−/+	+	+
1096	yel lq, 1.582[20]	+	r	−	−/+	−/+	+	+
1097	grn-yel lq, 1.644[20]	+	r	−	−/+	−/+	+	+
1098	cl lq, 1.834[20]	∞	+	r	r	r	+	+
1099	cl lq, 1.788[20]	∞	+	r	r	r	+	+
1100	cl lq, 1.650[20]	∞	+	r	r	r	+	+
1101	wh, 1.9	+	+	+	+	+	+	+
1102	cl lq, 1.79[20]	+	+	+	+	+	+	+
1103	wh, 2.13	r/+	i	r/+	r/+	r/+	+	+
1104	wh	+	r	+	+	+	+	+
1105	wh	+	r	+	+	+	+	+
1106	cl lq	+	+	+	+	+	+	+
1107	cl gas, 3.6643	r	...	r	r/+	r/+	+	+
1108	wh, 3.01	r	r	r	r	r	+	+
1109	wh, 2.95(2.63)	r	+	r/+	r	r	+	+
1110	red	r	...	r	r	r	+	+
1111	cl gas, 5.81	+	r	+	+	+	+	+
1112	wh, 3.05	d	i	d/+	d/+	d/+	+	+
1113	wh, 3.07	r	i	r	r	r	+	+
1114	cl gas, 0.17847	i	i	−	−	−	−	−
1115	wh, 13.29	−	−	−	−/+	−	−	−
1116	grey	−	i	−	−	−	−	−
1117	wh, 4.90	+	...	+	+	+	+	+
1118	grey, 12.68	−	...	−	−/+	−	−	−
1119	wh	−/+	r	−	−/+	+	−	−
1120	wh	+	r	+	+	+	+	+
1121	wh	r/+	i	+	+	+	+	+
1122	wh, 7.13	i/+	...	−	−	−	i	i
1123	wh	d	...	−	−	−	−/+	−
1124	wh, 2.80	r	...	−	−	−	−/+	−
1125	wh	+	r	+	+	+	+	+
1126	wh	r	...	−	+	−	+	+

No.	Formula and name	M_r	Phase transition temperature
1127	HfI$_4$ Hafnium(IV) iodide	686.11	subl 392, mp 455 (p)
1128	HfN Hafniun mononitride	192.50	mp >3000
1129	HfO$_2$ Hafnium(IV) oxide	210.49	mp 2780, bp ca. 5400
1130	HfO$_2 \cdot n$H$_2$O	–	140 → HfO(OH)$_2$
1131	HfO(OH)$_2$ Hafnium oxide-dihydroxide	228.50	1000 → HfO$_2$
1132	**Hg Mercury**	200.59	mp −38.862; bp 356.66
1133	Hg$_3$(AsO$_4$)$_2$ Mercury(II) arsenate	879.61	dec >300
1134	HgBr$_2$ Mercury(II) bromide	360.40	mp 238.1; bp 319
1135	Hg$_2$Br$_2$ Dimercury dibromide	560.99	subl ca. 390
1136	Hg(BrO$_3$)$_2$ (·2H$_2$O) Mercury(II) bromate	456.39	dec hydr >130
1137	Hg$_2$(BrO$_3$)$_2$ Dimercury dibromate	656.98	dec t
1138	Hg$_2$C$_2$ Dimercury acetylide	425.20	dec 100
1139	Hg(CH$_3$COO)$_2$ Mercury(II) acetate	318.68	mp ca. 180 dec
1140	Hg(CN)$_2$ Mercury(II) cyanide	252.63	dec 320
1141	Hg(CNO)$_2$ (·0.5H$_2$O) Mercury(II) fulminate	284.62	dec hydr t
1142	Hg$_2$CO$_3$ Dimercury carbonate	461.19	130 → Hg,HgO
1143	HgCl$_2$ Mercury(II) chloride	271.50	mp 280, bp 301.8
1144	Hg$_2$Cl$_2$ Dimercury dichloride	472.09	dec ca. 400
1145	Hg(ClO$_3$)$_2$ Mercury(II) chlorate	367.49	dec ca. 300
1146	Hg$_2$(ClO$_3$)$_2$ Dimercury dichlorate	568.08	dec 250
1147	HgF$_2$ (·2H$_2$O) Mercury(II) fluoride	238.59	mp 645, bp 647
1148	Hg$_2$F$_2$ Dimercury difluoride	439.18	mp 570
1149	α-HgI$_2$ Mercury(II) iodide	454.40	131 → β-HgI$_2$
1150	β-HgI$_2$	454.40	mp 256, bp 354
1151	Hg$_2$I$_2$ Dimercury diiodide	654.99	subl 140, mp 290
1152	[HgI$_4$],Ag$_2$ Silver(I) tetraiodomercurate(II)	923.94	dec >158
1153	[HgI$_4$],Ba (·5H$_2$O) Barium tetraiodomercurate(II)	845.53	dec hydr t
1154	[HgI$_3$],K (·H$_2$O) Potassium triiodomercurate(II)	620.40	−H$_2$O 60, mp 105
1155	[HgI$_4$],K$_2$ (·2H$_2$O) Potassium tetraiodomercurate(II)	786.40	dec hydr 400
1156	Hg(NCS)$_2$ Mercury(II) thiocyanate	316.76	dec 165
1157	Hg(NO$_3$)$_2$ (·H$_2$O) Mercury(II) nitrate	324.60	dec hydr >45
1158	Hg$_2$(NO$_2$)$_2$ Dimercury dinitrite	493.19	dec 100
1159	Hg$_2$(NO$_3$)$_2$ (·2H$_2$O) Dimercury dinitrate	525.19	mp hydr 70 dec
1160	HgO Mercury(II) oxide	216.59	dec 400–500
1161	α-HgS Mercury(II) sulfide	232.66	344 → β-HgS
1162	β-HgS	232.66	mp 820 (p)
1163	Hg(S$_2$) Mercury(II) disulfide(2−)	264.72	dec 390
1164	HgSO$_4$ Mercury(II) sulfate	296.65	dec >550
1165	Hg$_2$SO$_4$ Dimercury sulfate	497.24	dec 550–600
1166	HgSe$_{1 \pm x}$ Mercury(II) selenide	279.55	mp 799
1167	HgTe Mercury(II) telluride	328.19	mp 667
1168	**Ho Holmium**	164.930	mp 1470, bp 2707
1169	HoBr$_3$ (·6H$_2$O) Holmium(III) bromide	404.64	mp 919, bp 1336
1170	Ho$_2$(C$_2$O$_4$)$_3$ (·10H$_2$O) Holmium(III) oxalate	593.91	hydr 600 → Ho$_2$O$_3$
1171	HoCl$_3$ (·6H$_2$O) Holmium(III) chloride	271.29	mp 718, bp 1517
1172	HoF$_3$ Holmium(III) fluoride	221.92	mp 1143
1173	HoI$_3$ Holmium(III) iodide	545.64	mp 989, bp 1300

(Continued)

No.	Color and density	Solubility and reactivity						
		water	alcohol	HCl	H_2SO_4	HNO_3	NaOH	$NH_3 \cdot H_2O$
1127	yel-orange, 5.60	+	...	+	+	+	+	+
1128	dk-brn	−	...	−	−	−	−/+	−
1129	wh, 9.68	−	...	+	−/+	−	−	−
1130	yel	i	...	+	+	+	i/+	i
1131	wh	i	...	−/+	−/+	−/+	i/+	i
1132	lt-grey lq, 13.5461[20]	−	−	−	−/+	+	−	−
1133	yel	i	i	r	r	r	i	i
1134	wh, 6.11	d	r	−	−/+	−	+	...
1135	wh, 7.31	i	i	−	−/+	−/+	i	i
1136	wh hydr	d	...	+	−	−	+	...
1137	wh	+	...	+	+	−	+	+
1138	wh	i/+	...	+	+	+	−	+
1139	wh, 3.27	r/+	r	+	−	−	+	+
1140	wh, 3.40	r/+	r	−	−/+	−/+	−	+
1141	wh 4.42	i/r	r	−/+	−/+	−/+	+	+
1142	wh ($t \rightarrow$ grey)	i/+	i	+	+	+	−	−
1143	wh, 5.44	r	r	r	−	r	+	...
1144	wh, 7.15	i	i	i	−/+	−/+	−	−
1145	wh, 5.0	r/+	...	+	−	−/+	+	+
1146	wh, 6.41	r/+	r	+	+	r	+	+
1147	wh, 8.95	+	...	r	r	r	+	+
1148	lt-yel, 8.73	d/+	...	+	+	−	+	+
1149	red, 6.28	i	d	i	i/+	+	i	−
1150	yel, 6.27	i	d	i	i/+	+	i	−
1151	yel, 7.70	i	i	i	i/+	i/+	i	+
1152	yel, 6.02	i	i	+	+	+	−/+	−/+
1153	red hydr	r	r	−	−/+	+	−	+
1154	lt-yel hydr	r/+	r	−	−/+	+	−	+
1155	lt-yel hydr	r	r	−	−/+	+	−	+
1156	wh, 3.71	d/r	r	+	−/+	−/+	+	−
1157	wh hydr, 4.30	+	i	+	r	r	+	+
1158	yel, 7.33	+	...	+	−/+	−/+	+	+
1159	wh, 7.79(4.78)	+	...	+	r	r	+	+
1160	yel, 1.14	−/+	i	+	+	+	−	−
1161	red, 8.10	i/+	i	−	−/+	−/+	i	i
1162	blk, 7.65	i/+	i	−	−/+	−/+	i	i
1163	wh	i/+	...	+	+	+	−/+	−
1164	wh, 6.47	+	i	+	+	+	+	+
1165	wh, 7.56	i/+	...	+	+	+	+	...
1166	blk, 8.27	i	...	−	−/+	−/+	−	−
1167	blk	i	...	−	−	−/+	−	−
1168	wh, 8.799	psv/+	...	+	+	+	−	−
1169	yel	r	r	−	−/+	−	+	+
1170	yel hydr	i	i	i/+	i/+	i/+	i	i
1171	yel hydr, 3.715	r	r	r	−	−	+	+
1172	yel, 7.644	i	i	−	−	−	−	−
1173	yel	r	r	−	−/+	−/+	+	+

No.	Formula and name	M_r	Phase transition temperature
1174	Ho(NO$_3$)$_3$ (·5H$_2$O) Holmium(III) nitrate	350.94	dec hydr 560
1175	Ho$_2$O$_3$ Holmium(III) oxide	377.86	mp 2360
1176	Ho(OH)$_3$ Holmium(III) hydroxide	215.95	850 → Ho$_2$O$_3$
1177	Ho$_2$(SO$_4$)$_3$ (·8H$_2$O) Holmium(III) sulfate	618.05	dec hydr >850
1178	**I$_2$ Diiodine**	253.808	mp 113.5; bp 184.35
1179	[IAg$_2$]NO$_3$ Di{silver(I)}iodine(−I) nitrate	404.64	mp 94
1180	IBr Iodine monobromide	206.81	mp 40.5; bp 116 dec
1181	[I(Br)Cl],Cs Cesium bromochloroiodate(I)	375.17	mp 235, dec 290
1182	[I(Br)$_2$],Cs Cesium dibromoiodate(I)	419.62	mp 248, dec 320
1183	[I(Br)$_2$],K Potassium dibromoiodate(I)	325.81	mp 58, dec 180
1184	ICl Iodine monochloride	162.36	mp 27.19; bp 97.4 dec
1185	I$_2$Cl$_6$ Diiodine hexachloride	466.53	dec 77, mp 101 (*p*)
1186	[I(Cl)$_2$],Cs Cesium dichloroiodate(I)	330.72	mp 238, dec 290
1187	[ICl$_4$],Cs Cesium tetrachloroiodate(III)	401.62	mp 228 dec
1188	[I(Cl)$_2$],K Potassium dichloroiodate(I)	236.91	dec 215
1189	[ICl$_4$],K Potassium tetrachloroiodate(III)	307.81	mp 116 dec
1190	IF Iodine monofluoride	145.90	mp −14 dec
1191	IF$_3$ Iodine trifluoride	183.90	subl −28 dec
1192	IF$_5$ Iodine pentafluoride	221.89	mp 9.421; bp 104.48
1193	IF$_7$ Iodine heptafluoride	259.89	mp 6.4 (*p*), dec 530
1194	[I(I)$_2$],Cs Cesium diiodoiodate(I)	513.62	mp 215 dec
1195	[I(I)$_2$],K (·H$_2$O) Potassium diiodoiodate(I)	419.81	mp hydr 38, dec hydr 225
1196	[I(I)$_2$],NH$_4$ Ammonium diiodoiodate(I)	398.75	dec 175
1197	[I(I)$_2$],Rb Rubidium diiodoiodate(I)	466.18	mp 194 dec
1198	I$_3$N (·*n*NH$_3$) Triiodine nitride	394.72	dec hydr >20
1199	I$_2$O$_5$ Diiodine pentaoxide	333.80	dec 275–350
1200	IO$_2$F$_3$ Iodine dioxide-trifluoride	215.90	mp 42.5; bp 147
1201	[I(O)$_2$F$_4$],H Hydrogen dioxotetrafluoroiodate(VII)	235.90	mp 36, dec 130
1202	(IO)IO$_3$ Oxoiodine(III) iodate	317.80	dec 130
1203	**In Indium**	114.82	mp 156.634; bp 2024
1204	InAs Indium monoarsenide	189.74	mp 943
1205	InBr Indium(I) bromide	194.72	mp 285, bp 662
1206	InBr$_3$ (·5H$_2$O) Indium(III) bromide	354.53	subl 371, mp 419.7 (*p*)
1207	[In^{III}Br$_4$],In Indium(I) tetrabromoindate(III)	549.26	mp 198, bp 638
1208	InCl Indium(I) chloride	150.27	mp 225, bp 653
1209	InCl$_3$ (·4H$_2$O) Indium(III) chloride	221.18	subl 498, mp 583 (*p*)
1210	[In^{III}Cl$_4$],In Indium(I) tetrachloroindate(III)	371.45	mp 240, bp 655
1211	In(ClO$_4$)$_3$ (·8H$_2$O) Indium(III) perchlorate	413.17	mp hydr 80, dec 200
1212	InF$_3$ (·3H$_2$O) Indium(III) fluoride	171.81	mp 1172, bp >1200
1213	InI Indium(I) iodide	241.72	mp 365, bp 743
1214	InI$_3$ Indium(III) iodide	495.53	mp 210, bp 447
1215	[In^{III}I$_4$],In Indium(I) tetraiodoindate(III)	737.26	mp 225
1216	InN Indium mononitride	128.83	mp 1200
1217	In(NO$_3$)$_3$ (·5H$_2$O) Indium(III) nitrate	300.83	−H$_2$O 100, *t* → In$_2$O$_3$
1218	In$_2$O$_3$ Indium(III) oxide	277.64	mp 1910, bp ca. 3300
1219	In(OH)$_3$ Indium(III) hydroxide	165.84	ca. 150 → In$_2$O$_3$
1220	InP Indium monophosphide	145.79	mp 1062 (*p*)

(Continued)

No.	Color and density	Solubility and reactivity						
		water	alcohol	HCl	H_2SO_4	HNO_3	NaOH	$NH_3 \cdot H_2O$
1174	yel hydr	r	r	−	−	r	+	+
1175	yel, 8.24	−/+	−	+	+	+	−	−
1176	yel	i	i	+	+	+	i	i
1177	yel hydr	r/d	...	−	r	−	+	+
1178	viol-blk, 4.93	+	r	−	−/+	−/+	+	+
1179	wh	+	...	+	+	+	+	+
1180	blk-brn, 4.42	+	r	+	+	+	+	+
1181	yel-red	r	r	−	−/+	−/+	+	...
1182	red, 4.25	r	r	−	−/+	−/+	+	...
1183	red	r	r	−	−/+	−/+	+	...
1184	dk-red, 3.18	+	r/+	+	+	+	+	+
1185	orange-yel, 3.20	+	r	+	+	+	+	+
1186	orange, 3.86	r	r	−	−/+	−/+	+	...
1187	lt-orange, 3.37	d	r/+	−	−/+	−/+	+	+
1188	orange	r/+	r	−	−/+	−/+	+	...
1189	yel, 1.76	r/+	r	−	−/+	−/+	+	+
1190	red	+	+	+	+	+	+	+
1191	yel	+	+	+	+	+	+	+
1192	cl lq, 3.231[15]	+	+	+	+	+	+	+
1193	cl lq, 2.8[6]	+	+	+	+	+	+	+
1194	blk, 4.47	i	r	−	−/+	−/+	+	...
1195	dk-brn hydr	r	r	−	−/+	−/+	+	...
1196	dk-brn, 3.75	r/+	r	−	−/+	−/+	+	...
1197	blk, 4.03	r	r	−	−/+	−/+	+	...
1198	red-brn hydr	+	i	+	+	+	+	+
1199	wh, 4.80	+	i	+	+	+	+	+
1200	yel	+	+	+	+	+	+	+
1201	wh	+	+	+	+	+	+	+
1202	yel, 4.97	i/+	...	−	−/+	−/+	+	+
1203	wh, 7.30	−	...	+	+	+	−	−
1204	dk-grey, 5.666	−	i	+	+	+	−	−
1205	red, 4.96	+	...	+	+	+	+	+
1206	wh, 4.74	r	r	r	−/+	−	+	+
1207	wh, 4.22	+	...	+	+	+	+	+
1208	yel, 4.18	+	i	+	+	+	+	+
1209	wh, 3.46	r	r	r	−	−	+	+
1210	wh, 3.65	+	...	+	+	+	+	+
1211	wh hydr	r/+	r	−	−	−	+	+
1212	wh, 4.39	d/+	i	−	−	−	+	+
1213	dk-red, 5.32	+	i	+	+	+	+	+
1214	yel, 4.68	r	r	−	−/+	−/+	+	+
1215	red-brn, 4.71	+	...	+	+	+	+	+
1216	grey	−	i	+	−/+	−/+	−	−
1217	wh	r	r	−	−	r	+	+
1218	lt-yel, 7.18	−	...	+	+	+	−	−
1219	wh, 4.33	i	...	+	+	+	i/+	i
1220	dk-grey, 4.787	−	...	+	+	+	−	−

No.	Formula and name	M_r	Phase transition temperature
1221	In_2S Indium(I) sulfide	261.71	mp 653
1222	In_2S_3 Indium(III) sulfide	325.84	mp 1072
1223	$In_2(SO_4)_3$ ($\cdot 9H_2O$) Indium(III) sulfate	517.83	$-H_2O$ 200, dec 600
1224	InSb Indiumantimony	236.57	mp 546
1225	In_2Se_3 Indium(III) selenide	466.52	mp 900
1226	In_2Te_3 Indium(III) telluride	612.44	mp 670
1227	**Ir Iridium**	192.22	mp 2443, bp 4380
1228	$[IrBr_6],Na_3$ ($\cdot 12H_2O$) Sodium hexabromoiridate(III)	740.61	mp hydr 100, $-H_2O$ 150
1229	$[Ir(C_2O_4)_3],K_3$ ($\cdot 4H_2O$) Potassium trioxalatoiridate(III)	573.57	$-H_2O$ >240
1230	$[Ir_2(CO)_8]$ Octacarbonyldiiridium	608.52	subl 160
1231	$[Ir_4(CO)_{12}]$ Dodecacarbonyltetrairidium	1105.00	subl 210
1232	$IrCl_3$ ($\cdot nH_2O$) Iridium(III) chloride	298.58	>760 → Ir
1233	$IrCl_4$ Iridium(IV) chloride	334.03	dec ca. 700
1234	$[IrCl_6],K_2$ Potassium hexachloroiridate(IV)	483.13	dec ca. 600
1235	$[IrCl_6],K_3$ Potassium hexachloroiridate(III)	522.23	dec ca. 500
1236	$[IrCl_6],K_4$ Potassium hexachloroiridate(II)	561.33	dec >700
1237	$[IrCl_6],(NH_4)_2$ Ammonium hexachloroiridate(IV)	441.02	>200 → Ir,NH_4Cl
1238	$[IrCl_6],Na_2$ ($\cdot 6H_2O$) Sodium hexachloroiridate(IV)	450.92	dec hydr < 600
1239	$[IrCl_6],Na_3$ ($\cdot 12H_2O$) Sodium hexachloroiridate(III)	473.91	$-H_2O$ 50, dec 800
1240	IrF_3 Iridium(III) fluoride	249.21	dec 250
1241	IrF_4 Iridium(IV) fluoride	268.21	mp 106, dec 400
1242	IrF_5 Iridium(V) fluoride	287.21	mp 104.5
1243	IrF_6 Iridium(VI) fluoride	306.21	mp 44.1; bp 53.6
1244	$[Ir_2(NH_3)_{10}]$ Decaamminediiridium	554.75	dec 90
1245	$[Ir(NH_3)_6]Cl_3$ Hexaammineiridium(III) chloride	400.77	dec 500
1246	$[Ir(NH_3)_5Cl]Cl_2$ Pentaamminechloroiridium(III) chloride	383.73	dec 550
1247	$[Ir(NH_3)_6](NO_3)_3$ Hexaammineiridium(III) nitrate	480.42	dec >500
1248	IrO_2 Iridium(IV) oxide	224.22	dec >800
1249	Ir_2O_3 Iridium(III) oxide	432.44	dec 400–500
1250	$Ir_2O_3 \cdot nH_2O$	–	dec 300
1251	$Ir(OH)_4$ Iridium(IV) hydroxide	260.25	350 → IrO_2
1252	IrS_2 Iridium(IV) sulfide	256.35	dec 300
1253	Ir_2S_3 Iridium(III) sulfide	480.64	dec t
1254	$Ir_2(SO_4)_3$ ($\cdot nH_2O$) Iridium(III) sulfate	672.63	dec hydr >300
1255	**K Potassium**	39.098	mp 63.5; bp 760
1256	$KAg(NO_3)_2$ Potassium-silver(I) nitrate	270.97	mp 135; 300 → KNO_2,Ag
1257	$KAl(SO_4)_2$ ($\cdot 12H_2O$) Potassium-aluminum sulfate	258.20	mp hydr 92, $-H_2O$ 120
1258	$KAl_3(SO_4)_2(OH)_6$ Potassium-trialuminum bis(sulfate)-hexahydroxide	414.21	dec 800
1259	KBO_2 ($\cdot 1.33H_2O$) Potassiuim metaborate	81.91	$-H_2O$ 250, mp 940
1260	$K_2B_4O_7$ ($\cdot 8H_2O$) Potassium tetraborate	233.43	dec hydr >420, mp 815
1261	KBr Potassium bromide	119.00	mp 734, bp 1380
1262	$KBrO_3$ Potassium bromate	167.00	mp 434 dec
1263	$KBrO_4$ Potassium perbromate	183.00	275 → $KBrO_3$
1264	$K(CH_3COO)$ ($\cdot 1.5H_2O$) Potassium acetate	98.14	mp 310
1265	$K_2(C_4H_4O_6)$ ($\cdot 0.5H_2O$) Potassium tartrate	226.27	dec hydr >150

(Continued)

No.	Color and density	Solubility and reactivity						
		water	alcohol	HCl	H_2SO_4	HNO_3	NaOH	$NH_3 \cdot H_2O$
1221	blk, 5.87	i/+	i	+	+	+	−/+	−
1222	dk-red, 4.65	i	...	+	+	+	+	−
1223	wh hydr, 3.44	r	...	−	r	−	+	+
1224	grey, 5.755	−	−	+	+	+	−	−
1225	viol-blk, 5.67	i	...	−/+	−/+	−/+	−	−
1226	blk, 5.80	i	...	−/+	−/+	−/+	−	−
1227	wh, 22.421	−	−	−	−	−	−	−
1228	dk-grn hydr	r	...	−	−/+	+	+	+
1229	orange hydr, 2.51	r	i	−	−/+	−/+	−/+	−
1230	yel-grn	i/+	r	−	−/+	−/+	−	−
1231	yel	i/+	r	−	−/+	−	−/+	−
1232	dk-grn hydr, 5.30	i	i	i/+	i/−	i/−	−	i/−
1233	brn	r/+	r	+	−	r	+	+
1234	dk-red, 3.55	d	i	−	−	−	−/+	−
1235	dk-grn	r	i	−/+	−/+	−/+	−/+	−/+
1236	blk, 3.55	r	i	r	−/+	−/+	−/+	−
1237	red-blk, 2.86	d/r	i	−	−	−	+	+
1238	dk-red hydr	r	r	−	−	−	+	+
1239	dk grn	r	...	−	−/+	+	+	+
1240	blk	i	r	−/+	−	−	−/+	−/+
1241	yel	+	r	+	−	−	+	+
1242	yel	+	r/+	+	+	+	+	+
1243	yel, ca. 6	+	+	+	+	+	+	+
1244	yel	+	...	+	+	+	−/+	+/r
1245	wh, 2.43	r	...	+	+	+	...	r
1246	lt-yel, 2.68	d/r	...	i	+	+	−	d
1247	wh, 2.39	d	r	+	+	+	−	r
1248	blk-bl, 11.655	−	i	+	−	−	−	−
1249	bl-blk	−	...	+	+	+	+	+
1250	dk-grn	i	...	+	+	+	−/+	−/+
1251	bl-blk, 3.15	i	...	+	−/+	+	i	i
1252	brn-blk, 8.43	i	...	−	−/+	−/+	i	−
1253	brn-blk, 9.64	i	...	−	−/+	−/+	i	−
1254	yel hydr	r	r	−	r	−	+	+
1255	wh, 0.86	+	+	+	+	+	+	+
1256	wh, 3.22	r	r	+	+	r	+	+
1257	wh, 2.75(1.76)	r	i	r	r	r	+	+
1258	wh, 2.7	i	i	−/+	−/+	−/+	−/+	−
1259	wh	+	i	+	+	+	+	+
1260	wh, 1.74	r	...	−	−	−	−/+	−
1261	wh, 2.75	r	d	r	r/+	r	r	r
1262	wh, 3.27	r	d	r	r	r	r	r
1263	wh	r	...	−	−	−	−	...
1264	wh, 1.57	r	r	r	r	r	r	r
1265	wh hydr	i	...	−/+	−/+	−/+	−	−

No.	Formula and name	M_r	Phase transition temperature
1266	KCN Potassium cyanide	65.12	mp 634.5, bp 1625
1267	K_2CO_3 ($\cdot 1.5H_2O$) Potassium carbonate	138.20	mp 891, dec >1200
1268	$K_2C_2O_4$ ($\cdot H_2O$) Potassium oxalate	166.21	dec hydr 100–160
1269	$K_2Ca(SO_4)_2$ ($\cdot H_2O$) Dipotassium-calcium sulfate	310.40	mp 1004
1270	$K_2Cd(SO_4)_2$ ($\cdot 2H_2O$) Dipotassium-cadmium(II) sulfate	382.73	dec hydr 200
1271	KCl Potassium chloride	74.55	mp 770, bp 1430
1272	$KClO_3$ Potassium chlorate	122.55	mp 357. dec 400
1273	$KClO_4$ Potassium perchlorate	138.55	mp 525, 620 → KCl
1274	$K_2Co(SO_4)_2$ ($\cdot 6H_2O$) Dipotassium-cobalt(II) sulfate	329.25	dec hydr 210
1275	K_2CrO_4 Potassium chromate	194.19	mp 968.3
1276	$K_2Cr_2O_7$ Potassium dichromate	294.18	mp 397.5; dec ca. 600
1277	$K_2Cr_3O_{10}$ Potassium decaoxotrichromate(VI)	394.17	dec 243
1278	$K_2Cr_4O_{13}$ Potassium 13-oxotetrachromate(VI)	494.17	dec 210
1279	$K(CrO_3Cl)$ Potassium chlorochromate	174.54	dec >100
1280	$KCr(SO_4)_2$ ($\cdot 12H_2O$) Potassium-chromium(III) sulfate	283.22	mp hydr 89, $-H_2O$ 400
1281	KF Potassium fluoride	58.10	mp 857, bp 1505
1282	$KF\cdot 2H_2O$	94.13	mp 42, bp 156
1283	K_2FeO_4 Potassium ferrate	198.04	dec 700
1284	$KFe(SO_4)_2$ ($\cdot 12H_2O$) Potassium-iron(III) sulfate	287.07	mp hydr 33
1285	$K_2Fe(SO_4)_2$ ($\cdot 6H_2O$) Dipotassium-iron(II) sulfate	326.17	dec hydr ca. 200
1286	K_2GeO_3 Potassium germanate	198.80	mp 830
1287	KH Potassium hydride	40.11	mp 400 (p)
1288	KH_2AsO_4 Potassium dihydroarsenate	180.03	mp 288 dec
1289	K_2HAsO_4 Potassium hydroarsenate	218.12	dec 30
1290	$KHC_4H_4O_6$ Potassium hydrotartrate	188.18	dec >460
1291	$KHCO_3$ Potassium hydrocarbonate	100.11	dec 100–120
1292	KHC_2O_4 Potassium hydrooxalate	128.12	dec t
1293	$K(HCOO)$ Potassium formate	84.12	mp 168.7; dec >250
1294	$K(HF_2)$ Potassium hydrodifluoride	78.10	mp 238.7; >310 → KF
1295	$KH(PHO_3)$ Potassium hydrophosphonate	120.09	dec t
1296	KH_2PO_4 Potassium dihydroorthophosphate	136.08	mp 252.6 dec
1297	K_2HPO_4 ($\cdot 3H_2O$) Potasiium hydroorthophosphate	174.17	250 → $K_4P_2O_7$
1298	KHS ($\cdot 0.5H_2O$) Potassium hydrosulfide	72.17	$-H_2O$ t, mp 455
1299	$KHSO_3$ Potassium hydrosulfite	120.17	dec ca. 190
1300	$KHSO_4$ Potassium hydrosulfate	136.17	mp 218.6; dec 240–340
1301	KI Potassium iodide	166.00	mp 681, bp 1324
1302	KIO_3 Potassium iodate	214.00	mp 560 dec
1303	KIO_4 Potassium metaperiodate	230.00	mp 582 (p)
1304	$KMgCl_3$ ($\cdot 6H_2O$) Potassium-magnesium chloride	169.76	mp hydr 116, mp 487
1305	$K_2Mg(SO_4)_2$ ($\cdot 6H_2O$) Dipotassium-magnesium sulfate	294.63	dec hydr ca. 72
1306	$KMg(SO_4)Cl$ ($\cdot 3H_2O$) Potassium-magnesium sulfate-chloride	194.92	dec hydr >200
1307	$K_2Mg(SeO_4)_2$ ($\cdot 6H_2O$) Dipotassium-magnesium selenate	388.41	$-H_2O$ 180
1308	$KMnO_4$ Potassium permanganate	158.03	dec >240

(Continued)

No.	Color and density	Solubility and reactivity						
		water	alcohol	HCl	H_2SO_4	HNO_3	NaOH	$NH_3 \cdot H_2O$
1266	wh, 1.52	r	d	r	r	r	r	r
1267	wh, 2.43	r	i	+	+	+	r	r
1268	wh hydr, 2.13	r	i	−/+	−/+	−/+	r	r
1269	wh hydr, 2.57	d	i	−	+	−	+	+
1270	wh hydr, 2.92	r	i	−	r	−	+	+
1271	wh, 1.98	r	d	r/i	r/+	r	r	r
1272	wh, 2.32	r	r	−/+	−	r	−	−
1273	wh, 2.52	r	i	−	−	−	−	−
1274	red hydr, 2.22	r	i	−	r	i	+	+
1275	yel, 2.73	r	i	+	+	+	r	r
1276	orange-red, 2.68	r	i	r/+	r	r	+	+
1277	dk-red	+	...	−/+	−	−	+	+
1278	brn-red	+	...	−/+	−	−	+	+
1279	orange, 2.50	+	r	+	+	+	+	+
1280	dk-viol hydr, 1.84	r	i	−	r	−	+	+
1281	wh, 2.48	r	i	r	r/+	r	−	−
1282	wh, 2.45	r	i	r	r/+	r	−	−
1283	red-viol	r	...	+	+	+	−/+	+
1284	wh hydr, 1.83	r	i	−	r	−	+	+
1285	lt-grn hydr, 2.17	r	i	−	r	−/+	+	+
1286	wh, 3.40	r	...	+	+	+	−	−
1287	wh, 1.43	+	+	+	+	+	+	+
1288	wh, 2.87	r	i	r	r	r	+	+
1289	wh	r	i	r	r	r	+	+
1290	wh	d	i	−	−	−	+	+
1291	wh, 2.17	r	i	+	+	+	+	+
1292	wh, 2.04	r	i	−/+	−/+	−/+	+	+
1293	wh, 1.91	r	d	r	r/+	r/+	r	r
1294	wh, 2.37	r	i	−/+	−/+	−/+	+	+
1295	wh	r	i	−	−	−	+	+
1296	wh, 2.34	r	i	r	r	r	+	+
1297	wh hydr, 2.33	r	r	r	r	r	+	+
1298	wh, 1.68	r/+	r	+	+	+	+	+
1299	wh	r	i	−/+	−/+	−/+	+	+
1300	wh, 2.32	r	i	−	r	−	+	+
1301	wh, 3.12	r	r	r	r/+	r/+	r	r
1302	wh, 3.93	r	i	r/+	r	r	−	−
1303	wh, 3.62	d	d	d	d	d	d	d
1304	wh hydr, 1.60	r	...	r	−	−	+	+
1305	wh hydr, 2.03	r	...	−	r	−	+	+
1306	wh, 2.15	r	...	r	r	−	+	+
1307	wh hydr, 2.36	r	i	−/+	r/+	−	+	+
1308	dk-viol, 2.70	r	+	r/+	r/+	r/+	r/+	+

No.	Formula and name	M_r	Phase transition temperature
1309	K_2MnO_4 Potassium manganate	197.13	dec >500
1310	K_2MoO_4 ($\cdot nH_2O$) Potassium molybdate	238.13	mp 926, dec 1400
1311	KN_3 Potassium azide	81.12	mp 343, dec >355
1312	$K_3N(?)$ Tripotassium nitride	131.30	dec t
1313	KNCS Potassium thiocyanate	97.18	mp 173.2; bp 500 dec
1314	KNCSe Potassium selenocyanate	144.08	dec 100–158
1315	KNH_2 Potassium amide	55.12	mp 338 dec 700
1316	KNO_2 Potassium nitrite	85.10	mp 440 dec >900
1317	KNO_3 Potassium nitrate	101.10	mp 334.5; 400 $\rightarrow$ KNO_2
1318	$KNa(C_4H_4O_6)$ ($\cdot 4H_2O$) Potassium-sodium tartrate	210.16	mp hydr 75, $-H_2O$ 215
1319	$KNaCO_3$ ($\cdot 6H_2O$) Potassium-sodium carbonate	122.10	$-H_2O$ 100
1320	$K_2Ni(SO_4)_2$ ($\cdot 6H_2O$) Dipotassium-nickel(II) sulfate	329.01	dec hydr ca. 100
1321	KO_2 Potassium superoxide	71.10	mp 535 (p), dec 400
1322	KO_3 Potassium ozonide	87.10	dec 60
1323	K_2O Potassium oxide	94.20	350 $\rightarrow$ K_2O_2,K
1324	K_2O_2 Potassium peroxide	110.19	mp 545 (p), dec 500
1325	KOCN Potassium cyanate	81.12	dec >700
1326	KOH Potassium hydroxide	56.11	mp 404, bp 1324
1327	$KOH \cdot H_2O$	74.12	mp 145
1328	$KOH \cdot 2H_2O$	92.14	33 $\rightarrow$ $KOH \cdot H_2O$
1329	$K(PH_2O_2)$ Potassium phosphinate	104.09	dec t
1330	$K_2(PHO_3)$ Potassium phosphonate	158.18	dec t
1331	KPO_3 Potassium metaphosphate	118.07	mp 813, bp 1320
1332	K_3PO_4 ($\cdot 7H_2O$) Potassium orthophosphate	212.26	mp 1640
1333	$K_4P_2O_7$ ($\cdot 3H_2O$) Potassium diphosphate	330.33	$-H_2O$ 300
1334	$K_2(PO_3F)$ Potassium fluoroorthophosphate	176.17	mp 825
1335	$KReO_4$ Potassium perrhenate	289.30	mp 553, bp 1367
1336	$KRh(SO_4)_2$ ($\cdot 12H_2O$) Potassium-rhodium(III) sulfate	334.13	dec hydr >200
1337	$KRuO_4$ Potassium tetraoxoruthenate(VII)	204.16	dec >200
1338	K_2RuO_4 ($\cdot H_2O$) Potassium tetraoxoruthenate(VI)	243.26	$-H_2O$ 200
1339	K_2S Potassium sulfide	110.26	mp 912
1340	$K_2S \cdot 5H_2O$	200.34	mp 60, $-H_2O$ 150
1341	$K_2(S_2)$ Potassium disulfide(2−)	142.33	mp 520 dec
1342	$K_2(S_3)$ Potassium trisulfide(2−)	174.39	mp 292 dec
1343	$K_2(S_4)$ Potassium tetrasulfide(2−)	206.46	soft >159 dec
1344	$K_2(S_5)$ Potassium pentasulfide(2−)	238.53	mp 211 dec
1345	$K_2(S_6)$ Potassium hexasulfide(2−)	270.59	mp 196 dec
1346	K_2SO_3 ($\cdot H_2O$) Potassium sulfite	158.26	dec hydr 600
1347	K_2SO_4 Potassium sulfate	174.26	mp 1074
1348	$K_2S_2O_5$ Potassium pentaoxodisulfate(IV)	222.32	dec >190
1349	$K_2S_2O_6$ Potassium dithionate	238.32	dec >258
1350	$K_2S_2O_7$ Potassium disulfate	254.32	440 $\rightarrow$ K_2SO_4
1351	$K_2S_3O_6$ Potassium trithionate	270.39	dec 300–400
1352	$K_2S_4O_6$ Potassium tetrathionate	302.45	dec >500
1353	$K_2S_5O_6$ ($\cdot 1.5H_2O$) Potassium pentathionate	334.52	dec hydr t
1354	$K(SO_2F)$ Potassium fluorosulfite	122.16	dec 175
1355	$K(SO_3F)$ Potassium fluorosulfonate	138.16	mp 311

(Continued)

No.	Color and density	Solubility and reactivity						
		water	alcohol	HCl	H_2SO_4	HNO_3	NaOH	$NH_3 \cdot H_2O$
1309	dk-grn, 2.80	+	+	+	+	+	+/−	+/−
1310	wh, 2.91	r/+	i	+	+	+	−	−
1311	wh, 2.04	r	r	+	+	+	r	r
1312	grn-blk	+	...	+	+	+	+	+
1313	wh, 1.89	r	r	r	r/+	r/+	r	r
1314	wh, 2.35	r	r	+	+	+	−	−
1315	wh	+	+	+	+	+	+	+
1316	wh, 1.92	r	i/r	−/+	−/+	−/+	−	−
1317	wh, 2.11	r	i	r	r	r	r	r
1318	wh hydr, 1.79	r	d	−/+	−/+	−/+	−	−
1319	wh hydr, 1.63	r	...	+	+	+	r	r
1320	bl-grn hydr, 2.12	r	i	−	r	−	+	+
1321	orange-yel, 2.16	+	+	+	+	+	+	+
1322	red, 1.99	+	+	+	+	+	+	+
1323	wh, 2.33	+	+	+	+	+	+	+
1324	wh, 2.40	+	...	+	+	+	+	+
1325	wh, 2.06	r/+	i	−/+	−/+	−/+	r	r
1326	wh, 2.04	r	r	+	+	+	r	r
1327	wh	r	r	+	+	+	r	r
1328	wh	r	r	+	+	+	r	r
1329	wh	r	d	−	−/+	−/+	−	−
1330	wh	r	r	−	−	−	−	−
1331	wh, 2.39	r/+	i	−/+	−/+	−/+	+	+
1332	wh, 2.56	r	i	r	r	r	r	r
1333	wh hydr, 2.83	r/+	i	+	+	+	−/+	−
1334	wh	r	...	+	+	+	+	+
1335	wh, 4.89	d	d	−	−	−	−	−
1336	yel hydr, 2.23	r	i	−/+	r/+	−/+	+	+
1337	blk	d	...	−	−	−	+	+
1338	dk-grn	r	...	+	+	+	−	−
1339	wh, 1.74	r/+	r	+	+	+	r	r
1340	wh	r/+	r	+	+	+	r	r
1341	yel, 1.97	r/+	r	+	+	+	−	−
1342	yel-orange, 2.10	r/+	r	+	+	+	−	−
1343	orange-yel	r/+	r	+	+	+	−	−
1344	yel-brn, 2.13	r/+	d	+	+	+	−	−
1345	red-brn, 2.02	r/+	d	+	+	+	−	−
1346	wh hydr	r	d	+	+	+	r	r
1347	wh, 2.66	r	i	−	r	−	r	r
1348	wh, 2.34	r/+	d	+	+	+	−	−
1349	wh, 2.28	r	i	−	−/+	−/+	−	−
1350	wh, 2.27	r/+	i	+	+	+	+	+
1351	wh, 2.33	r/+	i	−	−/+	−/+	−	−
1352	wh, 2.29	r	i	−	−/+	−/+	−	−
1353	wh hydr, 2.11	r	i	−	−/+	−/+	+	+
1354	wh	+	...	+	+	+	+	+
1355	wh	r/+	...	−/+	−/+	−/+	−/+	...

No.	Formula and name	M_r	Phase transition temperature
1356	$K_2S_2O_6(O_2)$ Potassium peroxodisulfate	270.32	ca. 100 → $K_2S_2O_7$
1357	$K_2(SO_3S)$ (·1.67H_2O) Potassium thiosulfate	190.33	$-H_2O$ 200, dec >430
1358	K_3Sb Tripotassium stibide	239.04	mp 812
1359	K_2Se Potassium selenide	157.16	mp ca. 820
1360	K_2SeO_4 Potassium selenate	221.15	mp 1020 (p), dec 600
1361	K_2SiO_3 Potassium metasilicate	154.28	mp 976
1362	$K_2Si_2O_5$ Potassium pentaoxosilicate(IV)	214.36	mp 1045
1363	K_2Te Potassium telluride	205.80	mp ca. 1000
1364	K_2TeO_3 Potassium tellurite	253.79	mp 460–470
1365	$K_2UO_2(SO_4)_2$ (·2H_2O) Dipotassium-uranyl sulfate	540.35	$-H_2O$ 120
1366	KVO_3 Potassium metavanadate	138.04	mp 522
1367	K_3VO_4 Potassium orthovanadate	232.23	mp ca. 1300
1368	$KV(SO_4)_2$ (·12H_2O) Potassium-vanadium(III) sulfate	282.16	mp hydr 20, dec 300
1369	K_2WO_4 (·2H_2O) Potassium wolframate	326.04	mp 923
1370	**Kr Krypton**	83.80	mp −157.37; bp −153.22
1371	Kr·5.75 H_2O	187.39	dec −28
1372	KrF_2 Krypton difluoride	121.80	mp −77, dec >−40
1373	**Ku Kurchatovium** (1995 yr: Dubnium, Db)	261.109	...
1374	**La Lanthanum**	138.906	mp 920, bp 3450
1375	LaB_6 Lanthanum hexaboride	203.77	mp 2210
1376	$LaBr_3$ (·7H_2O) Lanthanum(III) bromide	378.62	mp 783, bp ca. 1700
1377	$La(BrO_3)_3$ (·9H_2O) Lanthanum(III) bromate	522.61	mp hydr 37.7; dec hydr t
1378	LaC_2 Lanthanum dicarbide	162.93	mp 2360
1379	$La_2(C_2O_4)_3$ (·10H_2O) Lanthanum(III) oxalate	541.87	hydr >600 → La_2O_3
1380	$LaCl_3$ Lanthanum(III) chloride	245.27	mp 862, bp 1710
1381	$LaCl_3·7H_2O$	371.37	mp 91 dec
1382	LaF_3 Lanthanum(III) fluoride	195.90	mp 1493, bp 2330
1383	LaI_3 Lanthanum(III) iodide	519.62	mp 779, bp 1580
1384	$La_2(MoO_4)_3$ Lanthanum(III) molybdate	757.62	mp 1015
1385	$La(NO_3)_3$ (·6H_2O) Lanthanum(III) nitrate	324.92	mp hydr 43, dec hydr t
1386	La_2O_3 Lanthanum(III) oxide	325.81	mp 2280, bp ca. 4200
1387	$La(OH)_3$ Lanthanum(III) hydroxide	189.93	1100 → La_2O_3
1388	LaS Lanthanum monosulfide	170.97	mp 2330
1389	La_2S_3 Lanthanum(III) sulfide	374.01	mp 2150
1390	$La_2(SO_4)_3$ (·9H_2O) Lanthanum(III) sulfate	566.00	$-H_2O$ 600, dec 1150
1391	$La_2(SeO_4)_3$ (·5H_2O) Lanthanum(III) selenate	706.68	$-H_2O$ 180–200
1392	**Li Lithium**	6.941	mp 180.5; bp 1336.6
1393	$LiAlO_2$ Lithium dioxoaluminate(III)	65.92	mp 1600
1394	Li_3AsO_4 Lithium arsenate	159.74	mp 1150
1395	$LiBO_2$ (·6H_2O) Lithium metaborate	49.75	$-H_2O$ 110, mp 849
1396	$Li_2B_4O_7$ (·5H_2O) Lithium tetraborate	169.12	$-H_2O$ 200, mp 918
1397	$LiBr$ (·2H_2O) Lithium bromide	86.85	mp 552, bp 1310
1398	Li_2C_2 Lithium acetylide	37.90	dec >750
1399	$Li(CH_3COO)$ (·2H_2O) Lithium acetate	65.99	mp hydr 54.5, dec >275
1400	Li_2CO_3 Lithium carbonate	73.89	mp 618, dec 730
1401	$Li_2C_2O_4$ Lithium oxalate	101.90	dec >400
1402	$LiCl$ Lithium chloride	42.39	mp 610, bp 1380

(*Continued*)

No.	Color and density	Solubility and reactivity						
		water	alcohol	HCl	H_2SO_4	HNO_3	NaOH	$NH_3 \cdot H_2O$
1356	wh, 2.48	r/+	i	−/+	−/+	−/+	−	−
1357	wh, 2.23(2.59)	r	i	+	+	+	−	−
1358	grn, 2.35	+	...	+	+	+	+	+
1359	wh, 2.29	r/+	r	+	+	+	r	r
1360	wh, 3.07	r	i	−/+	−/+	−	−	−
1361	wh	r	i	+	+	+	−	−
1362	wh, 2.46	r	i	+	+	+	+	+
1363	lt-yel, 2.52	r/+	...	+	+	+	r	r
1364	wh	r	...	−	−/+	−/+	−	−
1365	yel hydr, 3.36	r	i	−	r	−	+	+
1366	wh	d/r	i	+	+	+	−	−
1367	wh	+	i	+	+	+	−	−
1368	viol hydr, 1.78	r	i	−	r/+	−/+	+	+
1369	wh hydr, 3.11	r	i	+	+	+	−	−
1370	cl gas, 3.708	d/i	d/i	−	−	−	−	−
1371	wh	+	+	+	+	+	+	+
1372	wh	+	+	+	+	+	+	+
1373	...	−	...	−	...	−	...	...
1374	wh, 6.162	psv/+	...	+	+	+	+	+
1375	red-viol, 2.61	−	i	−/+	+	+	−	−
1376	wh, 5.06	r	r	r	+	−	+	+
1377	wh hydr	r	i	−	+	−	+	+
1378	yel, 5.02	+	...	+	+	+	+	+
1379	wh hydr	i	...	−/+	i/+	−/+	i	i
1380	wh, 3.84	r	r	r	+	−	+	+
1381	wh	r	r	r	+	−	+	+
1382	wh, 5.94	i	i	−	−	−	−	−
1383	grn-grey, 5.63	r	r	−	+	−/+	+	+
1384	lt-grey, 4.77	i	i	−/+	−/+	−	−/+	−
1385	wh hydr	r	r	r	+	r	+	+
1386	wh, 6.51	−/+	+	+	+	+	−	−
1387	wh	i	...	+	+	+	−	−
1388	yel	+	...	+	+	+	+	+
1389	red-yel, 4.91	i/+	...	+	+	+	+	+
1390	wh, 3.60(2.80)	d	d	−	d/r	−	+	+
1391	wh hydr	r	...	−/+	+	−	+	+
1392	wh, 0.534	+	+	+	+	+	+	+
1393	wh, 2.55	i/+	i	+	+	+	+	+
1394	wh, 3.07	i	...	+	+	+	−	−
1395	wh, 1.40(1.38)	+	...	+	+	+	+	+
1396	wh	r	i	+	+	+	r	r
1397	wh, 3.46	r	r	r	−/+	−	−	−
1398	wh, 1.65	+	...	+	+	+	+	+
1399	wh hydr	r	r	r	−	−	r	−
1400	wh, 2.11	r	i	+	+	+	r	r
1401	wh, 2.12	r	i	−	−/+	−/+	−	−
1402	wh, 2.07	r	r	r	−/+	−	r	−

No.	Formula and name	M_r	Phase transition temperature
1403	$LiCl \cdot H_2O$	60.41	mp 93.5, $-H_2O$ 98
1404	$LiClO_3$ Lithium chlorate	90.39	mp 129, dec >270
1405	$LiClO_3 \cdot 0.5H_2O$	99.40	mp 65, $-H_2O$ 90
1406	$LiClO_4$ Lithium perchlorate	106.39	mp 236.7; dec 400
1407	$LiClO_4 \cdot 3H_2O$	160.44	$-H_2O$ 90–150
1408	Li_2CrO_4 ($\cdot 2H_2O$) Lithium chromate	129.87	$-H_2O$ 150
1409	$Li_2Cr_2O_7$ ($\cdot 2H_2O$) Lithium dichromate	229.87	$-H_2O$ 130, dec 500
1410	LiD Lithium deuteride	8.96	mp 680
1411	LiF Lithium fluoride	25.94	mp 845.1; bp 1676
1412	Li_2GeO_3 Lithium germanate	134.49	mp 1239
1413	LiH Lithium hydride	7.95	mp 680, dec 850
1414	$Li(HCOO)$ Lithium formate	51.96	mp 273, dec 290–420
1415	$Li(HCOO) \cdot H_2O$	69.97	$-H_2O$ 94
1416	LiH_2PO_4 Lithium dihydroorthophosphate	103.93	mp >100
1417	$LiHSO_4$ Lithium hydrosulfate	104.01	mp 104 (p), dec >150
1418	LiI Lithium iodide	133.85	mp 469, bp 1170 dec
1419	$LiI \cdot 3H_2O$	187.89	mp 73, $-H_2O$ 300
1420	$LiMnO_4$ ($\cdot 3H_2O$) Lithium permanganate	125.88	mp hydr 105, dec hydr 190
1421	Li_2MoO_4 Lithium molybdate	173.82	mp 705
1422	LiN_3 Lithium azide	48.96	dec 115
1423	Li_3N Trilithium nitride	34.83	mp 813 (p)
1424	$LiNH_2$ Lithium amide	22.96	mp 374, dec >400
1425	Li_2NH Lithium imide	28.90	dec >500
1426	$LiNO_2$ ($\cdot 0.5H_2O$) Lithium nitrite	52.95	$-H_2O$ 100, mp 220
1427	$LiNO_3$ ($\cdot 3H_2O$) Lithium nitrate	68.95	mp 253.0; >600 $\rightarrow$ $LiNO_2$
1428	$(LiNb)O_3$ Lithium-niobium trioxide	147.84	mp 1253
1429	Li_2O Lithium oxide	29.88	mp 1453, bp ca. 2600
1430	Li_2O_2 Lithium peroxide	45.88	340 $\rightarrow$ Li_2O
1431	$LiOH$ ($\cdot H_2O$) Lithium hydroxide	23.95	mp 471, bp 925 (p)
1432	Li_3PO_4 ($\cdot 0.5H_2O$) Lithium orthophosphate	115.79	mp 837
1433	$Li_3PO_4 \cdot 2H_2O$	151.82	mp 100, dec 120
1434	Li_2S Lithium sulfide	45.95	mp 950
1435	Li_2SO_3 ($\cdot H_2O$) Lithium sulfite	93.95	$-H_2O$ 190, mp 455 dec
1436	Li_2SO_4 ($\cdot H_2O$) Lithium sulfate	109.94	$-H_2O$ 130, mp 859
1437	$Li(SO_3F)$ Lithium fluorosulfonate	106.00	mp 360
1438	Li_3Sb Trilithium stibide	142.57	mp >950
1439	Li_2SiO_3 Lithium metasilicate	89.97	mp 1202
1440	$Li_2Si_2O_5$ Lithium pentaoxodisilicate(IV)	150.05	mp 1033
1441	Li_4SiO_4 Lithium orthosilicate	119.85	mp 1225 dec
1442	$(LiTa)O_3$ Lithium-tantalum trioxide	235.89	mp 1650
1443	Li_2WO_4 Lithium wolframate	261.73	mp 740
1444	**Lr Lawrencium**	260.105	...
1445	**Lu Lutetium**	174.967	mp 1663, bp 3412
1446	$LuBr_3$ Lutetium(III) bromide	414.68	mp 960, bp 1410
1447	$Lu_2(C_2O_4)_3$ ($\cdot 6H_2O$) Lutetium(III) oxalate	613.99	hydr >600 $\rightarrow$ Lu_2O_3
1448	$LuCl_3$ Lutetium(III) chloride	281.33	mp 925, bp 1420
1449	LuF_3 Lutetium(III) fluoride	231.96	mp 1184, bp 2200

(Continued)

No.	Color and density	Solubility and reactivity						
		water	alcohol	HCl	H_2SO_4	HNO_3	NaOH	$NH_3 \cdot H_2O$
1403	wh, 1.78	r	r	r	−	−	r	−
1404	wh, 1.12	r	r	−/+	r/+	−	−	−
1405	wh	r	r	−/+	r/+	−	−	−
1406	wh, 2.43	r	r	−	−	r	−	−
1407	wh, 1.84	r	r	−	−	r	−	−
1408	yel	r	i	+	+	+	r	r
1409	red-orange, 2.34	r	i	r/+	r	r	+	+
1410	wh, 0.82	+	+	+	+	+	+	+
1411	wh, 2.64	d	i	i	−/+	−	−	i
1412	wh, 3.53	d	...	+	+	+	−/+	−
1413	wh, 0.82	+	+	+	+	+	+	+
1414	wh, 1.46	r	d	r	−/+	−/+	−	−
1415	wh	r	d	r	−/+	−/+	−	−
1416	wh, 2.46	r	i	r	r	r	+	+
1417	wh, 2.12	r	i	−	r	−	+	+
1418	wh, 4.06	r	r	r	r/+	r/+	r	r
1419	wh, 3.48	r	r	r	r/+	r/+	r	r
1420	viol hydr, 2.06	r	+	r/+	r	r	r	r/+
1421	wh, 2.66	r	i	+	+	+	−	−
1422	wh, 1.61	r	r	+	+	+	−	−
1423	dk-red, 1.28	+	...	+	+	+	+	+
1424	wh, 1.18	r/+	d/+	+	+	+	+	+
1425	wh, 1.48	+	d	+	+	+	+	+
1426	wh hydr, 1.62	r	r	−/+	−/+	−/+	−	−
1427	wh, 2.38	r	r	r	r	r/i	r	r
1428	wh	−	...	−/+	+	−/+	−	−
1429	wh, 2.01	+	...	+	+	+	+	+
1430	wh, 2.36	+	...	+	+	+	+	+
1431	wh, 1.46(1.51)	r	d	+	+	+	r	r
1432	wh, 2.54	d	...	r	r	r	−	−
1433	wh, 1.65	d	...	r	r	r	−	−
1434	lt-yel, 1.66	r	r	+	+	+	r	r
1435	wh	r	i	+	+	+	−	−
1436	wh, 2.22(2.06)	r	d	−	r	−	r	−
1437	wh	r/+	r	−/+	−/+	−/+	−/+	...
1438	dk-grey, 3.2	+	...	+	+	+	+	+
1439	wh, 2.52	i/+	...	+	+	+	−	−
1440	wh	i	...	+	+	+	−/+	−
1441	wh, 2.28	i/+	...	+	+	+	−	−
1442	wh	−	...	−/+	+	−/+	−	−
1443	wh, 3.71	r	i	+	+	+	−	−
1444	...	...	...	+	...	...	...	...
1445	wh, 9.835	psv/+	...	+	+	+	−	−
1446	wh	r	r	−	−/+	−	+	+
1447	wh hydr	i	i	i/+	−/+	−/+	i	−
1448	wh, 3.98	r	r	r	−	−	+	+
1449	wh, 8.29	i	i	−	i	−	i	−

No.	Formula and name	M_r	Phase transition temperature
1450	LuI_3 Lutetium(III) iodide	555.68	mp 1045, bp 1210
1451	$Lu(NO_3)_3 \cdot (4H_2O)$ Lutetium(III) nitrate	360.98	dec hydr t
1452	Lu_2O_3 Lutetium(III) oxide	397.93	mp 2450
1453	$Lu(OH)_3$ Lutetium(III) hydroxide	225.99	$1050 \rightarrow Lu_2O_3$
1454	$Lu_2(SO_4)_3$ $(\cdot 8H_2O)$ Lutetium(III) sulfate	638.12	$-H_2O$ 650, dec >850
1455	**Md Mendelevium**	258.099	...
1456	**Mg Magnesium**	24.305	mp 648, bp 1095
1457	$(MgAl_2)O_4$ Magnesium-dialuminum oxide	142.27	mp 2105
1458	Mg_3As_2 Trimagnesium diarsenide	222.76	mp 800
1459	MgB_2 Magnesium diboride	45.93	dec 1050
1460	$Mg_3(BO_3)_2$ Magnesium orthoborate	190.53	mp 1410
1461	Mg_3Bi_2 Trimagnesiumdibismuth	490.88	mp 821
1462	$MgBr_2$ Magnesium bromide	184.11	mp 710, bp ca. 1250
1463	$MgBr_2 \cdot 6H_2O$	292.20	mp 172.4; $-H_2O$ t
1464	$Mg(BrO_3)_2$ $(\cdot 6H_2O)$ Magnesium bromate	280.11	$-H_2O$ 200
1465	Mg_2C_3 Dimagnesium tricarbide	84.64	dec >600
1466	$[Mg(C_5H_5)_2]$ Bis(cyclopentadienyl)magnesium	154.50	mp 177, bp 221, dec >300
1467	$Mg(CH_3COO)_2$ Magnesium acetate	142.39	mp 323 dec
1468	$Mg(CH_3COO)_2 \cdot 4H_2O$	214.45	mp 80, $-H_2O$ 134
1469	$MgCN_2$ Magnesium cyanamide	64.33	dec 1370
1470	$MgCO_3$ $(\cdot 5H_2O)$ Magnesium carbonate	84.31	dec >400
1471	MgC_2O_4 $(\cdot 2H_2O)$ Magnesium oxalate	112.32	dec hydr 150
1472	$MgCl_2$ Magnesium chloride	95.21	mp 714, bp 1370
1473	$MgCl_2 \cdot 6H_2O$	203.30	mp 117, $-6H_2O$ 150
1474	$Mg(ClO_3)_2$ $(\cdot 6H_2O)$ Magnesium chlorate	191.21	mp hydr 35, dec hydr 120
1475	$Mg(ClO_4)_2$ Magnesium perchlorate	223.20	mp 246, dec 382
1476	$Mg(ClO_4)_2 \cdot 6H_2O$	331.29	mp 147, dec 180–190
1477	$MgCrO_4$ $(\cdot 5H_2O)$ Magnesium chromate	140.30	dec ca. 600
1478	$(MgCr_2)O_4$ Magnesium-chromium tetraoxide	192.29	mp 2350
1479	MgF_2 Magnesium fluoride	62.30	mp 1290, bp ca. 2270
1480	$(MgFe_2)O_4$ Magnesium-diiron tetraoxide	200.00	mp 1750
1481	$MgHAsO_4$ $(\cdot 7H_2O)$ Magnesium hydroarsenate	164.23	$-H_2O$ 190
1482	$MgHPO_4$ $(\cdot 3H_2O)$ Magnesium hydroorthophosphate	120.28	$-H_2O$ 200
1483	MgI_2 $(\cdot 8H_2O)$ Magnesium iodide	278.11	mp 633, bp 1014
1484	$Mg(IO_3)_2$ $(\cdot 4H_2O)$ Magnesium iodate	374.11	$-H_2O$ 210
1485	Mg_3N_2 Trimagnesium dinitride	100.93	dec 1500
1486	$MgNH_4AsO_4$ $(\cdot 6H_2O)$ Magnesium-ammonium arsenate	181.26	dec hydr 800 $(\rightarrow Mg_2P_2O_7)$
1487	$MgNH_4PO_4$ $(\cdot 6H_2O)$ Magnesium-ammonium orthophosphate	137.31	$t \rightarrow Mg_2P_2O_7$
1488	$Mg(NH_4)_2(SO_4)_2$ $(\cdot 6H_2O)$ Magnesium-diammonium sulfate	252.51	mp hydr ca. 120, dec 250
1489	$Mg(NO_3)_2$ Magnesium nitrate	148.31	$>300 \rightarrow MgO$
1490	$Mg(NO_3)_2 \cdot 6H_2O$	256.40	mp 89.9
1491	MgO Magnesium oxide	40.30	mp 2825, bp 3600
1492	$Mg(OH)_2$ Magnesium hydroxide	58.32	$>480 \rightarrow MgO$
1493	Mg_3P_2 Trimagnesium diphosphide	134.86	dec >1000

(Continued)

No.	Color and density	Solubility and reactivity						
		water	alcohol	HCl	H_2SO_4	HNO_3	NaOH	$NH_3·H_2O$
1450	brn	r	r	−	−/+	−/+	+	+
1451	wh hydr	r	r	−	−	r	+	+
1452	wh, 9.42	−/+	...	+	+	+	−	−
1453	wh	i	...	+	+	+	−	−
1454	wh hydr, 3.33	r/d	...	−	r	−	+	+
1455	...	...	...	...	...	...	...	...
1456	wh, 1.737	psv/+	...	+	+/psv	+	−	−
1457	wh, 3.55	−	i	−	−/+	−	−/+	−
1458	red-brn, 3.15	+	...	+	+	+	+	+
1459	blk	−/+	...	+	+	+	−	−
1460	wh, 2.99	i	...	+	+	+	−/+	−
1461	grey, 5.95	−	−	−/+	−	−/+	−/+	−
1462	wh, 3.72	r	r	−	−/+	−	+	+
1463	wh, 2.00	r	r	−	−/+	−	+	+
1464	wh hydr, 2.29	r	i	−/+	−/+	−/+	+	+
1465	lt-grey	+	−	+	+	+	+	+
1466	wh	+	r	−	+	+	+	+
1467	wh, 1.42	r	r	r	−	−	+	+
1468	wh, 1.45	r	r	r	−	−	+	+
1469	wh	+	i	+	+	+	+	+
1470	wh, 3.04	d/+	...	+	+	+	−	−
1471	wh hydr, 2.45	d	...	−/+	−/+	−/+	d	d
1472	wh, 2.32	r	r	r	−/+	−	+	+
1473	wh, 1.60	r	r	r	−/+	−	+	+
1474	wh hydr, 1.80	r	r	−	r	−	+	+
1475	wh, 2.21	r	r	−	−	−	+	+
1476	wh, 1.97	r	r	−	−	−	+	+
1477	yel hydr, 1.70	r	...	+	+	+	+	+
1478	dk-grn, 4.43	−	...	−	−/+	−	−	−
1479	wh, 3.177	d	i	d/r	d/+	d/r	−	−
1480	blk-brn, 4.52	−	...	+	−/+	−/+	−	−
1481	wh hydr, 1.94	d	...	−	−/+	−/+	+	+
1482	wh hydr, 2.10	d	...	−/+	−/+	−/+	+	+
1483	wh, 4.43	r	r	−	−/+	−/+	+	+
1484	wh hydr, 3.3	r	...	−/+	−	−	+	−/+
1485	yel-grn, 2.71	+	...	+	+	+	+	+
1486	wh hydr, 1.93	d	i	r	r	r	−	−
1487	wh hydr, 1.71	d	i	+	+	+	−	−
1488	wh hydr, 1.72	r	i	−	r	−	+	+
1489	wh	r	r	r	r	r	+	+
1490	wh, 1.64	r	r	r	r	r	+	+
1491	wh, 3.62	−	i	+	+	+	−	−
1492	wh, 2.39	i	...	+	+	+	−	−
1493	yel-grn, 2.05	+	i	+	−/+	−/+	+	+

No.	Formula and name	M_r	Phase transition temperature
1494	$Mg(PH_2O_2)_2$ ($\cdot 6H_2O$) Magnesium phosphinate	154.28	$-H_2O$ 180
1495	$Mg_2P_2O_7$ Magnesium diphosphate	222.55	mp 1395
1496	$Mg_3(PO_4)_2$ ($\cdot 8H_2O$) Magnesium orthophosphate	262.86	$-H_2O$ 400, mp 1357
1497	MgS Magnesium sulfide	56.37	mp 2000 dec
1498	$MgSO_3$ ($\cdot 6H_2O$) Magnesium sulfite	104.37	$-H_2O$ 200, dec t
1499	$MgSO_4$ ($\cdot H_2O$) Magnesium sulfate	120.37	mp 1137, dec 1168
1500	$MgSO_4 \cdot 7H_2O$	246.47	mp 54, $-H_2O$ 200
1501	$Mg(SO_3S)$ ($\cdot 6H_2O$) Magnesium thiosulfate	136.43	$-H_2O$ >240
1502	Mg_3Sb_2 Trimagnesiumdiantimony	316.42	mp 1250
1503	Mg_2Si Dimagnesium silicide	76.70	mp 1085
1504	Mg_2SiO_4 Magnesium orthosilicate	140.69	mp 1890
1505	$MgWO_4$ Magnesium wolframate	272.15	mp 1360
1506	**Mn Manganese**	54.938	mp 1245, bp 2080
1507	$(MnAl_2)O_4$ Manganese-dialuminum tetraoxide	172.90	mp 1560
1508	MnAs Manganese monoarsenide	129.86	mp 936
1509	MnB_2 Manganese diboride	76.56	mp 1990
1510	$MnBr_2$ ($\cdot 4H_2O$) Manganese(II) bromide	214.75	mp 698
1511	$Mn_{3+x}C$ Trimanganese carbide	176.83	mp 1520
1512	$[Mn(C_5H_5)_2]$ Bis(cyclopentadienyl)manganese	185.13	mp 173
1513	$Mn(CH_3COO)_2$ ($\cdot 4H_2O$) Manganese(II) acetate	173.03	mp hydr 80, dec hydr t
1514	$MnCO_3$ Manganese(II) carbonate	114.95	dec 100–300
1515	MnC_2O_4 ($\cdot 2H_2O$) Manganese(II) oxalate	142.96	$-H_2O$ 100
1516	$[Mn_2(CO)_{10}]$ Decacarbonyldimanganese	389.98	mp 154 (p), dec >110
1517	$[Mn(CO)_5Br]$ Pentacarbonylbromomanganese	274.89	>65 $\rightarrow [Mn_2(CO)_{10}]$
1518	$[Mn(CO)_3C_5H_5]$ Tricarbonyl(cyclopentadienyl)manganese	204.06	mp 77
1519	$[Mn(CO)_5Cl]$ Pentacarbonylchloromanganese	230.44	120 $\rightarrow [Mn_2(CO)_{10}]$
1520	$[Mn(CO)_5I]$ Pentacarbonyliodomanganese	321.89	>118 $\rightarrow [Mn_2(CO)_{10}]$
1521	$[Mn(CO)_4NO]$ Tetracarbonylnitrosylmanganese	196.98	mp ca. 1, subl 25
1522	$MnCl_2$ Manganese(II) chloride	125.84	mp 650, bp 1231
1523	$MnCl_2 \cdot 4H_2O$	197.90	mp 58, $-H_2O$ 198
1524	$(Mn_2Cu)O_4$ Dimanganese-copper tetraoxide	237.42	mp >1500
1525	MnF_2 ($\cdot 4H_2O$) Manganese(II) fluoride	92.93	mp 856, bp 1640
1526	MnF_3 Manganese(III) fluoride	111.93	600 $\rightarrow MnF_2$
1527	$(MnFe_2)O_4$ Manganese-diiron tetraoxide	230.63	mp 1570, dec >1800
1528	$[MnH(CO)_5]$ Hydropentacarbonylmanganese	196.00	mp -25, dec 100–150
1529	MnI_2 Manganese(II) iodide	308.75	mp 638
1530	$(Mn^{II}Mn_2^{III})O_4$ Manganese(II)-dimanganese(III) tetraoxide	228.81	mp 1705
1531	$Mn(NH_4)_2(SO_4)_2$ ($\cdot 6H_2O$) Manganese-diammonium sulfate	283.14	$-H_2O$ 180 dec
1532	$Mn(NO_3)_3$ ($\cdot 6H_2O$) Manganese(II) nitrate	178.95	mp hydr 25.3; dec hydr >195
1533	MnO_{1+x} ($0 \le x \le 0.13$) Manganese(II) oxide	70.94	mp 1780
1534	MnO_{2-x} ($\cdot nH_2O$) Manganese(IV) oxide	86.94	$-H_2O$ 250, dec >535
1535	Mn_2O_3 Manganese(III) oxide	157.87	940 $\rightarrow (Mn^{II}Mn_2^{III})O_4$
1536	$Mn_2O_3 \cdot nH_2O$	–	100 $\rightarrow MnO(OH)$

(Continued)

No.	Color and density	Solubility and reactivity						
		water	alcohol	HCl	H_2SO_4	HNO_3	NaOH	$NH_3 \cdot H_2O$
1494	wh hydr, 1.59	r	i	−	+	+	+	+
1495	wh, 2.56	i	i	+	+	+	−	−
1496	wh, 2.41(1.64)	i	...	r	r	r	i	−
1497	wh, 2.86	i/+	...	+	+	+	i/+	−
1498	wh hydr, 1.73	d	i	+	+	+	d	−
1499	wh, 2.66(2.57)	r	d	−	r	−	+	+
1500	wh, 1.68	r	d	−	r	−	+	+
1501	wh hydr, 1.82	r	i	−/+	−/+	−/+	+	+
1502	grey, 4.09	−	...	−/+	−/+	−/+	−/+	−
1503	dk-sk-bl, 1.94	−/+	...	+	+	+	−/+	−
1504	wh, 3.22	i	...	−/+	−/+	−/+	−	−
1505	wh, 5.66	i	i	+	+	+	−	−
1506	lt-grey, 7.44	psv/+	...	+	+	+	−	−
1507	red-blk, 4.04	−	i	−	−/+	−	−/+	−
1508	blk, 6.18	−	...	+	+	+	−	−
1509	grey-viol, 6.9	+	...	+	+	+	+	+
1510	pink, 4.39	r	r	−	−/+	−	+	+
1511	grey-blk, 6.89	+	...	+	+	+	+	+
1512	brn-red	−/+	r	−	−/+	+	−	−
1513	lt-pink hydr	r	r	−	−	−	+	+
1514	red-pink, 3.70	i/+	i	+	+	+	i	i
1515	lt-pink hydr	d	...	−	−/+	−/+	−/+	−
1516	yel, 1.75	−	r	−	−/+	−/+	+	−
1517	orange	−	+	−/+	−/+	−/+	−/+	−
1518	yel	−/+	r	−/+	−/+	−/+	−/+	−
1519	lt-yel	−	+	−/+	−/+	−/+	−/+	...
1520	red	−	+	−/+	−/+	−/+	−/+	...
1521	dk-red lq	+	∞	+	+	+	+	+
1522	pink, 2.98	r	r	r	−/+	−	+	+
1523	lt-pink, 2.01	r	r	r	−/+	−	+	+
1524	brn-blk, 5.01	−	i	−	−/+	−	−/+	−
1525	pink, 3.98	r/+	i	r	r/+	r	+	+
1526	red, 3.54	+	r	+	+	+	+	+
1527	dk-brn, 4.87	−	...	−/+	−/+	−	−/+	−
1528	cl lq	+	r	+	+	+	+	+
1529	pink, 5.01	r	r	−	−/+	−/+	+	+
1530	brn-blk, 4.84	−	...	+	−/+	−/+	−	−
1531	lt-pink hydr, 1.83	r	i	−	r	−	+	+
1532	lt-pink hydr, 1.82	r	r	r	−	r	+	+
1533	grey-grn, 5.18	−	...	+	+	+	−	−
1534	blk-brn, 5.03	−	...	−/+	−/+	−	−	−
1535	brn, 4.90	−	...	+	+	+	−	−
1536	brn	i	...	+	+	+	i	i

No.	Formula and name	M_r	Phase transition temperature
1537	Mn_2O_7 Manganese(VII) oxide	221.87	mp 5.9; dec >55 $\rightarrow Mn_2O_3$
1538	$Mn(OH)_2$ Manganese(II) hydroxide	88.95	>200 $\rightarrow$ MnO
1539	MnO(OH) Manganese metahydroxide	87.94	250 $\rightarrow Mn_2O_3$
1540	MnP Manganese monophosphide	85.91	mp 1147
1541	Mn_2P Dimanganese phosphide	140.85	mp 1327
1542	Mn_3P Trimanganese phosphide	195.79	mp 1105 dec
1543	Mn_3P_2 Trimanganese diphosphide	226.76	dec t
1544	$Mn_2P_2O_7$ Manganese(II) disphosphate	283.82	mp 1196
1545	$Mn_3(PO_4)_2$ ($\cdot 3H_2O$) Manganese(II) orthophosphate	354.75	$-H_2O$ 320
1546	MnS Manganese(II) sulfide	87.00	mp 1615
1547	$MnS \cdot nH_2O$	–	$-H_2O$ 175
1548	$Mn(S_2)$ Manganese(II) disulfide(2–)	119.07	dec >800
1549	$MnSO_4$ ($\cdot H_2O$) Manganese(II) sulfate	151.00	mp 700, dec >850
1550	$MnSO_4 \cdot 5H_2O$	241.08	$-H_2O$ 250
1551	$MnSO_4 \cdot 7H_2O$	277.11	$-H_2O$ 280
1552	$Mn(SO_4)_2$ Manganese(IV) sulfate	247.06	dec t
1553	$Mn_2(SO_4)_3$ ($\cdot H_2SO_4 \cdot 6H_2O$) Manganese(III) sulfate	398.06	dec hydr 160–300
1554	MnSe Manganese selenide	133.90	mp 1510
1555	MnSi Manganese monosilicide	83.02	mp 1270
1556	Mn_2Si Dimanganese silicide	137.96	mp 1316
1557	$MnSiO_3$ Manganese(II) metasilicate	131.02	mp 1323
1558	Mn_2SiO_4 Manganese(II) orthosilicate	201.96	mp 1327
1559	$MnWO_4$ Manganese(II) wolframate	302.78	mp >1400
1560	$(Mn_2Zn)O_4$ Dimanganese-zinc tetraoxide	239.26	mp >2000
1561	**Mo Molybdenum**	95.94	mp 2620, bp 4630
1562	MoB Molybdenum monoboride	106.75	mp 2550
1563	Mo_2B Dimolybdenum boride	202.69	mp 2270 dec
1564	Mo_2B_5 Dimolybdenum pentaboride	245.94	mp 2200 dec
1565	$MoBr_2$ Molybdenum(II) bromide	255.75	dec >700
1566	$MoBr_3$ Molybdenum(III) bromide	335.65	>500 $\rightarrow MoBr_2$
1567	$MoBr_4$ Molybdenum(IV) bromide	415.56	>110 $\rightarrow MoBr_3$
1568	MoC Molybdenum monocarbide	107.95	mp 2570
1569	Mo_2C Dimolybdenum carbide	203.89	mp 2690
1570	$[Mo(CN)_8]K_4$ ($\cdot 2H_2O$) Potassium octacyanomolybdate(IV)	460.48	$-H_2O$ 105, dec t
1571	$[Mo(CO)_6]$ Hexacarbonylmolybdenum	264.00	mp 148, bp 155 dec
1572	$[Mo_2(CO)_4(C_5H_5)_2]$ Tetracarbonylbis(cyclopentadienyl)dimolybdenum	434.11	mp 215–7 dec ($\rightarrow$ Mo)
1573	$MoCl_2$ Molybdenum(II) chloride	166.85	700 $\rightarrow Mo,MoCl_5$
1574	$MoCl_3$ Molybdenum(III) chloride	202.30	410 $\rightarrow MoCl_2,MoCl_4$
1575	$MoCl_4$ Molybdenum(IV) chloride	237.75	subl >180 dec
1576	$MoCl_5$ Molybdenum(V) chloride	273.21	mp 194, bp 268
1577	$MoCl_3N$ Molybdenum trichloride-nitride	216.31	subl 130
1578	$MoCl_2O_2$ Molybdenum dichloride-dioxide	198.84	mp 170
1579	$MoCl_3O$ Molybdenum trichloride-oxide	218.30	mp 308
1580	$MoCl_4O$ Molybdenum tetrachloride-oxide	253.75	mp 102
1581	MoF_5 Molybdenum(V) fluoride	190.93	mp 67, bp 214

(Continued)

No.	Color and density	Solubility and reactivity						
		water	alcohol	HCl	H_2SO_4	HNO_3	NaOH	$NH_3 \cdot H_2O$
1537	dk-grn lq, 2.396[20]	+	i	+	+	+	+	+
1538	wh, 3.26	i	i	+	+	+	i	−
1539	brn, 4.14	i	...	−/+	−/+	−/+	i	−
1540	dk-grey, 5.39	−/+	...	−/+	−/+	−/+	−/+	−
1541	dk-grey	−	...	−	−/+	−/+	−	−
1542	grey-blk	−	...	−	−	−/+	−	−
1543	dk-grey, 5.12	−	...	−	−	−/+	−	−
1544	brn-pink, 3.71	i	...	+	+	+	−	−
1545	lt-pink hydr, 3.10	d	...	+	+	+	−	−
1546	brn-blk, 3.99	i/+	r	+	+	+	i	i
1547	lt-pink hydr, 3.95	i/+	r	+	+	+	i	i
1548	blk, 3.46	i	i	+	+	+	−	−
1549	wh, 3.25	r	i	−	r	−	+	+
1550	pink, 2.38	r	i	−	r	−	+	+
1551	pink, 2.09	r	i	−	r	−	+	+
1552	blk	+	...	−	−	−	−	−
1553	grn, 3.24	+	...	+	r	−	+	+
1554	grey, 5.55	i	...	+	+	+	−	−
1555	grey, 5.90	−	...	−/+	−/+	−/+	−/+	−
1556	grey, 6.20	−	...	+	+	−/+	+	−
1557	pink, 3.72	i	...	−	−/+	−/+	−/+	−
1558	red, 4.1	i	...	−	−/+	−/+	−/+	−
1559	yel, 7.12	i	i	−/+	−/+	−/+	−	−
1560	dk-brn, 5.2	−	i	−	−/+	−	−/+	−
1561	lt-grey, 10.23	−	−	−/+	−/+	+	−	−
1562	grey, 8.65	−	i	−	−/+	−/+	−/+	−
1563	grey, 9.26	−	i	−	−/+	−/+	−/+	−
1564	grey, 7.12	−	i	−	−/+	−/+	−/+	−
1565	yel-brn, 4.88	i	r	+	+	+	+	+
1566	blk	i/+	d	−/+	−/+	−/+	+	+
1567	blk	+	r	+	+	+	+	+
1568	grey, 8.78	−	...	−/+	−/+	−/+	−	−
1569	dk-grey, 9.18	−	...	−	−/+	−/+	−	−
1570	yel hydr, 2.34	r	i	−/+	−/+	−/+	−	−
1571	wh, 1.96	−	d	−/+	−/+	−/+	−	−
1572	red-brn	−/+	r	−/+	−/+	−/+	−/+	...
1573	yel, 3.714	i	r	+	+	+	+	+
1574	dk-red, 3.58	i/+	d	−/+	−/+	−/+	+	+
1575	brn-blk	+	r	+	+	+	+	+
1576	blk-bl, 2.93	+	+	+	+	+	+	+
1577	red-brn	+	...	+	+	+	+	+
1578	lt-yel, 3.31	+	r	+	+	+	+	+
1579	brn-blk	+	r	+	+	+	+	+
1580	dk-grn	+	...	+	+	+	+	+
1581	yel	+	...	+	+	+	+	+

No.	Formula and name	M_r	Phase transition temperature
1582	MoF$_6$ Molybdenum(VI) fluoride	209.93	mp 17.58; bp 33.88
1583	[MoF$_6$],K$_3$ Potassium hexafluoromolybdate(III)	327.22	mp 734
1584	MoI$_2$ Molybdenum(II) iodide	349.75	mp 700
1585	MoI$_3$ Molybdenum(III) iodide	476.65	mp 900, $t \rightarrow$ MoI$_2$
1586	MoI$_4$ Molybdenum(IV) iodide	603.56	mp 400, dec t
1587	MoO$_2$ Molybdenum(IV) oxide	127.94	dec 1800
1588	MoO$_3$ Molybdenum(VI) oxide	143.94	mp 795, bp 1155
1589	MoO$_3$·H$_2$O	161.95	−H$_2$O 115
1590	MoO$_3$·2H$_2$O	179.97	70 $\rightarrow$ MoO$_3$·H$_2$O
1591	Mo(O)F$_4$ Molybdenum oxide-tetrafluoride	187.93	mp 97.2; bp 186
1592	MoO$_2$F$_2$ Molybdenum dioxide-difluoride	165.93	subl 270
1593	Mo(OH)$_3$ Molybdenum(III) hydroxide	143.96	dec t
1594	MoO(OH)$_2$ Molybdenum oxide-dihydroxide	145.95	900 $\rightarrow$ MoO$_2$
1595	MoO(OH)$_3$ Molybdenum oxide-trihydroxide	162.96	dec >800
1596	[Mo(PF$_3$)$_6$] Hexakis(trifluorophosporus)molybdenum	623.75	mp 196
1597	MoS$_2$ Molybdenum(IV) sulfide	160.07	subl 450, mp ca. 2100
1598	MoS$_3$ Molybdenum(VI) sulfide	192.14	>300 $\rightarrow$ MoS$_2$
1599	MoSe$_2$ Molybdenum(IV) selenide	253.86	dec 900
1600	MoSi$_2$ Molybdenum disilicide	152.11	mp ca. 2030
1601	**N$_2$ Dinitrogen**	28.014	mp −210.0; bp −195.802
1602	ND$_3$ Deuterioammonia	20.05	mp −74.36; bp −31.04
1603	ND$_4$Cl Tetradeuterioammonium chloride	57.52	mp >300
1604	NF$_3$ Nitrogen trifluoride	71.00	mp −206.78; bp −129
1605	trans-N$_2$F$_2$ Dinitrogen difluoride	66.01	mp −172, bp −111.4
1606	cis-N$_2$F$_2$	66.01	mp < −195, bp −105.7
1607	N$_2$F$_4$ Dinitrogen tetrafluoride	104.01	mp −161.5; bp −74.1
1608	NH$_3$ Ammonia	17.03	mp −77.75; bp −33.4
1609	NH$_3$·0.5H$_2$O	26.04	mp −78.2 dec 80−100
1610	NH$_3$·H$_2$O	35.05	mp −77 dec
1611	NH$_3$·2H$_2$O	53.06	mp −97, dec t
1612	N$_2$H$_4$ Hydrazine	32.05	mp 1.5; bp 113.5
1613	N$_2$H$_4$·H$_2$O	50.06	mp −51.6; bp 120.1
1614	(NH$_4$)$_2$B$_4$O$_7$ (·4H$_2$O) Ammonium tetraborate	191.32	−H$_2$O 87, dec 190
1615	NH$_4$Br Ammonium bromide	97.94	subl >394 dec
1616	NH$_4$(CH$_3$COO) Ammonium acetate	77.08	mp 114 dec
1617	N$_2$H$_5$(CH$_3$COO) Hydrazinium(1+) acetate	92.10	mp 101
1618	NH$_4$CN Ammonium cyanide	44.06	dec 36
1619	(NH$_4$)$_2$CO$_3$ Ammonium carbonate	96.09	dec 20−58
1620	(NH$_4$)$_2$C$_2$O$_4$ (·H$_2$O) Ammonium oxalate	124.10	dec hydr >120
1621	NH$_2$Cl Chloramine	51.48	mp −66, dec >−40
1622	NH$_4$Cl Ammonium chloride	53.49	subl >337.8 dec
1623	N$_2$H$_5$Cl Hydrazinium(1+) chloride	68.51	mp 89, dec 350
1624	N$_2$H$_6$Cl$_2$ Hydrazinium(2+) chloride	104.97	mp 198 dec
1625	NH$_4$ClO$_3$ Ammonium chlorate	101.49	dec ca. 100
1626	NH$_4$ClO$_4$ Ammonium perchlorate	117.49	dec 210−270
1627	N$_2$H$_5$ClO$_4$ (·0.5H$_2$O) Hydrazinium(1+) perchlorate	132.50	mp hydr 142.4; dec 220
1628	(NH$_4$)$_2$CrO$_4$ Ammonium chromate	152.07	mp 185 dec ($\rightarrow$ Cr$_2$O$_3$)

(*Continued*)

No.	Color and density	Solubility and reactivity						
		water	alcohol	HCl	H_2SO_4	HNO_3	NaOH	$NH_3 \cdot H_2O$
1582	cl lq, 2.543[19]	+	r	+	+	+	+	+
1583	brn	r	...	−/+	−/+	−/+	+	...
1584	brn-blk, 5.28	i/+	i	+	+	+	+	+
1585	blk	+	i	−	−/+	−/+	+	...
1586	blk	+	...	+	+	+	+	+
1587	brn-viol, 6.47	−	...	−	−/+	+	−	−
1588	grn-wh, 4.69	+	...	+	+	+	+	+
1589	wh, 3.11	d	i	+	+	+	+	+
1590	yel, 3.12	d	i	+	+	+	+	+
1591	wh, 3.0	+	r	+	+	+	+	+
1592	wh, 3.49	+	r	+	+	+	+	+
1593	blk	i	i	−/+	−/+	−	i	−
1594	dk-grn	i	i	−/+	−/+	+	+	−
1595	brn	i	...	+	+	+	+	...
1596	wh	i/+	...	−	−/+	−/+	+	−
1597	grey-sk-bl, 4.68–4.75	−	i	−	−/+	−/+	−	−
1598	dk-brn	i	...	−/+	−/+	−/+	+	+
1599	bl-brn	−	i	−	−	+	−	−
1600	dk-grey, 6.31	−	...	−	−	−/+	−	−
1601	cl gas, 1.2505	i	i	−	−	−	−	−
1602	cl gas	r	r	+	+	+	+	+
1603	wh	r	r	r	r	r	r/+	r
1604	cl gas	i/+	i	−	−/+	−/+	−/+	−
1605	cl gas	d/+	...	−	−	−	−	−
1606	cl gas	d/+	...	−	−	−	−	−
1607	cl gas	i/+	i	−	−/+	−/+	−/+	−
1608	cl gas, 0.771	r	r	+	+	+	r	r
1609	wh	r	r	+	+	+	+	+
1610	wh	r	r	+	+	+	r	r
1611	wh	r	r	+	+	+	+	+
1612	cl lq, 1.012[15]	∞	∞	+	+	+	r	r
1613	cl lq, 1.032[20]	∞	r	+	+	+	r	r
1614	wh	r	i	+	+	+	r/+	r/+
1615	wh, 2.43	r	r	r	r/+	r	r/+	r
1616	wh, 1.07	r/+	r	−	−	−	−/+	r
1617	wh	r/+	r	−	−	−	−/+	r
1618	wh	r/+	r	−	−	−	−/+	r
1619	wh	r/+	i	+	+	+	−/+	r
1620	wh hydr, 1.50	r	d	−	−/+	−/+	−/+	r
1621	cl lq	r/+	r	+	+	+	+	...
1622	wh, 1.53	r	r	r	r	r	r/+	r
1623	wh	r	d	r/+	−/+	−/+	−/+	−
1624	wh, 1.42	r/+	d	r	−	−	−/+	−
1625	wh, 1.80	r	d	−/+	−	−	−/+	−
1626	wh, 1.95	r	d	−	−	−	−/+	−
1627	wh hydr, 1.94	r	r	−	−/+	−/+	−/+	−
1628	yel, 1.91	r	d	+	+	+	r	r

No.	Formula and name	M_r	Phase transition temperature
1629	$(NH_4)_2Cr_2O_7$ Ammonium dichromate	252.06	dec ca. 168–185
1630	$(NH)F_2$ Difluoramine	53.01	mp –116.8; bp –23.6
1631	NH_4F Ammonium fluoride	37.04	$168 \rightarrow NH_4(HF_2)$
1632	NH_4HCO_3 Ammonium hydrocarbonate	79.06	dec 36–70
1633	$NH_4HC_2O_4$ ($\cdot 0.5H_2O$) Ammonium hydrooxalate	107.07	mp 220 dec
1634	$NH_4(HCOO)$ Ammonium formate	63.06	mp 119, dec 180
1635	$NH_4(HF_2)$ Ammonium hydrodifluoride	57.04	mp 126.2; dec 238
1636	$NH_4H(PHO_3)$ Ammonium hydrophosphonate	99.03	mp 123, dec 145
1637	$NH_4H_2PO_4$ Ammonium dihydroorthophosphate	115.03	mp 190 (p), dec 140
1638	$(NH_4)_2HPO_4$ Ammonium hydroorthophosphate	132.06	$70 \rightarrow NH_4H_2PO_4$
1639	NH_4HS Ammonium hydrosulfide	51.11	mp 120 (p), dec 20
1640	NH_4HSO_3 Ammonium hydrosulfite	99.11	dec 150
1641	NH_4HSO_4 Ammonium hydrosulfate	115.11	mp 251, bp 490
1642	NH_4I Ammonium iodide	144.94	subl 404.7 dec
1643	NH_4IO_3 Ammonium iodate	192.94	dec 150
1644	NH_4IO_4 Ammonium periodate	208.94	dec t
1645	NH_4MnO_4 Ammonium permanganate	136.97	dec 60–110 ($\rightarrow MnO_2$)
1646	$(NH_4)_6Mo_7O_{24}$ ($\cdot 4H_2O$) Ammonium 24-oxohepta-molybdate(VI)	1163.79	$-H_2O$ 90, dec 150
1647	NH_4N_3 Ammonium azide	60.06	mp 160, dec t
1648	NH_4NCS Ammonium thiocyanate	76.12	mp 149, dec 170
1649	$NH_4(NH_2COO)$ Ammonium carbamate	78.07	subl 60; mp 133.5
1650	NH_4NO_2 Ammonium nitrite	64.04	dec >60
1651	NH_4NO_3 Ammonium nitrate	80.03	mp 169.6; dec 210
1652	$N_2H_5NO_3$ Hydrazinium(1+) nitrate	95.06	mp 70.7; dec >300
1653	$N_2H_6(NO_3)_2$ Hydrazinium(2+) nitrate	158.07	mp 104 dec
1654	NH_4OCN Ammonium cyanate	60.06	$60 \rightarrow C(NH_2)_2O$
1655	NH_2OH Hydroxylamine	33.03	mp 32, dec >100
1656	$(NH_3OH)Cl$ Hydroxylaminium chloride	69.49	mp 159 dec
1657	$(NH_3OH)HSO_4$ Hydroxylaminium hydrosulfate	131.11	mp 57
1658	$(NH_3OH)NO_3$ Hydroxylaminium nitrate	96.04	mp 48, dec < 100
1659	$(NH_3OH)_2SO_4$ Hydroxylaminium sulfate	164.14	mp 170 dec
1660	$NH_4(PH_2O_2)$ Ammonium phosphinate	83.03	mp 200, dec 240
1661	NH_4ReO_4 Ammonium perrhenate	268.24	$400 \rightarrow ReO_2$
1662	$N_2H_6SO_4$ Hydrazinium(2+) sulfate	130.12	mp 254 dec
1663	$(NH_4)_2SO_3$ ($\cdot H_2O$) Ammonium sulfite	116.14	dec hydr ca. 60
1664	$(NH_4)_2SO_4$ Ammonium sulfate	132.14	dec 235–357
1665	$(NH_4)_2S_2O_6(O_2)$ Ammonium peroxodisulfate	228.20	$120 \rightarrow (NH_4)_2S_2O_7$
1666	$(NH_4)_2(SO_3S)$ Ammonium thiosulfate	148.21	dec 150
1667	$(NH_4)_2SeO_4$ Ammonium selenate	179.03	dec >500
1668	$(NH_4)_2TeO_4$ Ammonium metatellurate	227.67	dec >575
1669	$(NH_4)_2U_2O_7$ Ammonium heptaoxodiuranate(VI)	624.13	$350 \rightarrow UO_3$
1670	NH_4VO_3 Ammonium metavanadate	116.98	$100–150 \rightarrow V_2O_5$
1671	$(NH_4)_3VS_4$ Ammonium tetrathioorthovanadate	233.32	$60–150 \rightarrow V_2S_5$
1672	NO Nitrogen monooxide	30.01	mp –163.6; bp –151.6
1673	NO_2 Nitrogen dioxide	46.01	$< 20.7 \rightarrow N_2O_4$; 20.7–135 NO_2 and N_2O_4

(Continued)

No.	Color and density	Solubility and reactivity						
		water	alcohol	HCl	H_2SO_4	HNO_3	NaOH	$NH_3 \cdot H_2O$
1629	orange, 2.15	r	r	r/+	r	r	+	+
1630	cl gas	i/+	...	+	+	+	–/+	–
1631	wh, 1.01	r	r	–/+	–/+	–/+	–/+	r
1632	wh, 1.58	r/+	i	+	+	+	+	+
1633	wh	r	d	–	–/+	–/+	–/+	r
1634	wh, 1.27	r	r	–	–/+	–/+	–/+	r
1635	wh, 1.50	r	d	–	–/+	–/+	+	+
1636	wh	r	i	–	–/+	–/+	+	+
1637	wh, 1.80	r	i	r	r	r	+	+
1638	wh, 1.62	r	i	r	r	r	+	+
1639	wh, 1.17	r	r	–/+	–/+	–/+	+	+
1640	wh, 2.03	r	...	+	+	+	+	+
1641	wh, 1.78	r	d	–	–	–	+	+
1642	wh, 2.51	r	r	r	r/+	r/+	r/+	r
1643	wh, 3.31	r	i	r/+	r	r	–	–
1644	wh, 3.06	r	...	r	r	r	–	–
1645	viol, 2.22	r/+	+	r/+	r	r	–/+	+
1646	wh hydr, 2.50	r/+	i	+	+	+	+	–
1647	wh, 1.35	r	r	–	–/+	–/+	–/+	–
1648	wh, 1.31	r	r	–	–/+	–/+	–/+	–
1649	wh	+	...	+	+	+	+	+
1650	wh, 1.69	r/+	r	–/+	–/+	–/+	–/+	–
1651	wh, 1.72	r	r	r	r	r	r/+	r
1652	wh	r	d	–/+	–/+	–/+	–/+	–
1653	wh	r	d	–	–	–	–/+	–
1654	wh	r/+	d	–/+	–/+	–/+	–/+	–
1655	wh, 1.20	r	r	+	+	+	r	r
1656	wh, 1.67	r	r	r	–	–	–/+	–
1657	wh	r	d	–	r	–	–/+	+
1658	wh	r/+	r	–	–	r	–/+	–
1659	wh	r	i	–	r	–	–/+	–
1660	wh, 2.52	r	r	–	–/+	–/+	–/+	–
1661	wh, 3.97	r	r	–	–	–	–/+	–/+
1662	wh, 1.38	r	i	–	r	–	–/+	–
1663	wh hydr, 1.41	r	d	+	+	+	r	r
1664	wh, 1.77	r	i	–	r	–	r/+	r
1665	wh, 1.98	r/+	...	–/+	–/+	–/+	–	–
1666	wh, 1.68	r	i	+	+	+	–/+	–/+
1667	wh, 2.19	r	i	–	r	–	–/+	–
1668	wh, 3.01	+	–	+	+	+	+	+
1669	yel	i	...	–	–/+	–/+	–	–
1670	wh, 2.33	r	i	–/+	–/+	–/+	Na/+	–
1671	dk-viol, 1.62	r/+	i	+	+	+	–	–
1672	cl gas, 1.3402	d	r	–	–	–	–	–
1673	brn gas, 2.0527	+	+	+	+	+	+	+

No.	Formula and name	M_r	Phase transition temperature
1674	N_2O Dinitrogen oxide	44.01	mp −90.9; bp −88.6
1675	N_2O_3 Dinitrogen trioxide	76.01	$< -101 \rightarrow (NO)NO_2$, bp −40, dec >5
1676	N_2O_4 Dinitrogen tetraoxide	92.01	mp −11.2; bp 20.7 dec; 20.7–135 NO_2 and N_2O_4
1677	N_2O_5 Dinitrogen pentaoxide	108.01	$< 32 \rightarrow (NO_2)NO_3$
1678	(NO)Br Nitrosyl bromide	109.91	mp −55.5; dec >25
1679	(NO)Cl Nitrosyl chloride	65.46	mp −59.6; bp −5.4
1680	$(NO_2)Cl$ Nitroyl chloride	81.46	mp −141, bp −14.3
1681	$(NO)ClO_4$ ($\cdot H_2O$) Nitrosyl perchlorate	129.46	dec >108
1682	$(NO_2)ClO$ Nitryl hypochlorite	94.76	mp −107, bp 22.3
1683	(NO)F Nitrosyl fluoride	49.00	mp −132.5; bp −59.9
1684	$N(O)F_3$ Nitrogen oxide-trifluoride	87.00	mp −160, bp −85
1685	$(NO_2)F$ Nitryl fluoride	65.00	mp −166.0; bp −72.4
1686	$(NO)HSO_4$ Nitrosyl hydrosulfate	127.08	mp 73.5 dec
1687	$(NO_2)NH_2$ Nitryl amide	62.03	mp 75 dec
1688	$(NO)NO_2$ Nitrosyl nitrite	76.01	mp −101 ($\rightarrow N_2O_3$)
1689	$(NO_2)NO_3$ Nitryl nitrate	108.01	subl 32, mp 41 (p)
1690	$(NO_2)OF$ Nitryl fluorooxygenate(0)	81.00	mp −175, bp −45.0
1691	$(NO)_2S_2O_7$ Nitrosyl disulfate	236.14	mp 233, bp 360
1692	**Na Sodium**	22.990	mp 97.83; bp 886
1693	$NaAlO_2$ Sodium dioxoaluminate(III)	81.97	mp 1800
1694	$NaAl(SO_4)_2$ ($\cdot 12H_2O$) Sodium-aluminum sulfate	242.10	mp hydr 62.5; dec 800
1695	$NaAsO_2$ Sodium metaarsenite	129.91	dec 550
1696	$NaAsO_3$ Sodium trioxoarsenate(V)	145.91	mp 615
1697	Na_3AsO_4 ($\cdot 12H_2O$) Sodium arsenate	207.89	mp hydr 86.3; −H_2O 150
1698	$Na_4As_2O_7$ Sodium heptaoxodiarsenate(V)	353.80	mp 835, dec 1000
1699	Na_3AsS_4 ($\cdot 8H_2O$) Sodium tetrathioarsenate	272.16	dec 450–500
1700	$NaBO_2$ Sodium metaborate	65.80	mp 965, bp 1434
1701	$NaBO_2 \cdot 4H_2O$ see $[B(OH)_4]$,Na (No. 256)		
1702	NaB_5O_8 Sodium octaoxopentaborate(III)	205.04	mp 785 dec
1703	$NaB_5O_8 \cdot 5H_2O$	295.11	mp 117, −H_2O 350
1704	$Na_2B_4O_7$ Sodium tetraborate	201.22	mp 741, bp 1575 dec
1705	$Na_2B_4O_7 \cdot (4+5)H_2O$	−	−H_2O 320–340
1706	$Na_2B_4O_7 \cdot 10H_2O$	381.37	mp 64, −H_2O 380
1707	$NaBiO_3$ Sodium bismuthate	279.97	...
1708	Na_3BiO_4 Sodium tetraoxobismuthate(V)	341.95	...
1709	NaBr Sodium bromide	102.89	mp 755, bp 1390
1710	$NaBr \cdot 2H_2O$	138.92	−H_2O 50.2
1711	$NaBrO_3$ Sodium bromate	150.89	mp 384 dec
1712	Na_2C_2 Sodium acetylide	70.00	dec 800–825
1713	$Na(CH_3COO)$ Sodium acetate	82.03	mp 324
1714	$Na(CH_3COO) \cdot 3H_2O$	136.08	mp 58, −H_2O 120
1715	NaCN ($\cdot 2H_2O$) Sodium cyanide	49.01	mp 563.7; bp 1497
1716	Na_2CO_3 ($\cdot H_2O$) Sodium carbonate	105.99	mp 851, dec >1000
1717	$Na_2CO_3 \cdot 10H_2O$	286.14	mp 32.5; −H_2O 100
1718	$Na_2C_2O_4$ Sodium oxalate	134.00	mp 260, dec >400

(Continued)

No.	Color and density	Solubility and reactivity						
		water	alcohol	HCl	H_2SO_4	HNO_3	NaOH	$NH_3 \cdot H_2O$
1674	cl gas, 1.9778	d	r	−	−/+	−	−	−
1675	bl lq, 1.447²	+	+	+	+	+	+	+
1676	yel lq, 1.491⁰	+	+	+	+	+	+	+
1677	cl gas	+	+	+	+	+	+	+
1678	red gas	+	...	+	+	+	+	+
1679	orange-yel gas, 2.992	+	r	+	+	+	+	+
1680	cl gas, 2.57	+	+	+	+	+	+	+
1681	wh hydr, 2.17	+	+	+	+	+	+	+
1682	lt-yel lq	+	...	+	+	+	+	+
1683	cl gas, 2.335	+	...	+	+	+	+	+
1684	cl gas, lq 0.927⁻⁸⁸	i	+	−	−	−	−/+	...
1685	cl gas, 2.90	+	+	+	+	+	+	+
1686	wh	+	...	+	+/−	+	+	+
1687	wh	+	+	+	+	+	+	+
1688	sk-bl	+	+	+	+	+	+	+
1689	wh, 1.64	+	+	+	+	+	+	+
1690	cl gas	+	+	+	+	+	+	+
1691	wh	r/+	i	+	+	+	+	+
1692	wh, 0.97	+	+	+	+	+	+	+
1693	wh	+	i	+	+	+	+	+
1694	wh hydr, 1.68	r	i	−	r	−	+	+
1695	wh, 1.87	+	d	+	+	+	+	+
1696	wh, 2.30	+	...	+	+	+	+	+
1697	wh, 2.84(1.76)	r	r	r	r	r	r	r
1698	wh, 2.21	+	...	+	+	+	+	+
1699	lt-yel	r/+	i	+	+	+	r/+	r/+
1700	wh, 2.34	+	i	+	+	+	+	+
1701	see No.256							
1702	wh	r	...	+	+	+	r/+	...
1703	wh	r	...	+	+	+	r/+	...
1704	wh, 2.37	r	r	+	+	+	r/+	...
1705	wh, 1.88–1.91	r	r	+	+	+	r/+	...
1706	wh, 1.73	r	r	+	+	+	r/+	...
1707	yel	i	+	i/+	i/+	+	i	i
1708	brn	i	+	i/+	i/+	+	i	i
1709	wh, 3.21	r	r	r	r/+	r	r	r
1710	wh, 2.18	r	r	r	r/+	r	r	r
1711	wh, 3.34	r	i	−	−	−	−	−
1712	wh, 1.60	+	...	+	+	+	+	+
1713	wh, 1.53	r	r	r	r	r	r	r
1714	wh, 1.45	r	r	r	r	r	r	r
1715	wh, 1.60	r	d	r	r	r	r	r
1716	wh, 2.54(2.26)	r	i	+	+	+	r	r
1717	wh, 1.45	r	i	+	+	+	r	r
1718	wh, 2.34	r	i	−/+	−/+	r/+	r	r

No.	Formula and name	M_r	Phase transition temperature
1719	$Na_3CO_3HCO_3$ ($\cdot 2H_2O$) Sodium carbonate-hydrocarbonate	189.99	dec 135
1720	$Na_2Ca(CO_3)_2$ ($\cdot 5H_2O$) Disodium-calcium carbonate	206.07	$-H_2O$ 110–130
1721	$Na_2Ca(SO_4)_2$ ($\cdot 4H_2O$) Disodium-calcium sulfate	278.18	$-H_2O$ >160
1722	$NaCl$ ($\cdot 2H_2O$) Sodium chloride	58.44	mp 800.8; bp 1465
1723	$NaClO$ ($\cdot 5H_2O$) Sodium hypochlorite	74.44	mp hydr 24.5; dec >30
1724	$NaClO_2$ ($\cdot 3H_2O$) Sodium chlorite	90.44	$-H_2O$ 37.4; dec >180
1725	$NaClO_3$ Sodium chlorate	106.44	mp 262, dec 630
1726	$NaClO_4$ Sodium perchlorate	122.44	mp 482 dec
1727	$NaClO_4 \cdot H_2O$	140.45	mp 50.8; $-H_2O$ 130
1728	Na_2CrO_4 ($\cdot 10H_2O$) Sodium chromate	161.97	mp hydr 19.9; mp 792
1729	$Na_2Cr_2O_7$ Sodium dichromate	261.97	mp 320, dec >400
1730	$Na_2Cr_2O_7 \cdot 2H_2O$	298.00	mp 110 dec
1731	$NaCrS_2$ Sodium dithiochromate(III)	139.12	dec >850
1732	NaF Sodium fluoride	41.99	mp 997, bp 1700
1733	$NaFeO_2$ Sodium dioxoferrate(III)	110.84	mp 1350
1734	Na_2GeO_3 ($\cdot 7H_2O$) Sodium germanate	166.59	mp hydr 83, mp 1070
1735	NaH Sodium hydride	24.00	dec 400, mp 638 (p)
1736	NaH_2AsO_4 ($\cdot H_2O$) Sodium dihydroarsenate	163.92	$-H_2O$ 90, dec >230
1737	Na_2HAsO_4 ($\cdot 12H_2O$) Sodium hydroarsenate	185.91	mp hydr 20, $-H_2O$ 100
1738	$Na_2H_2As_2O_7$ Disodium-dihydrogen heptaoxodiarsenate(V)	309.83	$230 \rightarrow NaAsO_3$
1739	$NaHCO_3$ Sodium hydrocarbonate	84.01	dec 100–150
1740	$Na(HCOO)$ ($\cdot 2H_2O$) Sodium formate	68.01	mp 259, dec >300
1741	$Na(HF_2)$ Sodium hydrodifluoride	61.99	dec 270
1742	NaH_4IO_6 ($\cdot H_2O$) Sodium tetrahydroorthoperiodate	249.92	dec hydr 175
1743	$Na_3H_2IO_6$ Sodium dihydroorthoperiodate	293.88	mp 200 dec
1744	$NaH(PHO_3)$ ($\cdot 2.5H_2O$) Sodium hydrophosphonate	103.98	mp hydr 42, $-H_2O$ 100
1745	NaH_2PO_4 Sodium dihydroorthophosphate	119.98	$>160 \rightarrow Na_2H_2P_2O_7$
1746	$NaH_2PO_4 \cdot H_2O$	137.99	mp 57.4; $-H_2O$ 100
1747	$NaH_2PO_4 \cdot 2H_2O$	156.01	mp 40.8 dec
1748	Na_2HPO_4 ($\cdot 2H_2O$) Sodium hydroorthophosphate	141.96	$-H_2O$ 95, $300 \rightarrow Na_4P_2O_7$
1749	$Na_2HPO_4 \cdot 7H_2O$	268.06	$48.1 \rightarrow Na_2HPO_4 \cdot 2H_2O$
1750	$Na_2HPO_4 \cdot 12H_2O$	358.14	$35.1 \rightarrow Na_2HPO_4 \cdot 7H_2O$
1751	$Na_2H_2P_2^{IV}O_6$ ($\cdot 6H_2O$) Disodium-dihydrogen hexaoxodiphosphate(IV)	205.94	$-H_2O$ 100, mp 250
1752	$Na_2H_2P_2O_7$ ($\cdot 6H_2O$) Sodium dihydrodiphosphate	221.94	$220 \rightarrow NaPO_3$
1753	$NaHS$ ($\cdot nH_2O$) Sodium hydrosulfide	56.06	mp hydr 53, mp 350
1754	$NaHSO_3$ Sodium hydrosulfite	104.06	dec 25–100
1755	$NaHSO_4$ Sodium hydrosulfate	120.06	mp 186, dec 320
1756	$NaHSO_4 \cdot H_2O$	138.08	mp 58.5, dec 250
1757	$NaHSO_3(O_2)$ Sodium hydroperoxosulfate	136.06	dec t
1758	$NaHg$ Sodiummercury	223.58	mp 212
1759	$NaHg_2$ Sodiumdimercury	424.17	mp 354
1760	$NaHg_4$ Sodiumtetramercury	825.35	mp 136 dec
1761	Na_3Hg Trisodiummercury	269.56	mp 35 dec
1762	Na_3Hg_2 Trisodiumdimercury	470.15	mp 119 dec

(Continued)

No.	Color and density	Solubility and reactivity						
		water	alcohol	HCl	H_2SO_4	HNO_3	NaOH	$NH_3 \cdot H_2O$
1719	wh, 2.15	r	d	+	+	+	+	+
1720	wh hydr, 2.04	d	...	+	+	+	−	−
1721	wh, 2.80	d	...	−	−	−	−/+	−
1722	wh, 2.17	r	d	r/i	r/i	r	r/i	r
1723	wh hydr, 1.10	r/+	...	r/+	−/+	+	−	−
1724	wh	r/+	r	+	+	+	−	−
1725	wh, 2.49	r	r	r/+	r	−	−	−
1726	wh	r	r	−	−	−	−	−
1727	wh, 2.02	r	r	−	−	−	−	−
1728	yel, 2.72(1.48)	r	d	+	+	+	r	r
1729	orange	r	d	r/+	r	r	+	+
1730	orange, 2.52	r	d	r/+	r	r	+	+
1731	dk-grey-grn	+	i	+	+	+	+	+
1732	wh, 2.56	r	d	r	r/+	r	r	r
1733	grn or brn, 4.05	+	...	+	+	+	+	...
1734	wh, 3.31	r	...	+	+	+	−/+	−
1735	wh, 1.36	+	+	+	+	+	+	+
1736	wh hydr, 2.53	r	...	r	r	r	+	+
1737	wh hydr, 1.72	r	i	r	r	r	+	+
1738	wh	+	...	+	+	+	+	+
1739	wh, 2.24	r/+	r	+	+	+	+	+
1740	wh, 1.92	r	d	r	r/+	r/+	r	r
1741	wh, 2.08	r	...	−/+	−/+	−/+	+	+
1742	wh hydr	d	...	−	−	−	+	+
1743	wh	d	...	−	−	−	+	+
1744	wh	r	...	r	r	r	+	+
1745	wh	r	i	r	r	r	+	+
1746	wh, 2.04	r	i	r	r	r	+	+
1747	wh, 1.91	r	i	r	r	r	+	+
1748	wh, 2.07	r	i	r	r	r	+	+
1749	wh, 1.68	r	i	r	r	r	+	+
1750	wh, 1.52	r	i	r	r	r	+	+
1751	wh, 1.85	r	i	−/+	−/+	−/+	+	+
1752	wh, 1.86(1.85)	r	...	+	+	+	+	+
1753	wh, 1.79	r/+	r	+	+	+	+	+
1754	wh, 1.48	r	d	+	+	+	+	+
1755	wh, 2.74	r	+	−	r	−	+	+
1756	wh, 2.10	r	+	−	r	−	+	+
1757	wh	r/+	...	+	+	+	−/+	...
1758	grey	+	+	+	+	+	+	+
1759	grey	+	+	+	+	+	+	+
1760	grey	+	+	+	+	+	+	+
1761	grey	+	+	+	+	+	+	+
1762	grey	+	+	+	+	+	+	+

No.	Formula and name	M_r	Phase transition temperature
1763	Na_5Hg_2 Pentasodiumdimercury	516.13	mp 66 dec
1764	Na_7Hg_8 Heptasodiumoctamercury	1765.65	mp 222 dec
1765	NaI Sodium iodide	149.90	mp 662, bp 1304
1766	$NaI·2H_2O$	185.92	mp 68.9 dec
1767	$NaIO_3$ $(·H_2O)$ Sodium iodate	197.89	mp 422, dec >500
1768	$NaIO_4$ $(·3H_2O)$ Sodium metaperiodate	213.89	mp 300 dec
1769	Na_5IO_6 Sodium orthoperiodate	337.85	dec 800
1770	$Na_2Mg(CO_3)_2$ $(·H_2O)$ Disodium-magnesium carbonate	190.30	dec hydr >250
1771	$Na_2Mg(SO_4)_2$ $(·4H_2O)$ Disodium-magnesium sulfate	166.42	dec hydr >240
1772	$NaMnO_4$ $(·3H_2O)$ Sodiun permanganate	141.92	dec hydr 170
1773	Na_2MnO_4 $(·10H_2O)$ Sodium manganate	164.91	mp hydr 17
1774	Na_2MoO_4 $(·2H_2O)$ Sodium molybdate	205.92	$-H_2O$ 150, mp 688
1775	NaN_3 Sodium azide	65.01	mp ca. 200, dec 300
1776	$NaCNS$ $(·H_2O)$ Sodium thiocyanate	81.07	mp 307.5
1777	$NaNH_2$ Sodium amide	39.01	mp 210, bp 400
1778	$NaNO_2$ Sodium nitrite	69.00	mp 271, dec >520
1779	$NaNO_3$ Sodium nitrate	84.99	mp 306.5; dec 380
1780	$Na_2N_2O_2$ Sodium hyponitrite	105.99	dec 335
1781	$Na_2N_2O_4$ Tetrasodium tetraoxodinitrate	183.97	dec >100
1782	$(NaNb)O_3$ Sodium-niobium trioxide	163.89	mp 1365
1783	NaO_2 Sodium superoxide	54.99	$270 \rightarrow Na_2O_2$
1784	Na_2O Sodium oxide	61.98	mp 1132
1785	Na_2O_2 Sodium peroxide	77.98	mp 596 (p), 675 $\rightarrow Na_2O$
1786	$NaOCN$ Sodium cyanate	65.01	mp 550, dec >600
1787	$NaOH$ Sodium hydroxide	40.00	mp 321, bp 1390
1788	$NaOH·H_2O$	58.01	mp 64.3; $-H_2O$ 100
1789	$Na(PH_2O_2)$ $(·H_2O)$ Sodium phosphinate	87.98	dec hydr >200
1790	$Na_2(PHO_3)$ $(·5H_2O)$ Sodium phosphonate	125.96	mp hydr 53, dec hydr 120
1791	$NaPO_3$ Sodium metaphosphate	101.96	mp 627.6
1792	Na_3PO_4 Sodium orthophosphate	163.94	mp 1340
1793	$Na_3PO_4·12H_2O$	380.12	mp 73.4, $-H_2O$ 200
1794	$Na_4P_2^{IV}O_6$ $(·10H_2O)$ Sodium hexaoxodiphosphate(IV)	249.90	dec hydr >250
1795	$Na_4P_2O_7$ $(·10H_2O)$ Sodium diphosphate	265.90	mp hydr 79.5; mp 985
1796	$Na_5P_3O_{10}$ $(·6H_2O)$ Sodium decaoxotriphosphate(V)	367.86	$-H_2O$ 120, mp 692 dec
1797	$Na_2(PO_3F)$ Sodium fluoroorthophosphate	143.95	mp ca. 625
1798	$NaReO_4$ Sodium perrhenate	273.19	mp 300, dec 410
1799	Na_2S $(·9H_2O)$ Sodium sulfide	78.05	mp hydr 50, mp 1180
1800	$Na_2(S_2)$ Sodium disulfide(2−)	110.11	mp ca. 490
1801	$Na_2(S_4)$ Sodium tetrasulfide(2−)	174.24	mp 286 dec
1802	$Na_2(S_5)$ Sodium pentasulfide(2−)	206.31	mp 253
1803	Na_2SO_3 $(·7H_2O)$ Sodium sulfite	126.04	$-H_2O$ 150, mp 911 (p)
1804	Na_2SO_4 Sodium sulfate	142.04	mp 884, bp 1430
1805	$Na_2SO_4·10H_2O$	322.19	mp 32.4 dec
1806	$Na_2S_2O_4$ $(·2H_2O)$ Disodium tetraoxodisulfate	174.11	$-H_2O$ 50, dec>300
1807	$Na_2S_2O_5$ Disodium pentaoxodisulfate	190.11	dec 150

(Continued)

No.	Color and density	Solubility and reactivity						
		water	alcohol	HCl	H_2SO_4	HNO_3	NaOH	$NH_3 \cdot H_2O$
1763	grey	+	+	+	+	+	+	+
1764	grey	+	+	+	+	+	+	+
1765	wh, 3.67	r	r	r	r/+	r/+	r	r
1766	wh, 2.45	r	r	r	r/+	r/+	r	r
1767	wh, 4.28	r	i	r/+	r	r	r	r
1768	wh, 3.87(3.22)	r	...	r	r	r	–	–
1769	wh	d/+	...	d	–	d	d	–
1770	wh hydr, 2.41	i	...	+	+	+	–/+	–
1771	wh hydr, 2.25	r	...	r	r	r	+	+
1772	red-viol hydr, 2.46	r	+	r/+	r/+	r/+	r/+	+
1773	grn hydr	+	...	+	+	+	–	–
1774	wh, 3.78(3.28)	r	...	+	+	+	–	–
1775	wh, 1.85	r	d	+	+	+	r	r
1776	wh, 1.73	r	r	r	r/+	r/+	r	r
1777	wh, 1.39	+	+	+	+	+	+	+
1778	wh, 2.17	r	r	r/+	r/+	r/+	r	r
1779	wh, 2.26	r	d	r	r	r	r	r
1780	wh, 2.47	r/+	i	+	+	+	–	–
1781	yel	+	...	+	+	+	+	+
1782	wh	–	...	–/+	+	–/+	–	–
1783	dk-yel, 2.21	+	+	+	+	+	+	+
1784	wh, 2.36	+	+	+	+	+	+	+
1785	wh, 2.60	+	+	+	+	+	+	+
1786	wh, 1.94	r/+	i	r/+	r/+	r/+	r	r
1787	wh, 2.13	r	r	+	+	+	r	r
1788	wh	r	r	+	+	+	r	r
1789	wh hydr	r	r	–	–/+	–/+	–	–
1790	wh hydr	r	d	–	–	–	–	–
1791	wh, 2.48	r/+	...	r/+	r/+	r/+	+	+
1792	wh, 2.54	r	i	r	r	r	r	r
1793	wh, 1.62	r	i	r	r	r	r	r
1794	wh, 1.82	r/+	...	+	+	+	+	+
1795	wh, 2.37(1.82)	r	i	+	+	+	–/+	–
1796	wh, 2.52(2.12)	r/+	...	+	+	+	–/+	...
1797	wh	r	...	+	+	+	+	+
1798	wh, 5.24	r	r	–	–	–	–	–
1799	wh, 1.86(1.43)	r/+	d	+	+	+	r	r
1800	lt-yel	r/+	d	+	+	+	–	–
1801	orange-yel, 2.08	r/+	r	+	+	+	–	–
1802	yel-brn, 2.08	r/+	r	+	+	+	–	–
1803	wh, 2.63(1.56)	r/+	d	+	+	+	r	r
1804	wh, 2.66	r	d	–	r	–	r	r
1805	wh, 1.46	r	i	–	r	–	r	r
1806	wh hydr	r/+	i	+	+	+	–	–
1807	wh, 1.48	r/+	d	–	–/+	–/+	+	+

No.	Formula and name	M_r	Phase transition temperature
1808	$Na_2S_2O_6$ ($\cdot 2H_2O$) Sodium dithionate	206.11	$-H_2O$ 100, dec >200
1809	$Na_2S_2O_7$ Sodium disulfate	222.11	mp 405, dec 460
1810	$Na_2S_2O_6(O_2)$ Sodium peroxodisulfate	238.10	ca. 250 $\rightarrow$ $Na_2S_2O_7$
1811	$Na_2(SO_3S)$ Sodium thiosulfate	158.11	dec 200
1812	$Na_2(SO_3S)\cdot 5H_2O$	248.18	mp hydr 48.5; $-H_2O$ 100
1813	Na_3Sb Trisodium stibide	190.72	mp 1010
1814	$NaSbO_2$ ($\cdot 3H_2O$) Sodium dioxostibate(III)	176.74	dec t
1815	Na_2Se Sodium selenide	124.94	mp ca. 875
1816	Na_2SeO_3 ($\cdot 5H_2O$) Sodium selenite	172.94	$-H_2O$ 40, mp 710 dec
1817	Na_2SeO_4 ($\cdot 10H_2O$) Sodium selenate	188.94	mp 730
1818	Na_2SiO_3 Sodium metasilicate	122.06	mp 1089
1819	$Na_2SiO_3\cdot 9H_2O$	284.20	mp 47, $-H_2O$ >100
1820	$Na_2Si_2O_5$ Sodium pentaoxodisilicate(IV)	182.15	mp 874
1821	Na_4SiO_4 Sodium orthosilicate	184.04	mp 1120 dec
1822	$Na_6Si_2O_7$ Sodium heptaoxodisilicate(IV)	306.11	mp 1122
1823	Na_2Te Sodium telluride	173.58	mp 1035
1824	$NaVO_3$ ($\cdot 2H_2O$) Sodium metavanadate	121.93	mp 630
1825	Na_3VO_4 ($\cdot 10H_2O$) Sodium orthovanadate	183.91	mp 866
1826	$Na_4V_2O_7$ Sodium heptaoxodivanadate(V)	305.84	mp 654
1827	Na_2WO_4 ($\cdot H_2O$) Sodium wolframate	293.83	$-H_2O$ 150, mp 698
1828	$Na_6W_7O_{24}$ ($\cdot 16H_2O$) Sodium 24-oxoheptawolframate(VI)	1808.87	$-H_2O$ 300
1829	**Nb Niobium**	92.906	mp 2470, bp 4927
1830	Nb_3Al Triniobiumaluminum	305.70	mp 2040 dec
1831	NbB_2 Niobium diboride	114.53	mp 3050
1832	$NbBr_4$ Niobium(IV) bromide	412.52	subl 300
1833	$NbBr_5$ Niobium(V) bromide	492.43	mp 265.2; bp 361.6
1834	Nb_3Br_{8+x} ($0.01 \leq x \leq 1.12$) Triniobium octabromide	917.95	subl 400
1835	$NbBr_3O$ Niobium tribromide-oxide	348.62	subl t, dec >320
1836	NbC Niobium monocarbide	104.92	mp 3500 dec
1837	Nb_2C Diniobium carbide	197.82	mp 3080 dec
1838	$[Nb(C_5H_5)Cl_4]$ Cyclopentadienyltetrachloroniobium	299.81	subl 210
1839	$[Nb(C_5H_5)_2Cl_2]$ Bis(cylopentadienyl)dichloroniobium	294.00	dec >400
1840	$[Nb(CO)_4C_5H_5]$ Tetracarbonyl(cyclopentadienyl)niobium	270.04	mp 146.5
1841	$NbCl_4$ Niobium(IV) chloride	234.72	subl 275; dec >320
1842	$NbCl_5$ Niobium(V) chloride	270.17	mp 204.7; bp 247.5
1843	Nb_3Cl_{8+x} ($0.01 \leq x \leq 1.39$) Triniobium octachloride	562.34	800 $\rightarrow$ Nb_6Cl_{14}
1844	Nb_6Cl_{14} Hexaniobium 14-chloride	1053.78	>900 $\rightarrow$ $NbCl_4$,Nb
1845	$NbCl_3O$ Niobium trichloride-oxide	215.26	subl 332, mp 429
1846	NbF_5 Niobium(V) fluoride	187.90	mp 79.5; bp 234.5
1847	Nb_3Ga Triniobiumgallium	348.44	mp ca. 1900 dec
1848	Nb_3Ge Triniobiumgermanium	351.33	mp ca. 1970 dec
1849	NbH_{1-x} ($-0.05 \leq x \leq 0.3$) Niobium monohydride	93.91	600–800 $\rightarrow$ Nb
1850	NbH_2 Niobium dihydride	94.92	>400 $\rightarrow$ Nb
1851	NbI_4 Niobium(IV) iodide	600.52	subl >420

(Continued)

No.	Color and density	Solubility and reactivity						
		water	alcohol	HCl	H_2SO_4	HNO_3	NaOH	$NH_3 \cdot H_2O$
1808	wh hydr, 2.19	r	i	+	+	+	−	−
1809	wh, 2.66	r/+	...	+	+	+	+	+
1810	wh	r/+	i	−/+	−/+	−/+	−	−
1811	wh, 1.667	r	i	+	+	+	−	−
1812	wh, 1.72	r	i	+	+	+	−	−
1813	bl-blk, 2.67	+	...	+	+	+	+	+
1814	wh hydr, 2.86	i	...	−/+	−/+	+	−/+	−
1815	wh, 2.5	r/+	...	+	+	+	r	r
1816	wh	r/+	r	+	+	+	r	r
1817	wh, 3.21(1.61)	r	...	−/+	−	−	r	r
1818	wh, 2.4	r/+	i	+	+	+	+	+
1819	wh	r/+	i	+	+	+	+	+
1820	wh	r	i	+	+	+	+	+
1821	wh	r	i	+	+	+	r	r
1822	wh	r	i	+	+	+	+	+
1823	wh, 2.90	+	...	+	+	+	r	r
1824	wh hydr, 2.79	r	i	+	+	+	−/+	...
1825	wh	+	i	+	+	+	−	−
1826	wh	r/+	i	+	+	+	−/+	...
1827	wh, 4.18(3.25)	r	i	+	+	+	−	−
1828	wh hydr, 3.99	r/+	...	+	+	+	+	...
1829	lt-grey, 8.560	−	−	−	−/psv	psv	−	−
1830	lt-grey	−	i	+	+	+	+	...
1831	wh, 6.97	−	i	−	−	−	−	−
1832	brn-blk, 4.72	+	...	+	+	+	+	+
1833	dk-red, 4.36	+	r	+	+	+	+	+
1834	blk-brn, 5.4	+	...	+	+	+	+	+
1835	yel	+	r/+	+	+	+	+	+
1836	grey, 7.6	−	i	−	−	−/+	−	−
1837	dk-grey, 7.7	−	i	−	−	−/+	−	−
1838	dk-red	+	i	+	+	+	+	+
1839	brn	+	i	+	+	+	+	+
1840	red	−/+	r	−/+	−/+	−/+	−/+	...
1841	brn-blk, 3.14	+	...	+	+	+	+	+
1842	wh ($t \to$ yel), 2.75	+	+	+	+	+	+	+
1843	blk-grn, 3.75	+	...	+	+	+	+	...
1844	blk-grn, 3.78	−/+	...	−/+	−/+	−/+	+	...
1845	wh, 10.19	+	r/+	+	+	+	+	+
1846	wh, 3.29	+	+	+	+	+	+	+
1847	grey	−	i	+	+	+	+	...
1848	grey	−	i	+	+	+	−/+	...
1849	dk-grey	−	i	−	−/+	−	−	−
1850	grey	−	i	−	−/+	−	−	−
1851	grey, 5.64	+	...	+	+	+	+	+

No.	Formula and name	M_r	Phase transition temperature
1852	NbI_5 Niobium(V) iodide	727.43	mp 320 dec ($\rightarrow NbI_4$)
1853	Nb_3I_8 Triniobium octaiodide	1293.95	>800 $\rightarrow Nb_6I_{11}$
1854	Nb_6I_{11} Hexaniobium undecaiodide	1953.38	dec >950
1855	NbI_2O Niobium diiodide-oxide	362.71	dec >500
1856	NbI_3O Niobium triiodide-oxide	489.62	>150 $\rightarrow$ Nb
1857	NbN Niobium mononitride	106.91	dec >2300
1858	$Nb_{2-x}N$ (0 < x < 0.1) Diniobium nitride	199.82	mp 2420
1859	NbO Niobium(II) oxide	108.91	mp 1945
1860	NbO_2 Niobium(IV) oxide	124.90	mp 1915
1861	Nb_2O_5 Niobium(V) oxide	265.81	mp 1490–1500
1862	$Nb_2O_5 \cdot nH_2O$	–	$-H_2O$ >500
1863	NbS_2 Niobium disulfide	157.04	dec 1050
1864	$NbSi_2$ Niobium disilicide	149.08	mp ca. 1950
1865	Nb_3Sn Triniobiumtin	397.43	mp 2130 dec
1866	**Nd Neodymium**	144.24	mp 1024, bp ca. 3080
1867	NdB_6 Neodymium hexaboride	209.11	mp ca. 2600
1868	$NdBr_3$ Neodymium(III) bromide	383.95	mp 683, bp 1540
1869	$Nd(BrO_3)_3$ ($\cdot 9H_2O$) Neodymium(III) bromate	527.94	mp hydr 66.7; $-H_2O$ 150
1870	NdC_2 Neodymium dicarbide	168.26	mp 2300 dec
1871	$Nd_2(C_2O_4)_3$ ($\cdot 10H_2O$) Neodymium(III) oxalate	552.53	hydr >600 $\rightarrow Nd_2O_3$
1872	$NdCl_3$ Neodymium(III) chloride	250.60	mp 760, bp 1620
1873	$NdCl_3 \cdot 6H_2O$	358.69	mp 124, $-H_2O$ >160
1874	$Nd(ClO_4)_3$ ($\cdot 6H_2O$) Neodymium(III) perchlorate	442.59	dec hydr >170
1875	NdF_3 Neodymium(III) fluoride	201.23	mp 1377, bp 2300
1876	NdI_3 Neodymium(III) iodide	524.96	mp 787, bp 1350
1877	$Nd_2(MoO_4)_3$ Neodymium(III) molybdate	665.62	mp 1176
1878	$Nd(NO_3)_3$ ($\cdot 6H_2O$) Neodymium(III) nitrate	330.25	dec hydr t
1879	Nd_2O_3 Neodymium(III) oxide	336.48	mp 2320, bp ca. 4300
1880	$Nd(OH)_3$ Neodymium(III) hydroxide	195.26	>350 $\rightarrow Nd_2O_3$
1881	Nd_2S_3 Neodymium(III) sulfide	384.68	mp 2010 dec
1882	$Nd_2(S)O_2$ Dineodymium sulfide-dioxide	352.54	mp ca. 1990
1883	$Nd_2(SO_4)_3$ ($\cdot 8H_2O$) Neodymium(III) sulfate	576.67	$-H_2O$ 350, dec >700
1884	**Ne Neon**	20.180	mp -248.6; bp -246.048
1885	**Ni Nickel**	58.69	mp 1455, bp ca. 2900
1886	$(NiAl_2)O_4$ Nickel-dialuminum tetraoxide	176.65	mp 2110
1887	NiAs Nickel monoarsenide	133.61	mp 964
1888	$NiAs_2$ Nickel diarsenide	208.53	830 $\rightarrow$ NiAs
1889	$Ni_3(AsO_4)_2$ ($\cdot 8H_2O$) Nickel(II) arsenate	453.91	$-H_2O$ 200, dec 1000
1890	Ni(As)S Nickel arsenide-sulfide	165.68	700 $\rightarrow$ NiAs,NiS
1891	Ni_2B Dinickel boride	128.19	mp 1225
1892	$NiBr_2$ Nickel(II) bromide	218.50	subl 919, mp 963 (p)
1893	$NiBr_2 \cdot 6H_2O$	326.59	mp hydr 28.5; $-H_2O$ 150
1894	$Ni(BrO_3)_2$ ($\cdot 6H_2O$) Nickel(II) bromate	314.49	dec hydr t
1895	Ni_3C Trinickel carbide	188.08	dec 380–400
1896	$[Ni(C_5H_5)_2]$ Bis(cyclopentadienyl)nickel	188.88	dec 173, >200 $\rightarrow$ Ni
1897	$Ni(CH_3COO)_2$ ($\cdot 4H_2O$) Nickel(II) acetate	176.78	$-H_2O$ 90, dec 250
1898	$[Ni(C_4H_7O_2N_2)_2]$ Bis(dimethylglyoximato)nickel	288.91	subl 250

(Continued)

No.	Color and density	Solubility and reactivity						
		water	alcohol	HCl	H_2SO_4	HNO_3	NaOH	$NH_3 \cdot H_2O$
1852	yel-orange, 5.32	+	+	+	+	+	+	+
1853	blk, 5.89	+	...	+	+	+	+	+
1854	blk-brn, 5.4	+	...	+	+	+	+	+
1855	blk-brn	+	...	+	+	+	+	+
1856	blk	+	...	+	+	+	+	+
1857	yel-grey, 8.4	−	i	−	−	−/+	+	−
1858	grey	−	i	−	−	−	+	−
1859	dk-grey, 7.26	−	i	+	−/+	−	−/+	−/+
1860	bl-blk, 5.98	−	...	−	−/+	−	−/+	...
1861	wh, 4.47	−	i	−	−/+	−	−/+	−
1862	wh, 4.3	i	...	+	+	+	−/+	−
1863	blk	−	...	−	−/+	−	−/+	−
1864	grey	−	i	−	−	−	−/+	...
1865	grey	−	...	+	+	+	+	...
1866	wh, 7.01	psv/+	−	+	+	+	−	−
1867	bl	−	i	−	−	−	−	−
1868	viol	d	r	−	−/+	−	+	+
1869	red	r	...	−	−	−	+	+
1870	yel, 5.15	+	...	+	+	+	+	+
1871	pink hydr	i	i	i/+ −/+	−/+	i	i	i
1872	pink-viol, 4.13	r	r	r	−	−	+	+
1873	red, 2.28	r	r	r	−	−	+	+
1874	pink-viol hydr	r	...	−	−	−	+	+
1875	lt-viol	i	i	i	−	−	i	−
1876	grn	r	r	r	−/+	−/+	+	+
1877	red-viol, 5.14	i	i	−/+	−/+	−	−/+	−
1878	lt-viol hydr	r	r	−	−	r	+	+
1879	sk-bl, 7.24	−/+	...	+	+	+	−	−
1880	pink	i	...	+	+	+	i/+	i
1881	dk-grn, 5.18	i/+	...	+	+	+	i	i
1882	sk-bl-wh	−	...	+	+	+	−	−
1883	red hydr, 2.85	r	...	−	r	−	+	+
1884	cl gas, 0.90035	i	d	−	−	−	−	−
1885	wh, 8.91	−	−	+	+	+/psv	−	−
1886	sk-bl	−	i	−	−/+	−	−/+	−
1887	yel-red, 7.78	−	i	−/+	−/+	−/+	−	−
1888	lt-red, 7.1	−	i	−/+	−/+	−/+	−/+	−
1889	yel-grn hydr, 3.07	i	i	r	r	r	i	i/+
1890	grey, 5.9	−	i	−/+	−/+	−/+	−	−
1891	yel-grey, 4.64	−	i	−/+	−/+	+	−	−
1892	dk-yel, 5.10	r	r	r	−/+	−	+	+
1893	grn	r	r	−	−/+	−/+	+	+
1894	grn hydr, 2.58	r	...	−/+	−/+	r	+	+
1895	grey-blk, 7.97	−	i	+	+	+	−	−
1896	dk-grn	−/+	r	−	−/+	+	−	−
1897	grn hydr, 1.80	r	r	r	−	−	+	+
1898	red	i	r	+	+	+	−	−

No.	Formula and name	M_r	Phase transition temperature
1899	Ni(CN)$_2$ ($\cdot$2H$_2$O) Nickel(II) cyanide	110.73	$-$H$_2$O 180, dec 400
1900	[Ni(CN)$_4$],K$_2$ ($\cdot$H$_2$O) Potassium tetracyanonicco-late(II)	240.96	$-$H$_2$O 100
1901	[Ni(CN)$_4$],K$_4$ Potassium tetracyanonicccolate(0)	319.15	dec t
1902	NiCO$_3$ Nickel(II) carbonate	118.70	400 $\rightarrow$ NiO
1903	[Ni(CO)$_4$] Tetracarbonylnickel	170.73	mp $-$19.3; bp 42.3
1904	[Ni$_2$(CO)$_2$(C$_5$H$_5$)$_2$] Dicarbonylbis(cyclopentadie-nyl)dinickel	303.59	mp 139 dec ($\rightarrow$ Ni)
1905	NiCl$_2$ ($\cdot$6H$_2$O) Nickel(II) chloride	129.60	subl 970, mp 1009 (p)
1906	Ni(ClO$_4$)$_2$ ($\cdot$6H$_2$O) Nickel(II) perchlorate	257.59	mp hydr 186
1907	NiF$_2$ ($\cdot$4H$_2$O) Nickel(II) fluoride	96.69	subl 1474, mp 1160 (p)
1908	[NiF$_6$],K$_2$ Potassium hexafluoroniccolate(IV)	250.87	dec 400
1909	[NiF$_6$],K$_3$ Potassium hexafluoroniccolate(III)	289.97	dec ca. 500
1910	Ni(HCOO)$_2$ ($\cdot$2H$_2$O) Nickel(II) formate	148.72	dec hydr >260
1911	trans-[Ni(H$_2$O)$_4$Cl$_2$] ($\cdot$2H$_2$O) trans-Tetraaquadichlo-ronickel	201.66	mp hydr 28.8; dec hydr 140
1912	NiI$_2$ Nickel(II) iodide	312.50	>300 $\rightarrow$ Ni,I$_2$; mp 797 (p)
1913	NiI$_2\cdot$6H$_2$O	420.59	mp 43; $-$H$_2$O 140
1914	Ni(IO$_3$)$_2$ ($\cdot$4H$_2$O) Nickel(II) iodate	408.49	$-$H$_2$O ca. 100
1915	Ni$_3$N$_2$ Trinickel dinitride	204.08	585 $\rightarrow$ Ni
1916	Ni(NH$_2$)$_2$ Nickel(II) amide	90.74	ca. 80 $\rightarrow$ Ni$_3$N$_2$
1917	[Ni(NH$_3$)$_6$]Br$_2$ Hexaamminenickel(II) bromide	320.68	ca. 160 $\rightarrow$ NiBr$_2$
1918	[Ni(NH$_3$)$_6$]Cl$_2$ Hexaamminenickel(II) chloride	231.78	176.5 $\rightarrow$ NiCl$_2$
1919	[Ni(NH$_3$)$_6$]I$_2$ Hexaamminenickel(II) iodide	414.68	dec t
1920	Ni(NH$_4$)$_2$(SO$_4$)$_2$ ($\cdot$6H$_2$O) Nickel(II)-diammonium sulfate	286.89	$-$H$_2$O 130, dec >250
1921	Ni(NO$_3$)$_2$ ($\cdot$6H$_2$O) Nickel(II) nitrate	182.70	mp hydr 56.7; bp 136.7
1922	Ni$_{1-x}$O Nickel(II) oxide	74.69	mp 1955
1923	Ni$_2$O$_3$($\cdot n$H$_2$O) Nickel(III) oxide	165.38	hydr >200 $\rightarrow$ NiO(OH)
1924	Ni(OH)$_2$ Nickel(II) hydroxide	92.70	230 $\rightarrow$ NiO
1925	NiO(OH) Nickel metahydroxide	91.70	>250 $\rightarrow$ NiO
1926	Ni$_2$P Dinickel phosphide	148.35	mp 1110
1927	Ni$_3$P Trinickel phosphide	207.04	mp 970 dec
1928	Ni$_5$P$_2$ Pentanickel diphosphide	355.40	mp 1175
1929	[Ni(PF$_3$)$_4$] Tetrakis(trifluorophosphorus)nickel	410.56	mp $-$55, bp 70.5
1930	Ni$_2$P$_2$O$_7$ Nickel(II) diphosphate	291.32	mp 1395
1931	Ni$_3$(PO$_4$)$_2$ ($\cdot$7H$_2$O) Nickel(II) orthophosphate	366.01	$-$H$_2$O 110–260
1932	NiS Nickel(II) sulfide	90.76	mp 797
1933	Ni(S$_2$) Nickel(II) disulfide(2$-$)	122.82	mp 1010
1934	Ni$_3$S$_2$ Trinickel disulfide	240.20	mp 808 dec
1935	NiSO$_4$ ($\cdot$7H$_2$O) Nickel(II) sulfate	154.75	$-$H$_2$O 280, dec >700
1936	NiS$_2$O$_6$ ($\cdot$6H$_2$O) Nickel(II) dithionate	218.82	$t \rightarrow$ NiSO$_4$
1937	NiSb Nickelantimony	180.44	mp 1153
1938	(NiSb)S Nickel-antimony sulfide	212.51	$t \rightarrow$ NiSb,NiS
1939	Ni$_{1-x}$Se Nickel(II) selenide	137.65	mp ca. 1100
1940	Ni(Se$_2$) Nickel(II) diselenide(2$-$)	216.61	dec t
1941	NiSeO$_4$ ($\cdot$6H$_2$O) Nickel(II) selenate	201.65	dec hydr ca. 300

(Continued)

No.	Color and density	Solubility and reactivity						
		water	alcohol	HCl	H_2SO_4	HNO_3	NaOH	$NH_3 \cdot H_2O$
1899	lt-viol, 1.82	i	i	r	−	r	−	+
1900	orange-red hydr, 1.88	r/+	...	+	+	+	+	−
1901	dk-red	+	+	+	+	+	+	+
1902	yel-grn	i	...	+	+	+	−	−/+
1903	cl lq, 1.32[20]	d	r	−	−/+	+	−	−
1904	dk-grn	−/+	r	−/+	−/+	−/+	−/+	−/+
1905	yel, 3.55	r	r	r	−	−	+	+
1906	grn hydr	r	r	−	−	r	+	+
1907	yel-grn, 4.63	r	i	−	−	−	+	+
1908	dk-red	+	...	+	+	+	+	+
1909	viol	+	...	+	+	+	+	+
1910	lt-grn hydr	r	...	r	−/+	−/+	+	+
1911	grn hydr	+	r	−	−/+	−/+	+	+
1912	blk-grn, 5.83	r	r	r	−/+	−/+	+	+
1913	bl-grn	r	r	−	−/+	−/+	+	+
1914	yel, 5.07	d	...	+	−/+	+	+	+
1915	blk, 7.52	+	i	+	+	+	+	+
1916	red	+	+	+	+	+	+	+
1917	sk-bl-viol, 1.84	r/+	i	+	+	+	−/+	r/i
1918	sk-bl-viol, 1.47	r/+	i	+	+	+	−/+	r/i
1919	lt-sk-bl, 2.10	r/+	i	+	+	+	−/+	r/i
1920	grn-sk-bl, 1.92	r	i	−	r	r	+	+
1921	lt-grn hydr, 2.05	r	r	r	r	r	+	+
1922	yel ($t \rightarrow$ brn), 6.67	−	i	+	+	+	−	+
1923	grey-blk hydr, 4.83	i	i	+	+	+	i	+
1924	lt-grn, 3.65	i	i	+	+	+	i	+
1925	blk, 4.15	i	i	+	+	+	i	−
1926	bl-blk, 6.31	−	i	−	−/+	−/+	−	−
1927	dk-grey	−	...	−	−	−/+	−	−
1928	lt-blk	−	...	−	−	−/+	−	−
1929	cl lq	−/+	r	+	+	+	+	...
1930	yel	i	i	+	+	+	i	+
1931	grn hydr	i	i	r	r	r	i	+
1932	blk, 5.3–5.65	i	i	+	+	+	−	−
1933	viol-grey, 4.39	i	i	−	+	+	−	−/+
1934	yel, 5.82	−	i	+	+	+	−	−
1935	grn, 3.68(1.95)	r	d	r	r	r	+	+
1936	grn hydr, 1.91	r/+	...	−/+	−/+	−/+	+	+
1937	dk-red, 7.54	−	...	+	+	+	NaOH	−
1938	lt-grey, 6.65	−	...	+	+	+	−/+	−
1939	grey, 8.46	i	i	−	−	+	−	−/+
1940	dk-grey, 6.0	−	...	−	−/+	−/+	−/+	−/+
1941	grn hydr, 2.31	r	...	−	−	−	+	+

No.	Formula and name	M_r	Phase transition temperature
1942	NiSi Nickel monosilicide	86.78	mp 992
1943	Ni$_2$Si Dinickel silicide	145.47	mp 1290
1944	Ni(Te$_2$) Nickel(II) ditelluride(2−)	313.89	dec t
1945	**No Nobelium**	259.101	...
1946	**Np Neptunium**	237.048	mp 637, bp ca. 4100
1947	NpBr$_3$ Neptunium(III) bromide	476.76	mp 750
1948	NpBr$_4$ Neptunium(IV) bromide	556.66	mp 464
1949	Np(C$_2$O$_4$)$_2$ (·6H$_2$O) Neptunium(IV) oxalate	413.08	hydr >400 → NpO$_2$
1950	NpCl$_3$ Neptunium(III) chloride	343.41	mp 802, bp 1500
1951	NpCl$_4$ Neptunium(IV) chloride	378.86	mp 518, bp 850
1952	NpF$_3$ Neptunium(III) fluoride	294.04	mp 1425, bp 2223
1953	NpF$_4$ Neptunium(IV) fluoride	313.04	mp 830, bp 1480
1954	NpF$_5$ Neptunium(V) fluoride	332.04	dec >500
1955	NpF$_6$ Neptunium(VI) fluoride	351.04	mp and bp 54.76
1956	NpI$_3$ Neptunium(III) iodide	617.76	mp 767
1957	(Np$_2^V$Np^{VI})O$_8$ Dineptunium(V)-neptunium(VI) oxide	839.14	300 → Np$_2$O$_5$
1958	NpO$_2$ Neptunium(IV) oxide	269.05	mp 2560
1959	Np$_2$O$_5$ Neptunium(V) oxide	554.09	770 → NpO$_2$
1960	Np(O)F$_4$ Neptunium oxide-tetrafluoride	329.04	dec 100
1961	Np(OH)$_3$ Neptunium(III) hydroxide	288.07	dec t
1962	Np(OH)$_4$ Neptunium(IV) hydroxide	305.08	t → NpO$_2$
1963	NpO$_2$(OH) Neptunium dioxide-hydroxide	286.05	dec t
1964	NpO$_2$(OH)$_2$ (·H$_2$O) Neptunium dioxide-dihydroxide	303.06	−H$_2$O 90, dec >240
1965	NpO$_3$(OH) Neptunium trioxide-hydroxide	302.05	...
1966	**Ns Nielsbohrium** (1995 yr: Joliotium, Jl)	262.114	...
1967	**O$_2$ Dioxygen**	31.998	mp −218.7; bp −182.962
1968	**O$_3$ Ozone**	47.997	mp −192.7; bp −111,9
1969	OF$_2$ Oxygen difluoride	54.00	mp −223.8; bp −144.8
1970	O$_2$F$_2$ Dioxygen difluoride	69.99	mp −163, bp −57 dec
1971	**Osmium**	190.2	mp 3027, bp ca. 5000
1972	[Os(C$_5$H$_5$)$_2$] Bis(cyclopentadienyl)osmium	320.39	mp 229.5
1973	[Os(CO)$_5$] Pentacarbonylosmium	330.25	mp −15, dec >100
1974	[Os$_2$(CO)$_9$] Nonacarbonyldiosmium	632.49	dec >20, mp 64 (p)
1975	[Os$_3$(CO)$_{12}$] Dodecacarbonyltriosmium	906.72	subl 130, mp 224
1976	[Os(CO)$_3$Cl$_2$] Tricarbonyldichloroosmium	345.14	mp 271; >300 → Os
1977	[Os(CO)$_4$Cl$_2$] Tetracarbonyldichloroosmium	373.15	subl 220; >250 → Os
1978	OsCl$_3$ Osmium(III) chloride	296.56	subl 350, dec ca. 500
1979	OsCl$_4$ Osmium(IV) chloride	332.01	mp 323, t → OsCl$_3$
1980	[OsCl$_6$],K$_2$ Potassium hexachloroosmate(IV)	481.11	dec 600
1981	[OsCl$_6$],K$_3$ (·3H$_2$O) Potassium hexachloroosmate(III)	520.21	−H$_2$O 150, dec t
1982	[OsCl$_6$],(NH$_4$)$_2$ Ammonium hexachloroosmate(IV)	439.00	subl 170, dec >500
1983	[OsCl$_4$O$_2$],K$_2$ Potassium tetrachlorodioxoosmium(VI)	442.21	dec 200
1984	OsF$_4$ Osmium(IV) fluoride	266.19	mp 230
1985	OsF$_5$ Osmium(V) fluoride	285.19	mp 70, bp 233
1986	OsF$_6$ Osmium(VI) fluoride	304.19	mp 33.4; bp 47.5
1987	[OsH$_2$(CO)$_4$] Dihydrotetracarbonylosmium	304.26	mp 149

(Continued)

No.	Color and density	Solubility and reactivity						
		water	alcohol	HCl	H_2SO_4	HNO_3	NaOH	$NH_3 \cdot H_2O$
1942	grey-wh	−	i	−	−	−	−	−
1943	grey-wh, 7.2	−	i	−	−	−	−	−
1944	dk-grey, 7.3	−	...	−	−/+	−/+	−/+	−/+
1945	...	...	...	...	...	...	...	...
1946	wh, 20.48	psv/+	...	+	+/psv	+/psv	−	−
1947	grn, 6.62	r/+	...	−	−/+	−	+	+
1948	red-brn	r/+	...	−	−/+	−	+	+
1949	grn hydr	i	i	i	−/+	i/+	+	...
1950	grey-grn, 5.58	r/+	...	r	−	−	+	+
1951	red, 4.92	r/+	...	r	−	−	+	+
1952	dk-red, 9.12	i	...	−	−	−	−	−
1953	lt-grn, 6.80	i	i	−	−	−	−	−
1954	pink-sk-bl	i/+	...	−	−/+	−/+	−/+	−
1955	orange, 5.0	+	+	+	+	+	+	+
1956	brn, 6.82	r/+	...	−	−/+	−/+	+	+
1957	brn	−	...	−/+	−/+	+	−	−
1958	grn-brn, 11.1	−	i	−	−/+	+	−	−
1959	brn	−	...	+	+	+	−	−
1960	brn	+	...	+	+	+	+	+
1961	red	i	...	+	+	+	i	i
1962	grey-grn	i	...	+	+	+	i	i
1963	grn-viol	i	...	+	+	+	i	i
1964	brn	i	...	+	+	+	i	i
1965	grn-blk	i	...	+	+	+	i	i
1966	...	...	...	...	...	...	...	...
1967	cl gas, 1.42895	d	d	−	−	−	−	−
1968	bl gas, 2.144	d	+	−/+	+	+	+	+
1969	cl gas, 2.421	d/+	...	+	+	+	+	...
1970	dk-red lq, 1.45^{-58}	+	+	+	+	+	+	+
1971	lt-sk-bl, 22.61	−	−	−	−/+	−/+	−	−
1972	lt-yel	−	r	−	−/+	+	−	−
1973	cl lq	−/+	r	−	−/+	−/+	+	−
1974	orange	−	r	−	−	−/+	−/+	−
1975	yel	−	d	−	−	−/+	−	−
1976	wh	−/+	r	−/+	+	+	+	+
1977	wh	−/+	r	−/+	−/+	+	+	+
1978	brn	r	r	r/+	−/+	−/+	+	...
1979	brn	+	i	+	+	+	+	+
1980	red, 3.42	d/r	i	−	−/+	+	+	...
1981	dk-red hydr	r/+	r	−	−	+	+	...
1982	red-brn, 2.93	d/+	i	−	−/+	+	+	...
1983	red, 3.42	+	i	+	+	+	−	−
1984	yel	+	i	+	+	+	+	+
1985	sk-bl-grn	+	...	+	+	+	+	+
1986	yel-grn	+	r	+	+	+	+	+
1987	wh	+	...	+	+	+	+	+

No.	Formula and name	M_r	Phase transition temperature
1988	[Os(N)O$_3$],K Potassium nitridotrioxoosmate(VIII)	291.30	dec >180
1989	OsO$_2$ Osmium(IV) oxide	222.20	>500 → OsO$_4$
1990	OsO$_4$ Osmium(VIII) oxide	254.20	mp 40.6; bp 131.2
1991	OsOF$_5$ Osmium oxide-pentafluoride	301.19	mp 60, bp 100
1992	OsO$_3$F$_2$ Osmium trioxide-difluoride	276.19	mp 172
1993	Os(OH)$_4$ Osmium(IV) hydroxide	258.23	t → OsO$_2$
1994	[OsO$_2$(OH)$_4$],K Potassium dioxotetrahydroxoosmate(VI)	368.42	dec >140
1995	Os(S$_2$) Osmium(II) disulfide(2−)	254.33	dec >1000
1996	**P** (amor) **Phosphorus**, red	30.974	subl 416
1997	**P Phosphorus**, black	30.974	mp ca. 1000 (p)
1998	**P$_4$ Phosphorus**, white	123.896	mp 44.14; bp 287.3
1999	PBr$_3$ Phosphorus tribromide	270.69	mp −41.5; bp 173.3
2000	PBr$_5$ Phosphorus pentabromide	430.49	bp 106 dec
2001	P(Br)Cl$_2$O Phosphorus bromide-dichloride-oxide	197.78	mp 11; bp 136.5
2002	PBr$_2$(Cl)O Phosphorus dibromide-choride-oxide	242.23	mp 31; bp 165
2003	P(Br)Cl(O)F Phosphorus bromide-chloride-oxide-fluoride	181.33	bp 79
2004	P(Br)F$_2$ Phosphorus bromide-difluoride	148.87	mp −133.8; bp −16.1
2005	PBr$_2$F Phosphorus dibromide-fluoride	209.78	mp −115, bp 78.5
2006	PBr$_2$F$_3$ Phosphorus dibromide-trifluoride	247.78	mp −20, dec 15
2007	(PBr$_4$)F Tetrabromophosphonium fluoride	369.59	mp 87 dec
2008	PBr$_3$O Phosphorus tribromide-oxide	286.69	mp 55.7; bp 193
2009	PBr(O)F$_2$ Phosphorus bromide-oxide-difluoride	164.87	mp −84.8; bp 30.5
2010	PBr$_2$(O)F Phosphorus dibromide-oxide-fluoride	225.78	mp −117.2; bp 110.1
2011	PCl$_3$ Phosphorus trichloride	137.33	mp −90.34; bp 75.3
2012	PCl$_5$ Phosphorus pentachloride	208.24	subl 180, mp 166.8 (p)
2013	P$_2$Cl$_4$ Diphosphorus tetrachloride	203.76	mp −28, bp 180
2014	P(Cl)F$_2$ Phosphorus chloride-difluoride	104.42	mp −164.8; bp −47.3
2015	P(Cl)F$_4$ Phosphorus chloride-tetrafluoride	142.42	mp −132, bp −43.4
2016	PCl$_2$F Phosphorus dichloride-fluoride	120.88	mp −144, bp 13.85
2017	PCl$_2$F$_3$ Phosphorus dichloride-trifluoride	158.87	mp −125, bp 7.1
2018	PCl$_4$F Phosphorus tetrachloride-fluoride	191.78	mp −59, >30 → (PCl$_4$)F
2019	(PCl$_4$)F Tetrachlorophosphonium fluoride	191.78	subl 175, mp 177 (p)
2020	(PCl$_2$)$_3$N$_3$ Tris(dichlorophosphorus) trinitride	347.66	mp 114, bp 256
2021	(PCl$_2$)$_4$N$_4$ Tetrakis(dichlorophosphorus) tetranitride	463.55	mp 123.5; bp 328.5
2022	PCl$_3$O Phosphorus trichloride-oxide	153.33	mp 1.25; bp 105.8
2023	PCl(O)F$_2$ Phosphorus chloride-oxide-difluoride	120.42	mp −96.4; bp 3.1
2024	PCl$_2$(O)F Phosphorus dichloride-oxide-fluoride	136.88	mp −84.5; bp 54
2025	PF$_3$ Phosphorus trifluoride	87.97	mp −151.5; bp −101.8
2026	PF$_5$ Phosphorus pentafluoride	125.96	mp −93.75; bp −84.55
2027	[PF$_6$],H (·6H$_2$O) Hydrogen hexafluorophosphate(V)	145.97	mp hydr 32, dec hydr t
2028	[PF$_6$],K Potassium hexafluorophosphate(V)	184.06	mp 575
2029	[PF$_6$],NH$_4$ Ammonium hexafluorophosphate(V)	163.00	dec >200
2030	[PF$_6$],Na (·H$_2$O) Sodium hexafluorophosphate(V)	167.95	dec hydr 200
2031	[PF$_6$],PCl$_4$ Tetrahlorophosphonium hexafluorophosphate(V)	317.75	subl 135 dec

(Continued)

No.	Color and density	Solubility and reactivity						
		water	alcohol	HCl	H_2SO_4	HNO_3	NaOH	$NH_3 \cdot H_2O$
1988	yel-orange, 4.2	r	d	+	+	+	−	r
1989	brn-blk, 7.91	−	...	+	+	+	+	+
1990	lt-yel, 4.91	+	r	−/+	−/+	−/+	+	+
1991	grn	+	...	+	+	+	+	+
1992	orange	+	...	+	+	+	+	+
1993	blk	i	...	+	i	−/+	i	i
1994	viol-red	r/+	i	+	+	+	−	−
1995	brn-blk, 9.47	i	...	+	−/+	+	−	−
1996	red, 2.340	−	r	−	−/+	−/+	−	−
1997	blk, 2.7	−	i	−	−	−/+	−	−
1998	wh, 1.82	−	d	−/+	+	+	+	+
1999	cl lq, 2.852[15]	+	+	+	+	+	+	+
2000	orange	+	i	+	+	+	+	+
2001	cl lq, 2.104[14]	+	...	+	+	+	+	+
2002	yel	+	...	+	+	+	+	+
2003	cl lq	+	r	+	+	+	+	+
2004	cl gas	+	...	+	+	+	+	+
2005	cl gas	+	...	+	+	+	+	+
2006	yel-red lq	+	...	+	+	+	+	+
2007	lt-yel	+	...	+	+	+	+	+
2008	wh, 2.82	+	r/+	+	+	+	+	+
2009	cl lq	+	...	+	+	+	+	+
2010	cl lq	+	...	+	+	+	+	+
2011	cl lq, 1.5567[20]	+	r	+	+	+	+	+
2012	wh, 2.11	+	...	+	+	+	+	+
2013	cl lq	+	...	+	+	+	+	+
2014	cl gas	+	r	+	+	+	+	+
2015	cl gas	+	...	+	+	+	+	+
2016	cl gas, 1.590	+	...	+	+	+	+	+
2017	cl gas, 5.4	+	+	+	+	+	+	+
2018	cl lq	+	...	+	+	+	+	+
2019	wh	+	...	+	+	+	+	+
2020	wh, 1.98	−/+	r	−	−	−	−	−
2021	wh, 2.18	−/+	r	−	−	−	−	−
2022	cl lq, 1.645[20]	+	r/+	+	+	+	+	+
2023	cl gas	+	...	+	+	+	+	+
2024	cl lq	+	...	+	+	+	+	+
2025	cl gas, 4.19	+	r	+	+	+	+	+
2026	cl gas, 5.81	+	...	+	+	+	+	+
2027	wh hydr	r/+	...	−	−	−	+	+
2028	wh	r	...	−	−	−	−	−
2029	wh, 2.18	r/+	r	−/+	−/+	−/+	−/+	−
2030	wh, 2.51(2.37)	r	...	−	−	−	−	−
2031	wh	+	...	+	+	+	+	+

No.	Formula and name	M_r	Phase transition temperature
2032	PH_3 Phosphine	34.00	mp −133.8; bp −87.42
2033	P_2H_4 Diphosphan	65.98	mp −99.0; bp 63.2
2034	PH_4Br Phosphonium bromide	114.91	subl 38
2035	$(PH)F_2$ Difluorophosphine	69.98	mp −124, bp −65
2036	PH_4I Phosphonium iodide	161.91	mp 18.5; bp 80 dec
2037	PI_3 Phosphorus triiodide	411.69	mp 61.0; dec >200
2038	P_2I_4 Diphosphorus tetraiodide	569.56	mp 125.5
2039	PI_2Cl_3 Phosphorus diodide-trichloride	391.14	dec 259
2040	PI_3O Phosphorus triiodide-oxide	427.69	mp 50
2041	$[PMo_{12}O_{40}],H_3$ (·28H₂O) Hydrogen 40-oxododecamolybdophosphate(V)	1825.24	mp hydr 78
2042	$[PMo_{12}O_{40}],(NH_4)_3$ (·6H₂O) Ammonium 40-oxododecamolybdophosphate(V)	1876.33	...
2043	PN Phosphorus mononitride	44.98	dec 800
2044	$P(NH_2)_3O$ Phosphorus triamide-oxide	95.04	mp 160
2045	$PN(NH)$ Phosphorus nitride-imide	60.00	dec 400
2046	$PN(NH_2)_2$ Phosphorus nitride-diamide	77.03	mp 162
2047	P_4O_6 Tetraphosphorus hexaoxide	219.89	mp 23.8;bp 175.4
2048	P_4O_8 Tetraphosphorus octaoxide	251.89	subl 180
2049	P_4O_{10} Tetraphosphorus decaoxide	283.89	subl 359, mp 422 (p)
2050	POF_3 Phosphorus oxide-trifluoride	103.97	subl −39.8; mp −39.4
2051	P_2OF_6 Diphosphorus oxide-hexafluoride	191.94	bp −18, dec >20
2052	$[P(OH)_4]ClO_4$ Tetrahydroxophosphorus(V) perchlorate	198.45	mp 47
2053	P_4S_2 Tetraphosphorus disulfide	188.03	mp 46
2054	P_4S_3 Tetraphosphorus trisulfide	220.09	mp 172.5; bp 407.5
2055	P_4S_5 Tetraphosphorus pentasulfide	284.23	soft 170–220 dec
2056	P_4S_6 Tetraphosphorus hexasulfide	316.29	mp 232 dec
2057	P_4S_7 Tetraphosphorus heptasulfide	348.36	mp 312, bp 523
2058	P_4S_9 Tetraphosphorus nonasulfide	412.49	soft 240–270 dec
2059	P_4S_{10} Tetraphosphorus decasulfide	444.56	mp 288, bp 514
2060	$P(S)Br_3$ Phosphorus sulfide-tribromide	302.75	mp 38.2; bp 215 dec
2061	$PS(Br)Cl_2$ Phosphorus sulfide-bromide-dichloride	213.85	mp −30, bp 156
2062	$P(S)Br(Cl)F$ Phosphorus sulfide-bromide-chloride-fluoride	197.40	bp 98
2063	$PS(Br)F_2$ Phosphorus sulfide-bromide-difluoride	180.94	mp −136.9; bp 35.5
2064	$P(S)Br_2F$ Phosphorus sulfide-dibromide-fluoride	241.85	mp −75.2; bp 125.3
2065	$P(S)Cl_3$ Phosphorus sulfide-trichloride	169.40	mp −36.2; bp 125
2066	$PS(Cl)F_2$ Phosphorus sulfide-chloride-difluoride	136.49	mp −155.2; bp 6.3
2067	$P(S)Cl_2F$ Phosphorus sulfide-dichloride-fluoride	152.94	mp −96.0; bp 64.7
2068	$P(S)F_3$ Phosphorus sulfide-trifluoride	120.03	mp −149, bp −52
2069	$P(S)I_3$ Phosphorus sulfide-triiodide	443.75	mp 46
2070	$PS(NH_2)_3$ Phosphorus sulfide-triamide	111.11	mp 118–119
2071	$[PV_{12}O_{36}],(NH_4)_7$ Ammonium 36-oxododecavanadophosphate(V)	1344.52	dec >450
2072	$[PW_{12}O_{40}],H_3$ (·24H₂O) Hydrogen 40-oxododecawolframophosphate(V)	2880.16	mp hydr 80

(Continued)

No.	Color and density	Solubility and reactivity						
		water	alcohol	HCl	H_2SO_4	HNO_3	NaOH	$NH_3 \cdot H_2O$
2032	cl gas, 1.5294	d	r	−	−/+	−/+	−	−
2033	cl lq, 1.012[20]	i	r	−	−/+	−/+	−	−
2034	wh	+	...	+	+	+	+	+
2035	cl gas	+	...	+	+	+	+	+
2036	wh, 2.86	+	...	+	+	+	+	+
2037	dk-red, 4.18	+	r	+	+	+	+	+
2038	orange-red, 3.89	+	...	+	+	+	+	+
2039	red	+	...	+	+	+	+	+
2040	viol	+	r	+	+	+	+	+
2041	yel hydr, 2.53	r/+	d	−/+	−/+	−/+	+	+
2042	yel	i	i	−	−	−	+	−/+
2043	yel-brn	−	...	−	−	−/+	−	−
2044	wh	−/+	i	+	+	+	+	+
2045	wh	−	...	−	−	−	−	−
2046	wh	−	...	−/+	−/+	−/+	−	−
2047	wh, 2.14	+	...	+	+	+	+	+
2048	wh	+	i	+	+	+	+	+
2049	wh, 2.39	+	...	+	+	+	+	+
2050	cl gas, 4.8	+	r	+	+	+	+	+
2051	cl gas	+	...	+	+	+	+	+
2052	wh	+	...	+	+	+	+	+
2053	yel	−/+	...	−	−	−/+	−	−
2054	yel-grn, 2.03	−/+	...	−	−	+	−	−
2055	yel, 2.17	+	...	+	+	+	+	+
2056	dk-yel	+	r	+	+	+	+	+
2057	lt-yel, 2.19	+	i	+	+	+	+	+
2058	yel, 2.08	+	...	+	+	+	+	+
2059	dk-yel, 2.09	+	...	+	+	+	+	+
2060	yel, 2.85	+	r	+	+	+	+	+
2061	yel lq, 2.12[0]	+	...	+	+	+	+	+
2062	yel lq	+	...	+	+	+	+	+
2063	yel lq	+	...	+	+	+	+	+
2064	yel lq	+	...	+	+	+	+	+
2065	cl lq, 1.635[20]	+	...	+	+	+	+	+
2066	cl gas	+	...	+	+	+	+	+
2067	cl lq	+	...	+	+	+	+	+
2068	cl gas	+	...	+	+	+	+	+
2069	yel	+	...	+	+	+	+	+
2070	wh	−/+	i	−	−/+	−/+	+	+
2071	viol-red	r/+	r	−/+	−/+	−/+	−	−
2072	wh hydr	r/+	d	−/+	−/+	−/+	+	+

No.	Formula and name	M_r	Phase transition temperature
2073	**Pa Protactinium**	231.036	mp 1580, bp ca. 4500
2074	PaBr$_5$ Protactinium(v) bromide	630.56	mp 317
2075	PaCl$_4$ Protactinium(IV) chloride	372.85	mp 680, bp 850
2076	PaCl$_5$ Protactinium(v) chloride	408.30	mp 301, bp 420
2077	PaF$_4$ Protactinium(IV) fluoride	307.03	mp 1030, bp 1630
2078	PaF$_5$ Protactinium(v) fluoride	326.03	subl 500
2079	PaI$_5$ Protactinium(v) iodide	865.56	mp 300 dec
2080	PaO$_{2\pm x}$ Protactinium(IV) oxide	263.03	...
2081	Pa$_2$O$_5$ Protactinium(v) oxide	542.07	1550 → Pa$_2$O$_{2\pm x}$
2082	Pa(OH)$_5$ Protactinium(v) hydroxide	316.07	>500 → Pa$_2$O$_5$
2083	**Pb Lead**	207.2	mp 327.502; bp 1745
2084	Pb$_2$As$_2$O$_7$ Lead(II) heptaoxodiarsenate(v)	676.24	mp 802
2085	Pb$_3$(AsO$_4$)$_2$ Lead(II) arsenate	899.44	mp 1042 dec
2086	Pb(BO$_2$)$_2$ (·H$_2$O) Lead(II) metaborate	292.82	−H$_2$O 160, mp 600 dec
2087	PbBr$_2$ Lead(II) bromide	367.01	mp 373, bp 893
2088	Pb(BrO$_3$)$_2$ (·H$_2$O) Lead(II) bromate	463.00	dec hydr 180
2089	Pb(CH$_3$COO)$_2$ Lead(II) acetate	325.29	mp 280
2090	Pb(CH$_3$COO)$_2$·3H$_2$O	379.33	mp 75, −H$_2$O >75
2091	Pb(CH$_3$COO)$_4$ Lead(IV) acetate	443.38	mp 175–180 dec
2092	PbCO$_3$ Lead(II) carbonate	267.21	>300 → PbO
2093	PbC$_2$O$_4$ Lead(II) oxalate	295.22	dec 300
2094	Pb$_2$(CO$_3$)Cl$_2$ Lead(II) carbonate-dichloride	509.86	dec ca. 350
2095	PbCl$_2$ Lead(II) chloride	278.11	mp 501, bp 950
2096	PbCl$_4$ Lead(IV) chloride	349.01	mp −7, dec ca. 100
2097	Pb(Cl)F Lead(II) chloride-fluoride	261.65	mp 606
2098	[PbCl$_6$],K$_2$ Potassium hexachloroplumbate(IV)	498.11	dec >190
2099	[PbCl$_6$],(NH$_4$)$_2$ Ammonium hexachloroplumbate(IV)	456.00	dec >130
2100	Pb(ClO$_2$)$_2$ Lead(II) chlorite	342.10	dec 126
2101	Pb(ClO$_3$)$_2$ (·H$_2$O) Lead(II) chlorate	374.10	−H$_2$O 110, dec 230
2102	Pb(ClO$_4$)$_2$ (·3H$_2$O) Lead(II) perchlorate	406.10	dec hydr 100
2103	PbCl(OH) Lead(II) chloride-hydroxide	259.66	dec >140
2104	PbCrO$_4$ (·H$_2$O) Lead(II) chromate	323.19	mp 844, dec >850
2105	PbCr$_2$O$_7$ Lead(II) dichromate	423.19	dec t
2106	PbF$_2$ Lead(II) fluoride	245.20	mp 824, bp 1293
2107	PbF$_4$ Lead(IV) fluoride	283.19	mp ca. 600
2108	PbHAsO$_4$ Lead(II) hydroarsenate	347.13	ca. 400 → Pb$_2$As$_2$O$_7$
2109	Pb(HCOO)$_2$ Lead(II) formate	297.23	dec 190
2110	Pb(HSO$_4$)$_2$ (·H$_2$O) Lead(II) hydrosulfate	401.34	dec hydr t
2111	PbI$_2$ Lead(II) iodide	461.01	mp 402, bp 954
2112	[PbI$_3$],K (·2H$_2$O) Potassium triiodoplumbate(II)	627.01	−H$_2$O 30–97, dec 349
2113	Pb(IO$_3$)$_2$ Lead(II) iodate	557.00	dec 300
2114	PbMoO$_4$ Lead(II) molybdate	367.14	mp 1070
2115	Pb(N$_3$)$_2$ Lead(II) azide	291.24	dec 350
2116	Pb(NCS)$_2$ Lead(II) thiocyanate	323.37	dec >190
2117	Pb(NO$_3$)$_2$ Lead(II) nitrate	331.21	470 → PbO
2118	Pb(NO$_3$)OH Lead(II) nitrate-hydroxide	286.21	dec 180
2119	α-PbO Lead(II) oxide	223.20	488 → β-PbO
2120	β-PbO	223.20	mp 886, bp 1535

No.	Color and density	Solubility and reactivity						
		water	alcohol	HCl	H_2SO_4	HNO_3	NaOH	$NH_3 \cdot H_2O$
2073	lt-grey, 15.37	psv	−	+/psv	+	psv	−	−
2074	orange-red	+	r	+	+	+	+	+
2075	yel-grn, 4.68	d	...	+	+	+	+	+
2076	lt-yel	+	r	+	+	+	+	+
2077	red-brn	i/+	...	−	−	i	+	i
2078	wh	i	r	−	−	i	i	i
2079	blk	+	r	+	+	+	+	+
2080	bl (t → red)	−	i	−	−	−	−	−
2081	wh, 9.0	−	i	−	i/+	−	−	−
2082	wh, 13.43	i	i	+	+	+	−	−
2083	grey-sk-bl, 11.337	psv	−	psv	psv/+	+/psv	+	−
2084	wh	i/+	...	+	−	+	−	−
2085	wh	i	...	+	−/+	+	−	−
2086	wh	i	i	+	+	+	−	−
2087	wh, 6.66	d	d	−/+	−/+	−	−/+	−
2088	wh hydr, 5.53	d	...	−	−/+	−	+	−
2089	wh, 3.25	r	d	+	+	−	+	+
2090	wh, 2.55	r	i	+	+	−	+	+
2091	wh, 2.23	+	+	+	+	+	+	+
2092	wh, 6.55	i/+	i	+	+	+	+	−
2093	wh, 5.28	i	i	−/+	i/+	−/+	+	i
2094	wh, 6.1	i	i	−/+	−/+	−/+	−/+	−
2095	wh, 5.85	d	i	−/+	−/+	i	+	d
2096	yel lq, 3.18⁰	+	...	+	+/−	+	+	+
2097	yel-grn	i	i	−	−/+	d	+	d
2098	lt-yel	+	...	+	+	+	+	+
2099	lt-yel	+	...	+	+	+	+	+
2100	yel	d	...	−/+	−/+	−/+	+	−
2101	wh, 4.04(3.89)	r	r	−	−/+	−/+	+	+
2102	wh hydr, 2.6	r	r	+	+	−	+	+
2103	wh	i	...	+	+	+	−/+	−
2104	yel, 6.02	i	...	i	i	+	+	i
2105	red	+	...	+	+	r	+	+
2106	wh, 8.45	d	i	−	+	d	−/+	d
2107	wh, 6.7	+	...	+	+	+	+	+
2108	wh	i	...	−	−/+	+	+	+
2109	wh, 4.63	d	i	+	+	−/+	+	+
2110	wh hydr	d	...	−	d	−	+	+
2111	yel (t → brn), 6.16	d	i	d	d/+	d/+	+	−
2112	lt-yel	+	...	−/+	+	+	−/+	−
2113	wh	i	d	−/+	−/+	−	−	−
2114	lt-yel, 6.75	i	i	+	+	+	+	−
2115	wh, 4.71	i	...	+	+	−/+	+	−
2116	wh, 3.82	d	...	−/+	−/+	−/+	+	−
2117	wh, 4.53	r	d	+	+	r	+	+
2118	wh, 5.93	r	...	+	+	+	+	+
2119	red, 9.13	−	i	+	+	+	+	−
2120	yel, 9.45	−	i	+	+	+	+	−

No.	Formula and name	M_r	Phase transition temperature
2121	PbO$_2$ Lead(IV) oxide	239.20	dec >344
2122	Pb(OH)$_2$ Lead(II) hydroxide	241.21	100 → PbO
2123	[Pb(OH)$_6$],Na$_2$ Sodium hexahydroxoplumbate(IV)	355.22	dec 300
2124	Pb$_3$(PO$_4$)$_2$ Lead(II) orthophosphate	811.54	mp 1014
2125	Pb$_5$(PO$_4$)$_3$Cl Pentalead tris(orthophosphate)-chloride	1356.36	dec ca. 1100
2126	(Pb$_2^{II}$Pb^{IV})O$_4$ Dilead(II)-lead(IV) oxide	685.60	550 → PbO
2127	PbS$_{1+x}$ (0 ≤ x ≤ 0.005) Lead(II) sulfide	239.27	mp 1077, bp 1281
2128	PbSO$_4$ Lead(II) sulfate	303.26	mp 1170
2129	Pb(SO$_4$)$_2$ Lead(IV) sulfate	399.32	...
2130	PbSe Lead(II) selenide	286.16	mp 1065
2131	PbSeO$_4$ Lead(II) selenate	350.16	dec >1200
2132	PbSiO$_3$ Lead(II) metasilicate	283.28	mp 766
2133	PbTe Lead(II) telluride	334.80	mp 860
2134	Pb$_5$(VO$_4$)$_3$Cl Pentalead tris(orthovanadate)-chloride	1416.27	dec >1300
2135	PbWO$_4$ Lead(II) wolframate	455.05	mp 1123
2136	**Pd Palladium**	106.42	mp 1554, bp 2940
2137	PdBr$_2$ Palladium(II) bromide	266.23	mp 717
2138	[Pd(C$_4$H$_7$O$_2$N$_2$)$_2$] Bis(dimethylglyoximato)palladium	336.64	dec t
2139	Pd(CN)$_2$ Palladium(II) cyanide	158.46	dec 210
2140	Pd(CN)$_4$ Palladium(IV) cyanide	210.49	dec >20
2141	[Pd(CN)$_4$],K$_2$ (·3H$_2$O) Potassium tetracyanopalladate(II)	288.69	−H$_2$O 200, dec t
2142	PdCl$_2$ Palladium(II) chloride	177.33	mp 680, >800 → Pd
2143	[PdCl$_4$],K$_2$ Potassium tetrachloropalladate(II)	326.43	mp 525
2144	[PdCl$_6$],K$_2$ Potassium hexachloropalladate(IV)	397.33	380 → K$_2$[PdCl$_4$]
2145	[PdCl$_4$],(NH$_4$)$_2$ Ammonium tetrachloropalladate(II)	284.31	dec 430
2146	[PdCl$_6$],(NH$_4$)$_2$ Ammonium hexachloropalladate(IV)	355.22	>300 → (NH$_4$)$_2$[PdCl$_4$]
2147	PdF$_2$ Palladium(II) fluoride	144.42	subl >500 dec
2148	PdF$_4$ Palladium(IV) fluoride	182.41	...
2149	[Pd^{IV}F$_6$],Pd Palladium(II) hexafluoropalladate(IV)	326.83	dec t
2150	PdI$_2$ Palladium(II) iodide	360.23	dec 350
2151	trans-[Pd(NH$_3$)$_2$Cl$_2$] trans-Diamminedichloropalladium	211.39	210 → Pd,NH$_4$Cl
2152	trans-[Pd(NH$_3$)$_2$Cl$_4$] trans-Diamminetetrachloropalladium	282.29	260 → Pd,NH$_4$Cl
2153	[Pd(NH$_3$)$_4$]Cl$_2$ (·H$_2$O) Tetramminepalladium(II) chloride	245.45	dec hydr 120
2154	[Pd(NH$_3$)$_4$][Pd^{II}Cl$_4$] Tetraamminepalladium(II) tetrachloropalladate(II)	422.78	>180 → [Pd(NH$_3$)$_2$Cl$_2$]
2155	Pd(NO$_3$)$_2$ (·2H$_2$O) Palladium(II) nitrate	230.43	dec >350
2156	[Pd(NO$_2$)$_4$],K$_2$ Potassium tetranitropalladate(II)	368.64	dec 305
2157	PdO Palladium(II) oxide	122.42	dec 780
2158	PdO$_2$ (·nH$_2$O) Palladium(IV) oxide	138.42	hydr >200 → PdO
2159	Pd(OH)$_2$ Palladium(II) hydroxide	140.43	dec >500
2160	[Pd(OH)$_4$],Na$_2$ Sodium tetrahydroxopalladate(II)	220.43	dec >300
2161	PdS Palladium(II) sulfide	138.49	dec 950
2162	Pd(S$_2$) Palladium(II) disulfide(2−)	170.55	600 → PdS

(Continued)

No.	Color and density	Solubility and reactivity						
		water	alcohol	HCl	H_2SO_4	HNO_3	NaOH	$NH_3 \cdot H_2O$
2121	dk-brn, 9.38	−	i	+	−/+	−/+	−/+	−
2122	wh	i	...	+	+	+	+	i
2123	wh, 3.98	+	i	+	+	+	+/r	+
2124	wh, 6.9–7.3	i	i	−/+	−/+	+	+	−
2125	yel, 6.5–6.8	i	i	−/+	−/+	+	+	−
2126	orange-red, 9.07	−	i	+	+	+	−/+	−
2127	grey-blk, 7.58	i	i	−/+	−/+	+	i	i
2128	wh, 6.2–6.4	i	i	−/+	i	−/+	−/+	−
2129	yel	+	+	+	+/−	+	+	+
2130	grey-blk, 7.99	i	...	−/+	−/+	+	i/+	...
2131	wh, 6.37	i	...	−/+	−/+	−/+	−	−
2132	wh, 6.49	i	...	+	+	+	−/+	−
2133	dk-grey, 8.14	i	...	−/+	−/+	+	i/+	...
2134	red, 6.9	i	i	−/+	−/+	+	+	−
2135	wh, 8.10	i	i	+	+	+	+	−
2136	lt-grey, 12.02	−	−	−	−/+	−/+	−	−
2137	red-brn, 5.17	i	i	−/+	−/+	−	+	+
2138	yel	i	d/r	−	−	−	+	−
2139	wh	i	...	i/+	i/+	i/+	+	+
2140	pink	+	...	+	+	+	+	+
2141	wh	r	...	−/+	−/+	−/+	−	−
2142	viol-brn, 4.08	i/+	d	+	−	−	+	+
2143	yel-brn, 2.67	d/+	i/r	−	−	−	+	+
2144	red-brn, 2.74	d/+	i	−/+	−	−	+	+
2145	dk-grn, 2.17	r/+	i	−	−	−	+	+
2146	red-brn, 2.42	d/+	i	−	−	−	+	+
2147	lt-viol, 5.80	d/+	...	+	−	−	+	+
2148	red	+	...	+	+	+	+	+
2149	blk, 5.06	+	...	+	+	+	+	+
2150	blk, 6.00	i	i	−/+	−/+	−/+	+	+
2151	yel-orange, 2.5	d/+	i	+	+	+	+	+
2152	red-orange	r	...	+	+	+	+	−
2153	wh hydr, 1.91	r	...	+	+	+	−/+	r
2154	pink-red	i/+	...	+	+	+	+	i/+
2155	yel-brn hydr	r/+	...	+	−/+	r/d	+	+
2156	lt-yel	r	i	−/+	−	−	−/+	+
2157	blk, 8.7	−	i	−/+	−	−	−	−
2158	dk-red hydr	i	...	+	+	+	i/+	i/+
2159	brn	i	i	+	+	+	+	i/+
2160	yel	r	...	+	+	+	r	r
2161	brn-blk, 6.6	i	i	−	−/+	+	i	i
2162	blk-brn	i	i	−	−	−	−	−

No.	Formula and name	M_r	Phase transition temperature
2163	PdSO$_4$ ($\cdot$2H$_2$O) Palladium(II) sulfate	202.48	$-$H$_2$O 250
2164	PdSe Palladium(II) selenide	185.38	mp ca. 960
2165	Pd(Se$_2$) Palladium(II) diselenide(2$-$)	264.34	mp 1000
2166	PdSeO$_4$ Palladium(II) selenate	249.38	dec 600
2167	**Pm Promethium**	144.913	mp 1170, bp ca. 3000
2168	PmBr$_3$ Promethium(III) bromide	384.63	mp 680, bp 1530
2169	Pm$_2$(C$_2$O$_4$)$_3$ ($\cdot$10H$_2$O) Promethium(III) oxalate	553.88	hydr $t \rightarrow$ Pm$_2$O$_3$
2170	PmCl$_3$ Promethium(III) chloride	251.27	mp 740, bp 1670
2171	PmF$_3$ Promethium(III) fluoride	201.91	mp 1140, bp 2330
2172	PmI$_3$ Promethium(III) iodide	525.63	mp 800, bp 1370
2173	Pm(NO$_3$)$_3$ ($\cdot$6H$_2$O) Promethium(III) nitrate	330.93	dec hydr t
2174	Pm$_2$O$_3$ Promethium(III) oxide	337.82	mp ca. 2000
2175	Pm(OH)$_3$ Promethium(III) hydroxide	195.93	>350 $\rightarrow$ Pm$_2$O$_3$
2176	**Po Polonium**	208.982	mp 254, bp 962
2177	PoBr$_2$ Polonium(II) bromide	368.79	mp 275
2178	PoBr$_4$ Polonium(IV) bromide	528.60	mp 360
2179	PoCl$_2$ Polonium(II) chloride	279.89	subl 196
2180	PoCl$_4$ Polonium(IV) chloride	350.79	mp 300, bp 390
2181	PoI$_4$ Polonium(IV) iodide	716.60	dec t
2182	PoO$_{2\pm x}$ Polonium(IV) oxide	240.98	dec 500
2183	PoO(OH)$_2$ Polonium oxide-dihydroxide	259.00	dec t
2184	PoS Polonium(II) sulfide	241.05	...
2185	Po(SO$_4$)$_2$ Polonium(IV) sulfate	401.11	dec >550
2186	**Pr Praseodymium**	140.908	mp 931, bp 3510
2187	PrBr$_3$ Praseodymium(III) bromide	380.62	mp 693, bp 1550
2188	Pr(BrO$_3$)$_3$ ($\cdot$9H$_2$O) Praseodymium(III) bromate	524.61	mp hydr 58, $-$H$_2$O 130
2189	PrC$_2$ Praseodymium dicarbide	164.93	mp 2500 dec
2190	Pr$_2$(C$_2$O$_4$)$_3$ ($\cdot$10H$_2$O) Praseodymium(III) oxalate	545.87	dec hydr >600
2191	PrCl$_3$ Praseodymium(III) chloride	247.27	mp 786, bp 1630
2192	PrCl$_3$$\cdot$7H$_2$O	373.37	mp 115, dec t
2193	PrF$_3$ Praseodymium(III) fluoride	197.90	mp 1400, bp 2330
2194	PrF$_4$ Praseodymium(IV) fluoride	216.90	dec 90
2195	PrI$_3$ Praseodymium(III) iodide	521.62	mp 733, bp 1380
2196	Pr$_2$(MoO$_4$)$_3$ Praseodymium(III) molybdate	761.62	mp 1030
2197	Pr(NO$_3$)$_3$ ($\cdot$6H$_2$O) Praseodymium(III) nitrate	326.92	$-$H$_2$O 165, dec 400
2198	PrO$_2$ Praseodymium(IV) oxide	172.91	400 $\rightarrow$ Pr$_6$O$_{11}$
2199	Pr$_2$O$_3$ Praseodymium(III) oxide	329.81	mp >2000, bp ca. 4300
2200	Pr$_6$O$_{11}$ Hexapraseodymium undecaoxide	1021.44	...
2201	Pr(OH)$_3$ Praseodymium(III) hydroxide	191.93	dec >350
2202	PrS$_2$ Praseodymium(IV) oxide	205.04	mp 1780 dec
2203	Pr$_2$S$_3$ Praseodymium(III) sulfide	378.01	mp ca. 2000
2204	Pr$_2$(SO$_4$)$_3$ ($\cdot$8H$_2$O) Praseodymium(III) sulfate	570.00	$-$H$_2$O >600, dec >850
2205	**Pt Platinum**	195.08	mp 1772, bp ca. 3800
2206	PtAs$_2$ Platinum diarsenide	344.92	dec >800
2207	PtBr$_2$ Platinum(II) bromide	354.89	dec >250
2208	PtBr$_4$ ($\cdot$$nH_2$O) Platinum(IV) bromide	514.70	>180 $\rightarrow$ (Pt^{II}Pt^{IV})Br$_6$
2209	[PtBr$_4$],K$_2$ ($\cdot$2H$_2$O) Potassium tetrabromoplatinate(II)	592.89	mp hydr ca. 100

(Continued)

No.	Color and density	Solubility and reactivity						
		water	alcohol	HCl	H_2SO_4	HNO_3	NaOH	$NH_3 \cdot H_2O$
2163	red-brn hydr	r/+	...	+	d/r	−	+	+
2164	dk-grey	i	...	−/+	−/+	+	i/+	...
2165	grey-grn	−	−	−/+	−/+	−/+	−	−
2166	red-brn, 6.5	r	i	−/+	−	−	+	+
2167	wh, 7.26	−/+	...	+	+	+	−	−
2168	grn	d	r	−	−/+	−	+	+
2169	dk-pink hydr	i	i	−/+	−/+	−/+	i	i
2170	yel	r	r	r	−	−	+	+
2171	lt-viol	i	i	i	−	−	−	i
2172	dk-grey	r	r	−	−/+	−/+	+	+
2173	pink hydr	r	r	−	−	r	+	+
2174	lt-viol	−/+	...	+	+	+	−	−
2175	pink	i	...	+	+	+	i	i
2176	wh, 9.32	−	−	+	+	+	−	−
2177	brn	+	...	+	+	+	+	+
2178	red-brn	+	r	+	+	+	+	+
2179	red, 6.5	+	...	+	+	+	+	+
2180	yel	+	r	+	+	+	+	+
2181	blk	i/+	d	+	+	+	+	+
2182	red, 8.96	−	...	+	−/+	−/+	+	...
2183	lt-yel	i	...	+	+	+	+	...
2184	blk	i	i	−/+	−/+	−/+	i	ı
2185	wh	d/+	r	+	+	+	+	+
2186	yel-grey, 6.710	psv/+	−	+	+	+	−	−
2187	grn	d	r	−	−/+	−	+	+
2188	grn hydr	r	...	−	−	−	+	+
2189	yel, 5.10	+	...	+	+	+	+	+
2190	lt-grn hydr	i	i	−/+	−/+	−/+	i	i
2191	grn-sk-bl, 4.02	r	r	r	−	−	+	+
2192	grn, 2.25	r	r	r	−	−	+	+
2193	yel	i	i	−	−	i	i	−
2194	wh	i/+	...	+	+	+	−	−
2195	grn	r	r	−	−/+	−/+	+	+
2196	grn, 4.84	i	...	−	−	−	−/+	−
2197	grn hydr	r	r	−	−	r	+	+
2198	blk-brn, 6.82	−	...	+	+	+	−	−
2199	yel-grn, 6.97	−/+	...	+	+	+	−	−
2200	blk-brn	−	...	+	+	+	−	−
2201	grn, 3.73(2.83)	i	...	+	+	+	i	i
2202	dk-brn	i	...	−/+	−/+	−/+	i	i
2203	brn, 5.04	i/+	i	+	+	+	i	i/+
2204	grn hydr	r	...	−	r	−	+	+
2205	wh, 21.45	−	−	−	−	−	−	−
2206	grey, 10.6	−	...	−/+	−/+	−/+	−/+	−
2207	red-brn, 6.65	i	i	+	−/+	+	+	+
2208	blk-viol, 5.69	d/+	r	+	+	+	+	+
2209	blk hydr, 3.75	r	d	−	−/+	−/+	−/+	−/+

No.	Formula and name	M_r	Phase transition temperature
2210	$[PtBr_6],K_2$ Potassium hexabromoplatinate(IV)	752.70	dec >400
2211	$[PtBr_6],(NH_4)_2$ Ammonium hexabromoplatinate(IV)	704.53	dec 145
2212	$[PtBr_6],Na_2$ (·6H$_2$O) Sodium hexabromoplatinate(IV)	720.48	dec hydr 150
2213	$[Pt_2(C_2H_4)_2Cl_4]$ Diethylenetetrachlorodiplatinum	588.08	dec 200
2214	$[Pt(C_2H_4)Cl_3],K$ (·H$_2$O) Potassium ethylenetrichloroplatinate(II)	368.59	$-H_2O$ 100; >220 → PtCl$_2$
2215	$[Pt(CN)_4],Ba$ (·4H$_2$O) Barium tetracyanoplatinate(II)	436.48	$-H_2O$ 120, dec >420
2216	$[Pt(CN)_4],H_2$ (·5H$_2$O) Hydrogen tetracyanoplatinate(II)	301.17	mp hydr 100 dec
2217	$[Pt(CN)_4],K_2$ (·3H$_2$O) Potassium tetracyanoplatinate(II)	377.35	$-H_2O$ 100, dec >400
2218	$[Pt(CN)_6],K_2$ Potassium hexacyanoplatinate(IV)	429.38	dec 395
2219	$[Pt(CN)_4],Mg$ (·7H$_2$O) Magnesium tetracyanoplatinate(II)	323.46	$-H_2O$ 270
2220	$[Pt(CN)_4],(NH_4)_2$ Ammonium tetracyanoplatinate(II)	335.23	dec 200
2221	$[Pt(CN)_4],Na_2$ (·3H$_2$O) Sodium tetracyanoplatinate(II)	345.13	$-H_2O$ 125, dec 400
2222	$[Pt_2(CO)_2Br_4]$ Dicarbonyltetrabromodiplatinum	765.80	mp 177.7 dec
2223	$[Pt(CO)_2Cl_2]$ Dicarbonyldichloroplatinum	322.01	mp 103, dec 210
2224	$[Pt_2(CO)_2Cl_4]$ Dicarbonyltetrachlorodiplatinum	587.99	mp 194, dec >300
2225	$[Pt_2(CO)_3Cl_4]$ Tricarbonyltetrachlorodiplatinum	616.00	mp 130, dec 250
2226	$[Pt_2(CO)_2I_4]$ Dicarbonyltetraiododiplatinum	953.80	mp 140–150 dec
2227	$PtCl_2$ Platinum(II) chloride	265.99	dec 581
2228	$PtCl_4$ (·10H$_2$O) Platinum(IV) chloride	336.89	380 → (Pt^{II}Pt^{IV})Cl$_6$
2229	$[PtCl_4],Ba$ (·3H$_2$O) Barium tetrachloroplatinate(II)	474.22	$-H_2O$ 150
2230	$[PtCl_6],Ba$ (·6H$_2$O) Barium hexachloroplatinate(IV)	545.13	$-H_2O$ ca. 100
2231	$[PtCl_6],Cs_2$ Cesium hexachloroplatinate(IV)	673.61	dec 570
2232	$[PtCl_6],H_2$ (·6H$_2$O) Hydrogen hexachloroplatinate(IV)	409.81	mp hydr 60, dec hydr >115
2233	$[PtCl_4],K_2$ Potassium tetrachloroplatinate(II)	415.09	dec 475
2234	$[PtCl_6],K_2$ Potassium hexachloroplatinate(IV)	485.99	dec ca. 250
2235	$[PtCl_6],Li_2$ (·6H$_2$O) Lithium hexachloroplatinate(IV)	421.68	$-H_2O$ 180, dec t
2236	$[PtCl_6],Mg$ (·6H$_2$O) Magnesium hexachloroplatinate(IV)	432.10	$-H_2O$ ca. 90; dec >400
2237	$[PtCl_4],(NH_4)_2$ Ammonium tetrachloroplatinate(II)	372.97	dec ca. 140
2238	$[PtCl_6],(NH_4)_2$ Ammonium hexachloroplatinate(IV)	443.88	215 → Pt,NH$_4$Cl
2239	$[PtCl_4],Na_2$ (·4H$_2$O) Sodium tetrachloroplatinate(II)	382.87	mp hydr 100 dec
2240	$[PtCl_6],Na_2$ (·6H$_2$O) Sodium hexachloroplatinate(IV)	453.78	$-H_2O$ 100, dec < 270
2241	$[PtCl_6],Rb_2$ Rubidium hexachloroplatinate(IV)	578.73	dec 495
2242	PtF_4 Platinum(IV) fluoride	271.07	subl 300
2243	PtF_5 Platinum(V) fluoride	290.07	mp 80, dec 130
2244	PtF_6 Platinum(VI) fluoride	309.07	mp 61.3; bp 69.1
2245	$[PtF_6],K$ Potassium hexafluoroplatinate(V)	348.17	dec 750 (→ PtF$_4$)
2246	$[PtF_6],K_2$ Potassium hexafluoroplatinate(IV)	387.26	dec 800
2247	$[Pt^VF_6],O_2$ Dioxygenyl hexafuoroplatinate(V)	341.07	subl 100, mp 219 (p)
2248	$[Pt^VF_6],Xe$ Xenon(I) hexafluoroplatinate(V)	440.36	dec 165
2249	PtI_2 Platinum(II) iodide	448.89	dec 360
2250	PtI_4 Platinum(IV) iodide	702.70	270 → (Pt^{II}Pt^{IV})I$_6$

(Continued)

No.	Color and density	Solubility and reactivity						
		water	alcohol	HCl	H_2SO_4	HNO_3	NaOH	$NH_3 \cdot H_2O$
2210	red-brn, 4.66	d/+	i	−	−/+	−/+	−/+	−/+
2211	red-brn, 4.27	d/+	i	−	−/+	−/+	−/+	−/+
2212	red hydr, 3.32	r/+	r	−	−/+	−/+	−/+	−/+
2213	orange	+	r	+	+	+	+	+
2214	yel hydr, 2.88	r	r	−	−/+	−/+	+	−/+
2215	viol-yel, 2.08	d	i	−	+	−/+	−	−
2216	red hydr	r	r	−	+	−/+	+	+
2217	lt-yel, 2.46	r	d	−	+	−/+	−	−
2218	lt-yel	r	...	−	−/+	−/+	−	−
2219	red hydr, 2.18	r	r	−	−/+	−	+	+
2220	wh	r	...	−	+	−/+	−	−
2221	wh hydr, 2.65	r	r	−	+	−/+	−	−
2222	lt-red	+	r	+	+	+	+	+
2223	wh, 3.49	+	i	+	+	+	+	+
2224	yel-orange, 4.24	+	r	+	+	+	+	+
2225	orange-yel	+	+	+	+	+	+	+
2226	red	d	+	−	−/+	−/+	+	+
2227	grey-brn, 6.05	i	i	+	−	−	+	+
2228	red-brn, 4.30(2.43)	r	d	+	+	+	+	+
2229	dk-red hydr, 2.87	r	r	r	+	−	+	−/+
2230	orange-yel hydr, 2.86	r/+	r	r	+	−	+	−/+
2231	yel, 4.20	i/+	i	−	−	−	+	+
2232	orange-yel hydr, 2.43	r/+	r	r	+	+	+	+
2233	dk-red, 3.38	d/r	i	−	−	−	+	+
2234	yel, 3.50	d/+	i	−	−	−	+	+
2235	yel hydr	r/+	r	r	−	−	+	+
2236	yel hydr, 2.69	r/+	...	r	−	−	−/+	−/+
2237	red, 2.94	r	i	r	−	−	+	−/+
2238	yel, 3.07	d/+	i	−	−	−	+	+
2239	red hydr	r	...	r	−	−	+	−/+
2240	orange hydr, 2.50	r/+	r	r	−	−	+	+
2241	yel, 3.94	i/+	i	−	−	−	+	+
2242	brn	+	...	+	+	+	+	+
2243	dk-red	+	+	+	+	+	+	+
2244	dk-red	+	+	+	+	+	+	+
2245	yel-brn	+	+	+	+	+	+	+
2246	lt-yel	d/+	...	...	−	−	+	−/+
2247	orange	+	+	+	+	+	+	+
2248	red	+	+	+	+	+	+	+
2249	blk, 6.40	i	i	−	−/+	−/+	−	−
2250	blk, 6.55–6.70	i	r	+	+	+	+	−

No.	Formula and name	M_r	Phase transition temperature
2251	*trans*-[Pt(NH$_3$)$_2$Cl$_2$] *trans*-Diamminedichloroplatinum	300.05	>340 → Pt
2252	*cis*-[Pt(NH$_3$)$_2$Cl$_2$] *cis*-Diamminedichloroplatinum	300.05	275 → *trans*-isomer
2253	*trans*-[Pt(NH$_3$)$_2$Cl$_4$] *trans*-Diamminetetrachloroplatinum	370.95	>340 → Pt
2254	*cis*-[Pt(NH$_3$)$_2$Cl$_4$] *cis*-Diamminetetrachloroplatinum	370.95	>240 → Pt
2255	[Pt(NH$_3$)$_4$]Cl$_2$ (·H$_2$O) Tetraammineplatinum(II) chloride	334.11	−H$_2$O 110, 250 → *trans*-[Pt(NH$_3$)$_2$Cl$_2$]
2256	[Pt(NH$_3$)$_2$(NO$_2$)$_2$] Diamminedinitroplatinum	321.15	dec ca. 200
2257	[Pt(NH$_3$)$_4$][Pt^{II}Cl$_4$] Tetraammineplatinum(II) tetrachloroplatinate(II)	600.10	290 → *trans*-[Pt(NH$_3$)$_2$Cl$_2$]
2258	[Pt(NO$_2$)$_4$],K$_2$ Potassium tetranitroplatinate(II)	457.30	dec >150
2259	[Pt(NO$_2$)$_4$],Na$_2$ Sodium tetranitroplatinate(II)	425.08	dec hydr 100
2260	PtO Platinum(II) oxide	211.08	560 → Pt
2261	PtO$_2$ Platinum(IV) oxide	227.08	mp 450; 650 → Pt
2262	PtO$_2$·H$_2$O	245.09	dec 120
2263	PtO$_2$·2H$_2$O	263.11	dec 200
2264	PtO$_2$·3H$_2$O	281.12	dec 300
2265	PtO$_2$·4H$_2$O	299.14	dec 420
2266	PtO$_3$ Platinum(VI) oxide	243.08	>20 → PtO$_2$
2267	Pt(OH)$_2$ Platinum(II) hydroxide	229.09	dec >150
2268	[Pt(OH)$_6$],K$_2$ Potassium hexahydroxoplatinate(IV)	375.32	dec 160
2269	[Pt(OH)$_6$],Na$_2$ Sodium hexahydroxoplatinate(IV)	343.10	dec 150–170
2270	PtP$_2$ Platinum diphosphide	257.03	mp >1500
2271	[Pt(PF$_3$)$_4$] Tetrakis(trifluorophosphorus)platinum	546.95	mp −15, dec 90
2272	Pt$_2$O$_7$ Platinum(IV) diphosphate	369.02	dec 600
2273	(Pt^{II}Pt^{IV})Br$_6$ Platinum(II)-platinum(IV) bromide	869.58	>210 → PtBr$_2$
2274	(Pt^{II}Pt^{IV})Cl$_6$ Platinum(II)-platinum(IV) chloride	602.88	subl 350; 435 → PtCl2
2275	(Pt^{II}Pt^{IV})I$_6$ Platinum(II)-platinum(IV) iodide	1151.58	270 → PtI$_2$
2276	PtS Platinum(II) sulfide	227.15	800 → Pt
2277	PtS$_2$ Platinum(IV) sulfide	259.21	ca. 230 → Pt
2278	PtSe$_2$ Platinum(IV) selenide	353.00	>400 → Pt
2279	PtTe$_2$ Platinum(IV) telluride	450.28	mp ca. 1250
2280	**Pu Plutonium**	244.064	mp 640, bp ca. 3350
2281	PuBr$_3$ (·6H$_2$O) Plutonium(III) bromide	483.78	mp 681, bp 1531
2282	PuC$_{0.85}$ Plutonium 0.85-carbide	254.27	dec 1654
2283	Pu(C$_2$O$_4$)$_2$ (·6H$_2$O) Plutonium(IV) oxalate	420.10	hydr 380 → PuO$_2$
2284	PuCl$_3$ Plutonium(III) chloride	350.42	mp 760, bp 1770
2285	PuF$_3$ Plutonium(III) fluoride	301.06	mp 1426, bp 2300
2286	PuF$_4$ (·2.5H$_2$O) Plutonium(IV) fluoride	320.06	mp 1037, bp 1277
2287	PuF$_6$ Plutonium(VI) fluoride	358.05	mp 51.59; bp 62.16
2288	PuI$_3$ Plutonium(III) iodide	624.78	mp 777
2289	PuN Plutonium mononitride	258.07	mp 2585 (*p*)
2290	Pu(NO$_3$)$_4$ (·5H$_2$O) Plutonium(IV) nitrate	492.08	dec hydr 150–180
2291	PuO Plutonium(II) oxide	260.06	mp ca. 1900
2292	PuO$_{2+x}$ Plutonium(IV) oxide	276.06	mp 2390
2293	Pu(OH)$_3$ Plutonium(III) hydroxide	295.09	dec >350

(*Continued*)

No.	Color and density	Solubility and reactivity						
		water	alcohol	HCl	H_2SO_4	HNO_3	NaOH	$NH_3 \cdot H_2O$
2251	yel	i/d	...	−	−/+	+	−/+	+
2252	yel	i/d	...	−	−/+	+	−/+	+
2253	yel-grn, 3.51	d	...	−/r	−/+	−/+	−/+	r/+
2254	yel ($t \rightarrow$ grn), 3.42	d	...	−	+	+	−/+	+
2255	wh hydr, 2.74	r	i	+	+	+	−/+	r
2256	lt-yel	i	...	+	−/+	+	−/+	+
2257	dk-grn	d	...	+	+	+	+	−
2258	wh	d	i	−	−/+	−/+	−/+	−/+
2259	lt-yel	r	i	−	−/+	−/+	−/+	−/+
2260	viol-blk, 14.9	−	...	−	−	−	−	−
2261	blk, 10.2	−	...	−	−	−	−	−
2262	dk-brn	i	...	−	−/+	−	−/+	i
2263	brn	i	...	−	+	−/+	−/+	i
2264	yel	i	...	−/+	+	−/+	+	i
2265	wh	i	...	+	+	+	+	+
2266	red-brn	−	...	+	+	+	−	−
2267	blk	i	...	−/+	−/+	−/+	+	i
2268	yel-grn, 5.18	r	i	+	+	+	r	r
2269	yel	r	i	−/+	+	+	r	r
2270	grey, 9.01	−	...	−	−	−/+	−	−
2271	cl lq	i/+	...	+	+	+	+	+
2272	yel-grn	i	...	−	−/+	−/+	−/+	−
2273	blk	i/+	i	+	+	+	+	+
2274	dk-grn, 5.256	d/r	...	+	+	+	+	+
2275	blk	i	i	+	+	+	+	+
2276	blk, 8.85	i	...	−	−	−	−	−
2277	grey-brn, 7.66	−	...	−	−	−/+	−	−
2278	blk	−	...	−	−/+	−/+	−/+	−
2279	blk	−	...	−	−/+	−/+	−/+	−
2280	wh, 19.86	psv/+	...	+	+/psv	psv	−	−
2281	lt-grn, 6.69(9.07)	r	...	−	−/+	−	+	+
2282	blk, ca. 13.6	−/+	i	−	−/+	−/+	−	−
2283	yel-grn hydr	i	...	i	−/+	−/+	+	−/+
2284	bl-grn, 5.70	r	r	r	−	−	+	+
2285	viol-grn, 9.32	i/+	...	−	−	−	i	i
2286	pink-brn, 7.1(4.89)	i	i	−	i	−	−	−
2287	orange-red, 4.86	+	+	+	+	+	+	+
2288	lt-grn, 6.92	r	...	−	−/+	−/+	+	+
2289	blk, 14.25	+	i	+	+	+	−	−
2290	yel-grn hydr	r	...	−	−	r	+	+
2291	blk, 13.89	−	...	−/+	+	+	−	−
2292	dk-grn, 11.46	−	i	−	−/+	+	−	−
2293	grey-sk-bl	i	...	+	+	+	i	i

No.	Formula and name	M_r	Phase transition temperature
2294	Pu(OH)$_4$ Plutonium(IV) hydroxide	312.09	800 → PuO$_2$
2295	PuO$_2$(NO$_3$)$_2$ ($\cdot$6H$_2$O) Plutonyl nitrate	400.07	mp hydr 140, −H$_2$O >150
2296	PuS Plutonium(II) sulfide	276.13	mp 2350
2297	**Ra Radium**	226.025	mp 969, bp 1536
2298	RaBr$_2$ ($\cdot$2H$_2$O) Radium bromide	385.83	mp 728
2299	RaCO$_3$ Radium carbonate	286.03	dec >1500
2300	RaCl$_2$ ($\cdot$2H$_2$O) Radium chloride	296.93	mp ca. 900
2301	Ra(IO$_3$)$_2$ Radium iodate	575.83	dec t
2302	Ra(N$_3$)$_2$ Radium azide	310.07	dec t
2303	RaO Radium oxide	242.02	...
2304	Ra(OH)$_2$ Radium hydroxide	260.04	t → RaO
2305	RaS Radium sulfide	258.09	mp ca. 2000 dec
2306	RaSO$_4$ Radium sulfate	322.09	dec >1600
2307	**Rb Rubidium**	85.468	mp 39.3; bp 696
2308	RbAl(SO$_4$)$_2$ ($\cdot$12H$_2$O) Rubidium-aluminum sulfate	304.57	mp hydr 99, dec hydr t
2309	RbBr Rubidium bromide	165.37	mp 693, bp 1352
2310	RBrO$_3$ Rubidium bromate	213.37	mp 430
2311	Rb(CH$_3$COO) Rubidium acetate	144.51	mp 246
2312	Rb$_2$CO$_3$ ($\cdot$2H$_2$O) Rubidium carbonate	230.94	mp 873, dec 900
2313	RbCl Rubidium choride	120.92	mp 718, bp 1395
2314	RbClO$_3$ Rubidium chlorate	168.92	dec >500
2315	RbClO$_4$ Rubidium perchlorate	184.92	mp 597,900 → RbCl
2316	Rb$_2$CrO$_4$ Rubidiun chromate	286.93	mp 994
2317	Rb$_2$Cr$_2$O$_7$ Rubidium dichromate	386.92	mp 396
2318	RbCr(SO$_4$)$_2$ ($\cdot$12H$_2$O) Rubidium-chromium(III) sulfate	329.59	mp hydr 107, dec hydr t
2319	RbF ($\cdot$1.5H$_2$O) Rubidium fluoride	104.47	mp 798, bp 1430
2320	RbFe(SO$_4$)$_2$ ($\cdot$12H$_2$O) Rubidium-iron(III) sulfate	333.44	mp hydr 50, dec hydr t
2321	RbFe(SeO$_4$)$_2$ ($\cdot$12H$_2$O) Rubidium-iron(III) selenate	427.23	mp hydr 4.5; −H$_2$O 100
2322	RbH Rubidiun hydride	86.48	>200 → Rb
2323	RbHCO$_3$ ($\cdot$H$_2$O) Rubidium hydrocarbonate	146.48	dec >175
2324	Rb(HF$_2$) Rubidium hydrodifluoride	124.47	mp 204.5
2325	RbI Rubidium iodide	212.37	mp 656, bp 1327
2326	RuIO$_3$ Rubidium iodate	260.37	dec >700
2327	RuIO$_4$ Rubidium metaperiodate	276.37	dec >805
2328	RbMnO$_4$ Rubidium permanganate	204.40	dec ca. 300
2329	RbMn(SO$_4$)$_2$ ($\cdot$12H$_2$O) Rubidium-manganese(III) sulfate	332.53	mp hydr 22 dec
2330	RbN$_3$ Rubidium azide	127.49	mp 321, dec 395
2331	RbNH$_2$ Rubidium amide	101.49	mp 309
2332	RbNO$_2$ Rubidium nitrite	131.47	mp 422
2333	RbNO$_3$ Rubidium nitrate	147.47	mp 313; >600 → RbNO$_2$
2334	RbO$_2$ Rubidium superoxide	117.47	mp 540 (p), dec >400
2335	RbO$_3$ Rubidium ozonide	133.47	dec >70
2336	Rb$_2$O Rubidium oxide	186.94	mp 505 dec
2337	Rb$_2$O$_2$ Rubidium peroxide	202.93	mp 570, bp 1010 dec

(Continued)

No.	Color and density	Solubility and reactivity						
		water	alcohol	HCl	H_2SO_4	HNO_3	NaOH	$NH_3 \cdot H_2O$
2294	dk-grn	i	...	+	+	+	i	i
2295	pink-brn hydr	r	r	−	−	+	+	+
2296	yel-brn, 10.6	i	i	+	+	+	i	i
2297	wh, ca. 6	+	+	+	+	+	+	+
2298	lt-yel, 5.78	r	r	r	−/+	−	r	r
2299	wh	i	...	+	+	+	i	i
2300	wh, 4.91	r	r	r	−	−	r	−
2301	wh	i/d	...	i/d	−/+	−/+	i/d	−
2302	wh	r	r	−/+	−/+	−/+	−	−
2303	wh	+	...	+	+	+	+	+
2304	wh	r	...	+	+	+	r	r
2305	wh	+	...	+	+	+	r	r
2306	wh	i	...	−	i	−	i	i
2307	wh, 1.532	+	+	+	+	+	+	+
2308	wh hydr, 1.89	d/r	i	−	d/r	−	+	+
2309	wh, 3.35	r	d	r	−/+	−	r	r
2310	wh, 3.68	r	...	−	−	−	−	−
2311	wh	r	r	r	−	r	r	−
2312	wh	r	d	+	+	+	r	r
2313	wh, 2.76	r	d	r	−/+	−	−	−
2314	wh, 3.19	r	r	−/+	−	r	−	−
2315	wh, 2.9	d/r	d	−	−	−	−	−
2316	yel, 3.52	r	...	+	+	+	r	r
2317	orange, 3.02	r	...	r/+	r	r	+	+
2318	viol hydr, 1.95	r	...	−	r	−	+	+
2319	wh, 3.56	r	i	r	r	r	r	r
2320	wh hydr, 1.93	r	...	−	r	−	+	+
2321	wh hydr, 2.31	r	...	−/+	r	−	+	+
2322	wh, 2.59	+	...	+	+	+	+	+
2323	wh	r	r	+	+	+	+	+
2324	wh	r	i	−/+	−/+	−/+	+	+
2325	wh, 3.55	r	r	r	r/+	r/+	r	r
2326	wh, 4.33	d	...	−/+	d	d	−	−
2327	wh, 3.92	d	d	d	d	d	d	d
2328	red-viol, 3.24	d	...	−/+	−	r	−	−/+
2329	dk-red hydr	+	i	+	+	+	+	+
2330	wh, 2.48	r	...	+	+	+	−	−
2331	lt-grn	+	+	+	+	+	+	+
2332	wh	r	...	−/+	−/+	−/+	r	r
2333	wh, 3.11	r	d	−	−	r	r	r
2334	orange-yel, 3.06	+	...	+	+	+	+	+
2335	red, 2.75	+	+	+	+	+	+	+
2336	wh, 3.72	+	+	+	+	+	+	+
2337	yel, 3.80	+	...	+	+	+	+	+

No.	Formula and name	M_r	Phase transition temperature
2338	RbOH ($\cdot H_2O$) Rubidium hydroxide	102.48	mp hydr 145, mp 382
2339	RbOH$\cdot 2H_2O$	138.51	47 → PbOH·H_2O
2340	Rb_2S ($\cdot 4H_2O$) Rubidium sulfide	203.00	dec hydr 200, mp 530
2341	$Rb_2(S_5)$ Rubidium pentasulfide(2−)	331.27	mp 231
2342	Rb_2SO_4 Rubidium sulfate	267.00	mp 1066, bp ca. 1700
2343	$Rb_2S_2O_7$ Rubidium disulfate	347.06	>400 → Rb_2SO_4
2344	$Rb(SO_3F)$ Rubidium fluorosulfonate	184.53	mp 304
2345	Rb_2SeO_4 Rubidium selenate	313.89	mp 1050
2346	$RbV(SO_4)_2$ ($\cdot 12H_2O$) Rubidium-vanadium(III) sulfate	328.53	mp hydr 64
2347	**Re Rhenium**	186.207	mp 3190, bp ca. 5900
2348	$ReBr_3$ Rhenium(III) bromide	425.92	subl >400
2349	$ReBr_4$ Rhenium(IV) bromide	505.82	>250 → Re
2350	$Re(Br)O_3$ Rhenium bromide-trioxide	314.11	mp 39.5; bp 163
2351	$[Re(CN)_4O_2]$,K_3 Potassium tetracyanodioxo-rhenate(V)	439.57	dec >300
2352	$[Re_2(CO)_{10}]$ Decacarbonyldirhenium	652.51	mp 177, 250 → Re
2353	$[Re(CO)_5Br]$ Pentacarbonylbromorhenium	406.16	subl >60
2354	$[Re(CO)_3C_5H_5]$ Tricarbonyl(cyclopentadienyl)rhenium	335.33	mp 111
2355	$[Re(CO)_5Cl]$ Pentacarbonylchlororhenium	361.71	subl >50
2356	$[Re(CO)_5I]$ Pentacarbonyliodorhenium	453.16	subl ≤ 60, t → $[Re_2(CO)_{10}]$
2357	$ReCl_4$ Rhenium(IV) chloride	328.02	mp 22, 300 → $[Re_3Cl_9]$,$ReCl_5$
2358	$ReCl_5$ Rhenium(V) chloride	363.47	mp 278, bp 330 dec
2359	$[Re_3Cl_9]$ Nonachlorotrirhenium	877.70	mp 257, bp 327, dec >360
2360	$[Re(Cl)_6]$,K_2 Potassium hexachlororhenate(IV)	477.12	dec >300
2361	$[Re_2Cl_8]$,K_2 ($\cdot 2H_2O$) Potassium octachlorodirhenate(III)	734.23	−H_2O 130, dec >290
2362	$Re(Cl)O_3$ Rhenium chloride-trioxide	269.66	mp 4.5; bp 131
2363	$ReCl_4O$ Rhenium tetrachloride-oxide	344.02	mp 29.3; bp 228, dec >300
2364	$[ReCl_5O]$,Cs_2 Cesium pentachlorooxorhenate(V)	645.28	mp >300
2365	ReF_4 Rhenium(IV) fluoride	262.20	mp 125, dec 500
2366	ReF_5 Rhenium(V) fluoride	281.20	mp 48, bp 221
2367	ReF_6 Rhenium(VI) fluoride	300.20	mp 18.6; bp 33.7
2368	ReF_7 Rhenium(VII) fluoride	319.19	mp 48.3; bp 73.7
2369	$[ReH(CO)_5]$ Hydropentacarbonylrhenium	327.27	mp 13, dec t
2370	$[ReH_9]$,K_2 Potassium nonahydridorhenate(VII)	273.48	dec >300
2371	$[ReH_9]$,Na_2 Sodium nonahydridorhenate(VII)	241.26	dec 245
2372	ReI_3 Rhenium(III) iodide	566.92	800 → Re
2373	$[ReI_6]$,K_2 Potassium hexaiodorhenate(IV)	1025.83	dec t
2374	ReO_2 ($\cdot nH_2O$)] Rhenium(IV) oxide	218.21	−H_2O 500; >850 → Re_2O_7,Re
2375	ReO_3 Rhenium(VI) oxide	234.20	mp 160, >300 → Re_2O_7,ReO_2
2376	Re_2O_3 ($\cdot nH_2O$) Rhenium(III) oxide	420.41	dec hydr >25
2377	Re_2O_7 Rhenium(VII) oxide	484.41	mp 301.5; bp 358.5
2378	$Re(O)F_4$ Rhenium oxide-tetrafluoride	278.20	mp 108, bp 171.2

(Continued)

No.	Color and density	Solubility and reactivity						
		water	alcohol	HCl	H_2SO_4	HNO_3	NaOH	$NH_3 \cdot H_2O$
2338	wh, 3.20	r	r	+	+	+	r	r
2339	wh	r	r	+	+	+	−	−
2340	wh, 2.91	r/+	r	+	+	+	r	r
2341	red, 2.62	r/+	d	+	+	+	−	−
2342	wh, 3.61	r	d	−	r	−	r	−
2343	wh	r/+	...	+	+	+	+	+
2344	wh	r/+	...	−/+	−/+	−/+	−/+	...
2345	wh, 3.90	r	...	−/+	−/+	−	−	−
2346	yel hydr, 1.92	d	i	−	r	−/+	+	+
2347	lt-grey, 20.53	−	...	−	−/+	−/+	−/+	−
2348	blk	+	r	+	+	+	+	+
2349	blk, 5.47	+	...	+	+	+	+	+
2350	wh	+	...	+	+	+	+	+
2351	orange-brn, 2.70	r	d	−	−/+	−/+	−	−
2352	wh	−	r	−	−/+	−/+	+	−
2353	wh	−	r	−	−/+	−/+	−/+	...
2354	wh	−/+	r	−/+	−/+	−/+	−/+	...
2355	wh	−	r	−	−/+	−/+	−/+	...
2356	wh	−	r	−	−/+	−/+	−/+	...
2357	grn-blk	+	...	+	+	+	+	+
2358	brn-blk, 3.98	+	...	+	+	+	+	+
2359	red-viol	r/+	r	−	−	−	+	+
2360	yel-grn, 3.34	d/+	...	d/r	−	r/+	+	+
2361	dk-grn, 3.5	r	r	r/d	r/+	r/+	+	+
2362	cl lq, 3.87[20]	+	+	+	+	+	+	+
2363	brn-red, 3.76	+	+	+	+	+	+	+
2364	yel	+	...	+/r	+	+	+	+
2365	dk-grn, 5.38	+	...	+	+	+	+	+
2366	yel-grn	+	...	+	+	+	+	+
2367	yel lq, 3.58[22]	+	+	+	+	+	+	+
2368	yel	+	+	+	+	+	+	+
2369	cl lq	+	r	+	+	+	+	+
2370	wh, 3.07	r/+	d/+	+	+	+	r/+	r/+
2371	wh	r/+	d/+	+	+	+	r/+	r/+
2372	blk, 6.37	r	i	−/+	−/+	−/+	+	+
2373	blk, 4.4	+	d	+	+	+	+	+
2374	blk-brn, 11.4	−	...	−/+	−/+	−/+	−	−
2375	red, 7.43	−	...	−	−	+	−/+	...
2376	blk hydr	i	...	−	−	+	−	−
2377	lt-yel, 6.14	−/+	i	+	+	+	+	−
2378	sk-bl, 4.03	+	...	+	+	+	+	+

No.	Formula and name	M_r	Phase transition temperature
2379	Re(O)F$_5$ Rhenium oxide-pentafluoride	297.20	mp 41, bp 73
2380	ReO$_2$F$_2$ Rhenium dioxide-difluoride	256.20	mp 156
2381	ReO$_2$F$_3$ Rhenium dioxide-trifluoride	275.20	mp 90, bp 185
2382	ReO$_3$F Rhenium trioxide-fluoride	253.20	mp 147, bp 164
2383	Re(OH)$_4$ Rhenium(IV) hydroxide	254.24	>400 → ReO$_2$
2384	[Re$_2$(PF$_3$)$_{10}$] Decakis(trifluorophosphorus)dirhenium	1252.09	mp 182
2385	ReS$_2$ Rhenium(IV) sulfide	250.34	>1000 → Re
2386	Re$_2$S$_7$ Rhenium(VII) sulfide	596.88	250 → ReS$_2$
2387	**Rh Rhodium**	102.906	mp 1963, bp ca. 3700
2388	RhBr$_3$ Rhodium(III) bromide	342.62	>800 → Rh
2389	[Rh$_2$(CO)$_8$] Octacarbonyldirhodium	429.89	mp 76, dec >100
2390	[Rh$_4$(CO)$_{12}$] Dodecacarbonyltetrarhodium	747.74	mp 76; 150 → [Rh$_6$(CO)$_{16}$]
2391	[Rh$_6$(CO)$_{16}$] 16-Carbonylhexarhodium	1065.60	220 → Rh
2392	[Rh(CO)$_2$C$_5$H$_5$] Dicarbonyl(cyclopentadienyl)rhodium	224.02	mp −11, bp ca. 240
2393	[Rh$_2$(CO)$_3$(C$_5$H$_5$)$_2$] Tricarbonylbis(cyclopentadienyl)dirhodium	420.03	mp 139
2394	[Rh$_2$(CO)$_4$Cl$_2$] Tetracarbonyldichlorodirhodium	388.76	mp 121
2395	[Rh(C$_2$O$_4$)$_3$],K$_3$ (·4.5H$_2$O) Potassium tris(oxalato)rhodate(III)	484.25	−H$_2$O 190
2396	RhCl$_3$ (·nH$_2$O) Rhodium(III) chloride	209.27	−H$_2$O 180, 970 → Rh
2397	[RhCl$_6$],Cs$_2$ Cesium hexachlororhodate(IV)	581.43	dec >200
2398	[RhCl$_6$],K$_3$ (·3H$_2$O) Potassium hexachlororhodate(III)	432.92	dec ca. 600
2399	[RhCl$_6$],Na$_3$ (·12H$_2$O) Sodium hexachlororhodate(III)	384.59	dec hydr 550
2400	RhF$_3$ Rhodium(III) fluoride	159.90	subl >600
2401	RhF$_4$ Rhodium(IV) fluoride	178.90	subl 700
2402	Rh(HS)$_3$ Rhodium(III) hydrosulfide	202.13	t → Rh$_2$S$_3$
2403	RhI$_3$ Rhodium(III) iodide	483.62	>650 → Rh,I$_2$
2404	[Rh(NH$_3$)$_6$]Cl$_3$ Hexaamminerhodium(III) chloride	311.45	dec >200
2405	Rh(NO$_3$)$_3$ Rhodium(III) nitrate	288.92	>600 → Rh$_2$O$_3$
2406	[Rh(NO$_2$)$_6$],K$_3$ Potassium hexanitrorhodate(III)	496.23	dec 360–440
2407	[Rh(NO$_2$)$_6$],Na$_3$ Sodium hexanitrorhodate(III)	447.91	dec 440
2408	RhO$_2$ (·nH$_2$O) Rhodium(IV) oxide	134.90	hydr t → Rh$_2$O$_3$
2409	Rh$_2$O$_3$ Rhodium(III) oxide	253.81	dec >1100
2410	Rh(OH)$_3$ Rhodium(III) hydroxide	153.93	200 → Rh$_2$O$_3$
2411	Rh$_2$S$_3$ Rhodium(III) sulfide	302.01	dec >900
2412	**Rn Radon**	222.018	mp −71.0; bp −61.9
2413	**Ru Ruthenium**	101.07	mp 2607, bp ca. 4900
2414	RuBr$_3$ Ruthenium(III) bromide	340.78	subl >550
2415	[Ru(C$_5$H$_5$)$_2$] Bis(cyclopentadienyl)ruthenium	231.26	mp 199
2416	[Ru(C$_5$H$_7$O$_2$)$_3$] Tris(acetylacetonato)ruthenium	398.40	mp 76.5
2417	[Ru(CO)$_5$] Pentacarbonylruthenium	241.12	mp −22, 50 → [Ru$_3$(CO)$_{12}$]
2418	[Ru$_3$(CO)$_{12}$] Dodecacarbonyltriruthenium	639.33	dec 150, mp 155 (p)

(Continued)

No.	Color and density	Solubility and reactivity						
		water	alcohol	HCl	H_2SO_4	HNO_3	NaOH	$NH_3 \cdot H_2O$
2379	wh	+	...	+	+	+	+	+
2380	wh	+	...	+	+	+	+	+
2381	yel	+	...	+	+	+	+	+
2382	yel	+	...	+	+	+	+	+
2383	brn-blk	i	...	+	+	+	+	i
2384	wh	i/+	r	+	+	+	+	+
2385	blk, 7.51	i	i	−	−/+	+	i	...
2386	blk-brn, 4.87	i/+	...	−	−/+	−/+	+	−
2387	wh, 12.41	−	−	−	−/+	−	−	−
2388	red-brn, 5.56	r	i	−	−	−	+	+
2389	orange	−/+	r	−	−/+	−	−/+	...
2390	dk-red	−	d	−/+	−/+	−/+	−/+	...
2391	viol-brn	−	d	−	−/+	−	−/+	−
2392	yel lq	−/+	r	−/+	−/+	−/+	−/+	...
2393	red	−/+	r	−/+	−/+	−/+	−/+	...
2394	orange-red	+	r	+	+	+	−/+	+
2395	red hydr, 2.17	r	...	−	r	−/+	−/+	−
2396	red-brn, 3.11	r	r	+	−	−	+	+
2397	grn	+	...	+	+	+	+	+
2398	red hydr, 3.29	r/+	i	−/+	+	+	+	+
2399	dk-red hydr	r	i	−/+	+	+	+	+
2400	red, 5.38	i	...	−	−	−	−	−
2401	sk-bl	+	...	+	+	+	+	+
2402	brn	i	...	+	+	+	+	+
2403	blk, 6.40	r	i	−	−	−	+	+
2404	wh, 2.01	r	...	+	+	+	+	r
2405	yel-brn	r	i	+	−	r	+	+
2406	wh, 2.74	i/d	...	−/+	−/+	−/+	−/+	+
2407	wh	r	i	+	+	+	−/+	−
2408	dk-grn hydr	i	−	−/+	−	−	+	i
2409	grey-blk, 8.20	−	i	−	−	−	−	−
2410	yel	i	i	+	+	+	i/+	i
2411	blk, 6.40	i	...	−	−/+	−	i	i
2412	cl gas, 9.73	d	r	−	−	−	−	−
2413	wh, 12.3	−	−	+	−/+	−	−	−
2414	blk, 5.42	r	r	+	−/+	−/+	+	+
2415	yel	−	r	−	−	−	−	−
2416	dk-red	i	r	−	−/+	−/+	−	−
2417	cl lq	−	r	−	−/+	−/+	−	−
2418	orange	−	r	−	−/+	−/+	−/+	−

No.	Formula and name	M_r	Phase transition temperature
2419	[Ru$_2$(CO)$_4$(C$_5$H$_5$)$_2$] Tetracarbonylbis(cyclopentadienyl)diruthenium	444.37	mp 185
2420	[Ru$_2$(CO)$_6$Cl$_4$] Hexacarbonyltetrachlorodiruthenium	512.01	dec 315
2421	[Ru(CO)$_4$I$_2$] Tetracarbonyldiiodoruthenium	466.92	dec >140
2422	RuCl$_3$ Ruthenium(III) chloride	207.43	dec >500
2423	RuCl$_4$ (·5H$_2$O) Ruthenium(IV) chloride	242.88	dec hydr 100
2424	[RuCl$_6$],K$_2$ Potassium hexachlororuthenate(IV)	391.98	dec 520
2425	[RuCl$_6$],(NH$_4$)$_2$ Ammonium hexachlororuthenate(IV)	349.87	dec 360
2426	[Ru$_2$Cl$_{10}$O],K$_4$ (·H$_2$O) Potassium decachlorooxodiruthenate(IV)	729.06	dec hydr >200
2427	RuF$_5$ Ruthenium(V) fluoride	196.06	mp 86.5; bp 227
2428	RuF$_6$ Ruthenium(VI) fluoride	215.06	mp 54, dec >200
2429	[Ru(H$_2$O)Cl$_5$],K$_2$ Potassium aquapentachlororuthenate(III)	374.55	dec 180
2430	RuI$_3$ Ruthenium(III) iodide	481.78	>300 → Ru,I$_2$
2431	[Ru(NO$^+$)Cl$_5$],K$_2$ Potassium nitrosiliumpentachlororuthenate(II)	386.54	dec >570
2432	[Ru(NO$^+$)Cl$_5$],(NH$_4$)$_2$ Ammonium nitrosiliumpentachlororuthenate(II)	344.42	dec 440
2433	[Ru(NO$^+$)(H$_2$O)$_2$(OH)$_3$] Nitrosiliumdiaquatrihydroxoruthenium	218.13	dec t
2434	[Ru(NO$^+$)(NO$_2$)$_4$OH],Na$_2$ Sodium nitrosiliumtetranitrohydroxoruthenate(II)	378.08	dec t
2435	RuO$_2$ Ruthenium(IV) oxide	133.07	>1300 → RuO$_4$
2436	RuO$_2$·nH$_2$O	–	–H$_2$O 500
2437	RuO$_4$ Ruthenium(VIII) oxide	165.07	mp 25.4; >100 → RuO$_2$
2438	Ru(O)F$_4$ Ruthenium oxide-tetrafluoride	193.06	mp 115
2439	Ru(OH)$_3$ Ruthenium(III) hydroxide	152.09	dec t
2440	Ru(S$_2$) Ruthenium(II) disulfide(2−)	165.20	dec >1000
2441	**α-S$_8$ Octasulfur**	256.528	95.5 → β-S$_8$
2442	**β-S$_8$**	256.528	mp 119.3; bp 444.674
2443	S$_2$Br$_2$ Disulfur dibromide	223.94	mp −46, dec 90
2444	SBr$_2$O Sulfur dibromide-oxide	207.87	mp −49.5; bp 138 (p)
2445	S(Br)O$_2$F Sulfur bromide-dioxide-fluoride	162.97	mp −86, bp 40
2446	SCl$_2$ Sulfur dichloride	102.97	mp −121, bp 59.6
2447	SCl$_4$ Sulfur terachloride	173.88	mp −30, dec −15
2448	S$_2$Cl$_2$ Disulfur dichloride	135.04	mp −77, bp 138
2449	S(Cl)F$_5$ Sulfur chloride-pentafluoride	162.51	mp −64, bp −19
2450	SCl$_2$O Sulfur dichloride-oxide	118.97	mp −104.5; bp 75.6
2451	SCl$_2$O$_2$ Sulfur dichloride-dioxide	134.97	mp −54.1; bp 69.5
2452	S$_2$Cl$_4$O Disulfur tetrachloride-oxide	221.94	bp 60.5
2453	S(Cl)O(F) Sulfur chloride-oxide-fluoride	102.52	mp −139.5; bp 12.2
2454	S(Cl)O$_2$F Sulfur chloride-dioxide-fluoride	118.52	mp −124.7; bp 7.1
2455	SF$_4$ Sulfur tetrafluoride	108.06	mp −121.0; bp −37
2456	SF$_6$ Sulfur hexafluoride	146.05	mp −50, dec >800
2457	S$_2$F$_2$ Dissulfur difluoride	102.13	mp −133, >15 → S(S)F$_2$
2458	S$_2$F$_{10}$ Disulfur decafluoride	254.11	mp −60, bp 29

No.	Color and density	Solubility and reactivity						
		water	alcohol	HCl	H_2SO_4	HNO_3	NaOH	$NH_3 \cdot H_2O$
2419	orange	−/+	r	−/+	−/+	−/+	−/+	−/+
2420	wh ($t \rightarrow$ orange)	−/+	r	−/+	−/+	−/+	−/+	−/+
2421	yel	−/+	r	−/+	−/+	−/+	−/+	...
2422	blk-brn, 3.11	i/+	d	+	−/+	−/+	+	+
2423	red hydr	r	r	r	−	−	+	+
2424	blk-red	+	i	+	+	+	+	+
2425	red-brn	d/+	...	+	+	+	−/+	...
2426	brn-brn hydr	r/+	...	−/+	−/+	−/+	−/+	+
2427	dk-grn, 2.96	+	...	+	+	+	+	+
2428	dk-brn	+	...	+	+	+	+	+
2429	dk-red	+	...	+	−/+	−/+	−/+	+
2430	blk	d/+	d	+	−/+	−/+	+	+
2431	dk-red	r/+	i	r/+	−	−	+	+
2432	dk-orange	r/+	...	r/+	−	−	+	+
2433	brn	i	...	+	+	+	−	−
2434	red-orange	r	...	−	−	−	+	+
2435	blk-bl, 6.97	−	i	−	−	−	−	−
2436	blk	i	i	+	+	+	i	i
2437	orange-yel, 3.29	+	+	+	+	+	+	+
2438	wh	+	...	+	+	+	+	+
2439	dk-brn	i	...	+	+	+	i	i
2440	brn-blk, 6.99	i	...	−	−	−	i	i
2441	yel, 1.96	−	d	−	−/+	−/+	+	−/+
2442	yel, 2.07	−	d	−	−/+	−/+	+	−/+
2443	dk-red lq, 2.629[20]	+	...	+	+	+	+	+
2444	orange lq, 2.685[20]	+	...	+	+	+	+	+
2445	cl lq	+	...	+	+	+	+	+
2446	dk-red lq, 1.621[20]	+	+	+	+	+	+	+
2447	yel-brn lq	+	+	+	+	+	+	+
2448	lt-yel lq, 1.688[20]	+	r	+	+	+	+	+
2449	cl gas	+	+	+	+	+	+	+
2450	cl lq, 1.64[20]	+	+	+	+	+	+	+
2451	cl lq, 1.677[20]	+	r	+	+	+	+	+
2452	dk-red lq, 1.66[0]	+	+	+	+	+	+	+
2453	cl lq, 1.576[0]	+	...	+	+	+	+	+
2454	cl gas, 1.623	+	...	+	+	+	+	+
2455	cl lq, 1.919[−73]	+	+	+	+	+	+	+
2456	cl gas, 6.976	i	i	−	−	−	−	−
2457	cl lq, 1.5[−100]	+	r	+	+	+	+	+
2458	cl lq, 2.08[20]	−/+	...	−	−	−	−/+	−/+

No.	Formula and name	M_r	Phase transition temperature
2459	$(SF)_3(N)_3$ Tris(fluorosulfur) trinitride	195.21	mp 74, bp 93
2460	$(SF)_4(N)_4$ Tetrakis(fluorosulfur) tetranitride	260.28	mp 153 dec
2461	S_2I_2 Disulfur diiodide	317.94	dec 30
2462	S_4N_2 Tetrasulfur dinitride	156.28	mp 23 dec
2463	S_4N_4 Tetrasulfur tetranitride	184.29	mp 178, bp 185
2464	$S(N)F_3$ Sulfur nitride-trifluoride	103.07	bp −72.6, bp −27.1
2465	$S(NH)O$ Sulfur imide-oxide	63.08	mp −85, dec >20
2466	$S(NH_2)_2O_2$ Sulfur diamide-dioxide	96.11	mp 92.5; dec 140
2467	SO_2 Sulfur dioxide	64.06	mp −75.46; bp −10,1
2468	$SO_2·7.67H_2O$	202.24	mp 12, dec 60–80
2469	SO_3 Sulfur trioxide	80.06	mp16.8; bp 44.7
2470	S_2O Disulfur oxide	80.13	mp < −196
2471	S_3O_9 Trisulfur nonaoxide	240.19	soft >17, subl >43
2472	S_nO_{3n} Polysulfur tripolyoxide	–	soft >32
2473	$S(O)F_2$ Sulfur oxide-difluoride	86.06	mp −110.5; bp −43.7
2474	$S(O)F_4$ Sulfur oxide-tetrafluoride	124.06	mp −99.6; bp −48.5
2475	SO_2F_2 Sulfur dioxide-difluoride	102.06	mp −135.81; bp 55.37
2476	$S(S)F_2$ Sulfur sulfide-difluoride	102.13	mp −164.6; bp −10.6
2477	**Sb Antimony**	**121.75**	mp 630.74; bp 1634
2478	$SbBr_3$ Antimony(III) bromide	361.46	mp 96.6; bp 289
2479	$SbCl_3$ Antimony(III) chloride	228.11	mp 72.3; bp 221
2480	$SbCl_5$ Antimony(V) chloride	299.02	mp 2.8; bp 140 dec
2481	$SbCl_3F_2$ Antimony trichloride-difluoride	266.11	mp 55
2482	$[Sb^VCl_6],\{I(I)Cl\}$ Iodochloroiodine(I) hexachlorostibate(V)	623.73	subl 20, mp 70
2483	$Sb(Cl)O$ Antimony chloride-oxide	173.20	>320 → $SbCl_3,Sb_2O_3$
2484	SbF_3 Antimony(III) fluoride	178.74	mp 287, bp 319
2485	SbF_5 Antimony(V) fluoride	216.74	mp 8.3; bp 149.5
2486	$[Sb^VF_6],(ClO_2)$ Dioxochlorine(V) hexafluorostibate(V)	303.19	mp 235
2487	$[SbF_6],H\ (·H_2O)$ Hydrogen hexafluorostibate(V)	254.76	dec hydr 357
2488	$[SbF_6],Na$ Sodium hexafluorostibate(V)	258.73	mp >350
2489	SbH_3 Stibine	124.77	mp −94.2; bp −18.4
2490	SbI_3 Antimony(III) iodide	502.46	mp 170.5; bp 401.6
2491	α-Sb_2O_3 Antimony(III) oxide	291.50	460 → β-Sb_2O_3
2492	β-Sb_2O_3	291.50	mp 655, bp 1456
2493	$Sb_2O_3·nH_2O$	–	30 → Sb_2O_3
2494	Sb_2O_5 Antimony(V) oxide	323.50	>450 → $(Sb^{III}Sb^V)O_4$
2495	$Sb_2O_5·nH_2O$ ($n ≤ 3.5$)	–	1000 → $(Sb^{III}Sb^V)O_4$
2496	$[Sb(OH)_6],K$ Potassium hexahydroxostibate(V)	262.89	dec ca. 680
2497	$[Sb(OH)_6],Na$ Sodium hexahydroxostibate(V)	246.78	dec 600
2498	Sb_2S_3 Antimony(III) sulfide	339.70	mp 550.5; bp 1160
2499	$[SbS_3],Ag_3$ Silver(I) trithiostibate(III)	541.55	mp 486
2500	$[SbS_4],Ag_5$ Silver(I) tetrathiostibate(III)	789.35	dec 600
2501	$[SbS_3],Na_3\ (·9H_2O)$ Sodium trithiostibate(III)	286.92	−H_2O 150, mp 600
2502	$[SbS_4],Na_3\ (·9H_2O)$ Sodium tetrathiostibate(V)	318.98	mp hydr 87, dec hydr ca. 230

(Continued)

No.	Color and density	Solubility and reactivity						
		water	alcohol	HCl	H_2SO_4	HNO_3	NaOH	$NH_3 \cdot H_2O$
2459	wh	-/+	...	-	-/+	-/+	+	...
2460	wh	-/+	...	-	-/+	-/+	+	...
2461	brn	+	...	+	+	+	+	+
2462	red lq, 1.901[13]	-/+	d	-/+	-/+	-/+	...	...
2463	orange, 2.22	+	d	-/+	+	+	+	...
2464	cl gas	-/+	...	-	-	-	+	+
2465	cl lq	-/+	...	...	...	...	+	+
2466	wh	-/+	d/r	-	-	-	+	+
2467	cl gas, 2.9269	r	r	-	-	-/+	+	+
2468	wh	r	r	-	-	-/+	+	+
2469	wh, 1.97	+	...	+	+	+	+	+
2470	cl gas	+	r	+	+	+	+	+
2471	wh	+	...	+	+	+	+	+
2472	wh	+	...	+	+	+	+	+
2473	cl gas, 3.84	+	+	+	+	+	+	+
2474	cl gas	+	+	+	+	+	+	+
2475	cl gas, 3.992	d/+	r	-	-	-	+	+
2476	cl gas	+	r	+	+	+	+	+
2477	sk-bl-wh, 6.684	-	-	-	-/+	+	-	-
2478	wh, 4.15	+	r	+	+	+	+	+
2479	wh, 3.14	+	r	+	+	+	+	+
2480	cl lq, 2.346[20]	+	r	+	+	+	+	+
2481	wh	+	r	+	+	+	+	+
2482	dk-red	r/+	r	+	+	+	+	+
2483	wh	i/+	i	+	+	+	-	-
2484	wh, 4.38	r/+	r	r	r	r	+	+
2485	cl lq, 2.993[22]	+	...	+	+	+	+	+
2486	wh	r/+	r/+	+	+	+	+	+
2487	wh hydr	+	...	-	-	-	+	+
2488	wh, 3.38	r	r	r	r	r	-/+	-
2489	cl gas, 4.599	d	r	-/+	-/+	-/+	+	+
2490	red, 4.77	+	d	+	+	+	+	+
2491	wh, 5.19	-	...	+	+	+	+	+
2492	wh, 5.76	-	...	+	-/+	-/+	+	+
2493	wh	i	i	+	+	+	+	+
2494	yel, 3.78	-	i	-/+	...	-	+	+
2495	yel	i	i	-/+	...	i	+	+
2496	wh	d	r	+	+	+	-	-
2497	wh	i	r	+	+	+	i	i
2498	grey-orange, 4.63	i	...	-/+	...	+	+	+
2499	red, 5.85	i	...	-	-/+	+	-	-/+
2500	brn-blk, 6.25	i	...	-/+	-/+	-/+	-/+	+
2501	lt-yel	r	...	+	+	+	+	...
2502	yel-grn hydr, 1.84	r	i	+	+	+	+	...

No.	Formula and name	M_r	Phase transition temperature
2503	$Sb_2(SO_4)_3$ ($\cdot H_2O$) Antimony(III) sulfate	531.69	$-H_2O$ 20–30
2504	$(Sb^{III}Sb^V)O_4$ Antimony(III)-antimony(V) oxide	307.50	subl 1050
2505	Sb_2Se_3 Antimony(III) selenide	480.38	mp 617, bp 1031
2506	Sb_2Te_3 Antimony(III) telluride	626.30	mp 621.6; bp 1173
2507	**Sc Scandium**	44.956	mp 1541, bp ca. 2850
2508	ScB_2 Scandium diboride	66.58	mp 2250
2509	$ScBr_3$ Scandium(III) bromide	284.67	subl 929
2510	ScC Scandium monocarbide	56.97	mp 1800
2511	$Sc_2(C_2O_4)_3$ ($\cdot 5H_2O$) Scandium(III) oxalate	353.97	hydr 500 $\rightarrow$ Sc_2O_3
2512	$ScCl_3$ Scandium(III) chloride	151.32	mp 967 (p), bp 975
2513	$ScCl_3 \cdot 6H_2O$	259.41	dec >250
2514	ScF_3 Scandium(III) fluoride	101.95	mp 1552, bp 1607
2515	ScI_3 Scandium(III) iodide	425.67	subl 912
2516	ScN Scandium mononitride	58.96	mp 2650
2517	$Sc(NO_3)_3$ ($\cdot 4H_2O$) Scandium(III) nitrate	230.97	hydr 220 $\rightarrow$ Sc_2O_3
2518	Sc_2O_3 Scandium(III) oxide	137.91	mp ca. 2450
2519	$Sc(OH)_3$ Scandium(III) hydroxide	95.98	280 $\rightarrow$ ScO(OH)
2520	$[Sc(OH)_4]$,Na Sodium tetrahydroxoscandate(III)	135.97	dec >300
2521	$[Sc(OH)_6]$,Na$_3$ Sodium hexahydroxoscandate(III)	215.97	120 $\rightarrow$ $[Sc(OH)_4]$,Na
2522	$ScO(OH)$ Scandium metahydroxide	77.96	460 $\rightarrow$ Sc_2O_3
2523	$ScPO_4$ ($\cdot 2H_2O$) Scandium(III) orthophosphate	139.93	$-H_2O$ 200, mp 1780
2524	$Sc_2(SO_4)_3$ ($\cdot 5H_2O$) Scandium(III) sulfate	378.10	$-H_2O$ 400, dec >600
2525	**Se (amor) Selenium**, amorphous	78.96	>72 $\rightarrow$ Se_8
2526	**Se Selenium**, grey	78.96	mp 217, bp 685.3
2527	**Se$_8$ Selenium**, red	631.68	>130 $\rightarrow$ Se
2528	$SeBr_4$ Selenium tetrabromide	398.58	dec 75
2529	Se_2Br_2 Diselenium dibromide	317.73	dec >100
2530	$Se(Br)Cl_3$ Selenium bromide-trichloride	265.22	mp 190
2531	$SeBr_2O$ Selenium dibromide-oxide	254.77	mp 41.6; bp 217 dec
2532	SeC_2 see CSe_2 (No. 447)		
2533	$SeCl_4$ Selenium tetrachloride	220.77	subl ca. 196, mp 305 (p)
2534	Se_2Cl_2 Diselenium dichloride	228.83	mp -85, bp 127
2535	$SeCl_2O$ Selenium dichloride-oxide	165.87	mp 10.8; bp 177.6 dec
2536	SeF_4 Selenium tetrafluoride	154.95	mp -9.5; bp 107.7
2537	SeF_6 Selenium hexafluoride	192.95	subl -46.6; bp -34.8
2538	SeO_2 Selenium dioxide	110.96	subl 315, mp 340 (p)
2539	SeO_3 Selenium trioxide	126.96	mp 118.5; dec >185
2540	$Se(O)F_2$ Selenium oxide-difluoride	132.96	mp 15.5; bp 125
2541	Se_4S_4 Tetraselenium tetrasulfide	444.10	mp 113 dec
2542	**Si Silicon**	28.086	mp 1415, bp ca. 3250
2543	$SiBr_4$ Silicon tetrabromide	347.70	mp 5.4; bp 152.6
2544	SiC Silicon monocarbide	40.10	mp 2830 dec
2545	$SiCl_4$ Silicon tetrachloride	169.90	mp -68.8; bp 57.6
2546	SiF_4 Silicon tetrafluoride	104.08	subl -95.7, mp -90.3 (p)
2547	$[SiF_6]$,Ca ($\cdot 2H_2O$) Calcium hexafluorosilicate(IV)	182.15	dec hydr >100
2548	$[SiF_6]$,Co ($\cdot 6H_2O$) Cobalt(II) hexafluorosilicate(IV)	201.01	dec hydr >120
2549	$[SiF_6]$,Cs$_2$ Cesium hexafluorosilicate(IV)	407.88	dec 245–800

(Continued)

No.	Color and density	Solubility and reactivity						
		water	alcohol	HCl	H_2SO_4	HNO_3	NaOH	$NH_3 \cdot H_2O$
2503	wh, 3.63	i/+	...	+	+	+	+	...
2504	wh, 5.82	−	...	−/+	−/+	−	−	−
2505	grey	i	i	−/+	−/+	−/+	i/+	i
2506	grey, 6.50	i	...	+	+	+	i/+	i
2507	wh-yel, 3.02	+	−	+	+	+	+	+
2508	grey	−	i	−	−/+	−/+	−/+	−
2509	wh, 3.91	r	r	r	−/+	−	+	+
2510	blk	−	i	−	−/+	−	−	−
2511	wh hydr	i	...	−/+	−/+	i/+	+	+
2512	wh, 2.39	r	r	r/d	−/+	−	+	+
2513	wh	r	r	r	−	−	+	+
2514	wh	i	i	i	−	i	i	i
2515	wh	r	r	−	−/+	−/+	+	+
2516	dk-sk-bl, 4.2	−	i	−/+	−/+	−/+	−	−
2517	wh hydr	r	r	−	−	r	+	+
2518	wh, 3.86	−	i	−/+	−/+	−/+	−	−
2519	wh, 2.65	i	i	+	+	+	+	i
2520	wh	+	...	+	+	+	+	+
2521	wh	+	...	+	+	+	r	−
2522	wh	i	...	+	+	+	i/+	i
2523	wh	i	i	i/r	i	i	i	i
2524	wh, 2.58(2.52)	r	r	−	r	−	+	+
2525	red, 4.28	−	−	−	−/+	+	−/+	−
2526	grey, 4.79	−	−	−	−/+	+	−/+	−
2527	red, 4.46	−	−	−	−	−/+	−/+	−
2528	yel-orange	+	...	+	+	+	+	+
2529	dk-red lq, 3.604[15]	+	+	+	+	+	+	+
2530	yel-brn	+	...	+	+	+	+	+
2531	orange	+	...	+	+	+	+	+
2532	see No. 447							
2533	lt-yel, 3.80	+	...	+	+	+	+	+
2534	dk-red lq, 2.77[25]	+	+	+	+	+	+	+
2535	lt-yel lq, 2.43[20]	+	...	+	+	+	+	+
2536	cl lq, 2.72[25]	+	r	+	+	+	+	+
2537	cl gas, 2.917	+	+	+	+	+	+	+
2538	wh, 3.95	+	r	+	+	+	+	+
2539	wh, 2.75	+	+	+	+	+	+	+
2540	cl lq, 2.67[20]	+	r	+	+	+	+	+
2541	orange-yel, 3.06	−	...	−/+	−/+	−/+	−	−
2542	dk-grey, 2.33	−	−	−	−	−	−/+	...
2543	cl lq, 2.789[20]	+	...	+	+	+	+	+
2544	wh, 3.22	−	−	−	−	−	−	−
2545	cl lq, 1.483[20]	+	i	+	+	+	+	+
2546	cl gas, 4.684	+	r	+	+	+	+	+
2547	wh, 2.66	d	r	r/+	−/+	−	−/+	−/+
2548	lt-red hydr, 2.11	r	...	−	−	−	+	+
2549	wh, 3.37	d	i	−	−	−	−	−

No.	Formula and name	M_r	Phase transition temperature
2550	$[SiF_6]$,Cu ($\cdot 6H_2O$) Copper(II) hexafluorosilicate(IV)	205.62	dec hydr >150
2551	$[SiF_6]$,Fe ($\cdot 6H_2O$) Iron(II) hexafluorosilicate(IV)	197.92	dec hydr t
2552	$[SiF_6]$,H$_2$ ($\cdot 2H_2O$) Hydrogen hexafluorosilicate(IV)	144.09	mp hydr 19 dec
2553	$[SiF_6]$,K$_2$ Potassium hexafluorosilicate(IV)	220.27	dec 700, mp 873 (p)
2554	$[SiF_6]$,Li$_2$ ($\cdot 2H_2O$) Lithium hexafluorosilicate(IV)	155.96	$-H_2O$ 100
2555	$[SiF_6]$,Mg ($\cdot 6H_2O$) Magnesium hexafluorosilicate(IV)	166.38	dec hydr 120
2556	$[SiF_6]$,Mn ($\cdot 6H_2O$) Manganese(II) hexafluorosilicate(IV)	197.01	dec hydr >160
2557	$[SiF_6]$,(NH$_4$)$_2$ Ammonium hexafluorosilicate(IV)	178.15	dec 319
2558	$[SiF_6]$,Na$_2$ Sodium hexafluorosilicate(IV)	188.05	dec ca. 570
2559	$[SiF_6]$,Ni ($\cdot 6H_2O$) Nickel(II) hexafluorosilicate(IV)	200.76	dec hydr >185
2560	$[SiF_6]$,Rb$_2$ Rubidium hexafluorosilicate(IV)	313.01	dec 750
2561	$[SiF_6]$,Sr ($\cdot 2H_2O$) Strontium hexafluorosilicate(IV)	229.69	dec >110
2562	$[SiF_6]$,Tl$_2$ Thallium(I) hexafluorosilicate(IV)	550.84	dec hydr 165
2563	SiH_4 Monosilane	32.12	mp -185, bp -111.9
2564	Si_2H_6 Disilane	62.22	mp -132, bp -14.5
2565	Si_3H_8 Trisilane	92.32	mp -117.4; bp 52.9
2566	Si_4H_{10} Tetrasilane	122.42	mp -84.3; bp 107
2567	SiI_4 Silicon tetraiodide	535.70	mp 122, bp 290
2568	Si_3N_4 Trisilicon tetranitride	140.29	subl 1900
2569	$Si(NCO)_4$ Silicon tetracyanate-N	196.15	mp 26, bp 186
2570	SiO Silicon monooxide	44.09	dec 400–700
2571	SiO_2(amor) ($\cdot nH_2O$) Silicon dioxide	60.08	$-H_2O$ t, soft ca. 1300
2572	SiO_2 (α-quartz)	60.08	573 $\rightarrow$ β-quartz
2573	SiO_2 (β-quartz)	60.08	mp 1550, bp 2950
2574	SiO_2 (keatite)	60.08	...
2575	SiO_2 (coesite)	60.08	...
2576	SiO_2 (α-cristobalite)	60.08	470 $\rightarrow$ β-cristobalite
2577	SiO_2 (β-cristobalite)	60.08	mp 1720, bp 2950
2578	SiO_2 (melanophlogite)	60.08	>800 $\rightarrow$ β-cristobalite
2579	SiO_2 (stishovite)	60.08	...
2580	SiO_2 (α-tridymite)	60.08	163 $\rightarrow$ β-tridymite
2581	SiO_2 (β-tridymite)	60.08	1470 $\rightarrow$ β-cristobalite
2582	$Si(OCN)_4$ Silicon tetracyanate	196.15	mp 35, bp 245
2583	SiS_2 Silicon disulfide	92.22	mp 1090, bp 1130
2584	$[SiW_{12}O_{40}]$,H$_4$ ($\cdot 7H_2O$) Hydrogen 40-oxododeca-wolframosilicate(IV)	2878.28	dec hydr >100
2585	**Sm Samarium**	150.36	mp 1072, bp ca. 1800
2586	SmBr$_2$ Samarium(II) bromide	310.17	mp 669, bp 1880
2587	SmBr$_3$ ($\cdot 6H_2O$) Samarium(III) bromide	390.07	mp 640
2588	$Sm(BrO_3)_3$ ($\cdot 9H_2O$) Samarium(III) bromate	534.06	mp hydr 74.5; $-H_2O$ 150
2589	SmC$_2$ Samarium dicarbide	174.38	mp >2300
2590	$Sm_2(C_2O_4)_3$ ($\cdot 10H_2O$) Samarium(III) oxalate	564.77	hydr >600 $\rightarrow$ Sm_2O_3
2591	SmCl$_2$ Samarium(II) chloride	221.27	mp 859, bp 1950
2592	SmCl$_3$ ($\cdot 6H_2O$) Samarium(III) chloride	256.72	$-H_2O$ >110, mp 678
2593	SmF$_3$ Samarium(III) fluoride	207.35	mp 1305, bp 2330
2594	SmI$_2$ Samarium(II) iodide	404.17	mp 520, bp 1660

(Continued)

No.	Color and density	Solubility and reactivity						
		water	alcohol	HCl	H_2SO_4	HNO_3	NaOH	$NH_3 \cdot H_2O$
2550	sk-bl hydr, 2.21	r	...	−	−	−	+	+
2551	wh hydr, 1.96	r	...	−	−/+	−/+	+	+
2552	wh hydr	r	...	−	−	−	+	+
2553	wh, 2.67	d	i	−	−	−	−	−
2554	wh hydr, 2.33	r	r	−	−	−	−	−
2555	wh hydr, 1.79	r	i	−	−	−	+	+
2556	pink hydr, 1.90	r	r	−	−	−	+	+
2557	wh, 2.15	r	d	−	−	−	−	−
2558	wh, 2.68	d/r	i	−	−	−	−	−
2559	grn hydr, 2.13	r	...	−	−	r	+	+
2560	wh, 3.33	d	i	r	−/+	−/+	−	−
2561	wh hydr, 2.99	r	...	−/+	+.	−	−	−
2562	wh, 5.72	r	...	−	−/+	r	−/+	−
2563	cl gas, 1.44	i/+	r	−	−/+	+	+	+
2564	cl gas, 2.85	i/+	r	−	−/+	+	+	+
2565	cl lq, 0.739[20]	i/+	r	−	−/+	+	+	+
2566	cl lq, 0.79[0]	i/+	r	−	−/+	+	+	+
2567	wh	+	...	+	+	+	+	+
2568	wh, 3.44	+	i	+	+	+	+	+
2569	wh, 1.413	+	r	+	+	+	+	+
2570	brn-blk, 2.13	+	...	+	+	+	+	+
2571	wh, 2.203(2.175)	−	−	−	−	−	+	−
2572	wh, 2.648	−	−	−	−	−	−/+	−
2573	wh, 2.533	−	−	−	−	−	−/+	−
2574	wh, 2.503	−	−	−	−	−	−/+	−
2575	wh, 2.911	−	−	−	−	−	−/+	−
2576	wh, 2.334	−	−	−	−	−	−/+	−
2577	wh, 2.194	−	−	−	−	−	−/+	−
2578	wh, 2.05	−	−	−	−	−	−/+	−
2579	wh, 4.287	−	−	−	−	−	−/+	−
2580	wh, 2.265	−	−	−	−	−	−/+	−
2581	wh, 2.192	−	−	−	−	−	−/+	−
2582	wh, 1.41	+	r	+	+	+	+	+
2583	wh, 2.02	+	...	+	+	+	+	+
2584	wh	r	...	−/+	−/+	−/+	+	+
2585	wh, 7.47	+	−	+	+	+	−	−
2586	dk-brn, 5.1	r/+	...	r	−/+	−	+	+
2587	yel hydr, 2.97	r	r	r	−/+	−	+	+
2588	yel hydr	r	d	−	−	−	+	+
2589	yel, 5.86	+	...	+	+	+	+	+
2590	yel hydr	i	...	−/+	−/+	i/+	i	i
2591	red-brn, 4.56	r/+	i	r	−	−	+	+
2592	yel, 4.46(2.38)	r	r	r	−	r	+	+
2593	yel	i	i	i	−	i	−	i
2594	blk-grn	+	...	+	+	+	+	+

No.	Formula and name	M_r	Phase transition temperature
2595	SnI$_3$ Samarium(III) iodide	531.07	mp 816–824 dec
2596	Sm$_2$(MoO$_4$)$_3$ Samarium(III) molybdate	780.53	mp 1130
2597	Sm(NO$_3$)$_3$ (·6H$_2$O) Samarium(III) nitrate	336.37	mp hydr 79, dec hydr t
2598	Sm$_2$O$_3$ Samarium(III) oxide	348.72	mp 2270
2599	Sm(OH)$_3$ Samarium(III) hydroxide	201.38	$t \to$ Sm$_2$O$_3$
2600	SmS Samarium(II) sulfide	182.43	mp 1940
2601	Sm$_2$S$_3$ Samarium(III) sulfide	396.92	mp 1780
2602	Sm$_2$(S)O$_2$ Disamarium sulfide-dioxide	214.42	mp ca. 1980
2603	Sm$_2$(SO$_4$)$_3$ (·8H$_2$O) Samarium(III) sulfate	588.91	−H$_2$O 450, dec >900
2604	**α-Sn Tin**, grey	118.710	>13.2 → β-Sn
2605	**β-Sn Tin**, white	118.710	mp 231.9681; bp 2620
2606	SnBr$_2$ Tin(II) bromide	278.52	mp 232, bp 638
2607	SnBr$_4$ Tin(IV) bromide	438.33	mp 31, bp 202
2608	Sn(Br)Cl$_3$ Tin bromide-trichloride	304.97	mp −1; bp 50
2609	SnBr$_3$Cl Tin tribromide-chloride	393.88	mp 1, bp 73
2610	Sn(CH$_3$COO)$_2$ Tin(II) acetate	236.80	mp 182, bp 240
2611	Sn(CH$_3$COO)$_4$ Tin(IV) acetate	354.89	mp 253
2612	SnC$_2$O$_4$ Tin(II) oxalate	206.73	soft 280 dec
2613	SnCl$_2$ Tin(II) chloride	189.62	mp 247, bp 652
2614	SnCl$_4$ Tin(IV) chloride	260.52	mp −33, bp 114.1
2615	SnCl$_4$·5H$_2$O	350.60	mp hydr 56
2616	[SnCl$_6$],Cs$_2$ Cesium hexachlorostannate(IV)	597.24	dec t
2617	Sn(Cl)F Tin chloride-fluoride	173.16	mp 185–190
2618	[SnCl$_6$],H$_2$ (·6H$_2$O) Hydrogen hexachlorostan-nate(IV)	333.44	mp hydr 19.2; dec 30
2619	[SnCl$_6$],K$_2$ Potassium hexachlorostannate(IV)	409.62	dec t
2620	[SnCl$_6$],(NH$_4$)$_2$ Ammonium hexachlorostannate(IV)	367.51	dec >200
2621	SnF$_2$ Tin(II) fluoride	156.71	mp 215.05; bp 853
2622	SnF$_4$ Tin(IV) fluoride	194.70	subl 705
2623	[SnF$_6$],K$_2$ (·H$_2$O) Potassium hexafluorostan-nate(IV)	310.89	dec hydr >115
2624	SnH$_4$ Stannane	122.74	mp −146, bp −52
2625	[Sn(H$_2$O)Cl$_2$] (·H$_2$O) Aquadichlorotin	207.63	mp hydr 37.7; 80 → [Sn(H$_2$O)Cl$_2$]
2626	SnI$_2$ Tin(II) iodide	372.52	mp 320, bp 718
2627	SnI$_4$ Tin(IV) iodide	626.33	mp 144.5; bp 346
2628	SnI$_2$Br$_2$ Tin diiodide-dibromide	532.33	mp 50, bp 225
2629	SnI$_2$Cl$_2$ Tin diiodide-dichloride	443.42	bp 297
2630	Sn(NO$_3$)$_2$ (·20H$_2$O) Tin(II) nitrate	242.72	mp hydr 20, dec >150
2631	Sn(NO$_3$)$_4$ Tin(IV) nitrate	366.73	mp 91; 600 → SnO$_2$
2632	SnO Tin(II) oxide	134.71	mp 1040, bp 1425
2633	SnO$_2$ Tin(IV) oxide	150.71	mp 1630, bp 2500
2634	α-SnO$_2$·nH$_2$O (1 < n ≤ 2)	–	350 → SnO$_2$
2635	β-SnO$_2$·nH$_2$O (n ≤ 1)	–	600 → SnO$_2$
2636	Sn(OH)$_2$ Tin(II) hydroxide	152.72	120 → SnO
2637	[Sn(OH)$_6$],K$_2$ Potassium hexahydroxostannate(IV)	298.95	dec 185
2638	[Sn(OH)$_3$],Na Sodium trihydroxostannate(II)	192.72	dec >20

(Continued)

No.	Color and density	Solubility and reactivity						
		water	alcohol	HCl	H_2SO_4	HNO_3	NaOH	$NH_3 \cdot H_2O$
2595	orange-yel	r	r	r	−/+	−/+	+	+
2596	viol, 5.36	i	i	−/+	−/+	−	−/+	−
2597	lt-yel hydr, 2.38	r	...	−	−	r	+	+
2598	lt-yel, 8.35	−/+	...	+	+	+	−	−
2599	lt-yel	i	...	+	+	+	i	i
2600	brn	i	...	+	+	+	i	−
2601	yel-pink, 5.73	i/+	...	+	+	+	i	i
2602	dk-yel	−	...	+	+	+	−	−
2603	lt-yel hydr, 2.93	r	...	−	r	−	+	+
2604	grey, 5.75	−	−	+	+	+	−/+	−
2605	wh, 7.31	−	−	+	+	+	−/+	−
2606	yel, 4.92	+	r	+	+	+	+	+
2607	wh, 3.34	+	r	+	+	+	+	+
2608	cl lq, 2.51[20]	+	...	+	+	+	+	+
2609	cl lq	+	...	+	+	+	+	+
2610	lt-yel	+	...	+	+/−	+/−	+	+
2611	wh	+	...	+	+	+	+	+
2612	wh	i	...	+	+	+	−/+	i
2613	wh, 3.95	+	r	+	+	+	+	+
2614	cl lq, 2.2262[20]	+	+	+	+	+	+	+
2615	wh hydr	+	+	+	+	+	+	+
2616	wh, 3.33	d	...	r/d	−	−	+	+
2617	wh	+	...	+	+	+	+	+
2618	wh hydr, 1.93	r	...	−	−	−	+	+
2619	wh, 2.71	r	...	r/d	−	−	−/+	−/+
2620	wh, 2.4	r	...	r/d	−	−	−/+	−/+
2621	wh	r/+	r	−	−	−	+	+
2622	wh, 4.78	r/+	...	−	−	−	+	+
2623	wh, 3.05	r	i	−	−	r	−/+	−
2624	cl gas	i	...	−	−/+	−	−/+	−/+
2625	wh hydr, 2.71	r/+	r	+	+	+	+	+
2626	orange-red, 5.28	d/+	r	−/+	−/+	−/+	+	+
2627	yel, 4.5	+	r	+	+	+	+	+
2628	orange-red, 3.63	r/+	...	+	+	+	+	+
2629	red lq	r/+	...	+	+	+	+	+
2630	wh	+	...	+	+	+	+	+
2631	wh	+	...	+	+	+	+	+
2632	dk-bl, 6.25	−	...	+	+	+	−/+	−
2633	wh, 7.00	−	...	−	−/+	−	−/+	−
2634	wh	i	...	+	+	+	+	i
2635	wh	i	...	i	i	i	i	i
2636	wh	i	...	+	+	+	+	+
2637	wh, 3.20	+	i	+	+	+	+/r	+
2638	wh	+	...	+	+	+	+/r	+/r

No.	Formula and name	M_r	Phase transition temperature
2639	[Sn(OH)$_6$],Na$_2$ Sodium hexahydroxostannate(IV)	266.73	dec 140
2640	SnS Tin(II) sulfide	150.78	mp 880, bp 1230
2641	SnS$_2$ Tin(IV) sulfide	182.84	dec >500
2642	[SnS$_3$],K$_2$ ($\cdot$3H$_2$O) Potassium trithiostannate(IV)	293.10	$-$3H$_2$O 100
2643	SnSO$_4$ ($\cdot$2H$_2$O) Tin(II) sulfate	214.77	dec >360
2644	Sn(SO$_4$)$_2$ ($\cdot$2H$_2$O) Tin(IV) sulfate	310.83	dec hydr 50
2645	SnSe Tin selenide	197.67	mp 861
2646	SnSe$_2$ Tin(IV) selenide	276.63	mp 650
2647	SnTe Tin(II) telluride	246.31	mp 780
2648	**Sr Strontium**	87.62	mp 768, bp 1390
2649	SrB$_6$ Strontium hexaboride	152.49	mp 2235
2650	SrBr$_2$ Strontium bromide	247.43	mp 657, bp 1972
2651	SrBr$_2$$\cdot$6H$_2$O	355.52	$-$H$_2$O 345
2652	Sr(BrO$_3$)$_2$ ($\cdot$H$_2$O) Strontium bromate	343.42	$-$H$_2$O 120, dec 240
2653	SrC$_2$ Strontium acetylide	111.64	mp ca. 1700
2654	Sr(CH$_3$COO)$_2$ ($\cdot$0.5H$_2$O) Strontium acetate	205.71	dec 350–400
2655	SrCO$_3$ Strontium carbonate	147.63	mp 1497 (p), dec 1200
2656	SrC$_2$O$_4$ ($\cdot$H$_2$O) Strontium oxalate	175.64	$-$H$_2$O 150, 450 $\rightarrow$ SrCO$_3$
2657	SrCl$_2$ Strontium chloride	158.53	mp 873, bp 2040
2658	SrCl$_2$$\cdot$6H$_2$O	266.62	mp 115, $-$H$_2$O 250
2659	Sr(Cl)F Strontium chloride-fluoride	142.07	mp 962
2660	Sr(ClO$_3$)$_2$ ($\cdot$8H$_2$O) Strontium chlorate	254.52	dec hydr 120
2661	Sr(ClO$_4$)$_2$ ($\cdot n$H$_2$O) Strontium perchlorate	286.52	$-$H$_2$O 240, >415 $\rightarrow$ SrCl$_2$
2662	SrCrO$_4$ Strontium chromate	203.61	mp 1283 dec
2663	Sr$_2$Cr$_2$O$_7$ Strontium dichromate	303.61	dec hydr 110
2664	SrF$_2$ Strontium fluoride	125.62	mp 1473, bp ca. 2460
2665	SrH$_2$ Strontium hydride	89.64	mp 1050 dec
2666	SrHPO$_4$ Strontium hydroorthophosphate	183.60	dec ca. 900
2667	Sr(HS)$_2$ Strontium hydrosulfide	153.77	dec 150–200
2668	(SrHf)O$_3$ Strontium-hafnium trioxide	314.11	mp 2830
2669	SrI$_2$ Strontium iodide	341.43	mp 538, bp ca. 1900
2670	SrI$_2$$\cdot$6H$_2$O	449.52	dec 90
2671	Sr(IO$_3$)$_2$ ($\cdot$H$_2$O) Strontium iodate	437.42	dec >500
2672	Sr(MnO$_4$)$_2$ ($\cdot$3H$_2$O) Strontium permanganate	325.49	dec hydr 175
2673	SrMoO$_4$ Strontium molybdate	247.56	mp 1460
2674	Sr$_3$N$_2$ Tristrontium dinitride	290.87	dec >1000
2675	Sr(NCS)$_2$ ($\cdot$3H$_2$O) Strontium thiocyanate	203.79	$-$H$_2$O 100, dec >160
2676	SrN$_2$O$_2$ ($\cdot$5H$_2$O) Strontium hyponitrite	146.63	$-$H$_2$O 100
2677	Sr(NO$_2$)$_2$ ($\cdot$H$_2$O) Strontium nitrite	179.63	$-$H$_2$O ca. 100, dec 240
2678	Sr(NO$_3$)$_2$ ($\cdot$4H$_2$O) Strontium nitrate	211.63	$-$H$_2$O 100, dec 450
2679	SrO Strontium oxide	103.62	mp 2650, bp ca. 3000
2680	SrO$_2$ ($\cdot$8H$_2$O) Strontium peroxide	119.62	$-$H$_2$O 100, mp 215 dec
2681	Sr(OH)$_2$ Strontium hydroxide	121.63	mp 460, dec >500
2682	Sr(OH)$_2$$\cdot$8H$_2$O	265.75	$-$H$_2$O 100
2683	Sr$_3$(PO$_4$)$_2$ Strontium orthophosphate	452.80	mp 1650
2684	SrS Strontium sulfide	119.69	mp ca. 2000 dec
2685	SrSO$_4$ Strontium sulfate	183.68	mp 1500 (p), dec 1580

(Continued)

No.	Color and density	Solubility and reactivity						
		water	alcohol	HCl	H_2SO_4	HNO_3	NaOH	$NH_3 \cdot H_2O$
2639	wh	+	i	+	+	+	r	r
2640	brn, ca. 5.1	i	...	−/+	−/+	−/+	+	i
2641	yel ($t \to$ brn), 4.5	i	i	−/+	i	−/+	+	+
2642	dk-brn lq hydr, 1.18	r	i	−/+	−/+	−/+	−/+	−
2643	wh	r	...	+	−/r	−	+	+
2644	wh	r/+	...	+	−/+	−	+	+
2645	dk-grey, 6.18	i	...	+	+	+	i	i
2646	brn, 5.13	−	...	−/+	−/+	−/+	−	−
2647	grey, 6.48	i	...	+	+	+	i	i
2648	lt-yel, 2.630	+	+	+	+	+	+	+
2649	dk-grey, 3.39	−	i	+	−	+	−	−
2650	wh, 4.22	r	r	r	+	−	+	+
2651	wh, 2.36	r	r	r	+	−	+	+
2652	wh hydr, 3.77	r	...	−	+	−	+	+
2653	blk, 3.2	+	...	+	+	+	+	+
2654	wh, 2.10	r	i	r	+	r	+	+
2655	wh, 3.70	i	i	+	+	+	i	i
2656	wh	i	...	−/+	−/+	−/+	−	i
2657	wh, 3.05	r	d	r	+	r	+	+
2658	wh, 1.93	r	r	r	+	r	+	+
2659	wh, 4.18	i	d	i	−	−	i	i
2660	wh, 3.15	r	d	−/+	+	−	+	+
2661	wh hydr	r	r	−	+	r	+	+
2662	yel, 3.90	d	d	+	+	+	d	d
2663	red	r	...	−	−	−	+	+
2664	wh, 4.24	i/+	d	i	−	−	i	i
2665	wh, 3.27	+	+	+	+	+	+	+
2666	wh, 3.54	i	...	−/+	−/+	−/+	−/+	−/+
2667	wh	r/+	i	+	+	+	+	+
2668	wh	−	...	−	−/+	−	−	−
2669	wh, 4.55	r	r	r	−/+	−/+	+	+
2670	wh, 4.42	r	r	−	−/+	−/+	+	+
2671	wh, 5.05	i	...	−/+	+	−	−	−
2672	viol hydr, 2.75	r	r	−/+	+	r	−	−/+
2673	lt-grey, 4.54	i	...	+	+	+	−	−
2674	grey	+	...	+	+	+	+	+
2675	wh	r	r	−	+	−/+	+	+
2676	wh hydr, 2.17	d	...	−/+	+	d	−/+	−
2677	wh, 2.87(2.41)	r	d	−/+	+	r/+	+	+
2678	wh, 2.99(2.25)	r	d	r	+	r	+	+
2679	wh, 5.02	+	d	+	+	+	+	+
2680	wh, 4.56(1.95)	−/+	r	+	+	+	+	+
2681	wh, 3.63(1.9)	d	...	+	+	+	d	d
2682	wh	d	...	+	+	+	d	d
2683	wh	i	...	r	r	r	i	i
2684	wh, 3.65	r/+	r	+	+	+	+	+
2685	wh, 3.96	i	i	−	i	−	i	i

No.	Formula and name	M_r	Phase transition temperature
2686	SrS_2O_6 ($\cdot 4H_2O$) Strontium dithionate	247.75	$-H_2O$ 78, dec t
2687	SrS_4O_6 ($\cdot 6H_2O$) Strontium tetrathionate	311.88	$-H_2O$ >50
2688	$Sr(SO_3S)$ ($\cdot 5H_2O$) Strontium thiosulfate	199.75	$-H_2O$ 189
2689	SrSe Strontium selenide	166.58	mp 1600
2690	$SrSeO_4$ Strontium selenate	230.58	dec >1650
2691	$SrSiO_3$ ($\cdot 2H_2O$) Strontium metasilicate	163.70	$-H_2O$ 850, mp 1580
2692	Sr_2SiO_4 Strontium orthosilicate	267.32	mp 2325
2693	SrTe Strontium telluride	215.22	mp 1490
2694	$(SrTi)O_3$ Strontium-titanium trioxide	183.50	mp 2080
2695	$SrWO_4$ Strontium wolframate	335.47	mp 1535 dec
2696	$(SrZr)O_3$ Strontium-zirconium trioxide	226.84	mp 2750
2697	**T_2 Ditritium**	6.032	mp -252.52; bp -248.12
2698	T_2O Tritium oxide	22.03	mp 4.5
2699	**Ta Tantalum**	180.948	mp 3010, bp 5425
2700	TaB_2 Tantalum diboride	202.57	mp 3200
2701	$TaBr_4$ Tantalum(IV) bromide	500.56	subl >310
2702	$TaBr_5$ Tantalum(V) bromide	580.47	mp 265.8; bp 348.8
2703	Ta_6Br_{15} Hexatantalum 15-bromide	2284.25	subl >450
2704	TaC Tantalum monocarbide	192.96	mp 3780 dec
2705	Ta_2C Ditantalum carbide	373.91	mp 3500
2706	$[Ta(C_5H_5)Cl_4]$ Cyclopentadienyltetrachloro-tantalum	387.86	subl 230
2707	$TaCl_3$ Tantalum(III) chloride	287.31	dec >440
2708	$TaCl_4$ Tantalum(IV) chloride	322.76	subl 300 dec
2709	$TaCl_5$ Tantalum(V) chloride	358.21	mp 216.5; bp 239
2710	Ta_6Cl_{15} Hexatantalum 15-chloride	1617.48	subl >470
2711	$TaCl_3O$ Tantalum trichloride-oxide	303.31	soft 327 dec
2712	TaF_5 Tantlum(V) fluoride	275.94	mp 96.8; bp 229.5
2713	$[TaF_7],K_2$ Potassium heptafluorotantalate(V)	392.13	mp 775
2714	$[TaF_8],K_3$ Potassium octafluorotantalate(V)	450.23	mp 780
2715	$(Ta_2Fe)O_6$ Ditantalum-iron hexaoxide	513.74	dec >1900
2716	TaH_{1-x} ($0.12 \le x \le 0.4$) Tantalum monohydride	181.96	>800 $\rightarrow$ Ta
2717	TaI_4 Tantalum(IV) iodide	688.56	mp 398; $t \rightarrow Ta_6I_{14}$
2718	TaI_5 Tantalum(V) iodide	815.47	mp 380, bp 543
2719	Ta_6I_{14} Hexatantalum 14-iodide	2862.34	subl >535
2720	$(Ta_2Mn)O_6$ Ditantalum-manganese hexaoxide	512.83	dec >2000
2721	TaN Tantalum mononitride	194.96	mp ca. 3090 dec
2722	Ta_3N_5 Tritantalum pentanitride	612.88	$\ge 1000 \rightarrow$ TaN
2723	Ta_2O_5 Tantalum(V) oxide	441.89	mp 1890
2724	$Ta_2O_5 \cdot nH_2O$	–	$-H_2O$ >600
2725	$Ta(OH)_3$ Tantalum(III) hydroxide	231.97	dec t
2726	$TaSi_2$ Tantalum disilicide	237.12	mp ca. 2200
2727	**Tb Terbium**	158.925	mp 1356, bp 3073
2728	$Tb_2(C_2O_4)_3$ ($\cdot 10H_2O$) Terbiuum(III) oxalate	581.90	hydr >600 $\rightarrow Tb_2O_3$
2729	$TbCl_3$ ($\cdot 6H_2O$) Terbium(III) chloride	265.28	mp hydr 153, mp 588
2730	TbF_3 Terbium(III) fluoride	215.92	mp 1177, bp 2880
2731	TbF_4 Terbium(IV) fluoride	234.92	dec 180

(Continued)

No.	Color and density	Solubility and reactivity						
		water	alcohol	HCl	H_2SO_4	HNO_3	NaOH	$NH_3 \cdot H_2O$
2686	wh hydr, 2.37	r	i	+	+	+	+	+
2687	wh hydr, 2.15	r	...	r	−/+	−	−	−
2688	wh, 2.92(2.17)	d	i	−/+	−/+	+	−/+	−
2689	wh, 5.54	+	...	+	+	+	+	+
2690	wh, 4.23	d	...	−/+	−/+	−	−	−
2691	wh, 3.65	i	i	+	+	+	−	−
2692	wh, 3.84	i/+	i	+	+	+	−/+	−
2693	wh, 4.83	i/+	...	+	+	+	i	i
2694	wh	−	−	−/+	−/+	−/+	−	−
2695	wh, 6.19	i	i	+	+	+	−	−
2696	wh	−	−	+	+	+	−	−
2697	cl gas	i	i	−	−	−	−	−
2698	cl lq	∞	∞	r	r	r	r	r
2699	grey, 16.60	−	−	−	−	−	−	−
2700	wh, 12.38	−	i	−	−	−	−/+	−
2701	dk-grn, 5.77	+	...	+	+	+	+	+
2702	orange-yel, 4.67	+	r	+	+	+	+	+
2703	blk-grn, 6.29	+	...	+	+	+	+	+
2704	yel-grey, 14.5	−	i	−	−/+	−	−	−
2705	dk-grey, 15.0	−	i	−	−/+	−	−	−
2706	yel	+	i	+	+	+	+	+
2707	grn	r	...	r	−	−	+	+
2708	blk-brn, 4.35	+	...	+	+	+	+	+
2709	wh, 3.68	+	r	+	+	+	+	+
2710	blk-grn, 5.10	−	...	r	−/+	−/+	+	+
2711	wh	+	...	+	+	+	+	+
2712	wh, 4.74	+	d	+	+	+	+	+
2713	wh	d	...	+	+	+	+	+
2714	wh	+	...	+	+	+	+	+
2715	lt-brn, 7.33	−	...	−	−/+	−	−/+	−
2716	grey	−	i	+	+	+	+	+
2717	blk	+	...	+	+	+	+	+
2718	blk-brn, 5.80	+	...	+	+	+	+	+
2719	grn, 6.85	+	...	+	+	+	+	+
2720	blk, 7.03	−	...	−	−/+	−	−/+	−
2721	grey-sk-bl, 16.30	−	i	−	−/+	−/+	−	−
2722	red, 9.85	−	i	−	−/+	−/+	−	−
2723	wh, 8.24	−	i	−	−	−	−	−
2724	wh	i	i	+	+	+	i/+	i
2725	grn	i/+	...	+	+	+	i	i
2726	grey	−	i	−	−	−	−	−
2727	wh, 8.234	psv/+	−	+	+	+	−	−
2728	wh hydr	i	...	−/+	i/+	−/+	i	i
2729	wh, 4.35	r	r	r	−	−	+	+
2730	wh	i	i	−	−	i	i	−
2731	yel	i/+	...	i	i	i	+	+

No.	Formula and name	M_r	Phase transition temperature
2732	Tb(NO$_3$)$_3$ (·6H$_2$O) Terbium(III) nitrate	344.94	mp hydr 89.3; dec hydr t
2733	Tb$_2$O$_3$ Terbium(III) oxide	365.85	mp 2000, bp ca. 4300
2734	Tb(OH)$_3$ Terbium(III) hydroxide	209.95	dec t
2735	Tb$_2$(SO$_4$)$_3$ (·8H$_2$O) Terbium(III) sulfate	606.04	−H$_2$O 360, dec >850
2736	**Tc Technetium**	97.907	mp 2250, bp ca. 4600
2737	TcBr$_3$O Technetiumtribromide-oxide	353.62	subl 400
2738	[Tc$_2$(CO)$_{10}$] Dodecacarbonylditechnetium	475.91	mp 159–160
2739	TcCl$_4$ Technetium(IV) chloride	239.72	subl >300
2740	TcCl$_6$ Technetium(VI) chloride	310.63	dec >20
2741	[TcCl$_6$],K$_2$ Potassium hexachlorotechnetate(IV)	388.82	dec t
2742	Tc(Cl)O$_3$ Technetium chloride-trioxide	181.36	bp 25
2743	TcCl$_3$O Technetium trichloride-oxide	220.27	subl 900
2744	TcF$_5$ Technetium(V) fluoride	192.90	mp 50
2745	TcF$_6$ Technetium(VI) fluoride	211.90	mp 37, bp 55
2746	[TcF$_6$],K Potassium hexafluorotechnetate(V)	250.99	dec >150
2747	[TcF$_6$],K$_2$ Potassium hexafluorotechnetate(IV)	290.09	dec >225
2748	TcO$_2$ Technetium(IV) oxide	129.91	>1000 → Tc,Tc$_2$O$_7$
2749	Tc$_2$O$_7$ Technetium(VII) oxide	307.81	mp 119.5; dec 260
2750	TcOF$_4$ Technetium oxide-tetrafluoride	189.90	mp 134
2751	TcO$_3$F Technetium trioxide-fluoride	164.90	mp 18, bp 100
2752	Tc(OH)$_4$ Techetium(IV) hydroxide	165.94	400 → TcO$_2$
2753	Tc$_2$S$_7$ Technetium(VII) sulfide	420.28	dec >230
2754	**Te Tellurium**	127.60	mp 449.8; bp 990
2755	TeBr$_2$ Tellurium dibromide	287.41	mp 280, bp 339
2756	TeBr$_4$ Tellurium tetrabromide	447.22	mp 380, bp 421
2757	TeCl$_2$ Tellurium dichloride	198.51	mp 208, bp 328
2758	TeCl$_4$ Tellurium tetrachloride	269.41	mp 224, bp 380
2759	TeF$_4$ Tellurium tetrafluoride	203.59	mp 129.6
2760	TeF$_6$ Tellurium hexafluoride	241.59	subl −38.9; mp −37.6
2761	TeI$_4$ Tellurium tetraiodide	635.22	dec 100, mp 280 (p)
2762	[TeMo$_6$O$_{24}$],(NH$_4$)$_6$ (·7H$_2$O) Ammonium 24-oxohe-xamolybdotellurate(VI)	1195.45	dec hydr 550
2763	TeO$_2$ Tellurium dioxide	159.60	mp 733, bp 1257
2764	TeO$_3$ Tellurium trioxide	175.60	dec >400
2765	**Th Thorium**	232.038	mp 1750, bp ca. 4200
2766	ThB$_6$ Thorium hexaboride	296.90	mp 2195
2767	ThBr$_4$ (·4H$_2$O) Thorium(IV) bromide	551.65	mp 678, bp 857
2768	ThC Thorium monocarbide	244.05	mp 2625
2769	ThC$_2$ Thorium dicarbide	256.06	mp 2655, bp ca. 5000
2770	ThCl$_4$ Thorium(IV) chloride	373.85	mp 770, bp 922
2771	ThF$_4$ (·4H$_2$O) Thorium(IV) fluoride	308.03	mp 1110, bp 1680
2772	ThI$_4$ Thorium(IV) iodide	739.65	mp 566, bp 837
2773	ThN Thorium mononitride	246.05	mp 2630
2774	Th(NO$_3$)$_4$ (·5H$_2$O) Thorium(IV) nitrate	480.05	hydr >400 → ThO$_2$
2775	ThO$_2$ Thorium(IV) oxide	264.04	mp 3350, bp ca. 4400
2776	Th(OH)$_4$ Thorium(IV) hydroxide	300.07	>470 → ThO$_2$
2777	Th(PO$_3$)$_4$ Thorium(IV) metaphosphate	547.92	dec >1500

(*Continued*)

No.	Color and density	Solubility and reactivity						
		water	alcohol	HCl	H_2SO_4	HNO_3	NaOH	$NH_3 \cdot H_2O$
2732	wh hydr	r	r	r	–	r	+	+
2733	wh, ca. 7.5	–/+	...	+	+	+	–	–
2734	wh	i	...	+	+	+	i	i
2735	wh hydr	d	...	–	d	–	+	+
2736	wh, 11.49	–	–	–	–/+	+	–	–
2737	brn	–/+	...	–/+	+	+	+	+
2738	wh	i	d	–	–/+	–/+	+	–
2739	dk-red, 3.3	+	...	+	+	+	+	+
2740	grn	+	+	+	+	+	+	+
2741	yel, 3.6	+	...	+/r	+	+	+	+
2742	cl lq	+	...	+	+	+	+	+
2743	brn	–/+	...	+	+	+	+	+
2744	dk-yel	+	...	+	+	+	+	+
2745	yel, 3.02	+	+	+	+	+	+	+
2746	lt-yel	+	...	+	+	+	+	+
2747	lt-pink	r	...	–	–/+	–/+	–/+	–
2748	blk-brn, 6.9	–	...	+	+	+	–	–
2749	lt-yel, 3.5	+	r	+	+	+	–	+
2750	dk-yel	+	...	+	+	+	+	+
2751	yel lq	+	...	+	+	+	+	+
2752	brn-blk	i	...	+	+	+	+	i
2753	brn-blk	i/+	...	–	–/+	–/+	+	–
2754	grey, 6.13	–	–	–/+	+	+	–/+	–
2755	brn	+	...	+	+	+	+	+
2756	yel-orange, 4.31	+	...	+	+	+	+	+
2757	dk-grn, 7.05	+	...	–	–	–	+	...
2758	wh, 3.26	+	r	+	+	+	+	+
2759	wh	+	...	+	+	+	+	+
2760	cl gas	+	...	+	+	+	+	+
2761	dk-grey, 5.05	+	...	+	+	+	+	+
2762	wh hydr, 2.78	r	...	–	–	–	+	–/+
2763	wh ($t \rightarrow$ yel), 5.67	–	...	+	–/+	–/+	+	+
2764	grey, 5.64	–/+	i	–/+	–	–	–/+	–
2765	wh, 11.72	psv	–	–/+	psv	psv	–	–
2766	blk-viol, 7.31	–	...	–	–	+	–	–
2767	wh, 5.69(5.67)	r	...	–	–/+	–	+	+
2768	blk	+	...	+	+	+	+	+
2769	grey-blk, 8.96	+	...	+	+	+	+	+
2770	wh, 4.60	r/+	r	r	–	–	+	+
2771	wh, 5.71(3.36)	i	...	i	–	–	i	i
2772	yel, 6.00	r/+	r	–	–/+	–/+	+	+
2773	yel, 11.9	–	...	–	–/+	–/+	–/+	–
2774	wh hydr, 2.80	r/+	r	r	–	r	+	+
2775	wh, 9.7	–	i	–	–/+	–	–	–
2776	wh	i	...	+	+	+	i	i
2777	wh, 4.08	i	i	–	–/+	–	–/+	–

No.	Formula and name	M_r	Phase transition temperature
2778	ThS$_2$ Thorium(IV) sulfide	296.17	mp 1930 dec
2779	Th(SO$_4$)$_2$ (·9H$_2$O) Thorium(IV) sulfate	424.16	dec hydr >400
2780	Th(SeO$_4$)$_2$ (·9H$_2$O) Thorium(IV) selenate	517.95	−H$_2$O 220
2781	ThSi$_2$ Thorium disilicide	288.21	mp 1640 dec
2782	ThSiO$_4$ (·nH$_2$O) Thorium(IV) orthosilicate	324.12	mp 1975 dec
2783	**Ti Titanium**	47.88	mp 1668, bp 3260
2784	TiB$_2$ Titanium diboride	69.50	mp 2850
2785	TiBr$_2$ Titanium(II) bromide	207.69	dec >500
2786	TiBr$_3$ (·6H$_2$O) Titanium(III) bromide	287.59	mp hydr 115, dec >400
2787	TiBr$_4$ Titanium(IV) bromide	367.50	mp 38, bp 231
2788	TiC$_{1-x}$ Titanium monocarbide	59.89	mp 2781, bp 4300
2789	[Ti(C$_5$H$_5$)$_2$] Bis(cyclopentadienyl)titanium	178.07	mp 173
2790	[Ti(C$_5$H$_5$)Cl$_3$] Cyclopentadienyltrichlorotitanium	219.33	mp 216.75
2791	[Ti(C$_5$H$_5$)$_2$Cl] Bis(cyclopentadienyl)chlorotitanium	213.52	mp 283
2792	[Ti(C$_5$H$_5$)$_2$Cl$_2$] Bis(cyclopentadienyl)dichlorotitanium	248.98	mp 289–291
2793	[Ti(CO)$_2$(C$_5$H$_5$)$_2$] Dicarbonylbis(cyclopentadienyl)titanium	234.09	dec >90
2794	TiCl$_2$ Titanium(II) chloride	118.79	mp 1035
2795	TiCl$_3$ (·6H$_2$O) Tintanium(III) chloride	154.24	dec 440
2796	TiCl$_4$ Titanium(IV) chloride	189.69	mp −24.1, bp 136.3
2797	Ti(Cl)F$_3$ Titanium chloride-trifluoride	140.33	dec 125
2798	TiCl$_2$O Titanium dichloride-oxide	118.79	dec 180
2799	Ti(ClO$_4$)$_4$ Titanium(IV) perchlorate	445.68	mp 85, dec 110
2800	(TiCo$_2$)O$_4$ Titanium-dicobalt tetraoxide	229.74	dec t
2801	TiF$_3$ Titanium(III) fluoride	104.87	dec >950
2802	TiF$_4$ Titanium(IV) fluoride	123.87	subl 285.5
2803	[TiF$_6$],K$_2$ (·H$_2$O) Potassium hexafluorotitanate(IV)	240.06	−H$_2$O 32, mp 780
2804	(TiFe)O$_3$ Titanium-iron trioxide	151.72	mp 1367, dec >1900
2805	(TiFe$_2$)O$_4$ Titanium-diiron tetraoxide	223.57	dec >2100
2806	TiH$_{2-x}$ (0 ≤ x ≤ 0.5) Titanium dihydride	49.90	dec 400
2807	TiI$_2$ Titanium(II) iodide	301.69	mp 600, bp 1170
2808	TiI$_4$ Titanium(IV) iodide	555.50	mp 155, bp 379.5
2809	TiI$_2$O Titanium diodide-oxide	317.69	dec 105
2810	(TiMn)O$_3$ Titanium-manganese trioxide	150.82	mp 1404
2811	TiN Titanium mononitride	61.89	mp ca. 3000
2812	Ti(NO$_3$)$_4$ Titanium(IV) nitrate	295.90	mp 58
2813	TiO$_{1+x}$ (−0.23 ≤ x ≤ 0.3) Titanium(II) oxide	63.88	mp 1780, bp 3227
2814	α-TiO$_2$ Titanium(IV) oxide	79.88	650 → γ-TiO$_2$
2815	β-TiO$_2$	79.88	915 → γ-TiO$_2$
2816	γ-TiO$_2$	79.88	mp 1870, bp ca. 3000
2817	Ti$_2$O$_3$ Titanium(III) oxide	143.76	mp 1830 bp ca. 3000
2818	Ti(O)F$_2$ Titanium(IV) oxide-difluoride	101.88	dec 146
2819	Ti(OH)$_3$ Titanium(III) hydroxide	98.90	dec t
2820	TiO(OH)$_2$ Titanium oxide-hydroxide	97.89	dec 600–700
2821	TiP Titanium monophosphide	78.85	mp 1860
2822	(TiPb)O$_3$ Tutanium-lead trioxide	303.08	mp ca. 1290

(Continued)

No.	Color and density	Solubility and reactivity						
		water	alcohol	HCl	H_2SO_4	HNO_3	NaOH	$NH_3 \cdot H_2O$
2778	blk-brn, 7.36	i/+	...	+	+	+	i	i
2779	wh, 4.23(2.77)	d	...	−	d	−	+	+
2780	wh hydr, 3.03	d	...	−	i	−	−/+	−/+
2781	blk, 7.63	−	i	−/+	−	−	+	−
2782	wh, 4.6	i	...	−	−	−	−/+	−
2783	wh, 4.51	psv/+	−	−/+	psv/+	psv/+	−/+	−
2784	grey, 4.45	−	i	−	−/+	+	−	−
2785	blk, 4.41	+	r	+	+	+	+	+
2786	red-viol, 4.24	r	r	+	−/+	−	+	+
2787	yel, 3.25	+	r	+	+	+	+	+
2788	blk-grey, 4.93	−	...	−	+	+	−	−
2789	dk-grn	+	...	+	+	+	+	+
2790	yel	+	r	+	+	+	+	...
2791	brn	i/+	r	−	+	+	−/+	...
2792	red, 1.60	+	r	+	+	+	+	+
2793	red-brn	+	r	+	+	+	+	+
2794	blk, 3.13	+	r	+	+	+	+	+
2795	dk-viol, 2.64	r	r	r	−	−	+	+
2796	cl lq, 1.726[20]	+	r	+	+	+	+	+
2797	yel	+	...	+	+	+	+	+
2798	yel, 2.45	+	...	+	+	+	+	+
2799	wh	i/+	r	+	+	+	+	+
2800	grn-blk, 5.1	−	−	−/+	−/+	−	−	−
2801	dk-bl, 2.98	i	i	−	−	−	−	−
2802	wh, 2.80	+	r	+	+	+	+	+
2803	wh	d	...	+	+	+	+	−
2804	blk, 4.74	−	−	−/+	−/+	−	−	−
2805	dk-brn, 4.78	−	−	−/+	−/+	−	−	−
2806	grey, 3.91	−/+	+	+	+	+	+	+
2807	blk, 4.99	+	r	+	+	+	+	+
2808	red-brn, 4.40	+	...	+	+	+	+	+
2809	brn	+	...	+	+	+	+	+
2810	yel-red, 4.54	−	−	−/+	−/+	−/+	−/+	−
2811	yel-brn, 5.43	−	i	−	−	−	−/+	−
2812	wh, 2.16	+	+	+	+	+/−	+	+
2813	yel, 4.89	−	...	+	+	+	−	−
2814	wh, 4.14	−	...	−	−/+	−	−/+	−
2815	wh, 3.90	−	...	−	−/+	−	−/+	−
2816	wh, 4.85	−	...	−	−/+	−/+	−/+	−
2817	dk-viol, 4.59	−	...	−	+	−/+	−	−
2818	yel	+	...	+	+	+	+	+
2819	viol	i	...	+	+	+	i	i
2820	wh	i	i	+	+	+	i/+	i
2821	grey, 3.95	−	...	−	−/+	−/+	−/+	−
2822	yel, 7.52	−	...	−/+	−/+	−	−/+	−

No.	Formula and name	M_r	Phase transition temperature
2823	TiS Titanium(II) sulfide	79.95	mp 1930
2824	TiS$_2$ Titanium(IV) sulfide	112.01	...
2825	Ti$_2$S$_3$ Titanium(III) sulfide	191.96	dec 690–1200
2826	Ti$_2$(SO$_4$)$_3$ Titanium(III) sulfate	383.95	dec 500–600
2827	TiSO$_4$(OH)$_2$ Titanium sulfate-dihydroxide	177.96	mp 580 dec
2828	TiSi$_2$ Titanium disilicide	104.05	mp 1540
2829	**Tl Thallium**	204.383	mp 303.6; bp 1457
2830	TlAl(SO$_4$)$_2$ (·12H$_2$O) Thallium(I)-aluminum sulfate	423.49	mp hydr 91, dec hydr t
2831	Tl$_3$AsS$_4$ Thallium(I) tetrathioarsenate	816.34	mp 425
2832	TlBr Thallium(I) bromide	284.29	mp 461, bp 816
2833	TlC$_5$H$_5$ Thallium(I) cyclopentadienide	269.48	mp >230
2834	Tl(CH$_3$COO) Thallium(I) acetate	263.43	mp 131, dec ca. 200
2835	Tl(CH$_3$COO)$_3$ (·1.5H$_2$O) Thallium(III) acetate	381.52	mp 182 dec
2836	Tl$_2$CO$_3$ Thallium(I) carbonate	468.77	mp 272, 360 → Tl$_2$O
2837	Tl$_2$C$_2$O$_4$ Thallium(I) oxalate	496.78	dec >350
2838	TlCl Thallium(I) chloride	239.84	mp 431, bp 818
2839	TlCl$_3$ Thallium(III) chloride	310.74	dec 40, mp 155 (p)
2840	TlClO$_3$ Thallium(I) chlorate	287.83	dec 500
2841	TlClO$_4$ Thallium(I) perchlorate	303.83	mp 501, dec t
2842	Tl$_2$CrO$_4$ Thallium(I) chromate	524.76	mp 635
2843	Tl$_2$Cr$_2$O$_7$ Thallium(I) dichromate	624.75	mp 360 dec
2844	TlF Thallium(I) fluoride	223.38	mp 322, bp 840 dec
2845	TlF$_3$ Thallium(III) fluoride	261.38	mp 550 dec
2846	Tl(HCOO) Thallium(I) formate	249.40	mp 104, dec ca. 200
2847	TlH$_2$PO$_4$ Thallium(I) dihydroorthophosphate	301.37	mp ca. 190
2848	Tl$_2$HPO$_4$ (·2H$_2$O) Thallium(I) hydroorthophosphate	504.74	hydr 200 → Tl$_4$P$_2$O$_7$
2849	TlHSO$_4$ Thallium(I) hydrosulfate	301.45	mp 120 dec
2850	TlI Thallium(I) iodide	331.29	mp 442, bp 833 dec
2851	TlN$_3$ Thallium(I) azide	246.40	mp 334, dec >400
2852	TlNCS Thallium(I) thiocyanate	262.47	mp 234
2853	TlNO$_2$ Thallium(I) nitrite	250.39	mp 186
2854	TlNO$_3$ Thallium(I) nitrate	266.39	mp 206.5; dec >300
2855	Tl(NO$_3$)$_3$ (·3H$_2$O) Thallium(III) nitrate	390.40	mp hydr 102, dec 300
2856	Tl$_2$O Thallium(I) oxide	424.77	mp 303, bp ca. 1100
2857	Tl$_2$O$_3$ Thallium(III) oxide	456.76	500 → Tl$_2$O, mp 716 (p)
2858	Tl$_2$O$_3$·nH$_2$O	–	–H$_2$O 100
2859	TlOH (·2H$_2$O) Thallium(I) hydroxide	221.39	hydr 139 → Tl$_2$O
2860	Tl$_3$PO$_4$ Thallium(I) orthophosphate	708.12	dec ca. 250
2861	Tl$_4$P$_2$O$_7$ Thallium(I) diphosphate	991.47	dec >120
2862	Tl$_2$S Thallium(I) sulfide	440.83	mp 448.9; bp 1177
2863	Tl$_2$S$_3$ Thallium(III) sulfide	504.96	dec 260
2864	Tl$_2$SO$_4$ Thallium(I) sulfate	504.83	mp 632
2865	Tl$_2$S$_2$O$_6$ Thallium(I) dithionate	568.89	dec t
2866	Tl$_2$(SO$_3$S) Thallium(I) thiosulfate	520.90	dec 130
2867	Tl$_2$Se Thallium(I) selenide	487.73	mp 390

(Continued)

No.	Color and density	Solubility and reactivity						
		water	alcohol	HCl	H_2SO_4	HNO_3	NaOH	$NH_3 \cdot H_2O$
2823	brn	i	...	−	−/+	−/+	i	i
2824	yel, 3.28	−	...	−/+	−/+	+	+	−
2825	dk-grey, 3.58	i	i	−	−/+	+	−	−
2826	grn	i	i	+	+/i	i/r	i	i
2827	wh	r/+	...	+	+	+	+	+
2828	lt-grey	−	...	−/+	−/+	−	−/+	−
2829	wh, 11.84	−	−	−	+	+	−	−
2830	wh hydr, 2.32	r	...	+	r	−	+	+
2831	red	i	i	+	+	+	i/+	i/+
2832	lt-yel, 7.5	d	r	+	−/+	−	−	−
2833	lt-yel	−	d	+	+	+	−	−
2834	wh, 3.68	r	r	+	+	−	r	r
2835	wh hydr	+	...	+	−	r	+	+
2836	wh, 7.11	r	i	+	+	+	−	−
2837	wh, 6.31	d	...	+	+	−	−	−
2838	wh, 7.0	d	r	+	−	−	−	−
2839	wh hydr, 3.03	+	r	+	−	−	+	+
2840	wh, 5.05	r	d	+	−/+	r	−	−
2841	wh, 4.89	r	d	+	−	r	−	−
2842	yel, 6.91	i	...	+	+	+	i	i
2843	orange-red	i	...	+	i	i	i/+	...
2844	wh, 8.40	r	d	+	−	r	r	r
2845	lt-grn, 8.36	+	...	+	+	+	+	+
2846	wh	r	d	+	−/+	−/+	r	r
2847	wh, 4.73	d	i	−	−/+	−/+	−	...
2848	wh	r	...	−	−/+	−/+	−	−
2849	wh	r	...	+	r/d	−	+	+
2850	yel, 7.29	i	d	+	−/+	−/+	i	i
2851	yel	d	i	+	−/+	−/+	−	−
2852	wh, 4.96	d	i	+	−/+	−/+	−	−
2853	yel	r	...	+	−/+	−/+	r	r
2854	wh, 5.56	r	i	+	−	r	r	r
2855	wh hydr	+	...	+	−	r	+	+
2856	dk-brn, 9.52	+	r	+	+	+	+	+
2857	brn-blk, 10.11	−	...	+	+	+	−	−
2858	red-brn	i	...	+	+	+	i	i
2859	lt-yel, 7.44	r	r	+	+	+	r	r
2860	wh, 6.89	d	i	+	+	−	−	−
2861	wh, 6.79	r	i	+	+	−	−/+	−
2862	blk-bl, 8.39	i	r	+	+	+	i	i
2863	blk	i	...	−	−/+	−/+	i	i
2864	wh, 6.77	d/r	...	+	d/r	−	−	−
2865	wh, 5.57	r	r	+	+	−/+	−	−
2866	wh	d/r	...	+	+	+	−	−
2867	grey, 9.05	i	...	+	+	+	i	i

No.	Formula and name	M_r	Phase transition temperature
2868	Tl$_2$SeO$_4$ Thallium(I) selenate	551.72	mp >400
2869	TlVO$_3$ Thallium(I) metavanadate	303.32	mp 424
2870	**Tm Thulium**	168.934	mp 1545, bp 1947
2871	TmBr$_3$ Thulium(III) bromide	408.65	mp 954, bp 1440
2872	Tm$_2$(C$_2$O$_4$)$_3$ (·6H$_2$O) Thulium(III) oxalate	601.92	hydr >600 → Tm$_2$O$_3$
2873	TmCl$_3$ Thulim(III) chloride	275.29	mp 824, bp 1490
2874	TmCl$_3$·6H$_2$O	383.38	mp 154, dec t
2875	TmF$_3$ Thulium(III) fluoride	225.93	mp 1158, bp 2230
2876	TmI$_3$ Thulium(III) iodide	549.65	mp 1020, bp 1260
2877	Tm(NO$_3$)$_3$ (·5H$_2$O) Thulium(III) nitrate	354.95	dec hydr t
2878	Tm$_2$O$_3$ Thulium(III) oxide	385.87	mp 2000, bp ca. 4300
2879	Tm(OH)$_3$ Thulium(III) hydroxide	219.96	t → Tm$_2$O$_3$
2880	Tm$_2$(SO$_4$)$_3$ (·8H$_2$O) Thulium(III) sulfate	626.05	−H$_2$O 600, dec >850
2881	**U Uranium**	238.029	mp 1134, bp 4200
2882	UB$_2$ Uranium diboride	259.65	mp 2385
2883	UBr$_3$ Uranium(III) bromide	477.74	mp 730
2884	UBr$_4$ Uranium(IV) bromide	557.65	subl 519, bp 765
2885	UBr$_5$ Uranium(V) bromide	637.55	dec 80
2886	UC Uranium monocarbide	250.04	mp 2400
2887	UC$_{2-x}$ Uranium dicarbide	262.05	mp 2350, bp 4370
2888	[U(CO$_3$)$_3$O$_2$],(NH$_4$)$_4$ (·2H$_2$O) Ammonium tricarbonatodioxouranate(VI)	522.21	dec hydr 100
2889	UCl$_3$ Uranium(III) chloride	344.39	subl 835, bp 1780
2890	UCl$_4$ Uranium(IV) chloride	379.84	subl 590, bp 792
2891	UCl$_5$ Uranium(V) chloride	415.29	dec 120, mp 320 (p)
2892	UCl$_6$ Uranium(VI) chloride	450.75	mp 177 dec
2893	UF$_3$ Uranium(III) fluoride	295.02	mp 1495, bp 2300
2894	UF$_4$ Uranium(IV) fluoride	314.02	mp 1036, bp 1730
2895	UF$_4$·2.5H$_2$O	359.06	−H$_2$O 450
2896	α-UF$_5$ Uranium(V) fluoride	333.02	dec >150, mp 287 (p)
2897	β-UF$_5$	333.02	mp 348, bp 530
2898	UF$_6$ Uranium(VI) fluoride	352.02	subl 56.4; mp 64 (p)
2899	[UF$_5$],NH$_4$ Ammonium pentafluorouranate(IV)	351.06	dec 270
2900	UH$_{3-x}$ Uranium trihydride	241.05	dec 330
2901	UI$_3$ Uranium(III) iodide	618.74	mp 680, bp ca. 1750
2902	UI$_4$ Uranium(IV) iodide	745.65	dec 520
2903	UN Uranium mononitride	252.04	dec >1900, mp 2850 (p)
2904	UO$_{2+x}$ (0 ≤ x ≤ 0.67) Uranium(IV) oxide	270.03	subl 1400. mp 2850
2905	UO$_3$ Uranium(VI) oxide	286.03	dec >500
2906	UO$_2$(CH$_3$COO)$_2$ (·2H$_2$O) Uranyl acetate	388.12	−H$_2$O 110, 275 → UO$_3$
2907	(UO$_2$)CO$_3$ (·1.5÷2.0H$_2$O) Uranyl carbonate	330.04	dec 500
2908	(UO$_2$)C$_2$O$_4$ (·3H$_2$O) Uranyl oxalate	358.05	−H$_2$O 100, dec 500
2909	(UO$_2$)Cl$_2$ (·3H$_2$O) Uranyl chloride	340.93	−H$_2$O 400, mp 578
2910	UO$_2$(ClO$_4$)$_2$ (·6H$_2$O) Uranyl perchlorate	468.93	mp hydr 185, −H$_2$O 270
2911	U(O)F$_4$ Uranium oxide-tetrafluoride	330.02	dec 150
2912	(UO$_2$)F$_2$ Uranyl fluoride	308.02	dec 700

(Continued)

No.	Color and density	Solubility and reactivity						
		water	alcohol	HCl	H_2SO_4	HNO_3	NaOH	$NH_3 \cdot H_2O$
2868	wh, 6.88	r	i	+	+	r	−	−
2869	grey, 6.09	d	...	+	+	−/+	−/+	−/+
2870	wh, 9.332	psv/+	−	+	+	+	−	−
2871	wh	r	r	−	−/+	−	+	+
2872	lt-grn hydr	i	...	−/+	i/+	−/+	i	−
2873	lt-yel	r	r	r	−	−	+	+
2874	grn	r	r	r	−	−	+	+
2875	lt grn	i	i	−	i	−	i	−
2876	yel-grn	r	r	−	−/+	−/+	+	+
2877	lt-grn hydr	r	r	r	−	r	+	+
2878	lt-grn, ca. 9	−/+	...	+	+	+	−	−
2879	grn	i	...	+	+	+	i	i
2880	grn	r	...	−	r	−	+	+
2881	wh, 19.04	+	−	+	+	+/psv	−	−
2882	grey, 12.82	−	...	−	−	−/+	−	−
2883	dk-red, 5.98	+	r	+	+	+	+	+
2884	dk-brn, 5.35	r/+	i	−	−/+	−	+	+
2885	blk-brn	+	...	+	+	+	+	+
2886	grey-blk, ca. 13.6	+	...	+	+	+	−	−
2887	lt-grey, 11.68	+	i	+	+	+	+	+
2888	yel hydr	r	r	+	+	+	+	+
2889	grn ($t \rightarrow$ red), 5.51	+	...	+	+	+	+	+
2890	dk-grn, 4.87	r/+	r	−	−	−	+	+
2891	red-brn, 3.81	+	+	+	+	+	+	+
2892	dk-grn, 3.59	+	...	+	+	+	+	+
2893	red-viol, 8.97	i/+	...	+	+	+	i	i
2894	grn, 6.72	i	...	−	−	i	i/+	−
2895	grn-sk-bl, 4.71	i	...	−	−	i	i/+	−
2896	wh, 5.81	+	+	+	+	+	+	+
2897	lt-yel, 6.45	+	+	+	+	+	+	+
2898	wh, 5.06	+	r/+	+	+	+	+	+
2899	grn	i	...	−/+	−/+	i/+	−	−
2900	grey-blk, 10.92	+	...	+	+	+	+	+
2901	blk, 6.76	+	r	+	+	+	+	+
2902	blk, 5.6	+	r	+	+	+	+	+
2903	grey, 14.32	+	...	−	−	−/+	−	−
2904	brn-blk, 10.96	−	...	−	−/+	−/+	−	−
2905	yel-orange, 8.34	+	...	+	+	+	+	+
2906	yel-grn hydr, 2.89	r/+	r	r	−	r	+	+
2907	lt-yel, 5.24	i	r	+	+	+	i	i
2908	yel hydr, 3.07	d	i	d	−/+	−/+	+	+
2909	yel, 5.43	r	r	r	−	−	+	+
2910	yel hydr, 2.57	r	...	−	−	−	+	+
2911	orange	+	...	+	+	+	+	+
2912	lt-yel, 6.37	r	r	−	−	r	+	+

No.	Formula and name	M_r	Phase transition temperature
2913	$[UO_2F_5],(NH_4)_3$ Ammonium dioxopentafluorouranate(VI)	419.13	dec >185
2914	$U(OH)_4$ Uranium(IV) hydroxide	306.06	>350 → UO_{2+x}
2915	$UO_2(NO_3)_2$ ($\cdot 6H_2O$) Uranyl nitrate	394.04	mp hydr 59.5; dec hydr 300
2916	$UO_2(O_2)$ ($\cdot 2H_2O$) Uranyl peroxide	302.03	dec hydr 260
2917	$UO_2(OH)_2$ ($\cdot H_2O$) Uranyl hydroxide	304.04	400 → UO_3
2918	$(UO_2)S$ Uranyl sulfide	302.09	dec 40–50
2919	$(UO_2)SO_4$ ($\cdot 3H_2O$) Uranyl sulfate	366.09	$-H_2O$ 175; 720 → $(U_2^V U^{VI})O_8$
2920	UP Uranium monophosphide	269.00	mp 2850
2921	US Uranium(II) sulfide	270.10	mp 2462
2922	$U(S_2)$ Uranium(II) disulfide(2−)	302.16	mp 1680 dec
2923	$U(SO_4)_2$ ($\cdot 4H_2O$) Uranium(IV) sulfate	430.15	$-H_2O$ >300
2924	USi_2 Uranium disilicide	294.20	mp 1700
2925	U_3Si_2 Triuranium disilicide	770.26	mp 1665
2926	$USiO_4$ Uranium(IV) orthosilicate	330.11	mp ca. 1900
2927	$(U_2^V U^{VI})O_8$ Diuranium(V)-uranium(VI) oxide	842.08	dec >1500
2928	**V Vanadium**	50.942	mp 1920, bp 3450
2929	VB_2 Vanadium diboride	72.56	mp ca. 2400
2930	VBr_2 Vanadium(II) bromide	210.75	subl >1000
2931	VBr_3 Vanadium(III) bromide	290.65	subl >280; 500 → V
2932	$V(Br)O$ Vanadium bromide-oxide	146.85	dec 480
2933	VBr_2O Vanadium dibromide-oxide	226.75	subl ca. 600
2934	VBr_3O Vanadium tribromide-oxide	306.65	dec 180
2935	VC_{1-x} (0.1 ≤ x ≤ 0.3) Vanadium monocarbide	62.95	mp 2810, bp ca. 3900
2936	V_2C_{1-x} (0 ≤ x ≤ 0.2) Divanadium carbide	113.90	mp 2165
2937	$[V(C_5H_5)_2]$ Bis(cyclopentadienyl)vanadium	181.13	mp 167.5
2938	$[V(CO)_6]$ Hexacarbonylvanadium	219.00	dec >60–70
2939	$[V(CO)_4C_5H_5]$ Tetracarbonyl(cyclopentadienyl)vanadium	228.08	mp 138; >150 → V
2940	$[V(CO)_6],Na$ Sodium hexacarbonylvanadate(−I)	241.99	dec t
2941	VCl_2 Vanadium(II) chloride	121.85	subl 1000, mp ca. 1350
2942	VCl_3 Vanadium(III) chloride	157.30	dec 500
2943	VCl_4 Vanadium(IV) chloride	192.75	mp −20.5; bp 153 dec
2944	$V(Cl)O$ Vanadium chloride-oxide	102.39	ca. 600 → VCl_3, V_2O_3
2945	$V(Cl)O_2$ Vanadium chloride-dioxide	118.39	dec 150
2946	VCl_2O Vanadium dichloride-oxide	137.85	subl ca. 600
2947	VCl_3O Vanadium trichloride-oxide	173.30	mp −78, bp 126.7
2948	VF_3 Vanadium(III) fluoride	107.94	subl 800, mp 1410
2949	VF_4 Vanadium(IV) fluoride	126.93	dec >100
2950	VF_5 Vanadium(V) fluoride	145.93	mp 19.5; bp 47.9
2951	$[VF_6],K$ Potassium hexafluorovanadate(V)	204.03	dec 330
2952	$[VF_6],(NH_4)_3$ Ammonium hexafluorovanadate(III)	219.05	mp 1400
2953	$[V(H_2O)_4Cl_2]Cl$ ($\cdot 2H_2O$) Tetraaquadichlorovanadium(III) chloride	229.36	dec hydr t

(Continued)

No.	Color and density	Solubility and reactivity						
		water	alcohol	HCl	H_2SO_4	HNO_3	NaOH	$NH_3 \cdot H_2O$
2913	yel-grn	r	r	r	−	−	−/+	r
2914	lt-grn	i	...	+	+	+	−	−
2915	yel-grn hydr, 2.81	r	r	−	−	r	+	+
2916	lt-yel hydr	−	...	+	+	+	+	...
2917	yel, 5.93	i	...	+	+	+	i	i
2918	blk-brn	i	r	+	+	+	i	i
2919	yel hydr, 5.24	r	r	−	r	−	+	+
2920	grey-blk	−	...	−	−/+	−/+	−/+	−
2921	yel, 10.87	i	...	+	+	+	i	i
2922	grey-blk, 7.54	i/+	...	+	+	+	−	−
2923	grn hydr, 3.60	r/+	i	−	r	−	+	+
2924	lt-grey	−	...	−	−	−	−/+	−
2925	brn-blk	−	...	−	−	−	−/+	−
2926	grn, >5.1	i	...	−	−	−	−/+	−
2927	dk-grn, 8.39	−	...	−	−/+	+	−	−
2928	grey, 5.96	−	−	−	−/+	−/+	−	−
2929	grey, 5.10	−	i	−	−	−	−/+	−
2930	lt-brn, 4.58	r/+	i	+	−/+	+	+	+
2931	blk-grn, 4.44	d/+	r	+	−/+	−	+	+
2932	viol, 4.0	−	r	+	−/+	−	+	+
2933	dk-yel	−/+	r	+	−/+	−/+	+	...
2934	red lq, 2.933[15]	+	...	+	+	+	+	+
2935	blk, 5.46–5.80	−	−	−	−	−/+	−	−
2936	dk-grey	−	−	−	−	−/+	−	−
2937	viol-blk	−/+	r	−	−/+	+	−	−
2938	blk	−/+	r/+	−	−/+	−/+	−	−
2939	lt-red	−/+	r	−/+	−/+	−/+	−/+	...
2940	yel	i/+	...	+	+	+	−	−
2941	lt-grn, 3.09	r/+	i	+	−/+	+	+	+
2942	viol, 2.87	+	r	+	−/+	−/+	+	+
2943	red-brn lq, 1.83[20]	+	r	+	+	+	+	+
2944	yel-brn, 2.28	−	d	−	−/+	−/+	−	−
2945	orange-red, 2.29	+	r	+	+	+	+	+
2946	grn, 2.88	−/+	r	r/+	−	−/+	+	...
2947	yel lq, 1.829[20]	+	r	+	+	+	+	+
2948	yel-grn, 3.63	d/+	d	+	−/+	−	+	...
2949	grn, 2.98	+	d	+	+	+	+	+
2950	wh, 2.18	+	r	+	+	+	+	+
2951	wh	+	...	+	+	+	+	...
2952	yel-grn	i	...	−	−/+	−/+	−/+	−
2953	grn hydr	d	r	−	+/−	+/−	+	+

No.	Formula and name	M_r	Phase transition temperature
2954	$[V(H_2O)_4(SO_4)O]$ ($\cdot H_2O$) Tetraaquasulfatooxovanadium	235.06	dec hydr 280
2955	α-VI_2 Vanadium(II) iodide	304.75	subl 800
2956	β-VI_2	304.75	dec 1400
2957	VI_3 Vanadium(III) iodide	431.65	dec 280
2958	VN_{1-x} ($0 \le x \le 0.29$) Vanadium mononitride	64.95	mp ca. 2050
2959	V_2N_{1-x} ($0 \le x \le 0.26$) Divanadium nitride	115.89	mp ca. 2000
2960	$VNH_4(SO_4)_2$ ($\cdot 12H_2O$) Vanadium(III)-ammonium sulfate	261.11	mp hydr 45, dec t
2961	$VO_{1\pm x}$ ($x = 0.25$) Vanadium(II) oxide	66.94	mp 1830
2962	VO_2 Vanadium(IV) oxide	82.94	mp 1640, bp ca. 2700
2963	V_2O_3 Vanadium(III) oxide	149.88	mp 1970, bp ca. 3000
2964	V_2O_5 Vanadium(V) oxide	181.88	mp 690, dec >700
2965	$V_2O_5 \cdot nH_2O$ ($n = 1, 2, 3$)	–	dec t
2966	$(VO)Br_2$ ($\cdot 5H_2O$) Vanadyl bromide	226.75	dec hydr 140
2967	$(VO)Cl_2$ ($\cdot 5H_2O$) Vanadyl chloride	137.85	dec hydr 120
2968	$V(O)F_3$ Vanadium oxide-trifluoride	123.94	subl 109.5
2969	$V(OH)_2$ Vanadium(II) hydroxide	84.96	dec t
2970	$V(OH)_3$ Vanadium(III) hydroxide	101.96	$80 \rightarrow VO(OH)$
2971	$VO(OH)$ Vanadium metahydroxide	83.95	$300 \rightarrow V_2O_3$
2972	$VO(OH)_2$ Vanadyl hydroxide	100.96	dec 700
2973	$VS_{1\pm x}$ ($x = 0.17$) Vanadium(II) sulfide	83.01	dec ca. 1400
2974	V_2S_{3-x} ($-0.06 < x < 0.66$) Vanadium(III) sulfide	198.08	dec >600
2975	VS_{2+x} ($0.17 \le x \le 0.53$) Vanadium(IV) sulfide	115.07	dec >520
2976	V_2S_5 Vanadium(V) sulfide	262.21	dec >400
2977	VSO_4 ($\cdot 7H_2O$) Vanadium(II) sulfate	147.00	dec hydr t
2978	$V_2(SO_4)_3$ ($\cdot 9H_2O$) Vanadium(III) sulfate	390.07	dec >400
2979	V_3Si Trivanadium silicide	180.91	mp 1910 dec
2980	**W Tungsten**	**183.85**	mp 3387, bp ca. 5680
2981	WB_2 Tungsten diboride	205.47	mp 2900
2982	W_2B_5 Ditungsten pentaboride	421.76	mp 2370 dec
2983	WBr_2 Tungsten(II) bromide	343.66	dec 400
2984	WBr_3 Tungsten(III) bromide	423.56	$>180 \rightarrow WBr_2$
2985	WBr_4 Tungsten(IV) bromide	503.47	subl >240
2986	WBr_5 Tungsten(V) bromide	583.37	mp 295, bp 392
2987	WBr_6 Tungsten(VI) bromide	663.27	mp 309, dec >400 $\rightarrow WBr_5$
2988	WBr_2O Tungsten dibromide-oxide	359.66	subl 450
2989	WBr_2O_2 Tungsten dibromide-dioxide	375.66	subl >440
2990	WBr_3O Tungsten tribromide-oxide	439.56	dec >420
2991	WBr_4O Tungsten tetrabromide-oxide	519.45	mp 277, bp 331
2992	WC Tungsten monocarbide	195.86	mp 2780 dec
2993	W_2C_{1+x} Ditungsten carbide	379.71	mp 2750, bp ca. 6000
2994	$[W(CN)_8],K_2$ ($\cdot 2H_2O$) Potassium octacyanowolframate(IV)	548.39	$-H_2O$ 115
2995	$[W(CO)_6]$ Hexacarbonyltungsten	351.91	mp 169, bp 175 dec
2996	WCl_2 Tungsten(II) chloride	254.76	$>100 \rightarrow WCl_3, WCl_6$

(Continued)

No.	Color and density	Solubility and reactivity						
		water	alcohol	HCl	H_2SO_4	HNO_3	NaOH	$NH_3 \cdot H_2O$
2954	bl hydr	+	d	+	+	+	+	+
2955	blk, 5.44	r/+	i	+	−/+	+	+	+
2956	pink-red, 5.25	r/+	i	+	−/+	+	+	+
2957	blk, 5.2	d/+	r	+	−/+	−/+	+	+
2958	blk, 6.04	−	...	−	−	−	−	−
2959	blk	−	...	−	−	−	−	−
2960	bl-viol hydr, 1.69	r	i	+	d	−	−/+	+
2961	lt-grey, 5.6–5.75	−	...	+	+	+	−	−
2962	bl-blk, 4.34	−	...	+	+	+	+	+
2963	blk, 4.87	−	...	−	−	−/+	−	−
2964	orange, 3.36	−	r	+	+	+	+	+
2965	yel	i	...	+	+	+	+	+
2966	bl hydr	r	r	−/+	−/+	−	+	+
2967	bl hydr	r	r	r/+	−	−/+	+	+
2968	lt-yel, 2.46	+	...	+	+	+	+	+
2969	brn	i	...	+	+	+	−	−
2970	grn-yel	i	...	+	+	+	−	−
2971	blk	−	...	+	+	+	−	−
2972	yel	i	...	+	+	+	+	−
2973	blk, 4.51	i	...	−	−/+	−/+	i/+	−
2974	grey-blk, 3.7	i	...	−	−/+	+	i/+	i
2975	blk-grey, 2.5	−	...	−	+	+	−/+	−
2976	brn-blk, 3.0	−	i	−	−/+	+	+	+
2977	red-viol hydr	r/+	...	−	r	+	+	+
2978	yel-orange	i	i	+	i	−	−/+	+
2979	grey	−	i	−	−/+	−	−/+	−
2980	lt-grey, 19.35	−	−	−	−	−	−	−
2981	grey, 12.75	−	...	−	−	−	−/+	−
2982	grey	−	i	−	−/+	+	−	−
2983	yel-grn	i/+	...	+	+	+	+	+
2984	dk-grn	i/+	d	+	+	+	+	+
2985	blk	+	...	+	+	+	+	+
2986	dk-grey-grn	+	r	+	+	+	+	+
2987	blk-grey, 6.9	−/+	r	+	+	+	+	+
2988	dk-grey, 6.13	−	i	−	−/+	−/+	+	−
2989	red	−/+	i	−	−/+	−/+	−	−
2990	bl-blk, 5.87	−/+	i	−	−/+	−/+	−	−
2991	dk-brn	+	...	+	+	+	+	+
2992	grey-bl, 15.63	−	...	−	−	−	−	−
2993	dk-grey, 16.61	−	...	−	−	−	−	−
2994	lt-yel, 1.99	r	i	−	−/+	−/+	−	−
2995	wh, 2.65	−	d	−	−	−/+	−	−
2996	lt-grey, 5.44	i	r	+	+	+	+	+

No.	Formula and name	M_r	Phase transition temperature
2997	WCl_3 Tungsten(III) chloride	290.21	dec >200
2998	WCl_4 Tungsten(IV) chloride	325.66	>450 $\rightarrow$ WCl_2,WCl_5
2999	WCl_5 Tungsten(V) chloride	361.12	mp 248, bp 287
3000	WCl_6 Tungsten(VI) chloride	396.57	mp 275, bp 347
3001	WCl_2O Tungsten dichloride-oxide	270.76	subl 500
3002	WCl_3O Tungsten trichloride-oxide	306.21	290 $\rightarrow$ WCl_4O,WCl_2O
3003	WCl_2O_2 Tungsten dichloride-dioxide	286.75	mp 497 dec
3004	WCl_4O Tungsten tetrachloride-oxide	341.66	mp 204, bp 224
3005	WF_6 Tungsten(VI) fluoride	297.84	mp 2.0; bp 17.3
3006	WFe_2 Tungstendiiron	295.54	mp 1046
3007	$[WH_2(C_5H_5)_2]$ Dihydrobis(cyclopentadienyl)tungsten	316.06	mp 163
3008	WI_2 Tungsten(II) iodide	437.66	>800 $\rightarrow$ W
3009	WI_3 Tungsten(III) iodide	564.56	600 $\rightarrow$ WI_2
3010	$W(I)O_2$ Tungsten iodide-dioxide	342.75	subl >410
3011	WI_2O_2 Tungsten diiodide-dioxide	469.66	>400 $\rightarrow$ $W(I)O_2$
3012	WO_2 Tungsten(IV) oxide	215.85	mp ca. 1550, bp 1730
3013	WO_3 Tungsten(VI) oxide	231.85	mp 1473 (p), bp ca. 1700
3014	$WO_3 \cdot H_2O$	249.86	100 $\rightarrow$ WO_3
3015	$WO_3 \cdot 2H_2O$	267.88	70 $\rightarrow$ $WO_3 \cdot H_2O$
3016	$W(O)F_4$ Tungsten oxide-tetrafluoride	275.84	mp 106, bp 185.9
3017	WS_2 Tungsten(IV) sulfide	247.98	dec 1250
3018	WSe_2 Tungsten(IV) selenide	341.77	dec >1000
3019	WSi_{2+x} ($0 \leq x \leq 0.1$) Tungsten disilicide	240.02	mp 2165
3020	**Xe Xenon**	131.29	mp −111.85; bp −108.12
3021	$Xe \cdot 5.75H_2O$	234.88	dec −4
3022	$Xe(CF_3COO)_2$ Xenon bis(trifluoroacetate)	357.32	dec >20
3023	$XeCl_2$ Xenon dichloride	202.20	subl t, dec 80
3024	XeF_2 Xenon difluoride	169.29	mp 140, dec 600
3025	XeF_4 Xenon tetrafluoride	207.28	mp 135
3026	XeF_6 Xenon hexafluoride	245.28	mp 49.48; bp 75.57 dec
3027	$[XeF_8],(NO)_2$ Nitrosyl octafluoroxenonate(VI)	343.29	dec 400
3028	XeO_3 Xenon trioxide	179.29	subl 70 dec
3029	$Xe(O)F_2$ Xenon oxide-difluoride	185.29	0.5 $\rightarrow$ XeF_2,XeO_2F_2
3030	$Xe(O)F_4$ Xenon oxide-tetrafluoride	223.28	mp −41
3031	XeO_2F_2 Xenon dioxide-difluoride	201.28	mp 31 dec ($\rightarrow$ XeF_2)
3032	$Xe(OH)_4$ Xenon tetrahydroxide	199.32	mp 90, bp 115 dec
3033	**Y Yttrium**	88.906	mp 1528, bp 3322
3034	YB_6 Yttrium hexaboride	153.77	mp 2300
3035	YBr_3 ($\cdot 9H_2O$) Yttrium(III) bromide	328.62	mp 905, bp 1324
3036	$Y(BrO_3)_3$ ($\cdot 9H_2O$) Yttrium(III) bromate	472.61	mp hydr 74
3037	$Y_2(C_2O_4)_3$ ($\cdot 10H_2O$) Yttrium(III) oxalate	441.87	hydr >700 $\rightarrow$ Y_2O_3
3038	YCl_3 ($\cdot 6H_2O$) Yttrium(III) chloride	195.27	mp 721, bp 1482
3039	YF_3 ($\cdot 0.5H_2O$) Yttrium(III) fluoride	145.90	mp 1155, bp 2230
3040	YH_3 Yttrium(III) hydride	91.93	dec 900

(*Continued*)

No.	Color and density	Solubility and reactivity						
		water	alcohol	HCl	H_2SO_4	HNO_3	NaOH	$NH_3 \cdot H_2O$
2997	blk	i/+	d	−/+	−/+	−/+	+	+
2998	blk, 4.62	+	...	+	+	+	+	+
2999	blk-grn, 3.88	+	...	+	+	+	+	+
3000	blk-viol, 3.52	+	r	+	+	+	+	+
3001	yel-brn, 5.92	−/+	i	−/+	−/+	−/+	−	−
3002	dk-grn	−/+	i	−/+	−/+	−/+	+	−/+
3003	yel	+	i	+	+	+	+	+
3004	red, 3.95	+	...	+	+	+	+	+
3005	lt-yel lq, 3.44[20]	+	r	+	+	+	+	+
3006	grey	−	...	−	−/+	−/+	−	−
3007	yel	+	r	+	+	+	−/+	...
3008	lt-brn, 6.9	i	r	−/+	+	+	+	+
3009	blk	i/+	d	−/+	+	+	+	+
3010	bl-blk	−/+	i	−/+	−/+	−/+	−/+	−/+
3011	dk-grn, 6.39	+	i	+	+	+	+	+
3012	dk-brn, 11.05	−	...	−/+	−/+	+	−/+	−
3013	yel, 7.16	−	...	−	−	−	+	+
3014	yel, 5.5	i	...	−	−	−	+	+
3015	wh, 4.0	i	...	−	−	−	+	+
3016	wh	+	...	+	+	+	+	+
3017	dk-grey-sk-bl, 7.5	i	i	−	−	+	i	i
3018	dk-grey	i	...	−	−	+	−	−
3019	sk-bl-grey, 9.4	−	i	−	−	−	−/+	−
3020	cl gas, 5.85	d	r	−	−	−	−	−
3021	wh	+	+	+	+	+	+	+
3022	wh	r/+	r	−/+	+	+	+	+
3023	wh	+	+	+	...	...	+	+
3024	wh, 4.32	r	+	−	−	−	+	+
3025	wh, 4.10	+	+	+	+	+	+	+
3026	wh	+	r	+	+	+	+	+
3027	yel	+	+	+	+	+	+	+
3028	wh	r/+	+	−	−	−	+	+
3029	lt-yel	+	...	+	+	+	+	+
3030	cl lq, 3.17[20]	+	...	+	+	+	+	+
3031	wh	+	...	+	+	+	+	+
3032	wh	i	...	...	...	...	−	...
3033	wh, 4.45	psv/+	−	+	+	+	−/+	+
3034	bl-viol	−	i	−	−	−	−/+	−
3035	wh	r	r	r	−/+	−	+	+
3036	wh hydr	r	d	−	−	−	+	+
3037	wh hydr	i	...	i/+	−/+	i/+	i	i
3038	wh, 2.8(2.18)	r	r	r	−	r	+	+
3039	wh, 5.069(4.01)	d	i	−	−	−	d	d
3040	bl	−/+	i	+	+	+	−	−

No.	Formula and name	M_r	Phase transition temperature
3041	YI_3 Yttrium(III) iodide	469.62	mp 997, bp 1310
3042	$Y_2(MoO_4)_3$ ($\cdot 4H_2O$) Yttrium(III) molybdate	657.62	dec t
3043	$Y(NO_3)_3$ ($\cdot 5H_2O$) Yttrium(III) nitrate	274.92	dec hydr < 280
3044	Y_2O_3 Yttrium(III) oxide	225.81	mp 2430, bp ca. 4300
3045	$Y(OH)_3$ Yttrium(III) hydroxide	139.93	>200 $\rightarrow$ Y_2O_3
3046	YPO_4 Yttrium(III) orthophosphate	183.88	mp 1950
3047	$Y_2(S)O_2$ Diyttrium sulfide-dioxide	241.88	mp 2120
3048	$Y_2(SO_4)_3$ ($\cdot 8H_2O$) Yttrium(III) sulfate	466.00	$-H_2O$ 120, dec >900
3049	YSb Yttriumantimony	210.66	mp 2310
3050	**Yb Ytterbium**	173.04	mp 824, bp 1211
3051	$YbBr_2$ Ytterbium(II) bromide	332.85	mp 613, bp 1800
3052	$YbBr_3$ Ytterbium(III) bromide	412.75	mp 943, dec t
3053	$Yb(CH_3COO)_3$ ($\cdot 4H_2O$) Ytterbium(III) acetate	350.17	$-H_2O$ 100
3054	$Yb_2(C_2O_4)_3$ ($\cdot 6H_2O$) Ytterbium(III) oxalate	610.13	hydr >600 $\rightarrow$ Yb_2O_3
3055	$YbCl_2$ Ytterbium(II) chloride	243.95	mp 702, bp 2033
3056	$YbCl_3$ Ytterbium(III) chloride	279.40	mp 865, dec 1300
3057	$YbCl_3 \cdot 6H_2O$	387.49	mp 155, $-H_2O$ 180
3058	YbF_{2+x} Ytterbium(II) fluoride	211.04	mp 1407
3059	YbF_3 Ytterbium(III) fluoride	230.03	mp 1162, bp 2200
3060	YbI_2 Ytterbium(II) iodide	426.85	mp 772
3061	$Yb(NO_3)_3$ ($\cdot 5H_2O$) Ytterbium(III) nitrate	359.05	hydr >600 $\rightarrow$ Yb_2O_3
3062	Yb_2O_3 Ytterbium(III) oxide	394.08	mp >2000, bp ca. 4300
3063	$Yb(OH)_3$ Ytterbium(III) hydroxide	224.06	$t \rightarrow Yb_2O_3$
3064	$YbSO_4$ Ytterbium(II) sulfate	269.10	dec t
3065	$Yb_2(SO_4)_3$ ($\cdot 8H_2O$) Ytterbium(III) sulfate	634.27	$-H_2O$ 600, dec 900
3066	**Zn Zinc**	65.39	mp 419.5; bp 906.2
3067	$(ZnAl_2)O_4$ Zinc-dialuminum tetraoxide	183.35	mp 1950 dec
3068	Zn_3As_2 Trizinc diarsenide	346.01	mp 1015
3069	$Zn_3(AsO_4)_2$ ($\cdot 8H_2O$) Zinc(II) arsenate	474.01	$-H_2O$ 290
3070	$ZnBr_2$ ($\cdot 2H_2O$) Zinc(II) bromide	225.20	mp hydr 37, mp 394, bp 670
3071	$Zn(BrO_3)_2$ ($\cdot 6H_2O$) Zinc(II) bromate	321.19	mp hydr 100; $-H_2O$ 200
3072	$Zn(CH_3COO)_2$ Zinc(II) acetate	183.48	mp 242, dec t
3073	$Zn(CH_3COO)_2 \cdot 2H_2O$	219.51	dec 100
3074	$Zn(CN)_2$ Zinc(II) cyanide	117.43	dec 800
3075	$ZnCO_3$ Zinc(II) carbonate	125.40	200 $\rightarrow$ ZnO
3076	ZnC_2O_4 ($\cdot 2H_2O$) Zinc(II) oxalate	153.41	dec hydr 100
3077	$ZnCl_2$ ($\cdot 1.5H_2O$) Zinc(II) chloride	136.30	mp 293, bp 733
3078	$[ZnCl_4],(NH_4)_2$ Ammonium tetrachlorozincate(II)	243.28	mp 150
3079	$Zn(ClO_3)_2$ ($\cdot 4H_2O$) Zinc(II) chlorate	232.29	mp hydr 53.9; dec 60
3080	$Zn(ClO_4)_2$ ($\cdot 6H_2O$) Zinc(II) perchlorate	264.29	mp hydr 100, dec 200
3081	ZnF_2 ($\cdot 4H_2O$) Zinc(II) fluoride	103.39	mp 872, bp 1502
3082	$Zn(HCOO)_2$ ($\cdot 2H_2O$) Zinc(II) formate	191.45	$-H_2O$ 120, dec >213

(Continued)

No.	Color and density	Solubility and reactivity						
		water	alcohol	HCl	H_2SO_4	HNO_3	NaOH	$NH_3 \cdot H_2O$
3041	wh	r	r	−	−/+	−/+	+	+
3042	grey-yel hydr, 4.79	i	i	−	−	−	−/+	−
3043	wh hydr, 2.68	r	r	−	−	r	+	+
3044	wh, 5.01	−	i	+	+	+	−	−
3045	lt-yel	i	i	+	+	+	i	i
3046	wh, 4.25	i	i	i	i	i	i	i
3047	lt-grey	−	...	+	+	+	−	−
3048	wh, 2.52(2.56)	r	i	−	r	−	+	+
3049	grey	−	i	−/+	−/+	−/+	+	−
3050	wh, 6.760	psv/+	...	+	+	+	−	−
3051	lt-yel, 5.91	r/+	...	+	+	+	+	+
3052	wh	r	r	r	−/+	−	+	+
3053	wh hydr, 2.09	r	...	−	−/+	−/+	+	+
3054	wh hydr, 2.64	i	...	−/+	i/+	−/+	i	−
3055	wh, 5.08	r/+	...	+	+	+	+	+
3056	wh	r	r	r	−	r	+	+
3057	wh, 2.58	r	r	r	−	r	+	+
3058	grey	i	...	i	−	−	i	−
3059	wh, 8.168	i	i	r	r	r	−	i
3060	lt-yel, 5.40	r	...	+	+	+	+	+
3061	wh hydr, 2.68	r	r	−	−	r	+	+
3062	wh, 9.18	−/+	−	+	+	+	−	−
3063	wh	i	...	+	+	+	i	i
3064	yel-grn	d/+	...	+	+	+	+	+
3065	wh, 3.79(3.29)	r	...	−	r	−	+	+
3066	wh, 7.133	psv	...	+	+	+	+	+
3067	wh, 4.62	−	...	−	−	−	−/+	−
3068	grey, 5.60	−	i	+	+	+	−	−
3069	wh	i	...	−/+	−/+	−/+	+	+
3070	wh, 4.20	r	r	r	−/+	−	+	+
3071	wh hydr, 2.57	r	...	−/+	−/+	r	+	+
3072	wh, 1.84	r	r	−	−/+	−	+	+
3073	wh, 1.73	r	r	−	−/+	−	+	+
3074	wh, 1.85	i	i	i/+	i/+	i/+	+	−/+
3075	wh, 4.51	i/+	...	+	+	+	+	i
3076	wh hydr	i	...	−	−/+	−/+	+	+
3077	wh, 2.91	r	r	r	r/+	r	+	+
3078	wh, 1.88	+	...	+	+	+	+	+
3079	wh hydr, 2.15	r	r	−/+	−/+	−	+	+
3080	wh hydr, 2.25	r	r	−	−	r	+	+
3081	wh, 4.95(2.54)	d/+	i	d	−/+	d	+	+
3082	wh	r	i	−/+	−/+	−	+	+

No.	Formula and name	M_r	Phase transition temperature
3083	ZnI_2 Zinc(II) iodide	319.20	mp 446, bp 624 dec
3084	Zn_3N_2 Trizinc dinitride	224.18	dec 700
3085	$[Zn(NH_3)_2Cl_2]$ Diamminedichlorozinc	170.36	mp 210.8; dec 270
3086	$Zn(NO_3)_2$ ($\cdot 6H_2O$) Zinc(II) nitrate	189.40	mp hydr 36.4; dec hydr >130
3087	ZnO Zinc(II) oxide	81.39	subl 1725 dec
3088	$Zn(OH)_2$ Zinc(II) hydroxide	99.40	250 → ZnO
3089	$[Zn(OH)_4],Na_2$ Sodium tetrahydroxozincate(II)	179.40	dec >100
3090	Zn_3P_2 Trizinc diphosphide	258.12	mp 1193
3091	$Zn_3(PO_4)_2$ ($\cdot 4H_2O$) Zinc(II) orthophosphate	386.11	mp 1060
3092	α-ZnS Zinc(II) sulfide	97.46	1020 → β-ZnS
3093	β-ZnS	97.46	subl 1185, mp 1775
3094	$ZnSO_4$ Zinc(II) sulfate	161.45	dec >600
3095	$ZnSO_4\cdot 7H_2O$	287.56	mp 100, $-H_2O$ 280
3096	$ZnSb$ Zincantimony	187.14	mp 546 dec
3097	Zn_3Sb_2 Trizincdiantimony	439.67	mp 566
3098	$ZnSe$ Zinc(II) selenide	144.35	mp 1575
3099	$ZnSeO_4$ ($\cdot H_2O$) Zinc(II) selenate	208.35	dec hydr >50
3100	$ZnSiO_3$ Zinc(II) metasilicate	141.47	mp 1437
3101	Zn_2SiO_4 ($\cdot H_2O$) Zinc(II) orthosilicate	222.86	$-H_2O$ >350, mp 1509
3102	$ZnTe$ Zinc(II) telluride	192.99	mp 1238.5
3103	**Zr Zirconium**	**91.224**	**mp 1855, bp ca. 4340**
3104	ZrB_2 Zirconium diboride	112.85	mp 3200
3105	$ZrBr_4$ Zirconium(IV) bromide	410.84	subl 355, mp 450 (p)
3106	ZrC Zirconium monocarbide	103.24	mp 3530, bp ca. 5100
3107	$[Zr(C_5H_5)_2Cl_2]$ Bis(cyclopentadienyl)dichlorozirconium	292.32	mp 243.5; dec t
3108	$ZrCl_2$ Zirconium(II) chloride	162.13	mp 722 dec
3109	$ZrCl_3$ Zirconium(III) chloride	197.58	subl 650, dec 5
3110	$ZrCl_4$ Zirconium(IV) chloride	233.04	subl 331, mp 437 (p)
3111	$ZrCl_2O$ ($\cdot 8H_2O$) Zirconium dichloride-oxide	178.13	$-H_2O$ 150, dec 250
3112	ZrF_3 Zirconium(III) fluoride	148.22	dec >300
3113	ZrF_4 Zirconium(IV) fluoride	167.22	subl 600, mp 910 (p)
3114	$[ZrF_6],K_2$ Potassium hexafluorozirconate(IV)	283.41	mp 600 dec
3115	$[ZrF_7],(NH_4)_3$ Ammonium heptafluorozirconate(IV)	278.33	dec 365
3116	ZrH_2 Zirconium(II) hydride	93.24	dec 800
3117	$[Zr(H_2O)_4(SO_4)_2]$ Tetraaquabis(sulfato)zirconium	355.41	$-H_2O$ 340, dec 400
3118	ZrI_4 Zirconium(IV) iodide	598.84	subl 418, mp 500 (p)
3119	ZrN Zirconium mononitride	105.23	mp 2980
3120	$Zr(NCS)_4$ Zirconium(IV) thiocyanate	323.56	soft 40–50 dec
3121	ZrO_2 Zirconium(IV) oxide	123.22	mp 2700, bp ca. 4300
3122	$ZrO_2\cdot nH_2O$	–	140 → ZrO(OH)₂
3123	$ZrO(OH)_2$ Zirconium oxide-dihydroxide	141.24	1000 → ZrO₂
3124	ZrP_2O_7 Zirconium(IV) diphosphate	265.17	dec 1550
3125	ZrS_2 Zirconium(IV) sulfide	155.36	mp 1550
3126	$ZrSi_2$ Zirconium disilicide	147.40	mp 1520 dec
3127	$ZrSiO_4$ Zirconium(IV) orthosilicate	183.31	mp 1540, dec 1650

(*Continued*)

No.	Color and density	Solubility and reactivity						
		water	alcohol	HCl	H_2SO_4	HNO_3	NaOH	$NH_3 \cdot H_2O$
3083	lt-yel, 4.74	r	r	−	−/+	−/+	+	+
3084	grey, 6.22	+	...	+	+	+	+	+
3085	wh, 2.10	+	...	+	+	+	+	+
3086	wh hydr, 2.07	r	r	−	r	r	+	+
3087	wh ($t \rightarrow$ yel), 5.61	−	i	+	+	+	+	−/+
3088	wh, 3.03	i	...	+	+	+	+	i/+
3089	wh	+	r	+	+	+	+/r	+
3090	dk-grey, 4.55	−/+	i	+	+	+	−	−
3091	wh, 4.00(3.11)	i	i	i/+	i/+	i/+	+	+
3092	wh, 3.98	i	...	+	+	+	i	i
3093	wh, 4.14	i	...	+	+	+	i	i
3094	wh, 3.74	r	d	−	r	−	+	+
3095	wh, 1.96	r	d	−	r	−	+	+
3096	grey	−	i	−/+	−/+	+	−/+	−
3097	dk-grey, 6.33	−	i	−/+	−/+	−/+	−/+	−
3098	yel, 5.42	i	i	+	+	+	i	i
3099	wh	r	...	r	r	−	+	+
3100	wh, 3.42	i	...	−/+	−/+	−/+	−/+	−
3101	wh, 4.10	i	i	+	+	+	−/+	−
3102	red, 5.64	i/+	i	+	+	+	i	i
3103	wh, 6.50	−	−	−	−/+	−	−	−
3104	grey, 6.09	−	i	+	+	+	−	−
3105	wh, 3.98	+	...	+	+	+	+	+·
3106	grey, 6.62	−	...	−	−/+	−	−	−
3107	wh	+	r	+	+	+	+	+
3108	blk, 3.16	+	...	+	+	+	+	+
3109	blk-grn, 3.05	+	...	+	+	+	+	+
3110	wh, 2.80	+	r	+	+	+	+	+
3111	wh hydr, 1.91	r/+	i	+	+	+	+	+
3112	sk-bl-grey, 4.26	i	...	i	−	i	i	i
3113	wh, 4.6	+	...	+	+	+	+	+
3114	wh, 3.48	d	...	−	−	−	−/+	−
3115	wh, 1.43	d	...	−	−	−	−/+	−
3116	grey-blk	−	...	−/+	−/+	+	−/+	−
3117	wh, 2.80	r	i	+	+	+	+	+
3118	yel-orange	+	r	+	+	+	+	+
3119	yel, 7.3	−	−	−/+	−/+	−/+	−	−
3120	wh	+	...	+	+	+	+	+
3121	wh, 5.89	−	i	−	−/+	−	−	−
3122	yel	i	...	+	+	+	i	i
3123	wh, 3.25	i	...	+	+	+	i	i
3124	wh	i	...	−	−/+	−	−/+	−
3125	brn, 3.87	i	...	−	−/+	−/+	i/+	i
3126	grey, 4.88	−	...	−	−	−	−	−
3127	wh, 4.69	i	...	i	i	i	i	i

Formula Index

The Formula Index has been compiled on the assumption that all compounds may be regarded as a combination of two constituents, viz., a positively charged constituent (an arbitrary or real M^{v+} cation) and a negatively charged constituent (an arbitrary or real A^{v-} anion). The formulas of all M^{v+} cations and A^{v-} anions contained in the formulas of the compounds listed in this Section are presented separately.

The Formula Index is rubricated in terms of elements arranged in the alphabetical order. The first line of each rubric contains, after the symbol of an element, the numerals of the formulas wherein the symbol of this element stands in the first place in compliance with the system of formula presentation accepted in the above Table. Thus, the line "Ag 8–62" pertains to the formulas of the substances from No. 8 (Ag) to No. 62 (Ag_2TeO_3).

Other lines in this rubric embrace the formulas of all simple and multicomponent M^{v+} cations and A^{v-} anions that contain the element of the given rubric irrespective of the place said element occupies in the ion formula. For example, the "Ag^+" line cites not only the compounds Nos. 10–19, 21–30, 32, 33, 35–56, and 59–62 having the formulas with the Ag symbol in the first place, but likewise lists the numerals of the compounds wherein the Ag symbol occupies the third place, namely No. 1152 ($[HgI_4],Ag_2$), No. 2499 ($[SbS_3],Ag_3$), and No. 2500 ($[SbS_4],Ag_5$).

Moreover, the Formula Index lists the central atoms of the element and oxidation numbers thereof, e.g. in the "Ag^{+I} (complex)" line, and also the charged and neutral ligands, e.g. in the "Ag^+ (ligand)" line. The numerals correspond to the complexes formed by these central atoms and ligands.

A separate line lists the constituents of heteropolyanions. For example, the "B^{+III} (heteropolyanions)" line lists the compound comprising the $[BW_{12}O_{40}]^{5-}$ anion (No. 263) with the B^{+III} central atom, the same compound being listed in the "W^{+VI} (heteropolyanions)" and "O^{2-} (heteropolyanions)" lines.

The numerals pertaining to the formulas of M^{v+} cations and M^{v-} anions point to the compounds comprising only one species of the cations or anions, respectively. Thus, the "Ag^+" line lists the numerals that refer to the compounds containing one cation species only [silver(I) cations Ag^+], namely $Ag_3^+AsO_3^{3-}$ (No. 10), $Ag_3^+AsO_4^{3-}$ (No. 11), $Ag^+AsO_2^-$ (No. 12), etc.

In case a M^{v+} cation or a A^{v+} anion is contained in a compound comprising another cation or anion species, respectively, such an ion shall be placed on a separate line and designated by an upper left asterisk (*). Thus, the "*Ag^+" line lists the numerals of the compounds that contain, apart from the Ag^+ cation, other cation species, namely $Ag_2^+H_3^+IO_6^{5-}$ (No. 34, the other cation being H^+), $(Ag^+Sb^{3+})S_2^{2-}$ (No. 58, the other cation being Sb^{3+}), $K^+Ag^+(NO_3^-)_2$ (No. 1256, the other cation being K^+), etc.

If the charge of an ion can not be ascertained from the compound formula, the ion in question is listed in a separate line in the general form, viz., M^{v+} or A^{v-}. For example, the "$^*Ag^{v+}$" line corresponds to the compound $(Au^{v+}Ag^{v+})Te_4^{v-}$ (No. 169), wherein the charge of the silver ion Ag^{v+} is unknown. The same numeral is listed in the "$^*Au^{v+}$" and "Te^{v-}" lines.

The crystal water and water as a ligand are not included in the Formula Index, in so far as cystallohydrates and aquacomplexes are listed in the Table within the range of numerals included in the first line of the rubric of each element.

In the numerals listed one after another in an interval, all repeating figures are omitted(except for the first line) for the sake of brevity. For example, in the "Al^{3+}" line the intervals 64–6, 101–5, and 108–33 denote 64–66, 101–105, and 108–133, respectively, and should be read accordingly.

Pt^{5+} 2243
Pt^{+V} (complexes) 2245, 2247–8
Pt^{6+} 2244, 2266
Pt^{v+} 2206, 2270
Pu 2280–2296
Pu^{2+} 2291, 2296
Pu^{3+} 2281, 2284–5, 2288–9, 2293
Pu^{4+} 2283, 2286, 2290, 2292, 2294
Pu^{6+} 2287
Pu^{v+} 2282
Pu O$_2^{2+}$ 2295
Ra 2297–2306
Ra^{2+} 2298–306
Rb 2307–2346
Rb$^+$ 1197, 2241, 2309–17, 2319, 2322, 2324–8, 2330–45, 2560
˙Rb$^+$ 2308, 2318, 2320–1, 2323, 2329, 2346
Re 2347–2386
Re^{-I} (complexes) 2369
Re^0 (complexes) 2384
Re^{+I} (complexes) 2353–6
Re^{3+} 2348, 2372, 2376
Re^{+III} (complexes) 2359, 2361
Re^{4+} 2349, 2357, 2365, 2374, 2383, 2385
Re^{+IV} (complexes) 2360, 2373
Re^{5+} 2358, 2366
Re^{+V} (complexes) 2351, 2364
Re^{6+} 2363, 2367, 2375, 2378, 2380
Re^{7+} 2350, 2362, 2368, 2377, 2379, 2381–2, 2386
Re^{+VII} (complexes) 2370–1
Re O$_4^-$ 52, 317, 1335, 1661, 1798
Re$_2$O$_9^{4-}$ 1092
Rh 2387–2411
Rh^0 (complexes) 2389–91
Rh^{+I} (complexes) 2392–4
Rh^{3+} 2388, 2396, 2400, 2403, 2405, 2409–11
˙Rh^{3+} 1336, 2402
Rh^{+III} (complexes) 2395, 2398–9, 2404, 2406–7
Rh^{4+} 2401, 2408
Rh^{+IV} (complexes) 2397
Rn 2412
Ru 2413–2440
Ru^0 (complexes) 2417–8
Ru^{+I} (complexes) 2419
Ru^{2+} 2440
Ru^{+II} (complexes) 2415, 2420–1, 2431–4
Ru^{3+} 2414, 2422, 2430, 2439
Ru^{+III} (complexes) 2416, 2429
Ru^{4+} 2423, 2435–6

Ru^{+IV} (complexes) 2424–6
Ru^{5+} 2427
Ru^{6+} 2428, 2438
Ru^{8+} 2437
Ru O$_4^{2-}$ 1338
Ru O$_4^-$ 1337
S 2441–2476
S^{v-} 164–5, 443, 618, 1388, 1934, 2053–5, 2057–8, 2541
˙S^{v-} 1890
S^{2-} 53, 58, 120, 162–3, 197, 199, 260, 293, 318, 364, 395, 442, 484, 510, 553, 580, 678, 739–40, 783, 834–6, 851, 863, 876, 895, 942, 977, 1011, 1025, 1049–50, 1093, 1161–2, 1221–2, 1252–3, 1298, 1339–40, 1389, 1434, 1497, 1546–7, 1597–8, 1639, 1753, 1799, 1863, 1881, 1932, 2056, 2059, 2161, 2184, 2202–3, 2276–7, 2296, 2305, 2340, 2385–6, 2402, 2411, 2498, 2600–1, 2640–1, 2667, 2684, 2753, 2778, 2823–5, 2862–3, 2918, 2921, 2973–6, 3017, 3092–3, 3125
˙S^{2-} 444–6, 450–1, 582, 905, 1882, 2060–2, 2065–70, 2476, 2602, 3047
S^{2-} (ligand) 2499–502, 2642
S$^+$ 2443, 2448, 2457, 2461, 2470
S^{2+} 2446
S^{3+} 2452
S^{4+} 2444, 2447, 2450, 2453, 2455, 2465, 2467–8, 2473, 2476
S^{5+} 2458
S^{6+} 2445, 2449, 2451, 2454, 2456, 2464, 2466, 2469, 2471–2, 2474–5
S^{v+} 2462–3
S$_2^{2-}$ 198, 319, 679, 784, 978–9, 1094, 1163, 1341, 1548, 1800, 1933, 1995, 2162, 2440, 2922
S$_3^{2-}$ 320, 785, 1095, 1342
S$_4^{2-}$ 321, 1096, 1343, 1801
S$_5^{2-}$ 786, 1097, 1344, 1802, 2341
S$_6^{2-}$ 787, 1345
SF^{3+} 2459–60
S O$_3^{2-}$ 54, 322, 511, 1299, 1346, 1435, 1498, 1640, 1663, 1754, 1803
S O$_4^{2-}$ 55, 106–7, 121–4, 323, 365, 366, 396, 512–4, 546, 554–5, 573, 581, 583, 663–5, 680, 723, 741–4, 746, 757, 759–61, 770, 788, 792, 823, 837–8, 852, 864, 877–8, 896–8, 956–8, 980–2, 1003, 1012, 1026, 1098–100, 1164–5, 1177, 1223, 1254, 1257, 1269–70, 1280, 1284–5, 1300, 1305,

W^{6+} 2987, 2989, 2991, 3000, 3003–5, 3011, 3013–6
W^{+VI} (heteropolyanions) 263, 2072, 2584
W^{v+} 2981–2, 2992–3, 3006, 3019
WO_4^{2-} 334, 527, 585, 986, 1369, 1443, 1505, 1559, 1827, 2135, 2695
$W_7O_{24}^{6-}$ 1828
Xe 3020–3032
Xe^+ 2248
Xe^{2+} 3022–4
Xe^{4+} 3025, 3029, 3032
Xe^{6+} 3026, 3028, 3030–1
Xe^{+VI} (complexes) 3027
XeF_5^+ 220
XeO_6^{6-} 335
Y 3033–3049
Y^{3+} 3035–48
Y^{v+} 3034, 3049

Yb 3050–3065
Yb^{2+} 3051, 3055, 3058, 3060, 3064
Yb^{3+} 3052–4, 3056–7, 3059, 3061–3, 3065
Zn 3066–3102
Zn^{2+} 244, 3068–77, 3079–84, 3086–8, 3090–5, 3097–102
$^*Zn^{2+}$ 987, 1560, 3067
Zn^{+II} (complexes) 3078, 3085, 3089
Zn^{v+} 3096
Zr 3103–3127
Zr^{2+} 3108, 3116
Zr^{3+} 3109, 3112
Zr^{4+} 245, 3105, 3110–1, 3113, 3118, 3120–5, 3127
$^*Zr^{4+}$ 336, 528, 2696
Zr^{+IV} (complexes) 3107, 3114–5, 3117
Zr^{v+} 3104, 3106, 3119, 3126

Index of Minerals [4–10, 24–25, 27–29, 35–36]

Formula	Crystal system, hardness	Name	Formula	Crystal system, hardness	Name
Ag	cub, 2.7	Silver native	γ-AlO(OH)	rhomb, 3.5–4.0	Boehmite
$AgAsS_2$	mon	Smithite			
	trig	Trechmannite	$AlPO_4$	trig	Berlinite
Ag_3AsS_3	mon	Xanthoconite	$AlPO_4 \cdot 2H_2O$	rhomb, 4.5	Variscite
	trig, 2–2.5	Proustite	$Al_3(PO_4)_2(OH)_3$ $\cdot 5H_2O$	rhomb, 3.5–4.0	Wavellite
AgBr	cub, 1.5–2.5	Bromargyrite			
AgCl	cub, 2–3	Chlorargyrite	$Al_2(SO_4)_3$ $\cdot 13.5H_2O$	mon	Meta-aluno-gen
AgI	hex, 1.5	Iodargyrite			
	cub	Miersite	$Al_2(SO_4)_3 \cdot 18H_2O$	tricl, 1.5–2.0	Alunogen
Ag_2S	mon, 2	Acanthite	$Al(SO_4)OH$ $\cdot 5H_2O$	mon	Jurbanite
	rhomb, 2–2.5	Argentite			
$(AgSb)S_2$	mon, 2.5	Miargyrite		rhomb	Rostite
Ag_2Se	rhomb, 2.5	Naumannite	$Al_2SO_4(OH)_4$ $\cdot 7H_2O$	mon	Aluminite
Ag_2Te	mon, 2–3	Hessite			
$AlAsO_4 \cdot 2H_2O$	rhomb	Mansfieldite	α-$Al_2(SiO_4)O$	tricl, 5	Cyanite
$AlCl_3 \cdot 6H_2O$	trig	Chlor-aluminite	β-$Al_2(SiO_4)O$	rhomb, 6.5–7.5	Sillimanite
$AlCl(OH)_2 \cdot 4H_2O$	amor	Cadwaladerite	γ-$Al_2(SiO_4)O$	rhomb, 7.5	Andalusite
$[AlF_6], Na_3$	mon, 2.5	Cryolite	$Al_2Si_2O_7 \cdot 2H_2O$	mon	Dickite
Al_2O_3	trig, 9	Corundum		tricl, 2–2.5	Kaolinite
$Al_2O_3 \cdot 0.25H_2O$	hex	Akdalaite	$Al_2Si_2O_7 \cdot 6H_2O$	mon, 1–2	Halloysite
$Al(OH)_3$	mon, 2.5–3.5	Gibbsite	α-As	hex, 3.5	Arsenic native
	trig	Bayerite	As_4	rhomb	Arseno-lamprite
α-AlO(OH)	rhomb, 6.5–7.0	Diaspore	α-As_2O_3	mon, 2.5	Claudetite

(*Continued*)

Formula	Crystal system, hardness	Name	Formula	Crystal system, hardness	Name
$\beta\text{-}As_2O_3$	cub, 1.5	Arsenolite	$CaCO_3 \cdot H_2O$	hex	Monohydro-calcite
As_2S_3	mon, 1.5–2.0	Orpiment			
$\alpha\text{-}As_4S_4$	mon, 1.5–2.0	Realgar	$CaC_2O_4 \cdot H_2O$	mon, 2.5–3.0	Whewellite
$\beta\text{-}As_4S_4$	mon	Pararealgar	$CaCl_2\,(\cdot nH_2O?)$	rhomb	Hydrophilite
Au	cub, 2.5	Gold native	$CaCl_2 \cdot 2H_2O$	...	Sinjarite
$(AuAg)Te_4$	mon, 1.5	Sylvanite	$CaCl_2 \cdot 6H_2O$	trig	Antarcticite
$(AuAg_3)Te_2$	cub, 2.5–3.0	Petzite	$CaCrO_4$	tetr	Chromatite
$AuTe_2$	mon, 2.5	Calaverite	CaF_2	cub, 4	Fluorite
	rhomb, 2–3	Krennerite	$CaHPO_4$	tricl	Monetite
$[BF_4],K$	rhomb	Avogadrite	$CaHPO_4 \cdot 2H_2O$	mon	Brushite
$[BF_4],Na$	rhomb	Ferruccite	$Ca(IO_3)_2$	mon	Lautarite
$B(OH)_3$	tricl	Sassolite	$Ca(IO_3)_2 \cdot H_2O$	mon	Bruggenite
$BaCO_3$	rhomb, 3–3.5	Witherite	$CaMg(CO_3)_2$	trig, 3.5–4.0	Dolomite
$BaCa(CO_3)_2$	mon, 4	Barytocalcite	$CaMgSiO_4$	rhomb, 5.5	Monticellite
	tricl	Alstonite	$CaMg(SiO_3)_2$	mon, 5–6	Diopside
BaF_2	cub	Frankdick-sonite	$CaMoO_4$	tetr, 3.5	Powellite
			$Ca(NO_3)_2 \cdot 4H_2O$	mon	Nitrocalcite
$Ba(NO_3)_2$	cub	Nitrobarite	CaO	cub	Lime
$BaSO_4$	rhomb, 3–3.5	Barite	$Ca(OH)_2$	trig	Portlandite
			$Ca_3(PO_4)_2$	trig	Whitlockite
$(BeAl_2)O_4$	rhomb, 8.5	Chrysoberyl	$Ca_5(PO_4)_3F$	hex, 4.5–5.0	Fluorapatite
BeO	hex, 9	Bromellite	CaS	cub	Oldhamite
$Be(OH)_2$	rhomb	Behoite	$CaSO_4$	rhomb, 3–3.5	Anhydrite
Be_2SiO_4	trig, 7.5–8.0	Phenakite	$CaSO_4 \cdot 0.5H_2O$	trig	Bassanite
Bi	trig, 2–2.5	Bismuth native	$CaSO_4 \cdot 2H_2O$	mon, 2	Gypsum
$BiAsO_4$	mon	Rooseveltite	$CaSiO_3$	tricl, 4.5–5.0	Wollastonite
$Bi_2CO_3(OH)_4$	tetr, 4	Bismutite	$CaSiO_3 \cdot H_2O$	mon, ?	Suolunite
$Bi(Cl)O$	tetr	Bismoclite	Ca_2SiO_4	mon, 5.5	Larnite
Bi_2O_3	cub	Sillenite		rhomb	Bredigite
	mon	Bismite	$Ca_2SiO_4 \cdot H_2O$	mon	Hillebrandite
$Bi(O)F$	tetr	Zavaritskite	$Ca_3Si_2O_7$	mon	Rankinite
Bi_2S_3	rhomb, 2	Bismuthinite		rhomb	Kilchoanite
Bi_2Se_3	rhomb	Guanajuatite	$Ca_3Si_2O_7 \cdot 3H_2O$	mon	Afwillite
Bi_2Te_3	trig	Telluro-bismutite	$(CaTi)O_3$	rhomb, 5.5	Perovskite
			$CaWO_4$	tetr, 5	Scheelite
$\alpha\text{-}C$	hex, 1–2	Graphite	Cd	hex, 2	Cadmium native
$\beta\text{-}C$	cub, 10	Diamond			
$Ca_3(AsO_4)_2 \cdot 9H_2O$	trig	Machatsch-kiite	$CdCO_3$	trig	Otavite
			CdO	cub	Monteponite
$Ca(BO_2)_2$	rhomb	Calciborite	CdS	hex, 3–3.5	Greenockite
$Ca(BO_2)_2 \cdot 6H_2O$	hex	Hexahydro-borite		cub	Hawleyite
			$CdSe$	hex	Cadmoselite
$CaCO_3$	hex	Vaterite	CeF_3	hex	Fluocerite
	rhomb, 3.5–4.0	Aragonite	CeO_2	cub	Cerianite
	trig, 3	Calcite	$CoAs_2$	rhomb, 4–5	Safflorite
			$CoAs_{3-x}$	cub, 5.0–5.5	Smaltite

(Continued)

Formula	Crystal system, hardness	Name	Formula	Crystal system, hardness	Name
$CoAs_3$	cub, 5.5–6.0	Skutterudite	Fe_3C	rhomb	Cohenite
$Co_3(AsO_4)_2 \cdot 8H_2O$	mon, 1.5–2.5	Erythrite	$FeCO_3$	trig, 3.5–4.0	Siderite
$Co(As)S$	cub, 5.5	Cobaltite	$FeC_2O_4 \cdot 2H_2O$	mon	Humboldtine
$CoCO_3$	trig	Sphero-cobaltite	$[Fe(C_2O_4)_3],K_3$ $\cdot 3H_2O$	mon	Minguzzite
$CoCl_2 \cdot 6H_2O$	mon	Albrittonite	$FeCl_2$	trig	Lawrencite
$(Co^{II}Co_2^{III})S_4$	cub, 5.5	Linnaeite	$FeCl_2 \cdot 2H_2O$	mon	Rokühnite
$CoO(OH)$	hex; trig	Heterogenite	$FeCl_3$	trig	Molysite
CoS_{1+x}	hex	Jaipurite	$FeCl_3 \cdot 6H_2O$	mon	Hydromolysite
$Co(S_2)$	cub	Cattierite	$(FeCu)S_2$	tetr, 3.5–4.0	Chalcopyrite
$CoSO_4 \cdot 7H_2O$	mon	Bieberite	$(Fe^{II}Fe_2^{III})O_4$	cub, 6	Magnetite
$CoSe$	hex	Freboldite	$Fe_{2+x}N$	trig	Siderazot
$(Cr_2Fe)O_4$	cub, 5.5–7.5	Chromite	$Fe(NH_4)_2(SO_4)_2$ $\cdot 6H_2O$	mon	Mohrite
CrN	cub	Carlsbergite	$(Fe_2Ni)O_4$	cub	Trevorite
Cr_2O_3	trig	Eskolaite	$Fe_{1-x}O$	cub	Wüstite
$CrO(OH)$	rhomb	Bracewellite	Fe_2O_3	trig, 5.5–6.5	Hematite
	trig	Grimaldiite	$Fe_{2.67}O_4$	cub, 5	Maghemite
Cu	cub, 2.5–3.0	Copper native	$Fe(OH)_2$	trig	Amakinite
Cu_3As	cub; trig	Domeykite	$FeO(OH)$	hex	Feroxyhyte
Cu_3AsS_4	rhomb, 3	Enargite		rhomb, 5–5.5	Goethite
	tetr	Luzonite	Fe_3P	tetr	Schreibersite
$Cu_2CO_3(OH)_2$	mon, 3.5–4.0	Malachite	$FePO_4$	rhomb	Heterosite
$CuCl$	cub	Nantokite	$FePO_4 \cdot 2H_2O$	mon, 3.5–4.0	Phospho-siderite
$CuCl_2 \cdot 2H_2O$	rhomb	Eriochalcite		rhomb	Strengite
$CuCrO_2$	trig	McConnellite			
CuI	cub	Marshite	$Fe_3(PO_4)_2 \cdot 8H_2O$	mon, 1.5–2.0	Vivianite
$Cu(IO_3)_2 \cdot 0.67H_2O$	tricl	Bellingerite	$Fe_{1-x}S$	mon, 3.5–4.5	Pyrrhotine
CuO	mon, 3.5	Tenorite	FeS	hex	Troilite
Cu_2O	cub, 3.5–4.0	Cuprite	$\alpha\text{-}Fe(S_2)$	rhomb, 5.0–5.5	Marcasite
CuS	hex, 1.5–2 0	Covellite			
$Cu_{1.8}S$	cub, 2.5–3.0	Digenite	$\beta\text{-}Fe(S_2)$	cub, 6.0–6.5	Pyrite
$Cu_{1.94}S$	mon, 2.2–2.4	Djurleite	$FeSO_4 \cdot 5H_2O$	tricl	Siderotil
Cu_2S	rhomb, 2.5–3.0	Chalcosine	$FeSO_4 \cdot 7H_2O$	rhomb	Melanterite
$CuSO_4$	rhomb	Chalcocyanite	$Fe_2(SO_4)_3 \cdot 9H_2O$	trig	Coquimbite
$CuSO_4 \cdot 5H_2O$	tricl, 2.5	Chalcanthite	$FeSi$	cub	Fersilicite
$CuSe$	hex	Klockmannite	$FeSiO_3$	rhomb	Ferrosilite
Cu_2Se	cub	Berzelianite	Fe_2SiO_4	rhomb, 6.5	Fayalite
	tetr	Bellidoite	$FeWO_4$	mon, 4.5	Ferberite
Fe	cub, 4.5	Iron native	$(Fe_2Zn)O_4$	cub, 6	Franklinite
$(FeAl_2)O_4$	cub, 7.5–8.0	Hercynite	$Ga(OH)_3$	cub	Söhngeite
$FeAs$	rhomb	Westerveldite	HBO_2	cub	Metaborite
$FeAs_2$	rhomb, 5–5.5	Löllingite	H_2O	hex, 1.5	Ice
$FeAsO_4 \cdot 2H_2O$	rhomb, 3.5–4.0	Scorodite	Hg (lq)	–	Mercury native
$Fe(As)S$	mon, 5.5–6.0	Arsenopyrite	Hg_2Cl_2	tetr	Calomel

(Continued)

Formula	Crystal system, hardness	Name	Formula	Crystal system, hardness	Name
Hg_2I_2	tetr	Coccinite	$Mg(NH_4)_2(SO_4)_2$	mon	Boussin-
HgO	rhomb	Montroydite	$\cdot 6H_2O$		gaultite
α-HgS	trig, 2.0	Cinnabar	$Mg(NO_3)_2 \cdot 6H_2O$	mon	Nitromagne-
β-HgS	cub, 2.5	Metacinnabar			site
$HgSe_{1\pm x}$	cub, 2.5	Tiemannite	MgO	cub, 2.5	Periclase
$HgTe$	cub	Coloradoite	$Mg(OH)_2$	trig, 2.5	Brucite
In	tetr, < 1.2	Indium native	$Mg_3(PO_4)_2$	mon	Farringtonite
$In(OH)_3$	cub	Dzhalindite	$Mg_3(PO_4)_2 \cdot 8H_2O$	mon	Bobierrite
Ir	cub, 4.9–5.4	Iridium native	MgS	cub	Niningerite
$KAl(SO_4)_2 \cdot 12H_2O$	cub	Potash alum	$MgSO_4 \cdot H_2O$	mon, 3–3.5	Kieserite
$KAl_3(SO_4)_2(OH)_6$	trig, 4	Alunite	$MgSO_4 \cdot 7H_2O$	rhomb, 2–2.5	Epsomite
$K_2Ca(SO_4)_2 \cdot H_2O$	mon	Syngenite	Mg_2SiO_4	rhomb, 7	Forsterite
KCl	cub, 2	Sylvite	$(MnAl_2)O_4$	cub, 7.5–8.0	Galaxite
K_2CrO_4	rhomb	Tarapacaite	$MnAs$	hex	Kaneite
$K_2Cr_2O_7$	tricl	Lopezite	$MnCO_3$	trig, 3.5–4.0	Rhodochrosite
KF	cub	Carobbiite	$MnCl_2$	trig	Scacchite
$KFe(SO_4)_2 \cdot 12H_2O$	mon	Yavapaiite	$(Mn_2Cu)O_4$	mon, 4	Crednerite
$KHCO_3$	mon	Kalicinite	$(MnFe_2)O_4$	cub, 5.6	Jacobsite
$KHSO_4$	rhomb	Mercallite	$(Mn^{II}Mn_2^{III})O_4$	tetr, 5.5	Hausmannite
$KMgCl_3 \cdot 6H_2O$	rhomb, 2.0–2.5	Carnallite	MnO	cub, 5.6	Manganosite
			MnO_{2-x}	rhomb	Ramsdellite
$K_2Mg(SO_4)_2$ $\cdot 6H_2O$	mon	Picromerite		tetr, 2.5–3.0	Pyrolusite
$KMg(SO_4)Cl \cdot 3H_2O$	mon, 3	Kainite	Mn_2O_3	cub, 6–6.5	Bixbyite
				tetr, 6	Kurnakite
KNO_3	rhomb, 2	Niter	$Mn(OH)_2$	trig	Pyrochroite
K_2SO_4	rhomb	Arcanite	$MnO(OH)$	hex, 4	Feitknechtite
Li_3PO_4	rhomb	Lithiophos-		mon, 4	Manganite
		phate		rhomb, 3.5–4.0	Groutite
$(MgAl_2)O_4$	cub, 7.5–8.0	Spinel (precious)	MnS	cub, 3.5–4.0	Alabandine
			$Mn(S_2)$	cub	Hauerite
$Mg_3(BO_3)_2$	rhomb	Kotoite	$MnSO_4 \cdot H_2O$	mon	Szmikite
$MgCO_3$	trig, 3.7–4.3	Magnesite	$MnSO_4 \cdot 5H_2O$	tricl	Jokokuite
$MgCO_3 \cdot 5H_2O$	mon	Lansfordite	$MnSO_4 \cdot 7H_2O$	mon	Mallardite
$MgC_2O_4 \cdot 2H_2O$	mon	Glushinskite	$MnSiO_3$	tricl, 5.5–6.5	Rhodonite
$MgCl_2$	trig	Chloro-magnesite	Mn_2SiO_4	rhomb, 5.0–5.8	Tephroite
$MgCl_2 \cdot 6H_2O$	mon, 1–2	Bischofite	$MnWO_4$	mon, 4.5	Huebnerite
$(MgCr_2)O_4$	cub	Magnesio-chromite	$(Mn_2Zn)O_4$	tetr, 6	Hetaerolite
			MoO_2	mon	Tugarinovite
MgF_2	tetr	Sellaite	MoO_3	rhomb	Molybdite
$(MgFe_2)O_4$	cub	Magnesio-ferrite	MoS_2	amor, < 1	Jordisite
				hex; trig, 1–1.5	Molybdenite
$MgHAsO_4 \cdot 7H_2O$	mon	Rösslerite	$(NH_4)_2C_2O_4 \cdot H_2O$	rhomb	Oxammite
$MgHPO_4 \cdot 3H_2O$	rhomb	Newberyite	NH_4Cl	cub	Salammoniac
$MgNH_4PO_4 \cdot 6H_2O$	rhomb	Struvite			

(*Continued*)

Formula	Crystal system, hardness	Name	Formula	Crystal system, hardness	Name
NH_4HCO_3	rhomb	Teschemacherite	$NiCl_2 \cdot 6H_2O$	mon	Nickelbischofite
$(NH_4)_2HPO_4$	...	Phosphammite	$Ni_{1-x}O$ ·	cub	Bunsenite
NH_4NO_3	rhomb	Nitrammite	NiS	trig, 3–3.5	Millerite
$(NH_4)_2SO_4$	rhomb	Mascagnite	$Ni(S_2)$	cub	Vaesite
$NaAl(SO_4)_2$ $\cdot 12H_2O$	cub	Soda alum	Ni_3S_2	trig	Heazlewoodite
$NaB_5O_8 \cdot 5H_2O$	mon	Sborgite	$NiSO_4 \cdot 7H_2O$	rhomb	Morenosite
$Na_2B_4O_7 \cdot 4H_2O$	mon	Kernite	NiSb	hex, 5.5	Breithauptite
$Na_2B_4O_7 \cdot 5H_2O$	hex	Tincalconite	(NiSb)S	tricl, 5–5.5	Ullmannite
$Na_2B_4O_7 \cdot 10H_2O$	mon, 2–2.5	Tincal	$Ni_{0.95}Se$	hex	Sederholmite
$Na_2CO_3 \cdot H_2O$	rhomb	Thermonatrite	NiSe	trig	Mäkinenite
$Na_2CO_3 \cdot 10H_2O$	mon	Natron	$Ni(Se_2)$	rhomb, 2.5	Kullerudite
$Na_3CO_3(HCO_3)$ $\cdot 2H_2O$	mon, 2.5–3.0	Trona	$Ni(Te_2)$	trig, 1.0–1.5	Melonite
$Na_2Ca(CO_3)_2$	rhomb	Nyerereite	$Os(S_2)$	cub	Erlichmanite
$Na_2Ca(CO_3)_2$ $\cdot 5H_2O$	mon, 2.5–3.0	Gaylussite	Pb	cub, < 1	Lead native
$Na_2Ca(SO_4)_2$	mon, 2.5–3.0	Glauberite	$PbCO_3$	rhomb, 3–3.5	Cerussite
$Na_2Ca(SO_4)_2$ $\cdot 4H_2O$	mon	Wattevillite	$Pb_2(CO_3)Cl_2$	tetr, 2–3	Phosgenite
NaCl	cub, 2	Halite	$PbCl_2$	rhomb	Cotunnite
$NaCl \cdot 2H_2O$	mon	Hydrohalite	$Pb(Cl)F$	tetr	Matlockite
NaF	cub	Villiaumite	$Pb(Cl)OH$	rhomb	Laurionite
$NaHCO_3$	mon	Nahcolite	$PbCrO_4$	mon, 2.5–3.0	Crocoite
$Na_2HPO_4 \cdot 2H_2O$	rhomb	Dorfmanite	$PbCrO_4 \cdot H_2O$	tricl	Iranite
$NaHSO_4 \cdot H_2O$	mon	Matteuccite	$PbHAsO_4$	mon	Schultenite
$Na_2Mg(CO_3)_2$	hex	Eitelite	$PbMoO_4$	tetr, 2.75–3.0	Wulfenite
$Na_2Mg(CO_3)_2$ $\cdot H_2O$ (?)	...	Northupite	α-PbO	tetr, 2	Litharge
$Na_2Mg(SO_4)_2$ $\cdot 4H_2O$	mon, 2.5–3.0	Bloedite	β-PbO	rhomb, 2	Massicot
NaCl			PbO_2	tetr	Plattnerite
$NaNO_3$	trig, 1.5–2.0	Nitratine	$Pb_5(PO_4)_3Cl$	hex, 3.5–4.0	Pyromorphite
$(NaNb)O_3$	rhomb	Lueshite	$(Pb_2^{II}Pb^{IV})O_4$	tetr	Minium
Na_3PO_4	rhomb	Olympite	PbS_{1+x}	cub, 3	Galena
Na_2SO_4	rhomb, 2.5–3.0	Thenardite	$PbSO_4$	rhomb, 3	Anglesite
$Na_2SO_4 \cdot 10H_2O$	mon, 1.5–2.0	Mirabilite	PbSe	cub, 2.5	Clausthalite
$Na_2Si_2O_5$	mon	Natrosilite	$PbSeO_4$	rhomb	Kerstenite
Ni	cub, 4.3–4.5	Nickel native	$PbSiO_3$	mon	Alamosite
NiAs	hex, 5–5.5	Nickeline	PbTe	cub	Altaite
$NiAs_2$	rhomb, 5.5–6.0	Rammelsbergite	$Pb_5(VO_4)_3Cl$	hex, 2.75–3.0	Vanadinite
$Ni_3(AsO_4)_2$	mon	Xanthiosite	$PbWO_4$	mon	Raspite
$Ni_3(AsO_4)_2 \cdot 8H_2O$	mon, 1.5–2.5	Annabergite		tetr, 2.5–3.0	Stolzite
$Ni(As)S$	cub, 5.5	Gersdorffite	Pd	cub, 4–4.5	Palladium native
			PdO (?)	tetr	Palladinite
			Pt	cub, 4–4.5	Platinum native
			$PtAs_2$	cub, 6–7	Sperrylite
			PtS	tetr	Cooperite

(Continued)

Formula	Crystal system, hardness	Name	Formula	Crystal system, hardness	Name
Re	hex, 8	Rhenium native	β-Sn	tetr, < 1	Tin native
			SnO	tetr	Romarchite
Rh	cub, 4.8–5.3	Rhodium native	SnO_2	tetr, 6–7	Cassiterite
			SnS	rhomb, 2	Herzenbergite
Ru	hex, 5.4–5.9	Ruthenium native	SnS_2	hex; trig	Berndtite
			$SrCO_3$	rhomb, 3.5	Strontianite
$Ru(S_2)$	cub	Laurite	$SrSO_4$	rhomb, 3–3.5	Celestine
α-S	rhomb, 1.5–2.5	Sulfur native	$(Ta_2Fe)O_6$	rhomb	Ferrotantalite
			$(Ta_2Mn)O_6$	rhomb	Mangano-tantalite
β-S	mon	β-Sulfur			
γ-S	mon	Rosickyite	Te	trig, 2–2.5	Tellurium native
Sb	trig, 3–3.5	Antimony native			
			TeO_2	rhomb	Tellurite
α-Sb_2O_3	rhomb, 2.5–3.0	Valentinite	ThO_2	cub, 6.5	Thorianite
			$ThSiO_4$	mon	Huttonite
β-Sb_2O_3	cub	Senarmontite		tetr, 5	Thorite
Sb_2S_3	amor, < 1	Metastibnite	$ThSiO_4 \cdot nH_2O$	tetr	Orangite
	rhomb, 2	Stibnite	$(TiFe)O_3$	trig, 5.6	Ilmenite
$[SbS_3],Ag_3$	trig, 2.5	Pyrargyrite	$(TiFe_2)O_4$	cub	Ulvöspinel
$[SbS_4],Ag_5$	rhomb, 2–2.5	Stephanite	$(TiMn)O_3$	trig, 5.6	Pyrophanite
$(Sb^{III}Sb^V)O_4$	rhomb	Cervantite	TiN	cub	Osbornite
Sb_2Te_3	trig	Tellur-antimony	TiO_{1+x}	cub	Hongquiite
			α-TiO_2	rhomb, 5.5–6.0	Brookite
$ScPO_4 \cdot 2H_2O$	mon	Kolbeckite			
Se	trig, 2	Selenium native	β-TiO_2	tetr, 5.5–6.0	Anatase
			γ-TiO_2	tetr, 6.5	Rutile
SeO_2	tetr	Downeyite	$(TiPb)O_3$	tetr	Macedonite
SiC	hex	Moissanite	Tl_2O_3	cub	Avicennite
$[SiF_6],K_2$	cub	Hieratite	Tl_2S	trig	Carlinite
$[SiF_6],(NH_4)_2$	cub	Cryptohalite	UO_{2+x}	cub, 5.6	Uraninite
	trig, 2.5	Bararite	$(UO_2)CO_3$	rhomb	Rutherfordine
$[SiF_6],Na_2$	trig	Malladrite	$(UO_2)CO_3$ $\cdot(1.5 \div 2)H_2O$	rhomb	Joliotite
SiO_2	amor	Lechatelierite			
	hex	β-Quartz	$UO_2(OH)_2 \cdot H_2O$	rhomb, 2.5	Schoepite
	hex	β-Tridymite	$USiO_4$	tetr, 5–6	Coffinite
	cub	β-Cristobalite	$(U_2^VU^{VI})O_8$	cub, 4.5	Nasturan
	cub, 6.5–7.0	Melanophlogite	$[V(H_2O)_4(SO_4)O]$ $\cdot H_2O$	mon	Minasragrite
	mon	Coesite	VO_2	rhomb	Paramontro-seite
	rhomb, 7	α-Tridymite			
	tetr	Keatite	$VO_2 \cdot 0.67H_2O$	mon	Doloresite
	tetr, 6.5	α-Cristobalite	V_2O_3	trig	Karelianite
	tetr	Stishovite	V_2O_5	rhomb	Shcherbinaite
	trig, 7	α-Quartz	$V_2O_5 \cdot H_2O$	...	Alaite
$SiO_2 \cdot 0.33H_2O$	rhomb	Silhydrite	$V_2O_5 \cdot 3H_2O$	mon	Navajoite
$SiO_2 \cdot nH_2O$	amor	Opal	VO(OH)	rhomb, 4–4.2	Montroseite

(Continued)

Formula	Crystal system, hardness	Name	Formula	Crystal system, hardness	Name
$VO(OH)_2$	mon	Duttonite	ZnO	hex, 4	Zincite
VS_{2+x}	mon, 3.5	Patronite	$Zn_3(PO_4)_2 \cdot 4H_2O$	rhomb	Hopeite
$WO_3 \cdot H_2O$	rhomb, 2	Tungstite	α-ZnS	cub, 3.5–4.0	Sphalerite
$WO_3 \cdot 2H_2O$	amor, 2	Meymacite	β-ZnS	hex, 3.5–4.0	Wurtzite
	mon, 2.5	Hydrotungstite	$ZnSO_4$	rhomb	Zinkosite
WS_2	hex; trig	Tungstenite	$ZnSO_4 \cdot 7H_2O$	rhomb	Goslarite
YPO_4	tetr, 4–5	Xenotime	$ZnSe$	cub	Stilleite
Zn	hex, 2.5	Zinc native	Zn_2SiO_4	trig, 5.5	Willemite
$(ZnAl_2)O_4$	cub, 7.5–8.0	Gahnite	$Zn_2SiO_4 \cdot H_2O$	trig	Hemimorphite
$Zn_3(AsO_4)_2$ $\cdot 8H_2O$	mon	Köttigite	$[Zr(H_2O)_4(SO_4)_2]$	rhomb	Zircosulfate
			ZrO_2	mon, 6.5	Baddeleyite
$ZnCO_3$	trig, 4–4.5	Smithsonite	$ZrSiO_4$	tetr, 7.5	Zircon

Index of Trivial Names [4, 7–8, 10, 28, 31, 35]

Formula	Name	Formula	Name
$AgCl$	Horn silver	$BiNO_3(OH)_2$	Spanish white
Ag_3N	Fulminating silver	Bi_2O_3 (mon)	Bismuth ocher
Ag_2S (rhomb)	Silver glance	Bi_2S_3	Bismuth glance
$(AgSb)S_2$	Silver-antimony ore	CCl_2O	Phosgene
$[AlH_4],Li$	Lithium alanate	CH_4	Marsh (mine) gas
	Lithium-aluminum hydride	CH_3COOH (sld)	Glacial (acetic) acid
		CH_3OH	Wood alcohol
$AlNH_4(SO_4)_2 \cdot 12H_2O$	Ammonium alum	C_2H_5OH	Spirit of wine
$Al_2O_3 \cdot 0.25H_2O$	Alumogel		Rectificate
$Al_2O_3 \cdot nH_2O$	Bauxite	$C(NH_2)_2O$	Urea
$Al_2(SO_4)_3 \cdot 18H_2O$	Aluminum coagulant	CO	Suffocating gas
		CO_2	Carbonic acid gas
As_2O_3 (amor)	White arsenic	CO_2 (sld)	Dry ice
$[AuCl_4],H \cdot 4H_2O$	Gold chloride	$CSCl_2$	Thiophosgene
$[AuCl_4],Na \cdot 2H_2O$	Gold(en) salt	$CS(NH_2)_2$	Thiourea
$B_3H_6N_3$	Borazol	CaC_2	Calcium carbide
	Inorganic benzene	$Ca(CH_3COO)_2 \cdot 2H_2O$	Grey acetate of lime
α-BN	White graphite		Vinegar powder
β-BN	Borazon	$CaCO_3$ (trig)	Lime (Iceland) spar
$BaMnO_4$	Kassel green		Marble
	Manganese green	CaF_2	Fluorspar
$Ba(NO_3)_2$	Barium saltpeter	$CaHPO_4 \cdot 2H_2O$	Precipitate
$Ba(OH)_2$	Caustic baryta	$Ca(H_2PO_4)_2 \cdot H_2O$	Double superphosphate
$BaSO_4$	Baryta white	$CaMg(CO_3)_2$	Brown (bitter) spar
	Blanc fixe	$Ca(NO_3)_2 \cdot 4H_2O$	Lime (Norwegian) saltpeter
	Heavy spar		
$BiNO_3(OH)_2$	Pearl white		

(Continued)

Formula	Name	Formula	Name
CaO	Air-hardening lime	$Fe(As)S$	Arsenical pyrites
	Quicklime		Poisonous pyrites
	Unslaked (burned) lime	Fe_3C	Cementite
		$[Fe(C_5H_5)_2]$	Ferrocene
$Ca(OH)_2$	Slaked (caustic) lime	$[Fe(CN)_6],K_3$	Red prussiate of potash
	Powder lime		Gmelin's salt
$CaSO_4 \cdot 0.5H_2O$	Burned gypsum (stucco, gypsum wall plaster)		Potassium ferricyanide
$CaWO_4$	Tungstein	$[Fe(CN)_6],K_4 \cdot 3H_2O$	Yellow prussiate of potash
	Scheele's spar		Potassium ferrocyanide
CdS (hex)	Cadmium blende		
$(CoAl_2)O_4$	Thenard's blue	$[Fe^{II}(CN)_6],KFe$	Prussian (Paris) blue
$Co(As)S$	Cobalt glance		Turnbull's blue
$[Co(C_5H_5)_2]$	Cobaltocene	$FeCO_3$	Iron spar
$(Co^{II}Co_2^{III})O_4$	Cobalt pyrites	$(FeCu)S_2$	Copper pyrites
$[Co(NO_2)_6],K_3$	Fischer's salt	$(Fe^{II}Fe_2^{III})O_4$	Magnetic ironstone
$CoSO_4 \cdot 7H_2O$	Cobalt (pink-red) vitriol		Black iron ore
		$FeNH_4(SO_4)_2 \cdot 12H_2O$	Ferric ammonium alum
$[Cr(C_5H_5)_2]$	Chromocene		
$(Cr_2Fe)O_4$	Chrome ironstone	$Fe(NH_4)_2(SO_4)_2 \cdot 6H_2O$	Mohr's salt
$[Cr(NH_3)_2(NCS)_4]^-,$ $NH_4^+ \cdot H_2O$	Reinecke's salt	$[Fe(NO^+)(CN)_5],Na_2$	Sodium nitroprusside
$CrNH_4(SO_4)_2 \cdot 12H_2O$	Chrome-ammonium alum	Fe_2O_3	Iron glance
			Pyrite cinder
Cr_2O_3	Green crown		Red ironstone
$Cr_2O_3 \cdot nH_2O$	Guignet's green		Crocus
$CrSO_4 \cdot 7H_2O$	Light-blue (chromium) vitriol	$Fe_2O_3 \cdot nH_2O$	Brown (bog) ironstone
$CrTl(SO_4)_2 \cdot 12H_2O$	Chrome-thallium alum	$FeO(OH)$ (rhomb)	Acicular iron ore
$CsAl(SO_4)_2 \cdot 12H_2O$	Cesium alum	$Fe_3(PO_4)_2 \cdot 8H_2O$	Blue iron ore
$CsCr(SO_4)_2 \cdot 12H_2O$	Chrome-cesium alum	$Fe_{1-x}S$	Magnetic pyrites
		$\beta\text{-}Fe(S_2)$	Iron pyrites
$CsFe(SO_4)_2 \cdot 12H_2O$	Iron-cesium alum	$FeSO_4 \cdot 7H_2O$	Iron (green) vitriol
$CsGa(SO_4)_2 \cdot 12H_2O$	Gallium-cesium alum	$Fe_2(SO_4)_3 \cdot 9H_2O$	Ferric tanning agent
			Ferric coagulant
$CsMn(SO_4)_2 \cdot 12H_2O$	Manganese-cesium alum	$GaNH_4(SO_4)_2 \cdot 12H_2O$	Gallium-ammonium alum
$CsV(SO_4)_2 \cdot 12H_2O$	Vanadium-cesium alum	H_2SO_4	Monohydrate
		$H_2SO_3(O_2)$	Caro's acid
$Cu_3(AsO_4)_2 \cdot 4H_2O$	Veronese green	$Hg(CNO)_2 \cdot 0.5H_2O$	Fulminating mercury
Cu_2O	Red (cupric) oxide		
Cu_2S	Copper glance	$HgCl_2$	Corrosive sublimate
$CuSO_4 \cdot 5H_2O$	Blue (copper) vitriol	Hg_2Cl_2	Horn quicksilver ore
	Bluestone	$\beta\text{-}HgS$	Mercury black
D_2O	Heavy water	$[I(I_2)],K \cdot H_2O$	Johnson's salt
$(FeAl_2)O_4$	Iron spinel	$I_3N \cdot nNH_3$	Nitrogen iodide

(Continued)

Formula	Name	Formula	Name
$KAl(SO_4)_2$	Burnt alum	$Na_2B_4O_7$	Burnt borax
$KAl(SO_4)_2 \cdot 12H_2O$	Potash alum	$Na_2B_4O_7 \cdot 5H_2O$	Jewelry borax
	Alum tanning agent	$Na_2B_4O_7 \cdot 10H_2O$	Borax
$KAl_3(SO_4)_2(OH)_6$	Alumstone	NaCN	Soda cyanide
K_2CO_3	Potash	Na_2CO_3	Washing (calcined)
$KClO_3$	Berthollet's salt		soda
$K_2Cr_2O_7$	Potassium	$Na_2CO_3 \cdot 10H_2O$	Soda
	bichromate	NaCl	Rock (table,
$KCr(SO_4)_2 \cdot 12H_2O$	Chrome-potassium		common) salt
	alum	$NaClO_2 \cdot 3H_2O$	Texton
	Chrome tanning	$NaHCO_3$	Bicarbonate
	agent		Baking soda
$KFe(SO_4)_2 \cdot 12H_2O$	Ferric potassium	$NaNO_3$	Soda (Chile)
	alum		saltpeter
$K(HC_4H_4O_6)$	Potassium bitartrate	Na_2O	Soda (soda oxide)
	Tartar	NaOH	Caustic soda
$K(HC_2O_4)$	Potassium bioxalate	$Na_3PO_4 \cdot 12H_2O$	Trisoda-phosphate
$K_2Mg(SO_4)_2 \cdot 6H_2O$	Potash magnesia	Na_2S_n	Liver of sulfur
	Schoenite	Na_2SO_3	Sulfite
KNO_3	Potassium (Indian)	Na_2SO_4	Sulfate
	saltpeter	$Na_2SO_4 \cdot 10H_2O$	Glauber's salt
$KNa(C_4H_4O_6) \cdot 4H_2O$	Seignette's salt	$Na_2S_2O_4 \cdot 2H_2O$	Hydrosulfite
KOH	Caustic potash	$Na_2SO_3S \cdot 5H_2O$	Antichlor
$KRh(SO_4)_2 \cdot 12H_2O$	Rhodium-potassium		Hyposulfite
	alum		Fixing agent
$K_2S_2O_5$	Metabisulfite		Fixing salt
$KV(SO_4)_2 \cdot 12H_2O$	Vanadium-potas-	NiAs	Red (nickel) pyrites
	sium alum		Copper nickel
$MgCO_3$	Talc (bitter) spar	Ni_3C	Pyrophoren
$Mg(ClO_4)_2$	Anhydrone	$[Ni(C_5H_5)_2]$	Nickelocene
$Mg(NO_3)_2 \cdot 6H_2O$	Magnesium saltpeter	NiS	Hair (yellow) pyrites
MgO	Calcined magnesia		Nickel blende
$MgSO_4 \cdot 7H_2O$	Epsom (bitter) salt	$NiSO_4 \cdot 7H_2O$	Nickel (green)
$(MnAl_2)O_4$	Mangan-spinel		vitriol
$MnCO_3$	Scarlet spar	$[Os(C_5H_5)_2]$	Osmocene
$[Mn(CO)_3(C_5H_5)]$	Cymantrene	$Pb(CH_3COO)_2 \cdot 3H_2O$	Sugar of lead
$(Mn^{II}Mn_2^{III})O_4$	Lustrous pyrolusite	$PbCO_3$	White lead ore
MnO_{2-x} (tetr)	Grey (gougy)	$PbCrO_4$	Red lead ore
	manganese ore		Lead crown
MnS	Mangan-blende	$PbMoO_4$	Yellow lead ore
	Manganese glance		Lead molybdic spar
$Mn(S_2)$	Manganese pyrites	$(Pb_2^{II}Pb^{IV})O_4$	Minium
MoO_3	Molybdic ocher	PBS_{1+x}	Lead glance
MoS_2	Molybdenum glance	$PbSO_4$	Lead vitriol
NH_4NO_3	Ammonium	$PbWO_4$ (tetr)	Scheele's lead ore
	saltpeter	$[Pt(C_2H_4)Cl_3],K \cdot H_2O$	Zeise's salt
N_2O	Laughing gas	$[PtCl_6],(NH_4)_2$	Platinum
$NaAl(SO_4)_2 \cdot 12H_2O$	Soda alum		salammoniac

Formula	Name	Formula	Name
trans-[Pt(NH$_3$)$_2$Cl$_2$]	Reise base II chloride	SiO$_2$·nH$_2$O	Silica gel
		SnCl$_4$ (lq)	Oil of tin
cis-[Pt(NH$_3$)$_2$Cl$_2$]	Peyronet's salt	[SnCl$_6$],(NH$_4$)$_2$	Pink salt
trans-[Pt(NH$_3$)$_2$Cl$_4$]	Gérard's salt	[Sn(H$_2$O)Cl$_2$]·H$_2$O	Tin salt
cis-[Pt(NH$_3$)$_2$Cl$_4$]	Cleve's salt	SnO$_2$	Tinstone
[Pt(NH$_3$)$_4$]Cl$_2$·H$_2$O	Reise base I chloride	[Sn(OH)$_6$],Na$_2$	Preparing salt
		SnS$_2$	Mosaic (mock) gold
[Pt(NH$_3$)$_4$][Pt^{II}Cl$_4$]	Magnus salt	T$_2$O	Ultraheavy water
RbAl(SO$_4$)$_2$·12H$_2$O	Rubidium alum	[Ti(C$_5$H$_5$)$_2$]	Titanocene
RbCr(SO$_4$)$_2$·12H$_2$O	Chrome-rubidium alum	(TiFe)O$_3$	Titanic iron ore
		(TiFe$_2$)O$_4$	Titanium-iron spinel
RbFe(SO$_4$)$_2$·12H$_2$O	Ferric-rubidium alum	γ-TiO$_2$	Titanium white
		TlAl(SO$_4$)$_2$·12H$_2$O	Thallium alum
RbMn(SO$_4$)$_2$·12H$_2$O	Manganese-rubidium alum	UO$_{2+x}$	Pitchblende
		[V(C$_5$H$_5$)$_2$]	Vanadocene
RbV(SO$_4$)$_2$·12H$_2$O	Vanadium-rubidium alum	V(NH$_4$)(SO$_4$)$_2$·12H$_2$O	Vanadium-ammonium alum
[Ru(C$_5$H$_5$)$_2$]	Ruthenocene	VSO$_4$·7H$_2$O	Vanadium (violet) vitriol
SO$_2$	Sulfurous anhydride		
SbCl$_3$ (lq)	Oil of antimony	WO$_3$·H$_2$O	Tungstic ocher ·
Sb(Cl)O	Algaroth (powder)	YPO$_4$	Yittrium spar
Sb$_2$S$_3$	Antimonite	(ZnAl$_2$)O$_4$	Zinc spinel
	Grey antimony ore	ZnCO$_3$	Calamine, noble Zinc spar
	Antimony glance		
[SbS$_4$],Na$_3$·9H$_2$O	Schlippe's salt	ZnO	Zinc white
SiC	Carborundum	α-ZnS	Zinc blende
SiO$_2$ (amor)	White black	ZnSO$_4$·7H$_2$O	White (zinc) vitriol
	Diatomite	Zn$_2$SiO$_4$·H$_2$O	Silicic (common) calamine
	Infusorial earth		Silicon-zinc ore
	Kieselguhr		
SiO$_2$ (α-quartz)	Rock crystal	ZrO$_2$	Zirconium white
	Quartz sand		Zirconia

2

Atomic Properties

2.1 Relative Atomic Masses [1, 37]

The values of relative atomic masses (A_r) are derived from the 1987 International table and are given in the atomic mass units [the carbon scale, wherein $^{12}C = 12$ (exactly)], the accuracy of the last significant digit being indicated. Artificial elements are marked with a superscript *. The subscript left of the element symbol stands for the serial number of the element in the Periodic Table. The elements with serial numbers from 106 to 109 are listed at the end of this table.

Element	A_r	Element	A_r	Element	A_r
$_{89}^{*}$Ac	227.027750±3	$_{97}^{*}$Bk	247.070300±6	$_{29}$Cu	63.546±3
$_{47}$Ag	107.8682±2	$_{35}$Br	79.904±1	$_{66}$Dy	162.50±3
$_{13}$Al	26.981539±5	$_{6}$C	12.011±1	$_{68}$Er	167.26±3
$_{95}^{*}$Am	243.061375±3	$_{20}$Ca	40.078±4	$_{99}^{*}$Es	252.082944±23
$_{18}$Ar	39.948±1	$_{48}$Cd	112.411±8	$_{63}$Eu	151.965±9
$_{33}$As	74.92159±2	$_{58}$Ce	140.115±4	$_{9}$F	18.9984032±9
$_{85}^{*}$At	209.987126±12	$_{98}^{*}$Cf	251.079580±5	$_{26}$Fe	55.847±3
$_{79}$Au	196.96654±3	$_{17}$Cl	35.4527±9	$_{100}^{*}$Fm	257.095099±8
$_{5}$B	10.811±5	$_{96}^{*}$Cm	247.070347±5	$_{87}^{*}$Fr	223.019733±4
$_{56}$Ba	137.327±7	$_{27}$Co	58.93320±1	$_{31}$Ga	69.723±1
$_{4}$Be	9.012182±3	$_{24}$Cr	51.9961±6	$_{64}$Gd	157.25±3
$_{83}$Bi	208.98037±3	$_{55}$Cs	132.90543±5	$_{32}$Ge	72.61±2

(Continued)

Element	A_r	Element	A_r	Element	A_r
$_1$H	1.007825035±12	$_{102}^{\bullet}$No	259.100931±12	$_{62}$Sm	150.36±3
$_2$He	4.002602±2	$_{93}^{\bullet}$Np	237.0481678±23	$_{50}$Sn	118.710±7
$_{72}$Hf	178.49±2	$_{105}^{\bullet}$Ns	262.11376±16	$_{38}$Sr	87.62±1
$_{80}$Hg	200.59±3	$_8$O	15.9994±3	$_{73}$Ta	180.9479±1
$_{67}$Ho	164.93032±3	$_{76}$Os	190.2±1	$_{65}$Tb	158.92534±3
$_{53}$I	126.90447±3	$_{15}$P	30.973762±4	$_{43}^{\bullet}$Tc	97.907215±4
$_{49}$In	114.82±1	$_{91}^{\bullet}$Pa	231.03588±2	$_{52}$Te	127.60±3
$_{77}$Ir	192.22±3	$_{82}$Pb	207.2±1	$_{90}$Th	232.0381±1
$_{19}$K	39.0983±1	$_{46}$Pd	106.42±1	$_{22}$Ti	47.88±3
$_{36}$Kr	83.80±1	$_{61}^{\bullet}$Pm	144.912743±4	$_{81}$Tl	204.3833±2
$_{104}^{\bullet}$Ku	261.10869±22	$_{84}^{\bullet}$Po	208.982404±5	$_{69}$Tm	168.93421±3
$_{57}$La	138.9055±2	$_{59}$Pr	140.90765±3	$_{92}$U	238.0289±1
$_3$Li	6.941±2	$_{78}$Pt	195.08±3	$_{23}$V	50.9415±1
$_{103}^{\bullet}$Lr	260.105320±60	$_{94}^{\bullet}$Pu	244.064199±5	$_{74}$W	183.85±3
$_{71}$Lu	174.967±1	$_{88}^{\bullet}$Ra	226.025403±3	$_{54}$Xe	131.29±2
$_{101}^{\bullet}$Md	258.09857±12	$_{37}$Rb	85.4678±3	$_{39}$Y	88.90585±2
$_{12}$Mg	24.3050±6	$_{75}$Re	186.207±1	$_{70}$Yb	173.04±3
$_{25}$Mn	54.93805±1	$_{45}$Rh	102.90550±3	$_{30}$Zn	65.39±2
$_{42}$Mo	95.94±1	$_{86}^{\bullet}$Ra	222.017571±3	$_{40}$Zr	91.224±2
$_7$N	14.00674±7	$_{44}$Ru	101.07±2	$^{\bullet}$106	263.11822±3
$_{11}$Na	22.989768±6	$_{16}$S	32.066±6	$^{\bullet}$107	262.12293±45
$_{41}$Nb	92.90638±2	$_{51}$Sb	121.75±3	$^{\bullet}$108	265.13016±99
$_{60}$Nd	144.24±3	$_{21}$Sc	44.955910±9	$^{\bullet}$109	266.13764±45
$_{10}$Ne	20.1797±6	$_{34}$Se	78.96±3		
$_{28}$Ni	58.69±1	$_{14}$Si	28.0855±3		

2.2 Properties of Natural Isotopes [1, 37]

The table lists the relative atomic mass (A_r) and the content (molar fraction x) of naturally occurring stable and radioactive isotopes in a natural element. Radioactive isotopes are marked with a superscript $^{\bullet}$ (cf. Sec. 2.3).

Isotope	A_r, amu	x, % (mol)	Isotope	A_r, amu	x, % (mol)
^{107}Ag	106.9051	51.839	^{132}Ba	131.9050	0.101
^{109}Ag	108.9048	48.161	^{134}Ba	133.9045	2.417
^{27}Al	26.9815	100	^{135}Ba	134.9057	6.592
^{36}Ar	35.9675	0.337[a]	^{136}Ba	135.9046	7.854
^{38}Ar	37.9627	0.063[a]	^{137}Ba	136.9058	11.23
^{40}Ar	39.9624	99.600[a]	^{138}Ba	137.9052	71.70
^{75}As	74.9216	100	^{9}Be	9.0122	100
^{197}Au	196.9665	100	^{209}Bi	208.9804	100
^{10}B	10.0129	19.9	^{79}Br	78.9183	50.69
^{11}B	11.0093	80.1	^{81}Br	80.9193	49.31
^{130}Ba	129.9063	0.106	^{12}C	12.0000	98.90

Isotope	A_r, amu	x, % (mol)	Isotope	A_r, amu	x, % (mol)
^{13}C	13.0034	1.10	^{57}Fe	56.9354	2.2
^{40}Ca	39.9626	96.941	^{58}Fe	57.9333	0.28
^{42}Ca	41.9586	0.647	^{69}Ga	68.9256	60.1
^{43}Ca	42.9588	0.135	^{71}Ga	70.9247	39.9
^{44}Ca	43.9555	2.086	^{*152}Gd	151.9198	0.20
^{46}Ca	45.9537	0.004	^{154}Gd	153.9209	2.18
^{48}Ca	47.9525	0.187	^{155}Gd	154.9226	14.80
^{106}Cd	105.9065	1.25	^{156}Gd	155.9221	20.47
^{108}Cd	107.9042	0.89	^{157}Gd	156.9240	15.65
^{110}Cd	109.9030	12.49	^{158}Gd	157.9240	24.84
^{111}Cd	110.9042	12.80	^{160}Gd	159.9270	21.86
^{112}Cd	111.9028	24.13	^{70}Ge	69.9242	20.5
^{113}Cd	112.9044	12.22	^{72}Ge	71.9221	27.4
^{114}Cd	113.9034	28.73	^{73}Ge	72.9235	7.8
^{116}Cd	115.9048	7.49	^{74}Ge	73.9212	36.5
^{136}Ce	135.9071	0.19	^{76}Ge	75.9214	7.8
^{138}Ce	137.9060	0.25	^{1}H	1.0078	99.985[b]
^{140}Ce	139.9054	88.48	^{2}H	2.0141	0.015[b]
^{142}Ce	141.9092	11.08	^{3}He	3.0160	0.000138[a]
^{35}Cl	34.9689	75.77	^{4}He	4.0026	99.999862[a]
^{37}Cl	36.9658	24.23	^{*174}Hf	173.9400	0.162
^{59}Co	58.9332	100	^{176}Hf	175.9414	5.206
^{50}Cr	49.9460	4.345	^{177}Hf	176.9432	18.606
^{52}Cr	51.9405	83.789	^{178}Hf	177.9437	27.297
^{53}Cr	52.9407	9.501	^{179}Hf	178.9458	13.629
^{54}Cr	53.9389	2.365	^{180}Hf	179.9465	35.100
^{133}Cs	132.9054	100	^{196}Hg	195.9658	0.14
^{63}Cu	62.9296	69.17	^{198}Hg	197.9667	10.02
^{65}Cu	64.9278	30.83	^{199}Hg	198.9682	16.84
D	see ^{2}H		^{200}Hg	199.9683	23.13
^{156}Dy	155.9243	0.06	^{201}Hg	200.9703	13.22
^{158}Dy	157.9244	0.10	^{202}Hg	201.9706	29.80
^{160}Dy	159.9252	2.34	^{204}Hg	203.9735	6.85
^{161}Dy	160.9269	18.9	^{165}Ho	164.9303	100
^{162}Dy	161.9268	25.5	^{127}I	126.9045	100
^{163}Dy	162.9287	24.9	^{113}In	112.9041	4.3
^{164}Dy	163.9292	28.2	^{*115}In	114.9039	95.7
^{162}Er	161.9288	0.14	^{191}Ir	190.9606	37.3
^{164}Er	163.9292	1.61	^{193}Ir	192.9629	62.7
^{166}Er	165.9303	33.6	^{39}K	38.9637	93.2581
^{167}Er	166.9300	22.95	^{*40}K	39.9640	0.0117
^{168}Er	167.9324	26.8	^{41}K	40.9620	6.7302
^{170}Er	169.9355	14.9	^{78}Kr	77.9204	0.35[a]
^{151}Eu	150.9197	47.8	^{80}Kr	79.9164	2.25[a]
^{153}Eu	152.9212	52.2	^{82}Kr	81.9135	11.60[a]
^{19}F	18.9984	100	^{83}Kr	82.9141	11.50[a]
^{54}Fe	53.9396	5.8	^{84}Kr	83.9115	57.00[a]
^{56}Fe	55.9349	91.72	^{86}Kr	85.9106	17.30[a]

(*Continued*)

Isotope	A_r, amu	x, % (mol)	Isotope	A_r, amu	x, % (mol)
*^{138}La	137.9071	0.09	^{206}Pb	205.9744	24.1
^{139}La	138.9063	99.91	^{207}Pb	206.9759	22.1
^{6}Li	6.0151	7.5	^{208}Pb	207.9766	52.4
^{7}Li	7.0160	92.5	^{102}Pd	101.9056	1.020
^{175}Lu	174.9408	97.41	^{104}Pd	103.9040	11.14
*^{176}Lu	175.9427	2.59	^{105}Pd	104.9051	22.33
^{24}Mg	23.9850	78.99	^{106}Pd	105.9035	27.33
^{25}Mg	24.9858	10.00	^{108}Pd	107.9039	26.46
^{26}Mg	25.9826	11.01	^{110}Pd	109.9052	11.72
^{55}Mn	54.9380	100	^{141}Pr	140.9076	100
^{92}Mo	91.9068	14.84	*^{190}Pt	189.9599	0.01
^{94}Mo	93.9051	9.25	^{192}Pt	191.9610	0.79
^{95}Mo	94.9058	15.92	^{194}Pt	193.9627	32.9
^{96}Mo	95.9047	16.68	^{195}Pt	194.9648	33.8
^{97}Mo	96.9060	9.55	^{196}Pt	195.9649	25.3
^{98}Mo	97.9054	24.13	^{198}Pt	197.9679	7.2
^{100}Mo	99.9075	9.63	^{85}Rb	84.9118	72.165
^{14}N	14.0031	99.634	*^{87}Rb	86.9092	27.835
^{15}N	15.0001	0.366	^{185}Re	184.9530	37.40
^{23}Na	22.9898	100	*^{187}Re	186.9557	62.60
^{93}Nb	92.9064	100	^{103}Rh	102.9055	100
^{142}Nd	141.9077	27.13	^{96}Ru	95.9076	5.52
^{143}Nd	142.9098	12.18	^{98}Ru	97.9053	1.88
*^{144}Nd	143.9101	23.80	^{99}Ru	98.9059	12.7
^{145}Nd	144.9126	8.30	^{100}Ru	99.9042	12.6
^{146}Nd	145.9131	17.19	^{101}Ru	100.9056	17.0
^{148}Nd	147.9169	5.76	^{102}Ru	101.9043	31.6
^{150}Nd	149.9209	5.64	^{104}Ru	103.9054	18.7
^{20}Ne	19.9924	90.48[a]	^{32}S	31.9721	95.02
^{21}Ne	20.9938	0.27[a]	^{33}S	32.9715	0.75
^{22}Ne	21.9914	9.25[a]	^{34}S	33.9679	4.21
^{58}Ni	57.9353	68.27	^{36}S	35.9671	0.02
^{60}Ni	59.9308	26.10	^{121}Sb	120.9038	57.3
^{61}Ni	60.9311	1.13	^{123}Sb	122.9042	42.7
^{62}Ni	61.9283	3.59	^{45}Sc	44.9559	100
^{64}Ni	63.9280	0.91	^{74}Se	73.9225	0.9
^{16}O	15.9949	99.762	^{76}Se	75.9192	9.0
^{17}O	16.9991	0.038	^{77}Se	76.9199	7.6
^{18}O	17.9992	0.200	^{78}Se	77.9173	23.6
^{184}Os	183.9525	0.02	^{80}Se	79.9165	49.7
*^{186}Os	185.9538	1.58	^{82}Se	81.9167	9.2
^{187}Os	186.9557	1.6	^{28}Si	27.9769	92.23
^{188}Os	187.9558	13.3	^{29}Si	28.9765	4.67
^{189}Os	188.9581	16.1	^{30}Si	29.9738	3.10
^{190}Os	189.9584	26.4	^{144}Sm	143.9120	3.1
^{192}Os	191.9615	41.0	*^{147}Sm	146.9149	15.0
^{31}P	30.9738	100	*^{148}Sm	147.9148	11.3
^{204}Pb	203.9730	1.4	^{149}Sm	148.9172	13.8

(Continued)

Isotope	A_r, amu	x, % (mol)	Isotope	A_r, amu	x, % (mol)
^{150}Sm	149.9173	7.4	*^{234}U	234.0409	0.0055
^{152}Sm	151.9197	26.7	*^{235}U	235.0439	0.7200
^{154}Sm	153.9222	22.7	*^{238}U	238.0508	99.2745
^{112}Sn	111.9048	0.97	^{50}V	49.9472	0.250
^{114}Sn	113.9028	0.65	^{51}V	50.9440	99.750
^{115}Sn	114.9033	0.36	^{180}W	179.9467	0.13
^{116}Sn	115.9017	14.53	^{182}W	181.9482	26.3
^{117}Sn	116.9030	7.68	^{183}W	182.9502	14.3
^{118}Sn	117.9016	24.22	^{184}W	183.9509	30.67
^{119}Sn	118.9033	8.58	^{186}W	185.9544	28.6
^{120}Sn	119.9022	32.59	^{124}Xe	123.9059	0.10[a]
^{122}Sn	121.9034	4.631	^{126}Xe	125.9043	0.09[a]
^{124}Sn	123.9053	5.79	^{128}Xe	127.9035	1.91[a]
^{84}Sr	83.9134	0.56	^{129}Xe	128.9048	26.40[a]
^{86}Sr	85.9093	9.86	^{130}Xe	129.9035	4.10[a]
^{87}Sr	86.9089	7.00	^{131}Xe	130.9051	21.20[a]
^{88}Sr	87.9056	82.58	^{132}Xe	131.9041	26.90[a]
*^{180}Ta	179.9475	0.012	^{134}Xe	133.9054	10.40[a]
^{181}Ta	180.9480	99.988	^{136}Xe	135.9072	8.90[a]
^{159}Tb	158.9253	100	^{89}Y	88.9058	100
^{120}Te	119.9040	0.096	^{168}Yb	167.9339	0.13
^{122}Te	121.9031	2.60	^{170}Yb	169.9348	3.05
*^{123}Te	122.9043	0.908	^{171}Yb	170.9363	14.3
^{124}Te	123.9028	4.816	^{172}Yb	171.9364	21.9
^{125}Te	124.9044	7.14	^{173}Yb	172.9382	16.12
^{126}Te	125.9033	18.95	^{174}Yb	173.9389	31.8
^{128}Te	127.9045	31.69	^{176}Yb	175.9426	12.7
^{130}Te	129.9062	33.80	^{64}Zn	63.9291	48.6
*^{232}Th	232.0381	100	^{66}Zn	65.9260	27.9
^{46}Ti	45.9526	8.0	^{67}Zn	66.9271	4.1
^{47}Ti	46.9518	7.3	^{68}Zn	67.9249	18.8
^{48}Ti	47.9479	73.8	^{70}Zn	69.9253	0.6
^{49}Ti	48.9479	5.5	^{90}Zr	89.9047	51.45
^{50}Ti	49.9448	5.4	^{91}Zr	90.9056	11.22
^{203}Tl	202.9723	29.524	^{92}Zr	91.9050	17.15
^{205}Tl	204.9744	70.476	^{94}Zr	93.9063	17.38
^{169}Tm	168.9342	100	^{96}Zr	95.9083	2.80

[a] In air.
[b] In water.

2.3 Properties of Radioactive Isotopes [1, 10, 37]

The table includes the long-lived artificial radioactive isotopes of elements and also the isotopes used as tracers, and the natural radioactive isotopes (in bold letters). The isotopes of elements 106 and 107 are placed at the end of the Table.

This Table lists the relative atomic masses (A_r) and the half-life ($T_{1/2}$) of the isotopes.

Isotope	A_r, amu	$T_{1/2}$	Isotope	A_r, amu	$T_{1/2}$
^{227}Ac	227.0275	21.773 yr	^{159}Gd	158.9264	18.6 h
^{228}Ac	228.0310	6.13 h	^{68}Ge	67.9281	288 d
^{108}Ag	107.9060	127 yr	^{3}H (T)	3.0160	12.34 yr
^{110}Ag	109.9061	250 d	^{6}He	6.0189	0.808 s
^{26}Al	25.9869	$72 \cdot 10^5$ yr	^{174}**Hf**	173.9400	$2.0 \cdot 10^{15}$ yr
^{28}Al	27.9819	2.24 min	^{181}Hf	180.9491	42.4 d
^{243}Am	243.0614	$7.37 \cdot 10^3$ yr	^{182}Hf	181.9506	$9 \cdot 10^6$ yr
^{39}Ar	38.9643	2.69 yr	^{194}Hg	193.9654	260 yr
^{73}As	72.9238	80.3 d	^{203}Hg	202.9729	46.7 d
^{76}As	75.9224	26.3 h	^{166}Ho	165.9323	$1.2 \cdot 10^3$ yr
^{210}At	209.9871	8.1 h	^{129}I	128.9050	$1.6 \cdot 10^7$ yr
^{195}Au	194.9650	183 d	^{131}I	130.9061	8.054 d
^{198}Au	197.9682	2.696 d	^{114}In	113.9049	49.51 d
^{8}B	8.0246	0.769 s	^{115}**In**	114.9039	ca. $5 \cdot 10^{14}$ yr
^{133}Ba	132.9060	10.73 yr	^{192}Ir	191.9626	74.08 d
^{140}Ba	139.9088	12.79 d	^{194}Ir	193.9651	0.47 yr
^{7}Be	7.0169	53.2 d	^{40}**K**	39.9640	$1.28 \cdot 10^9$ yr
^{10}Be	10.0135	$1.6 \cdot 10^6$ yr	^{42}K	41.9630	12.36 h
^{210}Bi	209.9841	5.01 d	^{43}K	42.9607	22.3 h
^{247}Bk	247.0703	$1.4 \cdot 10^3$ yr	^{81}Kr	80.9166	$2.1 \cdot 10^5$ yr
^{77}Br	76.9214	57.0 h	^{85}Kr	84.9125	10.73 yr
^{82}Br	81.9168	35.34 h	^{261}Ku	261.1087	65 s
^{14}C	14.0032	$5.71 \cdot 10^3$ yr	^{137}La	136.9065	$6 \cdot 10^4$ yr
^{41}Ca	–	$1.03 \cdot 10^5$ yr	^{138}**La**	137.9071	$1.1 \cdot 10^{11}$ yr
^{45}Ca	44.9560	162.6 d	^{140}La	139.9095	40.24 h
^{115}Cd	114.9054	44.6 d	^{9}Li	9.0268	0.178 s
^{144}Ce	143.9137	284.4 d	^{260}Lr	260.1053	3 min
^{251}Cf	251.0796	$9.0 \cdot 10^2$ yr	^{174}Lu	173.9404	3.3 yr
^{36}Cl	35.9797	$3.07 \cdot 10^5$ yr	^{176}**Lu**	175.9427	$3.8 \cdot 10^{10}$ yr
^{247}Cm	247.0703	$1.58 \, 10^7$ yr	^{177}Lu	176.9438	6.71 d
^{60}Co	59.9338	5.272 yr	^{258}Md	258.0986	55 d
^{51}Cr	50.9450	27.703 d	^{28}Mg	27.9839	21.0 h
^{135}Cs	134.9059	$3 \cdot 10^6$ yr	^{53}Mn	52.9413	$3.7 \cdot 10^6$ yr
^{137}Cs	136.9071	30.17 yr	^{54}Mn	53.9402	312 d
^{64}Cu	63.9298	12.70 h	^{93}Mo	92.9068	$3.5 \cdot 10^3$ yr
^{67}Cu	66.9277	61.8 h	^{99}Mo	98.9077	66.02 h
^{154}Dy	153.9244	$1 \cdot 10^7$ yr	^{13}N	13.0057	9.96 min
^{159}Dy	158.9257	144.4 d	^{22}Na	21.9944	2.603 yr
^{169}Dy	168.9346	9.4 d	^{24}Na	23.9910	15.0 h
^{252}Es	252.0829	472 d	^{92}Nb	91.9072	$3.6 \cdot 10^7$ yr
^{150}Eu	149.9197	36 yr	^{95}Nb	94.9068	35 d
^{152}Eu	151.9218	13.60 yr	^{144}**Nd**	143.9101	$2.1 \cdot 10^{16}$ yr
^{18}F	18.0009	109.8 min	^{147}Nd	146.9161	10.98 d
^{59}Fe	58.9350	44.6 d	^{23}Ne	22.9945	37.6 s
^{60}Fe	59.9340	$3 \cdot 10^5$ yr	^{59}Ni	58.9344	$7.5 \cdot 10^4$ yr
^{257}Fm	257.0951	100.5 d	^{63}Ni	62.9297	100.1 yr
^{223}Fr	223.0197	22 min	^{259}No	259.1009	58 min
^{67}Ga	66.9282	78.26 h	^{237}Np	237.0482	$2.14 \cdot 10^6$ yr
^{72}Ga	71.9247	14.10 h	^{262}Ns	262.1138	34 s
^{150}Gd	149.9187	$1.8 \cdot 10^6$ yr	^{15}O	15.0031	122 s
^{152}**Gd**	151.9198	$1.1 \cdot 10^{14}$ yr	^{186}**Os**	185.9538	$2 \cdot 10^{15}$ yr

(Continued)

Isotope	A_r, amu	$T_{1/2}$	Isotope	A_r, amu	$T_{1/2}$
^{191}Os	190.9609	15.4 d	^{123}Sn	122.9057	129·2 d
^{194}Os	193.9652	6.0 yr	^{126}Sn	125.9077	$1 \cdot 10^5$ yr
^{32}P	31.9739	14.31 d	^{90}Sr	89.9077	28.7 yr
^{33}P	32.9717	25.4 d	T	see ^{3}H	
^{231}Pa	231.0359	$3.28 \cdot 10^4$ yr	^{179}Ta	178.9460	1.7 yr
^{233}Pa	233.0416	27.0 d	^{180}Ta	179.9475	$>1 \cdot 10^{13}$ yr
^{205}Pb	204.9745	$1.5 \cdot 10^7$ yr	^{182}Ta	181.9502	115.0 d
^{210}Pb	209.9842	27.1 d	^{157}Tb	156.9240	150 yr
^{103}Pd	102.9061	17.0 d	^{158}Tb	157.9254	150 yr
^{107}Pd	106.9051	$6.5 \cdot 10^6$ yr	^{160}Tb	159.9272	72.3 d
^{145}Pm	144.9127	18 yr	^{98}Tc	97.9072	$4.2 \cdot 10^6$ yr
^{147}Pm	146.9161	2.623 yr	^{99}Tc	98.9062	$2.13 \cdot 10^5$ yr
^{209}Po	208.9824	102 yr	^{121}Te	120.9050	154 d
^{210}Po	209.9829	138.38 d	^{123}Te	122.9043	$>1.25 \cdot 10^{13}$ yr
^{143}Pr	142.9108	13.57 d	^{127}Te	126.9052	109 d
^{188}Pt	187.9594	10.2 d	^{230}Th	230.0331	$7.53 \cdot 10^4$ yr
^{190}Pt	189.9599	$6 \cdot 10^{11}$ yr	^{232}Th	232.0381	$1.40 \cdot 10^{10}$ yr
^{193}Pt	192.9630	50 yr	^{234}Th	234.0436	24.1 d
^{244}Pu	244.0642	$8.2 \cdot 10^7$ yr	^{44}Ti	43.9596	47 yr
^{226}Ra	226.0254	1600 yr	^{45}Ti	44.9581	3.09 h
^{228}Ra	228.0311	5.76 yr	^{204}Tl	203.9739	3.78 yr
^{83}Rb	82.9152	86.2 d	^{170}Tm	169.9358	128.6 d
^{86}Rb	85.9112	18.8 d	^{171}Tm	170.9364	1.92 yr
^{87}Rb	86.9092	$4.8 \cdot 10$ yr	^{234}U	234.0409	$2.45 \cdot 10^5$ yr
^{186}Re	185.9544	$2 \cdot 10^5$ yr	^{235}U	235.0439	$7.038 \cdot 10^8$ yr
^{187}Re	186.9557	$4 \cdot 10^{10}$ yr	^{236}U	236.0456	$2.342 \cdot 10^7$ yr
^{101}Rh	100.9062	3.3 yr	^{237}U	237.0487	6.75 d
^{102}Rh	101.9043	206 d	^{238}U	238.0508	$4.468 \cdot 10^9$ yr
^{222}Rn	222.0176	3.824 d	^{48}V	47.9523	15.974 d
^{103}Ru	102.9063	39.4 d	^{49}V	48.9485	330 d
^{106}Ru	105.9073	367 d	^{181}W	180.9482	121.2 d
^{35}S	34.9690	87.4 d	^{185}W	184.9556	75.3 d
^{124}Sb	123.9059	60.20 d	^{127}Xe	126.9052	36.41 d
^{125}Sb	124.9053	2.7 yr	^{88}Y	87.9095	106.6 d
^{46}Sc	45.9550	83.8 d	^{90}Y	89.9072	64.3 h
^{79}Se	78.9185	$\leq 6.5 \cdot 10^4$ yr	^{169}Yb	168.9352	32.0 d
^{31}Si	30.9753	2.62 h	^{65}Zn	64.9292	243.9 d
^{32}Si	31.9740	ca. 650 yr	^{93}Zr	92.9065	$1.5 \cdot 10^6$ yr
^{146}Sm	145.9131	$3 \cdot 10^7$ yr	^{95}Zr	94.9080	64 d
^{147}Sm	146.9149	$1.06 \cdot 10^{11}$ yr	^{263}(106)	263.1182	0.9 s
^{148}Sm	147.9148	$7 \cdot 10^{15}$ yr	^{262}(107)	262.1229	0.115 s
^{153}Sm	152.9221	46.7 h			

2.4 Electronic Configurations of Atoms [28, 33, 36]

The table lists the electronic configurations of free atoms in the ground state. For the atoms of elements in the third and subsequent periods, the configurations are presented in abridged form by enclosing in brackets the symbol of the noble gas from the preceding period. For example, the electronic configuration of the sodium atom written as [Ne] $3s^1$ stands for $1s^2\,2s^2\,2p^6\,3s^1$.

Period	Element	Electronic configuration of atom	Period	Element	Electronic configuration of atom
1	$_1$H	$1s^1$	5	$_{41}$Nb	[Kr] $4d^45s^1$
	$_2$He	$1s^2$		$_{42}$Mo	[Kr] $4d^55s^1$
				$_{43}$Tc	[Kr] $4d^55s^2$
2	$_3$Li	$1s^22s^1$		$_{44}$Ru	[Kr] $4d^75s^1$
	$_4$Be	$1s^22s^2$		$_{45}$Rh	[Kr] $4d^85s^1$
	$_5$B	$1s^22s^22p^1$		$_{46}$Pd	[Kr] $4d^{10}$
	$_6$C	$1s^22s^22p^2$		$_{47}$Ag	[Kr] $4d^{10}5s^1$
	$_7$N	$1s^22s^22p^3$		$_{48}$Cd	[Kr] $4d^{10}5s^2$
	$_8$O	$1s^22s^22p^4$		$_{49}$In	[Kr] $4d^{10}5s^25p^1$
	$_9$F	$1s^22s^22p^5$		$_{50}$Sn	[Kr] $4d^{10}5s^25p^2$
	$_{10}$Ne	$1s^22s^22p^6$		$_{51}$Sb	[Kr] $4d^{10}5s^25p^3$
				$_{52}$Te	[Kr] $4d^{10}5s^25p^4$
3	$_{11}$Na	[Ne] $3s^1$		$_{53}$I	[Kr] $4d^{10}5s^25p^5$
	$_{12}$Mg	[Ne] $3s^2$		$_{54}$Xe	[Kr] $4d^{10}5s^25p^6$
	$_{13}$Al	[Ne] $3s^23p^1$			
	$_{14}$Si	[Ne] $3s^23p^2$	6	$_{55}$Cs	[Xe] $6s^1$
	$_{15}$P	[Ne] $3s^23p^3$		$_{56}$Ba	[Xe] $6s^2$
	$_{16}$S	[Ne] $3s^23p^4$		$_{57}$La	[Xe] $5d^16s^2$
	$_{17}$Cl	[Ne] $3s^23p^5$		$_{58}$Ce	[Xe] $4f^15d^16s^2$ or $4f^26s^2$
	$_{18}$Ar	[Ne] $3s^23p^6$		$_{59}$Pr	[Xe] $4f^36s^2$
				$_{60}$Nd	[Xe] $4f^46s^2$
4	$_{19}$K	[Ar] $4s^1$		$_{61}$Pm	[Xe] $4f^56s^2$
	$_{20}$Ca	[Ar] $4s^2$		$_{62}$Sm	[Xe] $4f^66s^2$
	$_{21}$Sc	[Ar] $3d^14s^2$		$_{63}$Eu	[Xe] $4f^76s^2$
	$_{22}$Ti	[Ar] $3d^24s^2$		$_{64}$Gd	[Xe] $4f^75d^16s^2$
	$_{23}$V	[Ar] $3d^34s^2$		$_{65}$Tb	[Xe] $4f^96s^2$ or $4f^85d^16s^2$
	$_{24}$Cr	[Ar] $3d^54s^1$		$_{66}$Dy	[Xe] $4f^{10}6s^2$
	$_{25}$Mn	[Ar] $3d^54s^2$		$_{67}$Ho	[Xe] $4f^{11}6s^2$
	$_{26}$Fe	[Ar] $3d^64s^2$		$_{68}$Er	[Xe] $4f^{12}6s^2$
	$_{27}$Co	[Ar] $3d^74s^2$		$_{69}$Tm	[Xe] $4f^{13}6s^2$
	$_{28}$Ni	[Ar] $3d^84s^2$		$_{70}$Yb	[Xe] $4f^{14}6s^2$
	$_{29}$Cu	[Ar] $3d^{10}4s^1$		$_{71}$Lu	[Xe] $4f^{14}5d^16s^2$
	$_{30}$Zn	[Ar] $3d^{10}4s^2$		$_{72}$Hf	[Xe] $4f^{14}5d^26s^2$
	$_{31}$Ga	[Ar] $3d^{10}4s^24p^1$		$_{73}$Ta	[Xe] $4f^{14}5d^36s^2$
	$_{32}$Ge	[Ar] $3d^{10}4s^24p^2$		$_{74}$W	[Xe] $4f^{14}5d^46s^2$
	$_{33}$As	[Ar] $3d^{10}4s^24p^3$		$_{75}$Re	[Xe] $4f^{14}5d^56s^2$
	$_{34}$Se	[Ar] $3d^{10}4s^24p^4$		$_{76}$Os	[Xe] $4f^{14}5d^66s^2$
	$_{35}$Br	[Ar] $3d^{10}4s^24p^5$		$_{77}$Ir	[Xe] $4f^{14}5d^76s^2$
	$_{36}$Kr	[Ar] $3d^{10}4s^24p^6$		$_{78}$Pt	[Xe] $4f^{14}5d^96s^1$
				$_{79}$Au	[Xe] $4f^{14}5d^{10}6s^1$
5	$_{37}$Rb	[Kr] $5s^1$		$_{80}$Hg	[Xe] $4f^{14}5d^{10}6s^2$
	$_{38}$Sr	[Kr] $5s^2$		$_{81}$Tl	[Xe] $4f^{14}5d^{10}6s^26p^1$
	$_{39}$Y	[Kr] $4d^15s^2$		$_{82}$Pb	[Xe] $4f^{14}5d^{10}6s^26p^2$
	$_{40}$Zr	[Kr] $4d^25s^2$		$_{83}$Bi	[Xe] $4f^{14}5d^{10}6s^26p^3$

(Continued)

Period	Element	Electronic configuration of atom	Period	Element	Electronic configuration of atom
6	$_{84}$Po	[Xe] $4f^{14}5d^{10}6s^26p^4$	7	$_{96}$Cm	[Rn] $5f^76d^17s^2$
	$_{85}$At	[Xe] $4f^{14}5d^{10}6s^26p^5$		$_{97}$Bk	[Rn] $5f^97s^2$ or $5f^86d^17s^2$
	$_{86}$Rn	[Xe] $4f^{14}5d^{10}6s^26p^6$		$_{98}$Cf	[Rn] $5f^{10}7s^2$
				$_{99}$Es	[Rn] $5f^{11}7s^2$
7	$_{87}$Fr	[Rn] $7s^1$		$_{100}$Fm	[Rn] $5f^{12}7s^2$
	$_{88}$Ra	[Rn] $7s^2$		$_{101}$Md	[Rn] $5f^{13}7s^2$
	$_{89}$Ac	[Rn] $6d^17s^2$		$_{102}$No	[Rn] $5f^{14}7s^2$
	$_{90}$Th	[Rn] $6d^27s^2$		$_{103}$Lr	[Rn] $5f^{14}6d^17s^2$
	$_{91}$Pa	[Rn] $5f^26d^17s^2$ or $5f^16d^27s^2$		$_{104}$Ku	[Rn] $5f^{14}6d^27s^2$
	$_{92}$U	[Rn] $5f^36d^17s^2$		$_{105}$Ns	[Rn] $5f^{14}6d^37s^2$
	$_{93}$Np	[Rn] $5f^46d^17s^2$ or $5f^57s^2$		106	[Rn] $5f^{14}6d^47s^2$
	$_{94}$Pu	[Rn] $5f^67s^2$		107	[Rn] $5f^{14}6d^57s^2$
	$_{95}$Am	[Rn] $5f^77s^2$			

2.5 Ionization Energy, Electron and Proton Affinity, and Electronegativity [11, 33]

The table lists the following quantities: ionization energy (the enthalpy of electron detachment or positive ionization I), electron affinity (the enthalpy of electron attachment or negative ionization A_e), proton affinity (the enthalpy of proton attachment A_p) for atoms; and the electronegativity of elements (χ). The values enclosed in parentheses were obtained by the semiempirical calculation method. The sign "−" denotes the absence of data.

Element	Ionization energy, eV			Affinity, eV		χ
	I_1	I_2	I_3	A_e	A_p	
Ac	5.12	12.06	20.00	−	−	1.00
Ag	7.58	21.49	34.83	−1.30	−	1.42
Al	5.99	18.83	28.45	−0.46	−	1.47
Am	5.99	−	−	−	−	(1.2)
Ar	15.76	27.63	40.74	(0.37)	−3.80	(3.20)
As	9.78	18.62	28.35	−1.07	−	2.11
At	9.20	20.10	−	−2.79	−	1.90
Au	9.23	20.50	30.50	−2.31	−	1.42
B	8.30	25.16	37.93	−0.30	−7.30	2.01
Ba	5.21	10.00	35.84	(0.48)	−	0.97
Be	9.32	18.21	153.90	(−0.38)	−	1.47
Bi	12.25	16.74	25.57	−0.95	−	1.67
Bk	6.30	−	−	−	−	(1.2)
Br	11.81	21.76	36.27	−3.37	−5.80	2.74
C	11.26	24.38	47.93	−1.27	−6.00	2.50
Ca	6.11	11.87	50.91	(1.93)	−	1.04

(Continued)

Element	Ionization energy, eV			Affinity, eV		χ
	I_1	I_2	I_3	A_e	A_p	
Cd	8.99	16.91	37.48	(0.27)	–	1.46
Ce	5.47	10.85	20.08	−0.52	–	1.08
Cf	6.41	–	–	–		(1.2)
Cl	12.97	23.81	39.61	−3.61	−5.30	2.83
Cm	6.09	–	–	–	–	(1.2)
Co	7.87	17.06	33.50	−0.94	–	1.70
Cr	6.77	15.50	30.96	−0.98	–	1.56
Cs	3.89	25.10	34.60	−0.39	–	0.86
Cu	7.73	20.29	36.83	−1.23	–	1.75
Dy	5.93	11.67	22.80	−0.52	–	1.10
Er	6.10	11.93	22.70	−0.52	–	1.11
Es	6.52	–	–	–	–	(1.2)
Eu	5.66	11.25	24.70	−0.52	–	1.01
F	17.42	34.99	62.71	−3.49	−3.70	4.10
Fe	7.89	16.18	30.65	−0.58	–	1.64
Fm	6.64	–	–	–	–	(1.2)
Fr	3.98	–	–	–	–	0.86
Ga	6.00	20.51	30.71	−0.39	–	1.82
Gd	6.16	12.10	20.60	−0.52	–	1.11
Ge	7.90	15.93	34.20	−1.74	–	2.02
H	13.60	–	–	−0.75	−2.69	2.10
He	24.59	54.42	–	(0.22)	−1.84	(5.50)
Hf	7.50	14.90	23.30	(0.63)	–	1.23
Hg	10.44	18.76	34.20	(0.19)	–	1.44
Ho	6.02	11.80	22.80	−0.52	–	1.10
I	10.45	19.10	33.00	−3.08	−6.30	2.21
In	5.79	18.87	28.03	−0.72	–	1.49
Ir	9.10	17.00	–	−1.97	–	1.55
K	4.34	31.62	45.72	−0.47	–	0.91
Kr	14.00	24.37	39.95	(0.42)	−4.50	(2.94)
Ku	–	–	–	–	–	(1.20)
La	5.58	11.06	19.18	−0.55	–	1.08
Li	5.39	75.64	122.42	−0.59	–	0.97
Lr	–	–	–	–	–	(1.20)
Lu	5.43	13.90	20.96	−0.52	–	1.14
Md	6.74	–	–	–	–	(1.2)
Mg	7.65	15.04	80.14	(0.22)	–	1.23
Mn	7.44	15.64	23.30	(0.97)	–	1.60
Mo	7.10	16.16	27.14	−1.18	–	1.30
N	14.53	29.60	47.45	(0.21)	−4.20	3.07
Na	5.14	47.30	71.65	−0.34	–	0.93
Nb	6.88	14.32	25.05	−1.13	–	1.23
Nd	5.49	10.72	22.10	−0.52	–	1.07
Ne	21.56	40.96	63.45	(0.22)	−2.08	(4.84)
Ni	7.63	18.15	35.17	−1.28	–	1.75
No	6.84	–	–	–	–	(1.2)
Np	6.20	–	–	–	–	1.22

(*Continued*)

Element	Ionization energy, eV			Affinity, eV		χ
	I_1	I_2	I_3	A_e	A_p	
O	13.62	35.12	54.90	−1.47	−4.90	3.50
Os	8.50	17.00	25.00	−1.44	−	1.52
P	10.49	19.73	30.16	−0.80	−	2.32
Pa	5.89	−	−	−	−	1.14
Pb	7.42	15.03	31.98	−1.14	−	1.55
Pd	8.34	19.43	32.95	(−1.02)	−	1.35
Pm	5.55	10.90	22.30	−0.52	−	1.07
Po	8.43	19.40	27.30	−1.87	−	1.76
Pr	5.42	10.55	21.63	−0.52	−	1.07
Pt	8.90	18.56	−	−2.13	−	1.44
Pu	6.06	−	−	−	−	1.22
Ra	5.28	10.15	−	−	−	0.97
Rb	4.18	27.50	40.00	(−0.42)	−	0.89
Re	7.88	16.60	−	−0.15	−	1.46
Rh	7.46	18.08	31.06	−1.24	−	1.45
Rn	10.75	21.40	29.40	(0.42)	−	(2.06)
Ru	7.37	16.76	28.47	−1.14	−	1.42
S	10.36	23.35	34.80	−2.08	−6.80	2.60
Sb	8.64	16.50	25.30	−1.05	−	1.82
Sc	6.56	12.80	24.76	(0.73)	−	1.20
Se	9.75	21.19	30.82	−2.02	−	2.48
Si	8.15	16.34	33.53	−1.38	−6.50	2.25
Sm	5.63	11.07	23.40	−0.52	−	1.07
Sn	7.34	14.63	30.50	−1.25	−	1.72
Sr	5.69	11.03	42.88	(1.51)	−	0.99
Ta	7.89	16.20	−	−0.62	−	1.33
Tb	5.85	11.52	21.90	−0.52	−	1.10
Tc	7.28	15.26	29.55	−0.73	−	1.36
Te	9.01	18.60	28.00	−1.96	−	2.02
Th	6.08	11.50	20.00	−	−	1.11
Ti	6.82	13.58	27.48	−0.39	−	1.32
Tl	6.11	20.43	29.83	−0.50	−	1.44
Tm	6.18	12.05	23.70	−0.52	−	1.11
U	6.19	11.60	19.80	−	−10.80	1.22
V	6.74	14.65	29.32	−0.64	−	1.45
W	7.98	17.70	−	−0.50	−	1.40
Xe	12.13	21.25	32.10	(0.45)	−5.40	(2.40)
Y	6.22	12.24	20.52	(0.4)	−	1.11
Yb	6.25	12.18	25.50	−0.52	−	1.06
Zn	9.39	17.96	39.72	(−0.09)	−	1.66
Zr	6.84	13.13	22.98	−0.45	−	1.22

3

Molecular Properties

3.1 Ionization Energy [11]

The values presented in this table are the ionization energies (the enthalpy of electron detachment or positive ionization, I) for molecules and radicals.

Substance	I, eV	Substance	I, eV	Substance	I, eV
AgF	11.40	$B(OH)_3$	13.50	C_2H_4	10.51
As_2	11.00	$BaCl_2$	9.20	C_6H_6	9.25
As_4	9.90	BaI_2	8.10	$CH_3C(H)O$	10.22
$AsCl_3$	11.70	BaO	6.97	CH_3CN	12.12
AsH_3	10.06	BeF_2	14.70	CH_3COOH	10.37
At_2	8.30	BeO	10.10	$CHCl_3$	11.35
BBr_3	10.72	Br_2	10.53	C_5H_5N	9.30
BCl_3	11.60	BrF	11.80	CH_3OH	10.85
BF_3	15.55	BrF_3	12.90	C_2H_5OH	10.47
B_2H_6	11.41	CBr_4	11.00	CN	14.20
B_4H_{10}	10.40	CCl_4	11.47	C_2N_2	13.37
B_5H_9	10.10	CF_4	15.56	(CN)Br	11.84
B_6H_{10}	9.50	CH_2	10.40	(CN)Cl	12.34
$B_{10}H_{14}$	10.70	CH_3	9.84	(CN)F	13.32
BI_3	9.40	CH_4	12.71	(CN)I	10.87
B_2O_3	14.00	C_2H_2	11.41	CO	14.01

(*Continued*)

Substance	I, eV	Substance	I, eV	Substance	I, eV
CO_2	13.79	HF	15.92	NO	9.27
CS	11.10	HI	10.38	NO_2	9.78
CS_2	10.07	HN_3	10.74	N_2O	12.89
CSO	11.18	HNCO	11.60	Na_2	4.90
$CaCl_2$	10.30	HNCS	10.05	NaCl	8.92
CaO	6.50	HNO_3	11.03	NaI	7.64
CeO_2	9.50	HO_2	11.53	$[Ni(CO)_4]$	8.28
Cl_2	11.48	H_2O	12.61	$NiCl_2$	11.20
ClF	12.70	H_2O_2	10.90	NiF_2	11.50
ClO	10.40	HS	10.40	NiO	9.50
ClO_2	11.10	H_2S	10.47	O_2	12.08
ClO_3	11.70	H_2S_2	10.20	O_3	12.52
ClO_3F	13.60	H_2Se	9.88	OF_2	13.70
CoO	9.00	H_2Te	9.14	OH	13.18
$[Co(CO)_4]$	8.30	$HgBr_2$	10.62	OsO_4	12.60
$[Co_2(CO)_8]$	8.20	$HgCl_2$	10.37	P_4	10.80
$[Cr(CO)_6]$	8.14	HgI_2	9.50	PCl_3O	13.10
CrF_2	10.10	I_2	9.28	PF_3	9.71
CrF_3	12.20	IBr	9.98	PH_3	9.98
CrO	8.40	ICl	10.31	P_2H_4	10.60
CrO_2	10.30	IF	10.50	$PbBr_2$	10.20
CrO_3	11.60	IF_2	9.70	$PbCl_2$	10.30
Cs_2	3.80	IF_5	13.50	PbF_2	11.60
CsBr	7.72	K_2	3.60	PbF_4	10.40
CsI	7.25	KI	8.20	PbO	9.00
Cs_2O	4.45	Li_2	5.15	PbS	8.60
CsOH	7.21	LiBr	9.40	PbTe	8.20
CuF_2	11.30	LiCl	10.10	PdO	9.10
D_2	15.47	LiF	11.30	PrO_2	9.60
DF	16.03	LiH	7.85	Rb_2	3.45
F_2	15.69	LiI	8.60	RbI	8.00
$FeBr_2$	10.70	Li_2O	6.80	RuO_2	10.60
$[Fe(C_5H_5)_2]$	6.74	$MgBr_2$	10.65	RuO_4	12.80
$[Fe(CO)_5]$	7.95	$MgCl_2$	11.60	S_2	9.36
$FeCl_2$	11.50	MgF_2	13.50	S_8	9.43
FeF_2	11.30	MgI_2	9.57	SF_6	16.15
FeF_3	12.50	$[Mn_2(CO)_{10}]$	8.50	SO_2	12.34
GaOF	9.50	MnF_2	11.50	SO_2F_2	13.30
$GeBr_4$	10.75	$[Mo(CO)_6]$	8.23	Sb_2	9.00
$GeCl_4$	12.12	MoO_2	9.40	Sb_4	8.40
GeH_4	10.50	MoO_3	12.00	$SbCl_3$	11.00
Ge_2H_6	10.00	N_2	15.58	SbH_3	9.58
Ge_3H_8	9.60	ND_3	10.37	Se_2	8.88
H_2	15.43	NF_3	13.20	SeO_2	11.50
HBr	11.62	N_2F_4	12.00	SeO_3	11.60
HCN	13.73	NH_2	11.40	$SiBr_4$	10.80
HCOOH	11.05	NH_3	10.15	SiC	9.20
HCl	12.74	N_2H_4	8.74	$SiCl_4$	11.64

(*Continued*)

Substance	I, eV	Substance	I, eV	Substance	I, eV
SiF_4	15.40	$SrCl_2$	9.70	UF_6	14.14
SiH_4	11.40	SrO	6.10	UO_2	5.50
SiO_2	11.70	T_2	15.49	UO_3	10.40
$SnBr_2$	10.00	Te_2	8.29	VCl_3	12.80
$SnBr_4$	11.00	TeO_2	11.00	$[W(CO)_6]$	8.24
$SnCl_2$	10.20	$TiBr_4$	10.56	WO_3	11.70
$SnCl_4$	12.10	$TiCl_4$	11.70	XeF_2	12.35
SnF_2	11.50	TiO	6.80	XeF_4	12.65
SnH_4	9.20	TiO_2	10.00	XeF_6	12.19
SnO	10.50	$TlBr$	9.14	$ZnBr_2$	10.40
SnS	9.70	$TlCl$	9.70	$ZnCl_2$	11.70
$SnSe$	9.70	TlF	10.50	ZrO_2	8.00
$SnTe$	9.10	TlI	8.47		

3.2 Electron Affinity [11]

This table lists the values of electron affinity (the enthalpies of electron attachment or negative ionization, A_e) for molecules and radicals.

Substance	A_e, eV	Substance	A_e, eV	Substance	A_e, eV
BF_3	−2.65	$[Cr(CO)_5]$	−1.20	LiF	≤−1.35
BF_4	<−3.40	CrO_3	−4.04	LiI	≤−1.12
BaO	−0.57	$CsCl$	≤−1.40	$[Mo(CO)_5]$	−1.80
BaS	−0.84	Cs_2O_2	6.65	MoO_3	−2.59
$BaSe$	−0.95	F_2	−3.08	N_2	≥2.80
$BaTe$	−1.43	$[Fe(CO)_4]$	−1.20	NCO	−1.56
Br_2	−2.51	$GeBr_4$	−1.60	NCS	−2.17
BrO_3	−3.22	$GeCl_4$	−0.87	$NCSe$	−2.64
CBr_4	−2.06	H_2	3.58	NH_2	−0.74
CCl_4	−2.12	HCN	≤−1.00	NO	−0.02
CH_2	0.95	$HCrO_4$	−2.40	NO_2	−2.50
CH_3	−1.08	HNO_3	−2.00	NO_3	−3.90
C_2H_4	1.81	HO_2	−3.04	N_2O	−0.70
C_6H_6	1.10	H_2O	−1.78	$NaCl$	≤−1.19
$CHCl_3$	−1.76	HS	−2.32	NaF	≤−1.35
CN	−3.82	HSe	−2.21	Na_2O_2	6.15
$(CN)I$	−0.90	I_2	−2.58	O_2	−0.44
CO_2	>0.00	IBr	−2.70	O_3	−1.99
CS	<−1.46	ICl	−1.70	OH	−1.83
CS_2	−1.13	IO_3	−5.50	$PbCl_2$	≤−3.20
Cl_2	−2.38	KCl	≤−1.27	PtF_6	<−6.77
ClO	−2.20	KF	≤−1.26	Rb_2O_2	6.47
ClO_2	−3.43	K_2O_2	6.3	ReO_4	≤−2.40
ClO_3	−2.83	$LiCl$	≤−1.28	S_2	≤−2.00

(*Continued*)

Substance	A_e, eV	Substance	A_e, eV	Substance	A_e, eV
S_8	< 0.00	SeF_4	−1.70	$SnCl_4$	−1.57
SF_4	−1.20	SeO_2	ca. −2.30	$TiCl_4$	−1.57
SF_6	−1.48	$SiBr_4$	−0.13	UF_6	−2.91
SO_2	−1.20	$SiCl_4$	0.30	VCl_3O	−2.03
Se_2	< 0.00	$SnBr_4$	−1.92	WF_4	≤ −2.30

3.3 Proton Affinity [11]

The values of proton affinity (the enthalpy of proton attachment A_p) for molecules and radicals are tabulated below.

Substance	A_p, eV	Substance	A_p, eV	Substance	A_p, eV
AsH_3	−7.50	CO	−6.21	H_2Se	−7.40
B_4H_{10}	−6.20	CO_2	−4.70	KOH	−11.40
B_5H_9	−7.24	CsOH	−11.65	LiOH	−10.40
B_6H_{10}	−8.46	GeH_4	−6.04	N_2	−5.20
CF_4	−5.20	H_2	−4.40	NCS	−6.50
CH_2	−8.65	HBr	−6.10	NF_3	−6.60
CH_3	−5.43	HCN	−7.40	NH_2	−8.00
CH_4	−5.30	HCOOH	−7.80	NH_3	−9.00
C_2H_2	−6.55	HCl	−6.10	NO	≥ −5.55
C_2H_4	−6.90	HF	−6.85	NO_2	−6.60
C_6H_6	−5.63	HI	−6.30	N_2O	−5.80
$CH_3C(H)O$	−8.04	HNO_3	−7.30	NaOH	−10.70
CH_3CN	−8.10	HO_2	−6.40	O_2	−4.17
CH_3COOH	−8.16	H_2O	−7.14	OH	−6.20
CH_3OH	−7.80	HS	−7.20	PH_3	−8.07
C_2H_5OH	−8.40	H_2S	−7.40	SiH_4	≥ −6.35

3.4 Dipole Moments [12]

The values of permanent dipole moments, $\vec{\mu}$, for molecules and radicals are tabulated below.

Substance	$\vec{\mu}$, D	Substance	$\vec{\mu}$, D	Substance	$\vec{\mu}$, D
Al_2	0	As_2	0	AsH_3	0.18
$AlBr_3$	0	As_4	0	AsI_3	1.83
$AlCl_3$	0	$AsBr_3$	2.90	As_4O_6	0.13
AlF_3	0	$AsCl_3$	1.97	BBr_3	0
AlI_3	0	AsF_3	2.82	BCl_3	0

(Continued)

Substance	$\vec{\mu}$, D	Substance	$\vec{\mu}$, D	Substance	$\vec{\mu}$, D
BF_3	0	C(Se)O	0.75	$HgBr_2$	0
B_2H_6	0	C(Te)S	0.17	$HgCl_2$	0
B_4H_{10}	0.56	Cl_2	0	HgI_2	0
B_5H_9	2.16	ClF	0.65	I_2	0
$B_{10}H_{14}$	3.52	Cl_2F_6	0.55	IBr	1.21
$B_3H_6N_3$	0	ClO_2	0.78	ICl	0.65
BI_3	0	Cl_2O	1.69	IF_5	2.28
B_2O_3	2.50	Cl_2O_7	0.72	$InBr_3$	0
$B(OH)_3$	0	$ClOF_3$	0.02	$InCl_3$	0
BaO	7.96	$CrCl_2O_2$	0.47	InF_3	0
BaS	10.90	CsCl	10.42	InI_3	0
$BeBr_2$	0	CsF	7.88	KBr	10.41
$BeCl_2$	0	CsI	12.1	KCl	11.05
BeF_2	0	D_2	0	KF	7.33
BeI_2	0	DBr	0.83	KI	9.20
Br_2	0	DCl	1.12	LiBr	6.20
BrCl	0.57	DI	0.45	LiF	6.60
BrF	1.29	D_2O	1.87	LiH	5.88
BrF_3	1.19	F_2	0	LiI	6.25
BrF_5	1.51	$GaBr_3$	0	N_2	0
CBr_4	0	$GaCl_3$	0	ND_3	1.50
CCl_4	0	GaF_3	0	NF_3	0.24
CCl_2O	1.18	GaI_3	0	N_2F_4	0.26
CD_4	0	$GeBr_4$	0	NH_3	1.46
CF_4	0	$GeCl_4$	0	N_2H_4	1.83
CH_4	0	GeF_4	0	NO	0.16
C_2H_2	0	GeH_4	0	NO_2	0.32
C_2H_4	0	Ge_2H_6	0	N_2O	0.17
C_6H_6	0	GeI_4	0	N_2O_4	0
$CH_3C(H)O$	2.69	GeO_2	0	N_2O_5	1.39
$CHCl_3$	1.06	H_2	0	(NO)Br	1.87
CH_3COOH	1.70	HBr	0.79	(NO)Cl	1.83
CH_3OH	1.71	HCN	2.96	$(NO_2)Cl$	0.53
C_2H_5OH	1.68	H_2CN_2	4.24	(NO)F	1.81
CI_4	0	HCNO	3.06	NOF_3	0.04
C_2N_2	0	HCOOH	1.35	$(NO_2)F$	0.47
(CN)Br	2.94	H_2CS_3	2.13	NaCl	10.00
(CN)Cl	2.80	HCl	1.08	NaI	8.50
(CN)F	1.68	HF	1.91	O_2	0
(CN)I	3.71	HI	0.42	O_3	0.53
CO	0.11	HN_3	0.85	OF_2	0.30
CO_2	0	HNCS	1.72	OH	1.65
COF_2	0.95	HNO_3	2.16	P_2	0
CS	1.97	H_2O	1.86	P_4	0
CS_2	0	H_2O_2	2.26	PBr_3	0.61
$CSCl_2$	0.28	H_2S	0.93	PCl_3	0.78
CSO	0.71	H_2S_2	1.17	PCl_5	0
CSe_2	0	H_2Se	0.24	PCl_2F_3	0.68

(*Continued*)

Substance	$\vec{\mu}$, D	Substance	$\vec{\mu}$, D	Substance	$\vec{\mu}$, D
PCl_4F	0.21	SF_4	0.63	$SiBr_4$	0
PCl_3O	2.39	SF_6	0	$SiCl_4$	0
PF_3	1.03	S_2F_2	1.45	SiF_2	1.23
PF_5	0	S_2F_{10}	0	SiF_4	0
PH_3	0.58	S_4N_4	0.72	SiH_4	0
$(PH)F_2$	1.32	SO_2	1.67	Si_2H_6	0
PI_3	0.34	SO_3	0	SiI_4	0
P_2I_4	0.45	S_2O	1.47	SiO_2	0
PI_3O	1.69	SOF_2	1.62	$SnBr_4$	0
PN	2.75	SO_2F_2	1.11	$SnCl_4$	0
P_4O_6	0	Sb_2	0	SnF_4	0
P_4O_{10}	0	Sb_4	0	SnH_4	0
POF_3	1.77	$SbBr_3$	3.28	SnI_4	0
$PSCl_3$	1.41	$SbCl_3$	3.93	$TaCl_5$	1.20
PSF_3	0.63	$SbCl_5$	0	Te_2	0
$PbCl_4$	0	SbH_3	0.12	$TeCl_4$	2.57
$RbBr$	10.00	SbI_3	0.40	TeF_6	0
RbF	8.80	Se_2	0	$TiBr_4$	0
S_2	0	Se_2Cl_2	2.10	$TiCl_4$	0
SBr_2O	1.47	$SeCl_2O$	2.62	$TlCl$	4.44
S_2Cl_2	0.92	SeF_4	1.78	$TlCl_3$	3.93
$S(Cl)F_5$	1.58	SeF_6	0	TlF	7.60
SCl_2O	1.44	SeO_2	2.70	XeF_2	0
SCl_2O_2	1.80	SeO_3	0	XeF_4	0
SF_2	1.05	$SeOF_2$	2.84		

3.5 Atomization Energy and Bond Length for Diatomic Molecules and Ions [4–5, 11, 13–14, 30, 33]

This table lists the values for atomization energy (bond dissociation energy E_b), which correspond to standard enthalpy at 298.15 K for the reactions of dissociation of diatomic molecules and radicals thereof into neutral atoms (or, in the case of ions, into neutral and charged atoms, e.g. $Ar_2^+ = Ar + Ar^+$) and also the values of bond lengths (l_b) that correspond to the internuclear distances. All substances are regarded as ideal gases whose particles (molecules, radicals, and ions) are in the ground (nonexcited) states. This table includes no molecules and radicals whose values of atomization energy (bond dissociation energy) may be evaluated by calculations using the data of Sec. 3.1. The sign "–" denotes the absence of data.

Substance	E_b kJ/mole	l_b pm	Substance	E_b kJ/mole	l_b pm	Substance	E_b kJ/mole	l_b pm
Ag_2	164	268	$AgBr$	292	239	AgF	359	198
Ag_2^+	184	–	$AgCl$	313	228	AgI	234	255

(Continued)

Substance	E_b kJ/mole	l_b pm	Substance	E_b kJ/mole	l_b pm	Substance	E_b kJ/mole	l_b pm
Al_2	175	247	CH	339	112	CsF	514	235
AlBr	430	229	CI	284	–	CsH	176	249
AlCl	503	213	CN^-	1004	111	CsI	336	332
AlF	675	165	CN	762	117	CsO	296	–
AlI	372	254	CN^+	477	129	Cu_2	200	222
AlN	359	179	CO	1076	113	CuBr	329	217
AlP	217	–	CO^+	811	112	CuCl	382	205
Ar_2^+	138	450	CS	714	154	CuF	431	174
ArF^+	297	–	CS^+	631	–	CuI	291	234
As_2	385	210	CSe	582	168	CuO	267	173
As_2^+	267	–	CaBr	322	281	CuS	284	205
AsH	267	153	CaCl	397	244	CuSe	271	211
At_2	121	–	CaF	535	193	CuTe	192	235
At_2^+	234	–	CaI	285	288	D_2	443	74.2
Au_2	230	247	CaO	423	182	DBr	367	141
B_2	280	159	CaS	338	232	DCl	433	128
BBr	434	188	CdBr	157	–	DF	572	92
BCl	548	172	CdCl	206	–	DI	299	161
BF	757	125	CdI	136	–	DS	354	134
BH	338	123	CdS	≤197	–	Dy_2	71	–
BI	380	214	Ce_2	242	306	DyF	527	–
BN	389	128	CeF	582	–	Er_2	75	–
BO	807	120	CeO	794	181	ErF	565	–
BP	347	–	Cl_2^-	124	–	Eu_2	51	370
BaBr	372	299	Cl_2	243	199	EuF	523	–
BaCl	448	268	Cl_2^+	392	189	EuO	468	–
BaF	589	216	ClF	251	163	EuS	326	233
BaI	275	320	ClN	259	161	F_2^-	121	–
BaO	565	194	ClO	269	157	F_2	159	141
BaS	396	251	Co_2	171	–	F_2^+	323	133
BeBr	381	195	CoBr	329	–	Fe_2	104	–
BeCl	389	180	CoCl	397	209	FeBr	247	–
BeF	569	136	CoF	435	–	FeCl	350	209
BeO	448	133	CoI	284	–	FeO	411	163
BeS	380	174	CoO	369	165	FeS	338	–
Bi_2	200	266	CoS	343	–	FrBr	380	314
Br_2^-	86	–	Cr_2	154	234	FrCl	430	297
Br_2	194	228	CrBr	331	–	FrF	506	239
Br_2^+	320	–	CrCl	367	–	FrI	322	338
BrCl	219	214	CrF	443	–	Ga_2	138	–
BrF	233	176	CrI	288	–	GaAs	213	208
C_2^-	786	127	CrO	457	163	GaCl	476	220
C_2	605	134	CrS	338	–	GaI	362	257
C_2^+	531	146	Cs_2	42	430	GaN	523	168
CBr	364	182	Cs_2^+	50	470	GaP	230	201
CCl	397	164	CsBr	395	307	GaSb	189	231
CF	545	127	CsCl	443	291	GdF	590	–

(Continued)

Substance	E_b kJ/mole	l_b pm	Substance	E_b kJ/mole	l_b pm	Substance	E_b kJ/mole	l_b pm
Ge_2	277	244	K_2	57	392	NS^-	389	–
GeH	313	159	K_2^+	92	418	NS	464	150
GeO	657	163	KBr	382	282	NS^+	506	144
GeS	551	201	KCl	425	267	Na_2	73	308
GeSe	490	214	KF	497	217	Na_2^+	98	343
GeTe	405	234	KH	182	224	NaBr	370	250
$^1H_2^-$	ca. 18	–	KI	325	305	NaCl	411	236
H_2	436	74	KO	280	–	NaF	480	193
$^1H_2^+$	260	108	Kr_2^+	100	–	NaH	200	189
HBr	366	141	KrF^+	188	175	NaI	290	271
HCl	432	128	La_2	246	–	NaO	255	205
HD	439	74	LaBr	406	263	NbO	773	169
HF	566	92	LaCl	464	247	Nd_2	<167	311
HI	298	161	LaF	600	203	NdF	573	–
HS^-	364	135	LaI	326	288	Ne_2^+	71	–
HS	349	134	Li_2^-	92	–	NeH^+	204	–
HS^+	359	138	Li_2	102	267	Ni_2	234	–
HSe^-	313	–	Li_2^+	125	314	NiBr	360	–
HSe	313	148	LiBr	423	217	NiCl	372	214
HT	437	–	LiCl	477	202	NiF	435	–
HTe	238 ?	170	LiF	577	156	NiI	292	–
He_2^+	234	108	LiH	236	160	NiO	349	–
HeF^+	142	–	LiI	353	239	NiS	341	–
HeH^+	182	78	LiO	343	239	O_2^{2-}	–	150
HfO	774	172	Lu_2	142	–	O_2^-	397	134
Hg_2^+	87	–	LuF	569	192	O_2	498	121
HgBr	71	–	MgBr	297	236	O_2^+	646	112
HgCl	99	233	MgCl	318	220	OD	435	97
HgF	129	200	MgF	464	175	OF	220	162
HgI	36	249	MgO	413	175	OH^-	463	96
HgO	≤272	184	MgS	309	214	OH	428	97
HgS	≤205	–	MnBr	312	–	OH^+	468	103
HgSe	≤166	–	MnCl	359	–	OT	437	97
HgTe	≤142	–	MnF	447	–	P_2	489	189
Ho_2	83	304	MnI	283	–	P_2^+	430	199
HoF	547	194	MnO	410	≤190	PCl	289	204
I_2^-	106	–	MnS	292	–	PF	464	159
I_2	153	267	MoO	503	–	PH	343	143
I_2^+	254	–	N_2^-	598	–	PN	731	149
IBr	179	247	N_2	945	110	PO	ca. 544	ca. 150
ICl	212	232	N_2^+	846	112	PS	ca. 335	186
IF	281	191	ND	318	–	Pb_2	100	308
In_2	104	–	NF	297	132	PbBr	240	255
InBr	387	254	NH	313	104	PbCl	300	218
InCl	430	240	NO^-	506	–	PbF	355	200
InI	336	275	NO	632	115	PbI	195	280
IrO	355	173	NO^+	1051	106	PbO	372	192

(Continued)

Substance	E_b kJ/mole	l_b pm	Substance	E_b kJ/mole	l_b pm	Substance	E_b kJ/mole	l_b pm
PbS	341	229	Si_2^+	380	–	TiO	666	162
PbSe	302	240	SiC	439	170	Tl_2	62	–
PbTe	245	260	SiCl	456	206	TlBr	332	262
Pd_2	74	257	SiF	540	160	TlCl	371	249
PdO	284	–	SiH	302	152	TlF	445	208
PmF	540	–	SiO	800	151	TlI	280	281
Po_2	186	–	Sm_2	54	–	Tm_2	54	–
Pr_2	129	–	SmF	536	–	TmF	569	–
PrF	580	–	SmO	619	–	U_2	221	300
PrO	753	196	SmS	389	–	UF	724	–
PtO	359	173	Sn_2	196	–	UO	757	–
PuF	540	–	SnBr	337	248	V_2	242	245
Rb_2	49	410	SnCl	413	229	VBr	439	–
Rb_2^+	75	444	SnF	472	194	VCl	477	–
RbBr	387	295	SnI	232	270	VO	612	159
RbCl	429	279	SnO	531	183	WCl	421	–
RbF	505	227	SnS	465	221	WF	548	183
RbH	165	237	SnSe	400	233	WO	678	181
RbI	337	318	SnTe	316	252	Xe_2^+	100	–
RbO	254	–	SrBr	331	282	XeF^+	–	184
RuO	490	170	SrCl	401	254	Y_2	159	–
S_2^-	348	–	SrF	542	208	YBr	483	–
S_2	426	189	SrI	263	303	YCl	527	241
S_2^+	522	183	SrO	431	192	YF	600	193
SCl	267	–	SrS	313	244	YI	433	–
SF	360	161	SrSe	247	–	YbF	570	202
SO	522	148	T_2	443	74	ZnBr	141	–
Sb_2	323	221	TS	356	–	ZnCl	22	–
Sb_2^+	267	–	TaF	603	–	ZnF	367	–
Sc_2	113	–	Tb_2	146	306	ZnI	136	–
ScBr	441	238	Te_2	263	256	ZnO	275	–
ScCl	493	223	Te_2^+	334	–	ZnS	205	–
ScF	599	179	TeO	391	183	ZnSe	138	–
ScI	400	261	Th_2	288	330	ZnTe	96	–
Se_2^-	301	–	Ti_2	129	265	ZrBr	418	247
Se_2	309	217	TiBr	438	250	ZrCl	523	230
Se_2^+	416	–	TiCl	494	230	ZrF	623	190
SeO	423	165	TiF	570	195	ZrI	359	290
Si_2	311	225	TiI	309	280	ZrO	758	171

3.6 Dissociation Energy of Polyatomic Molecules [10–11, 30, 33]

The tabulated values of dissociation energy (E_b) correspond to the standard enthalpy (298.15 K) for the dissociation reactions of molecules or radicals thereof into neutral fragments of a simpler composition (calculated for 1 mole of the starting compound). All the substances are regarded as ideal gases, the particles of which (molecules and radicals) are in the nonexcited electronic states. Many dissociation reactions involve not only the rupture of a particular bond, but changes in the electronic configuration, geometric parameters and/or the arrangement of atoms.

Dissociation reaction	E_b, kJ/mole	Dissociation reaction	E_b, kJ/mole
$AgF_2 = Ag + 2F$	ca. 536	$B(OH)_3 = B(OH)_2 + OH$	607
$AlBr_2 = AlBr + Br$	288	$B(OH)_3 = HBO_2 + H_2O$	207
$AlBr_3 = AlBr_2 + Br$	356	$BaBr_2 = BaBr + Br$	452
$Al_2Br_6 = 2AlBr_3$	121	$BaCl_2 = BaCl + Cl$	473
$AlCl_2 = AlCl + Cl$	361	$BaF_2 = BaF + F$	559
$AlCl_3 = AlCl_2 + Cl$	415	$BaI_2 = BaI + I$	444
$Al_2Cl_6 = 2AlCl_3$	129	$BaOH = Ba + OH$	467
$AlF_2 = AlF + F$	501	$BeBr_2 = BeBr + Br$	410
$AlF_3 = AlF_2 + F$	600	$BeCl_2 = BeCl + Cl$	539
$Al_2F_6 = 2AlF_3$	215	$BeF_2 = BeF + F$	710
$AlI_2 = AlI + I$	210	$BeOH = Be + OH$	573
$AlI_3 = AlI_2 + I$	272	$BeOH = BeO + H$	452
$Al_2I_6 = 2AlI_3$	96	$Be(OH)_2 = BeO + H_2O$	557
$AmCl_3 = Am + 3Cl$	ca. 1356	$Be(OH)_2 = BeOH + OH$	500
$AmF_3 = Am + 3F$	ca. 1716	$BiBr_3 = Bi + 3Br$	697
$As_4 = 2As_2$	243	$BiCl_3 = Bi + 3Cl$	824
$AsBr_3 = As + 3Br$	1375	$BiF_3 = Bi + 3F$	1179
$AsCl_3 = As + 3Cl$	965	$BiF_5 = Bi + 5F$	1485
$AsF_3 = As + 3F$	1452	$BiI_3 = Bi + 3I$	505
$AsF_5 = As + 5F$	2030	$BrF_2 = BrF + F$	152
$AsH_3 = AsH_2 + H$	310	$BrF_3 = BrF + F_2$	214
$AsI_3 = As + 3I$	600	$BrF_3 = BrF_2 + F$	222
$BBr_2 = BBr + Br$	281	$BrF_5 = BrF_3 + F_2$	173
$BBr_3 = BBr_2 + Br$	387	$CBr_2 = CBr + Br$	300
$BCl_2 = BCl + Cl$	318	$CBr_3 = CBr_2 + Br$	200
$BCl_3 = BCl_2 + Cl$	464	$CBr_4 = CBr_3 + Br$	208
$BF_2 = BF + F$	466	$CCl_2 = CCl + Cl$	259
$BF_3 = BF_2 + F$	715	$CCl_3 = CCl_2 + Cl$	261
$BH_2 = BH + H$	452	$CCl_4 = CCl_3 + Cl$	307
$BH_3 = BH_2 + H$	355	$CF_2 = CF + F$	494
$B_2H_6 = 2BH_2 + H_2$	381	$CF_3 = CF_2 + F$	385
$B_2H_6 = 2BH_3$	146	$CF_4 = CF_3 + F$	540
$BI_2 = BI + I$	176	$CH_2 = CH + H$	430
$BI_3 = BI_2 + I$	301	$CH_3 = CH_2 + H$	458
$BO_2 = BO + O$	544	$CH_4 = CH_3 + H$	435
$B_2O_3 = BO_2 + BO$	553	$C_2H_2 = 2CH$	962
$BOH = B + OH$	678	$C_2H_2 = C_2H + H$	502
$BOH = BO + H$	299	$C_2H_4 = 2CH_2$	712
$B(OH)_2 = BO + H_2O$	209	$C_2H_4 = C_2H_3 + H$	444
$B(OH)_2 = BOH + OH$	408	$C_6H_6 = C_6H_5 + H$	457
$B(OH)_2 = HBO_2 + H$	100	$CHCl_3 = CCl_3 + H$	402
$B(OH)_3 = 0.5B_2O_3 + 1.5H_2O$	229	$CHCl_3 = CHCl_2 + Cl$	323

(Continued)

Dissociation reaction	E_b, kJ/mole	Dissociation reaction	E_b, kJ/mole
$CI_4 = CI_3 + I$	185	$Cr(Cl)O_2 = Cr(Cl)O + O$	444
$C_2N_2 = 2CN$	547	$Cr(Cl)O_2 = CrO_2 + Cl$	372
$(CN)Br = CN + Br$	358	$CrCl_2O = CrCl_2 + O$	427
$(CN)Cl = CCl + N$	778	$CrCl_2O = Cr(Cl)O + Cl$	314
$(CN)Cl = CN + Cl$	416	$CrCl_2O_2 = Cr(Cl)O_2 + Cl$	339
$(CN)F = CF + N$	795	$CrCl_2O_2 = CrCl_2O + O$	469
$(CN)F = CN + F$	577	$CrCl_2O_2 = CrO_2 + Cl_2$	469
$(CN)I = CN + I$	315	$CrF_2 = CrF + F$	528
$CO_2 = CO + O$	532	$CrF_3 = CrF_2 + F$	572
$CS_2 = CS + S$	441	$CrI_2 = CrI + I$	210
$C(S)O = CO + S$	308	$CrO_2 = CrO + O$	498
$C(S)O = CS + O$	670	$CrO_3 = CrO_2 + O$	481
$CSe_2 = CSe + Se$	332	$Cs_2O = CsO + Cs$	256
$CaBr_2 = CaBr + Br$	471	$Cs_2(O)_2 = 2CsO$	295
$CaCl_2 = CaCl + Cl$	490	$Cs_2(O)_2 = Cs_2O + O$	336
$CaF_2 = CaF + F$	582	$CsOH = Cs + OH$	380
$CaI_2 = CaI + I$	339	$CsOH = CsO + H$	513
$CdBr_2 = CdBr + Br$	318	$CuBr_2 = CuBr + Br$	186
$CdCl_2 = CdCl + Cl$	343	$CuCl_2 = CuCl + Cl$	223
$CdI_2 = CdI + I$	249	$CuF_2 = CuF + F$	342
$CeF_2 = CeF + F$	607	$CuI_2 = Cu + 2I$	384
$CeF_3 = CeF_2 + F$	686	$D_2O = OD + D$	507
$CeO_2 = CeO + O$	649	$D_2S = DS + D$	391
$ClF_2 = ClF + F$	195	$D_2SO_4 = SO_3 + D_2O$	106
$ClF_3 = ClF + F_2$	115	$DyF_2 = DyF + F$	565
$ClF_3 = ClF_2 + F$	174	$DyF_3 = DyF_2 + F$	632
$ClF_5 = ClF_3 + F_2$	75	$ErF_2 = ErF + F$	561
$Cl_2N = ClN + Cl$	280	$ErF_3 = ErF_2 + F$	636
$Cl_3N = Cl_2N + Cl$	381	$EuF_2 = EuF + F$	577
$ClO_2 = ClO + O$	246	$EuF_3 = EuF_2 + F$	669
$ClO_2 = O_2 + Cl$	17	$FeBr_2 = FeBr + Br$	439
$ClO_3 = ClO + O_2$	−54	$FeBr_3 = FeBr_2 + Br$	193
$ClO_3 = ClO_2 + O$	199	$Fe_2Br_6 = 2FeBr_3$	149
$Cl_2O = Cl_2 + O$	173	$FeCl_2 = FeCl + Cl$	442
$Cl_2O = ClO + Cl$	144	$FeCl_3 = FeCl_2 + Cl$	242
$CoBr_2 = CoBr + Br$	321	$Fe_2Cl_6 = 2FeCl_3$	148
$[Co_2(CO)_8] = 2[Co(CO)_4]$	61	$FeF_2 = Fe + 2F$	962
$CoCl_2 = CoCl + Cl$	358	$FeF_3 = FeF_2 + F$	480
$CoCl_3 = CoCl_2 + Cl$	195*	$Fe_2F_6 = 2FeF_3$	136
$CoF_2 = CoF + F$	519	$FeI_2 = Fe + 2I$	558
$CoI_2 = CoI + I$	252	$FeI_3 = FeI_2 + I$	141
$Co(OH)_2 = CoO + H_2O$	318	$Fe_2I_6 = 2FeI_3$	145
$CrBr_2 = CrBr + Br$	349	$Fe(OH)_2 = FeO + H_2O$	347
$CrBr_3 = CrBr_2 + Br$	216	$GaCl_2 = GaCl + Cl$	278
$CrCl_2 = CrCl + Cl$	408	$GaCl_3 = GaCl_2 + Cl$	327
$CrCl_3 = CrCl_2 + Cl$	233	$Ga_2Cl_6 = 2GaCl_3$	88
$Cr(Cl)O = CrCl + O$	519	$Ga_2F_6 = 2GaF_3$	171
$Cr(Cl)O = CrO + Cl$	431	$GaI_2 = GaI + I$	128

Dissociation reaction	E_b, kJ/mole	Dissociation reaction	E_b, kJ/mole
$GaI_3 = GaI_2 + I$	247	$HgCl_2 = HgCl + Cl$	354
$Ga_2I_6 = 2GaI_3$	49	$HgF_2 = HgF + F$	385
$GdF_2 = GdF + F$	598	$HgI_2 = HgI + I$	254
$GdF_3 = GdF_2 + F$	628	$HoF_2 = HoF + F$	569
$GeBr_2 = Ge + 2Br$	651	$HoF_3 = HoF_2 + F$	632
$GeBr_4 = Ge + 4Br$	1104	$IF_3 = I + 3F$	ca. 816
$GeCl_2 = Ge + 2Cl$	ca. 770	$IF_5 = I + 5F$	1339
$GeCl_4 = Ge + 4Cl$	1396	$IF_7 = I + 7F$	1617
$GeF_2 = Ge + 2F$	481	$InBr_2 = InBr + Br$	201
$GeF_4 = Ge + 4F$	ca. 1808	$InBr_3 = InBr_2 + Br$	255
$GeH_4 = GeH_3 + H$	364	$InCl_2 = InCl + Cl$	251
$Ge_2H_6 = 2GeH_3$	314	$InCl_3 = InCl_2 + Cl$	297
$GeI_2 = Ge + 2I$	528	$InF_3 = In + 3F$	ca. 1332
$GeI_4 = Ge + 4I$	847	$InI_2 = InI + I$	138
$GeSe_2 = GeSe + Se$	205	$InI_3 = InI_2 + I$	201
$GeSe_2 = Se_2 + Ge$	385	$IrO_2 = IrO + O$	597
$HCN = CH + N$	933	$KCN = CN + K$	438
$HCN = CN + H$	511	$K_2O = KO + K$	218
$HClO_4 = HCl + 2O_2$	−97	$K_2(O)_2 = 2KO$	280
$H_2F_2 = 2HF$	25	$K_2(O)_2 = K_2O + O$	338
$HN_3 = N_3 + H$	402	$KOH = KO + H$	500
$HN_3 = NH + N_2$	83	$KOH = OH + K$	347
$HNO_2 = NO + OH$	209	$KrF_2 = Kr + 2F$	100
$HNO_2^{\cdot\cdot} = NO + OH$	208	$LaBr_2 = LaBr + Br$	439
$HNO_2 = NO_2 + H$	331	$LaBr_3 = LaBr_2 + Br$	473
$HNO_2^{\cdot\cdot} = NO_2 + H$	330	$LaCl_2 = LaCl + Cl$	498
$HNO_3 = NO_2 + OH$	208	$LaCl_3 = LaCl_2 + Cl$	527
$HNO_3 = NO_3 + H$	424	$LaF_2 = LaF + F$	628
$HO_2 = O_2 + H$	197	$LaF_3 = LaF_2 + F$	661
$HO_2 = OH + O$	268	$LaI_2 = LaI + I$	351
$H_2O = OH + H$	499	$LaI_3 = LaI_2 + I$	372
$H_2O_2 = 2OH$	214	$Li_2O = LiO + Li$	393
$(H_2O)_2 = 2H_2O$	22	$Li(O)_2 = 2Li$	356
$H_2S = HS + H$	385	$Li_2(O)_2 = Li_2O + O$	306
$H_2(S_2) = 2HS$	277	$Li_2(O)_2 = Li_2O + 0.5O_2$	57
$H_2(S_2) = H_2S + S$	241	$LiOH = LiO + H$	528
$H_2(S_3) = H_2S + S_2$	77	$LiOH = OH + Li$	442
$H_2(S_3) = H_2(S_2) + S$	262	$LuF_2 = LuF + F$	569
$H_2(S_4) = H_2S + S_3$	68	$LuF_3 = LuF_2 + F$	628
$H_2(S_4) = H_2(S_2) + S_2$	100	$MgBr_2 = MgBr + Br$	385
$H_2(S_4) = H_2(S_3) + S$	264	$MgCl_2 = MgCl + Cl$	464
$H_2(S_5) = H_2S + S_4$	59	$MgF_2 = MgF + F$	566
$H_2(S_5) = H_2(S_3) + S_2$	101	$MgOH = MgO + H$	418
$H_2(S_5) = H_2(S_4) + S$	264	$MgOH = OH + Mg$	403
$H_2SO_4 = SO_3 + H_2O$	106	$Mg(OH)_2 = MgO + H_2O$	311
$H_2SeO_4 = SeO_3 + H_2O$	84	$Mg(OH)_2 = MgOH + OH$	393
$HfO_2 = HfO + O$	594	$MnBr_2 = MnBr + Br$	371
$HgBr_2 = HgBr + Br$	300	$[Mn_2(CO)_{10}] = 2[Mn(CO)_5]$	108

(Continued)

Dissociation reaction	E_b, kJ/mole	Dissociation reaction	E_b, kJ/mole
$MnCl_2 = MnCl + Cl$	432	$(NO_2)Cl = (NO)Cl + O$	281
$MnCl_3 = MnCl_2 + Cl$	167*	$(NO)F = NF + O$	569
$MnF_2 = MnF + F$	528	$(NO)F = NO + F$	236
$MnF_3 = MnF_2 + F$	259	$(NO_2)F = NO_2 + F$	192
$MnI_2 = MnI + I$	257	$NaCN = Na + CN$	441
$MoCl_5 = MoCl_4 + Cl$	184	$Na_2O = NaO + Na$	251
$MoCl_4O = MoCl_3O + Cl$	196	$Na_2(O)_2 = 2NaO$	298
$MoO_2 = MoO + O$	660	$NaOH = NaO + H$	502
$MoO_3 = MoO_2 + O$	599	$NaOH = OH + Na$	329
$ND_2 = ND + D$	431	$NbCl_4 = NbCl_3 + Cl$	330
$ND_3 = ND_2 + D$	448	$NbCl_5 = NbCl_4 + Cl$	272
$NF_2 = NF + F$	296	$NbCl_3O = NbCl_3 + O$	658
$NF_3 = NF_2 + F$	245	$NbO_2 = NbO + O$	661
$N_2F_2 = 2NF$	410	$NdF_2 = NdF + F$	596
$N_2F_4 = 2NF_2$	85	$NdF_3 = NdF_2 + F$	690
$NH_2 = NH + H$	421	$NiBr_2 = NiBr + Br$	273
$NH_3 = NH_2 + H$	438	$NiCl_2 = NiCl + Cl$	365
$N_2H_2 = N_2 + H_2$	-205	$NiF_2 = NiF + F$	479
$N_2H_2 = 2NH$	548	$NiI_2 = NiI + I$	212
$N_2H_3 = NH_2 + NH$	356	$NpBr_3 = Np + 3Br$	ca. 1191
$N_2H_3 = N_2H_2 + H$	226	$NpCl_3 = Np + 3Cl$	ca. 1380
$N_2H_4 = 2NH_2$	253	$NpF_3 = Np + 3F$	ca. 1758
$N_2H_4 = N_2H_3 + H$	318	$NpF_4 = Np + 4F$	ca. 2244
$(NH)F_2 = NF_2 + H$	340	$NpF_5 = Np + 5F$	ca. 2595
$(NH)O_2 = NO + OH$	207	$NpF_6 = Np + 6F$	ca. 2862
$(NH)O_2 = NO_2 + H$	329	$NpI_3 = Np + 3I$	ca. 978
$NH_2OH = NH_2 + OH$	264	$O_3 = O_2 + O$	107
$NO_2 = 0.5N_2 + O_2$	-34	$OF_2 = OF + F$	165
$NO_2 = NO + O$	306	$OsO_4 = OsO_3 + O$	306
$NO_2 = O_2 + N$	439	$OsO_4 = OsO_3 + 0.5O_2$	57
$NO_3 = NO + O_2$	19	$P_4 = 4P$	804
$NO_3 = NO_2 + O$	212	$PBr_3 = P + 3Br$	792
$N_2O = N_2 + O$	167	$P(Br)F_2 = PF_2 + Br$	364
$N_2O = N_2 + 0.5O_2$	-82	$PBr_3O = PBr_3 + O$	506
$N_2O = NO + N$	481	$PCl_2 = PCl + Cl$	314
$N_2O_2 = 2NO$	12	$PCl_3 = PCl_2 + Cl$	356
$N_2O_2 = N_2O + O$	163	$P_2Cl_4 = 2PCl_2$	314
$N_2O_3 = NO_2 + NO$	41	$P(Cl)F_2 = PF_2 + Cl$	423
$N_2O_3 = N_2O_2 + O$	335	$PCl_2F = PCl_2 + F$	540
$N_2O_4 = 2NO_2$	58	$PCl_3O = PCl_3 + O$	527
$N_2O_4 = N_2O_3 + O$	323	$PF_2 = PF + F$	444
$N_2O_5 = NO_3 + NO_2$	93	$PF_3 = PF_2 + F$	607
$N_2O_5 = N_2O_4 + O$	247	$PH_2 = PH + H$	339
$(NO)Br = BrN + O$	479	$PH_3 = PH_2 + H$	284
$(NO)Br = NO + Br$	123	$P_2H_4 = 2PH_2$	238
$(NO)Cl = ClN + O$	531	$P_2H_4 = P_2H_3 + H$	222
$(NO)Cl = NO + Cl$	159	$PI_3 = P + 3I$	552
$(NO_2)Cl = NO_2 + Cl$	134	$P_2I_4 = 2PI_2$	259

Dissociation reaction	E_b, kJ/mole	Dissociation reaction	E_b, kJ/mole
$POF_3 = PF_3 + O$	544	$S_8 = S_6 + S_2$	130
$PSBr_3 = PBr_3 + S$	397	$S_8 = S_7 + S$	288
$PSCl_3 = PCl_3 + S$	406	$S_2Br_2 = S_2 + 2Br$	434 ?
$PSF_3 = PF_3 + S$	414	$SCl_2 = SCl + Cl$	274
$PbBr_2 = PbBr + Br$	284	$S_2Cl_2 = 2SCl$	279
$PbBr_4 = Pb + 4Br$	ca. 804	$S_2Cl_2 = SCl_2 + S$	272
$PbCl_2 = PbCl + Cl$	406	$S(Cl)F_5 = SF_5 + Cl$	193
$PbCl_4 = Pb + 4Cl$	ca. 972	$S(Cl)O = ClO + S$	484
$PbF_2 = PbF + F$	436	$S(Cl)O = SCl + O$	486
$PbF_4 = Pb + 4F$	ca. 1324	$S(Cl)O = SO + Cl$	227
$PbI_2 = PbI + I$	223	$SCl_2O = SCl_2 + O$	441
$PbI_4 = Pb + 4I$	ca. 568	$SCl_2O = S(Cl)O + Cl$	229
$PmF_2 = PmF + F$	590	$SCl_2O_2 = SCl_2O + O$	400
$PmF_3 = PmF_2 + F$	674	$SF_2 = F_2 + S$	582
$PrF_2 = PrF + F$	598	$SF_2 = SF + F$	381
$PrF_3 = PrF_2 + F$	686	$SF_4 = SF_2 + F_2$	464
$PrO_2 = PrO + O$	498	$SF_5 = SF_4 + F$	272
$PtO_2 = PtO + O$	522	$SF_6 = SF_4 + F_2$	437
$PuBr_3 = Pu + 3Br$	1163	$SF_6 = SF_5 + F$	324
$PuCl_3 = Pu + 3Cl$	1327	$SO_2 = SO + O$	551
$PuF_2 = PuF + F$	565	$SO_3 = SO + O_2$	400
$PuF_3 = PuF_2 + F$	615	$SO_3 = SO_2 + O$	348
$PuF_4 = Pu + 4F$	ca. 2144	$S_2O = SO + S$	383
$PuF_5 = Pu + 5F$	ca. 2425	$SOF_2 = SF_2 + O$	657
$PuF_6 = Pu + 6F$	ca. 2562	$SO_2F_2 = SO_2 + 2F$	720
$PuI_3 = Pu + 3I$	ca. 930	$SO_2F_2 = SOF_2 + O$	393
$Rb_2O = RbO + Rb$	218	$Sb_4 = 2Sb_2$	221
$Rb_2(O)_2 = 2RbO$	331	$SbBr_3 = Sb + 3Br$	779
$Rb_2(O)_2 = Rb_2O + O$	334	$SbBr_5 = Sb + 5Br$	920
$RbOH = OH + Rb$	356	$SbCl_3 = Sb + 3Cl$	944
$RbOH = RbO + H$	530	$SbCl_5 = Sb + 5Cl$	1243
$[Re_2(CO)_{10}] = 2[Re(CO)_5]$	187	$SbF_3 = Sb + 3F$	1320
$RhCl_3 = RhCl_2 + Cl$	180	$SbF_5 = Sb + 5F$	2010
$RuCl_4 = RuCl_3 + Cl$	271	$SbI_3 = Sb + 3I$	585
$RuO_2 = RuO + O$	490	$ScBr_2 = ScBr + Br$	381
$RuO_4 = RuO_3 + O$	355	$ScBr_3 = ScBr_2 + Br$	435
$S_3 = S_2 + S$	272	$ScCl_2 = ScCl + Cl$	439
$S_4 = 2S_2$	119	$ScCl_3 = ScCl_2 + Cl$	494
$S_4 = S_3 + S$	272	$ScF_2 = ScF + F$	594
$S_5 = S_3 + S_2$	134	$ScF_3 = ScF_2 + F$	632
$S_5 = S_4 + S$	287	$ScI_2 = ScI + I$	293
$S_6 = 2S_3$	163	$ScI_3 = ScI_2 + I$	326
$S_6 = S_4 + S_2$	163	$SeBr_2 = Se + 2Br$	ca. 402
$S_6 = S_5 + S$	301	$SeBr_4 = Se + 4Br$	ca. 604
$S_7 = S_5 + S_2$	143	$SeCl_2 = Se + 2Cl$	486
$S_7 = S_6 + S$	267	$SeCl_4 = Se + 4Cl$	768
$S_8 = 2S_4$	174	$SeF_2 = Se + 2F$	702
$S_8 = S_5 + S_3$	159	$SeF_4 = Se + 4F$	1240

(Continued)

Dissociation reaction	E_b, kJ/mole	Dissociation reaction	E_b, kJ/mole
$SeF_6 = Se + 6F$	1709	$ThF_4 = ThF_3 + F$	669
$SeI_2 = Se + 2I$	302	$TiBr_2 = TiBr + Br$	433
$SeO_2 = SeO + O$	426	$TiBr_3 = TiBr_2 + Br$	307
$SeO_3 = SeO_2 + O$	213	$TiBr_4 = TiBr_3 + Br$	289
$SiCl_2 = SiCl + Cl$	399	$TiCl_2 = TiCl + Cl$	456
$SiCl_3 = SiCl_2 + Cl$	363	$TiCl_3 = TiCl_2 + Cl$	424
$SiCl_4 = SiCl_3 + Cl$	377	$TiCl_4 = TiCl_3 + Cl$	343
$Si_2Cl_6 = 2SiCl_3$	184	$TiF_2 = TiF + F$	811
$SiF_2 = SiF + F$	690	$TiF_3 = TiF_2 + F$	515
$SiF_3 = SiF_2 + F$	460	$TiF_4 = TiF_3 + F$	442
$SiF_4 = SiF_3 + F$	695	$TiI_2 = TiI + I$	436
$SiH_2 = SiH + H$	247	$TiI_3 = TiI_2 + I$	195
$SiH_3 = SiH_2 + H$	345	$TiI_4 = TiI_3 + I$	245
$SiH_4 = SiH_3 + H$	395	$TiO_2 = TiO + O$	604
$Si_2H_6 = 2SiH_3$	344	$TmF_2 = TmF + F$	569
$SiO_2 = SiO + O$	473	$TmF_3 = TmF_2 + F$	628
$SmF_2 = SmF + F$	577	$UBr_3 = UBr_2 + Br$	289
$SmF_3 = SmF_2 + F$	669	$UBr_4 = UBr_3 + Br$	272
$SnBr_2 = SnBr + Br$	310	$UBr_5 = UBr_4 + Br$	130
$SnBr_4 = Sn + 4Br$	1091	$UCl_3 = UCl_2 + Cl$	368
$SnCl_2 = SnCl + Cl$	339	$UCl_4 = UCl_3 + Cl$	326
$SnCl_4 = Sn + 4Cl$	1292	$UCl_5 = UCl_4 + Cl$	272
$SnF_2 = SnF + F$	439	$UCl_6 = UCl_5 + Cl$	188
$SnF_4 = Sn + 4F$	1656	$U(Cl)O = UO + Cl$	598
$SnI_2 = SnI + I$	276	$U(Cl)O_2 = U(Cl)O + O$	598
$SnI_4 = Sn + 4I$	820	$U(Cl)O_2 = UO_2 + Cl$	515
$SrBr_2 = SrBr + Br$	479	$UCl_2O = UCl_2 + O$	598
$SrCl_2 = SrCl + Cl$	476	$UCl_2O = U(Cl)O + Cl$	414
$SrF_2 = SrF + F$	559	$UF_2 = UF + F$	565
$SrI_2 = SrI + I$	394	$UF_3 = UF_2 + F$	536
$T_2O = OT + T$	511	$UF_4 = UF_3 + F$	628
$TaCl_5 = TaCl_4 + Cl$	323	$UF_5 = UF_4 + F$	515
$TaF_2 = TaF + F$	603	$UF_6 = UF_5 + F$	188
$TaF_3 = TaF_2 + F$	603	$UI_3 = U + 3I$	1029
$TaF_4 = TaF_3 + F$	598	$(UO_2)Cl_2 = U(Cl)O_2 + Cl$	255
$TaF_5 = TaF_4 + F$	577	$(UO_2)Cl_2 = UCl_2O + O$	439
$[Tc_2(CO)_{10}] = 2[Tc(CO)_5]$	177	$VBr_2 = VBr + Br$	451
$TeBr_2 = Te + 2Br$	ca. 486	$VBr_3 = VBr_2 + Br$	217
$TeBr_4 = Te + 4Br$	ca. 704	$VBr_4 = VBr_3 + Br$	205
$TeCl_2 = Te + 2Cl$	ca. 568	$VCl_2 = VCl + Cl$	496
$TeCl_4 = Te + 4Cl$	ca. 1244	$VCl_3 = VCl_2 + Cl$	275
$TeF_2 = Te + 2F$	ca. 786	$VCl_3O = VCl_3 + O$	577
$TeF_4 = Te + 4F$	ca. 1340	$VI_3 = VI_2 + I$	131
$TeF_6 = Te + 6F$	1978	$VO_2 = VO + O$	634
$TeI_2 = Te + 2I$	ca. 384	$WBr_6 = WBr_5 + Br$	133
$TeI_4 = Te + 4I$	ca. 484	$WCl_2 = W + Cl_2$	862
$TeO_2 = TeO + O$	375	$WCl_3 = WCl + Cl$	683
$ThF_3 = ThF_2 + F$	590	$WCl_4 = WCl_2 + Cl_2$	349

Dissociation reaction	E_b, kJ/mole	Dissociation reaction	E_b, kJ/mole
$WCl_5 = WCl_4 + Cl$	198	$YI_3 = YI_2 + I$	372
$WCl_6 = WCl_4 + Cl_2$	158	$YbF_2 = YbF + F$	569
$WCl_6 = WCl_5 + Cl$	203	$YbF_3 = YbF_2 + F$	628
$WCl_6 = WCl_5 + 0.5Cl_2$	158	$ZnBr_2 = ZnBr + Br$	399
$WCl_2O_2 = WO_2 + Cl_2$	749	$ZnCl_2 = ZnCl + Cl$	414
$WCl_4O = WCl_4 + O$	521	$ZnF_2 = ZnF + F$	420
$WF_6 = WF_5 + F$	452	$ZnI_2 = ZnI + I$	273
$WO_2 = WO + O$	603	$ZrBr_2 = ZrBr + Br$	587
$WO_3 = WO_2 + O$	625	$ZrBr_3 = ZrBr_2 + Br$	368
$XeF_2 = F_2 + Xe$	108	$ZrBr_4 = ZrBr_3 + Br$	325
$XeF_4 = XeF_2 + 2F$	266	$ZrCl_2 = ZrCl + Cl$	508
$XeF_4 = XeF_2 + F_2$	107	$ZrCl_3 = ZrCl_2 + Cl$	462
$XeF_6 = XeF_4 + F_2$	79	$ZrCl_4 = ZrCl_3 + Cl$	469
$XeO_3 = Xe + 3O$	252	$ZrF_2 = ZrF + F$	697
$YBr_2 = YBr + Br$	439	$ZrF_3 = ZrF_2 + F$	632
$YBr_3 = YBr_2 + Br$	473	$ZrF_4 = ZrF_3 + F$	649
$YCl_2 = YCl + Cl$	498	$ZrI_2 = ZrI + I$	529
$YCl_3 = YCl_2 + Cl$	536	$ZrI_3 = ZrI_2 + I$	262
$YF_2 = YF + F$	649	$ZrI_4 = ZrI_3 + I$	239
$YF_3 = YF_2 + F$	649	$ZrO_2 = ZrO + O$	647
$YI_2 = YI + I$	351		

*At 1000 K.

**An equilibrium mixture comprising 0.691 mole HNO_2 and 0.309 mole $(NH)O_2$.

3.7 Hybridization Type and Geometric Form for Polyatomic Molecules and Ions with One Central Atom of an *sp*-Element [9, 29, 33–35]

Presented below are the polyatomic chemical particles (molecules and radicals or ions thereof) having the general molecular formula AB_n and, therefore, A-B bonds only.

The geometric form of AB_n was deduced by the method of valence bonds, i.e. from the stereochemical arrangement of the axes of valent hybrid orbitals of the central atom A and, consequently, of σ-bonds AB.

In hybridization types, there is retained the pattern of designating the valent orbitals of the central atom based on the energy diagram of the polyelectron atom A, viz., *ns-np-nd*, where *n* is the number of the element A period.

The number of hybrid orbitals includes the orbitals of the central atom A, both those employed for the formation of A-B bonds and those containing lone pairs, the latter being denoted by two dots in the electron-dot formula of the molecule and by a lobe in the spatial presentation. In case the AB_n particle is a radical, one electron instead of an electron pair will be present on the hybrid orbital.

This can be readily established by calculations. The NO_2^- particle (sp^2, $:AB_2$), for example, contains on one of the hybrid orbitals an electron pair (the oxidation state N^{III}), whereas the NO_2 particle on the same orbital contains one electron (the oxidation state N^{IV}).

The geometric form of AB_n particles ($n > 2$), provided the B atoms are the same and the orbitals having no lone pairs or unpaired electrons are absent, is classified as *regular*. For example, the NH_4^+ particle (sp^3, AB_4) is a regular tetrahedron, where in all of the H-N-H angles are equal to 109.5°, as corroborated experimentally (cf. Sec. 3.8).

The geometric shape of AB_n particles, provided the B atoms are dissimilar and the orbitals with lone pairs or unpaired electrons are absent, is designated as *irregular*. For example, the PCl_3O particle (sp^3, AB_4) is an irregular tetrahedron, wherein the Cl-P-Cl and Cl-P-O angles are distinct from 109.5°, this being also corroborated experimentally.

The geometric shape of AB_n particles having orbitals with lone pairs and unpaired electrons is classified as *incomplete*. Thus, the SO_3^{2-} particle (sp^3, $:AB_3$) is an incomplete tetrahedron, whereas the SO_4^{2-} particle (sp^3, AB_4) is a complete (regular) tetrahedron. The presence of orbitals having lone pairs or unpaired electrons invariably results in geometric form distortion. Indeed, in the SO_3^{2-} particle the O-S-O angles differ from the tetrahedric value, while in the SO_4^{2-} particle these angles are equal to 109.5° (cf. Sec. 3.8).

In the geometric forms corresponding to the sp^3d-, sp^3d^2- (for the $:AB_5$ particles) and sp^3d^3-hybridizations, the axial positions of the terminal B atoms are designated by an asterisk (B^*) to distinguish them from the equatorial positions (no asterisks).

The three-dimensional presentation of the geometric forms of particles is as follows:

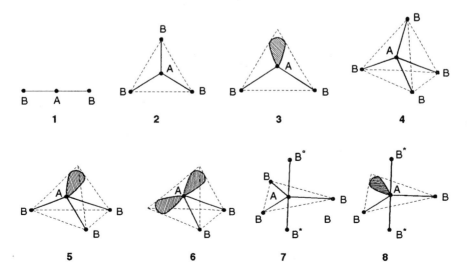

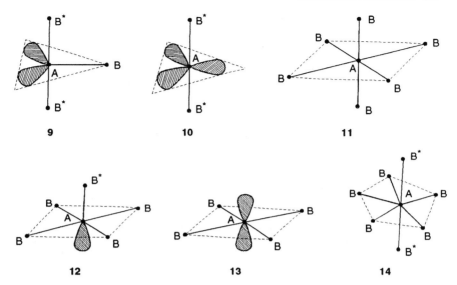

9   **10**   **11**

12   **13**   **14**

Hybridization type and geometric form	Particle	Examples
sp, linear (digonal), $\angle$ B–A–B 180°	AB$_2$ (**1**)	BO$_2^-$, BeF$_2$, CO$_2$, NO$_2^+$, N$_2$O, (CN)Cl, HCN, NCS$^-$, SnO$_2$
sp^2, triangular (trigonal), $\angle$ B–A–B 120°	AB$_3$ (**2**)	AlF$_3$, BO$_3^{3-}$, CO$_3^{2-}$, NO$_3^-$, SO$_3$, CCl$_2$O, COF$_2$, (NO$_2$)Cl
	:AB$_2$ (**3**)	NO$_2^-$, SO$_2$, SnF$_2$, O$_3$, NO$_2$, (NO)Cl
sp^3, tetrahedral, $\angle$ B–A–B 109.5°	AB$_4$ (**4**)	[BF$_4$]$^-$, CCl$_4$, ClO$_4^-$, NH$_4^+$, SO$_4^{2-}$, PCl$_3$O, SCl$_2$O$_2$, SO$_3$S^{2-}
	:AB$_3$ (**5**)	AsF$_3$, H$_3$O$^+$, NH$_3$, SO$_3^{2-}$, SOF$_2$
	::AB$_2$ (**6**)	[BrF$_2$]$^+$, ClO$_2^-$, H$_2$O, O$_3^-$, HOF
sp^3d, trigonal-bipyramidal, $\angle$ B–A–B 120°, $\angle$ B–A–B* 90°, $\angle$ B*–A–B* 180°	AB$_5$ (**7**)	AsF$_5$, PCl$_5$, PF$_5$, SbCl$_5$
	:AB$_4$ (**8**)	[BrF$_4$]$^+$, SF$_4$, TeCl$_4$, XeO$_2$F$_2$
	::AB$_3$ (**9**)	BrF$_3$, ClF$_3$, [XeF$_3$]$^+$, XeOF$_2$
	:::AB$_2$ (**10**)	[IBr$_2$]$^-$, KrF$_2$, XeF$_2$, [I(I)$_2$]$^-$
sp^3d^2, octahedral, $\angle$ B–A–B 90°	AB$_6$ (**11**)	[AlF$_6$]$^{3-}$, [AsF$_6$]$^-$, IO$_6^{5-}$, SF$_6$, XeO$_6^{4-}$, SClF$_5$
	:AB$_5$ (**12**)	BrF$_5$, [SbF$_5$]$^{2-}$, [XeF$_5$]$^+$
	::AB$_4$ (**13**)	[BrF$_4$]$^-$, [ICl$_4$]$^-$, XeF$_4$
sp^3d^3 pentagonal-bipyramidal, $\angle$ B–A–B 72°, $\angle$ B–A–B* 90°, $\angle$ B*–A–B* 180°	AB$_7$ (**14**)	IF$_7$

3.8 Bond Length and Valence Angles for Polyatomic Molecules and Ions with One Central Atom of an *sp*-Element [5, 7, 9, 13, 30, 33–34]

This table lists the polyatomic chemical particles (molecules, radicals thereof, and ions) having the general molecular formula AB_n and containing one central atom A, two or more terminal atoms (groups of atoms) and, therefore, involving the A-B bonds only. Where the central atom does not occupy the first place in the formula or is not obvious (in the formulas of complexes with polyatomic ligands), its symbol is listed in the "Hybridization" column in parentheses. The terminal atoms in axial positions are marked with an asterisk *. For spatial representations of geometric form of molecules in conformity with the type of hybridization and the molecular formula, see Sec. 3.7.

The experimental data for molecules and radicals pertain to the thermodynamic conditions of their formation, whereas the data for ions refer to the conditions of their existence in an aqueous solution or/and ionic crystals. In the latter case, the values of bond lengths may be given as an interval, in so far as the exact value depends on the actual counterion present. The dots (...) denote the absence of data.

Substance	Hybridization	Bond length (pm) and valence angles
$AlBr_3$	sp^2	Al–Br 227, Br–Al–Br 120°
$AlCl_3$	sp^2	Al–Cl 206, Cl–Al–Cl 120°
$[AlCl_4]^-$	sp^3	Al–Cl 209–214, Cl–Al–Cl 109.5°
AlF_3	sp^2	Al–F 163, F–Al–F 120°
$[AlF_4]^-$	sp^3	Al–F 169, F–Al–F 109.5°
$[AlF_6]^{3-}$	sp^3d^2	Al–F 181, F–Al–F 90°
$[AlH_4]^-$	sp^3	Al–H 155, H–Al–H 109.5°
$[Al(H_2O)_6]^{3+}$	sp^3d^2 (Al)	Al–O 188, O–Al–O 90°
AlI_3	sp^2	Al–I 244, I–Al–I 120°
$[Al(OH)_4]^-$	sp^3 (Al)	Al–O 175, O–Al–O 109.5°
$[Al(OH)_6]^{3-}$	sp^3d^2 (Al)	Al–O 185–198, O–Al–O 90°
$AsBr_3$	sp^3	As–Br 233, Br–As–Br 100°
$AsCl_3$	sp^3	As–Cl 216, Cl–As–Cl 99°
AsF_3	sp^3	As–F 171, F–As–F 96°
AsF_5	sp^3d	As–F 166, As–F* 171, F–As–F 120°, F–As–F* 90°
$[AsF_6]^-$	sp^3d^2	As–F 163–170, F–As–F 90°
AsH_3	sp^3 ?	As–H 151, H–As–H 92°
AsI_3	sp^3	As–I 256, I–As–I 100°
AsO_3^{3-}	sp^3	As–O 201, O–As–O ...
AsO_4^{3-}	sp^3	As–O 175, O–As–O 109.5°
AsS_4^{3-}	sp^3	As–S 223, S–As–S 109.5°
BBr_3	sp^2	B–Br 189, Br–B–Br 120°
BCl_3	sp^2	B–Cl 174, Cl–B–Cl 120°
BF_3	sp^2	B–F 131, F–B–F 120°
$[BF_4]^-$	sp^3	B–F 140, F–B–F 109.5°

Substance	Hybridization	Bond length (pm) and valence angles
$[BH_4]^-$	sp^3	B–H 126, H–B–H 109.5°
BI_3	sp^2	B–I 212, I–B–I 120°
$[B(NH_3)F_3]$	sp^3 (B)	B–N 160, B–F 158, N–B–F ...
BO_2^-	sp	B–O 127, O–B–O 180°
BO_3^{3-}	sp^2	B–O 132–140, O–B–O 120°
$B(OH)_3$	sp^2 (B)	B–O 136, O–B–O 120°
$[B(OH)_4]^-$	sp^3 (B)	B–O 143–150, O–B–O 109.5°
$BeBr_2$	sp	Be–Br 191, Br–Be–Br 180°
$BeCl_2$	sp	Be–Cl 175, Cl–Be–Cl 180°
BeF_2	sp	Be–F 140, F–Be–F 180°
$[BeF_4]^{2-}$	sp^3	Be–F 155, F–Be–F 109.5°
BeI_2	sp	Be–I 210, I–Be–I 180°
$BiBr_3$	sp^3	Bi–Br 263, Br–Bi–Br 100°
$BiCl_3$	sp^3	Bi–Cl 248, Cl–Bi–Cl 100°
BiF_3	sp^3	Bi–F 200, F–Bi–F 94°
BiI_3	sp^3	Bi–I 287, I–Bi–I 98°
$[Br(Br)_2]^-$	sp^3d	Br–Br* 250–290, Br*–Br–Br* 180°
$[BrF_2]^+$	sp^3	Br–F 169, F–Br–F 94°
BrF_3	sp^3d	Br–F 172, Br–F* 181, F–Br–F* 86°, F*–Br–F* 188°
$[BrF_4]^-$	sp^3d^2	Br–F 188, F–Br–F 90°
$[BrF_4]^+$	sp^3d	Br–F 170–190, Br–F* 220–250, F–Br–F ..., F–Br–F* ...
BrF_5	sp^3d^2	Br–F 177, Br–F* 169, F–Br–F 90°, F–Br–F* 85°
BrO_3^-	sp^3	Br–O 178, O–Br–O 112°
BrO_4^-	sp^3	Br–O 161, O–Br–O 109.5°
CBr_4	sp^3	C–Br 194, Br–C–Br 109.5°
CCl_4	sp^3	C–Cl 177, Cl–C–Cl 109.5°
CCl_2O	sp^2	C–Cl 175, C–O 117, Cl–C–Cl 111°, Cl–C–O 124°
CF_4	sp^3	C–F 132, F–C–F 109.5°
CH_4	sp^3	C–H 109, H–C–H 109.5°
$CHCl_3$	sp^3	C–H 110, C–Cl 176, H–C–Cl 108°, Cl–C–Cl 111°
CH_3OH	sp^3	C–H 110, C–O 143, O–H 96, H–C–H 109°, C–O–H 109°
CI_4	sp^3	C–I 215, I–C–I 109.5°
$(CN)Br$	sp	C–N 116, C–Br 179, N–C–Br 180°
$(CN)Cl$	sp	C–N 116, C–Cl 163, N–C–Cl 180°
$(CN)F$	sp	C–N 116, C–F 126, N–C–F 180°
$(CN)I$	sp	C–N 116, C–I 200, N–C–I 180°
CO_2	sp	C–O 117, O–C–O 180°
CO_3^{2-}	sp^2	C–O 129, O–C–O 120°
COF_2	sp^2	C–O 117, C–F 132, O–C–F 126°, F–C–F 108°
CS_2	sp	C–S 155, S–C–S 180°
CS_3^{2-}	sp^2	C–S 165, S–C–S 120°
$CSCl_2$	sp^2	C–S 160, C–Cl 173, S–C–Cl 124°, Cl–C–Cl 111°
CSO	sp	C–S 156, C–O 115, S–C–O 180°
CSe_2	sp	C–Se 205, Se–C–Se 180°
$C(Se)O$	sp	C–Se 171, C–O 116, Se–C–O 180°
$C(Se)S$	sp	C–Se 170, C–S 155, Se–C–S 180°
$C(Te)S$	sp	C–Te 190, C–S 156, Te–C–S 180°
CeC_2	sp (C)	C–Ce 260, C–C 128, Ce–C–C 180°

Substance	Hybridization	Bond length (pm) and valence angles
$[ClF_2]^+$	sp^3	Cl–F 156, F–Cl–F 100°
ClF_5	sp^3d^2	Cl–F 167, Cl–F* 158, F–Cl–F* 86°
Cl_3N	sp^3 (N)	N–Cl 176, Cl–N–Cl 108°
ClO_2^-	sp^3	Cl–O 157, O–Cl–O 111°
ClO_2	sp^3	Cl–O 147, O–Cl–O 118°
ClO_3^-	sp^3	Cl–O 145, O–Cl–O 106°
ClO_4^-	sp^3	Cl–O 148, O–Cl–O 109.5°
Cl_2O	sp^3 (O)	O–Cl 170, Cl–O–Cl 111°
ClO_2F	sp^3	Cl–O 148, Cl–F 167, O–Cl–O 114°, O–Cl–F 103°
ClO_3F	sp^3	Cl–O 140, Cl–F 162, O–Cl–O 117°, O–Cl–F 101°
D_2O	sp^3 (O)	O–D 97, D–O–D ...
D_2S	sp^3 ? (S)	S–D 135, D–S–D ...
$GaBr_3$	sp^2	Ga–Br 224, Br–Ga–Br 120°
$GaCl_3$	sp^2	Ga–Cl 209, Cl–Ga–Cl 120°
$[GaCl_4]^-$	sp^3	Ga–Cl 217, Cl–Ga–Cl 109.5°
$[GaH_4]^-$	sp^3	Ga–H 197, H–Ga–H 109.5°
GaI_3	sp^2	Ga–I 246, I–Ga–I 120°
$GeBr_4$	sp^3	Ge–Br 227, Br–Ge–Br 109.5°
$GeCl_2$	sp^2	Ge–Cl 211, Cl–Ge–Cl 107°
$GeCl_4$	sp^3	Ge–Cl 211, Cl–Ge–Cl 109.5°
$[GeCl_6]^{2-}$	sp^3d^2	Ge–Cl 235, Cl–Ge–Cl 90°
GeF_4	sp^3	Ge–F 167, F–Ge–F 109.5°
$[GeF_6]^{2-}$	sp^3d^2	Ge–F 177, F–Ge–F 90°
GeH_4	sp^3	Ge–H 153, H–Ge–H 109.5°
GeI_4	sp^3	Ge–I 250, I–Ge–I 109.5°
GeO_2	sp	Ge–O 163, O–Ge–O 180°
HCN	sp (C)	C–H 107, C–N 115, H–C–N 180°
HCO_3^-	sp^2 (C)	C–O 126, C–OH 135, O–C–O 125°, O–C–OH 118°
HClO	sp^3 (O)	O–H 96, O–Cl 169, H–O–Cl 103°
$HClO_4$	sp^3 (Cl)	Cl–O 141, Cl–OH 164, O–Cl–O 113°, O–Cl–OH 106°
HIO_3	sp^3 (I)	I–O 180, I–OH 190, O–I–O 98°, O–I–OH ...
HNC	sp (N)	N–H 100, N–C 119, H–N–C 180°
HNO_2	sp^2 (N)	N–O 117, N–OH 144, O–N–OH 111°, N–O–H 102°
HNO_3	sp^2 (N)	N–O 121, N–OH 141, O–N–O 130°, N–O–H 102°
H_2O	sp^3 (O)	O–H 96, H–O–H 105°
H_3O^+	sp^3 (O)	O–H 95, H–O–H 109°
HOF	sp^3 (O)	O–H 97, O–F 144, H–O–F 97°
H_3PO_4	sp^3 (P)	P–O 152, P–OH 157, O–P–OH 112°, HO–P–OH 106°
H_2S	sp^3 ? (S)	S–H 134, H–S–H 92°
H_3S^+	sp^3 (S)	S–H 135, H–S–H 96°
HSO_4^-	sp^3 (S)	S–O 143–146, S–OH 156, O–S–O ..., O–S–OH
H_2SO_4	sp^3 (S)	S–O 142, S–OH 153, O–S–O 125°, HO–S–OH 100°
H_2Se	sp^3 ? (Se)	Se–H 146, H–Se–H 91°
H_2SeO_3	sp^3 (Se)	Se–O 164, Se–OH 174, O–Se–OH ..., HO–Se–OH ...
H_2Te	sp^3 ? (Te)	Te–H 166, H–Te–H 90°
$[IBr_2]^-$	sp^3d	I–Br* 271, Br*–I–Br* 180°
$[I(Br)Cl]^-$	sp^3d	I–Br* 250, I–Cl* 238, Br*–I–Cl* 180°
$[ICl_2]^-$	sp^3d	I–Cl* 255, Cl*–I–Cl* 180°

Substance	Hybridization	Bond length (pm) and valence angles
[ICl$_2$]$^+$	sp^3	I–Cl 230, Cl–I–Cl 95°
[ICl$_4$]$^-$	sp^3d^2	I–Cl 242, 247, 260, 253, Cl–I–Cl 90°
IF$_5$	sp^3d^2	I–F 187, I–F$^•$ 184, F–I–F$^•$ 82°
IF$_7$	sp^3d^3	I–F 183, I–F$^•$ 194, F–I–F ..., F–I–F$^•$ ...
[I(I)$_2$]$^-$	sp^3d	I–I$^•$ 290, I$^•$–I–I$^•$ 180°
IO$_3^-$	sp^3	I–O 182, O–I–O 97°
IO$_4^-$	sp^3	I–O 178, O–I–O 109.5°
IO$_6^{5-}$	sp^3d^2	I–O 185, O–I–O 90°
InI$_3$	sp^2	In–I 262, I–In–I 120°
KCN	sp (C)	C–K 257, C–N 115, K–C–N 180°
KrF$_2$	sp^3d	Kr–F$^•$ 189, F$^•$–Kr–F$^•$ 180°
LaC$_2$	sp (C)	C–La 264, C–C 130, La–C–C 180°
N$_3^-$	sp	N–N 117, N–N–N 180°
NCS$^-$	sp (C)	C–N 117, C–S 160, N–C–S 180°
ND$_3$	sp^3	N–D 103, D–N–D 107°
NF$_2$	sp^3	N–F 135, F–N–F 103°
NF$_3$	sp^3	N–F 137, F–N–F 102°
NH$_2^-$	sp^3	N–H 103, H–N–H 104°
NH$_3$	sp^3	N–H 103, H–N–H 107°
NH$_4^+$	sp^3	N–H 104, H–N–H 109.5°
NH$_2$Cl	sp^3	N–H 102, N–Cl 175, H–N–H 107°, H–N–Cl 104°
(NH)F$_2$	sp^3	N–H 103, N–F 140, H–N–F 100°, F–N–F 103°
NH$_2$OH	sp^3 (N)	N–H 102, N–OH 145, H–N–H 107°, H–N–OH 103°
NH$_3$OH$^+$	sp^3 (N)	N–H 99, N–OH 141, H–N–H ..., H–N–OH ...
NO$_2^-$	sp^2	N–O 124, O–N–O 115°
NO$_2$	sp^2	N–O 120, O–N–O 134°
NO$_2^+$	sp	N–O 110, O–N–O 180°
NO$_3^-$	sp^2	N–O 124, O–N–O 120°
N$_2$O	sp (N)	N–N 113, N–O 118, N–N–O 180°
(NO)Br	sp^2	N–O 115, N–Br 214, O–N–Br 115°
(NO)Cl	sp^2	N–O 114, N–Cl 198, O–N–Cl 113°
(NO$_2$)Cl	sp^2	N–Cl 183, N–O 121, O–N–Cl 115°, O–N–O 130°
(NO)F	sp^2	N–O 114, N–F 151, O–N–F 110°
NOF$_3$	sp^3	N–O 116, N–F 143, O–N–F 117°, F–N–F 101°
(NO$_2$)F	sp^2	N–O 118, N–F 147, O–N–O 136°, O–N–F 112
NaCN	sp (C)	C–Na 226, C–N 115, Na–C–N 180°
O$_3^-$	sp^3	O–O 138, O–O–O 108°
O$_3$	sp^2	O–O 127, O–O–O 117°
OCN$^-$	sp (C)	C–O 118, C–N 120, O–C–N 180°
OF$_2$	sp^3	O–F 141, F–O–F 103°
PBr$_3$	sp^3	P–Br 222, Br–P–Br 101°
PBr$_4^+$	sp^3	P–Br 215, Br–P–Br 109.5°
PBr$_3$O	sp^3	P–Br 206, P–O 141, Br–P–Br 108°, Br–P–O 111°
PCl$_3$	sp^3	P–Cl 204, Cl–P–Cl 100°
PCl$_4^+$	sp^3	P–Cl 198, Cl–P–Cl 109.5°
PCl$_5$	sp^3d	P–Cl 202, P–Cl$^•$ 212, Cl–P–Cl 120°, Cl–P–Cl$^•$ 90°
[PCl$_6$]$^-$	sp^3d^2	P–Cl 207, Cl–P–Cl 90°
PCl$_2$F	sp^3	P–Cl 204, P–F–155, Cl–P–Cl 99°, Cl–P–F 102°

Substance	Hybridization	Bond length (pm) and valence angles
PCl_3O	sp^3	P–Cl 199, P–O 146, Cl–P–Cl 104°, Cl–P–O 115°
PF_3	sp^3	P–F 156, F–P–F 97°
PF_5	sp^3d	P–F 153, P–F* 158, F–P–F 120°, F–P–F* 90°
$[PF_6]^-$	sp^3d^2	P–F 159, F–P–F 90°
PH_3	sp^3	P–H 141, H–P–H 94°
PH_4^+	sp^3	P–H 142, H–P–H 109.5°
$(PH)F_2$	sp^3	P–H 141, P–F 158, H–P–F 96°, F–P–F 99°
PI_3	sp^3	P–I 243, I–P–I 100°
PO_4^{3-}	sp^3	P–O 154–156, O–P–O 109.5°
POF_3	sp^3	P–O 144, P–F 152, O–P–F 117°, F–P–F 101°
$[PO_2F_2]^-$	sp^3	P–O 147, P–F 158, O–P–O 122°, F–P–F 97°
PS_4^{3-}	sp^3	P–S 205, S–P–S 109.5°
PSF_3	sp^3	P–S 187, P–F 154, S–P–F 118°, F–P–F 100°
$PbBr_2$	sp^2	Pb–Br 260, Br–Pb–Br 95°
$PbCl_2$	sp^2	Pb–Cl 246, Cl–Pb–Cl 96°
$PbCl_4$	sp^3	Pb–Cl 243, Cl–Pb–Cl 109.5°
$[PbCl_6]^{2-}$	sp^3d^2	Pb–Cl 250, Cl–Pb–Cl 90°
PbF_2	sp^2 ?	Pb–F 213, F–Pb–F 90°
PbI_2	sp^2	Pb–I 279, I–Pb–I 95°
SBr_2O	sp^3	S–Br 227, S–O 145, Br–S–Br 96°, Br–S–O 108°
SCl_2	sp^3	S–Cl 201, Cl–S–Cl 103°
$SClF_5$	sp^3d^2	S–Cl 203, S–F 158, Cl–S–F 90°, F–S–F 90°
SCl_2O	sp^3	S–Cl 208, S–O 144, Cl–S–Cl 96°, Cl–S–O 106°
SCl_2O_2	sp^3	S–Cl 201, S–O 140, Cl–S–Cl 100, O–S–O 124°
SF_2	sp^3	S–F 159, F–S–F 98°
SF_3^+	sp^3	S–F 150, F–S–F 98°
SF_4	sp^3d	S–F 154, S–F* 164, F–S–F 104°, F–S–F* 89°
SF_6	sp^3d^2	S–F 156, F–S–F 90°
SO_2	sp^2	S–O 143, O–S–O 119°
SO_3^{2-}	sp^3	S–O 152, O–S–O 105°
SO_3	sp^2	S–O 142, O–S–O 120°
SO_4^{2-}	sp^3	S–O 147, O–S–O 109.5°
S_2O	sp^2 (S)	S–S 188, S–O 146, S–S–O 118°
SOF_2	sp^3	S–O 141, S–F 159, O–S–F 107°, F–S–F 93°
$[SOF_3]^+$	sp^3	S–O 135, S–F 144, O–S–F 116°, F–S–F 102°
SO_2F_2	sp^3	S–O 140, S–F 153, O–S–O 123°, F–S–F 97°
SO_3F^-	sp^3	S–O 142, S–F 157, O–S–O 113°, O–S–F 106°
SO_3S^{2-}	sp^3	S–O 148, S–S 199, O–S–O ..., O–S–S ...
$S(S)F_2$	sp^3	S–S 186, S–F 160, S–S–F 108°, F–S–F 93°
$SbBr_3$	sp^3	Sb–Br 249, Br–Sb–Br 98°
$SbCl_3$	sp^3	Sb–Cl 233, Cl–Sb–Cl 97°
$SbCl_5$	sp^3d	Sb–Cl 229, Sb–Cl* 234, Cl–Sb–Cl ..., Cl–Sb–Cl* ...
$[SbCl_6]^-$	sp^3d^2	Sb–Cl 221–244, Cl–Sb–Cl 90°
SbF_3	sp^3	Sb–F 188, F–Sb–F 95°
$[SbF_5]^{2-}$	sp^3d^2	Sb–F 202, F–Sb–F 90°
$[SbF_6]^-$	sp^3d^2	Sb–F 177–195, F–Sb–F 90°
SbH_3	sp^3 ?	Sb–H 171, H–Sb–H 92°
SbI_3	sp^3	Sb–I 272, I–Sb–I 99°

Substance	Hybridization	Bond length (pm) and valence angles
$[Sb(OH)_6]^-$	sp^3d^2 (Sb)	Sb–O 197, O–Sb–O 90°
$[SbS_4]^{3-}$	sp^3	Sb–S 238, S–Sb–S 109.5°
$SeCl_2O$	sp^3	Se–Cl 220, Se–O 161, Cl–Se–Cl 97°, Cl–Se–O 106°
SeF_4	sp^3d	Se–F 168, Se–F* 177, F–Se–F* 100°, F*–Se–F* 169°
SeF_6	sp^3d^2	Se–F 169, F–Se–F 90°
SeO_2	sp^2	Se–O 161, O–Se–O 113°
SeO_3^{2-}	sp^3	Se–O 170–176, O–Se–O 102°
SeO_3	sp^2	Se–O 169, O–Se–O 120°
SeO_4^{2-}	sp^3	Se–O 161, O–Se–O 109.5°
$SeOF_2$	sp^3	Se–O 158, Se–F 173, O–Se–F 105°, F–Se–F 92°
$SiBr_4$	sp^3	Si–Br 215, Br–Si–Br 109.5°
$SiCl_4$	sp^3	Si–Cl 202, Cl–Si–Cl 109.5°
SiF_2	sp^2	Si–F 159, F–Si–F 101°
SiF_4	sp^3	Si–F 156, F–Si–F 109.5°
$[SiF_6]^{2-}$	sp^3d^2	Si–F 170, F–Si–F 90°
SiH_4	sp^3	Si–H 148, H–Si–H 109.5°
SiI_4	sp^3	Si–I 243, I–Si–I 109.5°
SiO_2	sp	Si–O 155, O–Si–O 180°
SiO_4^{4-}	sp^3	Si–O 157, O–Si–O 109.5°
$SnBr_2$	sp^2	Sn–Br 255, Br–Sn–Br 95°
$SnBr_4$	sp^3	Sn–Br 244, Br–Sn–Br 109.5°
$[SnBr_6]^{2-}$	sp^3d^2	Sn–Br 259, Br–Sn–Br 90°
$SnCl_2$	sp^2	Sn–Cl 234, Cl–Sn–Cl 100°
$SnCl_4$	sp^3	Sn–Cl 228, Cl–Sn–Cl 109.5°
$[SnCl_6]^{2-}$	sp^3d^2	Sn–Cl 242, Cl–Sn–Cl 90°
SnF_2	sp^2	Sn–F 206, F–Sn–F 94°
SnF_4	sp^3	Sn–F 184, F–Sn–F 109.5°
SnH_4	sp^3	Sn–H 171, H–Sn–H 109.5°
SnI_2	sp^2	Sn–I 278, I–Sn–I 95°
SnI_4	sp^3	Sn–I 264, I–Sn–I 109.5°
$[SnI_6]^{2-}$	sp^3d^2	Sn–I 284, I–Sn–I 90°
SnO_2	sp	Sn–O 181, O–Sn–I 180°
$[Sn(OH)_6]^{2-}$	sp^3d^2 (Sn)	Sn–O 206, O–Sn–O 90°
$[SnS_4]^{4-}$	sp^3	Sn–S 237, S–Sn–S 109.5°
$TeBr_2$	sp^3	Te–Br 251, Br–Te–Br 98°
$TeCl_2$	sp^3	Te–Cl 236, Cl–Te–Cl ...
$TeCl_4$	sp^3d	Te–Cl 233, Te–Cl* >233, Cl–Te–Cl 120°, Cl–Te–Cl* 93°
TeF_6	sp^3d^2	Te–F 184, F–Te–F 90°
TeO_2	sp^2	Te–O 183, O–Te–O 110°
TeO_3^{2-}	sp^3	Te–O 188, O–Te–O 100°
TeO_6^{6-}	sp^3d^2	Te–O 198, O–Te–O 90°
XeF_2	sp^3d	Xe–F* 198, F*–Xe–F* 180°
$[XeF_3]^+$	sp^3d	Xe–F 183, Xe–F* 191, F–Xe–F* 80–82°, F*–Xe–F* 162°
XeF_4	sp^3d^2	Xe–F 194, F–Xe–F 90°
$[XeF_5]^+$	sp^3d^2	Xe–F 190, Xe–F* 177, F–Xe–F 90°, F–Xe–F* 79–83
XeO_3	sp^3	Xe–O 176, O–Xe–O 103°
XeO_4^{4-}	sp^3d^2	Xe–O 184–186, O–Xe–O 90°
$XeOF_2$	sp^3d	Xe–O 176, Xe–F* 195, O–Xe–F* 90°, F*–Xe–F* 180°

(Continued)

Substance	Hybridization	Bond length (pm) and valence angles
$XeOF_4$	sp^3d^2	Xe–O˙ 171, Xe–F 190, F–Xe–F ..., F–Xe–O˙ 91°
XeO_2F_2	sp^3d	Xe–O 171, Xe–F˙ 190, O–Xe–O 106°, O–Xe–F˙ 92°

3.9 Magnetic Moments of Atoms of *d*-Elements [6–9, 29, 33–34]

The theoretical magnetic moments of a chemical particle with an atom of the *d*-element comprises the sum of a purely spin (μ_{sp}) and an orbital (μ_{orb}) contribution of the magnetic moments of electrons in an atom of the *d*-element. In a definite field of terminal atoms or groups of atoms (ligands) the contribution μ_{orb} experiences a more or less pronounced killing, so the experimental values of magnetic moments deviate from the sum $\mu_{sp} + \mu_{orb}$.

One spin state only is possible for the atoms of *d*-elements, wherein in the particles the valence electron configurations are $d^0 - d^3$, d^9 and d^{10}. In the case of Period 4 elements, the atoms having the $d^4 - d^8$ configurations may be in two states. For example, for d^4 (C^{II},Mn^{III}) μ_{sp} = 4.9μ_B (high-spin state) and μ_{sp} = 2.38μ_B (low-spin state). For the elements of Periods 5 and 6, the atoms having the $d^4 - d^8$ configurations are in the low-spin state, wherein the μ_{orb} contribution is killed almost completely. For example, for d^4 (Os^{IV}, Re^{III}, Ru^{IV}) μ_{sp} = 2.83μ_B.

The table also lists the theoretical (computed) values of μ_{sp}, the sum $\mu_{sp} + \mu_{orb}$, and the interval of experimental values of magnetic moments (μ_{exp}) for the molecules, radicals and ions that contain an atom of the *d*-element. The values of μ_{exp} enclosed in parentheses pertain to the elements of Periods 5 and 6, provided they are distinct from the μ_{exp} values for the elements of Period 4 due to the antiferromagnetic spin-spin interaction of neighboring atoms in the crystalline lattice.

The exemplary atoms of *d*-elements with different oxidation numbers are listed for each spin state. The exact values of μ_{exp} for various particles are given in Sec. 3.11.

Configu-ration	Magnetic moment, μ_B			Examples
	μ_{sp}	$\mu_{sp} + \mu_{orb}$	μ_{exp}	
d^0	0	0	0	Cr^{VI}, Hf^{IV}, La^{III}, Mn^{VII}, Mo^{VI}, Nb^{V}, Os^{VII}, Re^{VII}, Ru^{VIII}, Sc^{III}, Ta^{V}, Tc^{VII}, Th^{IV}, Ti^{IV}, U^{VI}, V^{V}, W^{VI}, Zr^{IV}
d^1	1.73	3.00	1.7–1.9 (1.1–1.5)	Mn^{VI}, Mo^{V}, Re^{VI}, Ru^{VII}, Ti^{III}, V^{IV}, W^{V}
d^2	2.83	4.47	2.6–2.9 (0–2.1)	Fe^{VI}, Mn^{V}, Mo^{IV}, Os^{VI}, Ru^{VI}, Ti^{II}, V^{III}
d^3	3.87	5.20	3.3–4.1	Cr^{III}, Ir^{VI}, Mn^{IV}, Mo^{III}, Os^{V}, Re^{IV}, Ru^{V}, Tc^{IV}, V^{II}
d^4	4.90	5.48	4.7–5.0	Cr^{II}, Mn^{III}
	2.83	–	3.2–3.6 (0–3.1)	Cr^{II}, Mn^{III}, Os^{IV}, Re^{III}, Ru^{IV}

(Continued)

Configuration	Magnetic moment, μ_B			Examples
	μ_{sp}	$\mu_{sp} + \mu_{orb}$	μ_{exp}	
d^5	5.92	5.92	5.6–6.1	Fe^{III}, Mn^{II}
	1.73	–	1.6–2.5	Co^{IV}, Fe^{III}, Ir^{IV}, Mn^{II}, Os^{III}, Pt^V, Rh^{IV}, Ru^{III}, V^0
d^6	4.90	5.48	5.1–5.7	Co^{III}, Fe^{II}
	0	0	0	Co^{III}, Cr^0, Fe^{II}, Ir^{III}, Mo^0, Ni^{IV}, Os^{II}, Pd^{IV}, Pt^{IV}, Rh^{III}, Ru^{II}, W^0
d^7	3.87	5.20	4.3–5.2	Co^{II}, Fe^I
	1.73	–	1.7–2.0	Co^{II}, Ni^{III}
d^8	2.83	4.47	2.8–4.1	Ni^{II}
	0	0	0	Au^{III}, Fe^0, Mn^{-I}, Ni^{II}, Os^0, Pd^{II}, Pt^{II}, Ru^0
d^9	1.73	3.00	1.7–2.2	Ag^{II}, Cu^{II}
d^{10}	0	0	0	Ag^I, Au^I, Cd^{II}, Co^{-I}, Cu^I, Hg^{II}, Ni^0, Pd^0, Pt^0, Zn^{II}

3.10 Distortion of Bond Length in the Octahedral and Tetrahedral Field of Ligands [29, 33–34]

The crystalline field theory predicts that for the particle AB_6, where A is a *d*-element atom in the *octahedral* field of the ligands B, in the case of certain configurations and spin states of the atom A, there occurs a minor or a major alteration in the length of two axial bonds (tetragonal distortion) resulting in the formation of an octahedron either extended (more often) or flattened (less often) along the selected axis. A major tetragonal distortion that results in the formation of an extended octahedron would bring about an almost total disappearance of the effect of ligands disposed in axial positions. Here, the octahedral particle AB_6 with the dsp^3d-hybridization of the orbitals of the atom A will convert into the particle AB_4 having the dsp^2-hybridization of the orbitals of the atom A in the *square planar* field of ligands.

In the AB_4 particle, where the atom A in the *tetrahedral* field of ligands B, and under certain configurations of the atom A, there will occur a minor or a major alteration of bond lengths resulting in tetrahedron flattening along the selected axis.

The number of unpaired electrons in an atom of the *d*-element that characterizes its spin state is designated as N_e. The d^4–d^6 configurations in the octahedral field of ligands may be in the high- and low-spin states. For example, for the d^4 configuration the states in question appear when $N_e = 4$ and 2, respectively.

Configuration	Octahedral field			Tetrahedral field		
	N_e	Hybridization	Distortion	N_e	Hybridization	Distortion
d^0	0	d^2sp^3	none	0	d^3s	none
d^1	1	d^2sp^3	minor	1	d^3s	minor
d^2	2	d^2sp^3	minor	2	d^3s	none

(Continued)

Configuration	Octahedral field			Tetrahedral field		
	N_e	Hybridization	Distortion	N_e	Hybridization	Distortion
d^3	3	d^2sp^3	none	3	d^2sp	major
d^4	4	dsp^3d	major	4	sp^3	major
	2	d^2sp^3	minor			
d^5	5	sp^3d^2	none	5	sp^3	none
	1	d^2sp^3	minor			
d^6	4	sp^3d^2	minor	4	sp^3	minor
	0	d^2sp^3	none			
d^7	3	sp^3d^2	minor	3	sp^3	none
	1	dsp^3d	major			
d^8	2	sp^3d^2	none	2	sp^3	minor
	0	dsp^3d	major			
d^9	1*	dsp^3d	major	1	sp^3	major
d^{10}	0	sp^3d^2	none	0	sp^3	none

*$(n - 1)d^1 \rightarrow nd^1$

3.11 Valent State, Hybridization Type, Magnetic Moment and Bond Length for Polyatomic Molecules and Ions with One Central Atom of the *df*-Element [7, 9, 13, 29, 33–34]

This table lists polyatomic chemical particles (molecules, radicals, and ions) of the general formula AB_n containing one central atom of a *df*-element (an element of Group B in the Periodic Table) and two or more terminal atoms or groups of atoms B (ligands). The symbol of the central atom A always stands at the first place in the formula.

The valent state of the central atom is characterized by the oxidation number (v), the total number of *d*-electrons providing for this oxidation number (N_d), and the number of unpaired electrons (N_e).

The following geometric forms of the particles AB_n correspond to the hybridization types listed in the table:

sp	linear (digonal), particles AB_2
sp^2	triangular (trigonal), particles AB_3
d^2s	triangular (trigonal), particles AB_3
sp^3	tetrahedral, particles AB_4
d^3s	tetrahedral, particles AB_4
dsp^2	square planar, particles AB_4
dsp^3	trigonal-bipyramidal, particles AB_5
dsp^3	quadratic-pyramidal (central atom above the square base center), particles AB_5

d^2sp^2 quadratic-pyramidal (central atom above the square base center), particles AB_5

d^3sp trigonal-bipyramidal, particles AB_5

d^4s quadratic-pyramidal (central atom in the square base center), particles AB_5

sp^3d^2 octahedral, particles AB_6

d^2sp^3 octahedral, particles AB_6

dsp^3d octahedral, particles AB_6

d^3sp^3 pentagonal-bipyramidal, particles AB_7.

The hybridization type and geometric form of the particles AB_n are deduced from the method of valence bonds in combination with the crystalline field theory. It should be noted that actually there occurs mutual transition between the trigonal-bipyramidal and quadratic-pyramidal geometric forms, since the transition in question involves a relatively small alteration in the valence angles (dsp^3- and d^2sp^2-hybridization types).

The hybridization type and geometric form thus deduced are corroborated by experimental data, viz., the magnetic moment values of a particle and/or the length of chemical bonds A-B (in terminal groups of atoms the symbol B is ascribed to the coordinated atom in such a group of atoms).

The axial positions of the terminal atoms B (for the geometric forms corresponding to dsp^3-, d^2sp^2-, d^3sp-, d^4s-, sp^3d^2-, d^2sp^3-, dsp^3d- and d^3sp^3-hybridizations) are marked by an asterisk (B^*) in contrast to the equatorial positions (B without an asterisk). The known distorted values of valence angles between bonds are cited in footnotes.

Experimental data for molecules and radicals correspond to the thermodynamic conditions of their formation, whereas the data for ions pertain to the condition of ion existence in an aqueous solutions and/or in ionic crystals. The dots (...) denote the absence of data.

Full names of ligands listed in the abbreviated form are given in Sec. 6.3.1.

Substance	v	N_d/N_e	Hybridization	μ (μ_B)	Bond length (pm)
$[Ag(CN)_2]^-$	I	10/0	sp	0	Ag–C 213
$[Ag(NH_3)_2]^+$	I	10/0	sp	0	Ag–N 188
$[Ag(SO_3S)_2]^{3-}$	I	10/0	sp	0	...
$[AuBr_4]^-$	III	8/0	dsp^2	0	Au–Br 250
$[Au(CN)_2]^-$	I	10/0	sp	0	Au–C ...
$[Au(CN)_4]^-$	III	8/0	dsp^2	0	Au–C 197
$[AuCl_4]^-$	III	8/0	dsp^2	0	Au–Cl 224
$[Au(NH_3)_2]^+$	I	10/0	sp	0	Au–N ...
$[Au(NH_3)_4]^{3+}$	III	8/0	dsp^2	0	Au–N ...
$[Au(OH)_4]^-$	III	8/0	dsp^2	0	Au–O ...
$[Au(SO_3S')_2]^{3-}$	I	10/0	sp	0	Au–S' 227
$CdBr_2$	II	10/0	sp	0	Cd–Br 237

(Continued)

Substance	v	N_d/N_e	Hybridi-zation	μ (μ_B)	Bond length (pm)
$CdCl_2$	II	10/0	sp	0	Cd–Cl 221
$[CdCl_6]^{4-}$	II	10/0	sp^3d^2	0	Cd–Cl 253
CdF_2	II	10/0	sp	0	Cd–F 197
$[Cd(H_2O)_6]^{2+}$	II	10/0	sp^3d^2	0	Cd–O ...
CdI_2	II	10/0	sp	0	Cd–I 255
$[Cd(NH_3)_6]^{2+}$	II	10/0	sp^3d^2	0	Cd–N ...
$[Cd(OH)_6]^{4-}$	II	10/0	sp^3d^2	0	Cd–O ...
$CoBr_2$	II	7/3	sp	...	Co–Br 232
$[Co(C_5H_5)_2]$	II	7/1	–	...	Co–C 210
$[Co(bipy)_3]^{2+}$	II	7/3	sp^3d^2	4.86	Co–N ...
$[Co(CN)_5]^{3-}$	II	7/1	dsp^3	...	Co–C 189, Co–C* 201
$[Co(CN)_6]^{3-}$	III	6/0	d^2sp^3	0	Co–C 190
$[Co(CO)_4]^-$	–I	10/0	sp^3	0	Co–C 181
$[Co(CO)_3NO]$	0	9/1	sp^3	...	Co–C 183, Co–N 176
$CoCl_2$	II	7/3	sp	...	Co–Cl 212
$[CoCl_4]^{2-}$	II	7/3	sp^3	4.70	Co–Cl 234
CoF_2	II	7/3	sp	...	Co–F 172
$[CoF_6]^{3-}$	III	6/4	sp^3d^2	5.40	Co–F ...
$[Co(H_2O)_6]^{2+}$	II	7/3	sp^3d^2	4.90	Co–O ...
$[Co(H_2O)_6]^{3+}$	III	6/0	d^2sp^3	0	Co–O ...
$[Co(NCS)_4]^{2-}$	II	7/3	sp^3	4.30	Co–N ...
$[Co(NH_3)_6]^{2+}$	II	7/3	sp^3d^2	5.04	Co–N ...
$[Co(NH_3)_6]^{3+}$	III	6/0	d^2sp^3	0	Co–N ...
$[Co(NO_2)_6]^{4-}$	II	7/1	dsp^3d	1.90	Co–N ..., Co–N* ...
$[Co(NO_2)_6]^{3-}$	III	6/0	d^2sp^3	0	Co–N ...
$[Cr(C_5H_5)_2]$	II	4/2	–	...	Cr–C 217
$[Cr(C_6H_6)_2]$	0	6/0	–	0	Cr–C 214
$[Cr(en)_3]^{3+}$	III	3/3	d^2sp^3	3.93	Cr–N ...
$[Cr(bipy)_3]^{2+}$	II	4/2	d^2sp^3	3.27	Cr–N ...
$[Cr(bipy)_2(H_2O)_2]^{2+}$	II	4/4	dsp^3d	4.79	Cr–N ..., Cr–O* ...
$[Cr(CN)_6]^{4-}$	II	4/2	d^2sp^3	3.20	Cr–C ...
$[Cr(CN)_6]^{3-}$	III	3/3	d^2sp^3	3.72	Cr–C 207
$[Cr(CO)_6]$	0	6/0	d^2sp^3	0	Cr–C 192
$[Cr(Cl)O_3]^-$	VI	0/0	d^3s	0	Cr–Cl 219, Cr–O 161
$CrCl_2O_2$	VI	0/0	d^3s	0	Cr–Cl 212, Cr–O 158[a]
$[CrF_6]^{3-}$	III	3/3	d^2sp^3	...	Cr–F 190
$[Cr(H_2O)_6]^{2+}$	II	4/4	dsp^3d	4.80	Cr–O ..., Cr–O* ...
$[Cr(H_2O)_6]^{3+}$	III	3/3	d^2sp^3	3.85	Cr–O ...
cis-$[Cr(H_2O)_4Cl_2]^+$	III	3/3	d^2sp^3	...	Cr–O 201, Cr–Cl 228
$[Cr(NCS)_6]^{3-}$	III	3/3	d^2sp^3	3.72	Cr–N ...
$[Cr(NH_3)_6]^{3+}$	III	3/3	d^2sp^3	3.71	Cr–N ...
CrO_3	VI	0/0	d^2s	0	Cr–O 163
CrO_4^{2-}	VI	0/0	d^3s	0	Cr–O 166
$CrOF_4$	VI	0/0	d^2sp^2	0	Cr–F 174, Cr–O* 157[b]
CrO_2F_2	VI	0/0	d^3s	0	Cr–O 158, Cr–F 174[c]
$[Cu(bipy)_3]^{2+}$	II	9/1[d]	dsp^3d	...	Cu–N 203, Cu–N* 223, 245
$[Cu(acac)_2]$	II	9/1	sp^3	1.95	Cu–O ...

Substance	ν	N_d/N_e	Hybridization	μ (μ_B)	Bond length (pm)
$[Cu(CN)_2]^-$	I	10/0	sp	0	Cu–C 192
$[CuCl_2]^-$	I	10/0	sp	0	Cu–Cl ...
$[CuCl_4]^{2-}$	II	9/1	sp^3	...	Cu–Cl 223
$[Cu(H_2O)_6]^{2+}$	II	9/1^d	dsp^3d	1.80	Cu–O ..., Cu–O$^{\bullet}$ ...
$[Cu(H_2O)_2Cl_4]^{2-}$	II	9/1^d	dsp^3d	1.88	Cu–Cl 228, Cu–O$^{\bullet}$ 289
$[Cu(H_2O)_2(NH_3)_4]^{2+}$	II	9/1^d	dsp^3d	1.89	Cu–N 205, Cu–O$^{\bullet}$ 259, 337
$[Cu(NH_3)_2]^+$	I	10/0	sp	0	Cu–N ...
$[Cu(OH)_6]^{4-}$	II	9/1^d	dsp^3d	...	Cu–O 195, Cu–O$^{\bullet}$ 280
$FeBr_2$	II	6/4	sp	...	Fe–Br 224
$[Fe(C_5H_5)_2]$	II	6/0	–	...	Fe–C 205
$[Fe(en)_3]^{2+}$	II	6/4	sp^3d^2	5.45	Fe–N ...
$[Fe(bipy)_3]^{3+}$	III	5/1	d^2sp^3	2.20	Fe–N ...
$[Fe(phen)_3]^{3+}$	III	5/1	d^2sp^3	2.45	Fe–N ...
$[Fe(acac)_3]$	III	5/5	sp^3d^2	5.95	Fe–O ...
$[Fe(CN)_6]^{4-}$	II	6/0	d^2sp^3	0	Fe–C 189
$[Fe(CN)_6]^{3-}$	III	5/1	d^2sp^3	2.40	Fe–N ...
$[Fe(CO)_5]$	0	8/0	dsp^3	0	Fe–C 184, Fe–C$^{\bullet}$ 180
$[Fe(C_2O_4)_3]^{3-}$	III	5/5	sp^3d^2	5.75	Fe–O ...
$FeCl_2$	II	6/4	sp	...	Fe–Cl 217
$FeCl_3$	III	5/5	sp^2	...	Fe–Cl 213
$[FeCl_4]^{2-}$	II	6/4	sp^3	...	Fe–Cl 230
$[FeCl_4]^-$	III	5/5	sp^3	...	Fe–Cl 220
$[FeF_6]^{3-}$	III	5/5	sp^3d^2	5.98	Fe–F ...
$[Fe(H_2O)_6]^{2+}$	II	6/4	sp^3d^2	5.30	Fe–O ...
$[Fe(H_2O)_6]^{3+}$	III	5/5	sp^3d^2	5.90	Fe–O ...
$[Fe(NH_3)_6]^{2+}$	II	6/4	sp^3d^2	5.45	Fe–N ...
$[Fe(NO^+)(CN)_5]^{2-}$	II	6/0	d^2sp^3	0	Fe–N 163, Fe–C 190
FeO_3^{4-}	II	6/0	d^2s	0	Fe–O 180
FeO_4^{5-}	III	5/5	sp^3	...	Fe–O 189
FeO_4^{2-}	VI	2/2	d^3s	...	Fe–O 165
$HfBr_4$	IV	0/0	d^3s	0	Hf–Br 245
$HfCl_4$	IV	0/0	d^3s	0	Hf–Cl 233
HfF_4	IV	0/0	d^3s	0	Hf–F 191
HfI_4	IV	0/0	d^3s	0	Hf–I 266
$[Hf(OH)_6]^{2-}$	IV	0/0	d^2sp^3	0	Hf–O ...
$HgBr_2$	II	10/0	sp	0	Hg–Br 241
$[HgBr_4]^{2-}$	II	10/0	sp^3	0	Hg–Br ...
$[Hg(CN)_4]^{2-}$	II	10/0	sp^3	0	Hg–C ...
$HgCl_2$	II	10/0	sp	0	Hg–Cl 225
$[HgCl_4]^{2-}$	II	10/0	sp^3	0	Hg–Cl 229
$[Hg(H_2O)_6]^{2+}$	II	10/0	sp^3d^2	0	Hg–O 234–241
HgI_2	II	10/0	sp	0	Hg–I 259
$[HgI_3]^-$	II	10/0	sp^2	0	Hg–I 260–269
$[HgI_4]^{2-}$	II	10/0	sp^3	0	Hg–I 278
$[Hg(NCS)_4]^{2-}$	II	10/0	sp^3	0	Hg–S 256
$[Hg(NH_3)_4]^{2+}$	II	10/0	sp^3	0	Hg–N 211
$[IrCl_6]^{3-}$	III	6/0	d^2sp^3	0	Ir–Cl ...

(Continued)

Substance	ν	N_d/N_e	Hybridi-zation	μ (μ_B)	Bond length (pm)
$[IrCl_6]^{2-}$	IV	5/1	d^2sp^3	1.67	Ir–Cl 247
IrF_6	VI	3/3	d^2sp^3	3.30	Ir–F 183
$[Ir(H_2O)_6]^{3+}$	III	6/0	d^2sp^3	0	Ir–O ...
$LaBr_3$	III	0/0	d^2s	0	La–Br 274
LaI_3	III	0/0	d^2s	0	La–I 298
$[Mn(py)_6]^{2+}$	II	5/5	sp^3d^2	6.00	Mn–N ...
$[Mn(acac)_3]$	III	4/4	dsp^3d	4.95	Mn–O ..., Mn–O$^\bullet$ ...
$[Mn(CN)_6]^{4-}$	II	5/1	d^2sp^3	1.80	Mn–C ...
$[Mn(CN)_6]^{3-}$	III	4/2	d^2sp^3	3.18	Mn–C ...
$[Mn(CN)_6]^{2-}$	IV	3/3	d^2sp^3	3.94	Mn–C ...
$[Mn(CO)_5]^-$	–I	8/0	dsp^3	0	Mn–C 187, Mn–C$^\bullet$ 184
$[Mn(C_2O_4)_3]^{3-}$	III	4/4	dsp^3d	4.81	Mn–O ..., Mn–O$^\bullet$ ...
$[MnCl_4]^{2-}$	II	5/5	sp^3	5.95	Mn–Cl ...
$[MnCl_6]^{4-}$	II	5/5	sp^3d^2	...	Mn–Cl 258
$[MnCl_6]^{2-}$	IV	3/3	d^2sp^3	...	Mn–Cl 227
$Mn(Cl)O_3$	VII	0/0	d^3s	0	Mn–Cl 206, Mn–O 159
$[MnF_6]^{2-}$	IV	3/3	d^2sp^3	3.90	Mn–F ...
$[Mn(H_2O)_6]^{2+}$	II	5/5	sp^3d^2	5.96	Mn–O ...
$[Mn(H_2O)_6]^{3+}$	III	4/4	dsp^3d	4.90	Mn–O ..., Mn–O$^\bullet$ ...
$[Mn(NCS)_6]^{4-}$	II	5/5	sp^3d^2	6.06	Mn–N ...
MnO_4^{3-}	V	2/2	d^3s	...	Mn–O 171
MnO_4^{2-}	VI	1/1	d^3s	1.80	Mn–O 166
MnO_4^-	VII	0/0	d^3s	0	Mn–O 163
MnO_3F	VII	0/0	d^3s	0	Mn–O 159, Mn–F 172^e
$MoBr_4$	IV	2/2	d^3s	1.28	Mo–Br 239
$[MoBr_5O]^{2-}$	V	1/1	d^2sp^3	1.80	Mo–Br 255, Mo–Br$^\bullet$ 283, Mo–O$^\bullet$ 186
$[Mo(CN)_5]^-$	IV	2/0	d^4s	0	Mo–C ..., Mo–C$^\bullet$ ...
$[Mo(CO)_6]$	0	6/0	d^2sp^3	0	Mo–C 213
$MoCl_4$	IV	2/2	d^3s	2.10	Mo–Cl 223
$MoCl_5$	V	1/1	d^3sp	1.54	Mo–Cl 227, Mo–Cl$^\bullet$ 226
$[MoCl_6]^{3-}$	III	3/3	d^2sp^3	3.83	Mo–Cl 246
$MoCl_6$	VI	0/0	d^2sp^3	0	Mo–Cl 226
$MoCl_2O_2$	VI	0/0	d^3s	0	Mo–Cl 226, Mo–O 170^f
$MoCl_4O$	VI	0/0	d^2sp^2	0	Mo–Cl 228, Mo–O$^\bullet$ 166^g
$[MoCl_5O]^{2-}$	V	1/1	d^2sp^3	1.68	Mo–Cl 240, Mo–Cl$^\bullet$ 259, Mo–O$^\bullet$ 160
$[MoF_6]^{3-}$	III	3/3	d^2sp^3	...	Mo–F 200
$[MoF_6]^-$	V	1/1	d^2sp^3	1.51	Mo–F 174
MoF_6	VI	0/0	d^2sp^3	0	Mo–F 184
$[Mo(NCS)_6]^{3-}$	III	3/3	d^2sp^3	3.70	...
MoO_4^{2-}	VI	0/0	d^3s	0	Mo–O 177
$[MoOF_5]^{2-}$	V	1/1	d^2sp^3	...	Mo–F 185, Mo–F$^\bullet$ 203, Mo–O$^\bullet$ 166
$[MoOF_5]^-$	VI	0/0	d^2sp^3	0	Mo–F 185, Mo–F$^\bullet$ 203, Mo–O$^\bullet$ 166
MoO_2F_2	VI	0/0	d^3s	0	Mo–O 175, Mo–F 182^h
$[MoS_4]^{2-}$	VI	0/0	d^3s	0	Mo–S 218
$NbBr_5$	V	0/0	d^3sp	0	Nb–Br 245, Nb–Br$^\bullet$ 242
$NbCl_5$	V	0/0	d^3sp	0	Nb–Cl 228, Nb–Cl$^\bullet$ 225
$NbCl_3O$	V	0/0	d^3s	0	Nb–Cl 229, Nb–O 169^i

(Continued)

Substance	v	N_d/N_e	Hybridi-zation	μ (μ_B)	Bond length (pm)
NbF_5	V	0/0	d^3sp	0	Nb–F 188, Nb–F* 184
$[NbF_6]^-$	V	0/0	d^2sp^3	0	Nb–F 185
$NiBr_2$	II	8/2	sp	...	Ni–Br 221
$[Ni(C_5H_5)_2]$	II	8/2	–	...	Ni–C 211
$[Ni(en)_3]^{2+}$	II	8/2	sp^3d^2	3.16	Ni–N ...
$[Ni(bipy)_3]^{2+}$	II	8/2	sp^3d^2	2.88	Ni–N ...
$[Ni(CN)_4]^{2-}$	II	8/0	dsp^2	0	Ni–C 186
$[Ni(CO)_4]$	0	10/0	sp^3	0	Ni–C 184
$NiCl_2$	II	8/2	sp	...	Ni–Cl 206
$[Ni(H_2O)_6]^{2+}$	II	8/2	sp^3d^2	3.20	Ni–O ...
$[Ni(NCS)_4]^{2-}$	II	8/0	dsp^2	0	Ni–N ...
$[Ni(NH_3)_6]^{2+}$	II	8/2	sp^3d^2	3.11	Ni–N ...
$[OsBr_6]^{2-}$	IV	4/2	d^2sp^3	1.74	Os–Br 251
$[Os(C_5H_5)_2]$	II	6/0	–	0	Os–C 222
$[Os(en)_3]^{3+}$	III	5/1	d^2sp^3	1.60	Os–N ...
$[Os(acac)_3]$	III	5/1	d^2sp^3	1.81	Os–O ...
$[Os(CO)_5]$	0	8/0	dsp^3	0	Os–C ...
$[OsCl_6]^{2-}$	IV	4/2	d^2sp^3	1.66	Os–Cl 236
$[OsF_6]^{2-}$	IV	4/2	d^2sp^3	1.48	Os–F ...
$[OsF_6]^-$	V	3/3	d^2sp^3	3.25	Os–F 182
OsF_6	VI	2/2	d^2sp^3	...	Os–F 183
$[OsI_6]^{2-}$	IV	4/2	d^2sp^3	1.64	Os–I ...
$[Os(N_2)(NH_3)_5]^{2+}$	II	6/0	d^2sp^3	0	Os–N ..., Os–NN ...
$[Os(N)O_3]^-$	VIII	0/0	d^3s	0	Os–N 175, Os–O 175
OsO_4	VIII	0/0	d^3s	0	Os–O 171
$[Os(O)_2(OH)_4]^{2-}$	VI	2/2	d^2sp^3	0	Os–O 177, Os–OH 203
$[OsO_4(OH)_2]^{2-}$	VIII	0/0	d^2sp^3	0	Os–O ..., Os–OH ...
$[Pd(CN)_4]^{2-}$	II	8/0	dsp^2	0	Pd–C 200
$[Pd(C_2O_4)_2]^{2-}$	II	8/0	dsp^2	0	Pd–O 200
$[PdCl_4]^{2-}$	II	8/0	dsp^2	0	Pd–Cl 230
$[PdCl_6]^{2-}$	IV	6/0	d^2sp^3	0	Pd–Cl 230
$[PdF_6]^{2-}$	IV	6/0	d^2sp^3	0	Pd–F 210
$[Pd(NCS)_4]^{2-}$	II	8/0	dsp^2	0	Pd–S 230
$[Pd(NH_3)_4]^{2+}$	II	8/0	dsp^2	0	Pd–N ...
$[Pd(NH_3)_6]^{4+}$	IV	6/0	d^2sp^3	0	Pd–N ...
$[PtBr_4]^{2-}$	II	8/0	dsp^2	0	Pt–Br ...
$[PtBr_6]^{2-}$	IV	6/0	d^2sp^3	0	Pt–Br ...
$[Pt(CN)_4]^{2-}$	II	8/0	dsp^2	0	Pt–C 201
$[Pt(CN)_6]^{2-}$	IV	6/0	d^2sp^3	0	Pt–C ...
$[Pt(CO)(PPh_3)_3]$	0	10/0	sp^3	0	Pt–C 185, Pt–P 235
$[PtCl_4]^{2-}$	II	8/0	dsp^2	0	Pt–Cl 234
$[PtCl_6]^{2-}$	IV	6/0	d^2sp^3	0	Pt–Cl 232
$[PtF_6]^{2-}$	IV	6/0	d^2sp^3	0	Pt–F 191
$[PtF_6]^-$	V	5/1	d^2sp^3	...	Pt–F 186
$[Pt(H_2O)_2I_4]$	IV	6/0	d^2sp^3	0	Pt–O* 210, Pt–I 265
$[PtI_4]^{2-}$	II	8/0	dsp^2	0	Pt–I ...
$[PtI_6]^{2-}$	IV	6/0	d^2sp^3	0	Pt–I ...

Substance	v	N_d/N_e	Hybridi-zation	μ (μ_B)	Bond length (pm)
$[Pt(NH_3)_4]^{2+}$	II	8/0	dsp^2	0	Pt–N ...
$[Pt(NH_3)_6]^{4+}$	IV	6/0	d^2sp^3	0	Pt–N ...
cis-$[Pt(NH_3)_2Cl_2]$	II	8/0	dsp^2	0	Pt–N 200, Pt–Cl 230
trans-$[Pt(NH_3)_2Cl_2]$	II	8/0	dsp^2	0	Pt–N 205, Pt–Cl 232
$[Pt(OH)_6]^{2-}$	IV	6/0	d^2sp^3	0	Pt–O 196
$[Pt(PPh_3)_3]$	0	10/0	sp^2	0	Pt–P 226
$[Pt(PF_3)_4]$	0	10/0	sp^3	0	Pt–P 223
$[ReBr_6]^{2-}$	IV	3/3	d^2sp^3	3.80	Re–Br 250
$[Re(CN)_6]^{3-}$	III	4/2	d^2sp^3	0	Re–C ...
$[ReCl_6]^{2-}$	IV	3/3	d^2sp^3	3.80	Re–Cl 237
$Re(Cl)O_3$	VII	0/0	d^3s	0	Re–Cl 223, Re–O 170[j]
$ReCl_4O$	VI	1/1	d^2sp^2	1.48	Re–Cl 226, Re–O• 163[k]
$[ReF_6]^{2-}$	IV	3/3	d^2sp^3	3.50	Re–F ...
ReF_6	VI	1/1	d^2sp^3	...	Re–F 183
ReF_7	VII	0/0	d^3sp^3	0	Re–F 183, Re–F• 184
$[ReI_6]^{2-}$	IV	3/3	d^2sp^3	3.70	Re–I ...
$[Re(NCS)_6]^{2-}$	IV	3/3	d^2sp^3	3.90	...
ReO_4^-	VII	0/0	d^3s	0	Re–O 177
ReO_3F	VII	0/0	d^3s	0	Re–O 170, Re–F 186
$[Rh(CN)_6]^{3-}$	III	6/0	d^2sp^3	0	Rh–C ...
$[Rh(CO)Cl_5]^{2-}$	III	6/0	d^2sp^3	0	Rh–C ..., Rh–Cl ...
$[RhCl_6]^{3-}$	III	6/0	d^2sp^3	0	Rh–Cl ...
$[RhCl_6]^{2-}$	IV	5/1	d^2sp^3	...	Rh–Cl ...
$[RhF_6]^{3-}$	III	6/0	d^2sp^3	0	Rh–F ...
$[RhF_6]^{2-}$	IV	5/1	d^2sp^3	...	Rh–F 196
$[Rh(H_2O)_6]^{3+}$	III	6/0	d^2sp^3	0	Rh–O
$[Rh(NCS)_6]^{3-}$	III	6/0	d^2sp^3	0	Rh–S ...
$[Rh(NH_3)_6]^{3+}$	III	6/0	d^2sp^3	0	Rh–N ...
$[RuBr_6]^{2-}$	IV	4/2	d^2sp^3	2.80	Ru–Br ...
$[Ru(C_5H_5)_2]$	II	6/0	–	0	Ru–C 221
$[Ru(acac)_3]$	III	5/1	d^2sp^3	1.70	Ru–O ...
$[Ru(CO)_5]$	0	8/0	dsp^3	0	Ru–C ...
$[RuCl_6]^{3-}$	III	5/1	d^2sp^3	1.85	Ru–Cl 237
$[RuCl_6]^{2-}$	IV	4/2	d^2sp^3	2.90	Ru–Cl ...
$[RuF_6]^{2-}$	IV	4/2	d^2sp^3	3.08	Ru–F ...
$[RuF_6]^-$	V	3/3	d^2sp^3	3.70	Ru–F 187–191
$[Ru(H_2O)Cl_5]^{2-}$	III	5/1	d^2sp^3	1.76	Ru–O 210, Ru–Cl 234
$[Ru(H_2O)(NH_3)_5]^{3+}$	III	5/1	d^2sp^3	1.95	Ru–O ..., Ru–N ...
$[Ru(NH_3)_6]^{2+}$	II	6/0	d^2sp^3	0	Ru–N ...
$[Ru(NH_3)_6]^{3+}$	III	5/1	d^2sp^3	1.88	Ru–N ...
$[Ru(NH_3)_5Cl]^{2+}$	III	5/1	d^2sp^3	2.07	Ru–N ..., Ru–Cl ...
$[Ru(NH_3)_5(OH)]^{2+}$	III	5/1	d^2sp^3	1.91	Ru–N ..., Ru–O ...
$[Ru(N_2)(NH_3)_5]^{2+}$	II	6/0	d^2sp^3	0	Ru–NN 210, Ru–N ...
RuO_4^{2-}	VI	2/2	d^3s	...	Ru–O 179
RuO_4^-	VII	1/1	d^3s	...	Ru–O 179
RuO_4	VIII	0/0	d^3s	0	Ru–O 171
$[Ru(SO_2)(NH_3)_4Cl]^+$	II	6/0	d^2sp	0	Ru–S 207, Ru–N 217, Ru–Cl 241

Substance	v	N_d/N_e	Hybridization	μ (μ_B)	Bond length (pm)
$ScBr_3$	III	0/0	d^2s	0	Sc–Br 247
$[Sc(H_2O)_6]^{3+}$	III	0/0	d^2sp^3	0	Sc–O ...
ScI_3	III	0/0	d^2s	0	Sc–I 268
$[Sc(OH)_6]^{3-}$	III	0/0	d^2sp^3	0	Sc–O ...
$TaBr_5$	V	0/0	d^3sp	0	Ta–Br 244, Ta–Br$^\bullet$ 242
$TaCl_5$	V	0/0	d^3sp	0	Ta–Cl 227, Ta–Cl$^\bullet$ 227
$[TaCl_6]^-$	V	0/0	d^2sp^3	0	Ta–Cl 265
TaF_5	V	0/0	d^3sp	0	Ta–F 186, Ta–F$^\bullet$ 186
$TaCl_3O$	V	0/0	d^3s	0	Ta–Cl 227, Ta–O 169^l
$[TcBr_6]^{2-}$	IV	3/3	d^2sp^3	3.94	Tc–Br 250
$[TcCl_6]^{2-}$	IV	3/3	d^2sp^3	3.88	Tc–Cl ...
$[TcI_6]^{2-}$	IV	3/3	d^2sp^3	4.14	Tc–I ...
TcO_4^-	VII	0/0	d^3s	0	Tc–O ...
$ThBr_4$	IV	0/0	d^3s	0	Th–Br 272
$ThCl_4$	IV	0/0	d^3s	0	Th–Cl 258
ThF_4	IV	0/0	d^3s	0	Th–F 214
$TiBr_4$	IV	0/0	d^3s	0	Ti–Br 234
$TiCl_4$	IV	0/0	d^3s	0	Ti–Cl 231
$[TiCl_6]^{2-}$	IV	0/0	d^2sp^3	0	Ti–Cl 235
TiF_4	IV	0/0	d^3s	0	Ti–F 175
$[TiF_6]^{2-}$	IV	0/0	d^2sp^3	0	Ti–F 188
$[Ti(H_2O)_6]^{3+}$	III	1/1	d^2sp^3	1.75	Ti–O ...
TiI_4	IV	0/0	d^3s	0	Ti–I 255
$[UBr_4O_2]^{2-}$	VI	0/0	d^2sp^3	0	U–Br 280, U–O$^\bullet$ 170
$[UCl_4O_2]^{2-}$	VI	0/0	d^2sp^3	0	U–Cl 260, U–O$^\bullet$ 168
UF_6	VI	0/0	d^2sp^3	0	U–F 200
UO_2^{2+}	VI	0/0	sp	0	U–O 190
VBr_4	IV	1/1	d^3s	...	V–Br 230
$[V(C_5H_5)_2]$	II	3/3	–	...	V–C 228
$[V(CN)_6]^{4-}$	II	3/3	d^2sp^3	...	V–C 216
$[V(CN)_6]^{3-}$	III	2/2	d^2sp^3	...	V–C 215
$[V(CN)_5O]^{3-}$	IV	1/1	d^2sp^3	...	V–C 214, V–C$^\bullet$ 231, V–O$^\bullet$ 164
$[V(CO)_6]$	0	5/1	d^2sp^3	...	V–C 201
VCl_4	IV	1/1	d^3s	...	V–Cl 214
VCl_3O	V	0/0	d^3s	0	V–Cl 213, V–O 160^m
VF_5	V	0/0	d^3sp	0	V–F 171, V–F$^\bullet$ 171
$[VF_6]^{3-}$	III	2/2	d^2sp^3	2.79	V–F ...
$[VF_6]^{2-}$	IV	1/1	d^2sp^3	...	V–F 161
$[VF_6]^-$	V	0/0	d^2sp^3	0	V–F ...
$[V(H_2O)_6]^{2+}$	II	3/3	d^2sp^3	3.86	V–O ...
$[V(H_2O)_6]^{3+}$	III	2/2	d^2sp^3	2.76	V–O ...
$[V(H_2O)F_5]^{2-}$	III	2/2	d^2sp^3	2.74	V–O ..., V–F ...
$[V(H_2O)_5O]^{2+}$	IV	1/1	d^2sp^3	...	V–OH$_2$ 230, V–O 167
$[V(NH_3)_5(NH_2)]^{2+}$	III	2/2	d^2sp^3	2.74	V–N ..., V–NH$_2$ ...
VO_2^+	V	0/0	sp	0	V–O ...
VO_4^{3-}	V	0/0	d^3s	0	V–O 175
$[VS_4]^{3-}$	V	0/0	d^3s	0	V–S 215

(Continued)

Substance	v	N_d/N_e	Hybridization	μ (μ_B)	Bond length (pm)
WBr_5	V	1/1	d^3sp	1.14	W–Br 240, W–Br* 240
$[W(CO)_6]$	0	6/0	d^2sp^3	0	W–C 230
WCl_5	V	1/1	d^3sp	1.17	W–Cl 226, W–Cl* 226
WCl_6	VI	0/0	d^2sp^3	0	W–Cl 226
WCl_2O_2	VI	0/0	d^3s	0	W–Cl 227, W–O 171[n]
WCl_4O	VI	0/0	d^2sp^2	0	W–Cl 228, W–O* 169[o]
$[WCl_5O]^{2-}$	V	1/1	d^2sp^3	1.52	W–Cl ..., W–O ...
WF_6	VI	0/0	d^2sp^3	0	W–F 183
WO_4^{2-}	VI	0/0	d^3s	0	W–O 179
$[WS_4]^{2-}$	VI	0/0	d^3s	0	W–S 217
$ZnBr_2$	II	10/0	sp	0	Zn–Br 221
$[Zn(CN)_4]^{2-}$	II	10/0	sp^3	0	Zn–C 202
$ZnCl_2$	II	10/0	sp	0	Zn–Cl 205
$[ZnCl_4]^{2-}$	II	10/0	sp^3	0	Zn–Cl 225
ZnF_2	II	10/0	sp	0	Zn–F 181
$[Zn(H_2O)_6]^{2+}$	II	10/0	sp^3d^2	0	Zn–O ...
$[Zn(H_2O)_3(OH)_3]^-$	II	10/0	sp^3d^2	0	Zn–O 198
$[Zn(H_2O)_2(NH_3)_4]^{2+}$	II	10/0	sp^3d^2	0	ZnO* ..., Zn–N ...
ZnI_2	II	10/0	sp	0	Zn–I 238
$[Zn(NCS)_4]^{2-}$	II	10/0	sp^3	0	Zn–N ...
$[Zn(NH_3)_6]^{2+}$	II	10/0	sp^3d^2	0	Zn–N ...
ZnO_4^{6-}	II	10/0	sp^3	0	Zn–O 200
$[Zn(OH)_4]^{2-}$	II	10/0	sp^3	0	Zn–O ...
$ZrBr_4$	IV	0/0	d^3s	0	Zr–Br 247
$ZrCl_4$	IV	0/0	d^3s	0	Zr–Cl 232
$[ZrCl_6]^{2-}$	IV	0/0	d^2sp^3	0	Zr–Cl 245
ZrF_4	IV	0/0	d^3s	0	Zr–F 190
$[ZrF_6]^{2-}$	IV	0/0	d^2sp^3	0	Zr–F 204–212
ZrI_4	IV	0/0	d^3s	0	Zr–I 266
$[Zr(OH)_6]^{2-}$	IV	0/0	d^2sp^3	0	Zr–O ...

[a] Cl–Cr–Cl 113°, O–Cr–O 110°. [b] F–Cr–O* 100°. [c] O–Cr–O* 102°, F–Cr–F 119°. [d] $3d^9 \rightarrow 3d^84d^1$.
[e] O–Mn–O 110°, F–Mn–O 109°. [f] Cl–Mo–Cl 112°, O–Mo–O 104°. [g] Cl–Mo–Cl 87°, Cl–Mo–O* 103°.
[h] O–Mo–O 110°, F–Mo–F 113°. [i] Cl–Nb–Cl 107°, O–Nb–Cl 112°. [j] Cl–Re–O 108°, O–Re–O 111°.
[k] Cl–Re–O* 103°. [l] Cl–Ta–Cl 107°, Cl–Ta–O 112°. [m] Cl–V–Cl 112°, Cl–V–O 107°. [n] Cl–W–Cl 112°,
O–W–O 104°. [o] Cl–W–Cl 87°, Cl–W–O* 102°.

3.12 Geometric Forms, Bond Length and Valence Angles for Polyatomic Molecules and Ions with Two or More Central Atoms [7, 9, 10, 13, 28, 29, 33–36]

The table lists polyatomic chemical particles (molecules and ions) A_mB_n containing two or more atoms A and atoms B (groups of atoms), and also particles A_4. The spatial presentation of geometric forms demonstrates the position of atoms A and B and all σ-bonds A-A and B-B. The symbols of some atoms A and

B (in spatial presentation) and the symbols of some elements (in the table) are marked with asterisks for the purpose of identifying the position of specific atoms in spatial presentations. In some instances, the distances between unbonded atoms (AA) and the values of dihedral angles (α) are also listed. The dots (...) denote the absence of data.

The experimental data for molecules correspond to the thermodynamic conditions of their formation, while those for ions – to the conditions of their existence in aqueous solutions and/or ionic crystals.

The spatial presentations of the geometric forms of particles are as follows:

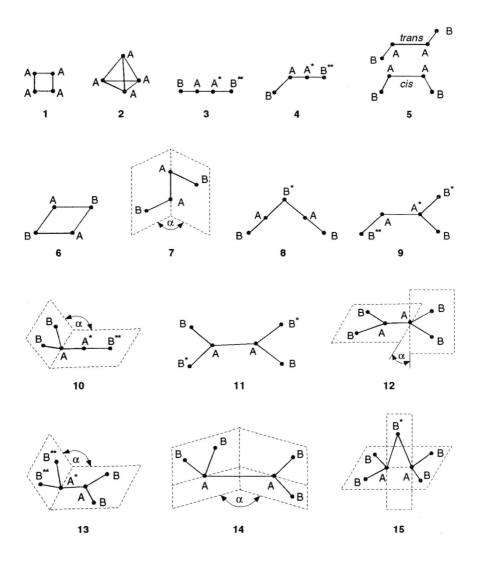

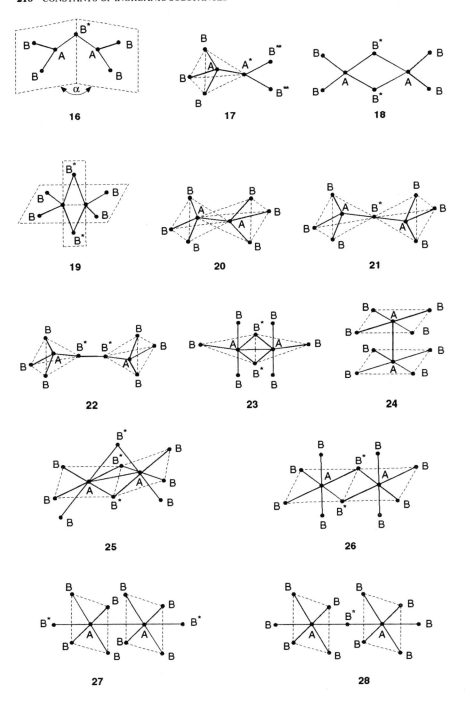

29

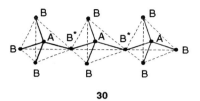

30

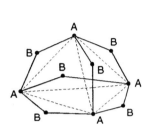

31

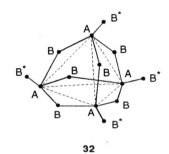

32

Substance	Form	Bond length (pm) and valence angles
Al_2Br_6	**19**	Al–Br 221, Al–Br• 238, Br–Al–Br 115°, Br•–Al–Br• 87°
$[Al_2Br_7]^-$	**21**	Al–Br 226, Al–Br• 243, Br–Al–Br 113°, Br–Al–Br• 104°
Al_2Cl_6	**19**	Al–Cl 208, Al–Cl• 230, Cl–Al–Cl 123°, Cl•–Al–Cl• 79°
Al_2I_6	**19**	Al–I 253, Al–I• 258, I–Al–I 112°, I•–Al–I• 102°
$[Al_2O(OH)_6]^{2-}$	**21**	Al–O• 175, Al–OH ..., Al–O•–Al 132°
As_4	**2**	As–As 244, As–As–As 60°
As_4O_6	**31**	As–O 178, O–As–O 99°, As–O–As 128°
$[As_2O_2F_8]^{2-}$	**26**	As–F 181, As–O• 181, AsAs 266, As–O•–As 95°
As_4S_6	**31**	As–S 225, S–As–S 114°, As–S–As 100°
Au_2Br_6	**18**	Au–Br(Br•) 359, AuAu 238–247, Br–Au–Br 90°, Br•–Au–Br• 86°
Au_2Cl_6	**18**	Au–Cl 224, Au–Cl• 233, Cl–Au–Cl 90°, Cl•–Au–Cl• 86°
B_2Cl_4	**12**	B–B 170, B–Cl 175, Cl–B–Cl 119°, α 90°
B_2H_6	**19**	B–H 120, B–H• 134, BB 178, H–B–H 122°, H•–B–H• 101°
B_2O_3	**8**	B–O 121, B–O• 131, O–B–O• 180°, B–O•–B 132°
B_2S_3	**8**	B–S 165, B–S• 181, S–B–S• 180°, B–S•–B 96°
$Be_2(O)_2$	**6**	Be–O 163, O–Be–O 90°
Bi_4	**2**	Bi–Bi 288, Bi–Bi–Bi 60°
C_2Br_2	**3**	C–C• 120, C(C•)–Br(Br••) 180, C–C•–Br(Br••) 180°
C_2Br_6	**20**	C–C 152, C–Br 193, C–C–Br 109.5°
$CBr_3(NO_2)$	**17**	C–N• 122, C–Br 192, N•–O•• 159, Br–C–Br 110°, Br–C–N• 108°, C–N•–O•• 111°, O••–N•–O•• 134°
C_2Cl_2	**3**	C–C• 120, C(C•)–Cl(Cl••) 164, C–C•–Cl(Cl••) 180°
C_2Cl_4	**11**	C–C 130, C–Cl(Cl•) 172, C–C•–Cl(Cl•) 114°
C_2Cl_6	**20**	C–C 157, C–Cl 174, Cl–C–Cl 109°
$CCl_3(NO_2)$	**17**	C–N• 159, C–Cl 173, N•–O•• 119, Cl–C–Cl 112°, Cl–C–N• 106°, C–N•–O•• 113°, O••–N•–O•• 132°
C_2F_4	**11**	C–C 131, C–F(F•) 132, C–C–F(F•) 124°
C_2F_6	**20**	C–C 156, C–F 132, F–C–F 109.5°

(Continued)

Substance	Form	Bond length (pm) and valence angles
C_2H_2	3	C–C· 120, C(C·)–H(H··) 106, C–C·–H(H··) 180°
C_2H_4	11	C–C 134, C–H(H·) 109, C–C–H(H·) 121°, H–C–H· 117°
C_2H_6	20	C–C 154, C–H 111, H–C–H 109.5°
C_2N_2	3	C–C· 138, C(C·)–N(N··) 116, C–C·–N(N··) 180°
$C(NO_2)F_3$	17	C–N· 156, C–F 133, N·–O·· 121, F–C–F 110°, F–C–N· 109°, C–N·–O·· 112°, O··–N·–O·· 132°
C_3O_2	29	C–C· 129, C–O 116, C–C·–C 180°, O–C–C· 180°
$C(PH_2)F_3$	17	C–P· 190, C–F 133, P·–H·· 143, H··–P·–H·· 97°, F–C–F 108°, C–P·–H·· 92°
Cl_2O_7	21	Cl–O 141, Cl–O· 172, O–Cl–O 97°, Cl–O·–Cl 115°
$[Co_2(CO)_8]$	23	Co–Co 252, Co–CO 180, Co–CO· 192, OC–Co–CO 90°
$Cr_2O_7^{2-}$	21	Cr–O 160, Cr–O· 185, Cr–O·–Cr 115–120°
$Cr_3O_{10}^{2-}$	30	Cr–O 160, Cr–O· 170–180, O–Cr–O 105–113°, Cr–O·–Cr 136°
Cs_2Br_2	6	Cs–Br 329, Br–Cs–Br 94°
Cs_2Cl_2	6	Cs–Cl 311, Cl–Cs–Cl 91°
Cs_2F_2	6	Cs–F 256, F–Cs–F 79°
Cs_2I_2	6	Cs–I 354, I–Cs–I 94°
Cu_2Cl_2	6	Cu–Cl 211, Cl–Cu–Cl 105°
$[Cu_2Cl_6]^{2-}$	18	Cu–Cl 220, Cu–Cl· 230, Cl–Cu–Cl ..., Cu–Cl·–Cu ...
$[Fe_2(CO)_9]$	25	Fe–CO 184, Fe–CO· 180, Fe–Fe 246, Fe–CO·–Fe 78°
Fe_2Cl_6	19	Fe–Cl 211, Fe–Cl· 228, Cl–Fe–Cl 128°, Cl·–Fe–Cl· 92°
$[Fe_2(NO)_4I_2]$	19	Fe–NO 167, Fe–I· 258, N–O 115, FeFe 305
Ga_2Br_6	19	Ga–Br 225, Ga–Br· 235, Br–Ga–Br 110°, Ga–Br·–Ga 93°
Ga_2Cl_6	19	Ga–Cl 209, Ga–Cl· 229, Cl–Ga–Cl 112°, Ga–Cl·–Ga 93°
Ge_2H_6	20	Ge–Ge 241, Ge–H 153, H–Ge–H 109°
$Ge_2(O)_2$	6	Ge–O 187, O–Ge–O 83°
H_2CN_2	10	N–C· 135, C·–N 116, N–H 100, N–C·–N·· 180°, H–N–H 113°, α 142°
HCNO	3	C–N· 117, C–H 103, N·–O·· 120, H–C–N· 180°, C–N·–O·· 180°
$H_2C_2O_4$	11	C–C 154, C–O 122, C–OH 137, C–C–O 122°, O–C–OH· 125°
HDS_2	7	S–S 206, S–H 133, S–D 132, S–S–H(D) 91°
HN_3	4	N–N· 124, N·–N·· 113, N–H 98, N–N·–N·· 180°, H–N–N· 115°
HNCO	4	N–C· 121, C·–O·· 117, N–H 99, N–C·–O·· 180°, H–N–C· 128°
HNCS	4	N–C· 122, C·–S·· 156, N–H 99, N–C·–S·· 180°, H–N–C· 135°
H_2O_2	7	O–O 145, O–H 97, H–O–O 100°, α 119°
H_2S_2	7	S–S 206, S–H 133, H–S–S 92°, α 91°
Hg_2Br_2	3	Hg–Hg· 249–269, Hg(Hg·)–Br(Br··) 271, Hg–Hg·–Br(Br··) 180°
Hg_2Cl_2	3	Hg–Hg· 249–269, Hg(Hg·)–Cl(Cl··) 243, Hg–Hg·–Cl(Cl··) 180°
Hg_2F_2	3	Hg–Hg· 249–269, Hg(Hg·)–F(F··) 214, Hg–Hg·–F(F··) 180°
$[Hg_2(H_2O)_2]^{2+}$	3	Hg–Hg· 240–270, Hg(Hg·)–O(O··) 215, Hg–Hg·–O(O··) 180°
Hg_2I_2	3	Hg–Hg· 249–269, Hg(Hg·)–I(I··) 269, Hg–Hg·–I(I··) 180°
I_2Cl_6	18	I–Cl 239, I–Cl· 270, Cl–I–Cl 94°, Cl·–I–Cl· 84°
I_2O_5	15	I–O 179, I–O· 194, O–I–O 100°, I–O·–I 139°
K_2Br_2	6	K–Br 302, Br–K–Br 101°
K_2Cl_2	6	K–Cl 286, Cl–K–Cl 98°
K_2F_2	6	K–F 235, F–K–F 88°
K_2I_2	6	K–I 326, I–K–I 104°
Li_2Br_2	6	Li–Br 235, Br–Li–Br 110°
Li_2Cl_2	6	Li–Cl 223, Cl–Li–Cl 108°

(*Continued*)

Substance	Form	Bond length (pm) and valence angles
Li_2F_2	6	Li–F 175, F–Li–F 105°
Li_2I_2	6	Li–I 254, I–Li–I 116°
$Li_2(O)_2$	6	Li–O 164, O–Li–O 53°
$[Mn_2(CO)_{10}]$	27	Mn–Mn 292, Mn–CO 179, Mn–CO$^\bullet$ 183, OC–Mn–CO 90°,
		OC–Mn–CO$^\bullet$ 90°, Mn–Mn–CO$^\bullet$ 180°
$[Mn_2(CO)_8Cl_2]$	26	Mn–CO 187, Mn–Cl$^\bullet$ ..., MnMn 374
$[Mo_2Cl_8]^{4-}$	24	Mo–Mo 214, Mo–Cl 244, Cl–Mo–Cl 90°, Cl–Mo–Mo 90°
cis-N_2F_2	5	N–N 121, N–F 141, N–N–F 114°
trans-N_2F_2	5	N–N 123, N–F 140, N–N–F 106°
N_2F_4	14	N–N 149, N–F 137, F–N–F 103°, N–N–F 101°, α 64°
cis-N_2H_2	5	N–N 123, N–H 101, N–N–H 100°
trans-N_2H_2	5	N–N 124, N–H 105, N–N–H 109°
N_2H_4	14	N–N 145, N–H 102, H–N–H 112°, N–N–H 110°, α 90°
$N_2H_6^{2+}$	20	N–N 146, N–H 98
$N(NH_2)O_2$	13	N–N$^\bullet$ 143, N–O 121, N$^\bullet$–H$^{\bullet\bullet}$ 101, O–N–O 130°,
		H$^{\bullet\bullet}$–N$^\bullet$–H$^{\bullet\bullet}$ 115°, α 128°
cis-$N_2O_2^{2-}$	5	N–N 175, N–O 115, N–N–O 90°
trans-$N_2O_2^{2-}$	5	N–N 125, N–O 141, O–N–N 120°
N_2O_3	9	N–N$^\bullet$ 186, N$^\bullet$–O 120, N$^\bullet$–O$^\bullet$ 122, N–O$^{\bullet\bullet}$ 114, O$^{\bullet\bullet}$–N–N$^\bullet$ 105°,
		N–N$^\bullet$–O$^\bullet$ 111°, N–N$^\bullet$–O 113°
N_2O_4	11	N–N 178, N–O 119, N–O$^\bullet$ 118, O–N–O$^\bullet$ 135°
N_2O_5	16	N–O 121, N–O$^\bullet$ 146, N–O$^\bullet$–N 109°, O–N–O 134°, α 95°
Na_2Br_2	6	Na–Br 269, Br–Na–Br 108°
Na_2Cl_2	6	Na–Cl 254, Cl–Na–Cl 105°
Na_2F_2	6	Na–F 208, F–Na–F 95°
Na_2I_2	6	Na–I 291, I–Na–I 111°
$Na_2(O)_2$	6	Na–O 190, O–Na–O 52°
O_2F_2	7	O–O 122, O–F 158, O–O–F 109.5°, α 88°
P_4	2	P–P 221, P–P–P 60°
P_2F_4	14	P–P 228, P–F 159, F–P–F 99°, P–P–F 95°
P_2H_4	14	P–P 222, P–H 145, H–P–H 91°, P–P–H 95°
P_2I_4	14	P–P 221, P–I 248, I–P–I 102°, P–P–I 90°
P_4O_6	31	P–O 165, O–P–O 99°, P–O–P 128°
P_4O_{10}	32	P–O 158, P–O$^\bullet$ 146, O–P–O 101°, P–O–P 124°, O–P–O$^\bullet$ 117°
$Pb_2(O)_2$	6	Pb–O 217, O–Pb–O 79°
Rb_2Br_2	6	Rb–Br 315, Br–Rb–Br 98°
Rb_2Cl_2	6	Rb–Cl 298, Cl–Rb–Cl 95°
Rb_2F_2	6	Rb–F 245, F–Rb–F 84°
Rb_2I_2	6	Rb–I 339, I–Rb–I 101°
$[Re_2(CO)_{10}]$	27	Re–Re 302, Re–CO 201, Re–CO$^\bullet$ 201, OC–Re–CO 90°,
		OC–Re–CO$^\bullet$ 90°, Re–Re–CO$^\bullet$ 180°
Re_2Cl_{10}	26	Re–Cl 225, Re–Cl$^\bullet$ 247
$[Re_2Cl_{10}O]^{4-}$	28	Re–O$^\bullet$ 186, Re–O$^\bullet$–Re 180°
Re_2O_7	21	Re–O 170, Re–O$^\bullet$ 180, Re–O$^\bullet$–Re 165°
$[RuCl_{10}O]^{4-}$	28	Ru–O$^\bullet$ 180, Ru–O$^\bullet$–Ru 180°
S_2Br_2	7	S–S 198, S–Br 224, S–S–Br 105°, α 84°
S_2Cl_2	7	S–S 193, S–Cl 206, S–S–Cl 108°, α 85°
S_2F_2	7	S–S 189, S–F 164, S–S–F 108°, α 93°

Substance	Form	Bond length (pm) and valence angles
S_2F_{10}	27	S–S 221, S–F 156, S–F˙ 156, F–S–F 90°, F–S–F˙ 90°, S–S–F˙ 180°
S_2N_2	6	N–S 162, N–S–N 85°
cis-S_2O_2	5	S–S 202, S–O 146, S–S–O 113°
$S_2O_6^{2-}$	20	S–S 215, S–O 143
$S_2O_6(O_2)^{2-}$	22	S–O˙ 150, O˙–O˙ 146, S–O˙–O˙ 122°
$[Sb_2F_{11}]^-$	28	Sb–F 180–190, Sb–F˙ 200–210, Sb–F˙–Sb 150°
Se_4^{2+}	1	Se–Se 228, Se–Se–Se 90°
$[Se_2(NCS)_6]^{2-}$	18	Se–SCN 230, Se–SCN˙ 300
Si_2Cl_6	20	Si–Si 234, Si–Cl 202, Cl–Si–Cl 109.5°
$(SiCl_3)_2O$	21	Si–O˙ 164, Cl–Si–Cl 110°, Cl–Si–O˙ 130°
Si_2F_6	20	Si–Si 234, Si–F 157, Si–Si–F 111°
Si_2H_6	20	Si–Si 232, Si–H 148
$Si_2(O)_2$	6	Si–O 171, O–Si–O 87°
$Sn_2(O)_2$	6	Sn–O 205, O–Sn–O 80°
$[Tc_2(CO)_{10}]$	27	Tc–Tc 302, Tc–CO 200, Tc–CO˙ 190, OC–Tc–CO 90°,
		OC–Tc–CO˙ 90°, Tc–Tc–CO˙ 180°
Te_4^{2+}	1	Te–Te 266, Te–Te–Te 90°
$V_2O_7^{4-}$	21	V–O 163, V–O˙ 182, O–V–O 93–123°, V–O˙–V 145°
$[W_2Cl_{10}O]^{4-}$	28	W–Cl 241, W–O˙ 187, W–O˙–W 180°
$[Xe_2F_3]^+$	8	Xe–F 190, Xe–F˙ 214, Xe–F˙–Xe 150°, F˙–Xe–F 178°
$[Xe_2F_{11}]^+$	28	Xe–F 184, Xe–F˙ 221, Xe–F˙–Xe 180°

4

Thermodynamic Properties

4.1 Enthalpy of Formation, Entropy, and Gibbs Energy of Formation at 298.15 K [2, 5, 14–17, 30, 36]

The table lists the thermodynamic properties, viz., the standard enthalpy of formation $\Delta H°$ the standard entropy $S°$, and the standard Gibbs energy of formation $\Delta G°$ for individual substances (atoms, molecules, ions, radicals and formula entities) in various states of aggregation (gaseous, liquid and solid states), and also for substances in aqueous solutions (in this case, the standard partial entropy is listed in column "$S°$"). For some substances, the listed values pertain to simplified formulas (more accurate formulas are given in footnotes). For the majority of solid substances, their specific crystal systems (syngonies) are indicated.

The sign "−" signifies the absence of data.

For strong electrolytes in an aqueous solution, thermodynamic properties are not cited, since the values in question can be calculated by summing up appropriate values listed for their ions.

Substance	State	$\Delta H°$ kJ/mole	$S°$ J/(K·mole)	$\Delta G°$ kJ/mole
Ac	gas	+385	+188	+348
	cub	0	+63	0
Ac_2O_3	hex	−2047	+134	−1958

(*Continued*)

Substance	State	$\Delta H°$ kJ/mole	$S°$ J/(K·mole)	$\Delta G°$ kJ/mole
Ag	gas	+285	+173	+246
	cub	0	+43	0
Ag^+	soln	+106	+73	+77
$(Ag^IAg^{III})O_2$	cub	−115	+28	−37
Ag_3AsO_4	cub	−634	+276	−545
AgBr	gas	+96	+257	−77
	cub	−101	+107	−97
$AgBrO_3$	tetr	−8	+152	+74
AgCN	trig	+146	+107	+157
$[Ag(CN)_2]^-$	soln	+269	+201	+302
$[Ag(CN)_2],K$	hex	−13	+142	+38
Ag_2CO_3	mon	−506	+167	−437
AgCl	gas	+92	+245	+70
	cub	−127	+96	−110
[AgCl]	soln	−73	+155	−73
$[AgCl_2]^-$	soln	−245	+231	−215
$AgClO_2$	tetr	+8	+136	+75
$AgClO_3$	tetr	−22	+150	+71
$AgClO_4$	cub	−32	+162	+88
Ag_2CrO_4	mon	−725	+218	−635
AgF	gas	+7	+236	−20
	cub	−206	+84	−188
[AgF]	soln	−240	+26	−205
$AgF·2H_2O$	sld	−802	+168	−670
[AgHS]	soln	−	−	+9
$[Ag(HS)_2]^-$	soln	−	−	−4
AgI	gas	+159	+264	+110
	hex	−62	+115	−66
$AgIO_3$	rhomb	−169	+149	−92
AgN_3	rhomb	+309	+99	+378
$[Ag(NH_3)_2]^+$	soln	−111	+246	−18
$AgNO_2$	rhomb	−45	+128	+19
$[AgNO_2]$	soln	−	+212	+40
$AgNO_3$	rhomb	−124	+141	−33
Ag_2O	cub	−31	+121	−11
[AgOH]	soln	−	−	−92
$[Ag(OH)_2]^-$	soln	−	−	−260
Ag_3PO_4	cub	−990	+258	−894
Ag_2S	mon	−33	+144	−41
	rhomb	−30	+151	−40
$AgSO_4^-$	soln	−797	+136	−674
Ag_2SO_4	rhomb	−715	+200	−618
$[Ag(SO_3S)]^-$	soln	−	−	−486
$[Ag(SO_3S)_2]^{3-}$	soln	−1279	+74	−1027
Ag_2Se	rhomb	−44	+151	−51
Ag_2SeO_4	rhomb	−431	+225	−338
Ag_2Te	mon	−36	+155	−42

(*Continued*)

Substance	State	$\Delta H°$ kJ/mole	$S°$ J/(K·mole)	$\Delta G°$ kJ/mole
Al	gas	+330	+164	+289
	cub	0	+28	0
Al$^+$	gas	+913	+150	+870
Al^{3+}	soln	−531	−300	−492
Al$_2$	gas	+462	+230	+410
AlAsO$_4$·2H$_2$O	sld	−	−	−1709
AlBr	gas	+16	+240	−25
AlBr$_2$	gas	−166	−	−
AlBr$_3$	gas	−422	+349	−450
	mon	−514	+180	−491
Al$_2$Br$_6$	gas	−971	+512	−971
Al$_4$C$_3$	trig	−208	+89	−196
AlCl	gas	−46	+228	−72
AlCl$_2$	gas	−319	+292	−331
AlCl$_3$	gas	−583	+322	−571
	mon	−704	+111	−629
AlCl$_3$·6H$_2$O	trig	−2693	+318	−2263
Al$_2$Cl$_6$	gas	−1293	+444	−1209
[AlCl$_4$],Na	rhomb	−1143	+188	−1042
AlF	gas	−264	+215	−289
AlF$_2$	gas	−687	−	−
AlF^{2+}	soln	−861	−165	−812
AlF$_2^+$	soln	−1191	−59	−1124
AlF$_3$	gas	−1211	+277	−1194
	trig	−1510	+66	−1431
[AlF$_3$]	soln	−1524	+8	−1427
[AlF$_4$]$^-$	soln	−1858	+40	−1721
[AlF$_5$]$^{2-}$	soln	−2193	+47	−2008
[AlF$_6$]$^{3-}$	soln	−2533	+21	−2291
[AlF$_6$],Na$_3$	mon	−3317	+238	−3153
AlH	gas	+258	+188	+230
AlH$_2$	gas	−289	−	−
AlH$_3$	hex	−12	+30	+46
[AlH$_4$],Li	mon	−117	+88	−48
[AlH$_4$],Na	sld	−114	+124	−49
AlI	gas	+64	+248	+16
AlI$_2$	gas	−38	−	−
AlI$_3$	gas	−208	+376	−260
	hex	−308	+190	−304
Al$_2$I$_6$	gas	−513	+585	−567
AlN	gas	+440	+227	+409
	hex	−318	+20	−287
Al(NH$_4$)(SO$_4$)$_2$	hex	−2352	+216	−2038
Al(NH$_4$)(SO$_4$)$_2$·12H$_2$O	cub	−5944	+697	−4938
Al(NO$_3$)$_3$·6H$_2$O	sld	−2851	+468	−2204
Al(NO$_3$)$_3$·9H$_2$O	mon	−3757	−	−
AlO	gas	+89	+218	+63

Substance	State	$\Delta H°$ kJ/mole	$S°$ J/(K·mole)	$\Delta G°$ kJ/mole
Al_2O_3	hex	−1675	+51	−1582
$Al_2O_3 \cdot H_2O$	amor	−2009	+97	−1821
$Al_2O_3 \cdot 3H_2O^{(1)}$	amor	−2552	+166	−2284
$AlOH^{2+}$	soln	−767	−157	−701
$Al(OH)_2^+$	soln	−	−	−907
$Al(OH)_3$	mon	−1295	+70	−1157
	trig	−1290	−	−
$[Al(OH)_3]$	soln	−	−	−1111
$[Al(OH)_4]^-$	soln	−1490	+144	−1305
$AlO(OH)$	rhomb (α)	−1000	+35	−921
	rhomb (γ)	−988	+48	−913
AlP	gas	ca. +420	−	−
$AlPO_4$	hex	−1734	+91	−1618
$AlPO_4 \cdot 2H_2O$	rhomb	−2353	+134	−2111
AlS	gas	+201	+230	+150
Al_2S_3	hex	−509	+96	−492
$[AlSO_4]^+$	soln	−1431	−192	−1253
$[Al(SO_4)_2]^-$	soln	−2337	−128	−2008
$Al_2(SO_4)_3$	hex	−3441	+239	−3100
$AlSe$	gas	ca. +230	−	−
Al_2Se_3	hex	−567	−	−
$AlTe$	gas	ca. +270	−	−
Al_2Te_3	sld	−326	−	−
Am	gas	+284	+195	+247
	hex	0	+72	0
Am^{3+}	soln	−	−159	−672
AmF_3	hex	−	−	−1518
AmO_2	cub	−1081	+84	−962
Am_2O_3	hex	−1761	+159	−1674
Ar	gas	0	+155	0
$[Ar]$	soln	−12	+59	+16
Ar^+	gas	+1514	+166	+1517
As	amor (β)	+14	−	−
	gas	+288	+174	+247
	hex (α)	0	+36	0
As_2	gas	+194	+242	+143
As_4	gas	+143	+330	+88
	rhomb	+8	−	−
$AsBr_3$	gas	−130	+364	−160
	rhomb	−193	+159	−162
$AsCl_3$	gas	−271	+326	−258
	lq	−315	+213	−268
AsF_3	gas	−921	+289	−906
	lq	−957	+181	−910
AsF_5	gas	−1238	+353	−1181
AsH_2	gas	+159	−	−
AsH_3	gas	+66	+223	+69

(Continued)

Substance	State	$\Delta H°$ kJ/mole	$S°$ J/(K·mole)	$\Delta G°$ kJ/mole
AsI_3	trig	−80	+164	−66
AsO	gas	+70	+233	+42
AsO_3^{3-}	soln	−664	−187	−448
AsO_4^{3-}	soln	−892	−165	−652
As_2O_3	cub	−666	+117	−588
	mon	−664	+127	−589
As_2O_5	amor	−925	+105	−782
As_4O_6	gas	−1209	+381	−1098
As_2S_3	mon	−96	+164	−95
As_2S_5	amor	−146	−	−
As_4S_4	mon (α)	−146	+254	−141
At^-	gas	−174	+175	−202
At	gas	+68	+187	+60
At_2	gas	+84	+276	+38
	rhomb	0	+121	0
Au	gas	+368	+180	+329
	cub	0	+48	0
Au^+	soln	+222	+128	+179
AuBr	sld	−14	+98	−6
$AuBr_3$	sld	−54	+155	−18
$[AuBr_4]^-$	soln	−	+314	−159
$[Au(CN)_2]^-$	soln	−	+123	+269
AuCl	rhomb	−35	+108	−20
$[AuCl_2]^-$	soln	−174	+257	−151
$AuCl_3$	mon	−118	+152	−49
$[AuCl_4]^-$	soln	−325	+258	−235
$[AuCl_3(OH)]^-$	soln	−	−	−306
AuF_3	sld	−368	+113	−297
AuI	sld	+1	+111	−1
Au_2O_3	cub	−3	+134	+77
$Au_2O_3 \cdot 3H_2O^{(1)}$	amor	−957	+243	−700
$[Au(OH)_3]$	soln	−	−	−318
$[Au(OH)_4]^-$	soln	−	−	−487
$AuTe_2$	mon	−12	+142	−10
B	amor	+4	+7	+4
	gas	+562	+152	+518
	hex	0	+6	0
B^+	gas	+1369	+138	+1323
B_2	gas	+846	+202	+789
BBr	gas	+239	+225	+196
BBr_2	gas	+70	+295	+29
BBr_3	gas	−205	+324	−232
	lq	−239	+230	−238
BCl	gas	+185	+213	+157
BCl_2	gas	−61	+273	−74
BCl_3	gas	−404	+290	−389
	lq	−427	+206	−387

(Continued)

Substance	State	$\Delta H°$ kJ/mole	$S°$ J/(K·mole)	$\Delta G°$ kJ/mole
B_2Cl_4	gas	−491	+357	−461
BF	gas	−117	+200	−145
BF_2	gas	−501	−248	−513
BF_3	gas	−1137	+254	−1119
$[BF_4]^-$	soln	−1574	+200	−1491
$[BF_4],K$	rhomb	−1886	+157	−1791
$[BF_4],Na$	rhomb	−1845	+120	−1743
BH_2	gas	+209	−	−
BH_3	gas	+92	+188	+96
B_2H_6	gas	+35	+232	+87
B_4H_{10}	gas	+68	−	−
B_5H_9	gas	+75	+276	+177
	lq	+44	+184	+174
B_6H_{10}	gas	+97	−	−
	lq	+58	−	−
$B_{10}H_{14}$	mon	−39	+176	+200
$[BH_4],K$	cub	−227	+107	−160
$[BH_4],Li$	rhomb	−191	+76	−125
$B_3H_6N_3$	gas	−510	+289	−388
	lq	−541	+193	−390
$[BH_4],Na$	cub	−192	+101	−127
BI	gas	+288	+233	+238
BI_2	gas	+220	+310	+164
BI_3	gas	+71	+349	+21
	sld	−37	+200	−43
BN	gas	+648	+212	+615
	hex (α)	−254	+15	−228
BO	gas	+6	+203	−22
BO_2	gas	−293	+229	−298
B_2O_3	amor	−1254	+78	−1182
	gas	−844	+280	−832
	hex	−1272	+54	−1193
$B(OH)_3$	gas	−1013	+303	−951
	tricl	−1094	+89	−969
$[B(OH)_3]$	soln	−1071	+165	−969
$[B(OH)_4]^-$	soln	−1343	+105	−1153
BP	sld	−101	+27	−95
BS	gas	+253	+216	+200
B_2S_3	gas	+67	+327	+2
	sld	−239	+106	−238
Ba	gas	+177	+170	+145
	α-cub	0	+61	0
Ba^+	gas	+706	+176	+666
Ba^{2+}	soln	−524	+9	−548
BaBr	gas	−77	−	−
$BaBr_2$	gas	−417	+348	−377
	rhomb	−751	+149	−732

Substance	State	$\Delta H°$ kJ/mole	$S°$ J/(K·mole)	$\Delta G°$ kJ/mole
$BaBr_2·2H_2O$	mon	−1356	+230	−1222
$Ba(BrO_3)_2·H_2O$	mon	−1003	+288	−772
BaC_2	tetr	−82	−	−
$BaCO_3$	rhomb	−1201	+112	−1123
$BaCl$	gas	−143	−	−
$BaCl_2$	gas	−495	−	−
	rhomb	−845	+124	−797
$BaCl_2·2H_2O$	mon	−1446	+203	−1283
$Ba(ClO_3)_2$	sld	−755	+234	−557
$Ba(ClO_3)_2·H_2O$	mon	−1048	−	−
$Ba(ClO_4)_2$	hex	−789	+255	−536
$BaCrO_4$	rhomb	−1426	+156	−1325
BaF	gas	−326	−	−
BaF_2	gas	−805	+299	−815
	cub	−1192	+96	−1142
BaH	gas	+184	+219	+156
BaH_2	rhomb	−190	+63	−152
$BaHPO_4$	rhomb	−1818	−	−
BaI	gas	+15	−	−
BaI_2	gas	−323	+362	−378
	rhomb	−610	+167	−607
$BaI_2·2H_2O$	rhomb	−1222	+247	−1104
$Ba(IO_3)_2$	mon	−1048	+249	−886
$Ba(IO_3)_2·H_2O$	mon	−1296	+323	−1087
$BaMoO_4$	tetr	−1538	+149	−1433
Ba_3N_2	sld	−335	−	−
$Ba(NO_2)_2$	hex	−756	−	−
$Ba(NO_3)_2$	cub	−979	+214	−784
BaO	gas	−131	+235	−152
	cub	−538	+70	−510
BaO_2	tetr	−644	+78	−588
$BaOH^+$	soln	−749	+56	−717
$Ba(OH)_2$	rhomb	−943	+100	−854
$Ba_3(PO_4)_2$	trig	−3960	+356	−3743
BaS	gas	+64	+249	+18
	cub	−442	+78	−437
$BaSO_3$	cub	−1030	+121	−947
$BaSO_4$	rhomb	−1458	+132	−1347
$[BaSO_4]$	soln	−	−	−1305
$BaSe$	cub	−364	+91	−360
$BaSeO_4$	rhomb	−1145	+133	−1032
$BaSiO_3$	rhomb	−1608	+110	−1525
$BaTe$	cub	−314	+57	−298
$(BaTi)O_3$	tetr	−1641	+108	−1554
$BaWO_4$	tetr	−1634	+152	−1529
$(BaZr)O_3$	cub	−2419	+125	−2335
Be	gas	+327	+136	+289

(Continued)

Substance	State	$\Delta H°$ kJ/mole	$S°$ J/(K·mole)	$\Delta G°$ kJ/mole
Be	hex	0	+10	0
Be$^+$	gas	+1233	+142	+1187
Be^{2+}	soln	−404	−197	−381
(BeAl$_2$)O$_4$	rhomb	−2299	+66	−2177
BeBr	gas	+50	−	−
BeBr$_2$	gas	−240	+273	−273
	rhomb	−368	+121	−356
Be$_2$C	cub	−91	+16	−88
Be(CH$_3$COO)$_2$	sld	−1302	−	−
BeCO$_3$	sld	−1024	+67	−948
BeCl	gas	+13	+218	+16
BeCl$_2$	gas	−360	+252	−366
	α-rhomb	−491	+83	−446
BeCl$_2$·4H$_2$O	sld	−1848	+243	−1573
BeF	gas	−208	+206	−236
BeF$^+$	soln	−738	−99	−695
BeF$_2$	amor	−1022	−	−
	gas	−784	+229	−789
	tetr	−1035	+53	−987
[BeF$_2$]	soln	−1066	−16	−998
[BeF$_3$]$^-$	soln	−	−	−1294
[BeF$_4$]$^{2-}$	soln	−	−	−1587
BeH	gas	+321	+177	+291
BeH$_2$	amor	−19	+25	+16
	gas	+126	+173	+116
BeI$_2$	gas	−85	+289	−134
	rhomb	−212	+130	−213
Be$_3$N$_2$	cub	−589	+50	−538
Be(NO$_3$)$_2$·4H$_2$O	sld	−1936	−	−
BeO	gas	+129	+198	+104
	hex	−607	+14	−578
Be(OH)$_2$	amor	−892	−	−
	gas	−671	−	−
	rhomb	−905	+53	−818
	tetr	−903	+55	−816
[Be(OH)$_3$]$^-$	soln	−	−	−959
[Be(OH)$_4$]$^{2-}$	soln	−	−	−1115
BeS	cub	−235	+34	−233
BeSO$_4$	tetr	−1205	+78	−1094
BeSO$_4$·2H$_2$O	sld	−1823	+163	−1598
BeSO$_4$·4H$_2$O	tetr	−2425	+233	−2081
BeSeO$_4$	sld	−947	−	−
BeSeO$_4$·4H$_2$O	rhomb	−2170	+247	−1828
Be$_2$SiO$_4$	trig	−2145	+64	−2030
Bi	gas	+209	+187	+170
	trig	0	+57	0
Bi$^+$	gas	+918	+175	+877

Substance	State	$\Delta H°$ kJ/mole	$S°$ J/(K·mole)	$\Delta G°$ kJ/mole
Bi$^{III(2)}$	soln	+81	−177	+92
Bi$_2$	gas	+225	+273	+178
BiAsO$_4$	tetr	−	−	−614
BiBr	gas	+53	+267	+13
BiBr$_3$	gas	−102	+385	−132
	sld	−259	−	−
BiCl	gas	+25	+255	−1
BiCl^{2+}	soln	−70	+1	−59
BiCl$_2^+$	soln	−	−	−199
BiCl$_3$	gas	−264	+356	−253
	cub	−378	+172	−313
[BiCl$_3$]	soln	−	−	−337
[BiCl$_4$]$^-$	soln	−	−	−479
Bi(Cl)O	tetr	−369	+103	−319
BiF	gas	−30	+244	−55
BiF$_3$	cub	−904	−	−
BiI	gas	+75	+275	+27
BiI$_3$	trig	−174	+234	−175
[BiI$_4$]$^-$	soln	−	−	−212
Bi(NO$_3$)$_3$·5H$_2$O	tricl	−2005	−	−
BiO	gas	+119	+246	+93
Bi$_2$O$_3$	mon	−575	+151	−494
Bi(OH)$_3$	amor	−715	+118	−583
BiPO$_4$	mon	−1173	+123	−1058
BiS	gas	+180	+285	+122
Bi$_2$S$_3$	rhomb	−156	+200	−153
Bi$_2$(SO$_4$)$_3$	sld	−2554	−	−2583
BiSe	gas	+176	+264	+127
Bi$_2$Se$_3$	rhomb	−140	+217	−133
BiTe	gas	+179	+273	+130
Bi$_2$Te$_3$	trig	−79	+251	−75
Br$^-$	gas	−219	+163	−239
	soln	−122	+83	−104
Br	gas	+112	+175	+82
Br$^+$	gas	+1260	+177	+1224
Br$_2$	gas	+31	+245	+3
	lq	0	+152	0
[Br$_2$]	soln	−1	+136	+4
BrCl	gas	+15	+240	−1
BrF	gas	−43	+229	−58
BrF$_2$	gas	−113	−	−
BrF$_3$	gas	−255	+292	−229
	lq	−303	+178	−243
BrF$_5$	gas	−429	+319	−350
	lq	−461	−	−
[BrO$_3^-$]	soln	−65	+164	+20
C$^-$	gas	+588	+151	+551

(Continued)

Substance	State	$\Delta H°$ kJ/mole	$S°$ J/(K·mole)	$\Delta G°$ kJ/mole
C	gas	+715	+158	+670
	hex (α)	0	+6	0
	cub (β)	+2	+2	+3
C^+	gas	+1810	+155	+1759
C_2	gas	+838	+199	+782
CBr_2	gas	+268	–	–
CBr_3	gas	+180	–	–
CBr_4	gas	+83	+358	+69
CCl	gas	+440	+224	+408
CCl_2	gas	+222	+266	+211
CCl_3	gas	+82	+300	+94
CCl_4	gas	–103	+310	–61
	lq	–135	+216	–65
CCl_2O	gas	–221	+284	–207
	lq	–244	+197	–204
CF	gas	+249	+213	+218
CF_2	gas	–164	+242	–174
CF_3	gas	–469	+265	–455
CF_4	gas	–933	+262	–888
CH	gas	+594	+183	+561
CH_2	gas	+376	+181	+363
CH_3	gas	+142	+193	+145
CH_4	gas	–75	+186	–51
C_2H_2	gas	+226	+200	+209
C_2H_4	gas	+55	+232	+68
C_6H_6	gas	+82	+269	+130
	lq	+48	+173	+124
$CH_3C(H)O$	gas	–166	+264	–133
CH_3COO^-	soln	–490	+88	–369
CH_3COOH	gas	–433	+282	–374
	lq	–484	+160	–389
$CHCl_3$	gas	–102	+296	–69
	lq	–132	+203	–72
C_5H_5N	gas	+139	+283	+190
	lq	+99	+178	+181
CH_3OH	gas	–202	+240	–163
	lq	–240	+127	–167
C_2H_5OH	gas	–235	+282	–168
	lq	–277	+161	–174
CI_3	gas	+385	–	–
CI_4	gas	+307	+392	+261
CN^-	soln	+146	+118	+172
C_2N_2	gas	+309	+244	+297
$(CN)Br$	gas	+186	+248	+165
$(CN)Cl$	gas	+138	+236	+131
$(CN)F$	gas	–12	+226	–19
$C(NH_2)_2O$	tetr	–333	+105	–197

Substance	State	$\Delta H°$ kJ/mole	$S°$ J/(K·mole)	$\Delta G°$ kJ/mole
$[C(NH_2)_2O]$	soln	−	+176	−203
(CN)I	gas	+225	+257	+196
CO	gas	−110	+198	−137
CO_2	gas	−393	+214	−394
$[CO_2]$	soln	−414	+121	−386
CO_3^{2-}	soln	−677	−57	−528
$C_2O_4^{2-}$	soln	−808	+80	−668
COF_2	gas	−635	+258	−619
CS	gas	+230	+210	+179
CS_2	gas	+117	+238	+67
	lq	+88	+151	+64
CSO	gas	−142	+231	−169
CSe_2	gas	+257	+263	+205
	lq	+219	−	−
Ca	gas	+178	+155	+144
	α-cub	0	+42	0
Ca^+	gas	+773	+161	+731
Ca^{2+}	soln	−543	−55	−553
$(CaAl_2)O_4$	rhomb	−2326	+114	−2209
$Ca(BO_2)_2$	rhomb	−2031	+105	−1924
CaBr	gas	−32	−	−
$CaBr_2$	gas	−391	+310	−426
	rhomb	−674	+134	−656
$CaBr_2·6H_2O$	trig	−2503	+305	−2119
CaC_2	tetr	−62	+70	−67
$CaCO_3$	rhomb	−1207	+88	−1127
	trig	−1207	+92	−1128
$[CaCO_3]$	soln	−	−	−1099
CaC_2O_4	mon	−1352	−	−
$CaC_2O_4·H_2O$	mon	−1670	+156	−1509
CaCl	gas	−98	+244	−125
$CaCl_2$	gas	−467	+286	−473
	rhomb	−794	+114	−749
$CaCl_2·2H_2O$	sld	−1389	−	−
$CaCl_2·6H_2O$	trig	−2625	+285	−2213
$Ca(ClO_4)_2$	sld	−725	−	−
$CaCrO_4$	tetr	−1379	+134	−1277
$(CaCr_2)O_4$	rhomb	−2049	+121	−1936
CaF	gas	−277	+230	−303
CaF_2	gas	−779	+274	−788
	cub	−1228	+69	−1176
$(CaFe_2)O_4$	tetr	−1521	+145	−1413
CaH	gas	+246	+202	+218
CaH_2	rhomb	−175	+42	−136
$Ca(HCO_3)_2$	sld	−2344	+155	−2152
$CaHPO_4$	tricl	−1813	+111	−1680
$CaHPO_4·2H_2O$	mon	−2403	+189	−2154

(Continued)

Substance	State	$\Delta H°$ kJ/mole	$S°$ J/(K·mole)	$\Delta G°$ kJ/mole
$Ca(H_2PO_4)_2$	sld	−3115	+190	−2812
$Ca(H_2PO_4)_2·H_2O$	tricl	−3410	+260	−3059
CaI	gas	ca. 0	−	−
CaI_2	gas	−232	+326	−282
	trig	−538	+145	−534
$Ca(IO_3)_2$	mon	−1008	+230	−846
$CaMg(CO_3)_2$	hex	−2315	+155	−2152
$CaMoO_4$	tetr	−1546	+123	−1439
$Ca(N_3)_2$	rhomb	+14	−	−
Ca_3N_2	hex	−432	+105	−369
$Ca(NO_3)_2$	cub	−938	+193	−743
$Ca(NO_3)_2·4H_2O$	mon	−2131	+339	−1701
CaO	gas	−60	+220	−82
	cub	−635	+40	−604
CaO_2	tetr	−659	+43	−598
$CaOH^+$	soln	−764	−15	−717
$Ca(OH)_2$	trig	−985	+83	−897
Ca_3P_2	sld	−494	−	−
$Ca(PO_3)_2$	amor	−1708	+147	−1532
$Ca_2P_2O_7$	tetr	−3333	+189	−3126
$Ca_3(PO_4)_2$	mon	−4105	+241	−3870
	trig	−4125	+236	−3889
$Ca_5(PO_4)_3F$	hex	−6888	+388	−6507
CaS	cub	−482	+56	−477
$CaSO_3·2H_2O$	hex	−1753	+184	−1555
$CaSO_4$	rhomb	−1434	+107	−1322
$CaSO_4·2H_2O$	mon	−2023	+194	−1797
$2CaSO_4·H_2O$	trig	−3203	+261	−2923
$[CaSO_4]$	soln	−1449	+21	−1310
CaSe	cub	−364	+69	−360
$CaSeO_4$	sld	−1120	−	−
$CaSeO_4·2H_2O$	rhomb	−1717	+185	−1486
$CaSiO_3$	tricl	−1635	+82	−1550
Ca_2SiO_4	mon	−2308	+128	−2193
	rhomb	−2318	+120	−2201
CaTe	cub	−293	+77	−289
$CaTeO_3$	sld	−1123	−	−
$CaTeO_3·2H_2O$	sld	−1436	−	−
$(CaTi)O_3$	mon	−1662	+94	−1577
$CaWO_4$	tetr	−1641	+126	−1534
$(CaZr)O_3$	rhomb	−2721	+100	−2635
Cd	gas	+112	+168	+77
	hex	0	+52	0
Cd^{2+}	soln	−76	−71	−78
Cd_3As_2	tetr	−42	+207	−36
CdBr	gas	+66	+263	+26
$CdBr_2$	gas	−140	+310	−172

Substance	State	$\Delta H°$ kJ/mole	$S°$ J/(K·mole)	$\Delta G°$ kJ/mole
$CdBr_2$	hex	−316	+139	−297
$CdBr_2·2H_2O$	sld	−1491	+311	−1245
$CdCO_3$	trig	−754	+97	−674
CdC_2O_4	sld	−922	−	−
$CdCl$	gas	+27	+251	+1
$CdCl^+$	soln	−240	+32	−220
$CdCl_2$	gas	−195	+286	−198
	trig	−392	+115	−344
$[CdCl_2]$	soln	−405	+110	−356
CdF_2	gas	−395	+265	−398
	cub	−700	+89	−651
CdI	gas	+82	+271	+34
CdI_2	gas	−60	+326	−107
	hex	−206	+161	−204
$Cd(IO_3)_2$	sld	−	−	−373
$CdNO_3^+$	soln	−305	−3	−187
$Cd(NO_3)_2$	sld	−453	−	−
$Cd(NO_3)_2·4H_2O$	rhomb	−2070	+393	−1653
CdO	gas	+81	+233	+58
	cub	−258	+55	−228
$CdOH^+$	soln	−	−	−272
$Cd(OH)_2$	trig	−563	+88	−474
$[Cd(OH)_2]$	soln	−	−	−446
CdS	hex	−152	+68	−147
$CdSO_4$	rhomb	−933	+123	−822
$3CdSO_4·8H_2O$	mon	−5188	+689	−4395
$[CdSO_4]$	soln	−981	+4	−835
$[Cd(SO_3S)]$	soln	−723	+59	−614
$[Cd(SO_3S)_2]^{2-}$	soln	−	−	−1142
$CdSe$	hex	−136	+97	−137
$CdTe$	cub	−108	+92	−98
Ce	gas	+423	−	−
	α-cub	0	+76	0
Ce^{3+}	soln	−698	−201	−676
$Ce^{IV(3)}$	soln	−577	−426	−508
CeC_2	tetr	−97	−	−
$CeCl_3$	gas	−717	−	−
	hex	−1060	+146	−981
CeF^{2+}	soln	−	−	−979
CeF_3	hex	−1680	+115	−1601
$Ce(NO_3)_3·6H_2O$	tricl	−3058	−	−
CeO_2	cub	−1090	+62	−1025
Ce_2O_3	trig	−1801	+148	−1708
$CeOH^{3+}$	soln	−812	−170	−749
$Ce(OH)_3$	amor	−1409	+110	−1269
$CePO_4$	mon	−1939	+115	−1816
$[CePO_4]$	soln	−	−	−1800

Substance	State	$\Delta H°$ kJ/mole	$S°$ J/(K·mole)	$\Delta G°$ kJ/mole
$CeSO_4^+$	soln	−1593	−59	−1440
$Ce(SO_4)_2$	sld	−2347	+201	−2121
$Ce_2(SO_4)_3$	mon	−3955	−	−
$Ce_2(SO_4)_3·8H_2O$	tricl	−6448	−	−
Cl^-	gas	−234	+153	−240
	soln	−167	+56	−131
Cl	gas	+122	+165	+106
Cl^+	gas	+1384	+167	+1361
Cl_2	gas	0	+223	0
$[Cl_2]$	soln	−23	+121	+7
ClF	gas	−50	+218	−51
ClF_5	gas	−239	+311	−147
Cl_2F_6	gas	−326	+562	−246
	lq	−262	+376	−126
ClN	gas	+335	−	−
Cl_2N	gas	+176	−	−
Cl_3N	gas	−84	−	−
	lq	+230	−	−
ClO^-	soln	−107	+42	−37
ClO	gas	+102	+227	+98
ClO_2^-	soln	−67	+101	−17
ClO_2	gas	+104	+257	+122
ClO_3	gas	+155	−	−
ClO_3^-	soln	−104	+164	−8
ClO_4^-	soln	−129	+183	−9
Cl_2O	gas	+76	+266	+94
Cl_2O_7	gas	+287	+565	+399
	lq	+251	−	−
ClO_3F	gas	−21	+279	+51
Cm	gas	+346	−	−
		(1600 K)		
Co	gas	+428	+179	+384
	hex	0	+30	0
Co^{2+}	soln	−59	−109	−56
Co^{3+}	soln	+25	−316	+78
$CoAs_2$	rhomb	−98	+100	−97
$Co_3(AsO_4)_2·8H_2O$	mon	−	−	−3530
CoBr	gas	+211	−	−
$CoBr_2$	gas	+2	+321	−39
	trig	−223	+134	−209
$CoBr_2·6H_2O$	mon	−2020	+375	−1660
$CoCO_3$	hex	−716	+89	−640
$[Co(CO_4)]$	gas	−6811	−	−
$[Co_2(CO)_8]$	gas	−1423	−	−
	sld	−1526	−	−
CoCl	gas	+152	−	−
$CoCl^+$	soln	−224	−48	−187

Substance	State	$\Delta H°$ kJ/mole	$S°$ J/(K·mole)	$\Delta G°$ kJ/mole
$CoCl_2$	gas	−84	+296	−97
	trig	−313	+109	−270
$CoCl_2·6H_2O$	mon	−2113	+349	−1724
$Co(ClO_4)_2·6H_2O$	hex	−2027	−	−
$(Co^{II}Co_2^{III})O_4$	cub	−892	+103	−774
$(Co^{II}Co_2^{III})S_4$	cub	−306	+274	−323
CoF	gas	+73	−	−
CoF_2	gas	−367	+270	−378
	tetr	−666	+82	−621
CoF_3	trig	−779	+93	−707
CoI	gas	+251	−	−
CoI_2	gas	+107	+340	+49
	trig	−96	+159	−100
$[Co(NH_3)_5]^{3+}$	soln	−	+318	−409
$[Co(NH_3)_6]^{2+}$	soln	−	−	−238
$[Co(NH_3)_6]^{3+}$	soln	−595	+334	−221
$[Co(NH_3)_6]Br_2$	cub	−909	−	−
$[Co(NH_3)_5Cl]^{2+}$	soln	−	+402	−361
$[Co(NH_3)_6]Cl_2$	cub	−999	−	−
$[Co(NH_3)_6]Cl_3$	mon	−1132	+255	−576
$[Co(NH_3)_5Cl]Cl_2$	rhomb	−1023	+366	−588
$[Co(NH_3)_6](NO_3)_3$	tetr	−1291	−	−
$Co(NO_3)_2$	sld	−422	+192	−230
$Co(NO_3)_2·6H_2O$	mon	−1656	+474	−1130
CoO	cub	−239	+53	−215
$Co_2O_3·3H_2O^{(1)}$	amor	−1451	+201	−1193
$Co(OH)_2$	trig	−541	+82	−456
$[Co(OH)_2]$	soln	−	−	−423
CoO(OH)	rhomb	−456	−	−
CoS	hex	−85	+62	−85
$Co(S_2)$	cub	−134	+103	−137
$CoSO_4$	rhomb	−889	+118	−784
$CoSO_4·7H_2O$	mon	−2980	+406	−2473
$[CoSO_4]$	soln	−966	−39	−813
$[Co(SO_3S)]$	soln	−	−	−582
CoSe	hex	−79	−	−
$CoSeO_4·7H_2O$	mon	−2367	−	−
CO_2SiO_4	rhomb	−1411	+159	−1313
Cr	gas	+398	+174	+353
	cub	0	+24	0
Cr^{2+}	soln	−180	+42	−164
Cr^{3+}	soln	−212	−316	−204
CrBr	gas	+179	−	−
$CrBr_2$	gas	−59	−	−
	mon	−298	+134	−285
$CrBr_3$	gas	−164	−	−
	trig	−400	+160	−373

Substance	State	$\Delta H°$ kJ/mole	$S°$ J/(K·mole)	$\Delta G°$ kJ/mole
$[Cr(CO)_6]$	gas	−1007	+488	−951
	rhomb	−1078	+314	−970
CrCl	gas	+152	−	−
$CrCl^{2+}$	soln	−	−	−339
$CrCl_2$	gas	−134	−	−
	rhomb	−395	+115	−356
$CrCl_2^+$	soln	−546	−87	−466
$CrCl_3$	gas	−246	−	−
	trig	−516	+123	−446
$CrCl_2O_2$	gas	−529	+330	−493
	lq	−597	+209	−525
	cub	−1432	+135	−1327
CrF	gas	+34	−	−
CrF_2	gas	−414	+266	−426
	mon	−754	+84	−711
CrF_3	gas	−906	−	−
	rhomb	−1112	+94	−1042
$(Cr_2Fe)O_4$	cub	−1454	+146	−1353
CrI_2	mon	−226	+155	−230
CrI_3	trig	−205	−	−
CrN	cub	−124	+53	−104
Cr_2N	sld	−129	+50	−101
$CrNH_4(SO_4)_2·12H_2O$	cub	−5562	+715	−4563
CrO	gas	+190	−	−
CrO_2	gas	−61	−	−
CrO_3	gas	−293	+271	−275
	rhomb	−584	+67	−505
CrO_4^{2-}	soln	−875	+46	−721
Cr_2O_3	trig	−1141	+81	−1059
$Cr_2O_3·3H_2O^{(1)}$	amor	−2026	+76	−1734
$Cr_2O_7^{2-}$	soln	−1479	+254	−1288
$CrOH^{2+}$	soln	−446	−153	−418
$Cr(OH)_2^{(4)}$	amor	−659	+81	−576
$Cr(OH)_2^+$	soln	−	−	−624
$Cr(OH)_3$	amor	−976	+95	−847
$[Cr(OH)_6]^{3-}$	soln	−	−	−312
CrS	hex	−176	+65	−179
$Cr_2(SO_4)_3$	hex	−3310	+288	−2986
Cs	gas	+78	+175	+51
	cub	0	+85	0
Cs^+	gas	+460	+170	+428
	soln	−258	+133	−292
Cs_2	gas	+111	+284	+77
CsBr	gas	−208	+267	−240
	cub	−392	+117	−379
$CsBrO_3$	hex	−376	+162	−285
Cs_2CO_3	sld	−1127	+189	−1039

(*Continued*)

Substance	State	$\Delta H°$ kJ/mole	$S°$ J/(K·mole)	$\Delta G°$ kJ/mole
CsCl	gas	−245	+256	−263
	α-cub	−443	+101	−415
CsClO$_3$	hex	−407	+163	−305
CsClO$_4$	rhomb	−443	+175	−314
Cs$_2$CrO$_4$	rhomb	−1429	+229	−1317
CsF	gas	−358	+243	−375
	cub	−556	+93	−528
CsH	gas	+121	+215	+102
	cub	−50	+79	−29
CsHCO$_3$	rhomb	−933	+130	−833
CsI	gas	−152	+275	−191
	cub	−349	+123	−343
CsIO$_3$	cub	−547	+167	−463
CsN$_3$	tetr	−20	+134	+51
CsNH$_2$	tetr	−119	−	−
CsNO$_3$	hex	−495	+146	−393
CsO$_2$	tetr	−278	+65	−211
Cs$_2$O	hex	−320	+126	−276
Cs$_2$O$_2$	rhomb	−404	+118	−327
CsOH	gas	−265	+255	−266
	sld	−406	+84	−356
Cs$_2$S	sld	−343	+146	−326
Cs$_2$SO$_4$	gas	−1104	+410	−1044
	rhomb	−1443	+212	−1324
Cs$_2$SeO$_4$	rhomb	−1136	+218	−1016
Cu	gas	+338	+166	+298
	cub	0	+33	0
Cu$^+$	gas	+1090	+161	+1046
	soln	+72	+42	+50
Cu^{2+}	soln	+66	−96	+65
Cu$_3$(AsO$_4$)$_2$·4H$_2$O	sld	−	−	−2732
CuBr	gas	+120	+248	+79
	cub	−105	+96	−101
CuBr$_2$	gas	+46	−	−
	mon	−143	+146	−131
CuBr$_2$·4H$_2$O	sld	−1325	+293	−1079
CuCN	mon	+98	+90	+111
[CuCO$_3$]	soln	−	−	−502
[Cu(CO$_3$)$_2$]$^{2-}$	soln	−	−	−1049
Cu$_2$CO$_3$(OH)$_2$	mon	−1048	+222	−901
CuCl	gas	+81	+237	+53
	cub	−136	+87	−119
CuCl$^+$	soln	−	−	−71
[CuCl$_2$]$^-$	soln	−278	+211	−244
CuCl$_2$	gas	−21	−	−
	mon	−216	+108	−172
CuCl$_2$·2H$_2$O	rhomb	−819	+191	−660

Substance	State	$\Delta H°$ kJ/mole	$S°$ J/(K·mole)	$\Delta G°$ kJ/mole
$Cu(ClO_4)_2 \cdot 6H_2O$	mon	−1915	−	−
CuF	gas	−13	−	−
CuF^+	soln	−262	−64	−222
CuF_2	gas	−276	+260	−283
	cub	−538	+69	−488
$CuF_2 \cdot 2H_2O$	mon	−1169	−	−
CuH	gas	+275	+196	+246
CuI	gas	+153	+256	+104
	cub	−68	+97	−70
$Cu(IO_3)_2$	mon	−411	−	−
$Cu(IO_3)_2 \cdot 0.67H_2O$	tricl	−	−	−513
$[Cu(NH_3)]^{2+}$	soln	−37	+18	−16
$[Cu(NH_3)_2]^+$	soln	−151	+264	−65
$[Cu(NH_3)_2]^{2+}$	soln	−140	+118	−31
$[Cu(NH_3)_3]^{2+}$	soln	−244	+204	−73
$[Cu(NH_3)_4]^{2+}$	soln	−347	+281	−112
$[Cu(NH_3)_4](NO_3)_2$	rhomb	−828	−	−
$Cu(NO_3)_2$	sld	−310	+192	−117
$Cu(NO_3)_2 \cdot 3H_2O$	sld	−1217	−	−
CuO	gas	+320	+231	+292
	mon	−156	+43	−128
Cu_2O	cub	−171	+92	−148
$CuOH^+$	soln	−	−	−130
$Cu(OH)_2$	amor	−444	+84	−359
$[Cu(OH)_3]^-$	soln	−	−	−492
$[Cu(OH)_4]^{2-}$	soln	−894	+43	−658
CuS	hex	−53	+67	−54
Cu_2S	rhomb	−79	+121	−86
$CuSO_4$	rhomb	−770	+112	−662
$CuSO_4 \cdot 5H_2O$	tricl	−2280	+301	−1880
$[CuSO_4]$	soln	−838	−18	−691
$CuSe$	hex	−40	+83	−42
Cu_2Se	α-cub	−55	+133	−62
$CuSeO_4 \cdot 5H_2O$	tricl	−1951	−	−
D	gas	+222	+123	+207
D_2	gas	0	+145	0
D_2O	gas	−250	+198	−235
	lq	−294	+76	−243
Dy	gas	+290	+196	+254
	hex	0	+75	0
Dy^{3+}	soln	−	−180	−682
$DyBr_3$	gas	−505	+414	−538
	hex	−777	−	−
$DyCl_3$	gas	−714	+383	−706
	mon	−999	+146	−920
DyF_3	gas	−1197	+347	−1187
	rhomb	−1720	+121	−1643

Substance	State	$\Delta H°$ kJ/mole	$S°$ J/(K·mole)	$\Delta G°$ kJ/mole
DyI_3	hex	−607	+218	−598
Dy_2O_3	α-cub	−1866	+150	−1774
$Dy(OH)_3$	amor	−1885	+151	−1757
e^-	gas	0	+21	0
$[e^-]$	soln	−164	−8	−157
Er	gas	+317	+194	+281
	hex	0	+73	0
Er^{3+}	soln	−664	−252	−589
$ErBr_3$	gas	−508	+412	−422
	hex	−781	−	−
$ErCl_3$	mon	−958	+158	−884
$ErCl_3·6H_2O$	mon	−2838	+399	−2418
ErF_3	gas	−954	+345	−944
	mon	−1723	+121	−1647
ErI_3	hex	−586	+218	+577
Er_2O_3	cub	−1898	+153	−1808
Eu	gas	+175	+189	+142
	cub	0	+78	0
Eu^{2+}	soln	−506	−77	−500
Eu^{3+}	soln	−581	−216	−554
$EuBr_2$	tetr	−745	+168	−727
$EuCl_2$	gas	−464	+323	−471
	rhomb	−812	+170	−773
$EuCl_3$	hex	−916	+147	−837
EuF_2	gas	−766	+300	−772
EuF_3	gas	−1356	+336	−1342
	rhomb	−1620	+109	−1538
Eu_2O_3	cub	−1628	+147	−1534
	α-mon	−1648	−	−
$Eu(OH)_3$	amor	−1317	+112	−1177
F^-	gas	−260	+145	−267
	soln	−334	−14	−280
F	gas	+79	+159	+62
F^+	gas	+1767	+162	+1743
F_2	gas	0	+203	0
Fe	gas	+417	+180	+371
	α-cub	0	+27	0
Fe^{2+}	soln	−93	−105	−92
Fe^{3+}	soln	−51	−279	−18
$(FeAl_2)O_4$	cub	−1975	+106	−1860
$FeAs_2$	rhomb	−44	+127	−52
$FeAsO_4·2H_2O$	rhomb	−	−	−1260
$Fe(As)S$	tricl	−106	+108	−110
FeBr	gas	+282	−	−
$FeBr^{2+}$	soln	−	−121	−113
$FeBr_2$	gas	−44	+321	−86
	trig	−250	+138	−238

(*Continued*)

Substance	State	$\Delta H°$ kJ/mole	$S°$ J/(K·mole)	$\Delta G°$ kJ/mole
$FeBr_3$	gas	−126	−	−
	hex	−267	+184	−246
Fe_3C	rhomb	+24	+108	+18
$[Fe(CN)_6]^{4-}$	soln	+458	+98	+696
$[Fe(CN)_6]^{3-}$	soln	+630	+269	+748
$[Fe(CN)_6],Ba_2·6H_2O$	mon	−2351	+115	−1741
$[Fe(CN)_6],K_3$	rhomb	−175	+420	−52
$[Fe(CN)_6],K_4$	tetr	−1188	+598	−1088
$[Fe(CN)_6],K_4·3H_2O$	mon	−1472	+593	−1172
$FeCO_3$	trig	−753	+96	−680
$[FeCO_3]$	soln	−	−187	−607
$[Fe(CO)_5]$	gas	−732	+445	−695
	lq	−764	+338	−695
$FeCl$	gas	+251	+257	+216
$FeCl^{2+}$	soln	−194	−116	−158
$FeCl_2$	gas	−149	+287	−160
	trig	−342	+118	−303
$FeCl_2·4H_2O$	mon	−1552	+246	−1272
$FeCl_2^+$	soln	−	−	−291
$FeCl_3$	gas	−253	+344	−248
	trig	−397	+142	−332
$FeCl_3·6H_2O$	mon	−2224	+352	−1804
$[FeCl_3]$	soln	−	−	−416
Fe_2Cl_6	gas	−654	+537	−599
$Fe(Cl)O$	rhomb	−410	−	−
$(FeCu)S_2$	tetr	−181	+118	−179
FeF	gas	+47	+240	+14
FeF^{2+}	soln	−370	−130	−332
FeF_2	gas	−390	+265	−400
	tetr	−661	+87	−618
FeF_3	gas	−821	+304	−813
	trig	−1042	+98	−972
$(Fe^{II}Fe_2^{III})O_4$	cub	−1118	+146	−1015
$Fe(HPO_4)^+$	soln	−1359	−158	−1170
FeI_2	gas	+57	+336	−1
	trig	−116	+170	−124
$FeMoO_4$	sld	−1074	+129	−974
$Fe(NH_4)_2(SO_4)_2·6H_2O$	mon	−3922	−	−
$(Fe_2Ni)O_4$	cub	−1077	+132	−969
$Fe_{0.95}O$	cub	−266	+57	−245
FeO	gas	+237	+236	+205
	cub	−271	+61	−251
FeO_4^{2-}	soln	−479	−	−
Fe_2O_3	trig	−824	+87	−742
$Fe_2O_3·3H_2O^{(1)}$	amor	−1653	+209	−1399
$FeOH^+$	soln	−328	+5	−291
$FeOH^{2+}$	soln	−293	−104	−243

(Continued)

Substance	State	$\Delta H°$ kJ/mole	$S°$ J/(K·mole)	$\Delta G°$ kJ/mole
$Fe(OH)_2$	trig	−574	+92	−493
$[Fe(OH)_2]$	soln	−	−	−459
$Fe(OH)_2^+$	soln	−	−	−454
$[Fe(OH)_3]^-$	soln	−	−	−621
$[Fe(OH)_3]$	soln	−	−	−677
$[Fe(OH)_4]^{2-}$	soln	−	−	−776
$[Fe(OH)_4]^-$	soln	−	−	−844
FeO(OH)	rhomb	−559	+67	−490
Fe_3P	tetr	−163	−	−
$FePO_4$	tetr	−1297	+101	−1185
$FePO_4 \cdot 2H_2O$	amor	−1857	+189	−1632
	mon	−1889	+171	−1658
FeS	α-hex	−101	+60	−101
$Fe(S_2)$	cub (β)	−174	+53	−163
	rhomb (α)	−151	−	−
$FeSO_4$	rhomb	−928	+108	−820
$FeSO_4 \cdot 7H_2O$	rhomb	−3015	+409	−2510
$[FeSO_4]$	soln	−1000	−37	−847
$FeSO_4^+$	soln	−934	−95	−785
$Fe_2(SO_4)_3$	rhomb	−2581	+259	−2247
$FeSiO_3$	rhomb	−1195	+94	−1118
Fe_2SiO_4	rhomb	−1480	+145	−1379
$FeWO_4$	mon	−1188	+132	−1087
$(Fe_2Zn)O_4$	cub	−1171	+152	−1066
Fr	gas	+73	+182	+47
	cub	0	+94	0
Ga	gas	+273	+169	+235
	rhomb	0	+41	0
Ga^+	gas	−21	+162	−76
Ga^{3+}	soln	−212	−321	−162
GaAs	cub	−72	+64	−68
$GaBr_3$	sld	−386	+167	−356
GaCl	gas	−82	+240	−108
$GaCl^{2+}$	soln	−	−	−290
$GaCl_2$	gas	−238	−	−
$GaCl_3$	gas	−444	−	−
	sld	−565	+133	−493
GaF^{2+}	soln	−538	−191	−477
GaF_2^+	soln	−864	−93	−783
GaF_3	sld	−1019	+84	−941
$[GaF_3]$	soln	−1208	−88	−1079
GaI_3	rhomb	−220	+209	−218
GaN	hex	−107	+42	−79
GaO	gas	+141	+231	+115
Ga_2O_3	gas	−516	−	−
	trig	−1104	+85	−1013
$GaOH^{2+}$	soln	−	−	−383

(*Continued*)

Substance	State	$\Delta H°$ kJ/mole	$S°$ J/(K·mole)	$\Delta G°$ kJ/mole
Ga(OH)$_2^+$	soln	–	–	−600
Ga(OH)$_3$	cub	−965	+113	−836
[Ga(OH)$_3$]	soln	–	–	−815
[Ga(OH)$_4$]$^-$	soln	–	–	−1013
GaP	cub	−102	+51	−93
GaPO$_4$	trig	−1413	+104	−1297
Ga$_2$S$_3$	cub	−510	+161	−505
GaSO$_4^+$	soln	−1101	−185	−922
[Ga(SO$_4$)$_2$]$^-$	soln	−2018	−148	−1679
GaSb	cub	−42	+76	−39
Ga$_2$Se$_3$	cub	−369	+179	−360
Ga$_2$Te$_3$	cub	−272	+198	−262
Gd	gas	+399	+194	+361
	hex	0	+68	0
Gd^{3+}	soln	−684	−201	−662
GdCl$_3$	gas	−690	–	–
	mon	−1007	+146	−931
GdCl$_3$·6H$_2$O	mon	−2863	+392	−2442
GdI$_3$	hex	−618	+218	−611
Gd(NO$_3$)$_3$·6H$_2$O	tricl	−3038	+577	−2411
Gd$_2$O$_3$	mon	−1816	+151	−1729
Ge	gas	+379	+168	+338
	cub	0	+31	0
Ge$^{IV(5)}$	soln	–	−657	−28
Ge$_2$	gas	+482	+253	+425
GeBr	gas	+146	+257	+101
GeBr$_2$	gas	−69	+318	−109
GeBr$_4$	gas	−305	+396	−323
	lq	−348	–	–
GeCl	gas	+91	+245	+60
GeCl$_2$	gas	−177	+295	−189
GeCl$_4$	gas	−495	+348	−457
	lq	−565	+251	−498
GeF	gas	−40	+234	−70
GeF$_2$	gas	−473	+271	−484
	rhomb	−586	–	–
GeF$_4$	gas	−1190	+303	−1150
[GeF$_5$]$^-$	soln	−1695	+125	−1552
[GeF$_6$]$^{2-}$	soln	−2026	+190	−1854
GeH	gas	ca. +280	–	–
GeH$_3$	gas	+238	–	–
GeH$_4$	gas	+90	+217	+113
Ge$_2$H$_6$	gas	+162	+297	+209
	lq	+138	–	–
Ge$_3$H$_8$	gas	+227	–	–
	lq	+192	–	–
GeI$_2$	gas	+74	+334	+18

Substance	State	$\Delta H°$ kJ/mole	$S°$ J/(K·mole)	$\Delta G°$ kJ/mole
GeI_2	trig	−78	−	−
GeI_4	gas	−38	+429	−87
	cub	−171	+268	−172
GeO_2	amor	−540	−	−
	tetr (α)	−580	+40	−522
	trig (β)	−555	+55	−501
$GeOH^{3+}$	soln	−	−	−264
$Ge(OH)_2^{2+}$	soln	−	−	−500
$Ge(OH)_3^+$	soln	−	−	−734
$[Ge(OH)_4]$	soln	−	−	−967
$[Ge(OH)_5]^-$	soln	−	−	−1154
$[Ge(OH)_6]^{2-}$	soln	−	−	−1319
GeS	gas	+93	+234	+42
	rhomb	−69	+66	−70
GeS_2	cub	−189	+87	−187
$GeSe$	gas	+110	−	−
	tetr	−84	+78	−85
$GeSe_2$	gas	+130	−	−
	rhomb	−63	+113	−62
$GeTe$	trig	−25	+91	−28
H^-	gas	+140	+109	+133
H	gas	+218	+115	+203
H^+	gas	+1536	+109	+1517
$[H^+]^{(6)}$	soln	0	0	0
H_2	gas	0	+131	0
	soln	−4	+58	+18
$HAsO_3^{2-}$	soln	−689	−15	−524
$HAsO_4^{2-}$	soln	−911	−5	−718
$H_2AsO_3^-$	soln	−721	+111	−593
$H_2AsO_4^-$	soln	−914	+117	−757
H_3AsO_3	soln	−749	+195	−646
H_3AsO_4	soln	−907	+183	−770
HBO_2	gas	−562	+240	−551
	cub	−804	+49	−736
	mon	−794	+38	−723
$(HBO_2)_3$	rhomb	−2368	+150	−2166
HBr	gas	−36	+199	−53
$[HBrO]$	soln	−113	+142	−82
HCN	gas	+134	+202	+124
	lq	+110	+113	+126
$[HCN]$	soln	+107	+127	+119
HCO_3^-	soln	−692	+91	−587
$[H_2CO_3]^{(7)}$	soln	−700	+187	−623
$HC_2O_4^-$	soln	−818	+117	−688
$H_2C_2O_4$	mon	−831	+120	−702
$HCOO^-$	soln	−426	+91	−352
$HCOOH$	gas	−379	+249	−351

(*Continued*)

Substance	State	$\Delta H°$ kJ/mole	$S°$ J/(K·mole)	$\Delta G°$ kJ/mole
HCOOH	lq	−425	+129	−362
[HCOOH]	soln	−426	+163	−373
HCl	gas	−92	+187	−95
[HClO]	soln	−121	+146	−80
[HClO$_2$]	soln	−52	+188	−28
HClO$_4$	gas	+5	+296	+92
	lq	−41	+188	+78
HClO$_4$·H$_2$O	sld	−382	−	−
HCrO$_4^-$	soln	−872	+180	−758
[H$_2$CrO$_4$]	soln	−844	+260	−752
HF	gas	−271	+174	−273
	lq	−303	−	−
[HF]	soln	−321	+91	−298
HF$_2^-$	soln	−657	+68	−581
[H$_2$GeO$_3$]	soln	−	+180	−730
HI	gas	+24	+206	−1
[HIO]	soln	−	+120	−99
HIO$_3$	rhomb	−236	+117	−142
H$_5$IO$_6$	mon	−762	−	−
HMoO$_4^-$	soln	−	−	−867
HN$_3$	gas	+294	+239	+328
	lq	+264	+141	+327
HNCO	gas	−117	+238	−107
	lq	−153	−	−
[HNCO]	soln	−	+134	−116
HNCS	gas	+127	+248	+113
HNO$_2$	gas	−80	+249	−45
[HNO$_2$]	soln	−119	+153	−56
HNO$_3$	gas	−135	+267	−75
	lq	−174	+156	−81
HN$_2$O$_2^-$	soln	−	+142	+84
[HNbO$_3$] ?	soln	−1039	+230	−985
H$_2$O	gas	−242	+189	−229
	hex	−292	+39	−234
	lq	−286	+70	−237
H$_2$O$_2$	gas	−137	+233	−106
	lq	−187	+110	−120
[H$_2$O$_2$]	soln	−191	+143	−134
H(PHO$_3$)$^-$	soln	−969	+80	−831
H(PH$_2$O$_2$)	sld	−615	−	−
H$_2$(PHO$_3$)	rhomb	−952	−	−
[H$_2$(PHO$_3$)]	soln	−965	+79	−847
HPO$_3$	amor	−977	−	−
HPO$_4^{2-}$	soln	−1292	−33	−1089
H$_2$PO$_4^-$	soln	−1296	+90	−1130
H$_2$P$_2$O$_7^{2-}$	soln	−2279	+176	−2015
H$_3$PO$_4$	lq	−1267	+201	−1134

(Continued)

Substance	State	$\Delta H°$ kJ/mole	$S°$ J/(K·mole)	$\Delta G°$ kJ/mole
H_3PO_4	mon	−1279	+110	−1119
$[H_3PO_4]$	soln	−1288	+158	−1143
$H_4P_2O_7$	sld	−2242	−	−
$[H_4P_2O_7]$	soln	−	+285	−2037
$H_4Re_2O_9$	sld	ca. −1800	ca. +370	ca. −1550
$[HS^-]$	soln	−18	+63	+12
HS	gas	+142	+196	+113
H_2S	gas	−21	+206	−34
$[H_2S]$	soln	−40	+121	−28
H_2S_2	gas	+15	+261	−5
H_2S_4	lq	−13	−	−
H_2S_5	gas	+33	−	−
HSO_3^-	soln	−626	+140	−528
HSO_4^-	soln	−889	+124	−755
H_2SO_4	gas	−744	+301	−663
	lq	−814	+157	−690
$H_2SO_4·H_2O$	lq	−1128	+212	−951
$H_2S_2O_7$	sld	−1272	−	−
HSO_3S^-	soln	−625	+163	−524
$[H_2SO_3S]$	soln	−604	+242	−527
HSe^-	soln	+16	+79	+44
H_2Se	gas	+30	+219	+16
$[H_2Se]$	soln	+19	+164	+22
$HSeO_3^-$	soln	−515	+134	−411
$HSeO_4^-$	soln	−576	+163	−451
H_2SeO_3	rhomb	−525	−	−
$[H_2SeO_3]$	soln	−507	+208	−426
H_2SeO_4	rhomb	−533	−	−
$H_2SiO_4^{2-}$	soln	−1397	−13	−1187
$H_3SiO_4^-$	soln	−1426	+113	−1254
$[H_4SiO_4]$	soln	−1462	+179	−1310
$[HTaO_3]$?	soln	−1182	+230	−1128
$HTcO_4$	sld	−703	+139	−593
HTe^-	soln	+69	+62	+105
H_2Te	gas	+99	+229	+85
$[H_2Te]$	soln	+78	+141	+90
$HTeO_3^-$	soln	−556	+139	−452
H_2TeO_3	rhomb	−613	−	−
$[H_2TeO_3]$	soln	−555	+218	−474
$H_4TeO_6^{2-}$	soln	−1128	+178	−866
$H_5TeO_6^-$	soln	−1158	+287	−928
H_6TeO_6	mon	−1287	−	−
$[H_6TeO_6]$	soln	−1182	+353	−972
$HV_{10}O_{28}^{5-}$	soln	−	−	−7707
$H_2V_{10}O_{28}^{4-}$	soln	−	−	−7728
HWO_4^-	soln	−	−	−952
$[H_2WO_4]$?	soln	−	−	−965

Substance	State	$\Delta H°$ kJ/mole	$S°$ J/(K·mole)	$\Delta G°$ kJ/mole
He	gas	0	+126	0
[He]	soln	−2	+56	+19
He$^+$	gas	+2379	+132	+2371
Hf	gas	+653	+187	+611
	hex	0	+46	0
Hf$^{IV(8)}$	soln	−	−	−569
HfC$_{0.94}$	cub	−219	+46	−217
HfCl$_4$	gas	−881	+381	−848
	cub	−991	+191	−901
HfF$_4$	gas	−1682	+325	−1644
	mon	−1931	+111	−1829
HfO$_2$	mon	−1145	+59	−1088
HfO$_2$·2H$_2$O$^{(9)}$	amor	−	−	−1362
Hg	gas	+61	+175	+31
	lq	0	+76	0
Hg^{2+}	soln	+171	−32	+164
Hg$_2^{2+}$	soln	+172	+85	+154
HgBr	gas	+103	+272	+67
HgBr$_2$	gas	−86	+312	−111
	rhomb	−169	+180	−155
Hg$_2$Br$_2$	tetr	−207	+218	−181
Hg(CN)$_2$	tetr	+269	−	−
Hg$_2$CO$_3$	sld	−553	+174	−466
HgCl	gas	+84	+260	+62
HgCl$^+$	soln	−19	+75	−5
HgCl$_2$	gas	−146	+295	−145
	rhomb	−225	+146	−179
[HgCl$_2$]	soln	−216	+155	−173
[HgCl$_3$]$^-$	soln	−389	+209	−309
Hg$_2$Cl$_2$	tetr	−266	+192	−211
HgF	gas	+3	+248	−18
HgF$_2$	gas	−294	+266	−290
	cub	−421	+116	−372
Hg$_2$F$_2$	tetr	−490	+175	−436
HgI	gas	+132	+281	+88
HgI$_2$	gas	−15	+329	−56
	tetr (α)	−105	+184	−103
[HgI$_4$]$^{2-}$	soln	−	−	−215
Hg$_2$I$_2$	tetr	−121	+235	−111
2Hg(NO$_3$)$_2$·H$_2$O	sld	−	−	−564
Hg$_2$(NO$_3$)$_2$·2H$_2$O	mon	−868	−	−
HgO	rhomb	−90	+70	−58
HgOH$^+$	soln	−85	+71	−52
HgS	cub (β)	−54	+88	−48
	trig (α)	−59	+82	−51
HgSO$_4$	rhomb	−704	+136	−590
Hg$_2$SO$_4$	mon	−743	+201	−626

(*Continued*)

Substance	State	$\Delta H°$ kJ/mole	$S°$ J/(K·mole)	$\Delta G°$ kJ/mole
HgSe	cub	−59	+99	−53
Ho	gas	+300	+195	+265
	hex	0	+76	0
Ho^{3+}	soln	−707	−197	−669
$HoCl_3$	mon	−979	+146	−900
HoF_3	gas	−1209	+347	−1199
	rhomb	−1714	+121	−1637
HoI_3	hex	−598	+218	−588
Ho_2O_3	cub	−1881	+158	−1791
I^-	gas	−195	+169	−222
	soln	−58	+103	−52
I	gas	+107	+181	+70
I^+	gas	+1115	+183	+1072
I_2	gas	+62	+261	+19
	lq	+22	+137	+16
	rhomb	0	+116	0
$[I_2]$	soln	+22	+135	+16
IBr	gas	+41	+259	+4
	rhomb	−10	−	−
ICl	gas	+17	+247	−6
	mon	−35	+98	−14
I_2Cl_6	tricl	−178	+334	−44
IF	gas	−94	+236	−117
IF_5	gas	−834	+329	−764
	lq	−883	+215	−778
IF_7	gas	−944	+346	−818
$[I(I)_2]^-$	soln	−53	+236	−52
$[I(I)_2],Cs$	rhomb	−361	+238	−355
$[I(I)_2],NH_4$	rhomb	−216	−	−
$[I(I)_2],Rb$	rhomb	−345	+226	−338
IO_3^-	soln	−220	+117	−126
I_2O_5	sld	−183	−	−
In	gas	+239	+174	+204
	tetr	0	+58	0
In^+	gas	+830	+168	+791
In^{3+}	soln	−136	−265	−98
In_2	gas	+392	+272	+346
InAs	cub	−58	+76	−53
InBr	gas	−37	+260	−75
	rhomb	−176	+112	−169
$InBr_3$	sld	−410	+167	−375
InCl	gas	−71	+248	−94
	cub	−186	+95	−164
$InCl^{2+}$	soln	−	−	−239
$InCl_2^+$	soln	−	−	−375
$InCl_3$	mon	−617	+138	−541
$[In^{III}Cl_4],In$	rhomb	−726	+244	−631

Substance	State	$\Delta H°$ kJ/mole	$S°$ J/(K·mole)	$\Delta G°$ kJ/mole
InF	gas	-192	$+236$	-215
InF^{2+}	soln	$-$	$-$	-404
InI	gas	$+83$	$+268$	$+38$
	rhomb	-116	$+124$	-118
InI$_2$	gas	-21	$-$	$-$
InI$_3$	gas	-116	$-$	$-$
	mon	-149	$+209$	-142
InO	gas	$+386$	$+236$	$+364$
In$_2$O$_3$	cub	-926	$+104$	-831
InOH^{2+}	soln	$-$	$-$	-315
In(OH)$_2^+$	soln	$-$	$-$	-528
In(OH)$_3$	cub	-916	$+107$	-780
[In(OH)$_3$]	soln	$-$	$-$	-736
InP	cub	-85	$+63$	-74
In$_2$S$_3$	cub	-427	$+164$	-413
[InSO$_4$]$^+$	soln	-1016	-92	-859
[In(SO$_4$)$_2$]$^-$	soln	-1932	-62	-1614
In$_2$(SO$_4$)$_3$	mon	-2726	$+302$	-2386
InSb	cub	-20	$+86$	-15
In$_2$Se$_3$	hex	-343	$+195$	-329
In$_2$Te$_3$	cub	-191	$+231$	-181
Ir	gas	$+669$	$+193$	$+622$
	cub	0	$+35$	0
IrCl$_2$	sld	-178	$+130$	-140
IrCl$_3$	gas	-103	$-$	$-$
	hex	-263	$+155$	-199
IrF$_6$	gas	-544	$+358$	-459
	rhomb	-580	$-$	$-$
IrO$_2$	tetr	-242	$+59$	-188
IrS$_2$	cub	-144	$+62$	-133
Ir$_2$S$_3$	rhomb	-245	$+97$	-224
K	gas	$+89$	$+160$	$+61$
	lq	$+2$	$+71$	0
	cub	0	$+65$	0
K$^+$	gas	$+514$	$+154$	$+481$
	soln	-252	$+101$	-283
K$_2$	gas	$+67$	$+250$	$+91$
KAl(SO$_4$)$_2$	trig	-2470	$+205$	-2240
KAl(SO$_4$)$_2$·12H$_2$O	cub	-6063	$+687$	-5142
KBO$_2$	trig	-982	$+80$	-924
K$_2$B$_4$O$_7$	sld	-3523	$+208$	-3325
KBr	gas	-182	$+250$	-214
	cub	-393	$+96$	-380
KBrO$_3$	trig	-358	$+149$	-269
KCH$_3$COO	mon	-723	$-$	$-$
KCN	gas	$+80$	$+253$	$+54$
	cub	-114	$+128$	-102

Substance	State	$\Delta H°$ kJ/mole	$S°$ J/(K·mole)	$\Delta G°$ kJ/mole
K_2CO_3	mon	-1150	$+156$	-1064
$K_2C_2O_4·H_2O$	mon	-1645	$-$	$-$
$K_2Ca(SO_4)_2·H_2O$	mon	-3177	$-$	$-$
KCl	gas	-214	$+239$	-233
	cub	-437	$+82$	-409
$KClO_3$	mon	-399	$+143$	-297
$KClO_4$	rhomb	-433	$+151$	-303
K_2CrO_4	rhomb	-1398	$+200$	-1290
$K_2Cr_2O_7$	tricl	-2068	$+291$	-1888
$KCr(SO_4)_2$	trig	-2352	$+240$	-2134
$KCr(SO_4)_2·12H_2O$	cub	-5784	$+708$	-4870
KF	gas	-326	$+226$	-344
	cub	-569	$+67$	-539
$KF·2H_2O$	rhomb	-1165	$+156$	-1023
KH	gas	$+123$	$+198$	$+103$
	cub	-58	$+50$	-34
KH_2AsO_4	tetr	-1187	$+155$	-1042
$KHCO_3$	mon	-966	$+109$	-867
$K(HF_2)$	tetr	-928	$+105$	-860
KH_2PO_4	rhomb	-1569	$+135$	-1416
K_2HPO_4	sld	-1776	$+179$	-1637
$KHSO_4$	α-rhomb	-1161	$+142$	-1033
KI	gas	-128	$+258$	-168
	cub	-331	$+104$	-325
KIO_3	mon	-500	$+151$	-417
KIO_4	tetr	-508	$+159$	-397
$KMgCl_3·6H_2O$	rhomb	-2945	$-$	$-$
$K_2Mg(SO_4)_2·6H_2O$	mon	-4538	$+536$	-3968
$KMnO_4$	rhomb	-829	$+172$	-729
K_2MnO_4	rhomb	-1180	$-$	$-$
K_2MoO_4	mon	-1499	$+174$	-1381
KN_3	tetr	-3	$+86$	$+77$
KNCS	rhomb	-202	$+124$	-180
KNO_2	mon	-354	$+117$	-280
KNO_3	rhomb	-495	$+133$	-395
KO_2	tetr	-284	$+117$	-238
KO_3	tetr	-261	$+105$	-181
K_2O	cub	-363	$+94$	-322
K_2O_2	rhomb	-440	$+120$	-376
KOH	gas	-232	$+244$	-235
	rhomb	-425	$+79$	-379
[KOH]	soln	-477	$+92$	-441
KPO_3	mon	-1246	$+108$	-1155
K_3PO_4	cub	-1988	$+212$	-1859
$K_4P_2O_7$	sld	-3202	$+318$	-2981
$K_4P_2O_7·3H_2O$	tetr	-4110	$+444$	-3718
$KReO_4$	tetr	-1098	$+168$	-995

(Continued)

Substance	State	$\Delta H°$ kJ/mole	$S°$ J/(K·mole)	$\Delta G°$ kJ/mole
K_2S	cub	−421	+113	−406
$K_2(S_3)$	rhomb	−474	+180	−460
$K_2(S_4)$	rhomb	−472	+210	−458
$K_2(S_5)$	tricl	−475	+257	−465
K_2SO_3	sld	−1119	+155	−1025
$K_2SO_3·H_2O$	mon	−1257	+157	−1025
$[KSO_4]^-$	soln	−1157	+147	−1031
K_2SO_4	gas	−1110	+365	−1048
	rhomb	−1438	+176	−1320
$K_2S_4O_6$	mon	−1766	+310	−1598
$K_2S_2O_6(O_2)$	tricl	−1918	+279	−1699
K_2Se	cub	−375	+120	−360
K_2SeO_4	rhomb	−1120	+180	−1000
K_2SiO_3	rhomb	−1542	+146	−1449
KVO_3	rhomb	−1170	−	−
K_2WO_4	mon	−1580	+170	−1460
Kr	gas	0	+164	0
[Kr]	soln	−15	+62	+15
Kr^+	gas	+1356	+175	+1347
KrF_2	gas	+63	−	−
La	gas	+419	+182	+382
	hex	0	+57	0
La^{3+}	soln	−707	−209	−686
LaB_6	cub	−130	+83	−127
LaC_2	tetr	−79	+73	−80
$LaCO_3^+$	soln	−	−	−1258
$LaCl_3$	gas	−695	+364	−687
	hex	−1071	+146	−998
$LaCl_3·7H_2O$	tricl	−3165	+463	−2699
LaF^{2+}	soln	−	−	−987
LaF_3	hex	−1732	+100	−1654
LaI_3	rhomb	−700	+213	−695
$La_2(MoO_4)_3$	mon	−4324	+390	−4014
La_2O_3	trig	−1793	+127	−1705
$La(OH)_3$	amor	−1410	+145	−1286
La_2S_3	rhomb	−1287	+134	−1264
$LaSO_4^+$	soln	−1602	−76	−1450
$La_2(SO_4)_3$	sld	−3932	+319	−3598
$La_2(SO_4)_3·9H_2O$	hex	−6655	−	−
Li	gas	+161	+139	+128
	lq	+2	+34	+1
	cub	0	+29	0
Li^+	gas	+687	+133	+650
	soln	−278	+11	−293
Li_2	gas	+211	+197	+170
$LiAlO_2$	hex	−1190	+53	−1128
$LiBO_2$	mon	−1018	+52	−962

Substance	State	$\Delta H°$ kJ/mole	$S°$ J/(K·mole)	$\Delta G°$ kJ/mole
$Li_2B_4O_7$	tetr	−3377	+158	−3186
LiBr	gas	−144	+225	−180
	cub	−350	+69	−339
$LiBr·2H_2O$	sld	−961	+113	−824
Li_2C_2	sld	−59	+59	−56
Li_2CO_3	mon	−1213	+90	−1129
LiCl	gas	−195	+213	−217
	cub	−408	+59	−384
$LiCl·H_2O$	cub	−712	+92	−628
$LiClO_4$	sld	−381	+126	−254
$LiClO_4·3H_2O$	hex	−1298	−	−
LiF	gas	−332	+200	−353
	cub	−616	+36	−588
LiH	gas	+135	+171	+112
	cub	−90	+20	−68
$LiHSO_4$	sld	−1143	+109	−1016
LiI	gas	−80	+232	−123
	cub	−271	+77	−268
Li_2MoO_4	hex	−1521	+115	−1407
LiN	gas	+163	+212	+137
LiN_3	sld	+4	+72	+77
Li_3N	hex	−198	+38	−155
$LiNH_2$	tetr	−180	−	−
$LiNO_2$	sld	−408	+84	−335
$LiNO_2·0.5H_2O$	sld	−674	−	−
$LiNO_3$	gas	−275	+279	−229
	trig	−488	+88	−385
$(LiNb)O_3$	trig	−	−	−1271
LiO	gas	+84	+211	+60
Li_2O	gas	−167	+229	−187
	cub	−599	+38	−562
$Li_2(O)_2$	gas	−225	−	−
Li_2O_2	hex	−636	+45	−571
LiOH	gas	−246	+218	−252
	tetr	−485	+43	−439
$LiOH·H_2O$	mon	−791	+71	−684
[LiOH]	soln	−509	−	−
Li_3PO_4	rhomb	−2092	+105	−1963
Li_2S	cub	−447	+63	−439
Li_2SO_4	mon	−1436	+114	−1321
$Li_2SO_4·H_2O$	mon	−1734	+146	−1559
Li_2SiO_3	rhomb	−1653	+76	−1561
Li_2WO_4	trig	−1604	+113	−1488
Lu	gas	+428	+185	+388
	hex	0	+51	0
Lu^{3+}	soln	−	−192	−653
$LuCl_3$	gas	−675	+359	−667

(Continued)

Substance	State	$\Delta H°$ kJ/mole	$S°$ J/(K·mole)	$\Delta G°$ kJ/mole
$LuCl_3$	mon	−950	+146	−879
LuF_3	gas	−1100	+321	−1090
	rhomb	−1701	+96	−1624
LuI_3	sld	−556	+213	−552
Lu_2O_3	cub	−1878	+110	−1789
Mg	gas	+147	+149	+112
	hex	0	+33	0
Mg^+	gas	+891	+154	+849
Mg^{2+}	soln	−462	−120	−455
$(MgAl_2)O_4$	cub	−2315	+81	−2190
Mg_3As_2	cub	−421	−	−
MgBr	gas	−39	−	−
$MgBr_2$	gas	−312	+296	−345
	trig	−517	+123	−499
$MgBr_2·6H_2O$	mon	−2408	+397	−2054
$MgCO_3$	trig	−1096	+66	−1012
$[MgCO_3]$	soln	−	−	−1002
$MgCO_3·5H_2O$	mon	−2567	+280	−2199
MgCl	gas	−44	+233	−70
$MgCl_2$	gas	−402	+271	−406
	trig	−641	+90	−592
$MgCl_2·6H_2O$	mon	−2500	+366	−2115
$Mg(ClO_4)_2$	sld	−587	+218	−331
$(MgCr_2)O_4$	cub	−1490	+119	−1379
MgF	gas	−222	+221	−248
MgF^+	soln	−776	−33	−746
MgF_2	gas	−724	+258	−731
	tetr	−1124	+57	−1071
$(MgFe_2)O_4$	α-cub	−1428	+124	−1317
MgI	gas	+53	−	−
MgI_2	gas	−170	+314	−219
	trig	−364	+134	−360
Mg_3N_2	cub	−461	+88	−401
$MgNH_4AsO_4·6H_2O$	rhomb	−3326	−	−
$MgNH_4(SO_4)_2·6H_2O$	mon	−4294	−	−
$Mg(NO_3)_2$	sld	−790	+164	−589
$Mg(NO_3)_2·2H_2O$	sld	−1409	−	−
$Mg(NO_3)_2·6H_2O$	mon	−2605	+453	−2072
MgO	gas	+4	+221	−21
	cub	−601	+27	−569
$MgOH^+$	soln	−	−	−628
$Mg(OH)_2$	trig	−925	+63	−834
Mg_3P_2	cub	−466	−	−
$Mg_2P_2O_7$	mon	−	+155	−
$Mg_3(PO_4)_2$	sld	−3790	+189	−3548
MgS	gas	+139	+240	+87
	cub	−347	+46	−341

Substance	State	$\Delta H°$ kJ/mole	$S°$ J/(K·mole)	$\Delta G°$ kJ/mole
$MgSO_3$	sld	−1013	+92	−929
$MgSO_3·6H_2O$	trig	−2824	+351	−2400
$MgSO_4$	rhomb	−1280	+92	−1166
$MgSO_4·H_2O$	mon	−1610	+126	−1436
$MgSO_4·7H_2O$	rhomb	−3388	+370	−2870
$[MgSO_4]$	soln	−1369	−52	−1212
Mg_3Sb_2	trig	−313	−	−
Mg_2Si	cub	−78	+82	−77
Mg_2SiO_4	rhomb	−2172	+95	−2053
$MgWO_4$	mon	−1535	+101	−1423
Mn	gas	+283	+174	+241
	α-cub	0	+32	0
	β-cub	+2	+35	+1
	γ-cub	−	+32	−
Mn^{2+}	soln	−220	−67	−230
Mn^{3+}	soln	−100	−213	−85
$(MnAl_2)O_4$	cub	−2098	−	−
MnAs	hex	−57	+60	−55
MnBr	gas	+84	−	−
$MnBr_2$	gas	−175	+318	−215
	trig	−379	+140	−366
$MnBr_2·4H_2O$	mon	−1539	+293	−1293
Mn_3C	rhomb	−17	+99	−16
$MnCO_3$	trig	−894	+86	−817
$[Mn(CO)_5]$	gas	−751	−	−
$[Mn_2(CO)_{10}]$	gas	−1610	−	−
$MnCl_2$	gas	−264	+295	−276
	trig	−481	+118	−440
$MnCl_2·4H_2O$	mon	−1686	+317	−1426
MnF_2	gas	−532	+270	−542
	tetr	−847	+93	−805
$(MnFe_2)O_4$	cub	−1226	−	−
MnI_2	trig	−263	+153	−264
$(Mn^{II}Mn_2^{III})O_4$	tetr	−1385	+149	−1279
$Mn(NH_4)_2(SO_4)_2·6H_2O$	mon	−4045	−	−
$Mn(NO_3)_2$	sld	−703	+167	−503
$Mn(NO_3)_2·6H_2O$	mon	−2372	−	−
MnO	cub	−385	+60	−363
MnO_2	tetr	−520	+53	−465
MnO_4^{2-}	soln	−	+191	−449
MnO_4^-	soln	−533	+196	−440
Mn_2O_3	cub	−958	+110	−880
$Mn_2O_3·3H_2O^{(1)}$	rhomb	−1775	+201	−1515
Mn_2O_7	lq	−726	+171	−544
$MnOH^+$	soln	−446	+3	−407
$Mn(OH)_2$	trig	−702	+82	−617
$MnO(OH)$	mon	−577	−	−

(Continued)

Substance	State	$\Delta H°$ kJ/mole	$S°$ J/(K·mole)	$\Delta G°$ kJ/mole
$Mn_3(PO_4)_2$	sld	−3066	+300	−2858
MnS	α-cub	−214	+78	−218
$Mn(S_2)$	cub	−245	+54	−232
$MnSO_4$	rhomb	−1066	+112	−958
$MnSO_4·H_2O$	mon	−1377	+156	−1213
$MnSO_4·7H_2O$	mon	−3139	−	−
$[MnSO_4]$	soln	−1127	+3	−987
$Mn_2(SO_4)_3$	amor	−2768	+259	−2431
MnSe	cub	−149	+91	−154
$MnSiO_3$	tricl	−1320	+89	−1240
Mn_2SiO_4	rhomb	−1730	+163	−1632
$MnWO_4$	mon	−1305	+135	−1204
Mo	gas	+659	+182	+613
	cub	0	+29	0
$MoBr_2$	sld	−121	+142	−109
$MoBr_3$	sld	−170	+163	−142
$MoBr_4$	sld	−189	+218	−155
$[Mo(CO)_6]$	gas	−914	+494	−859
	rhomb	−983	+327	−878
$MoCl_2$	amor	−288	+113	−247
$MoCl_3$	mon	−393	+138	−325
$MoCl_4$	gas	−385	+372	−354
	hex	−474	+201	−392
$MoCl_5$	gas	−446	+395	−389
	mon	−528	+225	−420
$MoCl_2O_2$	gas	−635	−	−
	sld	−717	−	−
MoF_6	gas	−1559	+350	−1473
	lq	−1586	+260	−1473
MoI_2	sld	−47	+167	−54
MoI_3	sld	−61	+209	−63
MoI_4	sld	−82	+268	−84
MoO_2	gas	−13	−	−
	mon	−590	+46	−534
MoO_3	gas	−363	+279	−346
	rhomb	−745	+78	−668
$MoO_3·H_2O$	hex	−1072	+159	−950
MoO_4^{2-}	soln	−998	+33	−838
$MoOF_4$	gas	−1255	−	−
	mon	−1310	−	−
MoO_2F_2	gas	−1121	−	−
$[MoO_2(OH)_2]$	soln	−	−	−877
MoS_2	hex	−234	+66	−226
N	gas	+473	+153	+456
N^+	gas	+1883	+162	+1857
N_2	gas	0	+192	0
$[N_2]$	soln	−11	+95	+18

Substance	State	$\Delta H°$ kJ/mole	$S°$ J/(K·mole)	$\Delta G°$ kJ/mole
N_2^+	gas	+1509	+197	+1501
N_3^-	soln	−	+108	+348
NCS^-	soln	+74	+146	+90
NF	gas	+257	+214	+252
NF_2	gas	+40	+250	+55
NF_3	gas	−133	+261	−91
N_2F_4	gas	−5	+313	+80
NH	gas	+340	+181	+334
NH_2	gas	+174	+195	+184
NH_3	gas	−46	+192	−16
	lq	−70	−	−
$[NH_3]$	soln	−	+111	−27
$NH_3·H_2O$	lq	−361	+166	−254
$[NH_3·H_2O]$	soln	−366	+179	−263
$NH_3·2H_2O$	lq	−650	+233	−493
$2NH_3·H_2O$	lq	−431	+268	−267
NH_4^+	gas	+664	+186	+709
	soln	−133	+111	−79
N_2H_4	gas	+95	+238	+159
	lq	+50	+122	+149
$[N_2H_4]^{(10)}$	soln	−	+138	+128
$N_2H_4·H_2O$	gas	−205	+264	−79
	lq	−243	−	−
NH_4Br	cub	−271	+113	−175
NH_4CH_3COO	sld	−615	−	−
NH_4Cl	α-cub	−315	+95	−203
$N_2H_6Cl_2$	cub	−364	−	−
NH_4ClO_4	rhomb	−295	+184	−88
$(NH_4)_2CrO_4$	mon	−1287	+173	−996
$(NH_4)_2Cr_2O_7$	mon	−1808	−	−
NH_4F	hex	−467	+72	−352
$NH_4(HCO_3)$	mon	−850	+121	−666
$NH_4(HF_2)$	rhomb	−807	+116	−655
$NH_4H_2PO_4$	tetr	−1446	+152	−1211
$(NH_4)_2HPO_4$	mon	−1566	−	−
NH_4HS	tetr	−158	+98	−51
$NH_4(HSO_3)$	rhomb	−769	−	−
$NH_4(HSO_4)$	rhomb	−1027	−	−
NH_4I	cub	−202	+117	−113
NH_4IO_3	rhomb	−384	−	−
NH_4N_3	rhomb	+115	+113	+274
NH_4NCS	mon	−58	+261	+11
NH_4NO_2	sld	−238	+254	−117
NH_4NO_3	rhomb	−366	+151	−184
$(NH)O_2$	gas	−78	+249	−43
NH_2OH	gas	−51	+236	−4
	rhomb	−115	+67	−17

(*Continued*)

Substance	State	$\Delta H°$ kJ/mole	$S°$ J/(K·mole)	$\Delta G°$ kJ/mole
[NH$_2$OH][11]	soln	−98	+167	−23
(NH$_4$)$_2$SO$_3$·H$_2$O	mon	−1185	−	−
(NH$_4$)$_2$SO$_4$	rhomb	−1181	+220	−901
(NH$_4$)$_2$S$_2$O$_6$(O$_2$)	mon	−1648	−	−
(NH$_4$)$_2$SeO$_4$	mon	−874	−	−
NH$_4$VO$_3$	rhomb	−1051	+141	−886
NO	gas	+91	+211	+87
NO$^+$	gas	+990	+198	+984
NO$_2^-$	gas	−360	+236	−334
	soln	−105	+140	−4
NO$_2$	gas	+33	+240	+51
NO$_3^-$	soln	−207	+147	−111
NO$_3$	gas	+71	+253	+116
N$_2$O	gas	+82	+220	+104
N$_2$O$_2^{2-}$	soln	−	+28	+146
N$_2$O$_3$	gas	+83	+312	+139
N$_2$O$_4$	gas	+9	+304	+98
	lq	−19	+209	+98
N$_2$O$_5$	gas	+11	+356	+115
	hex[12]	−43	+178	+114
(NO)Br	gas	+82	+274	+82
	lq	+52	−	−
(NO)Cl	gas	+53	+262	+67
	lq	+28	−	−
(NO$_2$)Cl	gas	+12	+272	+54
	lq	+21	+272	+63
(NO)F	gas	−66	+248	−51
(NO$_2$)F	gas	−108	+260	−66
(NO$_2$)OF	gas	−26	−	−
Na	gas	+108	+154	+77
	lq	+13	+58	+11
	cub	0	+51	0
Na$^+$	gas	+611	+148	+576
	soln	−240	+59	−262
Na$_2$	gas	+138	+230	+100
NaAlO$_2$	hex	−1133	+71	−1070
NaAl(SO$_4$)$_2$·12H$_2$O	cub	−5983	−	−
Na$_3$AsO$_4$	rhomb	−1535	+186	−1412
NaBO$_2$	trig	−976	+74	−920
Na$_2$B$_4$O$_7$	tricl	−3291	+190	−3096
NaBr	gas	−151	+241	−185
	cub	−361	+87	−349
NaBr·2H$_2$O	mon	−953	+176	−828
NaBrO$_3$	cub	−343	+131	−253
Na$_2$C$_2$	tetr	−19	+111	−18
NaCH$_3$COO	mon	−709	+124	−608
NaCH$_3$COO·3H$_2$O	mon	−1603	+262	−1334

(Continued)

Substance	State	$\Delta H°$ kJ/mole	$S°$ J/(K·mole)	$\Delta G°$ kJ/mole
NaCN	gas	+94	+243	+67
	cub	−90	+118	−80
Na_2CO_3	mon	−1132	+135	−1048
$Na_2CO_3 \cdot H_2O$	rhomb	−1432	+168	−1289
$Na_2CO_3 \cdot 10H_2O$	mon	−4083	+565	−3432
NaCl	gas	−181	+230	−201
	cub	−411	+72	−384
$NaClO_2$	sld	−306	+113	−230
$NaClO_3$	cub	−338	+223	−264
$NaClO_4$	rhomb	−382	+142	−254
$NaClO_4 \cdot H_2O$	mon	−678	−	−
Na_2CrO_4	rhomb	−1336	+180	−1230
$Na_2CrO_4 \cdot 10H_2O$	mon	−4277	−	−
$Na_2Cr_2O_7$	sld	−1967	+277	−1791
NaF	gas	−294	+217	−313
	cub	−577	+51	−547
$NaFeO_2$	trig	−698	+88	−640
NaH	gas	+124	+188	+103
	cub	−61	+40	−38
$NaHCO_3$	mon	−914	+102	−816
$Na(HF_2)$	trig	−922	+91	−854
NaH_2PO_4	sld	−1545	+127	−1394
Na_2HPO_4	sld	−1755	+150	−1615
$Na_2HPO_4 \cdot 2H_2O$	rhomb	−2346	+221	−2088
$Na_2HPO_4 \cdot 7H_2O$	mon	−3821	+434	−3279
$Na_2HPO_4 \cdot 12H_2O$	mon	−5298	+636	−4468
$Na_2H_2P_2O_7$	mon	−2764	+220	−2522
NaHS	trig	−236	−	−
$NaHSO_4$	tricl	−1132	+125	−1003
$NaHSO_4 \cdot H_2O$	mon	−1420	+170	−1235
NaI	gas	−76	+249	−118
	cub	−290	+99	−287
$NaI \cdot 2H_2O$	tricl	−886	+174	−766
$NaIO_3$	rhomb	−480	+135	−396
$NaMnO_4$	sld	−1682	+160	−1583
Na_2MoO_4	cub	−1454	+159	−1355
NaN_3	trig	+20	+71	+100
NaNCS	rhomb	−176	+113	−155
$NaNH_2$	rhomb	−124	+77	−64
$NaNO_2$	rhomb	−359	+105	−285
$NaNO_3$	gas	−257	+300	−211
	trig	−469	+117	−368
$(NaNb)O_3$	rhomb	−1325	+117	−1242
NaO_2	cub	−260	+116	−218
Na_2O	gas	−41	+272	−61
	α-cub	−418	+75	−379
Na_2O_2	hex	−510	+95	−447

Substance	State	$\Delta H°$ kJ/mole	$S°$ J/(K·mole)	$\Delta G°$ kJ/mole
NaOCN	trig	−120	+125	−81
NaOH	gas	−211	+236	−216
	rhomb	−425	+64	−379
[NaOH]	soln	−470	+48	−419
NaPO₃	rhomb	−1220	+96	−1130
Na₃PO₄	tetr	−1917	+174	−1789
Na₃PO₄·12H₂O	trig	−5480	+660	−4662
Na₄P₂O₇	rhomb	−3166	+270	−2947
Na₄P₂O₇·10H₂O	mon	−6125	+689	−5335
Na₂S	cub	−372	+77	−355
Na₂(S₄)	tetr	−411	−	−
Na₂SO₃	trig	−1090	+146	−1002
Na₂SO₃·7H₂O	mon	−3153	+450	−2668
Na₂SO₄	gas	−1012	+345	−953
	α-rhomb	−1388	+149	−1270
Na₂SO₄·10H₂O	mon	−4330	+586	−3647
Na₂S₂O₇	sld	−1938	−	−
Ma₂SO₃S	mon	−1117	+225	−1043
Ma₂SO₃S·5H₂O	mon	−2602	+314	−2207
Na₃Sb	hex	−227	−	−
Na₂Se	cub	−343	+105	−331
Na₂SeO₃	sld	−961	−	−
Na₂SeO₄	rhomb	−1101	+117	−971
Na₂SeO₄·10H₂O	mon	−4056	−	−
Na₂SiO₃	rhomb	−1588	+114	−1464
Na₄SiO₄	mon	−2106	+196	−1976
Na₂Te	cub	−353	+132	−347
NaVO₃	mon	−1147	+114	−1065
Na₃VO₄	cub	−1636	+190	−1516
Na₄V₂O₇	hex	−2919	+318	−2722
Na₂WO₄	cub	−1547	+161	−1433
Nb	gas	+722	+186	+677
	cub	0	+36	0
NbBr₅	mon	−556	+305	−523
NbCl₅	gas	−712	+401	−655
	mon	−797	+240	−692
NbF₅	mon	−1813	+160	−1699
NbO	gas	+199	+239	+169
	cub	−415	+48	−388
NbO₂	gas	−212	−	−
	tetr	−807	+55	−752
NbO₃⁻	soln	−1020	+151	−943
Nb₂O₅	mon	−1897	+137	−1764
Nd	gas	+327	+189	+292
	hex	0	+73	0
Nd³⁺	soln	−696	−210	−670
Nd(BrO₃)₂·9H₂O	hex	−3583	−	−

(Continued)

Substance	State	$\Delta H°$ kJ/mole	$S°$ J/(K·mole)	$\Delta G°$ kJ/mole
NdCl$_3$	gas	−721	+378	−712
	hex	−1041	+146	−963
NdCl$_3$·6H$_2$O	mon	−2863	+385	−2439
NdF$_3$	gas	−1293	+346	−1284
	hex	−1713	+121	−1637
NdI$_3$	rhomb	−671	+213	−661
Nd$_2$O$_3$	cub	−1808	+158	−1720
NdOH^{2+}	soln	−	−	−859
Nd(OH)$_3$	hex	−1413	+111	−1274
Nd$_2$S$_3$	rhomb	−1191	+134	−1159
Nd$_2$(SO$_4$)$_3$	sld	−3981	+289	−3628
Nd$_2$(SO$_4$)$_3$·8H$_2$O	mon	−6382	−	−
Ne	gas	0	+146	0
[Ne]	soln	−5	+67	+19
Ne$^+$	gas	+2087	+158	+2077
Ni	gas	+424	+182	+379
	cub	0	+30	0
Ni^{2+}	soln	−54	−129	−45
(NiAl$_2$)O$_4$	cub	−1939	+92	−1819
Ni$_3$(AsO$_4$)$_2$·8H$_2$O	mon	−	−	−3482
NiBr	gas	+181	−	−
NiBr$_2$	gas	+20	+317	−20
	trig	−220	+130	−205
Ni$_3$C	hex	+36	+106	+33
NiCO$_3$	rhomb	−689	+86	−612
[Ni(CO)$_4$]	gas	−586	+399	−567
	lq	−630	+313	−585
NiCl	gas	+178	−	−
NiCl$_2$	gas	−66	+294	−78
	trig	−305	+98	−259
NiCl$_2$·6H$_2$O	mon	−2109	+344	−1719
NiF	gas	+74	−	−
NiF$_2$	gas	−326	+269	−337
	tetr	−657	+74	−610
NiI$_2$	trig	−90	+141	−89
[Ni(NH$_3$)$_4$]$^{2+}$	soln	−	+238	−195
[Ni(NH$_3$)$_6$]$^{2+}$	soln	−	+356	−253
[Ni(NH$_3$)$_6$]Cl$_2$	cub	−997	−	−
[Ni(NH$_3$)$_6$]I$_2$	cub	−789	−	−
Ni(NO$_3$)$_2$	sld	−430	+192	−238
Ni(NO$_3$)$_2$·6H$_2$O	tricl	−2215	+511	−1700
NiO	gas	+248	+238	+217
	cub	−240	+38	−212
Ni$_2$O$_3$·3H$_2$O$^{(1)}$	amor	−1341	+192	−1080
NiOH$^+$	soln	−	−	−220
Ni(OH)$_2$	gas	−254	−	−
	trig	−544	+80	−459

(Continued)

Substance	State	$\Delta H°$ kJ/mole	$S°$ J/(K·mole)	$\Delta G°$ kJ/mole
[Ni(OH)$_2$]	soln	–	–	–418
NiS	trig	–83	+53	–80
Ni$_3$S$_2$	trig	–203	+134	–197
NiSO$_4$	rhomb	–873	+104	–763
NiSO$_4$·7H$_2$O	rhomb	–2977	+379	–2462
[NiSO$_4$]	soln	–961	–61	–802
NiSb	hex	–66	+73	–65
Np	gas	+464	+198	+420
	rhomb	0	+50	0
NpBr$_3$	hex	–725	+205	–703
NpBr$_4$	sld	–765	+243	–732
NpCl$_3$	hex	–904	+159	–837
NpCl$_4$	tetr	–992	+201	–904
NpF$_3$	hex	–1500	+109	–1427
NpF$_4$	mon	–1780	+151	–1690
NpI$_3$	rhomb	–499	+234	–502
NpO$_2$	cub	–1030	+83	–979
O$^-$	gas	+107	+158	+97
O	gas	+249	+161	+232
O$^+$	gas	+1569	+155	+1547
O$_2$	gas	0	+205	0
[O$_2$]	soln	–12	+111	+16
O$_2^+$	gas	+1172	+205	+1166
O$_3$	gas	+143	+239	+163
OCN$^-$	soln	–140	+107	–97
OD	gas	+36	+190	+32
OF	gas	+140	+214	+110
OF$_2$	gas	–22	+247	–5
OH$^-$	gas	–134	+171	–129
	soln	–230	–11	–157
OH	gas	+39	+184	+34
OH$^+$	gas	+1318	+183	+1307
OT	gas	+35	+193	+31
Os	gas	+790	+192	+743
	hex	0	+33	0
OsCl$_3$	cub	–191	–	–
OsCl$_4$	sld	–255	+155	–158
OsO$_4$	gas	–336	+294	–292
	mon	–391	+134	–299
Os(S$_2$)	cub	–145	+55	–132
P	amor	–17	+23	–12
	gas[13]	+316	+163	+280
	rhomb	–38	+23	–33
P$^+$	gas	+1337	+167	+1293
P$_2$	gas	+144	+218	+103
P$_4$	gas	+60	+280	+25
	cub	0	+164	0

Substance	State	$\Delta H°$ kJ/mole	$S°$ J/(K·mole)	$\Delta G°$ kJ/mole
PBr_3	gas	−139	+348	−163
	lq	−185	+240	−176
PBr_5	rhomb	−229	−	−
PBr_3O	gas	−390	+360	−387
	sld	−458	−	−
PCl	gas	+150	+237	+125
PCl_2	gas	−44	−	−
PCl_3	gas	−287	+312	−268
	lq	−319	+217	−272
PCl_5	gas	−375	+364	−305
	tetr	−445	+171	−318
PCl_3O	gas	−560	+325	−514
	lq	−597	+222	−521
PF	gas	−47	+225	−72
PF_2	gas	−431	−	−
PF_3	gas	−919	+273	−897
PF_5	gas	−1593	+293	−1517
PH	gas	+249	+196	+222
PH_2	gas	+126	−	−
PH_3	gas	+5	+210	+13
P_2H_4	gas	+21	−	−
PH_4Br	cub	−128	+110	−48
PH_4I	tetr	−69	+123	+2
$(PHO_3)^{2-}$	soln	−969	+17	−812
PI_3	hex	−46	−	−
P_2I_4	tricl	−83	−	−
PN	gas	+99	+211	+77
PO	gas	−27	+223	−51
PO_4^{3-}	soln	−1277	−220	−1019
$P_2O_7^{4-}$	soln	−	−105	−1924
P_4O_6	gas	+1594	+347	−1465
	lq	−2194	+284	−2046
	mon	−1640	−	−
P_4O_{10}	gas	−3199	+395	−2657
	rhomb	−3041	−	−
	trig	−2984	+229	−2698
POF_3	gas	−1198	+285	−1149
PS	gas	+125	+235	+77
P_4S_{10}	sld	−364	+382	−334
$PSBr_3$	gas	−256	+373	−277
$PSCl_3$	gas	−381	+337	−360
	lq	−331	−	−
PSF_3	gas	−1063	+299	−1040
Pa	tetr	0	+52	0
Pb	gas	+195	+175	+162
	cub	0	+65	0
Pb^+	gas	+917	+181	+876

(*Continued*)

Substance	State	$\Delta H°$ kJ/mole	$S°$ J/(K·mole)	$\Delta G°$ kJ/mole
Pb^{2+}	soln	−1	+11	−24
Pb_2	gas	+251	+272	+209
$Pb_3(AsO_4)_2$	sld	−	−	−1579
$Pb(BO_2)_2$	sld	−1550	+131	−1444
PbBr	gas	+49	+272	+10
$PbBr_2$	gas	−106	+338	−142
	rhomb	−278	+162	−262
$Pb(BrO_3)_2$	sld	−	−	−11
$Pb(CH_3COO)_2$	sld	−961	−	−
$Pb(CH_3COO)_2·3H_2O$	mon	−1849	−	−
$PbCO_3$	rhomb	−703	+131	−629
PbC_2O_4	sld	−852	+146	−750
$[Pb(CO_3)_2]^{2-}$	soln	−	−	−1127
PbCl	gas	+20	+259	−4
$PbCl^+$	soln	−	−	−165
$PbCl_2$	gas	−174	+316	−182
	rhomb	−359	+136	−314
$[PbCl_2]$	soln	−	−	−301
$PbCl_4$	gas	−314	+385	−276
	lq	−329	−	−
$Pb(Cl)F$	tetr	−534	+115	−485
$PbCl(OH)$	rhomb	−	−	−480
$PbCrO_4$	mon	−915	+169	−817
PbF	gas	−84	+249	−109
PbF^+	soln	−	−	−312
PbF_2	gas	−438	+293	−445
	rhomb	−679	+96	−628
$[PbF_2]$	soln	−	−	−597
$[PbF_3]^-$	soln	−	−	−884
PbF_4	gas	−778	+334	−737
	sld	−840	+151	−745
$Pb(HCOO)_2$	rhomb	−879	−	−
$[Pb(HS)_2]$	soln	−	−	−87
$[Pb(HS)_3]^-$	soln	−	−	−83
PbI	gas	+133	+280	+86
PbI_2	gas	−8	+354	−60
	trig	−178	+175	−176
$Pb(IO_3)_2$	sld	−489	+323	−348
$PbMoO_4$	tetr	−1049	+166	−948
$Pb(N_3)_2$	rhomb	+382	+150	+528
$Pb(NO_3)_2$	cub	−447	+213	−251
PbO	gas	+48	+240	+26
	rhomb (β)	−217	+69	−188
	tetr (α)	−219	+66	−189
PbO_2	tetr	−277	+72	−218
$PbOH^+$	soln	−	−	−226
$Pb(OH)_2$	hex	−545	+88	−452

Substance	State	$\Delta H°$ kJ/mole	$S°$ J/(K·mole)	$\Delta G°$ kJ/mole
[Pb(OH)$_2$]	soln	−	−	−401
[Pb(OH)$_3$]$^-$	soln	−	−	−572
[Pb(OH)$_4$]$^{2-}$	soln	−	0	−751
Pb$_3$(PO$_4$)$_2$	hex	−2655	+353	−2433
(Pb$_2^{II}$Pb^{IV})O$_4$	tetr	−723	+211	−606
PbS	gas	+122	+252	+76
	cub	−101	+91	−99
PbSO$_4$	rhomb	−920	+149	−813
[PbSO$_4$]	soln	−	−	−783
PbSe	gas	+117	+263	+71
	cub	−103	+103	−102
PbSeO$_4$	rhomb	−617	+142	−505
PbSiO$_3$	mon	−1146	+110	−1062
PbTe	gas	+166	+271	+120
	cub	−68	+118	−69
PbWO$_4$	tetr	−1121	+168	−1020
Pd	gas	+372	+167	+334
	cub	0	+38	0
Pd^{2+}	soln	−	−	+177
PdBr$_2$	gas	+109	−	−
	mon	−107	+140	−92
PdCl$^+$	soln	−	−	+11
PdCl$_2$	gas	+128	−	−
	rhomb	−163	−	−
[PdCl$_2$]	soln	−	−	−146
[PdCl$_3$]$^-$	soln	−	−	−292
[PdCl$_4$]$^{2-}$	soln	−	+297	−428
[PdCl$_6$]$^{2-}$	soln	−	+231	−441
PdI$_2$	mon	−63	−	−
[Pd(NH$_3$)$_2$Cl$_2$]	tetr	−429	−	−
PdO	tetr	−119	+36	−88
PdO$_2$·2H$_2$O$^{(9)}$	amor	−676	+113	−498
PdOH$^+$	soln	−	−	−53
Pd(OH)$_2$	amor	−386	+92	−302
[Pd(OH)$_2$]	soln	−	−	−287
PdS	tetr	−77	+56	−73
Pm	hex	0	+72	0
Pm^{3+}	soln	−	−172	−703
PmCl$_3$	sld	−1058	+140	−979
Po	gas	+147	+189	+109
	cub	0	+63	0
Po$^{II(14)}$	soln	−	−	+73
Po$_2$	gas	+145	+284	+98
PoO$_2$	cub	−251	+71	−192
Pr	gas	+356	+190	+321
	hex	0	+74	0
Pr^{3+}	soln	−665	−172	−713

Substance	State	$\Delta H°$ kJ/mole	$S°$ J/(K·mole)	$\Delta G°$ kJ/mole
$PrCl_3$	gas	−730	+379	−721
	hex	−1060	+144	−981
$PrCl_3·7H_2O$	tricl	−3184	+420	−2700
PrF_3	hex	−1762	+96	−1678
PrF_4	sld	−2003	+134	−1900
PrI_3	sld	−684	+213	−674
PrO_2	cub	−1044	+67	−920
Pr_2O_3	trig	−1833	+130	−1736
Pr_6O_{11}	α-cub	−5713	+488	−5390
$Pr(OH)_3$	hex	−1441	+96	−1297
Pt	gas	+566	+192	+521
	cub	0	+42	0
$PtBr_2$	cub	−100	+53	−58
$PtBr_4$	rhomb	−159	+164	−105
$PtCl_2$	rhomb	−106	+220	−93
$[PtCl_2]$	soln	−	−	−80
$[PtCl_3]^-$	soln	−	−	−228
$[PtCl_4]^{2-}$	soln	−503	+167	−369
$PtCl_4$	cub	−230	+268	−164
$[PtCl_6]^{2-}$	soln	−674	+220	−490
$[PtCl_4],K_2$	tetr	−1059	+180	−928
$[PtCl_6],K_2$	cub	−1252	+334	−1101
$[PtCl_4],(NH_4)_2$	tetr	−803	−	−
$[PtCl_6],(NH_4)_2$	cub	−984	−	−
PtI_2	sld	−63	−	−
PtI_4	amor	−57	+281	−
PtO	gas	+456	−	−
PtO_2	gas	+171	−	−
	hex	−88	+281	−98
$Pt(OH)_2$	amor	−352	+100	−269
$[Pt(OH)_4]^{2-}$	soln	−	−	−573
$(Pt^{II}Pt^{IV})Br_6$	sld	−261	+223	−167
$(Pt^{II}Pt^{IV})Cl_6$	hex	−336	+247	−185
$(Pt^{II}Pt^{IV})I_6$	sld	−168	−	−
PtS	tetr	−82	+55	−76
PtS_2	trig	−109	+75	−100
Pu	gas	+346	+177	+310
	α-mon	0	+55	0
Pu^+	gas	−	+183	−
Pu^{3+}	soln	−	−129	−590
Pu^{4+}	soln	−	−364	−
$PuBr_3$	rhomb	−740	+191	−713
$PuCl_3$	gas	−637	−	−
	hex	−961	+159	−892
PuF_3	hex	−1552	+113	−1479
PuF_4	gas	−1518	+334	−1480
	mon	−1734	+162	−1645

Substance	State	$\Delta H°$ kJ/mole	$S°$ J/(K·mole)	$\Delta G°$ kJ/mole
PuF_6	gas	−1705	+369	−1617
	sld	−1753	+223	−1622
PuI_3	rhomb	−560	+234	−561
PuO_2	gas	−473	−	−
		(1875 K)		
	cub	−1058	+82	−1005
Ra	gas	+162	+176	+131
	cub	0	+71	0
Ra^{2+}	soln	−530	+29	−556
$RaBr_2$	rhomb	−782	+155	−762
$RaCl_2$	rhomb	−869	+134	−821
$RaCl_2·2H_2O$	sld	−1468	+209	−1304
RaO	sld	−523	+75	−494
$RaSO_4$	rhomb	−1473	+146	−1364
Rb	gas	+82	+170	+54
	cub	0	+77	0
Rb^+	gas	+491	+164	+459
	soln	−251	+121	−284
Rb_2	gas	+115	+271	+80
RbBr	gas	−194	+261	−226
	cub	−394	+108	−381
$RbBrO_3$	hex	−364	+159	−454
Rb_2CO_3	mon	−1134	+172	−1046
RbCl	gas	−227	+250	−186
	cub	−436	+96	−408
$RbClO_3$	sld	−404	+152	−301
$RbClO_4$	rhomb	−437	+167	−308
Rb_2CrO_4	rhomb	−1411	+218	−1301
RbF	gas	−345	+237	−362
	cub	−554	+75	−523
RbH	cub	−59	+59	−34
$RbHCO_3$	rhomb	−958	+121	−858
$Rb(HF_2)$	sld	−920	−	−
RbI	gas	−149	+269	−109
	cub	−334	+118	−329
RbN_3	tetr	+4	+120	+77
$RbNH_2$	sld	−110	−	−
$RbNO_3$	trig	−490	+142	−389
RbO_2	tetr	−263	+59	−197
Rb_2O	cub	−337	+109	−293
Rb_2O_2	rhomb	−427	+105	−351
RbOH	gas	−236	+249	−237
	rhomb	−413	+79	−364
Rb_2S	cub	−354	+134	−339
Rb_2SO_4	gas	−1070	−	−
	rhomb	−1436	+197	−1317
Rb_2SeO_4	rhomb	−1124	+200	−1003

(Continued)

Substance	State	$\Delta H°$ kJ/mole	$S°$ J/(K·mole)	$\Delta G°$ kJ/mole
Re	gas	+778	+189	+733
	hex	0	+37	0
ReBr$_3$	sld	−165	+184	−141
[Re$_2$(CO)$_{10}$]	gas	−1583	−	−
	cub	−1666	+512	−1473
ReCl$_5$	mon	−361	+230	−252
[ReCl$_6$]$^{2-}$	soln	−763	+254	−628
[Re$_3$Cl$_9$]	hex	−789	+372	−567
[ReCl$_6$],K$_2$	cub	−1312	+372	−1174
ReF$_6$	gas	−1353	+364	−1269
	lq	−1382	+271	−1270
ReO$_2$	cub	−400	+65	−347
ReO$_3$	cub	−611	+81	−532
ReO$_4^-$	soln	−788	+204	−696
Re$_2$O$_7$	gas	−1115	−	−
	rhomb	−1263	+207	−1089
ReS$_2$	hex	−179	+96	−178
Re$_2$S$_7$	sld	−451	+204	−423
Rh	gas	+557	+186	+511
	cub	0	+32	0
RhCl$_3$	gas	+67	−	−
	sld	−229	+159	−167
[RhCl$_6$]$^{3-}$	soln	−	+209	−661
Rh$_2$O$_3$	trig	−356	+106	−277
Rh$_2$S$_3$	rhomb	−243	−	−
Rn	gas	0	+168	0
Ru	gas	+657	+186	+610
	hex	0	+29	0
RuCl$_3$	gas	+69	−	−
	hex	−266	+128	−196
RuF$_5$	gas	−744	−	−
	sld	−893	+161	−781
RuO$_2$	tetr	−306	+57	−253
RuO$_4$	gas	−184	+291	−140
	lq	−228	+177	−150
	mon	−240	+141	−151
Ru(S$_2$)	cub	−226	+48	−213
S^{2-}	soln	+33	−15	+86
S	gas	+279	+168	+238
	mon[15] (β)	ca. 0	+33	ca. 0
	rhomb[15] (α)	0	+32	0
S$^+$	gas	+1279	+164	+1233
S$_2^{2-}$	soln	+30	+28	+79
S$_2$	gas	+128	+228	+79
S$_3^{2-}$	soln	+26	+66	+74
S$_4^{2-}$	soln	+23	+103	+69
S$_5^{2-}$	soln	+21	+141	+66

(*Continued*)

Substance	State	$\Delta H°$ kJ/mole	$S°$ J/(K·mole)	$\Delta G°$ kJ/mole
S_6	gas	+100	+354	+52
S_8	gas	+102	+431	+50
S_2Br_2	lq	−15	−	−
SCl	gas	+126	+239	+98
SCl_2	gas	−20	+282	−28
	lq	−49	−	−
S_2Cl_2	gas	−19	+331	−32
	lq	−61	+167	−25
SCl_4	lq	−56	−	−
SCl_2O	gas	−213	+308	−198
	lq	−246	+279	−223
SCl_2O_2	gas	−364	+312	−320
	lq	−378	+216	−305
$S(Cl)O_2F$	gas	−564	−	−
SF	gas	+76	+228	+48
SF_2	gas	−137	+257	−144
SF_4	gas	−770	+290	−726
SF_6	gas	−1221	+291	−1117
S_2F_2[16]	gas	−228	+290	−235
SN	gas	+234	+222	+206
S_4N_4	mon	+460	−	−
SO	gas	+6	+222	−20
SO_2	gas	−297	+248	−300
	rhomb	−331	−	−
$[SO_2]$	soln	−323	+162	−301
$[SO_2 \cdot H_2O]$[17]	soln	−	+232	−538
SO_3^{2-}	soln	−636	−29	−487
SO_3	gas	−396	+257	−371
	lq	−468	+122	−373
	mon (α)	−455	+52	−369
SO_4^{2-}	soln	−909	+18	−744
S_2O	gas	−56	+267	−86
$S_2O_4^{2-}$	soln	−	+105	−598
$S_2O_5^{2-}$	soln	−	+105	−791
$S_2O_6^{2-}$	soln	−	+126	−967
$S_3O_6^{2-}$	soln	−	+138	−958
$S_4O_6^{2-}$	soln	−	+151	−1031
$S_5O_6^{2-}$	soln	−	+167	−958
SOF_2	gas	−565	+279	−548
SO_2F_2	gas	−858	+288	−813
$S_2O_6(O_2)^{2-}$	soln	−1339	+249	−1113
SO_3S^{2-}	soln	−652	+37	−514
Sb	gas	+268	+180	+228
	trig	0	+46	0
Sb^+	gas	+1109	+169	+1066
Sb_2	gas	+239	+257	+190
Sb_4	gas	+191	+351	+141

(Continued)

Substance	State	$\Delta H°$ kJ/mole	$S°$ J/(K·mole)	$\Delta G°$ kJ/mole
$SbBr_3$	gas	−194	+373	−224
	rhomb	−259	+207	−239
$SbCl_3$	gas	−314	+338	−301
	rhomb	−383	+184	−324
$[SbCl_4]^-$	soln	−	−	−476
$SbCl_5$	gas	−394	+402	−334
	lq	−440	+301	−350
$Sb(Cl)O$	mon	−374	−	−
SbF_3	rhomb	−907	+105	−834
SbH_3	gas	+145	+233	+148
SbI_3	gas	−8	+400	−62
	trig	−98	+213	−96
Sb_2O_3	cub (β)	−721	+132	−641
	rhomb (α)	−709	+141	−632
$Sb_2O_3 \cdot 3H_2O^{(1)}$	amor	−1543	+216	−1280
Sb_2O_5	cub	−1008	+125	−865
$Sb(OH)_2^+$	soln	−	−	−417
$[Sb(OH)_3]$	soln	−774	+126	−647
Sb_2S_3	rhomb	−158	+182	−156
$Sb_2(SO_4)_3$	sld	−2402	−	−
$(Sb^{III}Sb^V)O_4$	rhomb	−908	+127	−796
Sb_2Se_3	rhomb	−127	+197	−121
Sb_2Te_3	trig	−57	+234	−55
Sc	gas	+378	+175	+336
	hex	0	+35	0
Sc^{3+}	soln	−632	−264	−601
$ScBr_3$	sld	−752	+155	−720
$ScCl_3$	gas	−686	+337	−676
	trig	−941	+128	−869
ScF^{2+}	soln	−960	−124	−922
ScF_2^+	soln	−1296	−33	−1235
$[ScF_3]$	soln	−1635	+20	−1540
$[ScF_4]^-$	soln	−	−	−1836
Sc_2O_3	cub	−1909	+77	−1819
$ScOH_2^+$	soln	−	−	−812
$Sc(OH)_3$	cub	−1376	+93	−1243
Se^{2-}	soln	+64	−46	+129
Se	amor	+5	+48	+3
	gas	+228	+177	+188
	mon(18)	+7	−	−
	trig	0	+42	0
Se^+	gas	+1175	+175	+1129
Se_2^{2-}	soln	−	−	+114
Se_2	gas	+139	+252	+89
Se_6	gas	+149	+460	+87
$SeCl_2$	gas	−84	+293	−92
$SeCl_4$	mon	−191	+184	−100

(*Continued*)

Substance	State	$\Delta H°$ kJ/mole	$S°$ J/(K·mole)	$\Delta G°$ kJ/mole
Se_2Cl_2	lq	−85	+188	−50
$SeCl_2O$	gas	−25	+324	−12
SeF_6	gas	−1117	+314	−1017
SeO	gas	+55	+234	+28
SeO_2	gas	−127	+265	−132
	tetr	−226	+67	−172
SeO_3^{2-}	soln	−509	−7	−364
SeO_3	tetr	−167	+72	−84
SeO_4^{2-}	soln	−599	+54	−441
Si	gas	+452	+168	+408
	cub	0	+19	0
Si^+	gas	+1244	+163	+1195
Si_2	gas	+593	+231	+535
$SiBr$	gas	+181	+247	+136
$SiBr_2$	gas	−43	+303	−82
$SiBr_4$	gas	−420	+377	−436
	lq	−461	−	−
SiC	hex	−63	+16	−60
	cub	−65	+17	−63
$SiCl$	gas	+118	+238	+86
$SiCl_2$	gas	−160	+279	−171
$SiCl_3$	gas	−311	+316	−300
$SiCl_4$	gas	−657	+331	−617
	lq	−687	+240	−620
SiF	gas	+7	+226	−24
SiF_2	gas	−620	+256	−630
SiF_3	gas	−1000	−	−
SiF_4	gas	−1616	+282	−1573
$[SiF_4]^{2-}$	soln	−	−	−1612
$[SiF_6]^{2-}$	soln	−2396	+125	−2207
$[SiF_6],K_2$	cub	−2966	+236	−2810
$[SiF_6],Mg·6H_2O$	mon	−4579	−	−
$[SiF_6],(NH_4)_2$	cub	−2687	+284	−2371
	trig	−2689	+280	−2372
$[SiF_6],Na_2$	trig	−2918	+191	−2757
SiH	gas	+361	+198	+327
SiH_2	gas	+339	−	−
SiH_3	gas	+212	−	−
SiH_4	gas	+34	+205	+57
Si_2H_6	gas	+79	+274	+126
Si_3H_8	gas	+121	−	−
	lq	+92	−	−
SiN	gas	+486	+217	+456
Si_3N_4	hex	−744	+101	−643
SiO	gas	−99	+212	−126
SiO_2	amor	−902	+47	−849
	gas	−321	+229	−322

(Continued)

Substance	State	$\Delta H°$ kJ/mole	$S°$ J/(K·mole)	$\Delta G°$ kJ/mole
(coesite)	mon	−906	+40	−851
(α-tridymite)	rhomb	−909	+44	−855
(α-cristobalite)	tetr	−908	+43	−854
(stishovite)	tetr	−861	+28	−803
(α-quartz)	trig	−912	+41	−857
$SiO_2·H_2O$	amor	−1188	−	−
SiO_4^{4-}	soln	−1353	−402	−1027
SiS_2	lq	−157	+91	−159
	rhomb	−205	−	−
Sm	gas	+208	+183	+174
	trig	0	+70	0
Sm^{2+}	soln	−	−75	−515
Sm^{3+}	soln	−691	−212	−665
SmC_2	tetr	−71	+95	−75
$SmCl_2$	gas	−500	+316	−507
	rhomb	−816	+126	−866
$SmCl_3$	hex	−1053	+142	−975
$SmCl_3·6H_2O$	mon	−2873	+381	−2448
SmI_3	sld	−643	+220	−636
$Sm_2(MoO_4)_3$	rhomb	−4276	+446	−3975
Sm_2O_3	cub	−1826	−	−
	mon	−1845	+151	−1727
$Sm_2(SO_4)_3·8H_2O$	mon	−6299	+673	−5506
Sn	gas	+302	+168	+267
	cub (α)	−2	+44	ca. 0
	tetr (β)	0	+52	0
Sn^+	gas	+1018	+174	+975
$Sn^{2+(19)}$	soln	−24	−70	−27
$Sn^{IV(20)}$	soln	−2	−398	−2
$SnBr^+$	soln	−	+95	−138
$SnBr_2$	rhomb	−268	+146	−251
$SnBr_4$	gas	−314	+412	−331
	rhomb	−406	−	−
SnCl	gas	+43	+252	+17
$SnCl^+$	soln	−	+109	−167
$SnCl_2$	gas	−205	−	−
	rhomb	−331	+132	−288
$[SnCl_2]$	soln	−	−	−303
$[SnCl_3]^-$	soln	−	−	−433
$SnCl_4$	gas	−471	+366	−432
	lq	−511	+259	−440
$[SnCl_6],K_2$	cub	−1482	+371	−1339
$[SnCl_6],(NH_4)_2$	cub	−1234	+431	−934
SnF	gas	−150	+242	−117
SnF^+	soln	−	−	−335
SnF_2	gas	−452	−	−
	mon	−649	−	−

Substance	State	$\Delta H°$ kJ/mole	$S°$ J/(K·mole)	$\Delta G°$ kJ/mole
SnH	gas	+265	+213	+237
SnH_4	gas	+163	+229	+188
SnI_2	mon	−144	+167	−144
SnI_4	cub	−199	−	−
SnO	gas	+21	+232	−2
	tetr	−286	+56	−257
SnO_2	tetr	−581	+52	−520
$SnO_2·2H_2O^{(21)}$	amor (α)	−1116	+155	−946
$SnOH^+$	soln	−	+47	−252
$Sn(OH)_2$	amor	−561	+155	−492
$[Sn(OH)_2]$	soln	−	−	−464
$[Sn(OH)_3]^-$	soln	−	−	−646
SnS	gas	+110	+243	+63
	rhomb	−101	+77	−99
SnS_2	trig	−168	+87	−159
$[SnSO_4]^{2-}$	soln	−	−	−782
SnSe	gas	+123	+255	+75
	sld	−91	+92	−90
SnTe	gas	+201	+263	+153
	cub	−61	+103	−61
Sr	gas	+164	+165	+131
	α-cub	0	+53	0
Sr^+	gas	+719	+170	+678
Sr^{2+}	soln	−556	−27	−571
$SrBr_2$	rhomb	−729	+136	−708
$Sr(BrO_3)_2·H_2O$	mon	−1014	+283	−1244
SrC_2	tetr	−84	−	−
$SrCO_3$	rhomb	−1232	+97	−1152
SrCl	gas	−116	+250	−142
$SrCl_2$	gas	−471	+315	−459
	cub	−836	+126	−791
$SrCl_2·6H_2O$	trig	+2625	+350	−2230
$Sr(ClO_3)_2$	rhomb	−744	−	−
$SrCrO_4$	mon	−1435	+112	−1324
SrF_2	gas	−778	+290	−788
	cub	−1233	+82	−1181
SrH	gas	+220	+212	+192
SrH_2	rhomb	−180	+52	−141
$(SrHf)O_3$	mon	−1827	−	−
SrI_2	gas	−271	+341	−322
	rhomb	−568	+159	−565
$SrI_2·6H_2O$	trig	−2397	−	−
$Sr(IO_3)_2$	tricl	−1060	+243	−899
$Sr(IO_3)_2·H_2O$	sld	−1313	+292	−1097
$SrMoO_4$	tetr	−1559	+141	−1454
$Sr(NO_2)_2$	sld	−766	+176	−623
$Sr(NO_3)_2$	cub	−988	+195	−790

Substance	State	$\Delta H°$ kJ/mole	$S°$ J/(K·mole)	$\Delta G°$ kJ/mole
$Sr(NO_3)_2 \cdot 4H_2O$	mon	−2150	+364	−1724
SrO	gas	−60	+230	−82
	cub	−604	+54	−574
SrO_2	tetr	−637	−	−
$SrOH^+$	soln	−782	−6	−733
$Sr(OH)_2$	tetr	−957	+87	−867
$Sr(OH)_2 \cdot 8H_2O$	tetr	−3361	−	−
SrS	cub	−453	+68	−448
$SrSO_4$	rhomb	−1467	+111	−1353
$SrS_2O_6 \cdot 4H_2O$	hex	−2911	−	−
SrSe	cub	−387	−	−
$SrSeO_4$	mon	−1156	+113	−1039
$SrSiO_3$	hex	−1646	+97	−1562
Sr_2SiO_4	mon	−2329	+153	−2215
$(SrTi)O_3$	cub	−1684	+109	−1600
$SrWO_4$	tetr	−1656	+138	−1549
$(SrZr)O_3$	mon	−2634	+115	−2549
T	gas	+223	+128	+208
T_2	gas	0	+153	0
T_2O	gas	−252	+204	−237
Ta	gas	+782	+185	+739
	cub	0	+42	0
$TaCl_5$	gas	−763	−	−
	mon	−858	+236	−750
TaF_5	mon	−1904	+170	−1791
TaO_3^-	soln	−1151	+151	−1073
Ta_2O_5	rhomb	−2048	+143	−1913
Tb	gas	+389	+203	+350
	hex	0	+73	0
Tb^{3+}	soln	−698	−180	−690
$TbCl_3$	mon	−1045	+146	−937
$TbCl_3 \cdot 6H_2O$	mon	−2871	+404	−2452
$Tb(NO_3)_3 \cdot 6H_2O$	mon	−3044	−	−
Tb_2O_3	cub	−1865	+149	−1774
Tc	gas	+649	+181	+605
	hex	0	+33	0
TcO_2	mon	−433	+63	−381
TcO_4^-	soln	−724	−314	−630
Tc_2O_7	sld	−1117	+184	−938
Te^{2-}	soln	+101	−64	+174
Te	gas	+192	+183	+152
	trig	0	+50	0
Te^+	gas	+1067	+181	+1022
Te_2^{2-}	soln	−	−	+163
Te_2	gas	+167	+268	+117
$TeBr_4$	mon	−209	+71	−125
$TeCl_4$	gas	−207	−	−

Substance	State	$\Delta H°$ kJ/mole	$S°$ J/(K·mole)	$\Delta G°$ kJ/mole
$TeCl_4$	mon	−324	+209	−238
TeF_6	gas	−1369	+336	−1273
TeI_4	rhomb	−63	−	−
TeO	gas	+173	+241	+142
TeO_2	gas	−52	+273	−57
	rhomb	−323	+59	−265
TeO_3^{2-}	soln	−533	+13	−391
Th	gas	+576	+190	+535
	α-cub	0	+53	0
Th^{4+}	soln	−760	−330	−724
$ThBr_4$	gas	−766	+438	−790
	rhomb	−949	+234	−912
ThC_2	mon	−125	+69	−126
$ThCl^{3+}$	soln	−	−	−863
$ThCl_4$	gas	−973	+396	−942
	rhomb	−1194	+199	−1105
ThF^{3+}	soln	−	−	−1045
ThF_2^{2+}	soln	−	−	−1357
ThF_3^+	soln	−	−	−1666
ThF_4	gas	−1674	+346	−1640
	mon	−2096	+142	−2002
ThI_4	gas	−418	+470	−363
	mon	−552	+272	−548
ThO_2	gas	−510	−	−
	cub	−1227	+65	−1169
$ThOH^{3+}$	soln	−1021	−195	−939
$Th(OH)_2^{2+}$	soln	−1274	−150	−1152
$Th(OH)_3^+$	soln	−	−	−1354
$Th(OH)_4$	amor	−1772	+144	−1599
ThS_2	rhomb	−627	+96	−621
$Th(SO_4)_2$	sld	−2541	+148	−2306
Ti	gas	+471	+180	+427
	hex	0	+31	0
$TiBr_2$	cub	−385	+126	−368
$TiBr_3$	sld	−551	+177	−527
$TiBr_4$	gas	−551	+399	−570
	cub	−619	+241	−591
$TiCl_2$	gas	−303	+284	−312
	trig	−517	+106	−473
$TiCl_3$	gas	−541	+319	−527
	trig	−732	+140	−665
$TiCl_4$	gas	−761	+352	−724
	lq	−805	+252	−738
TiF_2	gas	−654	+261	−662
TiF_3	gas	−1164	+287	−1150
	rhomb	−1412	+103	−1343
TiF_4	amor	−1649	+134	−1559

(Continued)

Substance	State	$\Delta H°$ kJ/mole	$S°$ J/(K·mole)	$\Delta G°$ kJ/mole
TiF_4	gas	−1553	+312	−1516
$(TiFe)O_3$	trig	−1235	+106	−1158
TiH	gas	+215	+215	+180
TiH_2	cub	−144	+30	−105
TiI_2	gas	−62	+323	−10
	trig	−270	+148	−270
TiI_4	gas	−287	+433	−338
	cub	−386	+246	−381
$(TiMn)O_3$	trig	−1358	+105	−1279
TiN	cub	−338	+30	−309
TiO	gas	+63	+233	+33
	cub	−519	+35	−490
TiO_2	gas	−334	+236	−334
	tetr (β)	−938	+50	−883
	tetr (γ)	−944	+50	−889
Ti_2O_3	trig	−1521	+79	−1434
$Ti(OH)_3$	amor	−334	−	−
$TiO(OH)_2$	amor	−	−	−1059
TiP	hex	−283	−	−
TiS	hex	−269	+56	−267
TiS_2	trig	−425	+78	−430
Tl	gas	+181	+181	+146
	hex	0	+64	0
Tl^+	gas	+789	+175	+750
	soln	+6	+129	−32
Tl^{3+}	soln	+197	−192	+215
Tl_2	gas	+301	+287	+254
TlBr	gas	−39	+268	−77
	cub	−172	+123	−167
Tl_2CO_3	mon	−699	+159	−615
$Tl_2C_2O_4$	mon	−883	−	−
TlCl	gas	−69	+256	−93
	cub	−204	+111	−185
[TlCl]	soln	−167	+179	−168
$TlCl^{2+}$	soln	+7	−69	+41
$[TlCl_2]^-$	soln	−	−	−296
$TlCl_2^+$	soln	−179	+11	−116
$TlCl_3$	mon	−315	+325	−293
$[TlCl_3]$	soln	−351	+103	−263
$[TlCl_4]^-$	soln	−518	+202	−407
$TlClO_3$	sld	−134	+169	−40
$TlClO_4$	rhomb	−138	−	−
Tl_2CrO_4	rhomb	−957	+247	−863
TlF	gas	−154	+244	−208
	rhomb	−327	+96	−306
[TlF]	soln	−	−	−313
TlF_3	sld	−	−	−515

(Continued)

Substance	State	$\Delta H°$ kJ/mole	$S°$ J/(K·mole)	$\Delta G°$ kJ/mole
TlI	gas	+16	+277	−30
	rhomb	−123	+128	−125
TlN_3	tetr	−234	+126	+301
TlNCS	rhomb	+34	−	−
$TlNO_2$	cub	−142	−	−
$TlNO_3$	rhomb	−242	+161	−151
Tl_2O	gas	+9	+314	−16
	hex	−178	+126	−147
Tl_2O_3	cub	−411	+135	−321
$Tl_2O_3 \cdot 3H_2O^{(1)}$	amor	−1033	+204	−755
TlOH	gas	−100	+255	−107
	sld	−239	+88	−196
[TlOH]	soln	−222	+139	−194
$TlOH^{2+}$	soln	−	−	−15
$Tl(OH)_2^+$	soln	−	−	−243
$[Tl(OH)_3]$	soln	−	−	−470
Tl_2S	trig	−	−	−88
Tl_2S_3	amor	−116	+151	−94
$[TlSO_4]^-$	soln	−903	+170	−784
Tl_2SO_4	rhomb	−931	+231	−830
Tl_2SO_3S	rhomb	−	−	−610
Tl_2Se	tetr	−58	+172	−59
Tm	gas	+232	+190	+197
	hex	0	+74	0
Tm^{3+}	soln	−	−186	−661
$TmCl_3$	mon	−962	+146	−884
TmI_3	hex	−582	+218	−573
Tm_2O_3	cub	−1888	+139	−1794
U^-	gas	−	+203	−
U	gas	−524	+200	+479
	rhomb	0	+50	0
U^{3+}	soln	−491	−153	−489
U^{4+}	soln	−590	−346	−550
UBr_3	gas	−342	−	−
	hex	−711	+205	−689
UBr_4	mon	−822	+243	−789
UC_2	tetr	−88	+68	−90
$[U(CO_3)_2O_2]^{2-}$	soln	−	−	−2120
$[U(CO_3)_3O_2]^{4-}$	soln	−	−	−2673
UCl^{3+}	soln	−	−	−677
UCl_3	hex	−891	+159	−824
UCl_4	gas	−820	+390	−788
	tetr	−1049	+198	−960
UCl_5	mon	−1253	+243	−1144
UCl_6	trig	−1139	+286	−1010
$[U(Cl)O_2]^+$	soln	−	−	−1094
UF^{3+}	soln	−	−	−913

(Continued)

Substance	State	$\Delta H°$ kJ/mole	$S°$ J/(K·mole)	$\Delta G°$ kJ/mole
UF_2^{2+}	soln	–	–	−1223
UF_3	gas	−1096	+332	−1089
	hex	−1489	+117	−1418
UF_3^+	soln	–	–	−1528
UF_4	gas	−1591	+349	−1560
	mon	−1925	+152	−1834
$UF_4·2.5H_2O$	rhomb	−5367	+500	−4896
$[UF_4]$	soln	–	–	−1832
$[UF_5]^-$	soln	–	–	−2121
UF_5	gas	−2030	+377	−1976
	tetr (β)	−2039	+188	−1929
$[UF_6]^{2-}$	soln	–	–	−2414
UF_6	gas	−2137	+376	−2053
	rhomb	−2186	+228	−2058
UH_3	cub	−127	+64	−73
UI_3	rhomb	−478	+238	−482
UI_4	gas	−280	–	–
	rhomb	−529	+280	−528
UO_2	gas	ca. −430	–	–
	cub	−1085	+78	−1032
UO_2^+	soln	−1035	+50	−967
UO_2^{2+}	soln	−1024	−86	−961
$UO_{2.25}$	rhomb	−1131	+84	−1072
$UO_{2.33}$	cub	−1142	+84	−1081
UO_3	amor	−1209	–	–
	mon	−1225	+99	−1148
	rhomb	−1219	–	–
$UO_2(CH_3COO)_2·2H_2O$	rhomb	−2615	–	–
UO_2CO_3	rhomb	−1696	+146	−1570
$(UO_2)Cl_2$	gas	−992	–	–
	rhomb	−1252	+151	−1155
UO_2F^+	soln	−1367	−37	−1270
$(UO_2)F_2$	trig	−1650	+140	−1555
$[UO_2F_2]$	soln	–	–	−1574
$[UO_2F_3]^-$	soln	–	–	−1869
$[UO_2F_4]^{2-}$	soln	–	–	−2157
UOH^{3+}	soln	−827	−115	−786
$U(OH)_4$	amor	−1668	+67	−1473
$UO_2(NO_3)_2$	sld	−1377	+276	−1143
$UO_2(NO_3)_2·6H_2O$	rhomb	−3177	+505	−2594
$UO_2(OH)^+$	soln	−1258	+60	−1169
$UO_2(OH)_2$	rhomb	−1535	+142	−1401
$UO_2(OH)_2·H_2O$	rhomb	−1827	+192	−1639
$[UO_2(OH)_2]$	soln	−1515	+87	−1365
UO_2SO_4	sld	−1888	+155	−1726
$UO_2SO_4·3H_2O$	sld	−2790	+264	−2452
US	cub	−314	+78	−313

Substance	State	$\Delta H°$ kJ/mole	$S°$ J/(K·mole)	$\Delta G°$ kJ/mole
$U(S_2)$	rhomb	−533	+111	−532
	tetr	−525	+110	−524
$U(SO_4)_2$	sld	−2368	+161	−2138
$[U(SO_4)O_2]$	soln	−1913	+54	−1721
$[U(SO_4)_2O_2]^{2-}$	soln	−2818	+113	−2473
$(U_2^VU^{VI})O_8$	rhomb	−3574	+282	−3369
V	gas	+515	+182	+469
	cub	0	+29	0
V^{2+}	soln	−227	−135	−217
V^{3+}	soln	−263	−238	−242
VBr_2	sld	−	−	−389
VBr_3	hex	−446	+142	−412
$VC_{0.88}$	cub	−102	+28	−100
VCl_2	gas	−216	+284	−226
	hex	−443	+97	−397
VCl_3	gas	−369	+333	−360
	trig	−581	+131	−512
VCl_4	gas	−524	+364	−491
	lq	−576	+235	−504
$V(Cl)O$	rhomb	−609	+75	−559
VCl_3O	gas	−696	+343	−659
	lq	−736	+242	−669
VF_4	hex	−1421	+123	−1328
VF_5	gas	−1355	+331	−1294
	rhomb	−1481	+192	−1378
VI_2	gas	−22	−	−
	trig (α)	−256	−	−264
VN	cub	−217	+37	−191
VO	gas	+150	+231	+120
	cub	−432	+39	−404
VO^{2+}	soln	−489	−141	−446
VO_2	tetr	−713	+52	−659
VO_2^+	soln	−650	−42	−587
VO_4^{3-}	soln	−	−	−903
V_2O_3	trig	−1218	+99	−1139
V_2O_5	rhomb	−1550	+131	−1419
VOF^+	soln	−	−	−745
$[VOF_2]$	soln	−	−	−1038
$[VOF_3]^-$	soln	−	−	−1327
VOH^{2+}	soln	−	−	−462
$VO(OH)_2$	mon	−	−	−887
$[V(SO_4)O]$	soln	−	−	−1204
W	gas	+850	+174	+808
	cub	0	+33	0
WBr_2	sld	−80	+155	−71
WBr_3	sld	+104	+230	+113
WBr_5	gas	−223	+460	−237

(Continued)

Substance	State	$\Delta H°$ kJ/mole	$S°$ J/(K·mole)	$\Delta G°$ kJ/mole
WBr$_5$	sld	−313	+272	−271
WBr$_6$	gas	−245	+487	−244
	sld	−385	+301	−329
WBr$_4$O	tetr	−548	−	−
WC	hex	−41	+36	−40
[W(CO)$_6$]	gas	−881	+501	−826
	rhomb	−952	+332	−847
WCl$_5$	gas	−498	+432	−451
	mon	−518	+230	−411
WCl$_6$	gas	−551	+406	−463
	cub	−620	+301	−501
WCl$_2$O$_2$	gas	−701	+251	−638
	sld	−789	+132	−691
WCl$_4$O	gas	−612	+364	−547
	tetr	−689	+213	−579
WF$_6$	gas	−1722	+342	−1633
	lq	−1748	+253	−1632
	cub	−1750	+223	−1625
WI$_2$	amor	−5	+176	−13
WO	gas	+439	+256	+403
WO$_2$	gas	+88	+276	+77
	mon	−590	+50	−534
WO$_3$	gas	−293	+287	−277
	tricl	−843	+76	−764
WO$_3$·H$_2$O	rhomb	−1172	+117	−1036
WO$_4^{2-}$	soln	−1073	+97	−931
WOF$_4$	gas	−1372	−	−
	mon	−1445	−	−
WS$_2$	hex	−213	+68	−204
Xe	gas	0	+170	0
[Xe]	soln	−18	+66	+13
Xe$^+$	gas	+1176	+181	−1167
XeCl$_2$	gas	+161	−	−
XeF$_2$	tetr	−176	−	−
XeF$_4$	mon	−251	−	−
Y	gas	+425	+179	+385
	hex	0	+44	0
Y^{3+}	soln	−702	−202	−687
YCl^{2+}	soln	−	−	−825
YCl$_3$	gas	−770	+351	−762
	mon	−1025	+138	−953
YCl$_3$·6H$_2$O	mon	−2850	+372	−2431
YF^{2+}	soln	−	−	−994
YF$_2^+$	soln	−	−	−1295
YF$_3$	gas	−1263	+317	−1254
	rhomb	−1718	+110	−1647
[YF$_3$]	soln	−	−	−1596

Substance	State	$\Delta H°$ kJ/mole	$S°$ J/(K·mole)	$\Delta G°$ kJ/mole
YI_3	hex	-537	$+201$	-532
YO	gas	-42	$+236$	-69
Y_2O_3	cub	-1905	$+99$	-1817
YOH^{2+}	soln	$-$	$-$	-872
$Y(OH)_3$	hex	-1431	$+103$	-1298
$[YSO_4]^+$	soln	-1597	$+75$	-1449
$[Y(SO_4)_2]^-$	soln	-2498	-15	-2197
Yb	gas	$+152$	$+173$	$+118$
	α-cub	0	$+60$	0
Yb^{2+}	soln	-530	-75	-510
Yb^{3+}	soln	-671	-192	-657
$YbCl_2$	gas	-445	$+314$	-454
	rhomb	-772	$+134$	-728
$YbCl_3$	mon	-962	$+146$	-888
$YbCl_3 \cdot 6H_2O$	mon	-2486	$+383$	-2065
YbF_3	gas	-1355	$+331$	-1345
	rhomb	-1657	$+111$	-1581
Yb_2O_3	cub	-1815	$+133$	-1727
$Yb(OH)_3$	sld	$-$	$-$	-1281
Zn	gas	$+130$	$+161$	$+95$
	hex	0	$+42$	0
Zn^{2+}	soln	-154	-111	-147
$(ZnAl_2)O_4$	cub	-2067	$-$	$-$
$ZnBr$	gas	$+102$	$-$	$-$
$ZnBr_2$	gas	-186	$+300$	-218
	tetr	-330	$+136$	-313
$Zn(CH_3COO)_2$	tetr	-1076	$-$	$-$
$Zn(CH_3COO)_2 \cdot 2H_2O$	mon	-1671	$-$	$-$
$Zn(CN)_2$	cub	$+90$	$-$	$-$
$[Zn(CN)_4]^{2-}$	soln	-332	$+259$	-427
$ZnCO_3$	trig	-816	$+82$	-734
$ZnCl$	gas	$+6$	$+244$	-21
$ZnCl_2$	gas	-265	$+277$	-269
	tetr	-415	$+109$	-369
$Zn(ClO_4)_2 \cdot 6H_2O$	hex	-2121	$-$	$-$
ZnF	gas	-157	$+233$	-184
ZnF^+	soln	-478	-70	-434
ZnF_2	gas	-497	$+254$	-500
	tetr	-764	$+74$	-713
$[Zn(HS)_2]$	soln	$-$	$-$	-208
$[Zn(HS)_3]^-$	soln	$-$	$-$	-203
ZnI_2	gas	-65	$+317$	-112
	tetr	-212	$+157$	-212
Zn_3N_2	cub	-23	$-$	$-$
$[Zn(NH_3)_4]^{2+}$	soln	-677	$+298$	-536
$[Zn(NH_3)_2Cl_2]$	rhomb	-699	$-$	$-$
$Zn(NO_3)_2$	sld	-495	$+192$	-299

(*Continued*)

Substance	State	$\Delta H°$ kJ/mole	$S°$ J/(K·mole)	$\Delta G°$ kJ/mole
$Zn(NO_3)_2 \cdot 6H_2O$	tetr	−2305	+410	−1757
ZnO	gas	+105	+225	+81
	hex	−351	+44	−321
$ZnOH^+$	soln	−	−	−333
$Zn(OH)_2$	rhomb	−644	+75	−554
$[Zn(OH)_2]$	soln	−	−	−535
$[Zn(OH)_3]^-$	soln	−	−	−705
$[Zn(OH)_4]^{2-}$	soln	−	−	−871
Zn_3P_2	tetr	−195	−	−
$Zn_3(PO_4)_2$	mon	−2900	−	−
$Zn_3(PO_4)_2 \cdot 4H_2O$	rhomb	−	−	−3611
ZnS	gas	+204	+238	+155
	hex (β)	−195	+68	−193
	cub (α)	−209	+58	−204
$ZnSO_4$	rhomb	−980	+110	−869
$ZnSO_4 \cdot 7H_2O$	rhomb	−3077	+389	−2562
$[ZnSO_4]$	soln	−1060	−39	−905
ZnSe	gas	+218	+250	+169
	cub	−168	+83	−168
$ZnSeO_4$	sld	−674	−	−
$ZnSeO_4 \cdot H_2O$	sld	−993	−	−
$ZnSiO_3$	rhomb	−1232	+90	−1149
Zn_2SiO_4	trig	−1642	+131	−1528
ZnTe	cub	−120	+83	−117
Zr	gas	+608	+181	+566
	hex	0	+39	0
$ZrBr_4$	gas	−643	+415	−664
	cub	−759	+225	−724
ZrC	cub	−197	+33	−193
$ZrCl_2$	amor	−552	+113	−508
	gas	−326	+308	−340
$ZrCl_3$	gas	−603	+337	−592
	hex	−862	+131	−790
$ZrCl_4$	gas	−867	+369	−832
	mon	−980	+186	−891
$ZrCl_2O$	sld	−1083	+61	−993
$ZrCl_2O \cdot 8H_2O^{(22)}$	tetr	−3468	−	−
ZrF^{3+}	soln	−	−	−855
ZrF_2	gas	−614	+257	−618
ZrF_2^{2+}	soln	−	−	−1178
ZrF_3	gas	−1188	+303	−1176
ZrF_3^{3+}	soln	−	−	−1491
ZrF_4	gas	−1663	+329	−1628
	tetr	−1911	+105	−1810
$[ZrF_4]$	soln	−	−	−1800
$[ZrF_5]^-$	soln	−	−	−2106
$[ZrF_6]^{2-}$	soln	−	−	−2409

Substance	State	$\Delta H°$ kJ/mole	$S°$ J/(K·mole)	$\Delta G°$ kJ/mole
$[ZrF_7],(NH_4)_3$	cub	−3365	+337	−2922
ZrH_2	cub	−163	+35	−123
$[Zr(H_2O)_4(SO_4)_2]$	rhomb	−3647	−	−
ZrI_4	gas	−356	+447	−408
	cub	−485	+257	−481
ZrN	cub	−366	+39	−337
ZrO_2	gas	−295	+274	−304
	mon	−1101	+51	−1043
$ZrO_2·2H_2O^{(9)}$	amor	−1721	+130	−1548
$ZrOH^{3+}$	soln	−	−	−762
$Zr(OH)_2^{2+}$	soln	−1130	−69	−997
$Zr(OH)_3^+$	soln	−	−	−1232
$[Zr(OH)_4]$	soln	−	−	−1465
$ZrO(OH)_2$	amor	−1420	+92	−1305
ZrS_2	trig	−569	−	−
$ZrSO_4^{2+}$	soln	−	−	−1290
$Zr(SO_4)_2$	sld	−2410	−	−
$[Zr(SO_4)_2]$	soln	−	−	−2049
$[Zr(SO_4)_3]^{2-}$	soln	−	−	−2800
$ZrSiO_4$	tetr	−1990	+85	−1876

[1] $M_2O_3·nH_2O$ (M = Al, Au, Co, Cr, Fe, Mn, Ni, Sb, Tl). [2] $[Bi_6(OH)_{12}]^{6+}$.
[3] $[Ce_6(H_2O)_m(OH)_n]^{(24-n)+}$. [4] $[Cr(H_2O)_4(OH)_2]$. [5] $[GeO_2(OH)_2]^{2-}$. [6] $H_3O^+·3H_2O$.
[7] $CO_2·nH_2O + H_2CO_3$. [8] $[Hf_4(H_2O)_{16}(OH)_8]^{8+}$. [9] $MO_2·nH_2O$ (M = Hf, Pd, Zr).
[10] $N_2H_4·H_2O$. [11] $NH_2OH·H_2O$. [12] $(NO_2)NO_3$. [13] From white phosphorus P_4.
[14] Po^{2+}; $[PoCl_n]^{(n-2)-}$ ($n = 3, 4$). [15] S_8. [16] $S_2F_2 + S(S)F_2$. [17] $SO_2·nH_2O$.
[18] Se_8. [19] $[Sn_3(OH)_4]^{2+}$; $[Sn_4(OH)_6]^{2+}$. [20] $[Sn(H_2O)_{6-n}(OH)_n]^{(4-n)+}$; $[Sn(OH)_6]^{2-}$.
[21] α-$SnO_2·nH_2O$. [22] $[Zr_4(H_2O)_{16}(OH)_8]Cl_3·12H_2O$.

4.2 Enthalpy and Entropy of Phase Transitions [2, 5, 14–17, 29–30, 36]

Thermodynamic properties for phase transitions – the standard enthalpy ($\Delta H°$) and standard entropy ($\Delta S°$) at the specified temperatures – are tabulated below. For the majority of solid phases, the crystal systems (syngonies) they belong to are indicated. The sign "−" denotes the absence of data.

Formula	State		Tempera-ture, K	$\Delta H°$ kJ/mole	$\Delta S°$ J/(K·mole)
	Initial	Final			
Ac	cub	lq	1323	14.31	9.71
	lq	gas	3573	293.00	−
Ag	cub	lq	1235.08	11.95	9.67
	lq	gas	2443	254.30	104.35

(*Continued*)

Formula	State		Tempera-ture, K	$\Delta H°$ kJ/mole	$\Delta S°$ J/(K·mole)
	Initial	Final			
AgBr	cub	lq	705	9.16	13.05
AgCN	trig	lq	623	11.72	–
AgCl	cub	lq	728	12.89	17.70
	lq	gas	1823	182.84	–
AgI	α-cub	hex	409	–	–
	hex	β-cub	420	6.15	14.56
	β-cub	lq	827	9.41	11.34
	lq	gas	1799	143.93	–
$AgNO_3$	rhomb	hex	432.7	2.55	5.90
	hex	lq	482.9	11.55	23.85
Ag_2S	mon	cub	450	4.23	9.33
	rhomb	cub	453	–	–
	cub	lq	1118	14.06	12.59
Ag_2SO_4	rhomb	hex	700	7.95	11.72
	hex	lq	933	16.74	16.74
Ag_2Se	rhomb	cub	406	6.74	16.61
Ag_2Te	mon	cub	410	0.69	1.67
Al	cub	lq	933.52	10.92	11.72
	lq	gas	2773	290.78	–
$AlCl_3$	mon	trig	411	–	–
	trig	gas	456	111.71	–
	trig	lq	465.8	35.56	76.40
AlF_3	trig	sld	728	0.63	0.88
	sld	gas	1552	322.17	–
$[AlF_6],Na_3$	mon	lq	1300	115.65	88.95
AlN	hex	lq	2700	67.78	–
Al_2O_3	hex	lq	2326	117.15	50.88
	lq	gas	>3273	485.34	–
Am	hex	cub	>873	–	–
	cub	lq	1565	1437.00	–
Ar	cub	lq	83.8	1.17	–
	lq	gas	87.4	6.51	–
As	hex (α)	gas	888	31.80	35.82
AsH_3	cub	lq	156.23	2.34	–
	lq	gas	210.68	17.49	–
As_2O_3	mon	lq	587	18.41	–
	mon	gas	587	48.12	–
	cub	lq	551	24.89	–
	cub	gas	551	54.60	–
At_2	rhomb	lq	517	23.85	41.46
Au	cub	lq	1337.58	12.36	9.25
	lq	gas	3220	335.05	108.74
B	hex	lq	2348	22.13	9.62
	lq	gas	3973	519.90	127.58
BBr_3	lq	gas	363.0	30.54	–
B_4C	trig	lq	2623	104.60	–
BCl_3	lq	gas	285.7	23.85	–

| Formula | State | | Tempera-ture, K | $\Delta H°$ kJ/mole | $\Delta S°$ J/(K·mole) |
	Initial	Final			
BF_3	rhomb	lq	144.79	2.09	–
	rhomb	gas	144.79	23.85	–
	lq	gas	172.9	17.99	–
B_2H_6	mon	lq	107.7	4.44	–
	lq	gas	180.7	14.43	–
B_5H_9	lq	gas	333	32.22	–
BN	hex (α)	lq	3000	75.31	–
B_2O_3	hex	lq	723	23.01	31.84
	lq	gas	ca. 2273	366.31	–
Ba	α-cub	β-cub	643	0.63	0.96
	β-cub	lq	1000	7.66	7.82
	lq	gas	ca. 2133	149.37	–
$BaCO_3$	rhomb	hex	1079	18.79	17.41
	hex	cub	1241	3.05	2.47
$BaCl_2$	rhomb	cub	1198	16.95	14.14
	cub	lq	1234	15.98	12.93
	lq	gas	ca. 2323	238.49	–
BaF_2	cub	lq	1641	12.55	7.95
	lq	gas	2533	347.27	–
$Ba(NO_3)_2$	cub	lq	867	25.10	29.29
BaO	cub	gas	1650	372.38	–
	cub	lq	ca. 2293	57.74	26.28
$Ba(OH)_2$	rhomb	lq	681	19.20	27.82
$Ba_3(PO_4)_2$	trig	lq	1878	77.82	–
$BaSO_4$	rhomb	cub	1422	–	–
	cub	lq	1853	40.58	25.10
Be	hex	cub	1548	–	–
	cub	lq	1560	14.64	9.41
	lq	gas	2780	295.14	107.11
$(BeAl_2)O_4$	rhomb	lq	2143	175.73	–
$BeCl_2$	α-rhomb	β-rhomb	676	6.82	–
	β-rhomb	lq	688	8.66	–
	lq	gas	823	104.60	–
BeF_2	tetr	cub	403	–	–
	cub	lq	1076	16.74	15.56
Be_3N_2	cub	hex	1673	–	–
	hex	lq	2473	129.29	–
BeO	hex	tetr	2373	–	–
	tetr	lq	2853	80.75	31.25
Bi	trig	lq	544.59	11.05	20.29
	lq	gas	1837	176.98	97.07
$BiCl_3$	cub	lq	505	23.47	46.57
	lq	gas	714	72.59	–
Bi_2O_3	mon	cub	1003	–	–
	cub	lq	1098	28.45	26.11
Bi_2S_3	rhomb	lq	958	79.37	76.61
Bi_2Te_3	trig	lq	858	86.61	101.00

(Continued)

Formula	State		Tempera-ture, K	$\Delta H°$ kJ/mole	$\Delta S°$ J/(K·mole)
	Initial	Final			
Br_2	rhomb	lq	265.90	10.57	39.75
	lq	gas	332.97	29.54	88.87
BrF_3	rhomb	lq	282.0	12.03	42.68
	lq	gas	398.90	42.68	107.74
BrF_5	lq	gas	313.91	30.59	97.49
C	cub (β)	hex (α)	2073	1.88	—
CBr_4	mon	cub	320.1	6.28	—
	cub	lq	365.7	4.10	—
	lq	gas	462.7	43.51	—
CCl_4	mon	cub	225.5	4.56	—
	cub	lq	250.19	2.51	—
	lq	gas	349.90	30.00	85.77
CF_4	mon	cub	76.22	1.46	—
	cub	lq	89.6	0.70	—
	lq	gas	145	12.59	—
C_2N_2	sld	lq	245.32	8.11	—
	lq	gas	252.00	23.33	—
CO	cub	hex	61.52	0.63	—
	hex	lq	68.13	0.84	—
	lq	gas	81.7	6.04	—
CO_2	cub	gas	194.674	25.23	—
CSO	trig	lq	134.33	4.73	—
	lq	gas	222.91	18.49	—
CS_2	tetr	lq	161.3	4.39	—
	lq	gas	319.39	26.78	—
Ca	α-cub	β-cub	716	1.13	1.59
	β-cub	lq	1115	8.66	7.70
	lq	gas	1768	152.26	86.73
$Ca(BO_2)_2$	rhomb	lq	1435	73.93	51.51
CaC_2	tetr	cub	723	5.56	7.70
$CaCl_2$	rhomb	gas	934.6	225.94	—
	rhomb	lq	1055	28.37	26.87
CaF_2	cub	tetr	1423	4.77	3.35
	tetr	lq	1692	29.71	17.57
$(CaFe_2)O_4$	tetr	lq	1493	108.24	71.67
$Ca(NO_3)_2$	cub	lq	834	21.34	25.52
CaO	cub	lq	2887	79.50	27.80
$Ca_2P_2O_7$	tetr	mon	1413	6.69	4.73
	mon	lq	1631	100.83	62.01
$Ca_3(PO_4)_2$	trig	mon	1423	15.48	11.25
$CaSO_4$	rhomb	hex	1466	—	—
	hex	lq	1723	28.03	17.99
Ca_2SiO_4	mon	sld	970	1.81	1.88
	sld	rhomb	1710	14.18	8.28
$(CaTi)O_3$	mon	rhomb	1530	2.30	1.51
Cd	hex	lq	594.26	6.19	10.42
	lq	gas	1039.7	99.58	95.73

Formula	State		Tempera-ture, K	$\Delta H°$ kJ/mole	$\Delta S°$ J/(K·mole)
	Initial	Final			
$CdBr_2$	hex	gas	638	159.83	–
	hex	lq	838	20.92	–
$CdCl_2$	trig	lq	841.7	22.18	–
	lq	gas	1237	123.01	–
CdF_2	cub	lq	1345	22.59	16.32
	lq	gas	2026	234.30	–
CdI_2	hex	lq	661	33.47	–
$Cd(NO_3)_2 \cdot 4H_2O$	rhomb	lq	332.5	32.64	–
CdO	cub	gas	ca. 1173	225.10	–
CdS	hex	gas	958	215.06	–
Ce	α-cub	β-cub	1003	2.93	2.93
	β-cub	lq	1077	5.18	4.81
$CeCl_3$	hex	lq	1095	53.47	49.45
CeF_3	hex	lq	1703	55.23	31.88
Cl_2	tetr	lq	172.12	5.64	–
	lq	gas	239.1	20.40	–
Cl_2F_6	lq	gas	284.91	55.06	193.30
Cm	hex	lq	1618	14.64	–
Co	hex	cub	700	0.45	0.63
	cub	lq	1767	16.19	9.16
	lq	gas	3233	376.55	117.65
$CoCl_2$	trig	lq	1013	30.96	30.96
	lq	gas	1322	113.80	–
Cr	cub	lq	2163	16.93	7.95
	lq	gas	2953	344.31	116.90
$CrCl_2$	rhomb	lq	1097	32.22	–
	rhomb	gas	1097	251.46	–
$CrCl_2O_2$	lq	gas	390	35.15	–
Cs	cub	lq	301.9	2.18	7.20
	cub	gas	301.9	78.74	–
	lq	gas	940.8	68.28	–
CsCl	α-cub	β-cub	742.5	2.43	3.31
	β-cub	lq	918	20.75	22.59
	lq	gas	1575	149.33	–
CsF	cub	lq	976	10.25	–
$CsNO_3$	hex	cub	427	–	–
	cub	lq	687	13.60	–
CsOH	sld	lq	619	6.74	–
Cu	cub	lq	1357.7	13.05	9.62
	lq	gas	2813	304.36	–
CuCl	cub	hex	681	–	–
	hex	lq	703	10.96	15.61
CuO	mon	lq	1720	55.65	32.34
Cu_2O	cub	lq	1513	64.22	42.38
Cu_2S	rhomb	hex	376	3.85	10.25
	hex	cub	623	0.84	1.26
	cub	lq	1403	23.01	16.32

(Continued)

Formula	State		Tempera-ture, K	$\Delta H°$ kJ/mole	$\Delta S°$ J/(K·mole)
	Initial	Final			
Cu_2Se	α-cub	β-cub	383	4.85	12.68
D_2	hex	lq	18.7	0.20	–
D_2O	hex	lq	276.963	6.31	22.78
	hex	gas	276.963	52.77	190.54
	lq	gas	276.963	46.46	167.78
Dy	hex	cub	1657	4.16	2.51
	cub	lq	1682	11.06	6.57
	lq	gas	2860	230.09	81.17
Dy_2O_3	α-cub	β-cub	1590	0.92	0.59
Er	hex	lq	1795	19.90	11.09
	lq	gas	3130	261.35	83.35
Eu	cub	lq	1099	9.21	8.45
	lq	gas	1713	143.50	76.73
Eu_2O_3	α-mon	β-mon	895	0.54	0.59
F_2	mon	cub	45.55	0.73	–
	cub	lq	53.451	0.51	–
	lq	gas	89.9	6.54	–
Fe	α-cub	β-cub	1042	1.36	1.34
	β-cub	γ-cub	1185	0.90	0.76
	γ-cub	δ-cub	1667	0.69	0.41
	δ-cub	lq	1812	15.36	8.49
	lq	gas	ca. 3473	349.59	111.50
Fe_3C	rhomb	sld	463	0.75	1.63
$[Fe(CO)_5]$	mon	lq	253	13.60	–
	lq	gas	376	37.24	–
$FeCl_2$	trig	lq	947	43.01	45.69
	lq	gas	1296	124.81	–
$FeCl_3$	trig	lq	580.7	43.10	32.84
	trig	gas	580.7	69.04	–
	lq	gas	589	25.19	–
FeF_2	tetr	lq	1375	51.88	–
	lq	gas	2110	224.43	–
$(Fe^{II}Fe_2^{III})O_4$	cub	lq	1811	138.07	–
$(Fe_2Ni)O_4$	cub	sld	855	0.86	1.00
$Fe_{0.95}O$	cub	lq	1641	31.34	19.00
FeS	α-hex	β-hex	411	2.38	5.82
	β-hex	cub	598	0.50	0.84
	cub	lq	1468	32.34	22.01
Fe_2SiO_4	rhomb	lq	1490	92.17	61.84
Fr	cub	lq	294	2.07	6.90
Ga	rhomb	tetr	275.6	2.13	–
	tetr	lq	302.93	5.59	18.45
	lq	gas	2676	257.15	102.05
Gd	hex	cub	1535	3.91	2.55
	cub	lq	1585	10.05	6.36
	lq	gas	3545	359.38	101.55
Ge	cub	lq	1210	36.94	30.54

(Continued)

Formula	State		Tempera-ture, K	$\Delta H°$ kJ/mole	$\Delta S°$ J/(K·mole)
	Initial	Final			
Ge	lq	gas	ca. 3123	330.91	106.52
$GeCl_4$	lq	gas	356.3	33.05	–
GeH_4	sld	lq	107.4	0.84	–
	lq	gas	184.7	14.06	–
GeO_2	tetr (α)	trig (β)	1306	35.56	27.20
	trig (β)	lq	1389	43.93	–
H_2	hex	lq	13.96	0.12	–
	lq	gas	20.28	0.90	–
HD	hex	lq	16.7	0.16	–
HBr	cub	lq	186.24	2.41	–
	lq	gas	206.38	17.61	–
HCN	lq	gas	298.80	25.22	84.39
HF	lq	gas	292.67	7.49	25.61
HI	tetr	lq	222.3	2.87	–
	lq	gas	237.8	19.77	–
HN_3	lq	gas	308.9	29.71	–
HNO_3	mon	lq	231.6	10.47	45.23
H_2O	hex	lq	273.15	6.01	22.00
	hex	lq	273.16	6.01	22.00
	hex	gas	273.16	51.48	186.91
	lq	gas	273.16	45.05	164.92
	lq	gas	373.15	40.66	108.95
H_2O_2	tetr	lq	272.72	10.54	–
$H(PH_2O_2)$	sld	lq	299.7	9.71	–
$H_2(PHO_3)$	rhomb	lq	347	12.84	–
H_3PO_4	mon	lq	315.50	12.97	41.13
$H_4P_2O_7$	sld	lq	334	9.20	–
H_2S	cub	lq	187.61	2.38	–
	lq	gas	212.80	18.67	–
$H_2(S_2)$	sld	lq	183.6	7.53	–
	lq	gas	343.9	35.15	–
H_2SO_4	mon	lq	283.6	10.71	37.78
$H_2SO_4·H_2O$	mon	lq	281.63	19.37	–
$H_2S_2O_7$	sld	lq	308.37	13.35	–
H_2Se	cub	lq	207.43	2.51	–
	lq	gas	231.7	19.33	–
H_2SeO_4	rhomb	lq	335.6	14.43	–
$H_2SeO_4·H_2O$	sld	lq	299	19.87	–
H_2Te	sld	lq	222	4.18	–
	lq	gas	271.4	23.22	–
He	hex	lq	2.0	0.008	–
	lq	gas	4.22	0.081	–
Hf	hex	cub	2013	6.74	3.35
	cub	lq	2503	24.06	9.62
	lq	gas	ca. 4893	575.14	117.95
Hg	trig	lq	234.288	2.33	–
	lq	gas	629.81	59.30	94.15

Formula	State		Tempera-ture, K	$\Delta H°$ kJ/mole	$\Delta S°$ J/(K·mole)
	Initial	Final			
$HgBr_2$	tetr	gas	ca. 663	75.48	—
$HgCl_2$	rhomb	sld	428	—	—
	sld	lq	553	19.41	—
	lq	gas	575.0	58.91	—
HgI_2	tetr (α)	rhomb (β)	404	2.72	6.74
	rhomb (β)	lq	529	18.83	35.98
	lq	gas	627	59.66	—
HgS	trig (α)	cub (β)	617	4.18	6.28
Ho	hex	cub	1701	4.69	2.76
	cub	lq	1743	12.18	6.99
	lq	gas	2980	240.98	81.21
I_2	rhomb	lq	386.7	15.77	40.79
	lq	gas	457.50	41.80	91.13
ICl	mon	lq	300.34	11.13	37.07
	mon	gas	300.34	41.55	—
IF_5	mon	lq	282.571	15.90	56.07
In	tetr	lq	429.78	3.26	7.61
	lq	gas	2297	231.84	98.95
InBr	lq	gas	935	92.05	—
$InBr_3$	sld	gas	644	105.02	—
InCl	lq	gas	926	89.96	—
$InCl_3$	mon	gas	771	158.99	—
$[In^{III}Cl_4],In$	lq	gas	928	192.46	—
InI	lq	gas	1016	90.79	—
Ir	cub	lq	2716	26.36	9.66
	lq	gas	4653	612.20	131.34
IrF_6	lq	gas	326.8	35.98	—
K	cub	lq	336.7	2.33	6.90
	lq	gas	1033	80.16	—
$KAl(SO_4)_2·12H_2O$	cub	lq	365	28.03	—
KBO_2	trig	lq	1213	23.85	19.66
$K_2B_4O_7$	sld	lq	1088	104.18	—
KBr	cub	lq	1007	29.29	—
	lq	gas	1653	155.23	—
KCN	cub	lq	907.7	14.64	—
	lq	gas	1898	157.11	—
K_2CO_3	mon	hex	701	—	—
	hex	lq	1164	27.61	23.60
KCl	cub	lq	1043	25.52	24.47
	lq	gas	1703	120.08	—
$KClO_4$	rhomb	cub	574	13.77	—
K_2CrO_4	rhomb	hex	939	10.25	—
	hex	lq	1241.5	28.87	—
$K_2Cr_2O_7$	tricl	mon	530	—	—
	mon	lq	670.7	35.61	53.05
KF	cub	lq	1130	28.24	24.99
	lq	gas	1778	172.80	—

Formula	State		Tempera-ture, K	$\Delta H°$ kJ/mole	$\Delta S°$ J/(K·mole)
	Initial	Final			
$K(HF_2)$	tetr	cub	469.8	11.18	–
	cub	lq	511.9	6.62	–
$KHSO_4$	α-rhomb	β-rhomb	442	2.05	–
	β-rhomb	mon	634.2	0.40	–
$KMgCl_3$	sld	lq	760	53.09	69.87
KNCS	rhomb	tetr	414.6	0.13	–
	tetr	lq	446.4	10.46	–
KNO_3	rhomb	trig	402	5.86	14.60
	trig	lq	607.7	11.72	19.16
KOH	rhomb	cub	517	6.34	–
	cub	lq	677	9.37	–
	lq	gas	1597	133.89	–
K_2SO_4	rhomb	hex	857	8.95	10.46
	hex	lq	1347	37.91	28.24
Kr	cub	lq	115.78	1.64	–
	lq	gas	119.93	9.03	–
La	hex	α-cub	583	0.37	0.63
	α-cub	β-cub	1141	3.56	3.14
	β-cub	lq	1193	31.13	4.52
Li	cub	lq	453.7	3.00	6.61
	lq	gas	1609.8	14.31	–
$LiBO_2$	mon	lq	1122	33.85	30.67
$Li_2B_4O_7$	tetr	lq	1191	120.50	–
LiBr	cub	lq	825	12.13	–
	lq	gas	1583	148.11	–
Li_2CO_3	mon	lq	891	44.77	44.81
LiCl	cub	lq	883	19.54	22.13
	lq	gas	1653	150.62	–
LiF	cub	lq	1118.3	27.07	24.14
	lq	gas	1949	213.38	–
LiH	cub	lq	953	29.29	30.50
LiI	lq	gas	1443	170.71	–
$LiNO_3$	trig	lq	526.2	25.61	48.79
$(LiNb)O_3$	trig	lq	1526	70.88	46.02
Li_2O	cub	lq	1726	58.58	–
LiOH	tetr	lq	744	20.96	–
	lq	gas	1198	167.78	–
Li_2SO_4	mon	hex	773	–	–
	hex	cub	848	28.45	–
	cub	lq	1132	12.55	–
Li_2SiO_3	rhomb	lq	1475	30.12	–
Li_4SiO_4	mon	lq	1498	30.96	20.08
Lu	hex	lq	1936	19.25	9.62
	lq	gas	3685	355.91	97.03
Mg	hex	lq	921	8.95	9.71
	lq	gas	1368	127.40	93.47
$MgBr_2$	trig	lq	983	34.73	–

(*Continued*)

Formula	State		Temperature, K	$\Delta H°$ kJ/mole	$\Delta S°$ J/(K·mole)
	Initial	Final			
$MgCl_2$	trig	lq	987	43.10	43.68
	lq	gas	1643	156.23	–
$MgCl_2 \cdot 6H_2O$	mon	lq	390	34.31	–
MgF_2	tetr	lq	1563	58.12	37.87
	lq	gas	ca. 2543	271.96	–
$(MgFe_2)O_4$	α-cub	β-cub	1230	1.46	1.17
$Mg(NO_3)_2 \cdot 6H_2O$	mon	lq	363.1	41.00	–
MgO	cub	lq	3098	77.40	24.27
	lq	gas	3873	474.47	–
$Mg_2P_2O_7$	mon	lq	1668	134.47	76.44
$Mg_3(PO_4)_2$	mon	lq	1630	120.67	74.22
$MgSO_4$	rhomb	lq	1410	1464	10.46
Mg_2Si	cub	lq	1358	64.14	–
Mg_2SiO_4	rhomb	lq	2163	86.19	–
Mn	α-cub	β-cub	980	2.23	2.26
	β-cub	γ-cub	1360	2.12	1.55
	γ-cub	δ-cub	1410	1.88	1.34
	δ-cub	lq	1518	12.06	7.95
	lq	gas	2353	226.07	96.82
Mn_3C	rhomb	hex	1583	12.97	–
$MnCl_2$	trig	lq	923	37.53	40.67
	lq	gas	1504	120.50	–
$(Mn^{II}Mn_2^{III})O_4$	tetr	cub	1433	20.79	14.39
MnS	α-cub	β-cub	573	–	–
	hex	β-cub	593	–	–
	β-cub	lq	1888	26.11	14.48
$MnSiO_3$	tricl	lq	1596	40.17	25.73
Mn_2SiO_4	rhomb	lq	1600	89.62	55.35
Mo	cub	lq	2893	27.61	9.62
	lq	gas	4903	592.79	121.38
$MoCl_5$	mon	lq	467	7.53	–
	lq	gas	541	58.16	–
MoF_6	rhomb	cub	263.5	–	–
	cub	lq	290.73	4.33	14.88
MoO_3	rhomb	gas	973	271.96	–
	rhomb	lq	1068	48.91	45.52
	lq	gas	1428	138.07	–
N_2	cub	hex	35.61	0.23	–
	hex	lq	63.2	0.72	–
	lq	gas	77.348	5.58	–
NH_3	cub	lq	195.40	5.65	–
	lq	gas	239.8	23.35	–
N_2H_4	mon	lq	274.6	12.66	–
	lq	gas	386.7	40.58	–
NH_4Cl	α-cub	β-cub	457.6	4.18	–
N_2H_5Cl	cub	lq	362	15.23	–
NH_4NO_3	rhomb	α-cub	305.5	1.59	5.19

Formula	State		Temperature, K	$\Delta H°$ kJ/mole	$\Delta S°$ J/(K·mole)
	Initial	Final			
NH_4NO_3	α-cub	β-cub	357.6	1.34	3.77
	β-cub	tetr	398.6	4.23	10.63
	tetr	lq	442.8	5.44	12.26
NO	mon	lq	109.6	2.30	–
	lq	gas	121.5	13.78	–
N_2O	cub	lq	182.3	6.54	–
	lq	gas	184.6	16.55	–
N_2O_4	cub	lq	262.0	14.65	–
	lq	gas	293.9	38.12	–
(NO)Cl	lq	gas	267.8	25.10	–
$(NO_2)NO_3$	hex	gas (N_2O_5)	305	56.90	–
Na	cub	lq	370.98	2.60	6.99
	lq	gas	1159	86.37	–
$NaAlO_2$	hex	sld	740	1.30	1.76
$NaBO_2$	trig	lq	1238	36.23	–
$Na_2B_4O_7$	tricl	lq	1014	81.17	–
NaBr	cub	lq	1028	25.52	–
	lq	gas	1663	161.92	–
NaCN	cub	lq	836.9	8.79	–
	lq	gas	1770	148.07	–
Na_2CO_3	mon	hex	753	0.69	–
	hex	lq	1124	29.71	26.44
NaCl	cub	lq	1074.0	28.66	26.74
	lq	gas	1738	170.71	–
$NaClO_3$	cub	lq	535	22.59	42.80
NaF	cub	lq	1270	33.14	26.44
	lq	gas	1973	209.20	–
$NaFeO_2$	trig	tetr	1270	2.18	1.72
	tetr	lq	1623	49.20	30.38
NaI	cub	lq	934	21.76	–
	lq	gas	1577	159.83	–
Na_2MoO_4	cub	α-rhomb	723	61.09	–
	α-rhomb	β-rhomb	858	–	–
	β-rhomb	γ-rhomb	908	–	–
	γ-rhomb	lq	960	15.08	–
NaNCS	rhomb	lq	580.7	18.41	–
$NaNO_3$	trig	sld	549	3.39	6.15
	sld	lq	579.7	14.60	25.10
Na_2O	α-cub	β-cub	1023	–	–
	β-cub	γ-cub	1243	–	–
	γ-cub	lq	1405	41.84	–
Na_2O_2	hex	tetr	783	5.36	37.66
NaOH	rhomb	mon	516	–	–
	mon	cub	572.8	6.36	11.25
	cub	lq	594	6.36	10.75
	lq	gas	1663	158.57	–
Na_2S	cub	lq	1453	6.69	–

Formula	State		Tempera-ture, K	$\Delta H°$ kJ/mole	$\Delta S°$ J/(K·mole)
	Initial	Final			
Na_2SO_4	α-rhomb	β-rhomb	458	–	–
	β-rhomb	hex	514	7.49	14.56
	hex	lq	1157	23.72	20.54
$NaSO_3S·5H_2O$	mon	lq	321.7	23.43	–
Na_2SiO_3	rhomb	lq	1362	52.17	38.37
$Na_2Si_2O_5$	mon	rhomb	983	–	–
	rhomb	lq	1147	35.56	30.96
Na_2WO_4	cub	rhomb	861.4	35.15	–
	rhomb	lq	971	23.85	–
Nb	cub	lq	2743	26.38	9.62
	lq	gas	5200	718.22	135.94
$NbCl_5$	mon	sld	456	–	–
	sld	lq	477.9	35.56	–
	sld	gas	477.9	85.35	–
	lq	gas	520.7	52.72	–
NbF_5	mon	lq	352.7	12.22	34.85
	lq	gas	507.7	46.44	–
Nb_2O_5	mon	lq	1773	102.88	57.66
Nd	hex	cub	1128	3.03	2.68
	cub	lq	1297	7.14	5.52
	lq	gas	ca. 3353	273.04	81.71
Nd_2O_3	hex	cub	1395	0.59	0.42
Ne	cub	lq	24.6	0.34	–
	cub	gas	24.6	1.80	–
	lq	gas	27.102	1.84	–
Ni	cub	lq	1728	17.47	10.13
	lq	gas	ca. 3173	370.38	116.19
$NiCl_2$	trig	lq	1282	77.28	59.33
	trig	gas	1243	202.34	–
Ni_3S_2	trig	sld	823	–	–
	sld	lq	1081	24.27	23.01
Np	rhomb	lq	910	5.61	–
O_2	rhomb	hex	23.90	0.09	–
	hex	cub	43.76	0.74	–
	cub	lq	54.45	0.44	–
	lq	gas	90.188	6.82	–
O_3	lq	gas	161.25	10.84	–
Os	hex	lq	3300	31.76	9.62
	lq	gas	ca. 5273	750.00	–
OsO_4	mon	lq	313.8	14.27	–
	lq	gas	404.4	39.75	–
P (red)	amor	gas	689	29.83	42.51
P_4	cub	lq	317.29	2.63	6.79
	lq	gas	530	52.22	98.52
PBr_3	lq	gas	446.5	38.83	–
PBr_5	lq	gas	379	54.39	–
PBr_3O	lq	gas	466	37.99	–

Formula	State		Tempera-ture, K	$\Delta H°$ kJ/mole	$\Delta S°$ J/(K·mole)
	Initial	Final			
PCl_3	lq	gas	348.5	30.46	–
PCl_5	tetr	gas	453	67.36	–
PCl_3O	sld	lq	277.40	13.01	–
	lq	gas	379.0	34.35	–
PH_3	cub	lq	139.4	1.13	–
	lq	gas	185.73	14.60	–
P_4O_6	lq	gas	448.6	37.66	–
P_4O_{10}	rhomb	gas	632	73.64	116.69
	rhomb	lq	695	20.92	–
	trig	lq	853	48.12	–
	trig	gas	853	70.29	–
Pa	tetr	cub	1443	–	–
	cub	lq	1853	14.64	9.79
	lq	gas	ca. 4773	552.00	–
Pb	cub	lq	600.65	4.80	7.99
	lq	gas	2018	177.70	87.82
$PbBr_2$	rhomb	lq	646	20.92	–
	lq	gas	1166	115.90	–
$PbCl_2$	rhomb	lq	774	23.85	31.59
	lq	gas	1223	112.17	–
Pb(Cl)F	tetr	lq	879	36.74	41.80
PbF_2	rhomb	cub	720	2.47	2.93
	cub	lq	1097	11.92	10.88
	lq	gas	1566	167.36	–
PbI_2	trig	lq	675	21.76	–
	lq	gas	1227	103.76	–
PbO	tetr (α)	rhomb (β)	761	0.75	–
	rhomb (β)	lq	1159	29.29	25.27
	lq	gas	1808	207.23	–
PbS	cub	lq	1350	17.57	12.55
$PbSO_4$	rhomb	mon	1139	16.99	14.90
	mon	lq	1443	40.17	29.71
Pd	cub	lq	1827	17.57	9.62
	lq	gas	3213	332.00	109.91
Pm	hex	lq	1443	12.55	9.67
	lq	gas	ca. 3273	330.00	–
Po	cub	trig	327	–	–
	trig	lq	527	12.55	23.81
	lq	gas	1235	72.80	–
Pr	hex	cub	1069	3.17	2.97
	cub	lq	1204	6.89	5.69
	lq	gas	3783	296.78	78.41
Pr_6O_{11}	α-cub	β-cub	760	9.54	12.55
Pt	cub	lq	2045	19.66	9.62
	lq	gas	ca. 4073	509.74	124.43
Pu	α-mon	β-mon	395	3.36	8.57
	β-mon	rhomb	480	0.64	1.33

(Continued)

Formula	State		Tempera-ture, K	$\Delta H°$ kJ/mole	$\Delta S°$ J/(K·mole)
	Initial	Final			
Pu	rhomb	α-cub	588	0.52	0.90
	α-cub	β-cub	730	0.08	0.11
	β-cub	tetr	745	1.86	2.47
	tetr	lq	913	2.90	3.18
	lq	gas	ca. 3623	343.67	98.11
$PuCl_3$	hex	lq	1033	63.60	61.50
PuF_3	hex	lq	1699	54.39	32.01
PuF_4	mon	lq	1310	42.68	32.58
PuF_6	sld	lq	324.74	18.64	57.41
Ra	cub	lq	1242	9.41	9.67
	lq	gas	1809	162.00	–
Rb	cub	lq	312.5	2.26	7.24
	cub	gas	312.5	85.65	–
	lq	gas	969	75.77	–
RbBr	cub	lq	966	15.48	–
	lq	gas	1625	155.31	–
RbCl	cub	lq	991	18.41	–
	lq	gas	1668	154.47	–
RbF	cub	lq	1071	22.97	21.09
	lq	gas	1703	165.31	–
RbI	cub	lq	929	12.51	–
	lq	gas	1600	150.46	–
$RbNO_3$	trig	cub	437	–	–
	cub	α-rhomb	493	–	–
	α-rhomb	β-rhomb	564	–	–
	β-rhomb	lq	586	5.61	–
RbOH	rhomb	lq	655	6.78	–
Re	hex	lq	3463	33.05	9.62
	lq	gas	ca. 6173	744.00	–
ReF_6	rhomb	cub	269.7	–	–
	cub	lq	291.8	20.92	–
	lq	gas	306.9	28.87	–
Re_2O_7	rhomb	lq	574.7	64.02	–
	lq	gas	631.7	75.73	–
Rh	cub	lq	2236	21.55	9.62
	lq	gas	ca. 3973	496.06	123.85
Rn	cub	lq	202.2	2.90	–
	lq	gas	211.3	16.40	–
Ru	hex	lq	2880	25.98	9.62
	lq	gas	ca. 5173	591.49	134.68
S	lq	gas	717.82	9.20	12.84
S_8	rhomb (α)	mon (β)	368.7	3.21	8.70
	mon (β)	lq	392.5	13.74	35.38
S_2Cl_2	lq	gas	411	35.98	87.86
SCl_2O	lq	gas	348.8	31.80	91.21
SCl_2O_2	lq	gas	342.7	28.03	82.01
SF_6	sld	lq	223	5.02	–

(Continued)

Formula	State		Tempera- ture, K	$\Delta H°$ kJ/mole	$\Delta S°$ J/(K·mole)
	Initial	Final			
SO_2	rhomb	lq	197.69	7.40	–
	lq	gas	263.1	24.92	–
SO_3	rhomb	lq	290.0	5.61	19.33
	lq	gas	317.9	40.79	128.32
Sb	trig	cub	686	–	–
	cub	lq	903.89	19.62	21.71
	lq	gas	1907	124.39	65.23
$SbBr_3$	rhomb	lq	369.8	14.69	–
$SbCl_3$	rhomb	lq	345.5	12.68	–
	lq	gas	494	45.19	–
$SbCl_5$	hex	lq	276.0	10.04	–
	lq	gas	413	48.12	–
Sb_2O_3	rhomb (α)	cub (β)	733	6.78	8.03
	cub (β)	lq	928	61.71	66.48
Sb_2S_3	rhomb	lq	823.7	23.43	28.58
Sb_2Se_3	rhomb	lq	890	54.27	61.09
Sc	hex	cub	1609	4.01	2.51
	cub	lq	1814	14.10	7.78
	lq	gas	ca. 3123	314.19	101.21
Se	trig	lq	490	6.69	13.56
	lq	gas	958.5	29.29	30.54
$SeCl_2O$	sld	lq	284.0	4.18	14.64
	lq	gas	450.8	42.68	94.56
SeF_6	sld	gas	226.6	26.23	–
	lq	gas	238.4	18.33	–
SeO_2	tetr	gas	588	91.21	149.37
Si	cub	lq	1688	50.55	30.00
	lq	gas	ca. 3523	355.00	–
$SiBr_4$	lq	gas	425.8	38.07	–
$SiCl_4$	cub	lq	204.4	7.70	–
	lq	gas	330.8	29.29	88.70
SiF_4	cub	gas	177.5	25.73	–
	cub	lq	182.9	7.07	–
SiH_4	sld	lq	88	0.67	–
	lq	gas	161.3	12.13	–
SiO_2	lq	gas	3223	516.72	–
quartz	trig (α)	hex (β)	846	1.21	1.42
	hex (β)	lq	1823	8.54	4.52
cristobalite	tetr (α)	cub (β)	515	0.84	1.59
	cub (β)	lq	2001	7.70	3.85
tridymite	rhomb (α)	hex (β)	390	0.17	0.42
Sm	trig	cub	1190	3.11	2.64
	cub	lq	1345	8.62	6.40
	lq	gas	ca. 2073	166.41	80.63
Sm_2O_3	cub	mon	1148	1.05	0.88
Sn	cub	tetr	286.4	2.09	7.36
	tetr	lq	505.12	6.99	13.85

(Continued)

Formula	State		Tempera-ture, K	$\Delta H°$ kJ/mole	$\Delta S°$ J/(K·mole)
	Initial	Final			
Sn	lq	gas	2893	296.10	102.26
$SnBr_2$	rhomb	lq	505	7.11	–
	lq	gas	911	92.05	–
$SnBr_4$	mon	rhomb	288.5	–	–
	rhomb	lq	303	12.55	–
	lq	gas	475	41.84	–
$SnCl_2$	rhomb	lq	520	12.55	–
	lq	gas	925	87.86	–
$SnCl_4$	cub	lq	240	9.16	–
	lq	gas	387.3	34.73	–
SnI_2	lq	gas	991	100.42	–
SnI_4	cub	lq	417.7	18.74	–
	lq	gas	619	56.90	–
SnO_2	tetr	sld	698	1.88	–
SnS	rhomb	sld	875	0.67	0.79
	sld	lq	1153	31.59	27.41
Sr	α-cub	β-cub	878	0.84	0.96
	β-cub	lq	1041	10.04	9.62
	lq	gas	1663	141.42	–
$SrBr_2$	rhomb	lq	930	20.08	–
$SrCO_3$	rhomb	trig	1198	19.66	16.44
$SrCl_2$	cub	lq	1146	16.11	14.06
SrF_2	cub	lq	1746	17.99	10.88
SrO	cub	lq	2923	69.87	25.59
$Sr(OH)_2$	tetr	lq	733	22.97	28.45
Ta	cub	lq	3283	31.28	9.62
	lq	gas	5698	770.19	136.61
$TaBr_5$	mon	lq	539.0	44.77	–
	lq	gas	622.0	61.07	–
$TaCl_5$	mon	lq	489.7	46.44	–
	lq	gas	512	54.81	–
TaF_5	lq	gas	502.7	27.61	68.62
Tb	hex	cub	1563	5.02	3.22
	cub	lq	1629	10.79	6.65
	lq	gas	3346	330.89	94.64
Tc	hex	lq	2523	23.01	9.58
	lq	gas	ca. 4573	650.00	–
Te	trig	lq	723.0	17.49	24.18
	lq	gas	1263	51.04	40.42
$TeCl_4$	mon	lq	497	18.83	37.87
	lq	gas	653	76.99	–
TeF_6	sld	gas	234.3	27.07	–
	sld	lq	235.6	8.79	–
TeO_2	rhomb	lq	1006	29.06	51.51
Th	α-cub	β-cub	1636	2.74	1.67
	β-cub	lq	2023	16.12	7.95
	lq	gas	ca. 4473	514.46	101.67

| Formula | State | | Tempera- | $\Delta H°$ | $\Delta S°$ |
	Initial	Final	ture, K	kJ/mole	J/(K·mole)
$ThBr_4$	lq	gas	1130	127.19	–
ThI_4	lq	gas	1110	120.50	–
Ti	hex	cub	1155	4.26	3.68
	cub	lq	1941	17.15	8.79
	lq	gas	3533	421.03	118.20
TiB_2	hex	lq	3123	100.42	–
$TiBr_4$	mon	cub	258	–	–
	cub	lq	311	12.89	–
	lq	gas	504	44.35	–
$TiCl_4$	mon	lq	249.05	9.97	–
	lq	gas	409.6	35.98	–
$(TiFe)O_3$	trig	lq	1640	90.67	55.27
TiI_4	hex	cub	379	9.22	–
	cub	lq	428	19.83	–
	lq	gas	652.7	53.56	–
TiN	cub	lq	ca. 3273	66.94	–
TiO	mon	cub	1264	3.43	2.72
	cub	lq	2053	58.58	–
	lq	gas	3500	425.09	–
TiO_2	tetr (γ)	lq	2143	66.94	–
Ti_2O_3	mon	trig	433	0.90	1.88
	trig	lq	2103	129.70	–
Tl	hex	cub	507	0.38	0.75
	cub	lq	576.8	4.08	7.07
	lq	gas	1748	165.23	93.68
TlBr	cub	lq	734	150.62	–
	lq	gas	1089	102.93	–
TlCl	cub	lq	704	15.56	22.13
	lq	gas	1091	103.76	–
TlF	rhomb	tetr	355	0.38	1.09
	tetr	lq	595	13.87	23.30
TlI	rhomb	cub	451	–	–
	cub	lq	715	11.30	–
	lq	gas	1106	104.18	–
$TlNO_3$	rhomb	hex	348	1.00	3.01
	hex	cub	416.7	3.18	7.61
	cub	lq	479.7	9.58	–
Tl_2O	hex	lq	576	30.29	35.56
Tl_2S	trig	lq	722.1	12.55	–
Tl_2SO_4	rhomb	hex	773	–	–
	hex	lq	905	23.01	–
Tm	hex	lq	1818	16.84	9.29
	lq	gas	2220	190.67	85.90
U	rhomb	tetr	941	2.79	2.97
	tetr	cub	1048	4.76	4.56
	cub	lq	1407	8.52	6.07
	lq	gas	4473	464.07	105.31

(*Continued*)

Formula	State		Tempera-ture, K	$\Delta H°$ kJ/mole	$\Delta S°$ J/(K·mole)
	Initial	Final			
UBr_3	hex	lq	1003	46.02	–
	hex	gas	1003	287.02	–
UBr_4	mon	gas	792	175.31	–
	lq	gas	1038	129.70	–
UCl_3	hex	gas	1108	263.59	–
UCl_4	tetr	cub	820	–	–
	cub	gas	863	193.72	–
UF_4	mon	sld	1118	–	–
	sld	lq	1309	42.84	32.22
	lq	gas	2003	240.58	–
UF_6	rhomb	lq	337	19.20	56.94
	rhomb	gas	337	49.37	–
V	cub	lq	2193	20.93	9.62
	lq	gas	3723	451.89	122.72
VCl_4	lq	gas	426	37.66	88.70
VCl_3O	lq	gas	399.9	34.73	–
VO_2	mon	tetr	340	4.29	12.43
	tetr	lq	1913	56.92	31.32
V_2O_5	rhomb	lq	963	65.10	69.04
W	cub	lq	3660	35.15	9.62
	lq	gas	ca. 5953	823.91	141.38
WCl_5	mon	lq	521	20.92	–
	lq	gas	560	61.50	–
WCl_6	hex	cub	500.8	14.23	–
	cub	lq	548	10.88	–
	lq	gas	620	55.65	–
WCl_4O	tetr	lq	484	7.53	–
WF_6	cub	lq	275.2	2.09	7.53
	lq	gas	290.5	26.15	–
WO_3	tricl	mon	291	–	–
	mon	rhomb	603	–	–
	rhomb	tetr	1013	1.72	1.63
	tetr	lq	1746	73.43	42.09
Xe	cub	lq	161.30	0.55	–
	lq	gas	165.03	3.02	–
Y	hex	cub	1755	4.99	2.85
	cub	lq	1801	11.40	6.32
	lq	gas	3595	363.34	100.63
Yb	α-cub	β-cub	1033	1.75	1.67
	β-cub	lq	1097	7.66	6.99
	lq	gas	1484	128.87	87.82
Zn	hex	lq	692.7	7.38	10.67
	lq	gas	1179.4	115.56	97.61
$ZnBr_2$	tetr	gas	542	125.94	–
$ZnCl_2$	tetr	lq	566	10.25	17.41
	lq	gas	1006	129.29	–
ZnF_2	lq	gas	1775	180.33	–

(Continued)

Formula	State		Tempera-	ΔH°	ΔS°
	Initial	Final	ture, K	kJ/mole	J/(K·mole)
ZnI_2	tetr	gas	518	115.48	–
ZnS	cub (α)	hex (β)	1293	13.39	10.33
Zr	hex	cub	1136	3.94	3.47
	cub	lq	2128	20.92	10.04
	lq	gas	ca. 4613	582.04	124.31
ZrB_2	hex	lq	3473	104.60	–
$ZrBr_4$	cub	gas	630	107.95	–
$ZrCl_2$	amor	lq	995	26.78	–
$ZrCl_4$	mon	gas	604	105.86	–
	mon	lq	710	37.66	–
ZrF_4	tetr	mon	963	–	–
	mon	lq	1183	64.22	53.30
ZrI_4	cub	gas	691	121.34	–
ZrN	cub	lq	3253	67.36	–
ZrO_2	mon	tetr	1448	5.94	4.02
	tetr	lq	2973	87.03	29.50
	lq	gas	ca. 4573	624.25	–

4.3 Electrode Potentials
in Aqueous Solution at 298.15 K [15, 18, 29, 30, 33, 35, 36]

The electrochemical half-reaction of reduction of some oxidized form (an atom, molecule, ion or a formula entity)

$$Oxidized\ form\ (Of) + ne^- = Reduced\ form\ (Rf)$$

is characterized by the magnitude of standard potential E° for the Of/Rf pair measured with respect to the standard hydrogen electrode (pH = 0 and the pressure $H_{2(gas)}$ equals the standard atmospheric pressure).

The aqueous solution *medium*, provided it participates in the half-reaction (and, therefore, the value of E° pertains to the medium too) is designated as follows:

a) *acid* stands for the acid medium, viz., either H^+ ions

$$Of + hH^+ + ne^- = Rf$$

or an acid pair H^+/H_2O

$$Of + hH^+ + ne^- = Rf + mH_2O$$

b) *alk* stands for an alkaline medium, viz., either OH^- ions

$$Of + ne^- = Rf + gOH^-$$

or an alkaline pair OH^-/H_2O

$$Of + mH_2O + ne^- = Rf + gOH^-$$

c) *neutr* denotes a neutral medium (H_2O molecules)

$$Of + mH_2O + ne^- = Rf$$

In the table, the sign "−" denotes an medium that takes no part in the equation of reduction half-reaction.

If the value of E° is listed for a particular medium (either E°_{acid} at pH = 0 or E°_{alk} at pH = 14), then the value of E° in another medium can be calculated using the formula

$$E^\circ_{alk} = E^\circ_{acid} - 0.828h/n$$

where h and n are the coefficients for H^+ and e^- in the equation of the half-reaction for the acid medium.

To calculate the non-standard values of E in the $1 < pH < 14$ range, use is made of the formula

$$E = E^\circ_{acid} - 5.917 \cdot 10^{-2}(h \cdot pH/n)$$

The tabulated data make it possible to calculate the values of E° for the Of/Rf pairs not listed on one and the same line, but having the Of and Rf values listed on different lines. The value of E° for the Of/Rf pair of interest is obtained by the algebraic summation of the values of E° (provided these values refer to the medium of the same type) for the available pairs. For example, the value of E°_{acid} for the $2BrO_3^-/Br_2$ pair can be found by drawing the following scheme (*Latimer diagram*):

$$2BrO_3^- \xrightarrow[+1.495V]{8e^-} 2HBrO \xrightarrow[+1.574V]{2e^-} Br_2$$

$$\downarrow \qquad 10e^- \qquad \uparrow$$

$$E^\circ_{acid} = ?$$

and calculating therefrom the value of E°_{acid} for the $2BrO_3^-/Br_2$ pair as follows: $10E^\circ_{acid} = 8 \cdot 1.495V + 2 \cdot 1.574V$, so that $E^\circ_{acid} = +1.511V$.

All Ofs and Rfs in an aqueous solution are assumed to be hydrated. The state of aggregation for liquid and gaseous substances is indicated by (lq) and (gas), respectively, and is omitted for solid substances.

Oxidized form	$+ne^-$	Reduced form	E°, V	Medium
Ac^{3+}	3	Ac	−2.600	−
Ag^+	1	Ag	+0.799	−

(*Continued*)

Oxidized form	$+ne^-$	Reduced form	$E°$, V	Medium
$Ag^{II(1)}$	1	Ag^+	+1.980	–
$Ag^{III(1)}$	1	$Ag^{II(2)}$	+2.100	–
$(Ag^IAg^{III})O_2$	2	Ag_2O	+0.599	alk
$AgBr$	1	$Ag + Br^-$	+0.071	–
$AgBrO_3$	1	$Ag + BrO_3^-$	+0.558	–
$AgCH_3COO$	1	$Ag + CH_3COO^-$	+0.640	–
$AgCN$	1	$Ag + CN^-$	–0.040	–
$[Ag(CN)_2]^-$	1	$Ag + 2CN^-$	–0.430	–
$[Ag(CN)_3]^-$	1	$Ag + 3CN^-$	–0.510	–
Ag_2CO_3	2	$2Ag + CO_3^{2-}$	+0.471	–
$Ag_2C_2O_4$	2	$2Ag + C_2O_4^{2-}$	+0.465	–
$AgCl$	1	$Ag + Cl^-$	+0.222	–
Ag_2CrO_4	2	$2Ag + CrO_4^{2-}$	+0.477	–
AgI	1	$Ag + I^-$	–0.151	–
$AgIO_3$	1	$Ag + IO_3^-$	+0.355	–
AgN_3	1	$Ag + N_3^-$	+0.293	–
$AgNCS$	1	$Ag + NCS^-$	+0.090	–
$[Ag(NH_3)_2]^+$	1	$Ag + 2NH_{3(gas)}$	+0.164	–
$[Ag(NH_3)_2]^+$	1	$Ag + 2(NH_3 \cdot H_2O)$	+0.367	neutr
$AgNO_2$	1	$Ag + NO_2^-$	+0.564	–
Ag_2O	2	$2Ag$	+0.342	alk
Ag_2O_3	2	$(Ag^IAg^{III})O_2$	+0.740	alk
$AgOCN$	1	$Ag + OCN^-$	+0.410	–
$[Ag(OH)_2]^-$	1	Ag	+0.563	alk
Ag_2S (mon)	2	$2Ag + S^{2-}$	–0.655	–
Ag_2S (mon)	2	$2Ag + HS^-$	–0.687	alk
Ag_2S (mon)	2	$2Ag + H_2S$	–0.066	acid
Ag_2S (mon)	2	$2Ag + H_2S_{(gas)}$	–0.037	acid
$[Ag(SO_3)_2]^{3-}$	1	$Ag + 2SO_3^{2-}$	+0.430	–
Ag_2SO_4	2	$2Ag + SO_4^{2-}$	+0.653	–
$[Ag(SO_3S)_2]^{3-}$	1	$Ag + 2SO_3S^{2-}$	+0.004	–
Al^{3+}	3	Al	–1.700	–
$[AlF_6]^{3-}$	3	$Al + 6F^-$	–2.112	–
$AlOH^{2+}$	3	Al	–1.601	acid
$Al(OH)_2^+$	3	Al	–1.496	acid
$Al(OH)_3$	3	Al	–1.538	acid
			–2.366	alk
$Al(OH)_3$ (amor)	3	Al	–1.538	acid
			–2.316	alk
$[Al(OH)_4]^-$	3	Al	–2.336	alk
$\alpha\text{-AlO(OH)}$	3	Al	–1.544	acid
$\gamma\text{-AlO(OH)}$	3	Al	–1.516	acid
Am^{3+}	3	Am	–2.320	–
$Am^{IV(1)}$	1	Am^{3+}	+2.250	–
AmO_2^+	2	Am^{3+}	+1.720	acid
AmO_2^+	1	$Am^{IV(2)}$	+1.150	acid
AmO_2^{2+}	1	AmO_2^+	+1.640	–

(Continued)

Oxidized form	$+ne^-$	Reduced form	$E°$, V	Medium
$Am(OH)_3$	3	Am	−2.700	alk
As	3	AsH_3	−1.370	alk
As	3	$AsH_{3(gas)}$	−0.238	acid
			−1.066	alk
AsO_2^-	3	As	−0.680	alk
AsO_3^{3-}	3	As	−1.572	alk
AsO_4^{3-}	2	AsO_2^-	−0.710	alk
AsO_4^{3-}	2	AsO_3^{3-}	−0.658	alk
As_2S_3	6	$2As + 3S^{2-}$	−0.609	−
As_2S_3	6	$2As + 3HS^-$	−0.641	alk
As_2S_3	6	$2As + 3H_2S$	−0.020	acid
As_2S_3	6	$2As + 3H_2S_{(gas)}$	+0.009	acid
AsS_3^{3-}	3	$As + 3S^{2-}$	−0.750	−
AsS_4^{3-}	2	$AsS_3^{3-} + S^{2-}$	−0.600	−
$HAsO_2$	3	As	+0.234	acid
H_3AsO_3	3	As	+0.226	acid
H_3AsO_4	2	$HAsO_2$	+0.560	acid
H_3AsO_4	2	H_3AsO_3	+0.586	acid
$2At^+$	2	At_2	+0.700	−
At_2	2	$2At^-$	+0.200	−
$2AtO^-$	2	At_2	±0.000	alk
AtO_3^-	4	At^+	+1.400	acid
AtO_3^-	4	AtO^-	+0.500	alk
Au⁺	1	Au	+1.691	−
$Au^{III(1)}$	3	Au	+1.498	−
$[AuBr_2]^-$	1	$Au + 2Br^-$	+0.960	−
$[AuBr_4]^-$	2	$[AuBr_2]^- + 2Br^-$	+0.800	−
$[Au(CN)_2]^-$	1	$Au + 2CN^-$	−0.764	−
AuCl	1	$Au + Cl^-$	+1.157	−
$[AuCl_2]^-$	1	$Au + 2Cl^-$	+1.154	−
$[AuCl_4]^-$	3	$Au + 4Cl^-$	+1.002	−
$[AuCl_4]^-$	2	$AuCl + 3Cl^-$	+0.924	−
$[AuCl_4]^-$	2	$[AuCl_2]^- + 2Cl^-$	+0.926	−
$[AuCl_3(OH)]^-$	3	$Au + 3Cl^-$	+1.123	acid
$[Au(NCS)_2]^-$	1	$Au + 2NCS^-$	+0.690	−
$[Au(NCS)_4]^-$	2	$[Au(NCS)_2]^- + 2NCS^-$	+0.645	−
Au_2O_3	6	2Au	+0.535	alk
Au_2O_3	4	Au_2O ?	+1.640	acid
$[Au(OH)_4]^-$	3	Au	+0.490	alk
AuO(OH)	3	Au	+1.253	acid
			+0.622	alk
$[BF_4]^-$	3	$B\ (amor) + 4F^-$	−1.295	−
$[BF_4]^-$	3	$B + 4F^-$	−1.283	−
$B(OH)_3$	3	$B\ (amor)$	−0.900	acid
$B(OH)_3$	3	B	−0.888	acid
$[B(OH)_4]^-$	3	$B\ (amor)$	−1.822	alk
$[B(OH)_4]^-$	3	B	−1.810	alk

(*Continued*)

Oxidized form	$+ne^-$	Reduced form	$E°$, V	Medium
$[B(OH)_4]^-$	8	$[BH_4]^-$	-1.240	alk
Ba^{2+}	2	Ba	-2.905	—
Ba^{2+}	2	Ba (amalgam)	-1.570	—
Be^{2+}	2	Be	-1.847	—
$[BeF_4]^{2-}$	2	Be + 4HF	-2.042	acid
$Be(OH)_2$	2	Be	-2.599	alk
$[Be(OH)_4]^{2-}$	2	Be	-2.520	alk
Bi$^{III(1)}$	3	Bi	$+0.317$	—
$[BiCl_4]^-$	3	Bi + 4Cl$^-$	$+0.168$	—
$Bi(Cl)O$	3	Bi + Cl$^-$	$+0.158$	acid
$Bi(OH)_3$	3	Bi	-0.383	alk
$[Bi_6(OH)_{12}]^{6+}$	18	6Bi	$+0.215$	acid
$BiO(OH)$	3	Bi	-0.460	alk
Bi_2S_3	6	$2Bi + 3S^{2-}$	-0.709	—
Bi_2S_3	6	$2Bi + 3HS^-$	-0.741	alk
Bi_2S_3	6	$2Bi + 3H_2S$	-0.120	acid
Bi_2S_3	6	$2Bi + 3H_2S_{(gas)}$	-0.091	acid
$NaBiO_3$	2	$Bi^{III(2)} + Na^+$	$+1.808$	acid
$NaBiO_3$	2	$Bi(OH)_3 + Na^+$	$+0.370$	alk
Bk^{4+}	1	Bk^{3+}	$+1.600$	—
Br$_2$	2	$2Br^-$	$+1.087$	—
$Br_{2(lq)}$	2	$2Br^-$	$+1.065$	—
$[Br(Br)_2]^-$	2	$3Br^-$	$+1.050$	—
$2BrO^-$	2	Br_2	$+0.434$	alk
BrO_3^-	6	Br^-	$+1.440$	acid
			$+0.612$	alk
$2BrO_3^-$	10	Br_2	$+1.511$	acid
			$+0.517$	alk
BrO_3^-	4	BrO^-	$+0.538$	alk
BrO_3^-	4	$HBrO$	$+1.495$	acid
BrO_4^-	2	BrO_3^-	$+1.763$	acid
			$+0.935$	alk
$2HBrO$	2	Br_2	$+1.574$	acid
$2HBrO$	2	$Br_{2(lq)}$	$+1.595$	acid
CH_3CHO	2	C_2H_5OH	$+0.190$	acid
CH_3COOH	2	CH_3CHO	-0.118	acid
$C_2N_{2(gas)}$	2	$2CN^-$	-0.236	—
$C_2N_{2(gas)}$	2	$2HCN$	$+0.241$	acid
$CO_{(gas)}$	6	$CH_{4(gas)}$	$+0.260$	acid
$CO_{2(gas)}$	2	$CO_{(gas)}$	-0.104	acid
$2CO_{2(gas)}$	2	$H_2C_2O_4$	-0.470	acid
$CO_{2(gas)}$	2	$HCOOH$	-0.199	acid
CO_3^{2-}	2	$CO_{(gas)}$	-1.223	alk
$2CO_3^{2-}$	2	$C_2O_4^{2-}$	-1.210	alk
CO_3^{2-}	2	$HCOO^-$	-1.015	alk
$HC(H)O$	2	CH_3OH	$+0.232$	acid
$2HCO_3^-$	2	$C_2O_4^{2-}$	-0.993	alk

(*Continued*)

Oxidized form	$+ne^-$	Reduced form	$E°$, V	Medium
$2H_2CO_3$	2	$H_2C_2O_4$	−0.386	acid
$2HCO_3^- + N_{2(gas)}$	10	$2CN^-$	+0.925	alk
$2HNCO$	2	$C_2N_{2(gas)}$	−0.285	acid
$2NCO^-$	2	$C_2N_{2(gas)}$	−1.750	alk
$(SCN)_2$	2	$2NCS^-$	+0.770	−
$\mathbf{Ca^{2+}}$	2	Ca	−2.864	−
$Ca(OH)_2$	2	Ca	−3.018	alk
$\mathbf{Cd^{2+}}$	2	Cd	−0.404	−
Cd^{2+}	2	Cd (amalgam)	−0.352	−
$[Cd(CN)_4]^{2-}$	2	$Cd + 4CN^-$	−1.028	−
$[Cd(NH_3)_4]^{2+}$	2	$Cd + 4(NH_3 \cdot H_2O)$	−0.597	neutr
$CdOH^+$	2	Cd	−0.178	acid
$Cd(OH)_2$	2	Cd	−0.825	alk
CdS	2	$Cd + S^{2-}$	−1.208	−
CdS	2	$Cd + HS^-$	−1.239	alk
CdS	2	$Cd + H_2S$	−0.619	acid
CdS	2	$Cd + H_2S_{(gas)}$	−0.589	acid
$\mathbf{Ce^{3+}}$	3	Ce	−2.483	−
Ce^{3+}	3	Ce (amalgam)	−1.437	−
$Ce^{IV(1)}$	1	Ce^{3+}	+1.743	−
CeO_2	1	Ce^{3+}	+1.293	acid
$\mathbf{Cf^{3+}}$	1	Cf^{2+}	−1.600	−
$Cf^{IV(1)}$	1	Cf^{3+}	+3.200	−
$\mathbf{Cl_2}$	2	$2Cl^-$	+1.396	−
$Cl_{2(gas)}$	2	$2Cl^-$	+1.358	−
ClO^-	2	Cl^-	+0.920	alk
$2ClO^-$	2	$Cl_{2(gas)}$	+0.482	alk
ClO_2^-	2	ClO^-	+0.681	alk
ClO_2	1	ClO_2^-	+0.954	−
$ClO_{2(gas)}$	1	ClO_2^-	+1.032	−
$ClO_{2(gas)}$	1	$HClO_2$	+1.270	acid
$2ClO_3^-$	10	$Cl_{2(gas)}$	+1.470	acid
			+0.476	alk
ClO_3^-	2	ClO_2^-	+0.268	alk
ClO_3^-	1	$ClO_{2(gas)}$	+1.160	acid
			−0.496	alk
ClO_3	2	$HClO_2$	+1.215	acid
ClO_4^-	8	Cl^-	+1.386	acid
			+0.558	alk
ClO_4^-	2	ClO_3^-	+1.189	acid
			+0.361	alk
HClO	2	Cl^-	+1.494	acid
2HClO	2	$Cl_{2(gas)}$	+1.630	acid
$HClO_2$	2	HClO	+1.645	acid
$\mathbf{Cm^{4+}}$	1	Cm^{3+}	+3.500	−
$\mathbf{Co^{2+}}$	2	Co	−0.277	−
Co^{3+}	1	Co^{2+}	+1.380	−

(*Continued*)

Oxidized form	$+ne^-$	Reduced form	$E°$, V	Medium
$[Co(CN)_6]^{3-}$	1	$[Co(CN)_5]^{3-} + CN^-$	-0.830	$-$
$(Co^{II}Co_2^{III})O_4$	2	$3Co(OH)_2$	-0.160	alk
$(Co^{II}Co_2^{III})O_4$	2	$3Co(OH)_2$ (amor)	-0.220	alk
$[Co(NH_3)_6]^{2+}$	2	$Co + 6(NH_3 \cdot H_2O)$	-0.411	neutr
$[Co(NH_3)_6]^{3+}$	1	$[Co(NH_3)_6]^{2+}$	$+0.178$	$-$
$Co(OH)_2$	2	Co	-0.739	alk
$3CoO(OH)$	1	$[Co^{II}Co_2^{III}]O_4$	$+0.887$	alk
$CoO(OH)$	1	$Co(OH)_2$	$+0.189$	alk
CoS	2	$Co + S^{2-}$	-0.882	$-$
CoS	2	$Co + HS^-$	-0.914	alk
CoS	2	$Co + H_2S$	-0.294	acid
CoS	2	$Co + H_2S_{(gas)}$	-0.264	acid
Cr^{2+}	2	Cr	-0.852	$-$
Cr^{3+}	1	Cr^{2+}	-0.409	$-$
$[Cr(CN)_6]^{3-}$	1	$[Cr(CN)_6]^{4-}$	-1.280	$-$
$CrCl^{2+}$	1	$Cr^{2+} + 2Cl^-$	-0.403	$-$
CrO_4^{2-}	3	$Cr(OH)_3$	-0.125	alk
CrO_4^{2-}	3	$[Cr(OH)_6]^{3-}$	-0.165	alk
$Cr_2O_7^{2-}$	6	$2Cr^{3+}$	$+1.333$	acid
$CrOH_2^+$	3	Cr	-0.626	acid
$Cr(OH)_2$	2	Cr	-1.355	alk
$Cr(OH)^{2+}$	3	Cr	-0.516	acid
$Cr(OH)_3$	1	$Cr(OH)_2$	-1.175	alk
$[Cr(OH)_6]^{3-}$	1	$Cr(OH)_2$	-1.057	alk
$HCrO_4^-$	3	Cr^{3+}	$+1.363$	acid
Cs^+	1	Cs	-2.923	$-$
Cu^+	1	Cu	$+0.518$	$-$
Cu^{2+}	2	Cu	$+0.338$	$-$
Cu^{2+}	2	Cu (amalgam)	$+0.345$	$-$
Cu^{2+}	1	Cu^+	$+0.158$	$-$
$Cu^{2+} + Br^-$	1	$CuBr$	$+0.657$	$-$
$Cu^{2+} + 2CN^-$	1	$[Cu(CN)_2]^-$	$+1.105$	$-$
$Cu^{2+} + Cl^-$	1	$CuCl$	$+0.551$	$-$
$Cu^{2+} + 2Cl^-$	1	$[CuCl_2]^-$	$+0.486$	$-$
$Cu^{2+} + I^-$	1	CuI	$+0.860$	$-$
$Cu^{2+} + 2I^-$	1	$[CuI_2]^-$	$+0.690$	$-$
$2Cu^{2+}$	2	Cu_2O	$+0.214$	acid
$[Cu(CN)_2]^-$	1	$Cu + 2CN^-$	-0.429	$-$
$CuCl$	1	$Cu + Cl^-$	$+0.124$	$-$
$[CuCl_2]^-$	1	$Cu + 2Cl^-$	$+0.190$	$-$
CuI	1	$Cu + I^-$	-0.184	$-$
$[CuI_2]^-$	1	$Cu + 2I^-$	±0.000	$-$
$CuNCS$	1	$Cu + NCS^-$	-0.270	$-$
$[Cu(NH_3)_2]^+$	1	$Cu + 2NH_{3(gas)}$	$+0.030$	$-$
$[Cu(NH_3)_2]^+$	1	$Cu + 2(NH_3 \cdot H_2O)$	-0.120	neutr
$[Cu(NH_3)_4]^{2+}$	2	$Cu + 4NH_{3(gas)}$	-0.232	$-$
$[Cu(NH_3)_4]^{2+}$	2	$Cu + 4(NH_3 \cdot H_2O)$	-0.065	neutr

(*Continued*)

Oxidized form	$+ne^-$	Reduced form	$E°$, V	Medium
$2CuO$	2	Cu_2O	+0.669	acid
			−0.159	alk
Cu_2O	2	$2Cu$	+0.462	acid
			−0.366	alk
$CuOH^+$	2	Cu	+0.555	acid
$Cu(OH)_2$	2	Cu	−0.226	alk
$2Cu(OH)_2$	2	Cu_2O	−0.086	alk
$[Cu(OH)_4]^{2-}$	2	Cu	−0.132	alk
CuS	2	$Cu + S^{2-}$	−0.722	−
CuS	2	$Cu + HS^-$	−0.754	alk
CuS	2	$Cu + H_2S$	−0.133	acid
CuS	2	$Cu + H_2S_{(gas)}$	−0.104	acid
Cu_2S	2	$2Cu + S^{2-}$	−0.891	−
Cu_2S	2	$2Cu + HS^-$	−0.923	alk
Cu_2S	2	$2Cu + H_2S$	−0.302	acid
Cu_2S	2	$2Cu + H_2S_{(gas)}$	−0.273	acid
$2D^+$	2	$D_{2(gas)}$	−0.044	−
$2D_2O$	2	$D_{2(gas)} + 2OD^-$	−0.870	−
$\mathbf{Dy^{3+}}$	3	Dy	−2.356	−
$Dy(OH)_3$	3	Dy	−2.780	alk
$\mathbf{Er^{3+}}$	3	Er	−2.296	−
$Er(OH)_3$	3	Er	−2.830	alk
$\mathbf{Es^{3+}}$	1	Es^{2+}	−1.500	−
$\mathbf{Eu^{2+}}$	2	Eu	−3.395	−
Eu^{3+}	3	Eu	−2.406	−
Eu^{3+}	1	Eu^{2+}	−0.550	−
$Eu(OH)_3$	3	Eu	−2.830	alk
$\mathbf{F}_{(gas)}$	1	F^-	+2.850	−
$F_{(gas)}$	1	HF	+3.030	acid
$F_{2(gas)}$	2	$2F^-$	+2.866	−
$F_{2(gas)}$	2	$2HF$	+3.090	acid
$F_{2(gas)}$	2	HF_2^-	+3.013	acid
$\mathbf{Fe^{2+}}$	2	Fe	−0.441	−
Fe^{3+}	3	Fe	−0.037	−
Fe^{3+}	1	Fe^{2+}	+0.771	−
$Fe^{3+} + HS^-$	1	FeS	+1.809	alk
$[Fe(phen)_3]^{3+}$	1	$[Fe(phen)_3]^{2+}$	+1.140	−
$[Fe(CN)_6]^{4-}$	2	$Fe + 6CN^-$	−1.728	−
$[Fe(CN)_6]^{3-}$	1	$[Fe(CN)_6]^{4-}$	+0.543	−
$[Fe(C_2O_4)_3]^{3-}$	1·	$[Fe(C_2O_4)_2]^{2-} + C_2O_4^{2-}$	+0.020	−
$[FeF_6]_3$	1	$Fe^{2+} + 6F^-$	+0.400	−
$(Fe^{II}Fe_2^{III})O_4$	8	$3Fe$	−0.086	acid
			−0.914	alk
FeO_4^{2-}	3	Fe^{3+}	+1.900	acid
FeO_4^{2-}	3	$FeO(OH)$	+0.720	alk
Fe_2O_3	2	$2Fe(OH)_2$	−0.794	alk
$FeOH^+$	2	Fe	−0.279	acid

Oxidized form	$+ne^-$	Reduced form	$E°$, V	Medium
$FeOH^{2+}$	3	Fe	−0.019	acid
$Fe(OH)_2$	2	Fe	−0.875	alk
$Fe(OH)_2^+$	3	Fe	+0.072	acid
$[Fe(OH)_6]^{4-}$	2	Fe	−0.762	alk
$FeO(OH)$	1	$Fe(OH)_2$	−0.666	alk
FeS	2	$Fe + S^{2-}$	−0.965	−
FeS	2	$Fe + HS^-$	−0.997	alk
Fm^{3+}	1	Fm^{2+}	−1.000	−
Ga^{3+}	3	Ga	−0.560	−
$GaOH^{2+}$	3	Ga	−0.505	acid
$Ga(OH)_2^+$	3	Ga	−0.435	acid
$Ga(OH)_3$	3	Ga	−1.258	alk
$[Ga(OH)_4]^-$	3	Ga	−1.326	alk
Gd^{3+}	3	Gd	−2.397	−
$Gd(OH)_3$	3	Gd	−2.822	alk
Ge^{2+} ?	2	Ge	−0.240	−
GeO ?	2	Ge	+0.156	acid
GeO_2 (trig)	4	Ge	−0.069	acid
GeO_2 (trig)	2	Ge^{2+} ?	+0.052	acid
GeO_2 (trig)	2	GeO ?	−0.293	acid
$[Ge(OH)_6]^{2-}$	4	Ge	−0.972	alk
H_2GeO_3	4	Ge	−0.129	acid
H_2GeO_3	2	Ge^{2+} ?	−0.068	acid
$H_{(gas)}$	1	H^-	−2.230	−
H^+	1	$H_{(gas)}$	−2.107	−
$2H^+$	2	$H_{2(gas)}$	±0.000	−
$2H^+$ (pH = 7)	2	$H_{2(gas)}$	−0.414	−
$H_{2(gas)} + Ca^{2+}$	2	CaH_2	−2.157	−
$2H_{2(gas)} + Li^+ + Al^{3+}$	4	$Li[AlH_4]$	−1.896	−
$2H_{2(gas)} + SiO_2$	4	SiH_4	−1.138	acid
H_2O	1	$H_{(gas)}$	−2.935	alk
$2H_2O$	2	$H_{2(gas)}$	−0.828	alk
$2NH_4^+$	2	$H_{2(gas)} + 2(NH_3·H_2O)$	−0.547	neutr
$Hf^{IV(I)}$	4	Hf	−1.474	−
HfO_2	4	Hf	−1.570	acid
$Hf(OH)_2^{2+}$	4	Hf	−1.680	acid
$HfO(OH)_2$	4	Hf	−2.600	alk
Hg^{2+}	2	$Hg_{(lq)}$	+0.852	−
Hg_2^{2+}	2	$2Hg_{(lq)}$	+0.796	−
$2Hg^{2+}$	2	Hg_2^{2+}	+0.908	−
$[HgBr_4]^{2-}$	2	$Hg_{(lq)} + 4Br^-$	+0.223	−
Hg_2Br_2	2	$2Hg_{(lq)} + 2Br^-$	+0.142	−
$[Hg(CN)_4]^{2-}$	2	$Hg_{(lq)} + 4CN^-$	−0.370	−
Hg_2CO_3	2	$2Hg_{(lq)} + CO_3^{2-}$	+0.306	−
$2HgCl_2$	2	$Hg_2Cl_2 + 2Cl^-$	+0.657	−
$[HgCl_4]^{2-}$	2	$Hg_{(lq)} + 4Cl^-$	+0.438	−
Hg_2Cl_2	2	$2Hg_{(lq)} + 2Cl^-$	+0.268	−

(Continued)

Oxidized form	$+ne^-$	Reduced form	$E°$, V	Medium
$[HgI_4]^{2-}$	2	$Hg_{(liq)} + 4I^-$	-0.038	–
Hg_2I_2	2	$2Hg_{(liq)} + 2I^-$	-0.041	–
HgO	2	$Hg_{(liq)}$	$+0.099$	alk
$2HgO$	2	Hg_2^{2+}	-0.599	alk
$2HgO + CO_3^{2-}$	2	Hg_2CO_3	-0.109	alk
HgS (cub)	2	$Hg_{(liq)} + S^{2-}$	-0.692	–
HgS (cub)	2	$Hg_{(liq)} + HS^-$	-0.724	alk
HgS (cub)	2	$Hg_{(liq)} + H_2S$	$+0.103$	acid
HgS (cub)	2	$Hg_{(liq)} + H_2S_{(gas)}$	$+0.073$	acid
$2HgS$ (cub)	2	$Hg_2^{2+} + 2S^{2-}$	-2.199	–
$2HgS$ (cub)	2	$Hg_2^{2+} + 2HS^-$	-2.262	alk
$2HgS$ (cub)	2	$Hg_2^{2+} + 2H_2S$	-1.021	acid
$2HgS$ (cub)	2	$Hg_2^{2+} + 2H_2S_{(gas)}$	-0.962	acid
Hg_2SO_4	2	$2Hg_{(liq)} + SO_4^{2-}$	$+0.615$	–
Ho^{3+}	3	Ho	-2.319	–
$Ho(OH)_3$	3	Ho	-2.770	alk
I_2	2	$2I^-$	$+0.535$	–
$2[I(Br)_2]^-$	2	$I_2 + 4Br^-$	$+0.870$	–
$[I(I)_2]^-$	2	$3I^-$	$+0.534$	–
IO^-	2	I^-	$+0.484$	alk
$2IO^-$	2	I_2	$+0.433$	alk
IO_3^-	6	I^-	$+1.081$	acid
			$+0.253$	alk
$2IO_3^-$	10	I_2	$+1.190$	acid
			$+0.196$	alk
IO_3^-	4	HIO	$+1.128$	acid
IO_3^-	4	IO^-	$+0.137$	alk
IO_4^-	2	IO_3^-	$+1.653$	acid
HIO	2	I^-	$+0.987$	acid
$2HIO$	2	I_2	$+1.439$	acid
$H_3IO_6^{2-}$	2	IO_3^-	$+0.700$	alk
H_5IO_6	2	IO_3^-	$+1.601$	acid
In^{3+}	3	In	-0.338	–
$InOH^{2+}$	3	In	-0.268	acid
$In(OH)_2^+$	3	In	-0.184	acid
$In(OH)_3$	3	In	-1.066	alk
In_2S_3	6	$2In + 3H_2S$	-0.568	acid
In_2S_3	3	$2In + 3H_2S_{(gas)}$	-0.539	acid
Ir^{3+}	3	Ir	$+1.000$	–
$[IrBr_6]^{2-}$	1	$[IrBr_6]^{3-}$	$+0.947$	–
$[IrCl_6]^{3-}$	3	$Ir + 6Cl^-$	$+0.720$	–
$[IrCl_6]^{2-}$	1	$[IrCl_6]^{3-}$	$+1.017$	–
$[IrI_6]^{2-}$	1	$[IrI_6]^{3-}$	$+0.480$	–
IrO_2	1	Ir^{3+}	$+0.720$	acid
Ir_2O_3	6	$2Ir$	$+0.098$	alk
K^+	1	K	-2.924	–
La^{3+}	3	La	-2.522	–

Oxidized form	$+ne^-$	Reduced form	$E°$, V	Medium
$La(OH)_3$	3	$La + 3OH^-$	-2.760	–
Li^+	1	Li	-3.045	–
Lu^{3+}	3	Lu	-2.255	–
$Lu(OH)_3$	3	Lu	-2.720	alk
Md^{3+}	1	Md^{2+}	-0.150	–
Mg^{2+}	2	Mg	-2.370	–
$MgOH^+$	2	Mg	-2.023	acid
$Mg(OH)_2$	2	Mg	-2.689	alk
Mn^{2+}	2	Mn	$-.1.192$	–
Mn^{3+}	1	Mn^{2+}	$+1.499$	–
$[Mn(CN)_6]^{3-}$	1	$[Mn(CN)_6]^{4-}$	-0.244	–
$(Mn^{II}Mn_2^{III})O_4$	2	$3Mn(OH)_2$	-0.324	alk
MnO_2	2	Mn^{2+}	$+1.239$	acid
$3MnO_2$	4	$(Mn^{II}Mn_2^{III})O_4$	$+0.098$	alk
$2MnO_2$	2	Mn_2O_3	$+0.139$	alk
MnO_2	2	$Mn(OH)_2$	-0.043	alk
MnO_2	1	$MnO(OH)$	-0.260	alk
MnO_4^{2-}	2	MnO_2	$+2.308$	acid
			$+0.652$	alk
MnO_4^-	7	Mn	$+0.753$	acid
			-0.194	alk
MnO_4^-	5	Mn^{2+}	$+1.531$	acid
MnO_4^-	3	MnO_2	$+1.725$	acid
			$+0.621$	alk
MnO_4^-	1	MnO_4^{2-}	$+0.558$	–
MnO_4^-	5	$Mn(OH)_2$	$+0.355$	alk
Mn_2O_3	2	$2Mn(OH)_2$	-0.225	alk
$MnOH^+$	2	Mn	-0.878	acid
$Mn(OH)_2$	2	Mn	-1.566	alk
$MnO(OH)$	1	$Mn(OH)_2$	$+0.174$	alk
MnS	2	$Mn + S^{2-}$	-1.575	–
MnS	2	$Mn + HS^-$	-1.607	alk
MnS	2	$Mn + H_2S$	-0.986	acid
MnS	2	$Mn + H_2S_{(gas)}$	-0.956	acid
Mo^{3+}	3	Mo	-0.200	–
$[Mo(CN)_8]^{3-}$	1	$[Mo(CN)_8]^{4-}$	$+0.726$	–
$[Mo(CN)_7]^{4-}$	1	$[Mo(CN)_7]^{5-}$	$+0.730$	–
MoO_2	4	Mo	-0.983	alk
MoO_3	2	MoO_2	$+0.537$	acid
$MoO_3 (\cdot 2H_2O)$	6	Mo	±0.000	acid
MoO_4^{2-}	6	Mo	-0.913	alk
MoO_4^{2-}	2	MoO_2	-0.772	alk
$3N_{2(gas)}$	2	$2HN_3$	-3.100	acid
$N_{2(gas)}$	6	$2NH_{3(gas)}$	-0.771	alk
$N_{2(gas)}$	6	$2NH_4^+$	$+0.272$	acid
$N_{2(gas)}$	4	$N_2H_5^+$	-0.227	acid
$N_{2(gas)}$	6	$2(NH_3 \cdot H_2O)$	-0.737	alk

(Continued)

Oxidized form	$+ne^-$	Reduced form	$E°$, V	Medium
$N_{2(gas)}$	4	$N_2H_4 \cdot H_2O$	-1.119	alk
$N_{2(gas)}$	2	$2NH_3OH^+$	-1.871	acid
$N_{2(gas)}$	2	$2(NH_2OH \cdot H_2O)$	-3.043	alk
$N_2H_5^+$	2	$2NH_4^+$	$+1.270$	acid
NH_3OH^+	2	NH_4^+	$+1.344$	acid
$2NH_3OH^+$	2	$N_2H_5^+$	$+1.417$	acid
$NH_2OH \cdot H_2O$	2	$NH_3 \cdot H_2O$	$+0.416$	alk
$2(NH_2OH \cdot H_2O)$	2	$N_2H_4 \cdot H_2O$	$+0.805$	alk
$N_2H_4 \cdot H_2O$	2	$2(NH_3 \cdot H_2O)$	$+0.027$	alk
$2NO_{(gas)}$	2	$N_2O_{(gas)}$	$+1.600$	acid
			$+0.772$	alk
$2NO_{(gas)}$	2	$N_2O_2^{2-}$	$+0.100$	–
$2NO_2^-$	6	$N_{2(gas)}$	$+0.410$	alk
NO_2^-	1	$NO_{(gas)}$	-0.453	alk
$2NO_2^-$	4	$N_2O_{(gas)}$	$+0.160$	alk
$2NO_2^-$	4	$N_2O_2^{2-}$	-0.177	alk
$NO_{2(gas)}$	2	$NO_{(gas)}$	$+1.046$	acid
$NO_{2(gas)}$	1	HNO_2	$+1.088$	acid
$NO_{2(gas)}$	1	NO_2^-	$+0.889$	–
$2NO_3^-$	10	$N_{2(gas)}$	$+1.244$	acid
			$+0.250$	alk
NO_3^-	2	HNO_2	$+0.930$	acid
NO_3^-	8	$NH_{3(gas)}$	-0.133	alk
NO_3^-	8	NH_4^+	$+0.880$	acid
$2NO_3^-$	14	$N_2H_5^+$	$+0.824$	acid
NO_3^-	8	$NH_3 \cdot H_2O$	-0.120	alk
$2NO_3^-$	14	$N_2H_4 \cdot H_2O$	-0.141	alk
NO_3^-	6	NH_3OH^+	$+0.725$	acid
NO_3^-	6	$NH_2OH \cdot H_2O$	-0.299	alk
NO_3^-	3	$NO_{(gas)}$	$+0.955$	acid
NO_3^-	2	NO_2^-	$+0.010$	alk
NO_3^-	1	$NO_{2(gas)}$	$+0.772$	acid
			-0.884	alk
$2NO_3^-$	8	$N_2O_{(gas)}$	$+1.116$	acid
$2NO_3^-$	2	$N_2O_{4(lq)}$	$+0.810$	acid
			-0.846	alk
$N_2O_{(gas)}$	2	$N_{2(gas)}$	$+1.756$	acid
$N_2O_{(gas)}$	8	$2NH_4^+$	$+0.644$	acid
$N_2O_{4(lq)}$	2	$2HNO_2$	$+1.051$	acid
$N_2O_{4(lq)}$	4	$2NO$	$+1.028$	acid
$N_2O_{4(lq)}$	2	$2NO_2^-$	$+0.866$	–
HN_3	8	NH_4^+	$+0.660$	acid
$2HNO_2$	4	$H_2N_2O_2$	$+0.855$	acid
$2HNO_2$	6	$N_{2(gas)}$	$+1.453$	acid
HNO_2	1	$NO_{(gas)}$	$+1.004$	acid
$2HNO_2$	4	$N_2O_{(gas)}$	$+1.302$	acid
$H_2N_2O_2$	2	$N_{2(gas)}$	$+2.650$	acid

(*Continued*)

Oxidized form	$+ne^-$	Reduced form	$E°$, V	Medium
Na$^+$	1	Na	-2.711	$-$
Nb$^{V(1)}$	5	Nb	-0.522	$-$
NbO$_3^-$	5	Nb	-0.480	acid
			-1.474	alk
Nb$_2$O$_5$	10	2Nb	-0.599	acid
Nd^{3+}	3	Nd	-2.431	$-$
Nd(OH)$_3$	3	Nd	-2.540	alk
Ni^{2+}	2	Ni	-0.234	$-$
[Ni(NH$_3$)$_6$]$^{2+}$	2	Ni + 6(NH$_3$·H$_2$O)	-0.490	neutr
NiOH$^+$	2	Ni	$+0.089$	acid
Ni(OH)$_2$	2	Ni	-0.749	alk
NiO(OH)	1	Ni^{2+}	$+2.252$	acid
NiO(OH)	1	Ni(OH)$_2$	$+0.784$	alk
NiS	2	Ni + S^{2-}	-0.856	$-$
NiS	2	Ni + HS$^-$	-0.888	alk
NiS	2	Ni + H$_2$S	-0.268	acid
NiS	2	Ni + H$_2$S$_{(gas)}$	-0.238	acid
No^{3+}	1	No^{2+}	$+1.400$	$-$
Np^{3+}	3	Np	-1.856	$-$
Np^{4+}	1	Np^{3+}	$+0.155$	$-$
NpO$_2^+$	1	Np^{4+}	$+0.739$	$-$
NpO$_2^{2+}$	1	NpO$_2^+$	$+1.130$	$-$
Np(OH)$_3$	3	Np	-2.248	alk
Np(OH)$_4$	1	Np(OH)$_3$	-1.756	alk
NpO$_2$(OH)	1	Np(OH)$_4$	$+0.391$	alk
NpO$_2$(OH)$_2$	1	NpO$_2$(OH)	$+0.482$	alk
O$_{(gas)}$ + 2H$^+$ (pH = 7)	2	H$_2$O	$+0.815$	$-$
O$_{(gas)}$	2	H$_2$O	$+2.430$	acid
O$_{(gas)}$	2	OH$^-$	$+1.602$	alk
O$_2^-$	1	HO$_2^-$	$+0.408$	alk
O$_{2(gas)}$	2	HO$_2^-$	-0.076	alk
O$_{2(gas)}$	4	2H$_2$O	$+1.229$	acid
O$_{2(gas)}$	2	H$_2$O + O$_{(gas)}$	$+0.028$	acid
O$_{2(gas)}$	2	H$_2$O$_2$	$+0.694$	acid
O$_{2(gas)}$	1	O$_2^-$	-0.560	$-$
O$_{2(gas)}$	4	2OH$^-$	$+0.401$	alk
O$_{3(gas)}$	6	3H$_2$O	$+1.511$	acid
O$_{3(gas)}$	2	H$_2$O + O$_{2(gas)}$	$+2.075$	acid
O$_{3(gas)}$	6	3OH$^-$	$+0.683$	alk
O$_{3(gas)}$	2	OH$^-$ + O$_{2(gas)}$	$+1.247$	alk
OF$_{2(gas)}$	4	H$_2$O + 2F$^-$	$+2.176$	acid
OF$_{2(gas)}$	4	OH$^-$ + 2F$^-$	$+1.762$	alk
OH	1	H$_2$O	$+2.814$	acid
OH	1	OH$^-$	$+1.985$	$-$
BaO$_2$	2	2H$_2$O + Ba^{2+}	$+2.369$	acid
CaO$_2$	2	2H$_2$O + Ca^{2+}	$+2.222$	acid
HO$_2^-$	2	2OH$^-$	$+0.878$	alk

(Continued)

Oxidized form	$+ne^-$	Reduced form	$E°$, V	Medium
HO_2^-	1	OH^-	-0.240	alk
HO_2	1	H_2O_2	$+1.500$	acid
H_2O_2	2	$2H_2O$	$+1.764$	acid
Na_2O_2	2	$2H_2O + 2Na^+$	$+2.860$	acid
Na_2O_2	2	$4OH^- + 2Na^+$	$+1.204$	neutr
$S_2O_6(O_2)^{2-}$	2	$2SO_4^{2-}$	$+1.961$	–
$[OsBr_6]^{2-}$	1	$[OsBr_6]^{3-}$	$+0.349$	–
$[OsCl_6]^{3-}$	3	$Os + 6Cl^-$	$+0.597$	–
$[OsCl_6]^{2-}$	1	$[OsCl_6]^{3-}$	$+0.449$	–
OsO_4 (mon)	8	Os	$+0.780$	acid
OsO_4	8	Os	$+0.838$	acid
			$+0.010$	alk
$OsO_4 + 6Cl^-$	4	$[OsCl_6]^{2-}$	$+1.000$	acid
OsO_4	4	$Os(OH)_4$	$+0.170$	alk
$Os(OH)_4$	4	Os	-0.150	alk
$[OsO_2(OH)_4]^{2-}$	2	$Os(OH)_4$	$+0.100$	alk
$[OsO_4(OH)_2]^{2-}$	8	Os	$+0.025$	alk
$[OsO_4(OH)_2]^{2-}$	2	$[OsO_2(OH)_4]^{2-}$	$+0.300$	alk
P (amor)	3	$PH_{3(gas)}$	-0.915	alk
P_4	12	$4PH_{3(gas)}$	-0.046	acid
			-0.874	alk
$4(PHO_3)^{2-}$	12	P_4	-1.650	alk
$(PHO_3)^{2-}$	2	$(PH_2O_2)^-$	-1.565	alk
$(PH_2O_2)^-$	1	P (amor)	-1.697	alk
$4(PH_2O_2)^-$	4	P_4	-1.820	alk
$4PO_4^{3-}$	20	P_4	-1.510	alk
PO_4^{3-}	2	$(PHO_3)^{2-}$	-1.300	alk
PO_4^{3-}	4	$(PH_2O_2)^-$	-1.432	alk
$H(PH_2O_2)$	1	P (amor)	-0.387	acid
$4H(PH_2O_2)$	4	P_4	-0.510	acid
$H_2(PHO_3)$	3	P (amor)	-0.454	acid
$4H_2(PHO_3)$	12	P_4	-0.495	acid
$H_2(PHO_3)$	2	$H(PH_2O_2)$	-0.488	acid
HPO_4^{2-}	5	P (amor)	-1.447	alk
$4HPO_4^{2-}$	20	P_4	-1.475	alk
$H_2PO_4^-$	5	P (amor)	-0.358	acid
$4H_2PO_4^-$	20	P_4	-0.377	acid
H_3PO_4	5	P (amor)	-0.383	acid
H_3PO_4	20	P_4	-0.408	acid
H_3PO_4	4	$H(PH_2O_2)$	-0.382	acid
H_3PO_4	2	$H_2(PHO_3)$	-0.276	acid
Pa$^{V(1)}$	5	Pa	-1.000	–
Pa$^{V(1)}$	1	Pa^{4+}	-0.100	–
Pb^{2+}	2	Pb	-0.126	–
Pb^{2+}	2	Pb (amalgam)	-0.121	–
$PbBr_2$	2	$Pb + 2Br^-$	-0.277	–
$PbBr_2$	2	$Pb(Hg)_{(lq)} + 2Br^-$	-0.275	–

(*Continued*)

Oxidized form	$+ne^-$	Reduced form	$E°$, V	Medium
$PbCl_2$	2	$Pb + 2Cl^-$	-0.267	–
$PbCl_2$	2	Pb (amalgam) $+ 2Cl^-$	-0.262	–
PbF_2	2	$Pb + 2F^-$	-0.350	–
PbF_2	2	Pb (amalgam) $+ 2F^-$	-0.344	–
PbI_2	2	$Pb + 2I^-$	-0.379	–
PbI_2	2	Pb (amalgam) $+ 2I^-$	-0.358	–
PbO_2	2	Pb^{2+}	$+1.455$	acid
PbO_2	2	$Pb(OH)_2$	$+0.386$	alk
PbO_2	2	$[Pb(OH)_4]^{2-}$	$+0.210$	neutr
$PbO_2 + SO_4^{2-}$	2	$PbSO_4$	$+1.685$	acid
$PbOH^+$	2	Pb	$+0.056$	acid
$Pb(OH)_2$	2	Pb	-0.714	alk
$[Pb(OH)_4]^{2-}$	2	Pb	-0.538	alk
$[Pb(OH)_6]^{2-}$	2	$[Pb(OH)_4]^{2-}$	$+0.305$	alk
$[Pb_2^{II}Pb^{IV}]O_4$	2	$3Pb^{2+}$	$+2.156$	acid
$(Pb_2^{II}Pb^{IV})O_4$	2	$3Pb(OH)_2$	$+0.587$	alk
$(Pb_2^{II}Pb^{IV})O_4$	2	$3[Pb(OH)_4]^{2-}$	$+0.076$	alk
PbS	2	$Pb + S^{2-}$	-0.956	–
PbS	2	$Pb + HS^-$	-0.988	alk
PbS	2	$Pb + H_2S$	-0.367	acid
PbS	2	$Pb + H_2S_{(gas)}$	-0.338	acid
$PbSO_4$	8	PbS	$+0.304$	acid
PbS_4	2	$Pb + SO_4^{2-}$	-0.356	–
$PbSO_4$	2	Pb (amalgam) $+ SO_4^{2-}$	-0.350	–
Pd^{2+}	2	Pd	$+0.915$	–
$[PdBr_4]^{2-}$	2	$Pd + 4Br^-$	$+0.600$	–
$[PdBr_6]^{2-}$	2	$[PdBr_4]^{2-} + 2Br^-$	$+0.993$	–
$[PdCl_4]^{2-}$	2	$Pd + 4Cl^-$	$+0.640$	–
$[PdCl_6]^{2-}$	2	$[PdCl_4]^{2-} + 2Cl^-$	$+1.288$	–
$[PdI_4]^{2-}$	2	$Pd + 4I^-$	$+0.180$	–
$[PdI_6]^{2-}$	2	$[PdI_4]^{2-} + 2I^-$	$+0.482$	–
PdO_2	2	$Pd(OH)_2$	$+0.639$	alk
$PdOH^+$	2	Pd	$+0.954$	acid
$Pd(OH)_2$	2	Pd	$+0.065$	alk
Pm^{3+}	3	Pm	-2.428	–
$Pm(OH)_3$	3	Pm	-2.836	alk
Po	2	Po^{2-}	-1.400	–
$Po^{IV(1)}$	4	Po	$+0.765$	–
$Po^{IV(1)}$	2	$Po^{II(2)}$	$+0.879$	–
PoO_3^{2-}	4	Po	-0.490	alk
$PoO(OH)_2$	4	Po	-0.098	alk
Pr^{3+}	3	Pr	-2.462	–
$Pr^{IV(1)}$	1	Pr^{3+}	$+2.900$	–
$Pr(OH)_3$	3	Pr	-2.850	alk
$Pt^{II(1)}$	2	Pt	$+0.963$	–
$[PtBr_4]^{2-}$	2	$Pt + 4Br^-$	$+0.580$	–
$[PtBr_6]^{2-}$	2	$[PtBr_4]^{2-} + 2Br^-$	$+0.630$	–

(*Continued*)

Oxidized form	$+ne^-$	Reduced form	$E°$, V	Medium
$[PtCl_4]^{2-}$	2	$Pt + 4Cl^-$	+0.811	–
$[PtCl_6]^{2-}$	2	$[PtCl_4]^{2-} + 2Cl^-$	+0.734	–
$[PtI_6]^{2-}$	2	$[PtI_4]^{2-} + 2I^-$	+0.393	–
PtO_2	2	$Pt(OH)_2$	+0.272	alk
$Pt(OH)_2$	2	Pt	+0.237	alk
PtS	2	$Pt + S^{2-}$	–0.839	–
PtS	2	$Pt + HS^-$	–0.871	alk
PtS	2	$Pt + H_2S$	–0.250	acid
PtS	2	$Pt + H_2S_{(gas)}$	–0.221	acid
PtS_2	2	$PtS + H_2S_{(gas)}$	+0.052	acid
Pu^{3+}	3	Pu	–2.031	–
Pu^{4+}	1	Pu^{3+}	+0.967	–
PuO_2^+	1	Pu^{4+}	+1.156	acid
PuO_2^+	1	PuO_2	+1.580	–
PuO_2^{2+}	3	Pu^{3+}	+1.017	acid
PuO_2^{2+}	2	Pu^{4+}	+1.042	acid
PuO_2^{2+}	1	PuO_2^+	+0.928	–
PuO_2^{2+}	2	$P(OH)_4$	+0.935	neutr
PuO_3	2	$Pu(OH)_4$	–0.331	alk
$Pu(OH)_3$	3	Pu	–2.420	alk
$Pu(OH)_4$	1	$Pu(OH)_3$	–0.963	alk
Ra^{2+}	2	Ra	–2.916	–
Rb^+	1	Rb	–2.925	–
$Re^{III(1)}$	3	Re	+0.300	–
$Re^{IV(1)}$	4	Re	+0.264	–
$Re^{IV(1)}$	1	$Re^{III(2)}$	+0.157	–
$[ReCl_6]^{2-}$	4	Re	+0.414	–
$2[ReCl_6]^{2-}$	2	$[Re_2Cl_8]^{2-} + 4Cl^-$	+0.756	–
ReO_2	4	Re	+0.264	acid
			–0.564	alk
ReO_3	2	$Re^{IV(2)}$	+0.399	acid
ReO_3	2	$Re(OH)_4$	–0.429	alk
ReO_4^{2-}	2	$Re(OH)_4$	–0.541	alk
ReO_4^-	7	Re	+0.369	acid
			–0.577	alk
ReO_4^-	4	$Re^{III(2)}$	+0.421	acid
ReO_4^-	3	$Re^{IV(2)}$	+0.510	acid
ReO_4^-	3	ReO_2	+0.510	acid
			–0.594	alk
ReO_4^-	1	ReO_3	–0.924	alk
ReO_4^-	1	$ReO_4{}^{2-}$	–0.700	–
Rh^{3+}	3	Rh	+0.800	–
$Rh^{IV(1)}$	1	Rh^{3+}	+1.430	–
$[RhCl_6]^{3-}$	3	$Rh + 6Cl^-$	+0.437	–
$[RhCl_6]^{2-}$	1	$[RhCl_6]^{3-}$	+1.200	–
RhO_2	1	$Rh(OH)_3$	–0.256	alk
RhO_4^{2-} ?	3	$Rh(OH)_3$	–0.768	alk

Oxidized form	$+ne^-$	Reduced form	$E°$, V	Medium
$Rh(OH)_3$	3	Rh	±0.000	alk
Ru^{2+}	2	Ru	+0.450	−
Ru^{3+}	1	Ru^{2+}	+0.230	−
$Ru^{IV(1)}$	1	Ru^{3+}	+0.490	−
$RuCl_3$	3	$Ru + 3Cl^-$	+0.682	−
RuO_2	4	Ru	−0.254	alk
RuO_4^-	1	RuO_4^{2-}	+0.590	−
RuO_4	1	RuO_4^-	+1.000	−
S	2	HS^-	−0.476	alk
S	2	H_2S	+0.144	acid
S	2	$H_2S_{(gas)}$	+0.174	acid
			−0.654	alk
S	2	S^{2-}	−0.444	−
2S	2	S_2^{2-}	−0.411	−
3S	2	S_3^{2-}	−0.381	−
4S	2	S_4^{2-}	−0.357	−
5S	2	S_5^{2-}	−0.340	−
$2SO_2$	2	$HS_2O_4^-$	−0.080	acid
SO_2	4	S	+0.450	acid
$SO_{2(gas)}$	4	S	+0.451	acid
$3SO_2$	2	$S_3O_6^{2-}$	+0.291	−
$4SO_2$	6	$S_4O_6^{2-}$	+0.512	acid
$5SO_2$	10	$S_5O_6^{2-}$	+0.418	acid
$2SO_2$	4	SO_3S^{2-}	+0.338	acid
$2SO_{2(gas)}$	4	SO_3S^{2-}	+0.390	acid
SO_3^{2-}	4	S	−0.659	alk
$2SO_3^{2-}$	2	$S_2O_4^{2-}$	−1.142	alk
$2SO_3^{2-}$	4	SO_3S^{2-}	−0.589	alk
SO_4^{2-}	8	HS^-	−0.682	alk
SO_4^{2-}	8	H_2S	+0.302	acid
SO_4^{2-}	8	$H_2S_{(gas)}$	+0.309	acid
SO_4^{2-}	2	HSO_3^-	+0.110	acid
SO_4^{2-}	8	S^{2-}	+0.154	acid
			−0.674	alk
SO_4^{2-}	6	S	+0.354	acid
SO_4^{2-}	2	SO_2	+0.161	acid
SO_4^{2-}	2	$SO_{2(gas)}$	+0.159	acid
SO_4^{2-}	2	SO_3^{2-}	−0.104	acid
			−0.932	alk
$2SO_4^{2-}$	2	$S_2O_6^{2-}$	−0.244	acid
$2SO_4^{2-}$	8	SO_3S^{2-}	+0.275	acid
			−0.760	alk
$SO_4^{2-} + Cu^{2+}$	8	CuS	+0.419	acid
$S_2O_6^{2-}$	2	$2SO_2$	+0.566	acid
$S_4O_6^{2-}$	2	$2SO_3S^{2-}$	+0.015	−
SO_3S^{2-}	4	2S	+0.512	acid
			−0.730	alk

Oxidized form	$+ne^-$	Reduced form	$E°$, V	Medium
HSO_3^{3-}	4	S	+0.476	acid
Sb$^{III(1)}$	3	Sb	+0.240	–
Sb	3	$SbH_{3(gas)}$	–0.510	acid
$[SbCl_4]^-$	3	$Sb + 4Cl^-$	+0.170	–
$[SbCl_6]^-$	2	$[SbCl_4]^- + 2Cl^-$	+0.750	–
$2[Sb(H_2O)(OH)_5]$	10	2Sb	+0.333	acid
$2[Sb(H_2O)(OH)_5]$	4	Sb_2O_3 (amor)	+0.629	acid
$[Sb(H_2O)(OH)_5]$	2	$Sb(OH)_2^+$	+0.533	acid
Sb_2O_3 (amor)	6	2Sb	+0.135	acid
Sb_2O_3 (cub)	6	2Sb	+0.122	acid
Sb_2O_5	4	Sb_2O_3 (cub)	+0.649	acid
$Sb(OH)_2^+$	3	Sb	+0.199	acid
$[Sb(OH)_6]^{3-}$	3	Sb	–0.649	alk
$[Sb(OH)_6]^-$	2	$[Sb(OH)_6]^{3-}$	–0.590	–
$[SbS_3]^{3-}$	3	$Sb + 3HS^-$	–0.850	alk
Sb_2S_3	6	$2Sb + 3S^{2-}$	–0.714	–
Sb_2S_3	6	$2Sb + 3HS^-$	–0.746	alk
Sb_2S_3	6	$2Sb + 3H_2S$	–0.125	acid
Sb_2S_3	6	$2Sb + 3H_2S_{(gas)}$	–0.096	acid
Sc^{3+}	3	Sc	–2.077	–
$ScOH^{2+}$	3	Sc	–1.986	acid
$Sc(OH)_3$	3	Sc	–2.663	alk
Se	2	HSe^-	–0.642	alk
Se	2	H_2Se	–0.115	acid
Se	2	$H_2Se_{(gas)}$	–0.082	acid
Se	2	Se^{2-}	–0.670	–
2Se	4	Se_2^{2-}	–0.592	–
SeO_3^{2-}	2	Se	–0.341	alk
SeO_4^{2-}	2	$HSeO_3^-$	+1.073	acid
SeO_4^{2-}	2	H_2SeO_3	+1.150	acid
SeO_4^{2-}	2	SeO_3^{2-}	–0.001	alk
$HSeO_3^-$	4	Se	+0.778	acid
H_2SeO_3	4	Se	+0.741	acid
Si	4	$SiH_{4(gas)}$	–0.148	acid
$[SiF_6]^{2-}$	4	$Si + 6F^-$	–1.366	–
SiO_2 (α-quartz)	4	Si	–0.990	acid
SiO_4^{4-}	4	Si	–1.859	alk
H_4SiO_4	4	Si	–0.936	acid
Sm^{3+}	3	Sm	–2.414	–
Sm^{3+}	1	Sm^{2+}	–1.561	–
$Sm(OH)_3$	3	Sm	–2.830	alk
Sn^{2+}	2	Sn	–0.141	–
$Sn^{IV(1)}$	2	Sn^{2+}	+0.154	–
$[SnCl_3]^-$	2	$Sn + 3Cl^-$	–0.201	–
$[SnCl_6]^{2-}$	2	$[SnCl_3]^- + 3Cl^-$	+0.139	–
$[SnF_6]^{2-}$	4	$Sn + 6F^-$	–0.250	–
SnO_2	4	Sn	–0.118	acid

(*Continued*)

Oxidized form	$+ne^-$	Reduced form	$E°$, V	Medium
SnO_2	2	Sn^{2+}	−0.094	acid
$SnOH^+$	2	Sn	−0.078	acid
$Sn(OH)_2$	2	Sn	−0.917	alk
$[Sn(OH)_3]^-$	2	Sn	−0.790	alk
$[Sn(OH)_6]^{2-}$	4	Sn	−0.875	alk
$[Sn(OH)_6]^{2-}$	2	$[Sn(OH)_3]^-$	−0.960	alk
SnS	2	$Sn + S^{2-}$	−0.955	−
SnS	2	$Sn + HS^-$	−0.987	alk
SnS	2	$Sn + H_2S$	−0.366	acid
SnS	2	$Sn + H_2S_{(gas)}$	−0.337	acid
Sr^{2+}	2	Sr	−2.888	−
Sr^{2+}	2	Sr (amalgam)	−1.793	−
Ta_2O_5	10	2Ta	−0.753	acid
Tb^{3+}	3	Tb	−2.391	−
$Tb^{IV(1)}$	1	Tb^{3+}	+2.900	−
$Tb(OH)_3$	3	Tb	−2.790	alk
TcO_4^-	7	Tc	+0.472	acid
TcO_4^-	3	TcO_2	+0.770	acid
TcO_4^-	3	$Tc(OH)_4$	−0.366	alk
Te	2	HTe^-	−0.956	alk
Te	2	H_2Te	−0.464	acid
Te	2	$H_2Te_{(gas)}$	−0.441	acid
Te	2	Te^{2-}	−0.902	−
2Te	2	Te_2^{2-}	−0.845	−
$[TeCl_6]^{2-}$	4	$Te + 6Cl^-$	+0.630	−
TeO_3^{2-}	4	Te	−0.412	alk
H_2TeO_3	4	Te	+0.543	acid
$H_4TeO_6^{2-}$	2	TeO_3^{2-}	+0.401	alk
H_6TeO_6	2	H_2TeO_3	+1.249	acid
Th^{4+}	4	Th	−1.875	−
$ThOH^{3+}$	4	Th	−1.818	acid
$Th(OH)_2^{2+}$	4	Th	−1.755	acid
$Th(OH)_3^+$	4	Th	−1.663	acid
$Th(OH)_4$	4	Th	−2.513	alk
Ti^{2+} ?	2	Ti	−1.628	−
Ti^{3+}	3	Ti	−1.208	−
Ti^{3+}	1	Ti^{2+} ?	−0.368	−
$Ti^{IV(1)}$	1	Ti^{3+}	+0.092	−
$[TiF_6]^{2-}$	4	$Ti + 6F^-$	−1.190	−
TiO_2 (rutile)	4	Ti	−1.075	acid
$Ti(OH)_2^{2+}$	1	Ti^{3+}	+0.099	acid
$TiO(OH)_2$	4	Ti	−0.860	acid
Tl^+	1	Tl	−0.336	−
Tl^+	1	Tl (amalgam)	−0.334	−
Tl^{3+}	2	Tl^+	+1.280	−
$Tl^{3+} + Cl^-$	2	TlCl	+1.390	−
TlBr	1	$Tl + Br^-$	−0.653	−

(Continued)

Oxidized form	$+ne^-$	Reduced form	$E°$, V	Medium
TlCl	1	$Tl + Cl^-$	-0.556	$-$
$[TlCl_4]^-$	2	$TlCl + 3Cl^-$	$+0.810$	$-$
$[TlCl_4]^-$	2	$[TlCl_2]^- + 2Cl^-$	$+0.786$	$-$
TlI	1	$Tl + I^-$	-0.761	$-$
Tl_2O_3	4	$2Tl^+$	-0.063	alk
Tl_2S	2	$2Tl + S^{2-}$	-1.040	alk
Tm^{3+}	3	Tm	-2.278	$-$
$Tm(OH)_3$	3	Tm	-2.740	alk
U^{3+}	3	U	-1.690	$-$
U^{4+}	1	U^{3+}	-0.631	$-$
UO_2^+	1	U^{4+}	$+0.593$	acid
UO_2^+	1	UOH^{3+}	$+0.581$	acid
UO_2^{2+}	3	U^{3+}	$+0.007$	acid
UO_2^{2+}	1	UO_2^+	$+0.062$	$-$
UOH^{3+}	4	U	-1.422	acid
UO_2OH^+	2	UOH^{3+}	$+0.474$	acid
$UO_2(OH)_2$	2	$U(OH)_4$	-0.455	alk
V^{2+}	2	V	-1.125	$-$
V^{3+}	1	V^{2+}	-0.255	$-$
VO^{2+}	1	V^{3+}	$+0.361$	acid
VO^{2+}	1	VOH^{2+}	$+0.164$	acid
VO_2^+	5	V	-0.229	acid
VO_2^+	1	VO^{2+}	$+0.999$	acid
$4VO_4^{3-}$	4	$V_4O_9^{2-}$	-0.740	alk
V_2O_5	2	$2VO^{2+}$	$+0.958$	acid
$V(OH)_3$	1	$V(OH)_2$	-1.313	alk
$VO(OH)_2$	1	$V(OH)_3$	-0.436	alk
$H_2V_{10}O_{28}^{4-}$	10	$10VO^{2+}$	$+1.042$	acid
$[W(CN)_8]^{3-}$	1	$[W(CN)_8]^{4-}$	$+0.457$	$-$
WO_2	4	W	-0.154	acid
WO_3	6	W	-0.091	acid
WO_4^{2-}	6	W	-1.055	alk
$HW_6O_{21}^{5-}$	36	6W	-0.074	acid
XeF_2	2	$Xe_{(gas)} + 2F^-$	$+2.640$	$-$
XeO_3	6	$Xe_{(gas)}$	$+2.120$	acid
$HXeO_4^-$	6	$Xe_{(gas)}$	$+1.260$	alk
$HXeO_6^{3-}$	2	$HXeO_4^-$	$+0.940$	alk
H_4XeO_6	2	XeO_3	$+2.360$	acid
Y^{3+}	3	Y	-2.372	$-$
YOH^{2+}	3	Y	-2.193	acid
$Y(OH)_3$	3	Y	-2.855	alk
Yb^{3+}	3	Yb	-2.269	$-$
Yb^{3+}	1	Yb^{2+}	-1.518	$-$
$Yb(OH)_3$	3	Yb	-2.730	alk
Zn^{2+}	2	Zn	-0.763	$-$
$[Zn(CN)_4]^{2-}$	2	$Zn + 4CN^-$	-1.260	$-$
$[Zn(NH_3)_4]^{2+}$	2	$Zn + 4(NH_3 \cdot H_2O)$	-1.030	neutr

Oxidized form	+ne^-	Reduced form	$E°$, V	Medium
ZnOH$^+$	2	Zn	−0.498	acid
Zn(OH)$_2$	2	Zn	−1.243	alk
Zn(OH)$_2$ (amor)	2	Zn	−1.222	−
[Zn(OH)$_3$]$^-$	2	Zn	−1.207	alk
[Zn(OH)$_4$]$^{2-}$	2	Zn	−1.255	alk
ZnS (cub)	2	Zn + S^{2-}	−1.500	−
ZnS (cub)	2	Zn + HS$^-$	−1.532	alk
ZnS (cub)	2	Zn + H$_2$S$_{(sol)}$	−0.912	acid
ZnS (cub)	2	Zn + H$_2$S$_{(gas)}$	−0.882	acid
[ZrF$_6$]$^{2-}$	4	Zr + 6HF$_{(sol)}$	−1.608	acid
ZrO$_2$	4	Zr	−1.473	acid
Zr(OH)$_2^{2+}$	4	Zr	−1.355	acid
ZrO(OH)$_2$	4	Zr	−2.225	alk

4.4 Acidity Constants in Aqueous Solutions at 298.15 K [4, 5, 7–10, 19, 20, 22, 23, 29, 33]

The equilibrium state of the reversible reaction protolysis of HA acid (molecule, ion) in a dilute aqueous solution (0.1–0.001 M)

$$HA + H_2O \rightleftarrows A^- + H_3O^+$$

is characterized by the *acidity constant*:

$$K_a = [A^-][H_3O^+]/[HA] = f(T)$$

The value of K_a determines the strength of HA acid in an aqueous solution.

In an analogous manner, the equilibrium state of the reversible protolysis of A$^-$ base (molecule, ion)

$$A^- + H_2O \rightleftarrows HA + OH^-$$

is characterized by the *basicity constant*:

$$K_b = [HA][OH^-]/[A^-] = f(T)$$

The value of K_b determines the strength of A$^-$ base in an aqueous solution.

For each HA acid – A$^-$ base pair the values of K_a and K_b are interrelated as follows:

$$K_a \cdot K_b = K_w$$

where K_w is the ion product constant for water (at 25°C $K_w = 1.008 \cdot 10^{-14}$).

The table lists the values of K_a and also the values of exponents $pK_a = -\log K_a$ and $pK_b = -\log K_b$.

In the case of HBr, HBrO$_4$, HCl, HClO$_3$, HClO$_4$, HI, HMnO$_4$, H$_2$SO$_4$, and H$_2$SeO$_4$, protolysis in a dilute solution is irreversible, e.g.

$$H_2SO_4 + H_2O = HSO_4^- + H_3O^+$$

and, therefore, can not be characterized by the value of K_a.

The Formula Index for bases is given at the end of this Section.

Acid	K_a	pK_a	Base	pK_b
Ag$^+$·H$_2$O	1.02·10^{-12}	11.99	AgOH	2.01
AgOH·H$_2$O	9.74·10^{-13}	12.01	[Ag(OH)$_2$]$^-$	1.99
Al^{3+}·H$_2$O	9.55·10^{-6}	5.02	AlOH^{2+}	8.98
AlOH^{2+}·H$_2$O	4.68·10^{-6}	5.33	Al(OH)$_2^+$	8.67
Al(OH)$_2^+$·H$_2$O	1.35·10^{-6}	5.87	Al(OH)$_3$	8.13
Al(OH)$_3$·H$_2$O	3.16·10^{-8}	7.50	[Al(OH)$_4$]$^-$	6.50
Am^{3+}·H$_2$O	1.20·10^{-6}	5.92	AmOH^{2+}	8.08
AsO$^+$·H$_2$O	2.00	−0.30	HAsO$_2$	14.30
Au(OH)$_3$·H$_2$O	1.78·10^{-12}	11.75	[Au(OH)$_4$]$^-$	2.25
B(OH)$_3$·H$_2$O	5.75·10^{-10}	9.24	[B(OH)$_4$]$^-$	4.76
Ba^{2+}·H$_2$O	4.37·10^{-14}	13.36	BaOH$^+$	0.64
Be^{2+}·H$_2$O	2.00·10^{-6}	5.70	BeOH$^+$	8.30
BeOH$^+$·H$_2$O	3.16·10^{-8}	7.50	Be(OH)$_2$	6.50
Be(OH)$_2$·H$_2$O	1.23·10^{-11}	10.91	[Be(OH)$_3$]$^-$	3.09
[Be(OH)$_3$]$^-$·H$_2$O	3.55·10^{-14}	13.45	[Be(OH)$_4$]$^{2-}$	0.55
Bi^{3+}·H$_2$O	2.69·10^{-2}	1.57	BiOH^{2+}	12.43
BiOH^{2+}·H$_2$O	1.00·10^{-2}	2.00	Bi(OH)$_2^+$	12.00
Bi(OH)$_2^+$·H$_2$O	3.24·10^{-3}	2.49	Bi(OH)$_3$	11.51
Bk^{3+}·H$_2$O	2.19·10^{-6}	5.66	BkOH^{2+}	8.34
CH$_3$COOH	1.74·10^{-5}	4.76	CH$_3$COO$^-$	9.24
C(NH$_3$)(NH$_2$)O$^+$	6.61·10^{-1}	0.18	C(NH$_2$)$_2$O	13.82
CS(NH$_3$)(NH$_2$)$^+$	3.02	−0.48	CS(NH$_2$)$_2$	14.48
Ca^{2+}·H$_2$O	1.70·10^{-13}	12.77	CaOH$^+$	1.23
Cd^{2+}·H$_2$O	2.40·10^{-8}	7.62	CdOH$^+$	6.38
CdOH$^+$·H$_2$O	1.23·10^{-11}	10.91	Cd(OH)$_2$	3.09
Cd(OH)$_2$·H$_2$O	1.70·10^{-14}	13.77	[Cd(OH)$_3$]$^-$	0.23
[Cd(OH)$_3$]$^-$·H$_2$O	3.72·10^{-15}	14.43	[Cd(OH)$_4$]$^{2-}$	−0.43
Ce^{3+}·H$_2$O	1.00·10^{-9}	9.00	CeOH^{2+}	5.00
Ce^{4+}·H$_2$O	6.61	−0.82	CeOH^{3+}	14.82
Cf^{3+}·H$_2$O	2.40·10^{-6}	5.62	CfOH^{2+}	8.38
Cm^{3+}·H$_2$O	1.20·10^{-6}	5.92	CmOH^{2+}	8.08
Co^{2+}·H$_2$O	1.26·10^{-9}	8.90	CoOH$^+$	5.10
CoOH$^+$·H$_2$O	1.26·10^{-10}	9.90	Co(OH)$_2$	4.10
Cr^{2+}·H$_2$O	2.51·10^{-7}	6.60	CrOH$^+$	7.40
Cr^{3+}·H$_2$O	1.12·10^{-4}	3.95	CrOH^{2+}	10.05
CrOH^{2+}·H$_2$O	2.82·10^{-6}	5.55	Cr(OH)$_2^+$	8.45
Cu^{2+}·H$_2$O	4.57·10^{-8}	7.34	CuOH$^+$	6.66
CuOH$^+$·H$_2$O	1.51·10^{-7}	6.82	Cu(OH)$_2$	7.18
Cu(OH)$_2$·H$_2$O	1.91·10^{-13}	12.72	[Cu(OH)$_3$]$^-$	1.28

Acid	K_a	pK_a	Base	pK_b
$[Cu(OH)_3]^-·H_2O$	$1.38·10^{-14}$	13.86	$[Cu(OH)_4]^{2-}$	0.14
$Dy^{3+}·H_2O$	$2.69·10^{-8}$	7.57	$DyOH^{2+}$	6.43
$Er^{3+}·H_2O$	$3.47·10^{-8}$	7.46	$ErOH^{2+}$	6.54
$Es^{3+}·H_2O$	$7.24·10^{-6}$	5.14	$EsOH^{2+}$	8.86
$Eu^{3+}·H_2O$	$1.66·10^{-8}$	7.78	$EuOH^{2+}$	6.22
$Fe^{2+}·H_2O$	$1.82·10^{-7}$	6.74	$FeOH^+$	7.26
$Fe^{3+}·H_2O$	$6.76·10^{-3}$	2.17	$FeOH^{2+}$	11.83
$FeOH^+·H_2O$	$8.32·10^{-13}$	12.08	$Fe(OH)_2$	1.92
$FeOH^{2+}·H_2O$	$5.50·10^{-4}$	3.26	$Fe(OH)_2^+$	10.74
$Fe(OH)_2^+·H_2O$	$2.09·10^{-4}$	3.68	$Fe(OH)_3$	10.32
$Fe(OH)_3·H_2O$	$4.27·10^{-13}$	12.37	$[Fe(OH)_4]^-$	1.63
$Fm^{3+}·H_2O$	$1.58·10^{-4}$	3.80	$FmOH^{2+}$	10.20
$Ga^{3+}·H_2O$	$1.55·10^{-3}$	2.81	$GaOH^{2+}$	11.19
$GaOH^{2+}·H_2O$	$3.09·10^{-4}$	3.51	$Ga(OH)_2^+$	10.49
$Ga(OH)_2^+·H_2O$	$9.55·10^{-5}$	4.02	$Ga(OH)_3$	9.98
$Ga(OH)_3·H_2O$	$1.58·10^{-7}$	6.80	$[Ga(OH)_4]^-$	7.20
$Gd^{3+}·H_2O$	$1.51·10^{-8}$	7.82	$GdOH^{2+}$	6.18
$HAsO_2$	$6.03·10^{-10}$	9.22	AsO_2^-	4.78
$HAsO_3^{2-}$	$3.89·10^{-14}$	13.41	AsO_3^{3-}	0.59
$HAsO_4^{2-}$	$3.02·10^{-12}$	11.52	AsO_4^{3-}	2.48
$H_2AsO_3^-$	$7.41·10^{-13}$	12.13	$HAsO_3^{2-}$	1.87
$H_2AsO_4^-$	$1.07·10^{-7}$	6.97	$HAsO_4^{2-}$	7.03
H_3AsO_3	$5.89·10^{-10}$	9.23	$H_2AsO_3^-$	4.77
H_3AsO_4	$5.50·10^{-3}$	2.26	$H_2AsO_4^-$	11.74
$HBrO$	$2.06·10^{-9}$	8.69	BrO^-	5.31
$HBrO_3$	$2.00·10^{-1}$	0.70	BrO_3^-	13.30
HCN	$4.93·10^{-10}$	9.31	CN^-	4.69
HCO_3^-	$4.68·10^{-11}$	10.33	CO_3^{2-}	3.67
$HC_2O_4^-$	$6.17·10^{-5}$	4.21	$C_2O_4^{2-}$	9.79
H_2CO_3	$4.27·10^{-7(1)}$	6.37	HCO_3^-	7.63
$H_2C_2O_4$	$6.46·10^{-2}$	1.19	$HC_2O_4^-$	12.81
$HCOOH$	$1.78·10^{-4}$	3.75	$HCOO^-$	10.25
$HClO$	$2.82·10^{-8}$	7.55	ClO^-	6.45
$HClO_2$	$1.07·10^{-2}$	1.97	ClO_2^-	12.03
$HCrO_4^-$	$3.16·10^{-7}$	6.50	CrO_4^{2-}	7.50
$HCr_2O_7^-$	$2.29·10^{-2}$	1.64	$Cr_2O_7^{2-}$	12.36
H_2CrO_4	9.55	−0.98	$HCrO_4^-$	14.98
HF	$6.67·10^{-4}$	3.18	F^-	10.82
$2HF$	$2.63·10^{-3}$	2.58	HF_2^-	11.42
$H[Fe(CN)_6]^{3-}$	$6.76·10^{-5}$	4.17	$[Fe(CN)_6]^{4-}$	9.83
$H_2[Fe(CN)_6]^{2-}$	1.0010^{-3}	3.00	$H[Fe(CN)_6]^{3-}$	11.00
$HGeO_3^-$	$1.91·10^{-13}$	12.72	GeO_3^{2-}	1.28
H_2GeO_3	$1.86·10^{-9}$	8.73	$HGeO_3^-$	5.27
HIO	$2.29·10^{-11}$	10.64	IO^-	3.36
HIO_3	$1.69·10^{-1}$	0.77	IO_3^-	13.23
HIO_4	$2.30·10^{-2}$	1.64	IO_4^-	12.36

(*Continued*)

Acid	K_a	pK_a	Base	pK_b
$H_3IO_6^{2-}$	$1.05 \cdot 10^{-15}$	14.98	$H_2IO_6^{3-}$	-0.98
$H_4IO_6^{-}$	$5.37 \cdot 10^{-9}$	8.27	$H_3IO_6^{2-}$	5.73
H_5IO_6	$2.82 \cdot 10^{-2}$	1.55	$H_4IO_6^{-}$	12.45
$HMnO_4^{-}$	$7.08 \cdot 10^{-11}$	10.15	MnO_4^{2-}	3.85
$H_3Mo_8O_{26}^{-}$	$9.77 \cdot 10^{-6}$	5.01	$H_2Mo_8O_{26}^{2-}$	8.99
$H_4Mo_8O_{26}$	$1.45 \cdot 10^{-2}$	1.84	$H_3Mo_8O_{26}^{-}$	12.16
HN_3	$1.90 \cdot 10^{-5}$	4.72	N_3^{-}	9.28
$HNCO$	$3.47 \cdot 10^{-4}$	3.46	OCN^{-}	10.54
$HNCS$	$1.41 \cdot 10^{-1}$	0.85	NCS^{-}	13.15
HNO_2	$5.13 \cdot 10^{-4}$	3.29	NO_2^{-}	10.71
HNO_3	$2.69 \cdot 10^{1}$	-1.43	NO_3^{-}	15.43
$HN_2O_2^{-}$	$2.00 \cdot 10^{-12}$	11.70	$N_2O_2^{2-}$	2.30
$H_2N_2O_2$	$2.00 \cdot 10^{-8}$	7.70	$HN_2O_2^{-}$	6.30
$HNbO_3$?	$4.07 \cdot 10^{-8}$	7.39	NbO_3^{-}	6.61
HO_2	$1.58 \cdot 10^{-5}$	4.80	O_2^{-}	9.20
H_2O	$1.816 \cdot 10^{-16}$	15.741	OH^{-}	-1.744
H_2O_2	$2.40 \cdot 10^{-12}$	11.62	HO_2^{-}	2.38
H_3O^{+}	$5.551 \cdot 10^{1}$	-1.744	H_2O	15.741
$H(PHO_3)^{-}$	$2.57 \cdot 10^{-7}$	6.59	PHO_3^{2-}	7.41
$H(PH_2O_2)$	$7.94 \cdot 10^{-2}$	1.10	$PH_2O_2^{-}$	12.90
$H_2(PHO_3)$	$1.00 \cdot 10^{-2}$	2.00	$H(PHO_3)^{-}$	12.00
HPO_4^{2-}	$4.57 \cdot 10^{-13}$	12.34	PO_4^{3-}	1.66
$HP_2^{IV}O_6^{3-}$	$9.55 \cdot 10^{-11}$	10.02	$P_2^{IV}O_6^{4-}$	3.98
$HP_2O_7^{3-}$	$6.03 \cdot 10^{-9}$	8.22	$P_2O_7^{4-}$	5.78
$H_2PO_4^{-}$	$6.17 \cdot 10^{-8}$	7.21	HPO_4^{2-}	6.79
$H_2P_2^{IV}O_6^{2-}$	$5.50 \cdot 10^{-8}$	7.26	$HP_2^{IV}O_6^{3-}$	6.74
$H_2P_2O_7^{2-}$	$1.70 \cdot 10^{-6}$	5.77	$HP_2O_7^{3-}$	8.23
H_3PO_4	$7.24 \cdot 10^{-3}$	2.14	$H_2PO_4^{-}$	11.86
$H_3P_2^{IV}O_6^{-}$	$1.55 \cdot 10^{-3}$	2.81	$H_2P_2^{IV}O_6^{2-}$	11.19
$H_3P_2O_7^{-}$	$3.24 \cdot 10^{-2}$	1.49	$H_2P_2O_7^{2-}$	12.51
$H_4P_2^{IV}O_6$	$6.46 \cdot 10^{-3}$	2.19	$H_3P_2^{IV}O_6^{-}$	11.81
$H_4P_2O_7$	$1.41 \cdot 10^{-1}$	0.85	$H_3P_2O_7^{-}$	13.15
HPO_3F^{-}	$1.58 \cdot 10^{-5}$	4.80	$PO_3F_2^{-}$	9.20
H_2PO_3F	$2.82 \cdot 10^{-1}$	0.55	HPO_3F^{-}	13.45
$HPO_2(NH_2)_2$	$1.00 \cdot 10^{-5}$	5.00	$PO_2(NH_2)_2^{-}$	9.00
$HPO_3(NH_2)^{-}$	$1.00 \cdot 10^{-8}$	8.00	$PO_3(NH_2)^{2-}$	6.00
$H_2PO_4(NH_2)$	$1.00 \cdot 10^{-3}$	3.00	$HPO_3(NH_2)^{-}$	11.00
HS^{-}	$1.23 \cdot 10^{-13}$	12.91	S^{2-}	1.09
H_2S	$1.05 \cdot 10^{-7}$	6.98	HS^{-}	7.02
HSO_3^{-}	$6.31 \cdot 10^{-8}$	7.20	SO_3^{2-}	6.80
HSO_4^{-}	$1.12 \cdot 10^{-2}$	1.95	SO_4^{2-}	12.05
$HS_2O_4^{-}$	$3.55 \cdot 10^{-3}$	2.45	$S_2O_4^{2-}$	11.55
$HS_2O_6^{-}$	$3.98 \cdot 10^{-4}$	3.40	$S_2O_6^{2-}$	10.60
$H_2S_2O_4$	$5.00 \cdot 10^{-1}$	0.30	$HS_2O_4^{-}$	13.70
$H_2S_2O_6$	$6.31 \cdot 10^{-1}$	0.20	$HS_2O_6^{-}$	13.80
$HSO_3(O_2)^{-}$	$4.00 \cdot 10^{-10}$	9.40	$SO_3(O_2)^{2-}$	4.60

Acid	K_a	pK_a	Base	pK_b
HSO_3S^-	$1.91 \cdot 10^{-2}$	1.72	SO_3S^{2-}	12.28
H_2SO_3S	$2.51 \cdot 10^{-1}$	0.60	HSO_3S^-	13.40
HSe^-	$1.00 \cdot 10^{-11(2)}$	11.00	Se^{2-}	3.00
H_2Se	$1.55 \cdot 10^{-4}$	3.81	HSe^-	10.19
$HSeO_3^-$	$4.79 \cdot 10^{-9}$	8.32	SeO_3^{2-}	5.68
$HSeO_4^-$	$2.19 \cdot 10^{-2}$	1.66	SeO_4^{2-}	12.34
H_2SeO_3	$2.45 \cdot 10^{-3}$	2.61	$HSeO_3^-$	11.39
$HSiO_4^{3-}$	$2.00 \cdot 10^{-14}$	13.70	SiO_4^{4-}	0.30
$H_2SiO_4^{2-}$	$1.00 \cdot 10^{-12}$	12.00	$HSiO_4^{3-}$	2.00
$H_3SiO_4^-$	$1.86 \cdot 10^{-12}$	11.73	$H_2SiO_4^{2-}$	2.27
H_4SiO_4	$1.58 \cdot 10^{-10}$	9.80	$H_3SiO_4^-$	4.20
$HTaO_3$?	$2.51 \cdot 10^{-10}$	9.60	TaO_3^-	4.40
HTe^-	$6.76 \cdot 10^{-13}$	12.17	Te^{2-}	1.83
H_2Te	$2.29 \cdot 10^{-3}$	2.64	HTe^-	11.36
$HTeO_3^-$	$2.00 \cdot 10^{-11}$	10.70	TeO_3^{2-}	3.30
H_2TeO_3	$1.35 \cdot 10^{-4}$	3.87	$HTeO_3^-$	10.13
$H_4TeO_6^{2-}$	$1.00 \cdot 10^{-15}$	15.00	$H_3TeO_6^{3-}$	-1.00
$H_5TeO_6^-$	$1.10 \cdot 10^{-11}$	10.96	$H_4TeO_6^{2-}$	3.04
H_6TeO_6	$1.90 \cdot 10^{-8}$	7.72	$H_5TeO_6^-$	6.28
HVO_4^{2-}	$7.41 \cdot 10^{-12}$	11.13	VO_4^{3-}	2.87
$H_2VO_4^-$	$1.12 \cdot 10^{-9}$	8.95	HVO_4^{2-}	5.05
H_3VO_4 ?	$3.24 \cdot 10^{-5}$	4.49	$H_2VO_4^-$	9.51
HWO_4^-	$2.19 \cdot 10^{-4}$	3.66	WO_4^{2-}	10.34
H_2WO_4 ?	$6.31 \cdot 10^{-3}$	2.20	HWO_4^-	11.80
$H_2XeO_6^{2-}$	$1.00 \cdot 10^{-11}$	11.00	$HXeO_6^{3-}$	3.00
$H_3XeO_6^-$	$1.00 \cdot 10^{-6}$	6.00	$H_2XeO_6^{2-}$	8.00
H_4XeO_6	$1.00 \cdot 10^{-2}$	2.00	$H_3XeO_6^-$	12.00
$Hf(OH)_2^{2+} \cdot H_2O$	$1.58 \cdot 10^{-1}$	0.80	$Hf(OH)_3^+$	13.20
$Hf(OH)_3^+ \cdot H_2O$	$3.16 \cdot 10^{-2}$	1.50	$Hf(OH)_4$	12.50
$Hg^{2+} \cdot H_2O$	$2.63 \cdot 10^{-4}$	3.58	$HgOH^+$	10.42
$Hg_2^{2+} \cdot H_2O$	$3.31 \cdot 10^{-5}$	4.48	Hg_2OH^+	9.52
$HgOH^+ \cdot H_2O$	$8.91 \cdot 10^{-5}$	4.05	$Hg(OH)_2$	9.95
$Hg(OH)_2 \cdot H_2O$	$1.41 \cdot 10^{-15}$	14.85	$[Hg(OH)_3]^-$	-0.85
$Ho^{3+} \cdot H_2O$	$3.09 \cdot 10^{-8}$	7.51	$HoOH^{2+}$	6.49
$I^+ \cdot H_2O$?	$3.02 \cdot 10^{-10}$	9.52	HIO	4.48
$In^{3+} \cdot H_2O$	$2.63 \cdot 10^{-4}$	3.58	$InOH^{2+}$	10.42
$InOH^{2+} \cdot H_2O$	$5.75 \cdot 10^{-5}$	4.24	$In(OH)_2^+$	9.76
$In(OH)_2^+ \cdot H_2O$	$7.59 \cdot 10^{-6}$	5.12	$In(OH)_3$	8.88
$In(OH)_3 \cdot H_2O$	$6.92 \cdot 10^{-10}$	9.16	$[In(OH)_4]^-$	4.84
$K^+ \cdot H_2O$	$3.47 \cdot 10^{-15}$	14.46	KOH	-0.46
$La^{3+} \cdot H_2O$	$2.00 \cdot 10^{-11}$	10.70	$LaOH^{2+}$	3.30
$Li^+ \cdot H_2O$	$2.29 \cdot 10^{-14}$	13.64	$LiOH$	0.36
$Lu^{3+} \cdot H_2O$	$4.68 \cdot 10^{-8}$	7.33	$LuOH^{2+}$	6.67
$Mg^{2+} \cdot H_2O$	$3.80 \cdot 10^{-12}$	11.42	$MgOH^+$	2.58
$Mn^{2+} \cdot H_2O$	$2.57 \cdot 10^{-11}$	10.59	$MnOH^+$	3.41
NH^{4+}	$5.75 \cdot 10^{-10}$	9.24	$NH_3 \cdot H_2O$	4.76

(Continued)

Acid	K_a	pK_a	Base	pK_b
$N_2H_5^+$	$5.89 \cdot 10^{-9}$	8.23	$N_2H_4 \cdot H_2O$	5.77
$N_2H_6^{2+}$	$1.58 \cdot 10^1$	-1.20	$N_2H_5^+$	15.20
NH_3OH^+	$9.35 \cdot 10^{-7}$	6.03	$NH_2OH \cdot H_2O$	7.97
$Na^+ \cdot H_2O$	$6.61 \cdot 10^{-15}$	14.18	$NaOH$	-0.18
$NbO_2^+ \cdot H_2O$	3.98	-0.60	$HNbO_3$?	14.60
$Nd^{3+} \cdot H_2O$	$1.26 \cdot 10^{-8}$	7.90	$NdOH^{2+}$	6.10
$Ni^{2+} \cdot H_2O$	$1.20 \cdot 10^{-11}$	10.92	$NiOH^+$	3.08
$Np^{3+} \cdot H_2O$	$3.72 \cdot 10^{-8}$	7.43	$NpOH^{2+}$	6.57
$Np^{4+} \cdot H_2O$	$5.01 \cdot 10^{-3}$	2.30	$NpOH^{3+}$	11.70
$NpO_2^+ \cdot H_2O$	$1.26 \cdot 10^{-9}$	8.90	$NpO_2(OH)$	5.10
$NpO_2^{2+} \cdot H_2O$	$6.76 \cdot 10^{-6}$	5.17	$NpO_2(OH)^+$	8.83
$[Os(H_2O)_2O_4]$	$7.95 \cdot 10^{-13}$	12.10	$[Os(H_2O)O_4(OH)]^-$	1.90
$[Os(H_3O)(H_2O)O_4]^+$	$1.00 \cdot 10^1$	-1.00	$[Os(H_2O)_2O_4]$	15.00
$[Os(H_2O)O_4(OH)]^-$	$1.00 \cdot 10^{-15}$	15.00	$[OsO_4(OH)_2]^{2-}$	-1.00
$Pa^{4+} \cdot H_2O$	$7.24 \cdot 10^{-1}$	0.14	$PaOH^{3+}$	13.86
$PaOH^{3+} \cdot H_2O$	$4.17 \cdot 10^{-1}$	0.38	$Pa(OH)_2^{2+}$	13.62
$Pa(OH)_2^{2+} \cdot H_2O$	$5.62 \cdot 10^{-2}$	1.25	$Pa(OH)_3^+$	12.75
$Pa(OH)_3^{2+} \cdot H_2O$	$8.91 \cdot 10^{-2}$	1.05	$Pa(OH)_4^+$	12.95
$Pa(OH)_4^+ \cdot H_2O$	$3.02 \cdot 10^{-5}$	4.52	$Pa(OH)_5$	9.48
$Pb^{2+} \cdot H_2O$	$7.08 \cdot 10^{-7}$	6.15	$PbOH^+$	7.85
$PbOH^+ \cdot H_2O$	$2.00 \cdot 10^{-11}$	10.70	$Pb(OH)_2$	3.30
$Pb(OH)_2 \cdot H_2O$	$9.55 \cdot 10^{-12}$	11.02	$[Pb(OH)_3]^-$	2.98
$Pd^{2+} \cdot H_2O$	$4.90 \cdot 10^{-2}$	1.31	$PdOH^+$	12.69
$Pr^{3+} \cdot H_2O$	$9.55 \cdot 10^{-9}$	8.02	$PrOH^{2+}$	5.98
$Pu^{3+} \cdot H_2O$	$1.12 \cdot 10^{-7}$	6.95	$PuOH^{2+}$	7.05
$Pu^{4+} \cdot H_2O$	$3.09 \cdot 10^{-2}$	1.51	$PuOH^{3+}$	12.49
$PuO_2^+ \cdot H_2O$	$2.00 \cdot 10^{-10}$	9.70	$PuO_2(OH)$	4.30
$PuO_2^{2+} \cdot H_2O$	$3.98 \cdot 10^{-4}$	3.40	$PuO_2(OH)^+$	10.60
$SO_2 \cdot H_2O$	$1.66 \cdot 10^{-2}$	1.78	HSO_3^-	12.22
$Sb(OH)_5 \cdot H_2O$	$3.98 \cdot 10^{-5}$	4.40	$[Sb(OH)_6]^-$	9.60
$Sb(OH)_2^+ \cdot H_2O$	$6.92 \cdot 10^{-2}$	1.16	$Sb(OH)_3$	12.84
$Sb(OH)_3 \cdot H_2O$	$1.62 \cdot 10^{-12}$	11.79	$[Sb(OH)_4]^-$	2.21
$Sc^{3+} \cdot H_2O$	$2.45 \cdot 10^{-5}$	4.61	$ScOH^{2+}$	9.39
$ScOH^{2+} \cdot H_2O$	$2.00 \cdot 10^{-6}$	5.70	$Sc(OH)_2^+$	8.30
$Sc(OH)_2^+ \cdot H_2O$	$3.16 \cdot 10^{-7}$	6.50	$Sc(OH)_3$	7.50
$Sm^{3+} \cdot H_2O$	$1.55 \cdot 10^{-8}$	7.81	$SmOH^{2+}$	6.19
$Sn^{2+} \cdot H_2O$	$7.94 \cdot 10^{-3}$	2.10	$SnOH^+$	11.90
$SnOH^+ \cdot H_2O$	$2.88 \cdot 10^{-5}$	4.54	$Sn(OH)_2$	9.46
$Sn(OH)_2 \cdot H_2O$	$3.02 \cdot 10^{-10}$	9.52	$[Sn(OH)_3]^-$	4.48
$Sn(OH)_2^{2+} \cdot H_2O$	$4.68 \cdot 10^{-1}$	0.33	$Sn(OH)_3^+$	13.67
$Sn(OH)_3^+ \cdot H_2O$	$6.03 \cdot 10^{-2}$	1.22	$Sn(OH)_4$	12.78
$Sn(OH)_4 \cdot H_2O$	$5.75 \cdot 10^{-10}$	9.24	$[Sn(OH)_5]^-$	4.76
$[Sn(OH)_5]^- \cdot H_2O$	$1.29 \cdot 10^{-12}$	11.89	$[Sn(OH)_6]^{2-}$	2.11
$Sr^{2+} \cdot H_2O$	$6.76 \cdot 10^{-14}$	13.17	$SrOH^+$	0.83
$TaO_2^+ \cdot H_2O$	$9.55 \cdot 10^{-2}$	1.02	$HTaO_3$?	12.98
$Tb^{3+} \cdot H_2O$	$2.34 \cdot 10^{-8}$	7.63	$TbOH^{2+}$	6.37

Acid	K_a	pK_a	Base	pK_b
$TeO(OH)^+ \cdot H_2O$	$3.63 \cdot 10^{-4}$	3.44	H_2TeO_3	10.56
$Th^{4+} \cdot H_2O$	$4.37 \cdot 10^{-3}$	2.36	$ThOH^{3+}$	11.64
$ThOH^{3+} \cdot H_2O$	$5.01 \cdot 10^{-4}$	3.30	$Th(OH)_2^{2+}$	10.70
$Th(OH)_2^{2+} \cdot H_2O$	$4.17 \cdot 10^{-4}$	3.38	$Th(OH)_3^+$	10.62
$Th(OH)_3^+ \cdot H_2O$	$2.82 \cdot 10^{-4}$	3.55	$Th(OH)_4$	10.45
$Ti^{3+} \cdot H_2O$	$5.62 \cdot 10^{-3}$	2.25	$TiOH^{2+}$	11.75
$Ti(OH)_2^{2+} \cdot H_2O$	$3.24 \cdot 10^{-2}$	1.49	$Ti(OH)_3^+$	12.51
$Ti(OH)_3^+ \cdot H_2O$	$9.77 \cdot 10^{-4}$	3.01	$Ti(OH)_4$	10.99
$Tl^+ \cdot H_2O$	$6.61 \cdot 10^{-14}$	13.18	$TlOH$	0.82
$Tl^{3+} \cdot H_2O$	$5.75 \cdot 10^{-2}$	1.24	$TlOH^{2+}$	12.76
$TlOH^{2+} \cdot H_2O$	$2.09 \cdot 10^{-2}$	1.68	$Tl(OH)_2^+$	12.32
$Tl(OH)_2^+ \cdot H_2O$	$1.51 \cdot 10^{-2}$	1.82	$Tl(OH)_3$	12.18
$Tm^{3+} \cdot H_2O$	$3.80 \cdot 10^{-8}$	7.42	$TmOH^{2+}$	6.58
$U^{3+} \cdot H_2O$	$1.0 \cdot 10^{-7}$	7.00	UOH^{2+}	7.00
$U^{4+} \cdot H_2O$	$3.02 \cdot 10^{-8}$	1.52	UOH^{3+}	12.48
$UO_2^+ \cdot H_2O$	$1.00 \cdot 10^{-8}$	8.00	$UO_2(OH)$	6.00
$UO_2^{2+} \cdot H_2O$	$6.92 \cdot 10^{-6}$	5.16	$UO_2(OH)^+$	8.84
$UOH^{3+} \cdot H_2O$	$3.80 \cdot 10^{-3}$	2.42	$U(OH)_2^{2+}$	11.58
$U(OH)_2^{2+} \cdot H_2O$	$7.24 \cdot 10^{-4}$	3.14	$U(OH)_3^+$	10.86
$UO_2(OH)^+ \cdot H_2O$	$5.25 \cdot 10^{-8}$	7.28	$UO_2(OH)_2$	6.72
$V^{2+} \cdot H_2O$	$3.47 \cdot 10^{-7}$	6.46	VOH^+	7.54
$V^{3+} \cdot H_2O$	$1.20 \cdot 10^{-3}$	2.92	VOH^{2+}	11.08
$VO^{2+} \cdot H_2O$	$2.24 \cdot 10^{-3}$	2.65	$VO(OH)^+$	11.35
$VOH^{2+} \cdot H_2O$	$3.02 \cdot 10^{-4}$	3.52	$V(OH)_2^+$	10.48
$VO(OH)^+ \cdot H_2O$	$8.32 \cdot 10^{-4}$	3.08	$VO(OH)_2$	10.92
$Y^{3+} \cdot H_2O$	$8.32 \cdot 10^{-10}$	9.08	YOH^{2+}	4.92
$Yb^{3+} \cdot H_2O$	$4.27 \cdot 10^{-8}$	7.37	$YbOH^{2+}$	6.63
$Zn^{2+} \cdot H_2O$	$2.04 \cdot 10^{-8}$	7.69	$ZnOH^+$	6.31
$ZnOH^+ \cdot H_2O$	$7.59 \cdot 10^{-10}$	9.12	$Zn(OH)_2$	4.88
$Zn(OH)_2 \cdot H_2O$	$1.29 \cdot 10^{-12}$	11.89	$[Zn(OH)_3]^-$	2.11
$[Zn(OH)_3]^- \cdot H_2O$	$4.27 \cdot 10^{-13}$	12.37	$[Zn(OH)_4]^{2-}$	1.63
$Zr(OH)_2^{2+} \cdot H_2O$	$6.17 \cdot 10^{-1}$	0.21	$Zr(OH)_3^+$	13.79
$Zr(OH)_3^+ \cdot H_2O$	$1.55 \cdot 10^{-1}$	0.81	$Zr(OH)_4$	13.19

[1] The quantity listed is the apparent value of K_a (and also pK_a and pK_b) that corresponds to the overall content of H_2CO_3 and $CO_2 \cdot H_2O$ in solution. The true value of K_a for H_2CO_3 equals $1.32 \cdot 10^{-4}$ ($pK_a = 3.38$; $pK_b = 10.12$).

[2] The quantity listed is the apparent value of K_a (and also pK_a and pK_b) that corresponds to the overall content of Se^{2-} and Se_n^{2-} ($n > 1$) ions in solution. The true value of K_a for the HSe^-/Se^{2-} pair equals $1.12 \cdot 10^{-15}$ ($pK_a = 14.95$; $pK_b = -0.95$).

Formula Index of Bases

The formula index is intended for facilitating the search of those acid-base pairs wherein the formulas of acids and bases start with the symbols of dissimilar elements. The pairs are written in an inverted form (Base/Acid). For the sake of brevity, in the formulas of bases and acids reference to hydration is omitted.

$AsO_2^-/HAsO_2$	$H_2TeO_3/TeO(OH)^+$	$PO_2((NH_2)_2^-/HPO_2(NH_2)_2$
$AsO_3^{3-}/HAsO_3^{2-}$	IO^-/HIO	$PO_3((NH_2)^{2-}/HPO_3(NH_2)^-$
$AsO_4^{3-}/HAsO_4^{2-}$	IO_3^-/HIO_3	S^{2-}/HS^-
$BrO^-/HBrO$	IO_4^-/HIO_4	SO_3^{2-}/HSO_3^-
$BrO_3^-/HBrO_3$	$MnO_4^{2-}/HMnO_4^-$	SO_4^{2-}/HSO_4^-
CN^-/HCN	N_3^-/HN_3	$S_2O_4^{2-}/HS_2O_4^-$
CO_3^{2-}/HCO_3^-	$NCS^-/HNCS$	$S_2O_6^{2-}/HS_2O_6^-$
$C_2O_4^{2-}/HC_2O_4^-$	NO_2^-/HNO_2	$SO_3(O_2)^{2-}/HSO_3(O_2)^-$
$ClO^-/HClO$	NO_3^-/HNO_3	SO_3S^{2-}/HSO_3S^-
$ClO_2^-/HClO_2$	$N_2O_2^{2-}/HN_2O_2^-$	Se^{2-}/HSe^-
$CrO_4^{2-}/HCrO_4^-$	$NbO_3^-/HNbO_3$	$SeO_3^{2-}/HSeO_3^-$
$Cr_2O_7^{2-}/HCr_2O_7^-$	O_2^-/HO_2	$SeO_4^{2-}/HSeO_4^-$
F^-/HF	$OCN^-/HNCO$	$SiO_4^{4-}/HSiO_4^{3-}$
$[F(CN)_6]^{4-}/H[Fe(CN)_6]^{3-}$	OH^-/H_2O	$TaO_3^-/HTaO_3$
$GeO_3^{2-}/HGeO_3^-$	$PHO_3^{2-}/H(PHO_3)^-$	Te^{2-}/HTe^-
$HAsO_2/AsO^+$	$PH_2O_2^-/H(PH_2O_2)$	$TeO_3^{2-}/HTeO_3^-$
HIO/I^+	PO_4^{3-}/HPO_4^{2-}	VO_4^{3-}/HVO_4^{2-}
$HNbO_3/NbO_2^+$	$P_2^{IV}O_6^{4-}/HP_2^{IV}O_6^{3-}$	WO_4^{2-}/HWO_4^-
HSO_3^-/SO_2	$P_2O_7^{4-}/HP_2O_7^{3-}$	
$HTaO_3/TaO_2^+$	PO_3F^{2-}/HPO_3F^-	

4.5 Stability Constants of Complexes in Aqueous Solution at 298.15 K [21, 22]

The equilibrium state for a reversible reaction of mononuclear complex $[ML_n]$ formation (here and herein below the complex charge is not cited) in an aqueous solution from a central atom M and ligands L

$$M + nL \rightleftharpoons [ML_n]$$

is characterized by the *overall stability constant*:

$$\beta_n = [ML_n]/[M][L]^n$$

The value of the constant β_n determines the overall degree of $[ML_n]$ complex stability in an aqueous solution.

The equilibrium state for the reversible reaction of the same $[ML_n]$ complex formation from the $[ML_{n-1}]$ complex having one ligand less and the ligand L

$$[ML_{n-1}] + L \rightleftharpoons [ML_n]$$

is characterized by the *stepwise stability constant*

$$K_n = [ML_n]/[ML_{n-1}][L]$$

The value of the constant K_n determines the degree of $[ML_n]$ complex stability as compared to the stability of the initial $[ML_{n-1}]$ complex.

The values of constants K_n and β_n for the $[ML_n]$ complex are interrelated as follows:

$$\text{for } n = 1 \qquad K_n = \beta_n$$
$$\text{for } n \geq 2 \qquad K_n = \beta_n/\beta_{n-1}$$

where β_{n-1} is the overall stability of the $[ML_{n-1}]$ complex, i.e.

$$\beta_{n-1} = [ML_{n-1}]/[M][L]^{n-1}.$$

The table lists the values of $\log \beta_n$; to calculate the values of $\log K_n$ (for $n \geq 2$), use should be made of the following expression: $\log K_n = \log \beta_n - \log \beta_{n-1}$. The complexes of various M atoms are grouped in terms of the L ligand present.

List of Ligands

Br^-	$C_2O_4^{2-}$	H_3PO_4	NO_2^-
CH_3COO^-	$CS(NH_2)_2$	HS^-	NO_3^-
$C_2H_8N_2$ (en)	Cl^-	I^-	OH^-
C_5H_5N (py)	F^-	IO_3^-	$P_2O_7^{4-}$
$(C_5H_4N)_2$ (bipy)	HCO_3^-	NCS^-	SO_3^{2-}
CN^-	HPO_4^{2-}	NH_3	SO_4^{2-}
CO_3^{2-}	$H_2PO_4^-$	N_2H_4	SO_3S^{2-}

M	n	$\lg \beta_n$	M	n	$\lg \beta_n$	M	n	$\lg \beta_n$
Br^- Ligand			Ce^{III}	1	0.38	Pb^{II}	4	3.00
			Co^{II}	2	−0.42		3	3.30
Ag^I	4	8.73		1	−0.13		2	1.92
	3	8.00	Cu^I	2	5.92		1	1.15
	2	7.34	Cu^{II}	2	5.89	Pd^{II}	4	13.10
	1	4.38		1	0.30	Pt^{II}	4	20.50
Au^I	2	12.46	Fe^{III}	2	0.82	Sn^{II}	5	1.98
Au^{III}	4	31.50		1	0.55		4	1.66
Bi^{III}	6	9.52	Hg^{II}	4	20.01		3	2.13
	5	9.42		3	18.75		2	1.73
	4	7.84		2	17.33		1	0.90
	3	6.33		1	9.05	Tl^I	4	−0.20
	2	4.45	In^{III}	4	−1.25		3	0.60
	1	2.26		3	0.67		2	1.01
Cd^{II}	4	3.70		2	1.89		1	0.93
	3	3.32		1	1.30	Tl^{III}	6	26.20
	2	2.34	Ni^{II}	4	−8.12		5	25.50
	1	1.75		2	−3.24		4	23.90

(*Continued*)

M	n	$\lg \beta_n$	M	n	$\lg \beta_n$	M	n	$\lg \beta_n$
Tl^{III}	3	21.20	Cd^{II}	3	12.29	Ni^{II}	2	2.82
	2	16.60		2	10.22		1	1.78
	1	9.70		1	5.63	Ni^{III}	3	3.14
U^{IV}	1	0.18	Co^{II}	3	13.82	Zn^{II}	4	1.93
UO_2^{2+}	1	−0.30		2	10.72		3	1.61
Zn^{II}	4	−2.50		1	5.89		2	1.41
	3	−2.90	Co^{III}	3	48.69		1	1.11
	2	−2.20	Cu^{I}	2	10.80			
	1	−0.80	Cu^{II}	2	20.13		$(C_5H_4N)_2$ (bipy) Ligand	
				1	10.76	Ag^{I}	1	6.80
	CH_3COO⁻ Ligand		Fe^{II}	3	9.70	Cd^{II}	3	10.47
				2	7.65		2	8.00
Ag^{I}	2	0.64		1	4.34		1	4.50
	1	0.74	Hg^{II}	3	23.42	Cu^{II}	3	17.85
Ba^{II}	1	1.15	Mg^{II}	1	0.37		2	14.20
Ca^{II}	1	0.77	Mn^{II}	3	5.67	Fe^{II}	3	17.58
Cd^{II}	2	3.15		2	4.79		2	9.21
	1	1.93		1	2.73		1	4.21
Ce^{III}	3	3.23	Ni^{II}	3	19.11	Mg^{II}	1	0.50
	2	2.65		2	14.08	Pb^{II}	1	3.00
	1	1.68		1	7.60	Zn^{II}	3	13.50
Cu^{II}	2	3.63	Tl^{I}	1	0.30		2	9.80
	1	2.23	Zn^{II}	3	12.08		1	5.40
In^{III}	6	10.30		2	10.37			
	5	9.23		1	5.71		CN⁻ Ligand	
	4	9.08						
	3	7.90		C_5H_5N (py) Ligand		Ag^{I}	4	19.42
	2	5.95					3	20.55
	1	3.51	Ag^{I}	2	4.11		2	19.85
Mg^{II}	1	1.25		1	1.97	Au^{I}	2	38.30
Mn^{II}	1	1.40	Cd^{II}	4	2.49	Au^{III}	4	56.00
Ni^{II}	2	2.12		2	2.14	Cd^{II}	4	17.11
	1	1.43		1	1.27		3	13.92
Pb^{II}	4	8.58	Co^{II}	2	1.54		2	9.60
	3	6.48		1	1.14		1	5.18
	2	4.08	Cu^{I}	4	8.70	Co^{II}	6	19.09
	1	2.68		3	7.90	Co^{III}	6	64.00
Sr^{II}	1	1.19		2	6.60	Cu^{I}	4	30.30
Tl^{I}	1	−0.11		1	3.90		3	28.59
Tl^{III}	4	18.30	Cu^{II}	4	6.54		2	24.00
UO_2^{2+}	3	6.30		3	5.69	Fe^{II}	6	36.90
	2	4.90		2	4.38		5	18.60
	1	2.61		1	2.52	Fe^{III}	6	43.90
Zn^{II}	2	2.38	Fe^{II}	4	6.70	Hg^{II}	4	38.97
	1	1.57		1	0.71		3	36.31
			Hg^{II}	3	10.40		2	32.75
	$C_2H_8N_2$ (en) Ligand			2	10.00		1	17.00
Ag^{I}	2	7.84		1	5.10	Ni^{II}	4	31.00
	1	4.70	Ni^{II}	3	3.13	Tl^{III}	4	35.00

(Continued)

M	n	$\lg \beta_n$	M	n	$\lg \beta_n$	M	n	$\lg \beta_n$
Zn^{II}	4	19.62	Ni^{II}	2	7.64	Cd^{II}	1	2.00
	3	16.05		1	>5.30	Ce^{III}	1	0.22
	2	11.07	NpO_2^{2+}	2	7.07	Cm^{III}	1	1.17
CO_3^{2-} Ligand				1	3.30	Co^{II}	1	0.14
			Pb^{II}	2	6.54	Cr^{III}	2	−0.11
Ca^{II}	1	3.20		1	4.90		1	0.60
Cd^{II}	3	6.22	Sr^{II}	2	1.90	Cu^I	3	5.63
Cu^{II}	2	10.16		1	1.25		2	5.54
	1	6.91	Th^{IV}	4	29.6	Cu^{II}	3	−2.10
La^{III}	1	7.72	Tl^I	1	2.03		2	−0.52
Mg^{II}	1	3.40	Yb^{III}	3	>12.90		1	0.95
Pb^{II}	2	8.20		2	11.89	Fe^{II}	2	0.40
UO_2^{2+}	3	22.36		1	7.30		1	0.36
	2	18.09	Zn^{II}	3	8.34	Fe^{III}	4	−0.85
$C_2O_4^{2-}$ Ligand				2	7.55		3	0.78
				1	4.85		2	1.78
Al^{III}	3	16.30					1	1.48
	2	13.00	$CS(NH_2)_2$ Ligand			Ga^{III}	4	−6.80
Ba^{II}	1	2.31	Ag^I	3	13.14		3	−4.50
Be^{II}	2	5.91	Bi^{III}	6	11.94		2	−2.30
	1	4.08	Cd^{II}	3	2.92		1	0.60
Cd^{II}	2	5.37		2	2.63	Hg^{II}	4	15.22
	1	3.52		1	1.58		3	13.99
Ce^{III}	3	11.30	Cu^{II}	4	15.39		2	13.16
	2	10.48		3	12.82		1	6.76
	1	6.52	Hg^{II}	4	26.30	In^{III}	3	3.23
Co^{II}	3	9.70		3	24.60		2	2.64
	2	6.70		2	21.90		1	1.72
	1	4.70	Pb^{II}	3	1.77	Ir^{III}	6	14.00
Cu^{II}	2	10.30	Cl^- Ligand			La^{III}	1	−0.15
	1	6.70				Mn^{III}	1	0.95
Fe^{II}	3	5.22	Ag^I	4	5.92	Mo^{VI}	3	−2.69
	2	4.52		3	5.40		2	−0.80
Fe^{III}	3	20.20		2	5.25		1	−0.30
	2	16.20		1	3.31	NpO_2^{2+}	1	0.21
	1	9.40	Am^{III}	1	1.17	Pb^{II}	3	1.85
La^{III}	3	10.30	Au^I	2	11.79		2	2.44
	2	7.90	Au^{III}	4	21.30		1	1.62
	1	4.30	Bi^{III}	6	8.14	Pd^{II}	4	15.10
Mn^{II}	2	5.80		5	6.11		3	13.10
	1	3.89		4	4.98		2	10.60
Mn^{III}	3	19.42		3	3.42		1	6.00
	2	16.57		2	3.10	Pt^{II}	4	16.00
	1	9.98		1	2.43		3	14.48
Nd^{III}	3	>13.50	Cd^{II}	6	2.58	Pu^{III}	1	1.17
	2	11.51		4	2.90	Pu^{IV}	1	−0.25
	1	7.21		3	2.11	PuO_2^{2+}	2	−0.35
Ni^{II}	3	ca. 14.00		2	2.70		1	0.10

(Continued)

M	n	$\lg \beta_n$	M	n	$\lg \beta_n$	M	n	$\lg \beta_n$
Sn^{II}	3	2.45	Cr^{III}	2	8.54	U^{IV}	1	14.57
	2	2.24		1	5.20	UO_2^{2+}	4	13.25
	1	1.51	Cu^{II}	1	1.23		3	11.89
Sn^{IV}	1	0.82	Fe^{III}	4	15.74		2	9.33
Th^{IV}	4	−0.51		3	13.74		1	4.93
	3	0.23		2	10.74	VO^{2+}	3	7.15
	2	0.38		1	6.04		2	5.46
	1	1.38	Ga^{III}	3	13.49		1	3.30
Tl^{I}	3	−0.80		2	10.72	Y^{III}	3	12.14
	2	0.25		1	6.19		2	8.54
	1	0.68	Gd^{III}	1	3.46		1	4.81
Tl^{III}	5	17.47	Ge^{IV}	6	25.57	Zn^{II}	1	1.26
	4	17.00		5	21.84	Zr^{IV}	6	35.86
	3	14.75	H^{I}	2	3.77		5	31.85
	2	12.00		1	3.18		4	27.17
	1	7.50	Hg^{II}	1	1.56		3	22.20
U^{IV}	1	−0.85	In^{III}	3	10.23		2	16.35
UO_2^{2+}	3	−2.62		2	7.41		1	8.89
	2	−0.92		1	4.63	HCO_3^- Ligand		
	1	0.21	La^{III}	1	3.56	Ca^{II}	1	1.26
VO^{2+}	1	0.04	Mg^{II}	1	1.82	Na^{I}	1	−0.25
Y^{III}	1	1.26	Mn^{III}	1	5.76	Mg^{II}	1	1.16
Zn^{II}	4	−1.52	Ni^{II}	1	0.66	Mn^{II}	1	1.80
	3	−1.40	Pb^{II}	3	3.42	Pb^{II}	3	5.19
	2	0.18		2	2.27		2	4.77
	1	−0.19		1	1.26	HPO_4^{2-} Ligand		
Zr^{IV}	4	1.20	Pu^{III}	1	7.94	Ca^{II}	1	2.77
	3	1.50	Pu^{IV}	1	6.77	Fe^{III}	1	10.89
	2	1.30	Sc^{III}	4	20.18	Li^{I}	1	0.72
	1	0.90		3	17.33	Mg^{II}	1	2.91
F^- Ligand				2	12.88	Ni^{II}	1	2.08
Ag^{I}	1	0.38		1	7.08	Pu^{IV}	5	52.00
Al^{III}	6	20.83	Sn^{II}	3	10.00		4	43.20
	5	20.36		1	4.85		3	33.40
	4	19.03	Sn^{IV}	6	ca. 25.00		2	23.70
	3	16.65	Th^{IV}	3	17.96		1	12.90
	2	12.60		2	13.27	UO_2^{2+}	2	18.57
	1	6.98		1	7.28		1	8.43
Be^{II}	4	14.97	Ti^{IV}	4	20.38	Zn^{II}	1	2.40
	3	12.71		3	16.32	$H_2PO_4^-$ Ligand		
	2	9.94		2	11.74	Al^{III}	3	7.60
	1	6.00		1	6.65		2	5.30
Cd^{II}	3	1.20	Tl^{I}	1	0.10		1	3.00
	2	0.53	U^{IV}	6	32.21	Ca^{II}	1	1.41
	1	0.30		5	29.91	Fe^{III}	4	9.15
Ce^{III}	1	3.99		4	28.32			
Cr^{III}	3	11.02		3	24.05			
				2	19.82			

(Continued)

M	n	$\lg \beta_n$	M	n	$\lg \beta_n$	M	n	$\lg \beta_n$
Th^{IV}	2	8.15	Hg^{II}	3	27.60	Bi^{III}	6	4.23
	1	4.30		2	23.82		4	3.41
UO_2^{2+}	3	7.33		1	12.87		2	2.26
	2	5.43	In^{III}	3	2.48		1	1.15
	1	3.00		2	2.56	Cd^{II}	6	−0.08
				1	1.64		4	2.91
H_3PO_4 Ligand			Pb^{II}	4	3.92		3	2.30
Pu^{IV}	1	2.30		3	3.42		2	2.40
Th^{IV}	2	3.86		2	2.80		1	1.74
	1	1.89		1	1.26	Co^{II}	4	2.30
UO_2^{2+}	3	5.23	Tl^{I}	4	1.60		3	0
	2	3.88		3	2.00		2	−0.7
	1	<1.88		2	1.82		1	−0.04
				1	1.41	Cr^{III}	6	3.80
HS^- Ligand			Tl^{III}	4	31.82		5	5.40
Ag^{I}	2	18.45		3	27.60		4	6.10
	1	14.05		2	20.88		3	5.80
Cd^{II}	4	20.86		1	11.41		2	4.80
	3	18.49	Zn^{II}	4	−0.51		1	3.08
	2	16.57		3	1.26	Cu^{I}	6	9.27
	1	9.41		2	−1.53		5	9.59
Co^{II}	2	8.77		1	−0.47		4	10.05
	1	5.67					3	9.90
Cu^{II}	3	25.90	IO_3^- Ligand				2	12.11
Fe^{II}	3	10.97	Ag^{I}	2	1.90	Cu^{II}	4	6.52
I	2	8.94		1	0.63		3	5.19
Hg^{II}	2	37.72	Ba^{II}	1	1.05		2	3.65
Pb^{II}	3	16.52	Ca^{II}	1	0.89		1	2.30
	2	15.25	Cu^{II}	1	0.82	Fe^{II}	4	4.53
Zn^{II}	3	16.10	K^{I}	1	−0.30		1	1.33
	2	14.90	Mg^{II}	1	0.72	Fe^{III}	6	3.23
			Sr^{II}	1	0.98		5	4.23
I^- Ligand			Th^{IV}	6	11.02		4	4.53
Ag^{I}	4	13.10		3	7.18		3	4.63
	3	13.68		2	4.81		2	4.33
	2	11.74		1	2.88		1	3.03
	1	6.58	Tl^{I}	1	0.50	Hg^{II}	4	21.20
Bi^{III}	6	19.10					3	20.40
	5	16.80	NCS^- Ligand				2	17.60
	4	14.95	Ag^{I}	4	9.67	In^{III}	3	4.63
	1	2.89		3	9.45		2	3.00
Cd^{II}	6	6.00		2	8.23		1	2.58
	4	5.41		1	4.75	Ni^{II}	3	1.81
	3	5.00	Al^{III}	1	0.42		2	1.64
	2	3.92	Au^{I}	2	23.00		1	1.18
	1	2.28	Au^{III}	6	42.04	Pb^{II}	4	0.85
Cu^{I}	2	8.76		4	42.00		3	1.90
Hg^{II}	4	29.83		2	23.00		2	2.52

(Continued)

M	n	$\lg \beta_n$	M	n	$\lg \beta_n$	M	n	$\lg \beta_n$
Pb^{II}	1	1.09	Co^{III}	4	25.70	Cd^{II}	4	3.89
Ru^{III}	1	1.78		3	20.10		3	2.78
Th^{IV}	3	1.78		2	14.00		2	2.40
	1	1.08		1	7.30		1	2.25
Ti^{IV}	1	1.70	Cu^{I}	2	10.86	Ni^{II}	6	11.99
Tl^{I}	4	0		1	5.93		5	10.75
	3	0.20	Cu^{II}	6	8.90		4	9.20
	2	0.65		5	11.43		3	7.35
	1	0.80		4	12.90		2	5.20
U^{IV}	3	2.18		3	10.72		1	2.76
	2	2.11		2	7.82	Zn^{II}	4	3.88
	1	1.49		1	4.27		3	3.78
UO_2^{2+}	2	1.90	Fe^{II}	2	2.20		2	3.70
	1	1.50		1	1.40		1	3.40
V^{II}	1	2.00	Hg^{II}	4	7.47			
VO^{2+}	2	3.68		3	6.40	NO_2^- Ligand		
	1	2.32		2	4.80	Ag^{I}	2	2.83
Zn^{II}	4	3.70		1	2.68		1	1.88
	3	2.20	Li^{I}	1	−0.30	Cd^{II}	4	3.10
	2	2.10	Mg^{II}	6	−3.29		3	3.81
	1	1.62		5	−1.99		2	3.01
Zr^{IV}	6	7.90		4	−1.04		1	1.80
	5	6.90		3	−0.34	Cu^{II}	2	1.65
	4	5.80		2	0.08		1	1.30
	3	4.70		1	0.23	Hg^{II}	4	13.54
	2	3.40	Mn^{II}	6	9.00			
	1	2.00		2	1.30	NO_3^- Ligand		
				1	0.80	Ag^{I}	1	−0.29
NH_3 Ligand			Ni^{II}	6	8.31	Ba^{II}	1	0.92
Ag^{I}	2	7.21		5	7.89	Bi^{III}	1	1.26
	1	3.37		4	7.04	Ca^{II}	1	0.28
Au^{I}	2	27.00		3	5.81	Cd^{II}	1	−0.31
Au^{III}	4	30.00		2	4.26	Ce^{III}	2	1.51
Ca^{II}	1	−0.20		1	2.36		1	1.04
Cd^{II}	6	5.14	Tl^{I}	1	−0.92	Cs^{I}	1	0.01
	5	6.80	Tl^{III}	4	13.00	Fe^{III}	1	1.00
	4	7.37		3	11.60	Hf^{IV}	4	2.08
	3	6.37		2	9.30		3	1.89
	2	4.92		1	4.60		2	1.51
	1	2.74	Zn^{II}	4	8.62		1	0.92
Co^{II}	6	4.39		3	6.92	Hg^{II}	2	0.01
	5	5.13		2	4.91		1	0.35
	4	5.07		1	2.59	Hg_2^{2+}	2	−0.24
	3	4.43					1	0.08
	2	3.50	N_2H_4 Ligand			K^{I}	1	−0.14
	1	1.99	Ca^{II}	3	−1.91	La^{III}	1	−0.26
Co^{III}	6	35.21		2	−0.80	Li^{I}	1	−1.45
	5	30.80		1	−0.16	Mg^{II}	1	0.00

(Continued)

M	n	$\lg \beta_n$	M	n	$\lg \beta_n$	M	n	$\lg \beta_n$
Na^I	1	−0.59	Cr^{III}	6	14.41	Pb^{II}	2	10.83
Pb^{II}	1	1.18		5	21.19		1	7.83
Pu^{IV}	1	1.80		4	27.97	Pd^{II}	2	26.10
Sr^{II}	1	0.54		2	18.45		1	12.68
Th^{IV}	4	0.74		1	10.02	Pt^{II}	4	22.67
	3	1.00	Cu^{II}	4	15.89	Pu^{III}	1	7.05
	2	1.11		3	15.00	Pu^{IV}	1	9.12
	1	0.78		2	13.68	Sb^{III}	4	38.30
Tl^I	1	0.33		1	6.65		3	36.70
Tl^{III}	3	1.10	Fe^{II}	4	9.60		2	24.30
	2	0.12		3	10.00		1	6.07
	1	0.90		2	9.17	Sc^{III}	1	9.12
U^{IV}	4	0.18		1	7.25	Sn^{II}	3	11.93
	3	0.42	Fe^{III}	4	34.49		2	9.01
	2	0.47		3	32.87		1	4.45
	1	0.36		2	21.23	Sn^{IV}	6	63.00
UO_2^{2+}	1	−1.40		1	11.83	Sr^{II}	1	0.82
Zr^{IV}	4	−0.82	Ga^{III}	6	40.30	Th^{IV}	6	38.70
	3	−0.26		4	38.84		3	27.65
	2	0.11		3	31.65		2	19.87
	1	0.34		2	21.68		1	10.12
				1	11.19	Tl^I	1	0.82
OH⁻ Ligand			Ge^{IV}	6	60.91	Tl^{III}	2	25.37
Ag^I	3	5.20		5	59.63		1	12.86
	2	3.99		4	54.37	U^{IV}	2	25.7
	1	2.00		3	41.09		1	13.78
Al^{III}	4	32.50		2	27.61	UO_2^{2+}	4	32.40
	3	26.00		1	13.87		2	15.56
	2	17.86	Hg^{II}	3	21.20		1	8.84
	1	8.98		2	21.70	V^{II}	1	7.53
Ba^{II}	1	2.22		1	10.30	V^{III}	2	21.54
Be^{II}	4	18.42	Hg_2^{2+}	1	9.00		1	11.07
	3	18.63	In^{III}	4	33.88	Y^{III}	1	4.92
	1	7.48		3	29.04	Zn^{II}	4	16.63
Bi^{III}	4	35.20		2	20.17		3	15.00
	2	15.80		1	10.41		2	12.90
	1	12.43	La^{III}	1	3.30		1	5.04
Ca^{II}	1	1.22	Li^I	1	0.17	Zr^{IV}	4	54.51
Cd^{II}	4	9.25	Lu^{III}	1	7.40		3	41.32
	3	9.68	Mg^{II}	1	2.64		2	27.75
	2	9.46	Mn^{II}	3	7.81		1	14.04
	1	6.37		1	3.41			
Ce^{III}	1	4.60	Na^I	1	−0.48	$P_2O_7^{4-}$ Ligand		
Ce^{IV}	2	27.06	Nd^{III}	1	5.50	Ba^{II}	1	4.64
	1	14.81	Ni^{II}	3	12.96	Ca^{II}	1	5.60
Co^{II}	3	10.49		2	10.21	Cd^{II}	1	8.70
	2	9.19		1	3.08	Ce^{III}	1	17.15
	1	4.40	Pb^{II}	3	13.92	Co^{II}	1	6.10

(Continued)

M	n	$\lg \beta_n$	M	n	$\lg \beta_n$	M	n	$\lg \beta_n$
Cu^I	2	26.72	Co^{II}	1	2.36	VO^{2+}	1	2.48
Cu^{II}	2	12.45	Co^{III}	1	1.34	Y^{III}	2	3.99
	1	7.60	Cr^{III}	1	1.60		1	3.34
Fe^{III}	2	5.55	Cu^{II}	1	2.26	Yb^{III}	1	3.58
K^I	1	2.30	Er^{III}	1	3.58	Zn^{II}	1	2.38
La^{III}	2	18.57	Fe^{II}	1	2.20	Zr^{IV}	3	7.64
	1	16.72	Fe^{III}	2	5.38		2	6.54
Li^I	1	3.10		1	4.15		1	3.74
Mg^{II}	1	7.20	Ga^{III}	2	5.06			
Na^I	1	2.22		1	2.77		$SO_3S_2^-$ Ligand	
Ni^{II}	2	7.19	Gd^{III}	1	3.66	Ag^I	2	13.46
	1	5.82	Hg^{II}	2	2.44		1	8.60
Pb^{II}	2	9.40		1	1.34	Ba^{II}	1	2.33
	1	6.40	Hg_2^{2+}	2	2.40	Ca^{II}	1	1.98
Sn^{II}	2	16.40		1	1.30	Cd^{II}	2	6.46
Sr^{II}	1	5.40	Ho^{III}	1	3.58		1	3.94
Tl^I	2	1.87	In^{III}	2	5.00	Co^{II}	1	2.05
	1	1.69		1	3.04	Cu^I	3	13.71
Zn^{II}	2	11.00	K^I	1	0.75		2	12.27
	1	8.70	La^{III}	1	3.50		1	10.35
			Li^I	1	0.64	Cu^{II}	3	13.84
	SO_3^{2-} Ligand		Mg^{II}	1	2.23		2	12.22
			Mn^{II}	1	2.26		1	10.27
Ag^I	3	9.00	Na^I	1	0.72	Fe^{II}	1	2.17
	2	8.68	Nd^{III}	1	3.64	Fe^{III}	1	3.25
	1	5.60	Ni^{II}	1	2.32	Hg^{II}	4	33.61
Cd^{II}	2	4.19	Np^{IV}	2	3.47		3	32.26
Ce^{III}	1	8.04		1	2.43		2	30.80
Cu^I	3	9.36	Pb^{II}	2	3.47		1	29.27
	2	8.70		1	2.62	K^I	1	0.92
	1	7.85	Pr^{III}	1	3.62	La^{III}	1	2.99
Hg^{II}	3	24.96	Pu^{III}	2	1.62	Mg^{II}	1	1.79
	2	24.07		1	1.00	Mn^{II}	1	1.95
	1	22.66	Pu^{IV}	1	3.66	Na^I	1	1.08
Tl^{III}	4	34.00	Sm^{III}	1	3.66	Ni^{II}	1	2.06
			Th^{IV}	2	5.70	Pb^{II}	4	7.20
	SO_4^{2-} Ligand			1	3.32		3	6.35
			Ti^{IV}	1	2.40		2	5.13
Ag^I	1	1.30	Tl^I	1	1.27		1	2.70
Al^{III}	2	4.90	U^{IV}	2	5.42	Sr^{II}	1	2.04
	1	3.01		1	3.24	Tl^I	1	1.91
Ba^{II}	1	2.35	UO_2^{2+}	3	3.40	Tl^{III}	4	41.00
Ca^{II}	1	2.43		2	2.72	Zn^{II}	2	4.59
Cd^{II}	1	2.29		1	1.48		1	2.29
Ce^{III}	1	3.48						
Ce^{IV}	1	1.78						

4.6 Solubility Products in Aqueous Solution at 298.15 K [5, 15, 22]

The state of heterogeneous equilibrium between the saturated aqueous solution of a difficultly soluble strong electrolyte $M_m A_a$ and the precipitate of said substance

$$M_m A_{a(sld)} \rightleftharpoons M_m A_a = mM^{v+} + aA^{v-}$$

is characterized by the solubility product (SP):

$$L = [M^{v+}]^m [A^{v-}]^a = f(T)$$

The table lists the values of solubility products and exponents thereof ($pL = -\log L$); the values of log L with a superscript * (and, accordingly, the values of L) correspond to the room temperature (18–25°C).

Discrepancy between the composition of the precipitate of the dissolving substance and the formula entity of the solute in the solution, as well as the unobvious formula compositions of the cations and anions are stated in footnotes at the end of the table.

Solubility of difficultly soluble electrolytes in water, S, is expressed in terms of the analytical concentration of the solute in the saturated solution (mole/l). This solubility is evaluated using the following formulas derived from the mathematical expression of L:

for $m{:}a = 1{:}1$ $S = \sqrt{L}$ for $m{:}a = 1{:}3$ or $3{:}1$ $S = \sqrt[4]{L/27}$

     $m{:}a = 2{:}1$ or $1{:}2$ $S = \sqrt[3]{L/4}$ $m{:}a = 2{:}3$ or $3{:}2$ $S = \sqrt[3]{L/108}$

     $m{:}a = 2{:}2$ $S = \sqrt[4]{L}$ $m{:}a = 1{:}4$ $S = \sqrt[5]{L/256}$

If the electrolyte yields in the solution three ion species A, B and C (ion charges are not specified), then the formula for calculating will be as follows:

for $a{:}b{:}c = 1{:}1{:}1$, $S = \sqrt[3]{L}$ for $a{:}b{:}c = 1{:}2{:}1$, $S = \sqrt[4]{L/4}$, etc.

where a, b, and c stand for the coefficients of the ions A, B, and C, respectively.

Where there exists a discrepancy between the formula composition of the substance in solution and precipitate, the value of S should be recalculated for the amount of the dissolving substance, $M_m A_a$, in compliance with the reaction equations given in footnotes.

The solubilities of strong electrolytes difficultly soluble in water determined by the technique described above are estimated values. To obtain accurate values, it is essential to account for protolysis of cations and/or anions in an aqueous solution and complex formation therefrom (the magnitude of these phenomena exceeds the solubilities of difficultly soluble electrolytes). The techniques of calculations in question is described in handbooks on chemical analysis (for example, H. Laitinen and W. Harris, Chemical Analysis. New York/Toronto/ London, McGraw-Hill, 1976).

Because of the same factors, the experimental data on solubility of substances (Sec. 5.2) may differ significantly from solubility values assessed using the L values.

Substance	L	pL	Solubility	
			mole/l	g/l
$Ac(OH)_3$	$1.30 \cdot 10^{-21}$	20.89	$2.6 \cdot 10^{-6}$	$7.3 \cdot 10^{-4}$
$Ac(OH)_3$ (amor)	$2.10 \cdot 10^{-19}$	18.68*	$9.4 \cdot 10^{-6}$	$2.6 \cdot 10^{-3}$
Ag_3AsO_3	$4.50 \cdot 10^{-19}$	18.35	$1.1 \cdot 10^{-5}$	$5.1 \cdot 10^{-3}$
Ag_3AsO_4	$1.00 \cdot 10^{-22}$	22.00*	$1.4 \cdot 10^{-6}$	$6.4 \cdot 10^{-4}$
$AgBr$	$5.01 \cdot 10^{-13}$	12.30	$7.1 \cdot 10^{-7}$	$1.3 \cdot 10^{-4}$
$AgBrO_3$	$5.77 \cdot 10^{-5}$	4.24	$7.6 \cdot 10^{-3}$	1.79
$AgCH_3COO$	$4.40 \cdot 10^{-3}$	2.36	$6.6 \cdot 10^{-2}$	11.07
$AgCN$	$7.00 \cdot 10^{-15}$	14.16	$8.4 \cdot 10^{-8}$	$1.1 \cdot 10^{-5}$
Ag_2CN_2	$7.00 \cdot 10^{-11}$	10.16	$2.6 \cdot 10^{-4}$	$6.6 \cdot 10^{-2}$
Ag_2CO_3	$8.70 \cdot 10^{-12}$	11.06	$1.3 \cdot 10^{-4}$	$3.6 \cdot 10^{-2}$
$Ag_2C_2O_4$	$1.10 \cdot 10^{-11}$	10.96	$1.4 \cdot 10^{-4}$	$4.3 \cdot 10^{-2}$
$AgCl$	$1.82 \cdot 10^{-10}$	9.74	$1.3 \cdot 10^{-5}$	$1.9 \cdot 10^{-3}$
$AgClO_2$	$2.00 \cdot 10^{-4}$	3.70*	$1.4 \cdot 10^{-2}$	2.48
Ag_2CrO_4	$1.16 \cdot 10^{-12}$	11.94	$6.6 \cdot 10^{-5}$	$2.2 \cdot 10^{-2}$
$Ag_2Cr_2O_7$	$2.00 \cdot 10^{-7}$	6.70	$3.7 \cdot 10^{-3}$	1.59
AgI	$2.30 \cdot 10^{-16}$	15.64	$1.5 \cdot 10^{-8}$	$3.6 \cdot 10^{-6}$
$AgIO_3$	$3.20 \cdot 10^{-8}$	7.49	$1.8 \cdot 10^{-4}$	$5.1 \cdot 10^{-2}$
$AgMnO_4$	$1.60 \cdot 10^{-3}$	2.80*	$4.0 \cdot 10^{-2}$	9.07
AgN_3	$2.90 \cdot 10^{-9}$	8.54*	$5.4 \cdot 10^{-5}$	$8.1 \cdot 10^{-3}$
$AgNCS$	$1.57 \cdot 10^{-12}$	11.80	$1.3 \cdot 10^{-6}$	$2.1 \cdot 10^{-4}$
$AgNO_2$	$3.07 \cdot 10^{-10}$	9.51	$1.8 \cdot 10^{-5}$	$2.7 \cdot 10^{-3}$
$Ag_2O^{(1)}$	$2.00 \cdot 10^{-8}$	7.70	$7.1 \cdot 10^{-5}$	$1.6 \cdot 10^{-2}$
$AgOCN$	$2.30 \cdot 10^{-7}$	6.64*	$4.8 \cdot 10^{-4}$	$7.2 \cdot 10^{-2}$
Ag_3PO_4	$1.80 \cdot 10^{-18}$	17.75	$1.6 \cdot 10^{-5}$	$6.7 \cdot 10^{-3}$
$AgReO_4$	$7.95 \cdot 10^{-5}$	4.10*	$8.9 \cdot 10^{-3}$	3.19
Ag_2S	$7.24 \cdot 10^{-50}$	49.14	$2.6 \cdot 10^{-17}$	$6.5 \cdot 10^{-15}$
Ag_2SO_3	$1.50 \cdot 10^{-14}$	13.82*	$1.6 \cdot 10^{-5}$	$4.6 \cdot 10^{-3}$
Ag_2SO_4	$1.15 \cdot 10^{-5}$	4.94	$1.4 \cdot 10^{-2}$	4.43
Ag_2Se	$2.49 \cdot 10^{-59}$	58.60	$1.8 \cdot 10^{-20}$	$5.4 \cdot 10^{-18}$
Ag_2SeO_4	$1.23 \cdot 10^{-9}$	8.91	$6.7 \cdot 10^{-4}$	$2.4 \cdot 10^{-1}$
Ag_2Te	$4.71 \cdot 10^{-52}$	51.33	$4.9 \cdot 10^{-18}$	$1.7 \cdot 10^{-15}$
$AlAsO_4$	$1.60 \cdot 10^{-16}$	15.80*	$1.3 \cdot 10^{-8}$	$2.1 \cdot 10^{-6}$
$[AlF_6],K_3$	$1.60 \cdot 10^{-9}$	8.80*	$2.8 \cdot 10^{-3}$	$7.2 \cdot 10^{-1}$
$[AlF_6],Na_3$	$7.80 \cdot 10^{-14}$	13.11	$2.3 \cdot 10^{-4}$	$4.9 \cdot 10^{-2}$
$Al(OH)_3$	$5.70 \cdot 10^{-32}$	31.24	$6.8 \cdot 10^{-9}$	$5.3 \cdot 10^{-7}$
$Al(OH)_3^{(2)}$	$3.70 \cdot 10^{-15}$	14.43	$6.1 \cdot 10^{-8}$	$4.7 \cdot 10^{-6}$
$AlPO_4$	$1.69 \cdot 10^{-19}$	18.77	$4.1 \cdot 10^{-10}$	$5.0 \cdot 10^{-8}$
$Am(OH)_3$	$2.70 \cdot 10^{-20}$	19.57*	$5.6 \cdot 10^{-6}$	$1.7 \cdot 10^{-3}$
$Am(OH)_4$	$1.00 \cdot 10^{-56}$	56.00*	$2.1 \cdot 10^{-12}$	$6.5 \cdot 10^{-10}$
$AuBr$	$5.00 \cdot 10^{-17}$	16.30*	$7.1 \cdot 10^{-9}$	$2.0 \cdot 10^{-6}$
$AuBr_3^{(3)}$	$4.00 \cdot 10^{-36}$	35.40*	$2.0 \cdot 10^{-18}$	$8.7 \cdot 10^{-16}$

(Continued)

Substance	L	pL	Solubility	
			mole/l	g/l
AuCl	$1.79 \cdot 10^{-12}$	11.75	$1.3 \cdot 10^{-6}$	$3.1 \cdot 10^{-4}$
$AuCl_3$[3]	$3.20 \cdot 10^{-25}$	24.50*	$5.7 \cdot 10^{-13}$	$1.7 \cdot 10^{-10}$
$[AuCl_4]$,Cs	$1.00 \cdot 10^{-3}$	3.00*	$3.2 \cdot 10^{-2}$	14.92
AuI	$1.60 \cdot 10^{-23}$	22.80*	$4.0 \cdot 10^{-12}$	$1.3 \cdot 10^{-9}$
AuI_3[3]	$1.00 \cdot 10^{-46}$	46.00*	$1.0 \cdot 10^{-23}$	$5.8 \cdot 10^{-21}$
Au_2O_3[4]	$8.50 \cdot 10^{-46}$	45.07	$2.4 \cdot 10^{-12}$	$1.0 \cdot 10^{-9}$
$[BF_4]$,Cs	$2.00 \cdot 10^{-5}$	4.70*	$4.5 \cdot 10^{-3}$	$9.8 \cdot 10^{-1}$
$[BF_4]$,K	$2.00 \cdot 10^{-3}$	2.70*	$4.5 \cdot 10^{-2}$	5.63
$[BH_4]$,Cs	$2.50 \cdot 10^{-7}$	6.60*	$5.0 \cdot 10^{-4}$	$7.4 \cdot 10^{-2}$
$[BH_4]$,K	$1.30 \cdot 10^{-3}$	2.89*	$3.6 \cdot 10^{-2}$	1.94
$Ba_3(AsO_4)_2$	$7.80 \cdot 10^{-51}$	50.11*	$3.7 \cdot 10^{-11}$	$2.6 \cdot 10^{-8}$
$Ba(BrO_3)_2$	$3.30 \cdot 10^{-5}$	4.48	$2.0 \cdot 10^{-2}$	7.94
$BaCO_3$	$4.90 \cdot 10^{-9}$	8.31	$7.0 \cdot 10^{-5}$	$1.4 \cdot 10^{-2}$
BaC_2O_4	$1.10 \cdot 10^{-7}$	6.96	$3.3 \cdot 10^{-4}$	$7.5 \cdot 10^{-2}$
$BaCrO_4$	$1.06 \cdot 10^{-10}$	9.97	$1.0 \cdot 10^{-5}$	$2.6 \cdot 10^{-3}$
BaF_2	$1.73 \cdot 10^{-6}$	5.76	$7.6 \cdot 10^{-3}$	1.33
$Ba(IO_3)_2$	$6.50 \cdot 10^{-10}$	9.19	$5.5 \cdot 10^{-4}$	$2.7 \cdot 10^{-1}$
$BaMnO_4$	$2.50 \cdot 10^{-10}$	9.60	$1.6 \cdot 10^{-5}$	$4.1 \cdot 10^{-3}$
$BaMoO_4$	$4.88 \cdot 10^{-9}$	8.31	$7.0 \cdot 10^{-5}$	$2.1 \cdot 10^{-2}$
$Ba_2P_2O_7$	$3.00 \cdot 10^{-11}$	10.52*	$2.0 \cdot 10^{-4}$	$8.8 \cdot 10^{-2}$
$Ba_3(PO_4)_2$	$6.03 \cdot 10^{-39}$	38.22*	$8.9 \cdot 10^{-9}$	$5.4 \cdot 10^{-6}$
$BaSO_3$	$8.00 \cdot 10^{-7}$	6.10*	$8.9 \cdot 10^{-4}$	$1.9 \cdot 10^{-1}$
$BaSO_4$	$1.82 \cdot 10^{-10}$	9.74	$1.3 \cdot 10^{-5}$	$3.1 \cdot 10^{-3}$
$Ba(SO_3S)$[5]	$1.60 \cdot 10^{-5}$	4.80*	$4.0 \cdot 10^{-3}$	1.00
$BaSeO_4$	$3.12 \cdot 10^{-8}$	7.51	$1.8 \cdot 10^{-4}$	$5.0 \cdot 10^{-2}$
$BaWO_4$	$1.77 \cdot 10^{-9}$	8.75	$4.2 \cdot 10^{-5}$	$1.6 \cdot 10^{-2}$
$Be(OH)_2$	$8.02 \cdot 10^{-22}$	21.10	$5.9 \cdot 10^{-8}$	$2.5 \cdot 10^{-6}$
$Be(OH)_2$[6]	$6.59 \cdot 10^{-21}$	20.18	$8.1 \cdot 10^{-11}$	$3.5 \cdot 10^{-9}$
$BiAsO_4$	$4.37 \cdot 10^{-10}$	9.36	$2.1 \cdot 10^{-5}$	$7.3 \cdot 10^{-3}$
$Bi_2(C_2O_4)_3$	$4.00 \cdot 10^{-36}$	35.40*	$3.3 \cdot 10^{-8}$	$2.2 \cdot 10^{-5}$
$Bi(Cl)O$[7]	$3.37 \cdot 10^{-36}$	35.47	$9.6 \cdot 10^{-10}$	$2.5 \cdot 10^{-7}$
BiI_3	$8.10 \cdot 10^{-19}$	18.09*	$1.3 \cdot 10^{-5}$	$7.8 \cdot 10^{-3}$
$Bi(OH)_3$	$2.99 \cdot 10^{-36}$	35.52	$5.8 \cdot 10^{-10}$	$1.5 \cdot 10^{-7}$
$BiPO_4$	$1.05 \cdot 10^{-23}$	22.98	$3.2 \cdot 10^{-12}$	$9.8 \cdot 10^{-10}$
Bi_2S_3	$8.91 \cdot 10^{-105}$	104.05	$6.1 \cdot 10^{-22}$	$3.1 \cdot 10^{-19}$
$Ca_3(AsO_4)_2$	$6.80 \cdot 10^{-19}$	18.17*	$9.1 \cdot 10^{-5}$	$3.6 \cdot 10^{-2}$
$CaCO_3$	$4.37 \cdot 10^{-9}$	8.36	$6.6 \cdot 10^{-5}$	$6.6 \cdot 10^{-3}$
CaC_2O_4	$2.30 \cdot 10^{-9}$	8.64*	$4.8 \cdot 10^{-5}$	$6.1 \cdot 10^{-3}$
CaF_2	$3.95 \cdot 10^{-11}$	10.40	$2.1 \cdot 10^{-4}$	$1.7 \cdot 10^{-2}$
$CaHPO_4$[5]	$2.18 \cdot 10^{-7}$	6.66	$4.7 \cdot 10^{-4}$	$6.4 \cdot 10^{-2}$
$Ca(H_2PO_4)_2$[8]	$1.00 \cdot 10^{-3}$	3.00*	$6.3 \cdot 10^{-2}$	14.74
$Ca(IO_3)_2$	$1.93 \cdot 10^{-6}$	5.71	$7.8 \cdot 10^{-3}$	3.06
$CaMg(CO_3)_2$[9]	$3.63 \cdot 10^{-16}$	15.44	$9.8 \cdot 10^{-5}$	$1.8 \cdot 10^{-2}$
$CaMoO_4$	$3.15 \cdot 10^{-9}$	8.50	$5.6 \cdot 10^{-5}$	$1.1 \cdot 10^{-2}$
$Ca(OH)_2$	$6.30 \cdot 10^{-6}$	5.20	$1.2 \cdot 10^{-2}$	$8.6 \cdot 10^{-1}$
$Ca_3(PO_4)_2$	$1.00 \cdot 10^{-25}$	25.00	$3.9 \cdot 10^{-6}$	$1.2 \cdot 10^{-3}$
$Ca(PO_3F)$[5]	$4.00 \cdot 10^{-3}$	2.40*	$6.3 \cdot 10^{-2}$	8.73

(Continued)

Substance	L	pL	Solubility	
			mole/l	g/l
$CaSO_3$	$3.20 \cdot 10^{-7}$	6.49^*	$5.7 \cdot 10^{-4}$	$6.8 \cdot 10^{-2}$
$CaSO_4$	$3.72 \cdot 10^{-5}$	4.43	$6.1 \cdot 10^{-3}$	$8.3 \cdot 10^{-1}$
$CaWO_4$	$1.58 \cdot 10^{-9}$	8.80	$4.0 \cdot 10^{-5}$	$1.1 \cdot 10^{-2}$
$Cd(CN)_2$	$1.00 \cdot 10^{-8}$	8.00^*	$1.4 \cdot 10^{-3}$	$2.2 \cdot 10^{-1}$
$CdCO_3$	$2.50 \cdot 10^{-14}$	13.60	$1.6 \cdot 10^{-7}$	$2.7 \cdot 10^{-5}$
CdC_2O_4	$1.53 \cdot 10^{-8}$	7.81^*	$1.2 \cdot 10^{-4}$	$2.5 \cdot 10^{-2}$
$Cd(OH)_2$	$4.29 \cdot 10^{-15}$	14.37	$1.0 \cdot 10^{-5}$	$1.5 \cdot 10^{-3}$
$Cd(OH)_2$ (amor)	$2.20 \cdot 10^{-14}$	13.66^*	$1.8 \cdot 10^{-5}$	$2.6 \cdot 10^{-3}$
CdS	$6.46 \cdot 10^{-28}$	27.19	$2.5 \cdot 10^{-14}$	$3.7 \cdot 10^{-12}$
$CdSe$	$1.06 \cdot 10^{-33}$	32.98	$3.3 \cdot 10^{-17}$	$6.2 \cdot 10^{-15}$
$CdTe$	$8.71 \cdot 10^{-35}$	34.06	$9.3 \cdot 10^{-18}$	$2.2 \cdot 10^{-15}$
$Ce_2(C_2O_4)_3$	$2.50 \cdot 10^{-29}$	28.60	$7.5 \cdot 10^{-7}$	$4.1 \cdot 10^{-4}$
$CeO_2^{(10)}$	$9.11 \cdot 10^{-23}$	22.04	$1.3 \cdot 10^{-5}$	$2.2 \cdot 10^{-3}$
$Ce(OH)_3$	$6.38 \cdot 10^{-22}$	21.20	$2.2 \cdot 10^{-6}$	$4.2 \cdot 10^{-4}$
$CePO_4$	$5.10 \cdot 10^{-22}$	21.29	$2.3 \cdot 10^{-11}$	$5.3 \cdot 10^{-9}$
$Co_3(AsO_4)_2$	$7.60 \cdot 10^{-29}$	28.12^*	$9.3 \cdot 10^{-7}$	$4.2 \cdot 10^{-4}$
$CoCO_3$	$1.48 \cdot 10^{-10}$	9.83	$1.2 \cdot 10^{-5}$	$1.4 \cdot 10^{-3}$
CoC_2O_4	$6.30 \cdot 10^{-8}$	7.20^*	$2.5 \cdot 10^{-4}$	$3.7 \cdot 10^{-2}$
$Co(IO_3)_2$	$1.00 \cdot 10^{-4}$	4.00^*	$2.9 \cdot 10^{-2}$	11.95
$[Co(NO_2)_6],K_3$	$4.30 \cdot 10^{-10}$	9.37^*	$2.0 \cdot 10^{-3}$	$9.0 \cdot 10^{-1}$
$Co(OH)_2$	$1.56 \cdot 10^{-15}$	14.81	$7.3 \cdot 10^{-6}$	$6.8 \cdot 10^{-4}$
$CoO(OH)^{(11)}$	$2.50 \cdot 10^{-43}$	42.60	$9.8 \cdot 10^{-12}$	$9.0 \cdot 10^{-10}$
CoS	$1.82 \cdot 10^{-20}$	19.74	$1.3 \cdot 10^{-10}$	$1.2 \cdot 10^{-8}$
$Co(S_2)^{(5)}$	$7.28 \cdot 10^{-29}$	28.14	$8.5 \cdot 10^{-15}$	$1.1 \cdot 10^{-12}$
$[Cr(H_2O)_4(OH)_2]^{(12)}$	$9.70 \cdot 10^{-18}$	17.01	$1.3 \cdot 10^{-6}$	$2.1 \cdot 10^{-4}$
$Cr(OH)_3$	$1.09 \cdot 10^{-30}$	29.96	$1.4 \cdot 10^{-8}$	$1.5 \cdot 10^{-6}$
$Cr(OH)_3^{(2)}$	$1.03 \cdot 10^{-16}$	15.99	$1.0 \cdot 10^{-8}$	$1.0 \cdot 10^{-6}$
$CrPO_4$	$1.00 \cdot 10^{-17}$	17.00^*	$3.2 \cdot 10^{-9}$	$4.6 \cdot 10^{-7}$
$CsClO_4$	$4.00 \cdot 10^{-3}$	2.40^*	$6.3 \cdot 10^{-2}$	14.70
$CsIO_3$	$1.00 \cdot 10^{-2}$	2.00^*	$1.0 \cdot 10^{-1}$	30.78
$CsIO_4$	$4.40 \cdot 10^{-3}$	2.36^*	$6.6 \cdot 10^{-2}$	21.48
$CsMnO_4$	$9.10 \cdot 10^{-5}$	4.04^*	$9.5 \cdot 10^{-3}$	2.40
$Cu_3(AsO_4)_2$	$7.60 \cdot 10^{-36}$	35.12^*	$3.7 \cdot 10^{-8}$	$1.7 \cdot 10^{-5}$
$CuBr$	$6.61 \cdot 10^{-9}$	8.18	$8.1 \cdot 10^{-5}$	$1.2 \cdot 10^{-2}$
$CuCN$	$3.20 \cdot 10^{-20}$	19.50^*	$1.8 \cdot 10^{-10}$	$1.6 \cdot 10^{-8}$
CuC_2O_4	$2.87 \cdot 10^{-8}$	7.54	$1.7 \cdot 10^{-4}$	$2.6 \cdot 10^{-2}$
$CuCl$	$2.24 \cdot 10^{-7}$	6.65	$4.7 \cdot 10^{-4}$	$4.7 \cdot 10^{-2}$
CuI	$1.10 \cdot 10^{-12}$	11.96	$1.0 \cdot 10^{-6}$	$2.0 \cdot 10^{-4}$
$Cu(IO_3)_2$	$1.40 \cdot 10^{-7}$	6.85	$3.3 \cdot 10^{-3}$	1.35
$CuNCS$	$4.80 \cdot 10^{-15}$	14.32^*	$6.9 \cdot 10^{-8}$	$8.4 \cdot 10^{-6}$
$Cu_2O^{(1)}$	$1.15 \cdot 10^{-15}$	14.94	$1.7 \cdot 10^{-8}$	$2.4 \cdot 10^{-6}$
$Cu(OH)_2$	$5.60 \cdot 10^{-20}$	19.25	$2.4 \cdot 10^{-7}$	$2.4 \cdot 10^{-5}$
CuS	$1.41 \cdot 10^{-36}$	35.85	$1.2 \cdot 10^{-18}$	$1.1 \cdot 10^{-16}$
Cu_2S	$2.29 \cdot 10^{-48}$	47.64	$8.3 \cdot 10^{-17}$	$1.3 \cdot 10^{-14}$
$CuSe$	$1.62 \cdot 10^{-27}$	26.79	$4.0 \cdot 10^{-14}$	$5.7 \cdot 10^{-12}$
Cu_2Se	$1.11 \cdot 10^{-51}$	50.95	$6.5 \cdot 10^{-18}$	$1.3 \cdot 10^{-15}$
$Eu(OH)_3$	$2.84 \cdot 10^{-27}$	26.55	$1.0 \cdot 10^{-7}$	$2.1 \cdot 10^{-5}$

Substance	L	pL	Solubility mole/l	Solubility g/l
FeAsO$_4$	$5.80 \cdot 10^{-21}$	20.24*	$7.6 \cdot 10^{-11}$	$1.5 \cdot 10^{-8}$
[Fe(CN)$_6$],Ba$_2$	$3.00 \cdot 10^{-8}$	7.52*	$2.0 \cdot 10^{-3}$	$9.5 \cdot 10^{-1}$
[Fe(CN)$_6$],KFe[13]	$9.72 \cdot 10^{-20}$	19.01*	$4.6 \cdot 10^{-7}$	$1.4 \cdot 10^{-4}$
FeCO$_3$	$2.88 \cdot 10^{-11}$	10.54	$5.4 \cdot 10^{-6}$	$6.2 \cdot 10^{-4}$
FeC$_2$O$_4$	$2.10 \cdot 10^{-7}$	6.68	$4.6 \cdot 10^{-4}$	$6.6 \cdot 10^{-2}$
Fe(OH)$_2$	$7.94 \cdot 10^{-16}$	15.10	$5.8 \cdot 10^{-6}$	$5.2 \cdot 10^{-4}$
FeO(OH)[11]	$2.24 \cdot 10^{-42}$	41.65	$1.7 \cdot 10^{-11}$	$1.5 \cdot 10^{-9}$
FeO(OH)[11] (amor)	$6.30 \cdot 10^{-38}$	37.20*	$2.2 \cdot 10^{-10}$	$2.0 \cdot 10^{-8}$
FePO$_4$	$1.08 \cdot 10^{-26}$	25.97	$1.0 \cdot 10^{-13}$	$1.6 \cdot 10^{-11}$
FeS	$3.39 \cdot 10^{-17}$	16.47	$5.8 \cdot 10^{-9}$	$5.1 \cdot 10^{-7}$
Fe(S$_2$)[5]	$5.37 \cdot 10^{-27}$	26.27	$7.3 \cdot 10^{-14}$	$8.8 \cdot 10^{-12}$
Ga(OH)$_3$	$4.12 \cdot 10^{-36}$	35.39	$6.3 \cdot 10^{-10}$	$7.5 \cdot 10^{-8}$
Ga(OH)$_3$[2]	$2.85 \cdot 10^{-11}$	10.55	$5.3 \cdot 10^{-6}$	$6.4 \cdot 10^{-4}$
GaPO$_4$	$3.80 \cdot 10^{-21}$	20.42	$6.2 \cdot 10^{-11}$	$1.0 \cdot 10^{-8}$
[GeF$_6$],K$_2$	$3.00 \cdot 10^{-5}$	4.52*	$2.0 \cdot 10^{-2}$	5.18
β-GeO$_2$[10]	$1.70 \cdot 10^{-56}$	55.77	$2.3 \cdot 10^{-12}$	$2.4 \cdot 10^{-10}$
GeS	$3.00 \cdot 10^{-35}$	34.52*	$5.5 \cdot 10^{-18}$	$5.7 \cdot 10^{-16}$
[HfF$_6$],K$_2$	$2.00 \cdot 10^{-3}$	2.70*	$7.9 \cdot 10^{-2}$	29.42
HfO$_2$[14]	$4.00 \cdot 10^{-26}$	25.40*	$2.2 \cdot 10^{-9}$	$4.5 \cdot 10^{-7}$
Hg$_2$Br$_2$[8]	$7.87 \cdot 10^{-23}$	22.10	$2.7 \cdot 10^{-8}$	$1.5 \cdot 10^{-5}$
Hg$_2$CO$_3$[5]	$8.90 \cdot 10^{-17}$	16.05	$9.4 \cdot 10^{-9}$	$4.4 \cdot 10^{-6}$
Hg$_2$Cl$_2$[8]	$1.47 \cdot 10^{-18}$	17.83	$7.2 \cdot 10^{-7}$	$3.4 \cdot 10^{-4}$
Hg$_2$I$_2$[8]	$5.37 \cdot 10^{-29}$	28.27	$2.4 \cdot 10^{-10}$	$1.6 \cdot 10^{-7}$
HgO[15]	$3.34 \cdot 10^{-26}$	25.48	$2.0 \cdot 10^{-9}$	$4.4 \cdot 10^{-7}$
HgO + Hg[16]	$2.69 \cdot 10^{-24}$	23.57	–	–
HgS (cub)	$1.39 \cdot 10^{-45}$	44.86	$3.7 \cdot 10^{-23}$	$8.7 \cdot 10^{-21}$
HgS (cub) + Hg[17]	$1.12 \cdot 10^{-43}$	42.95	–	–
Hg$_2$SO$_4$[5]	$6.20 \cdot 10^{-7}$	6.21	$7.9 \cdot 10^{-4}$	$3.9 \cdot 10^{-1}$
HgSe	$1.00 \cdot 10^{-59}$	59.00*	$3.2 \cdot 10^{-30}$	$8.8 \cdot 10^{-28}$
In(OH)$_3$	$1.32 \cdot 10^{-37}$	36.88	$2.6 \cdot 10^{-10}$	$4.4 \cdot 10^{-8}$
In$_2$S$_3$	$9.05 \cdot 10^{-84}$	83.04	$9.7 \cdot 10^{-18}$	$3.1 \cdot 10^{-15}$
In$_2$Se$_3$	$5.59 \cdot 10^{-92}$	91.25	$2.2 \cdot 10^{-19}$	$1.0 \cdot 10^{-16}$
[IrCl$_6$],K$_2$	$6.80 \cdot 10^{-5}$	4.17*	$2.6 \cdot 10^{-2}$	12.42
[IrCl$_6$],(NH$_4$)$_2$	$3.00 \cdot 10^{-5}$	4.52*	$2.0 \cdot 10^{-2}$	8.63
Ir$_2$O$_3$[18]	$2.00 \cdot 10^{-48}$	47.70*	$2.6 \cdot 10^{-13}$	$1.1 \cdot 10^{-10}$
Ir(OH)$_4$[19]	$1.60 \cdot 10^{-72}$	71.80*	$1.4 \cdot 10^{-15}$	$3.8 \cdot 10^{-13}$
IrS$_2$[20]	$1.00 \cdot 10^{-75}$	75.00*	$6.3 \cdot 10^{-26}$	$1.6 \cdot 10^{-23}$
KClO$_4$	$1.00 \cdot 10^{-2}$	2.00*	$1.0 \cdot 10^{-1}$	13.86
KIO$_4$	$8.30 \cdot 10^{-4}$	3.08*	$2.9 \cdot 10^{-2}$	6.63
KReO$_4$	$1.90 \cdot 10^{-3}$	2.72*	$4.4 \cdot 10^{-2}$	12.61
La$_2$(C$_2$O$_4$)$_3$	$2.00 \cdot 10^{-28}$	27.70*	$1.1 \cdot 10^{-6}$	$6.1 \cdot 10^{-4}$
La$_2$(MoO$_4$)$_3$	$8.20 \cdot 10^{-22}$	21.09	$2.4 \cdot 10^{-5}$	$1.8 \cdot 10^{-2}$
La(OH)$_3$	$3.62 \cdot 10^{-23}$	22.44	$1.1 \cdot 10^{-6}$	$2.0 \cdot 10^{-4}$
La(OH)$_3$ (amor)	$6.50 \cdot 10^{-20}$	19.19*	$7.0 \cdot 10^{-6}$	$1.3 \cdot 10^{-3}$
La$_2$S$_3$	$2.00 \cdot 10^{-13}$	12.70*	$1.1 \cdot 10^{-3}$	$4.2 \cdot 10^{-1}$
La$_2$(SO$_4$)$_3$	$3.00 \cdot 10^{-5}$	4.52*	$4.9 \cdot 10^{-2}$	27.64
Li$_2$CO$_3$	$1.94 \cdot 10^{-3}$	2.71	$7.9 \cdot 10^{-2}$	5.81

(Continued)

Substance	L	pL	Solubility	
			mole/l	g/l
LiF	$1.46 \cdot 10^{-3}$	2.84	$3.8 \cdot 10^{-2}$	$9.9 \cdot 10^{-1}$
Li_3PO_4	$3.20 \cdot 10^{-9}$	8.49*	$3.3 \cdot 10^{-3}$	$3.8 \cdot 10^{-1}$
$Lu(OH)_3$	$1.00 \cdot 10^{-26}$	26.00	$1.4 \cdot 10^{-7}$	$3.1 \cdot 10^{-5}$
$MgCO_3$	$7.94 \cdot 10^{-6}$	5.10	$2.8 \cdot 10^{-3}$	$2.4 \cdot 10^{-1}$
MgC_2O_4	$8.60 \cdot 10^{-5}$	4.07*	$9.3 \cdot 10^{-3}$	1.04
MgF_2	$6.40 \cdot 10^{-9}$	8.19	$1.2 \cdot 10^{-3}$	$7.3 \cdot 10^{-2}$
$Mg(IO_3)_2$	$3.00 \cdot 10^{-3}$	2.52	$9.1 \cdot 10^{-2}$	33.99
$MgNH_4PO_4$[21]	$2.50 \cdot 10^{-13}$	12.60	$6.3 \cdot 10^{-5}$	$8.7 \cdot 10^{-3}$
$Mg(OH)_2$	$6.76 \cdot 10^{-12}$	11.17	$1.2 \cdot 10^{-4}$	$6.9 \cdot 10^{-3}$
$Mg(OH)_2$ (amor)	$6.00 \cdot 10^{-10}$	9.22*	$5.3 \cdot 10^{-4}$	$3.1 \cdot 10^{-2}$
$Mg_3(PO_4)_2$	$3.88 \cdot 10^{-26}$	25.41	$3.2 \cdot 10^{-6}$	$8.5 \cdot 10^{-4}$
$MgSO_3$	$3.00 \cdot 10^{-3}$	2.52*	$5.5 \cdot 10^{-2}$	5.72
$MgWO_4$	$3.61 \cdot 10^{-7}$	6.44	$6.0 \cdot 10^{-4}$	$1.6 \cdot 10^{-1}$
$MnCO_3$	$4.90 \cdot 10^{-11}$	10.31	$7.0 \cdot 10^{-6}$	$8.0 \cdot 10^{-4}$
MnC_2O_4	$5.00 \cdot 10^{-6}$	5.30	$2.2 \cdot 10^{-3}$	$3.2 \cdot 10^{-1}$
MnO_2[10]	$1.00 \cdot 10^{-50}$	50.00*	$3.3 \cdot 10^{-11}$	$2.9 \cdot 10^{-9}$
Mn_2O_3[4]	$1.67 \cdot 10^{-42}$	41.78	$7.9 \cdot 10^{-12}$	$1.2 \cdot 10^{-9}$
$Mn(OH)_2$	$2.27 \cdot 10^{-13}$	12.64	$3.8 \cdot 10^{-5}$	$3.4 \cdot 10^{-3}$
MnS	$1.12 \cdot 10^{-13}$	12.95	$3.3 \cdot 10^{-7}$	$2.9 \cdot 10^{-5}$
MnS ($\cdot nH_2O$)	$2.50 \cdot 10^{-10}$	9.60*	$1.6 \cdot 10^{-5}$	$1.4 \cdot 10^{-3}$
MnSe	$5.44 \cdot 10^{-10}$	9.26	$2.3 \cdot 10^{-5}$	$3.1 \cdot 10^{-3}$
$MnWO_4$	$3.16 \cdot 10^{-8}$	7.50	$1.8 \cdot 10^{-4}$	$5.4 \cdot 10^{-2}$
$MoO(OH)_2$[22]	$1.00 \cdot 10^{-50}$	50.00*	$3.3 \cdot 10^{-11}$	$4.8 \cdot 10^{-9}$
$NaIO_4$	$3.00 \cdot 10^{-3}$	2.52	$5.5 \cdot 10^{-2}$	11.72
$Nd_2(C_2O_4)_3$	$5.87 \cdot 10^{-29}$	28.23	$8.9 \cdot 10^{-7}$	$4.9 \cdot 10^{-4}$
$Nd(OH)_3$	$7.79 \cdot 10^{-24}$	23.11	$7.3 \cdot 10^{-7}$	$1.4 \cdot 10^{-4}$
$Ni_3(AsO_4)_2$	$3.10 \cdot 10^{-26}$	25.51*	$3.1 \cdot 10^{-6}$	$1.4 \cdot 10^{-3}$
$[Ni(C_4H_7O_2N_2)_2]$[23]	$2.30 \cdot 10^{-25}$	24.64*	$3.9 \cdot 10^{-9}$	$1.1 \cdot 10^{-6}$
$Ni(CN)_2$	$3.00 \cdot 10^{-23}$	22.52*	$2.0 \cdot 10^{-8}$	$2.2 \cdot 10^{-6}$
$NiCO_3$	$1.32 \cdot 10^{-7}$	6.88	$3.6 \cdot 10^{-4}$	$4.3 \cdot 10^{-2}$
$Ni(IO_3)_2$	$1.40 \cdot 10^{-8}$	7.85*	$1.5 \cdot 10^{-3}$	$6.2 \cdot 10^{-1}$
$Ni(OH)_2$	$1.60 \cdot 10^{-14}$	13.80	$1.6 \cdot 10^{-5}$	$1.5 \cdot 10^{-3}$
NiS	$9.33 \cdot 10^{-22}$	21.03	$3.1 \cdot 10^{-11}$	$2.8 \cdot 10^{-9}$
$NpO_2(OH)_2$[8]	$2.50 \cdot 10^{-22}$	21.60*	$4.0 \cdot 10^{-8}$	$1.2 \cdot 10^{-5}$
$Pb_3(AsO_4)_2$	$3.97 \cdot 10^{-36}$	35.40	$3.3 \cdot 10^{-8}$	$2.9 \cdot 10^{-5}$
$PbBr_2$	$5.00 \cdot 10^{-5}$	4.30	$2.3 \cdot 10^{-2}$	8.52
$Pb(BrO_3)_2$	$1.60 \cdot 10^{-4}$	3.80	$3.4 \cdot 10^{-2}$	15.83
$PbCO_3$	$3.63 \cdot 10^{-14}$	13.44	$1.9 \cdot 10^{-7}$	$5.1 \cdot 10^{-5}$
PbC_2O_4	$7.30 \cdot 10^{-11}$	10.14	$8.5 \cdot 10^{-6}$	$2.5 \cdot 10^{-3}$
$PbCl_2$	$1.70 \cdot 10^{-5}$	4.77	$1.6 \cdot 10^{-2}$	4.50
$Pb(Cl)F$[24]	$2.45 \cdot 10^{-9}$	8.61	$1.3 \cdot 10^{-3}$	$3.5 \cdot 10^{-1}$
$Pb(Cl)OH$[24]	$2.00 \cdot 10^{-14}$	13.70*	$2.7 \cdot 10^{-5}$	$7.0 \cdot 10^{-3}$
$PbCrO_4$	$2.83 \cdot 10^{-13}$	12.55	$5.3 \cdot 10^{-7}$	$1.7 \cdot 10^{-4}$
PbF_2	$2.69 \cdot 10^{-8}$	7.57	$1.9 \cdot 10^{-3}$	$4.6 \cdot 10^{-1}$
PbI_2	$8.70 \cdot 10^{-9}$	8.06	$1.3 \cdot 10^{-3}$	$6.0 \cdot 10^{-1}$
$Pb(IO_3)_2$	$2.60 \cdot 10^{-13}$	12.59	$4.0 \cdot 10^{-5}$	$2.2 \cdot 10^{-2}$
$PbMoO_4$	$8.51 \cdot 10^{-16}$	15.07	$2.9 \cdot 10^{-8}$	$1.1 \cdot 10^{-5}$

(Continued)

Substance	L	pL	Solubility	
			mole/l	g/l
$Pb(N_3)_2$	$2.60 \cdot 10^{-9}$	8.59[*]	$8.7 \cdot 10^{-4}$	$2.5 \cdot 10^{-1}$
$Pb(NCS)_2$	$2.00 \cdot 10^{-5}$	4.70[*]	$1.7 \cdot 10^{-2}$	5.53
$PbO_2^{(10)}$	$3.00 \cdot 10^{-66}$	65.52[*]	$2.6 \cdot 10^{-14}$	$6.2 \cdot 10^{-12}$
$Pb(OH)_2$	$5.50 \cdot 10^{-16}$	15.26	$5.2 \cdot 10^{-6}$	$1.2 \cdot 10^{-3}$
$Pb(OH)_2^{(6)}$	$2.06 \cdot 10^{-14}$	13.64	$1.4 \cdot 10^{-7}$	$3.5 \cdot 10^{-5}$
$Pb_3(PO_4)_2$	$7.90 \cdot 10^{-43}$	42.10[*]	$1.5 \cdot 10^{-9}$	$1.2 \cdot 10^{-6}$
PbS	$8.71 \cdot 10^{-29}$	28.06	$9.3 \cdot 10^{-15}$	$2.2 \cdot 10^{-12}$
$PbSO_4$	$1.74 \cdot 10^{-8}$	7.76	$1.3 \cdot 10^{-4}$	$4.0 \cdot 10^{-2}$
$PbSe$	$6.52 \cdot 10^{-37}$	36.19	$8.1 \cdot 10^{-19}$	$2.3 \cdot 10^{-16}$
$PbSeO_4$	$1.45 \cdot 10^{-7}$	6.84	$3.8 \cdot 10^{-4}$	$1.3 \cdot 10^{-1}$
$PbTe$	$4.13 \cdot 10^{-39}$	38.38	$6.4 \cdot 10^{-20}$	$2.2 \cdot 10^{-17}$
$PbWO_4$	$4.47 \cdot 10^{-12}$	11.35	$2.1 \cdot 10^{-6}$	$9.6 \cdot 10^{-4}$
$[PdCl_4],K_2$	$1.60 \cdot 10^{-5}$	4.80[*]	$1.6 \cdot 10^{-2}$	5.18
$[PdCl_6],K_2$	$5.97 \cdot 10^{-6}$	5.22	$1.1 \cdot 10^{-2}$	4.54
$PdO_2^{(10)}$	$6.50 \cdot 10^{-71}$	70.19[*]	$3.0 \cdot 10^{-15}$	$4.2 \cdot 10^{-13}$
$Pd(OH)_2$	$1.00 \cdot 10^{-24}$	24.00	$6.3 \cdot 10^{-9}$	$8.8 \cdot 10^{-7}$
$PoO(OH)_2^{(22)}$	$6.3 \cdot 10^{-52}$	51.20[*]	$1.9 \cdot 10^{-11}$	$4.9 \cdot 10^{-9}$
PoS	$5.00 \cdot 10^{-29}$	28.30[*]	$7.1 \cdot 10^{-15}$	$1.7 \cdot 10^{-12}$
$Po(SO_4)_2^{(20)}$	$2.60 \cdot 10^{-7}$	6.59[*]	$4.0 \cdot 10^{-3}$	1.61
$PtBr_4^{(19)}$	$3.00 \cdot 10^{-41}$	40.52[*]	$2.6 \cdot 10^{-9}$	$1.3 \cdot 10^{-6}$
$[Pt(CN)_4],Ba$	$4.00 \cdot 10^{-3}$	2.40[*]	$6.3 \cdot 10^{-2}$	27.61
$PtCl_4^{(19)}$	$8.00 \cdot 10^{-29}$	28.10[*]	$7.9 \cdot 10^{-7}$	$2.7 \cdot 10^{-4}$
$[PtCl_6],Cs_2$	$3.00 \cdot 10^{-8}$	7.52[*]	$2.0 \cdot 10^{-3}$	1.32
$[PtCl_6],K_2$	$1.10 \cdot 10^{-5}$	4.96[*]	$1.4 \cdot 10^{-2}$	6.81
$[PtCl_6],(NH_4)_2$	$9.00 \cdot 10^{-6}$	5.05[*]	$1.3 \cdot 10^{-2}$	5.82
$[PtCl_6],Rb_2$	$9.00 \cdot 10^{-8}$	7.06[*]	$2.8 \cdot 10^{-3}$	1.63
$[PtF_6],K_2$	$2.90 \cdot 10^{-5}$	4.54[*]	$1.9 \cdot 10^{-2}$	7.50
$PtO_2^{(10)}$	$1.60 \cdot 10^{-72}$	71.80[*]	$1.4 \cdot 10^{-15}$	$3.3 \cdot 10^{-13}$
$Pt(OH)_2$	$1.00 \cdot 10^{-25}$	25.00	$2.9 \cdot 10^{-9}$	$6.7 \cdot 10^{-7}$
PtS	$1.24 \cdot 10^{-61}$	60.92	$3.5 \cdot 10^{-31}$	$8.0 \cdot 10^{-29}$
$Pu(OH)_3$	$2.00 \cdot 10^{-20}$	19.70[*]	$5.2 \cdot 10^{-6}$	$1.5 \cdot 10^{-3}$
$Pu(OH)_4$	$1.00 \cdot 10^{-52}$	52.00[*]	$1.3 \cdot 10^{-11}$	$4.1 \cdot 10^{-9}$
$Ra(IO_3)_2$	$8.80 \cdot 10^{-10}$	9.06[*]	$6.0 \cdot 10^{-4}$	$3.5 \cdot 10^{-1}$
$RaSO_4$	$4.25 \cdot 10^{-11}$	10.37	$6.5 \cdot 10^{-6}$	$2.1 \cdot 10^{-3}$
$RbClO_4$	$2.50 \cdot 10^{-3}$	2.60	$5.0 \cdot 10^{-2}$	9.25
$RbIO_4$	$5.50 \cdot 10^{-4}$	3.26[*]	$2.3 \cdot 10^{-2}$	6.48
$RbMnO_4$	$2.90 \cdot 10^{-3}$	2.54[*]	$5.4 \cdot 10^{-2}$	11.01
$Rh(OH)_3$	$2.00 \cdot 10^{-48}$	47.70[*]	$5.2 \cdot 10^{-13}$	$8.0 \cdot 10^{-11}$
$RuO_2^{(10)}$	$1.00 \cdot 10^{-49}$	49.00[*]	$5.2 \cdot 10^{-11}$	$7.0 \cdot 10^{-9}$
$Ru(OH)_3$	$1.00 \cdot 10^{-38}$	38.00[*]	$1.4 \cdot 10^{-10}$	$2.1 \cdot 10^{-8}$
$Sb_2O_3^{(4)}$	$4.00 \cdot 10^{-42}$	41.40[*]	$9.8 \cdot 10^{-12}$	$2.9 \cdot 10^{-9}$
$Sb_2O_3^{(25)}$	$1.35 \cdot 10^{-17}$	16.87	$1.8 \cdot 10^{-9}$	$5.4 \cdot 10^{-7}$
$[Sb(OH)_6],Na$	$4.00 \cdot 10^{-8}$	7.40[*]	$2.0 \cdot 10^{-4}$	$4.9 \cdot 10^{-2}$
Sb_2S_3	$2.23 \cdot 10^{-90}$	89.65	$4.6 \cdot 10^{-19}$	$1.6 \cdot 10^{-16}$
$Sc(OH)_3$	$8.73 \cdot 10^{-28}$	27.06	$7.5 \cdot 10^{-8}$	$7.2 \cdot 10^{-6}$
$[SiF_6],Ca$	$8.10 \cdot 10^{-4}$	3.09[*]	$2.8 \cdot 10^{-2}$	5.18
$[SiF_6],Cs_2$	$1.26 \cdot 10^{-5}$	4.90[*]	$1.5 \cdot 10^{-2}$	5.98

(Continued)

Substance	L	pL	Solubility mole/l	Solubility g/l
$[SiF_6],K_2$	$2.19 \cdot 10^{-7}$	6.66	$3.8 \cdot 10^{-3}$	$8.4 \cdot 10^{-1}$
$[SiF_6],Na_2$	$3.08 \cdot 10^{-5}$	4.51	$2.0 \cdot 10^{-2}$	3.71
$[SiF_6],Rb_2$	$5.00 \cdot 10^{-7}$	6.30*	$5.0 \cdot 10^{-3}$	1.57
$[SnCl_6],Cs_2$	$3.60 \cdot 10^{-8}$	7.44*	$2.1 \cdot 10^{-3}$	1.24
SnI_2	$8.30 \cdot 10^{-6}$	5.08*	$1.3 \cdot 10^{-2}$	4.75
$SnO_2^{(10)}$	$4.84 \cdot 10^{-58}$	57.32	$1.1 \cdot 10^{-12}$	$1.7 \cdot 10^{-10}$
$Sn(OH)_2$	$5.46 \cdot 10^{-27}$	26.26	$1.1 \cdot 10^{-9}$	$1.7 \cdot 10^{-7}$
$Sn(OH)_2^{(6)}$	$3.72 \cdot 10^{-15}$	14.43	$6.1 \cdot 10^{-8}$	$9.3 \cdot 10^{-6}$
SnS	$2.95 \cdot 10^{-28}$	27.53	$1.7 \cdot 10^{-14}$	$2.6 \cdot 10^{-12}$
$SnSe$	$1.86 \cdot 10^{-34}$	33.73	$1.4 \cdot 10^{-17}$	$2.7 \cdot 10^{-15}$
$SnTe$	$3.61 \cdot 10^{-37}$	36.44	$6.0 \cdot 10^{-19}$	$1.5 \cdot 10^{-16}$
$SrCO_3$	$5.25 \cdot 10^{-10}$	9.28	$2.3 \cdot 10^{-5}$	$3.4 \cdot 10^{-3}$
SrC_2O_4	$5.60 \cdot 10^{-8}$	7.25*	$2.4 \cdot 10^{-4}$	$4.2 \cdot 10^{-2}$
$SrCrO_4$	$2.71 \cdot 10^{-5}$	4.57	$5.2 \cdot 10^{-3}$	1.06
SrF_2	$2.51 \cdot 10^{-9}$	8.60	$8.6 \cdot 10^{-4}$	$1.1 \cdot 10^{-1}$
$Sr(IO_3)_2$	$3.30 \cdot 10^{-7}$	6.48*	$4.4 \cdot 10^{-3}$	1.90
$SrMoO_4$	$1.39 \cdot 10^{-8}$	7.86	$1.2 \cdot 10^{-4}$	$2.9 \cdot 10^{-2}$
$Sr(OH)_2$	$3.20 \cdot 10^{-4}$	3.49*	$4.3 \cdot 10^{-2}$	5.24
$Sr_3(PO_4)_2$	$1.00 \cdot 10^{-31}$	31.00*	$2.5 \cdot 10^{-7}$	$1.1 \cdot 10^{-4}$
$SrSO_4$	$2.14 \cdot 10^{-7}$	6.67	$4.6 \cdot 10^{-4}$	$8.5 \cdot 10^{-2}$
$SrSeO_4$	$2.50 \cdot 10^{-5}$	4.60	$5.0 \cdot 10^{-3}$	1.15
$SrWO_4$	$7.07 \cdot 10^{-9}$	8.15	$8.4 \cdot 10^{-5}$	$2.8 \cdot 10^{-2}$
$TeO_2^{(10)}$	$3.00 \cdot 10^{-54}$	53.52*	$6.5 \cdot 10^{-12}$	$1.0 \cdot 10^{-9}$
$Th(OH)_4$	$7.67 \cdot 10^{-44}$	43.12	$7.9 \cdot 10^{-10}$	$2.4 \cdot 10^{-7}$
$Th(SO_4)_2$	$4.00 \cdot 10^{-3}$	2.40*	$1.0 \cdot 10^{-1}$	42.42
$[TiF_6],K_2$	$5.00 \cdot 10^{-4}$	3.30*	$5.0 \cdot 10^{-2}$	12.00
$TiO_2^{(14)}$	$1.00 \cdot 10^{-29}$	29.00*	$1.4 \cdot 10^{-10}$	$1.1 \cdot 10^{-8}$
$TiO(OH)_2$	$6.30 \cdot 10^{-52}$	51.20*	$1.9 \cdot 10^{-11}$	$1.9 \cdot 10^{-9}$
$TlBr$	$4.31 \cdot 10^{-6}$	5.37	$2.1 \cdot 10^{-3}$	$5.9 \cdot 10^{-1}$
Tl_2CO_3	$4.00 \cdot 10^{-3}$	2.40*	$1.0 \cdot 10^{-1}$	46.88
$Tl_2C_2O_4$	$2.00 \cdot 10^{-4}$	3.70*	$3.7 \cdot 10^{-2}$	18.30
$TlCl$	$1.90 \cdot 10^{-4}$	3.72	$1.4 \cdot 10^{-2}$	3.31
Tl_2CrO_4	$1.02 \cdot 10^{-12}$	11.99	$6.3 \cdot 10^{-5}$	$3.3 \cdot 10^{-2}$
TlI	$6.56 \cdot 10^{-8}$	7.18	$2.6 \cdot 10^{-4}$	$8.5 \cdot 10^{-2}$
TlN_3	$2.20 \cdot 10^{-4}$	3.66*	$1.5 \cdot 10^{-2}$	3.65
$TlNCS$	$5.80 \cdot 10^{-4}$	3.24	$2.4 \cdot 10^{-2}$	6.32
$Tl_2O_3^{(18)}$	$2.01 \cdot 10^{-44}$	43.70	$2.6 \cdot 10^{-12}$	$1.2 \cdot 10^{-9}$
Tl_3PO_4	$6.70 \cdot 10^{-8}$	7.17*	$7.1 \cdot 10^{-3}$	5.00
Tl_2S	$1.00 \cdot 10^{-24}$	24.00	$6.3 \cdot 10^{-9}$	$2.8 \cdot 10^{-6}$
Tl_2SO_4	$1.52 \cdot 10^{-4}$	3.82	$3.4 \cdot 10^{-2}$	16.97
$Tl_2SO_3S^{(26)}$	$2.00 \cdot 10^{-7}$	6.70*	$3.7 \cdot 10^{-3}$	1.92
Tl_2Se	$2.39 \cdot 10^{-22}$	21.62	$3.9 \cdot 10^{-8}$	$1.9 \cdot 10^{-5}$
Tl_2SeO_4	$9.95 \cdot 10^{-5}$	4.00	$2.9 \cdot 10^{-2}$	16.11
$UO_2CO_3^{(5)}$	$6.31 \cdot 10^{-15}$	14.20	$7.9 \cdot 10^{-8}$	$2.6 \cdot 10^{-5}$
$UO_2C_2O_4^{(5)}$	$2.00 \cdot 10^{-4}$	3.70*	$1.4 \cdot 10^{-2}$	5.06
$U(OH)_4$	$1.00 \cdot 10^{-45}$	45.00*	$3.3 \cdot 10^{-10}$	$1.0 \cdot 10^{-7}$
$UO_2(OH)_2^{(8)}$	$2.00 \cdot 10^{-15}$	14.70	$7.9 \cdot 10^{-6}$	$2.4 \cdot 10^{-3}$

Substance	L	pL	Solubility	
			mole/l	g/l
$V_2O_5^{(27)}$	$1.60 \cdot 10^{-15}$	14.80^*	$2.0 \cdot 10^{-8}$	$3.6 \cdot 10^{-6}$
$VO(OH)_2^{(8)}$	$7.06 \cdot 10^{-23}$	22.15	$2.6 \cdot 10^{-8}$	$2.6 \cdot 10^{-6}$
$WO_2^{(10)}$	$1.00 \cdot 10^{-50}$	50.00^*	$3.3 \cdot 10^{-11}$	$7.1 \cdot 10^{-9}$
$Y(OH)_3$	$3.21 \cdot 10^{-25}$	24.49	$3.3 \cdot 10^{-7}$	$4.6 \cdot 10^{-5}$
$Zn_3(AsO_4)_2$	$1.30 \cdot 10^{-27}$	26.89^*	$1.6 \cdot 10^{-6}$	$7.8 \cdot 10^{-4}$
$Zn(CN)_2$	$2.60 \cdot 10^{-13}$	12.59^*	$4.0 \cdot 10^{-5}$	$4.7 \cdot 10^{-3}$
$ZnCO_3$	$5.25 \cdot 10^{-11}$	10.28	$7.2 \cdot 10^{-6}$	$9.1 \cdot 10^{-4}$
ZnC_2O_4	$1.40 \cdot 10^{-9}$	8.85	$3.7 \cdot 10^{-5}$	$5.7 \cdot 10^{-3}$
$ZnO^{(15)}$	$2.20 \cdot 10^{-17}$	16.66	$1.8 \cdot 10^{-6}$	$1.4 \cdot 10^{-4}$
$Zn(OH)_2$	$3.03 \cdot 10^{-16}$	15.52	$4.2 \cdot 10^{-6}$	$4.2 \cdot 10^{-4}$
$Zn(OH)_2^{(6)}$	$8.42 \cdot 10^{-11}$	10.08	$9.2 \cdot 10^{-6}$	$9.1 \cdot 10^{-4}$
$Zn_3(PO_4)_2$	$9.10 \cdot 10^{-33}$	32.04^*	$1.5 \cdot 10^{-7}$	$5.9 \cdot 10^{-5}$
α-ZnS	$1.20 \cdot 10^{-25}$	24.92	$3.5 \cdot 10^{-13}$	$3.4 \cdot 10^{-11}$
β-ZnS	$7.94 \cdot 10^{-24}$	23.10	$2.8 \cdot 10^{-12}$	$2.7 \cdot 10^{-10}$
ZnSe	$4.70 \cdot 10^{-27}$	26.33	$6.9 \cdot 10^{-14}$	$9.9 \cdot 10^{-12}$
ZnTe	$7.00 \cdot 10^{-26}$	25.15	$2.6 \cdot 10^{-13}$	$5.1 \cdot 10^{-11}$
$[ZrF_6],K_2$	$5.00 \cdot 10^{-4}$	3.30^*	$5.0 \cdot 10^{-2}$	14.17
$ZrO_2^{(14)}$	$1.11 \cdot 10^{-36}$	35.96	$6.5 \cdot 10^{-13}$	$8.0 \cdot 10^{-11}$
$ZrO(OH)_2$	$7.90 \cdot 10^{-55}$	54.10	$5.0 \cdot 10^{-12}$	$7.0 \cdot 10^{-10}$

(1) $0.5M_2O_{(sld)} + 0.5H_2O \rightleftharpoons M^+ + OH^-$ (M = Ag, Au, Cu)

(2) $M(OH)_{3(sld)} + 2H_2O \rightleftharpoons [M(OH)_4]^- + H_3O^+$ (M = Al, Cr, Ga)

(3) $AuA_{3(sld)} + 2H_2O \rightleftharpoons [AuA_3(OH)]^- + H_3O^+$ (A = Br, Cl, I)

(4) $0.5M_2O_{3(sld)} + 1.5H_2O \rightleftharpoons M^{III} + 3OH^-$ (M = Au, Mn, Sb)

(5) $MA_{(sld)} \rightleftharpoons M^{2+} + A^{2-}$ (M^{2+} = Ba^{2+}, Ca^{2+}, Co^{2+}, Fe^{2+}, Hg_2^{2+}, UO_2^{2+};
    A^{2-} = CO_3^{2-}, $C_2O_4^{2-}$, HPO_4^{2-}, PO_3F^{2-}, S^{2-}, SO_4^{2-}, SO_3S^{2-})

(6) $M(OH)_{2(sld)} + 2H_2O \rightleftharpoons [M(OH)_3]^- + H_3O^+$ (M = Be, Pb, Sn, Zn)

(7) $Bi(Cl)O_{(sld)} + H_2O \rightleftharpoons Bi^{III} + Cl^- + 2OH^-$

(8) $MA_{2(sld)} \rightleftharpoons M^{2+} + 2A^-$ (M^{2+} = Ca^{2+}, Hg_2^{2+}, NpO_2^{2+}, UO_2^{2+}, VO^{2+};
    A^- = Br^-, Cl^-, $H_2PO_4^-$, I^-, OH^-)

(9) $CaMg(CO_3)_{2(sld)} \rightleftharpoons Ca^{2+} + Mg^{2+} + 2CO_3^{2-}$

(10) $MO_{2(sld)} + 2H_2O \rightleftharpoons M^{IV} + 4OH^-$ (M = Ce, Ge, Mn, Pb, Pd, Pt, Ru, Sn, Te, W)

(11) $MO(OH)_{(sld)} + H_2O \rightleftharpoons M^{III} + 3OH^-$ (M = Co, Fe)

(12) $[Cr(H_2O)_4(OH)_2]_{(sld)} + 2H_2O \rightleftharpoons [Cr(H_2O)_6]^{2+} + 2OH^-$

(13) $KFe[Fe^{II}(CN)_6]_{(sld)} \rightleftharpoons K^+ + Fe^{3+} + [Fe^{II}(CN)_6]^{4-}$

(14) $MO_{2(sld)} + 2H_2O \rightleftharpoons M(OH)_2^{2+} + 2OH^-$ (M = Hf, Ti, Zr)

(15) $MO_{(sld)} + H_2O \rightleftharpoons M^{2+} + 2OH^-$ (M = Hg, Zn)

(16) $HgO_{(sld)} + Hg_{(lq)} + H_2O \rightleftharpoons Hg_2^{2+} + 2OH^-$

(17) $HgS_{(sld)} + Hg_{(lq)} \rightleftharpoons Hg_2^{2+} + S^{2-}$

(18) $0.5M_2O_{3(sld)} + 1.5H_2O \rightleftharpoons M^{3+} + 3OH^-$ (M = Ir, Tl)

(19) $MA_{4(sld)} \rightleftharpoons M^{IV} + 4A^-$ (M = Ir, Pt; A^- = Br^-, Cl^-, OH^-)

(20) $MA_{2(sld)} \rightleftharpoons M^{IV} + 2A^{2-}$ (M = Ir, Po; A^{2-} = S^{2-}, SO_4^{2-})

(21) $MgNH_4PO_{4(sld)} \rightleftharpoons Mg^{2+} + NH_4^+ + PO_4^{3-}$

(22) $MO(OH)_{2(sld)} + H_2O \rightleftharpoons M^{IV} + 4OH^-$ (M = Mo, Po, Ti, Zr)

(23) $[Ni(C_4H_7O_2N_2)_2]_{(sld)} \rightleftharpoons Ni^{2+} + 2C_4H_7O_2N_2^-$

(24) $Pb(Cl)A_{(sld)} \rightleftharpoons Pb^{2+} + Cl^- + A^-$ (A^- = F^-, OH^-)

(25) $0.5Sb_2O_{3(sld)} + 3.5H_2O \rightleftharpoons [Sb(OH)_4]^- + H_3O^+$

(26) $Tl_2(SO_3S)_{(sld)} \rightleftharpoons 2Tl^+ + SO_3S^{2-}$

(27) $0.5V_2O_{5(sld)} + 0.5H_2O \rightleftharpoons VO^{2+} + OH^-$

5

Solubility

5.1 Composition and Preparation
of Aqueous Solutions [4, 10, 22, 34, 35]

The table lists the names, designations and definitions (in the form of mathematical formulas) of the methods for expressing the composition of aqueous solutions in terms of the content therein of the solute having the formula unity B.

In the formulas, use is made of the following physical quantities and units:

m_B — mass of solute B, grams;

m_{H_2O} — mass of water (solvent), grams (except for the definition of C_{mB}, where the mass of water is expressed in kilograms);

$m_{(soln)}$ — mass of solution, grams;

V_{H_2O} — volume of water at a given temperature, liters;

ρ_{H_2O} — density of water at a given temperature, grams per liter (cf. Sec. 5.4);

$V_{(soln)}$ — volume of solution at a given temperature, liters;

$\rho_{(soln)}$ — density of solution at a given temperature, grams per given (cf. Sec. 5.5);

M_B — molar mass of solute M, grams per mole;

M_{H_2O} — molar mass of water equal to 18.02 grams per mole;

n_{H_2O} — formula quantity of water as solvent, moles;

n_B – formula quantity of solute, moles;

z_B – equivalent number that is equal to the number of substance B equivalents (eqv. B) in one formula unity of the same substance (f.u. B) in a given reaction; z_B = 1 f.u. B/1 eqv. B;

$n_{eqv. B}$ – equivalent quantity of solute B in a given reaction in moles.

These physical quantities are interrelated as follows:

$$m_{H_2O} = \rho_{H_2O} \cdot V_{H_2O} = n_{H_2O} \cdot M_{H_2O}$$

$$m_{(soln)} = m_B + m_{H_2O} = \rho_{(soln)} \cdot V_{(soln)}$$

$$m_B = n_B \cdot M_B \qquad n_B = n_{eqv. B}/z_B$$

$$C_{eqv. B} = n_{eqv. B}/V_{(soln)} = m_B z_B/(V_{(soln)} \cdot M_B) \qquad [7]$$

The formulas ([1]–[7]) are employed in the practice of preparing aqueous solutions having the desired composition.

To prepare a solution with the desired mass fraction, solubility coefficient, mole fraction or molal concentration, use is made of formula [1], [2], [4] or [5], respectively, to calculate either the substance mass m_B if a definite water volume V_{H_2O} is to be used), or the water volume V_{H_2O} (if a definite substance sample having the mass m_B is employed), in which the sample m_B should be dissolved.

In order to prepare a solution having the desired mass, molar or equivalent concentration, recourse is made to formula [3], [6] or [7], respectively, to calculate the substance mass m_B, followed by placing the obtained sample m_B into a preliminarily selected measuring flask having the volume $V_{(soln)}$, gradually pouring water into the flask until the sample dissolves completely, and thereafter bringing the solution volume to the mark.

If instead of an anhydrous substance its crystallohydrate $B \cdot n_{H_2O}$ is employed, water content in the crystallohydrate (hydr) should be accounted for. To do so, after calculating, according to formulas [1]–[7], the values of m_B and V_{H_2O} required to prepare a solution of the desired concentration, these values should be recalculated to account for the crystal water. The crystallohydrate mass m_{hydr} is determined using the formula

$$m_{hydr} = m_B M_{hydr}/M_B$$

where M_{hydr} is the molar mass of the crystallohydrate (in gram per mole).

The formula

$$V_{ad} = V_{H_2O} - (m_{hydr} - m_B)/\rho_{H_2O}$$

yields the volume of water (V_{ad}) to be mixed with the crystallohydrate of the mass m_{hydr} in order to obtain the solution having the desired composition.

No.	Method of expressing composition of solution	Mass fraction w_B	Solubility coefficient $k_{s,B}$
1	Mass fraction $w_B = m_B/m_{soln}$	[1] $\dfrac{m_B}{m_B + \rho_{H_2O} V_{H_2O}}$	$\dfrac{k_{s,B}}{1+k_{s,B}}$
2	Solubility coefficient $k_{s,B} = m_B/m_{H_2O}$	$\dfrac{w_B}{1-w_B}$	[2] $\dfrac{m_B}{\rho_{H_2O} V_{H_2O}}$
3	Mass concentration $\rho_B = m_B/V_{(soln)}$, g/l	$\rho_{(soln)} w_B$	$\dfrac{\rho_{(soln)} k_{s,B}}{1+k_{s,B}}$
4	Molar fraction $x_B = n_B/(n_B + n_{H_2O})$	$\dfrac{w_B M_{H_2O}}{w_B M_{H_2O} + (1-w_B) M_B}$	$\dfrac{k_{s,B} M_{H_2O}}{M_B + k_{s,B} M_{H_2O}}$
5	Molal concentration $C_{m,B} = n_B/m_{H_2O}$, mole/kg	$\dfrac{1000 w_B}{(1-w_B) M_B}$	$\dfrac{1000 k_{s,B}}{M_B}$
6	Molar concentration $C_B = n_B/V_{(soln)}$, mole/l	$\dfrac{\rho_{(soln)} w_B}{M_B}$	$\dfrac{\rho_{(soln)} k_{s,B}}{(1+k_{s,B}) M_B}$

(Continued)

No.	Method of expressing composition of solution	Mass concentration ρ_B	Molar fraction x_B	Molal concentration $C_{m,B}$	Molar concentration C_B
1	Mass fraction $w_B = m_B/m_{soln}$	$\dfrac{\rho_B}{\rho_{(soln)}}$	$\dfrac{x_B M_B}{x_B M_B + (1-x_B)M_B}$	$\dfrac{C_{m,B}M_B}{1000 + C_{m,B}M_B}$	$\dfrac{C_B M_B}{\rho_{(soln)}}$
2	Solubility coefficient $k_{s,B} = m_B/m_{H_2O}$	$\dfrac{\rho_B}{\rho_{(soln)} - \rho_B}$	$\dfrac{x_B M_B}{(1-x_B)M_{H_2O}}$	$\dfrac{C_{m,B}M_B}{1000}$	$\dfrac{C_B M_B}{\rho_{(soln)} - C_B M_B}$
3	Mass concentration $\rho_B = m_B/V_{(soln)}$, g/l	[3] $\dfrac{m_B}{V_{(soln)}}$	$\dfrac{\rho_{(soln)}x_B M_B}{x_B M_B + (1-x_B)M_{H_2O}}$	$\dfrac{\rho_{(soln)}C_{m,B}M_B}{1000 + C_{m,B}M_B}$	$C_B M_B$
4	Molar fraction $x_B = n_B/(n_B + n_{H_2O})$	$\dfrac{\rho_B M_{H_2O}}{\rho_B M_{H_2O} + (\rho_{(soln)} - \rho_B)M_B}$	[4] $\dfrac{m_B M_{H_2O}}{m_B M_{H_2O} + \rho_{H_2O}V_{H_2O}M_B}$	$\dfrac{C_{m,B}M_{H_2O}}{1000 + C_{m,B}M_{H_2O}}$	$\dfrac{C_B M_{H_2O}}{\rho_{(soln)} + C_B(M_{H_2O} - M_B)}$
5	Molal concentration $C_{m,B} = n_B/m_{H_2O}$, mole/kg	$\dfrac{1000\rho_B}{(\rho_{(soln)} - \rho_B)M_B}$	$\dfrac{1000 x_B}{(1-x_B)M_{H_2O}}$	[5] $\dfrac{1000 m_B}{\rho_{H_2O}V_{H_2O}M_B}$	$\dfrac{1000 C_B}{\rho_{(soln)} - C_B M_B}$
6	Molar concentration $C_B = n_B/V_{(soln)}$, mole/l	$\dfrac{\rho_B}{M_B}$	$\dfrac{\rho_{(soln)}x_B}{x_B M_B + (1-x_B)M_{H_2O}}$	$\dfrac{\rho_{(soln)}C_{m,B}}{1000 + C_{m,B}M_B}$	[6] $\dfrac{m_B}{V_{(soln)}M_B}$

5.2 Solubility of Solid Substances in Water [5–8, 10, 17, 23, 36]

The table lists the experimental values for the solubility of substances that are solids at the temperature of dissolution.

Solubility is expressed in terms of the solubility coefficient k_s multiplied by 100 ($100k_s$), e.g. in grams of a anhydrous solute per 100 g of water. As a rule, solubility is listed for 20°C and 80°C; the temperatures other than those cited above are noted by superscripts (asterisk* denotes room temperature of 18° to 25°C).

Also tabulated is the solid phase composition, i.e. the number of water molecules (N_{H_2O}) per one formula unit of the substance in the solid phase in stable equilibrium with the saturated solution under standard atmospheric pressure. The sign "−" denotes that the substance does not crystallize out from aqueous solution and is occasionally decomposed by water at a given temperature.

The dots "..." denote the absence of data.

Changes in the formula composition of substances (on dissolution or crystallization) as a result of a chemical reaction between the solute and water are presented in the footnotes at the end of the table in the form of heterogeneous equilibria between precipitates and saturated solutions.

Solubility of practically insoluble substances is listed in Sec. 4.6.

Substance	Solubility, $100k_s$			
	20°C	N_{H_2O}	80°C	N_{H_2O}
$AgBrO_3$	0.159	0	0.936	0
$AgCH_3COO$	1.04	0	2.52	0
$[Ag(CN)_2],K$	25	0	100	0
$AgClO_3$	18.03[25]	0	50	0
$AgClO_4$	526.6	1	792.9[99]	0
AgF	172	2	216[50]	0
$AgMnO_4$	0.91[18]	0	1.69[29]	0
$AgNO_2$	0.34	0	1.36[60]	0
$AgNO_3$	227.9	0	635.3	0
$AgReO_4$	0.32	0	2.71[50]	0
Ag_2SO_4	0.79	0	1.30	0
Ag_2SeO_4	0.118	0	...	...
$AlCl_3$	45.9	6	48.6	6
AlF_3	0.50[25]	1	0.89[75]	3
$[AlF_6],K_3$	0.072*	0	...	...
$AlNH_4(SO_4)_2$	7.74	12	53.9	12
$Al(NO_3)_3$	73.9	9	132.6	6
$Al_2(SO_4)_3$	36.4	18	73.1	18
$As_2O_3 (amor)^{(1)}$	3.7	0	10.1[100]	0
$\beta\text{-}As_2O_3^{(1)}$	1.85	0	6.16	0
$As_2O_5^{(2)}$	65.8	−	75.1	1.67

(Continued)

Substance	Solubility, $100k_s$			
	20°C	N_{H_2O}	80°C	N_{H_2O}
[AuBr$_4$],K	18.3[15]	2	191.6[67]	2
[Au(CN)$_2$],K	14˙	0	200	0
AuCl$_3$[(3)]	68˙	2	...	...
[AuCl$_4$],Cs	0.8	0	19.5	0
[AuCl$_4$],K	61.8	0.5	405[60]	0.5
[AuCl$_4$],Na	151.2	2	900.0[60]	2
[BF$_4$],Cs	1.6[17]	0	...	...
[BF$_4$],K[(4)]	0.44	0	6.27[100]	−
[BF$_4$],NH$_4$[(4)]	25[16]	0	97[100]	−
[BF$_4$],Na[(4)]	108[27]	0	210[100]	−
[BH$_4$],Na	55	0	...	...
B$_2$O$_3$[(5)]	2.2	−	9.5	−
B(OH)$_3$[(5)]	4.87	0	23.54	0
[B$_2$(O$_2$)$_2$(OH)$_4$],Na$_2$	0.76	6	...	...
Ba$_3$(AsO$_4$)$_2$	0.055˙	0	...	...
BaBr$_2$	98.0	2	120.7[75]	2
Ba(BrO$_3$)$_2$	0.656	1	3.65	1
Ba(CH$_3$COO)$_2$	71.4[25]	3	73.8	1
BaCl$_2$	36.2	2	52.2	2
Ba(ClO$_3$)$_2$	33.81	1	84.84	1
Ba(ClO$_4$)$_2$	289.1	3	495.2	3
BaF$_2$	0.161	0	0.162[30]	0
BaHAsO$_4$	0.055˙	1	...	...
BaHPO$_4$	0.015˙	0	...	...
Ba(HS)$_2$[(6)]	48.8	4	63.9	−
BaI$_2$	203.1	7.5	257.1	2
Ba(IO$_3$)$_2$	0.033[18]	1	0.197[100]	1
Ba(MnO$_4$)$_2$	62.5[11]	0	75.4[25]	0
BaMoO$_4$	...	...	0.0058[23]	0
Ba(N$_3$)$_2$	17.3[17]	6	...	...
Ba(NCS)$_2$	...	...	169.3[25]	3
Ba(NO$_2$)$_2$	70.4	1	213.5	1
Ba(NO$_3$)$_2$	9.05	0	26.64	0
BaO[(7)]	1.5	−	90.8	−
BaO$_2$[(8)]	0.091˙	8	...	...
Ba(OH)$_2$[(7)]	3.89	8	101.4	8
Ba(PH$_2$O$_2$)$_2$	28[15]	1	31[100]	1
Ba(ReO$_4$)$_2$	8.13[30]	0	21.51[50]	0
BaS[(9)]	7.86	−	49.91	−
Ba(S$_4$)	38.4[15]	2	...	...
BaSO$_3$	0.02	0	0.002	0
BaS$_2$O$_6$	22.1[18]	2	81.1[100]	2
BaS$_2$O$_6$(O$_2$)	42.83[0]	4	...	...
Ba(SO$_3$S)	0.2	1	...	...
BaSeO$_4$	0.0155[25]	0	0.138[100]	0

(Continued)

Substance	Solubility, $100k_s$			
	20°C	N_{H_2O}	80°C	N_{H_2O}
$BeCO_3$	0.36^0	4	...	...
$BeCl_2$	72.8	4	77.0^{30}	4
$[BeF_4],K_2$	2	0	5.3^{100}	0
$Be(NO_3)_2$	106.6	4	177.8^{60}	4
$BeSO_4$	39.1	4	67.2	2
$Ca_3(AsO_4)_2$	...	...	0.013^{25}	9
$Ca(BO_2)_2^{(10)}$	0.13	6	...	...
$CaBr_2$	143	6	295	3
$Ca(CH_3COO)_2$	34.7	2	33.5	2
$CaCl_2$	74.5	6	147.0	2
$Ca(ClO_3)_2$	195.9	2	354.5^{93}	0
$Ca(ClO_4)_2$	189.0	4	226.8^{50}	4
$CaCrO_4$	13.2	2	14.8^{45}	2
$Ca(HCOO)_2$	16.60	2	17.95	0
$CaHPO_4$	0.02^{25}	2	0.11^{60}	2
$Ca(H_2PO_4)_2$	1.17^*	1	1.7^{30}	1
CaI_2	208.6	6	354	0
$Ca(IO_3)_2$	0.25	1	0.67	0
$CaMg(CO_3)_2$	0.032^{18}	0	...	...
$Ca(MnO_4)_2$	331^{14}	4	338^{25}	4
$Ca(N_3)_2$	38.1^0	0	45^{15}	0
$Ca(NO_2)_2$	88.0	4	151.8	1
$Ca(NO_3)_2$	129.3	4	358.7	0
$Ca(OH)_2$	0.160	0	0.092	0
$Ca(PH_2O_2)_2$	15.4^{25}	0	12.5^{100}	0
$Ca(PO_3F)$	0.87^*	2	...	...
$CaSO_4$	0.206	2	0.102	0
$Ca(SO_3S)$	49.3	6	67.9^{40}	6
$CaSeO_4$	8.3^{18}	2	6^{60}	2
$CaWO_4$	0.2^{18}	0	...	...
$CdBr_2$	98.4	2	157.1	0
$Cd(CN)_2$	1.7^{15}	0	...	...
$[Cd(CN)_4],K_2$	...	...	100^{100}	0
$CdCl_2$	113.4	2.5	140.4	1
$Cd(ClO_3)_2$	2.64^0	2	431.4^{65}	2
CdF_2	4.35^2	0	4.5^{25}	0
CdI_2	84.8	0	112.8	0
$Cd(NO_3)_2$	149.4	4	666.3	0
$CdSO_4$	76.4	2.67	67.2	1
$CeNH_4(SO_4)_2$	5.33	4	1.05^{90}	4
$Ce(NO_3)_3$	175.5^{25}	6	282.8^{50}	6
$Ce_2(SO_4)_3$	9.66	8	1.09	4
$Ce_2(SeO_4)_3$	40.0^0	8	2.5^{100}	4

(Continued)

Substance	Solubility, $100k_s$			
	20°C	N_{H_2O}	80°C	N_{H_2O}
$CoBr_2$	113.2	6	240.1	2
$[Co(C_2H_8N_2)_3]Br_3$	4.53[16]	3	...	...
$CoCl_2$	52.9	6	97.6	2
$Co(ClO_3)_2$	135.0[0]	6	318.6[61]	4
$Co(ClO_4)_2$	103.8	6	115[45]	6
CoF_2	...	...	1.5[25]	4
$Co(HCOO)_2$	5.03	2	...	...
CoI_2	187.4	6	400.0	6
$Co(IO_3)_2$	0.46	6	0.75	2
$[Co(NH_3)_6]Cl_3$	6.95	0	12.04[50]	0
trans-$[Co(NH_3)_4Cl_2]Cl$	0.329[0]	1	...	...
$[Co(NH_3)_5Cl]Cl_2$	0.4[25]	0	1.03[47]	0
$[Co(NH_3)_3(NO_2)_3]^{(11)}$	0.177[16]	0	0.28[25]	0
$[Co(NH_3)_6](NO_3)_3$	1.81	0	2.44[30]	0
$[Co(NH_3)_5(NO_2)]Cl_2$	2.87	0	...	...
$Co(NH_4)_2(SO_4)_2$	14.9	6	33.0	6
$[Co(NH_3)_6]_2(SO_4)_3$	1.2[17]	5	...	...
$Co(NO_3)_2$	97.3	6	211	3
$[Co(NO_2)_6],K_3$	0.9[17]	1.5	...	...
$[Co(NO_2)_6],K_2Na$	...	...	0.07[25]	1
$CoSO_4$	36.3	7	49.3	1
$CoSeO_4$	54.99[15]	7	31.47[90]	1
$[Cr(CN)_6],K_3$	30.96	0	...	...
$CrCl_3$	...	...	34.9[25]	6
$[Cr(H_2O)_4Cl_2]Cl$	...	...	50.6[25]	2
$[Cr(NCS)_6],K_3$	138.89*	4	...	...
$[Cr(NH_3)_5Cl]Cl_2$	0.65[16]	0	...	...
$CrNH_4(SO_4)_2$	2.1[0]	12	15.7[40]	12
$CrO_3^{(12)}$	167	0	190	0
$CrSO_4$	21.0[0]	7	...	...
$Cr_2(SO_4)_3$	...	...	64[25]	18
$CrTl(SO_4)_2$	...	...	110.5[25]	12
$CsAl(SO_4)_2$	0.40	12	5.4	12
$CsBr$	123.3[25]	0	214	0
$CsBrO_3$	3.7[25]	0	5.3[35]	0
$CsCH_3COO$	10[11]	0	1134[60]	0
Cs_2CO_3	260.5[15]	0	...	...
$CsCl$	186.5	0	250	0
$CsClO_3$	6.2	0	45	0
$CsClO_4$	1.6	0	14.4	0
Cs_2CrO_4	71.4[13]	0	...	...
$CsCr(SO_4)_2$	...	...	6.0[25]	12
CsF	572.9[25]	1.5	599.3[50]	0
$Cs_2Fe(SO_4)_2$	...	...	78.7[25]	6
$CsGa(SO_4)_2$	...	...	0.8[25]	12

(Continued)

Substance	Solubility, $100k_s$			
	20°C	N_{H_2O}	80°C	N_{H_2O}
$CsHCO_3$	209.3[15]	0	...	...
CsI	85.6[25]	0	170.8[75]	0
$CsIO_3$	...	...	2.6[24]	0
$CsIO_4$	2.2[15]	0	...	...
$CsMnO_4$	0.23[19]	0	1.27[59]	0
CsN_3	22[40]	0	307.4[16]	0
$CsNO_3$	23.0	0	134	0
$CsOH$	385.6[15]	1	303[30]	1
Cs_2SO_4	178.7	0	210.3	0
Cs_2SeO_4	245[12]	0	...	...
$CsV(SO_4)_2$	0.29[10]	12	...	...
$CuBr_2$	126.8	4	131.5[50]	4
$CuCl_2$	72.7	2	96.1	2
$Cu(ClO_3)_2$	164.4[18]	6	332[70]	4
CuF_2	3.5	2	...	...
$Cu(HCOO)_2$	12.5*	4	...	...
$Cu(IO_3)_2$	0.15	0.67	...	...
$[Cu(NH_3)_4]SO_4$	...	...	16.9[22]	1
$Cu(NO_3)_2$	124.7	6	207.7	3
$CuSO_4$	20.5	5	55.5	5
$CuSeO_4$	16.9	5	53.7	5
$Er_2(SO_4)_3$	13	8	5.3[40]	8
$Eu_2(SO_4)_3$	2.1	8	1.54[40]	8
$FeBr_2$	116.0	6	159.7[75]	4
$FeBr_3$	...	...	455[25]	6
$[Fe(CN)_6],Ba_2$	0.14[15]	6	0.7[100]	6
$[Fe(CN)_6],H_4$	13[14]	0	...	...
$[Fe(CN)_6],K_3$	46.0	0	81.8	0
$[Fe(CN)_6],K_4$	28.0	3	67.0	3
$[Fe(CN)_6],Na_4$	20	10	62.4	10
$[Fe(CN)_6],Tl_4$	0.36[18]	2	3.8[100]	2
FeC_2O_4	...	...	0.078[25]	2
$[Fe(C_2O_4)_3],K_3$	4.2[0]	3	105[100]	0
$FeCl_2$	68.5	4	90.7	2
$FeCl_3^{(13)}$	91.9	6	526	–
$Fe(ClO_4)_2$	80.1[0]	6	106.6[60]	6
$Fe(ClO_4)_3$	97.0[0]	10	142.0[60]	9
$FeNH_4(SO_4)_2$	68[25]	12	221[100]	12
$Fe(NH_4)_2(SO_4)_2$	26.4	6	52[70]	6
$Fe(NO_3)_2$	83.03	6	166[60]	6
$Fe(NO_3)_3$	82.5	9	104.8[40]	9
$FeSO_4$	26.3	7	43.7	1
$Fe_2(SO_4)_3$	440*	9	...	...

(*Continued*)

Substance	Solubility, $100k_s$			
	20°C	N_{H_2O}	80°C	N_{H_2O}
$GaNH_4(SO_4)_2$	...	...	14.5[25]	12
$Gd_2(SO_4)_3$	2.3	8	1.8[40]	8
$[GeF_6],K_2$	0.575[18]	0	2.93[100]	0
β-GeO_2[14]	0.43	0	1.0[100]	0
H_3AsO_4	see As_2O_5	0.5	see As_2O_5	—
$H_2C_2O_4$	9.52	2	84.5	2
HIO_3	see I_2O_5	0	see I_2O_5	0
$H_2(PHO_3)$	309[0]	0	694[30]	0
H_3PO_4	548	0	...	...
$H_4P_2O_7$	...	...	709[23]	0
$HSO_3(NH_2)$	14.7[0]	0	47	0
H_2SeO_3	see SeO_2	0	see SeO_2	0
H_2SeO_4	566.6	1	2753[50]	...
H_6TeO_6	50.05[30]	0	106.4	0
$[HfF_7],(NH_4)_3$	15.5*	0	...	...
$HgBr_2$	0.55	0	2.8	0
$Hg(BrO_3)_2$	0.14*	2	1.5[80]	2
$Hg(CH_3COO)_2$[15]	25[10]	0	100[100]	—
$Hg(CN)_2$	...	...	11.3[25]	0
$HgCl_2$	6.59	0	24.2	0
$Hg(NCS)_2$	...	...	0.07[25]	0
Hg_2SO_4[16]	0.04[25]	0	0.09[100]	—
I_2[17]	0.03	0	0.22	0
I_2O_5[18]	253.4	—	360.8	—
$InBr_3$	414.1	5	618.9[60]	0
$InCl_3$	195.1[22]	3	373.7	2.5
InI_3	1308[22]	0	2024[70]	0
$In_2(SO_4)_3$	117	9	...	...
$[IrCl_6],K_2$	1.25[19]	0	1.12[20]	0
$[IrCl_6],K_3$	8.0	0	...	...
$[IrCl_6],(NH_4)_2$	1.09[25]	0	4.38	0
$[IrCl_6],Na_3$	21.33[15]	12	...	...
$[Ir(NH_3)_6](NO_3)_3$	1.7[14]	0	...	...
$KAl(SO_4)_2$	5.9	12	71.0	12
$K_2B_4O_7$	...	...	16.5[30]	8
KBr	65.2	0	94.6	0
$KBrO_3$	6.87	0	34.28	0

(Continued)

Substance	Solubility, $100k_s$			
	20°C	N_{H_2O}	80°C	N_{H_2O}
$KBrO_4$	...	...	4.21^{25}	0
$K(CH_3COO)$	255.6	1.5	380.1	0.5
$K_2(C_4H_4O_6)$	0.015	0.5	...	...
KCN	69.9	0	99.8	0
K_2CO_3	111.0	1.5	139.2	1.5
$K_2C_2O_4$	36.4	1	63.6	1
$K_2Ca(SO_4)_2$	0.2^{10}	1	...	...
$K_2Cd(SO_4)_2$	39.2^{16}	2	42.3^{40}	1.5
KCl	34.4	0	51.1	0
$KClO_3$	7.3	0	37.6	0
$KClO_4$	1.68	0	13.4	0
$K_2Co(SO_4)_2$	19.2^0	6	...	...
K_2CrO_4	63.0	0	75.1	0
$K_2Cr_2O_7$	12.48	0	73.01	0
$KCr(SO_4)_2$	...	...	12.51^{25}	12
KF	94.93	2	150.1	0
$KFe(SO_4)_2$	11.4^{13}	12	...	...
$K_2Fe(SO_4)_2$	36.5^{25}	6	64.2^{70}	6
KH_2AsO_4	19^6	0	...	...
K_2HAsO_4	18.86^6	0	...	...
$KHCO_3$	33.3	0	68.3^{70}	0
$K(HC_4H_4O_6)$	0.57	0	6.9^{100}	0
$K(HC_2O_4)$	...	...	16.7^{100}	0
$K(HCOO)$	331.0^{18}	0	657.6^{90}	0
$K(HF_2)$	39.2	0	114	0
$KH(PHO_3)$	220	0	...	...
KH_2PO_4	22.6	0	70.4	0
K_2HPO_4	159.8	3	267.5^{63}	0
$KHSO_3$	49	0	115^{100}	0
$KHSO_4$	51.4	0	121.6^{100}	0
KI	144.5	0	190.7	0
KIO_3	8.1	0	24.8	0
KIO_4	0.42	0	4.44	0
$KMgCl_3$	39.4^{19}	6	...	...
$K_2Mg(SO_4)_2$	18.3	6	43.8^{75}	6
$KMnO_4$	6.36	0	25^{65}	0
K_2MoO_4	...	...	184.6^{25}	n
KN_3	49.7^{17}	0	105.7^{100}	0
KNCS	217	0	408^{67}	0
KNO_2	306.7	0	376	0
KNO_3	31.6	0	168.8	0
$KNa(C_4H_4O_6)$	19.4^0	4	49^{25}	4
$K_2Ni(SO_4)_2$	5.3^0	6	45.8^{75}	6
KOCN	...	...	75^{25}	0
KOH	112.4	2	162.5	1
$K(PH_2O_2)$	...	...	200^{25}	0
$K_2(PHO_3)$	170	0	...	...

(*Continued*)

Substance	Solubility, $100k_s$			
	20°C	N_{H_2O}	80°C	N_{H_2O}
K_3PO_4	98.5	7	178.5[60]	3
$KReO_4$	1.20[25]	0	7.47[75]	0
K_2SO_3	107.0	1	111.5	1
K_2SO_4	11.1	0	21.4	0
$K_2S_2O_5^{(19)}$	44.5	0	107.9	–
$K_2S_2O_6$	6.64	0	63.3[100]	0
$K_2S_3O_6$	8.86[0]	0	22.59[20]	0
$K_2S_4O_6$	14.42[0]	0	30.17[20]	0
$K_2S_5O_6$	18.34[0]	1.5	32.94[20]	1.5
KSO_3F	6.9[19]	0	...	...
$K_2S_2O_6(O_2)$	4.7	0	11.0[40]	0
$K_2(SO_3S)$	155.4	1.67	312[90]	0
K_2SeO_4	111.0	0	118.8[60]	0
$La(BrO_3)_3$	149.0	9	230.4[35]	9
$LaCl_3$	97.2[25]	7	170.3[92]	7
$La(NO_3)_3$	...	...	113.4[25]	6
$La_2(SO_4)_3$	2.14[25]	9	0.96[75]	9
$LiBr$	160.4	2	244.8	1
$LiCH_3COO$	45.5[26]	2	98.2[51]	2
Li_2CO_3	1.33	0	0.85[75]	0
$Li_2C_2O_4$	8.0	0	...	...
$LiCl$	78.5	1	112.3	1
$LiClO_3$	313.5[18]	0.5	...	...
$LiClO_4$	56.12	3	123.0	3
Li_2CrO_4	110[18]	2	...	...
$Li_2Cr_2O_7$	186.5[30]	2	242	2
LiF	0.27[18]	0	0.135[35]	0
Li_2GeO_3	...	...	0.85[25]	0
$Li(HCOO)$	39.5	1	93.8	1
LiI	165	3	435	1
$LiMnO_4$	49.97[16]	3	...	...
LiN_3	197.71[16]	1	...	...
$LiNO_2$	73.5[18]	0.5	...	...
$LiNO_3$	74.5	3	194.1[70]	0
$LiOH$	12.8	1	15.3	1
Li_3PO_4	0.03	2	...	...
Li_2SO_4	34.6	1	30.0	1
$MgBr_2$	101.1	6	112[60]	6
$Mg(BrO_3)_2$	27.3[18]	6	...	...
$Mg(CH_3COO)_2$	65.6[25]	4	97.9[55]	4
$MgCO_3$	0.18	5	...	...
MgC_2O_4	0.05[16]	2	0.06[100]	2
$MgCl_2$	54.8	6	65.8	6
$Mg(ClO_3)_2$	129.9[18]	6	280.3[93]	2

(Continued)

Substance	Solubility, $100k_s$			
	20°C	N_{H_2O}	80°C	N_{H_2O}
$Mg(ClO_4)_2$	99.2	6	109.2[50]	6
$MgCrO_4$	54.8[25]	5	66.5[75]	5
$MgHPO_4$	0.15[*]	3	0.09	3
MgI_2	139.8	8	185.7	6
$Mg(IO_3)_2$	8.6	4	16.2[100]	4
$MgNH_4PO_4$	0.03	6	0.011	6
$Mg(NH_4)_2(SO_4)_2$	17.96	6	48.32	6
$Mg(NO_3)_2$	73.3	6	110.1	6
$Mg(PH_2O_2)$	...	...	11.8[25]	6
$MgSO_3$	0.65[25]	6	0.67[75]	3
$MgSO_4$	35.1	7	54.8	1
$MnBr_2$	146.9	4	224.7	2
MnC_2O_4	...	...	0.03[25]	2
$MnCl_2$	73.9	4	112.7	2
MnF_2	1.06	4	...	...
$Mn(NH_4)_2(SO_4)_2$	...	...	37.1[25]	6
$Mn(NO_3)_2$	132.3	6	499	1
$MnSO_4$	62.9	5	45.6	1
$MoO_3^{(20)}$	0.138	0	2.107	0
NH_4Br	74.2	0	119.3	0
NH_4CH_3COO	148[4]	0	...	...
$(NH_4)_2CO_3$	100[15]	0	...	...
$(NH_4)_2C_2O_4$	4.45	1	22.4	1
NH_4Cl	37.2	0	65.6	0
$N_2H_6Cl_2$	...	...	270.4[23]	0
NH_4ClO_3	28.7[0]	0	115[75]	0
NH_4ClO_4	24.8[25]	0	74.2[85]	0
$(NH_4)_2CrO_4$	37.0[25]	0	70.1[75]	0
$(NH_4)_2Cr_2O_7$	35.6	0	115	0
NH_4F	82.6	0	117.6	0
NH_4HCO_3	21.7	0	...	...
$NH_4(HCOO)$	143	0	531	0
$NH_4(HF_2)$	60.15	0	292.7	0
$NH_4H(PHO_3)$	17[10]	0	260[31]	0
$NH_4H_2PO_4$	35.3	0	118.3	0
$(NH_4)_2HPO_4$	69	0	106[70]	0
NH_4HS	128.1[0]	0	...	...
NH_4HSO_3	71.8[0]	0	84.7[60]	0
NH_4HSO_4	100[*]	0	...	...
NH_4I	172.3	0	228.8	0
NH_4IO_3	2.6[15]	0	14.5[100]	0
NH_4IO_4	2.7[16]	0	...	...
NH_4MnO_4	7.9[15]	0	...	...
NH_4N_3	20.2[30]	0	27.07[40]	0

(*Continued*)

Substance	Solubility, $100k_s$			
	20°C	N_{H_2O}	80°C	N_{H_2O}
NH_4NCS	170	0	431[70]	0
NH_4NO_2	180.1	0	300[34]	0
NH_4NO_3	192.0	0	580.0	0
$N_2H_5NO_3$	266	0	2135[60]	0
$(NH_3OH)Cl$	83[17]	0	...	...
$(NH_3OH)_2SO_4$	63.7[25]	0	68.5[90]	0
NH_4ReO_4	6.23	0	32.34	0
$N_2H_6SO_4$	2.87	0	14.39	0
$(NH_4)_2SO_3$	28.0[0]	1	...	...
$(NH_4)_2SO_4$	75.4	0	94.1	0
$(NH_4)_2S_2O_6(O_2)$	58.2[0]	0	74.8[16]	0
$(NH_4)_2SO_3S$	...	...	103.3[100]	0
$(NH_4)_2SeO_4$	122[12]	0	197[100]	0
NH_4VO_3	4.8	0	30.5[70]	0
$NaAl(SO_4)_2$	39.72	12	40.8[45]	12
Na_3AsO_4	19.1[16]	12	23.4[30]	12
NaB_5O_8	10.18[0]	5	...	...
$Na_2B_4O_7$	2.5	10	24.3	4
$NaBr$	90.8	2	118.3	0
$NaBrO_3$	36.4	0	75.7	0
$NaCH_3COO$	46.5	3	153	0
$NaCN$	58.2	2	82.5[55]	0
Na_2CO_3	21.8	10	45.1	1
$Na_2C_2O_4$	3.41	0	5.71	0
$NaCl$	35.9	0	38.1	0
$NaClO$	53.4	5	129.9[50]	2.5
$NaClO_2$	64[17]	3	122[60]	0
$NaClO_3$	95.9	0	203.9[100]	0
$NaClO_4$	211[25]	1	300[75]	0
Na_2CrO_4	84.6[25]	10	124.7	0
$Na_2Cr_2O_7$	180	2	355	2
NaF	4.28	0	4.69	0
Na_2GeO_3	23.6	7	49.2[50]	7
NaH_2AsO_4	179	1	...	...
Na_2HAsO_4	51	12	65[30]	12
$NaHCO_3$	9.59	0	20.2	0
$Na(HCOO)$	81.2	2	122.2[60]	0
$Na(HF_2)$	3.25	0	7.5[90]	0
NaH_2PO_4	85.2	2	207.3	0
Na_2HPO_4	7.66	12	92.4	2
$NaHSO_4$	28.6[0]	1	50[100]	0
NaI	179.3	2	296	0
$NaIO_3$	9.5[25]	1	26.6	0
$NaIO_4$	10.2	3	38.8[50]	0
$NaMnO_4$	144	3	733[70]	3
Na_2MoO_4	64.7[15]	2	69[50]	2

(*Continued*)

Substance	Solubility, $100k_s$			
	20°C	N_{H_2O}	80°C	N_{H_2O}
NaN_3	40.8	0	55[100]	0
NaNCS	142.6[25]	1	200	0
$NaNO_2$	82.9	0	135.5	0
$NaNO_3$	87.6	0	149	0
NaOH	108.7	1	314	0
$Na(PH_2O_2)$	83[25]	1	554[100]	1
$Na_2(PHO_3)$	419[0]	5	...	...
Na_3PO_4	14.5[25]	12	68.0	6
$Na_4P_2O_7$	5.50	10	54.2[82]	0
$NaReO_4$	145.3[30]	0	173[50]	0
Na_2S	18.6	9	49.2	6
Na_2SO_3	26.1	7	29.0	0
Na_2SO_4	19.2	10	43.3	0
$Na_2S_2O_4$	24.1	2	...	...
$Na_2S_2O_5$[(19)]	65.3	0	88.7	–
$Na_2S_2O_6$	15.12	2	49.26	2
$Na_2(SO_3S)$	70.1	5	229	0
Na_2SeO_3	...	...	60[37]	5
Na_2SeO_4	57.2[25]	10	74.8[75]	0
Na_2WO_4	73.0	2	90.1	2
$Nd(BrO_3)_3$	75.5	9	132[45]	9
$NdCl_3$	98	6	127	6
$Nd(NO_3)_3$	153[25]	6	185[50]	6
$Nd_2(SO_4)_3$	7.1	8	2.0[85]	8
$NiBr_2$	131	6	154	3?
$Ni(CH_3COO)_2$	16.6*	4	...	...
$NiCl_2$	64.0	6	86.2[75]	2
$Ni(ClO_4)_2$	156.8[0]	6	192.9[45]	6
NiF_2	2.56	4	2.58[90]	4
NiI_2	148.1	6	187.3	4
$Ni(IO_3)_2$	1.1[30]	4	1.0[90]	4
$Ni(NH_4)_2(SO_4)_2$	6.5	6	21.8	6
$Ni(NO_3)_2$	94.2	6	172[70]	4
$NiSO_4$	38.4	7	66.7	6
$NiSeO_4$	33.44[15]	6	74.79	6
OsO_4[(21)]	6.44	0	...	...
$[PF_6],NH_4$	74.8	0	...	...
$PbBr_2$	0.97[25]	0	3.34	0
$Pb(BrO_3)_2$	1.33	1	...	...
$Pb(CH_3COO)_2$	44.3	3	221[50]	3
$PbCl_2$	0.978	0	2.62	0
$Pb(Cl)F$	0.035*	0	...	...

(Continued)

Substance	Solubility, $100k_s$			
	20°C	N_{H_2O}	80°C	N_{H_2O}
$Pb(ClO_2)_2$	0.095	0	0.42^{100}	0
$Pb(ClO_3)_2$	144.3^{18}	1	163	1
$Pb(ClO_4)_2$	...	...	441^{25}	3
PbF_2	0.064	0	...	...
$Pb(HCOO)_2$	1.6^{16}	0	20^{100}	0
PbI_2	0.076^{25}	0	0.3	0
$Pb(N_3)_2$	0.023^{18}	0	0.09^{70}	0
$Pb(NCS)_2$	0.05	0	0.2^{100}	0
$Pb(NO_3)_2$	52.2	0	107.4	0
$Pb(NO_3)OH$	19.4^{19}	0	...	...
$PbSeO_4$	...	...	0.013^{25}	0
$PbWO_4$	0.03*	0	...	...
trans-$[Pd(NH_3)_2Cl_2]$	0.304^{16}	0	...	...
$Pr(BrO_3)_3$	91.5	9	163.9^{45}	9
$Pr_2(C_2O_4)_3$	...	...	0.02^{25}	10
$PrCl_3$	98.4^{25}	7	141.6	7
$Pr(NO_3)_3$	144^{15}	6	302^{56}	6
$Pr_2(SO_4)_3$	12.6	8	3.5	8
$PtBr_4^{(22)}$	0.41	n	...	...
$[PtBr_6],K_2^{(23)}$	2.02	0	10^{100}	–
$[PtBr_6],(NH_4)_2^{(23)}$	0.59	0	0.36^{100}	–
$[Pt(CN)_4],Ba$	2.8*	4	...	...
$[Pt(CN)_4],K_2$	33.8	3	139^{60}	2
$PtCl_4^{(22)}$	142.1^{25}	10	367	3
$[PtCl_6],Cs_2^{(23)}$	0.024^0	0	0.377^{100}	–
$[PtCl_6],H_2$	140^{18}	6	...	...
$[PtCl_4],K_2$	0.93^{16}	0	5.3^{100}	0
$[PtCl_6],K_2^{(23)}$	0.774	0	3.71	–
$[PtCl_6],(NH_4)_2^{(23)}$	0.5	0	1.25^{100}	–
$[PtCl_6],Na_2$	53^{16}	6	...	...
$[PtCl_6],Rb_2^{(23)}$	0.014^0	0	0.33^{100}	–
cis-$[Pt(NH_3)_2Cl_2]$	0.26^{25}	0	3^{100}	0
trans-$[Pt(NH_3)_2Cl_2]$	0.036^{25}	0	0.7^{100}	0
cis-$[Pt(NH_3)_2Cl_4]$	0.33^0	0	1.54^{100}	0
trans-$[Pt(NH_3)_2Cl_4]$	0.14^0	0	3.0^{100}	0
$[Pt(NH_3)_4]Cl_2$	20^{10}	1	...	...
$[Pt(NO_2)_4],K_2$	3.8^{15}	0	...	...
$RaBr_2$	70.6	2	...	...
$RaCl_2$	24.5	2	...	...
$Ra(IO_3)_2$	0.018^0	0	0.17^{100}	0
$RbAl(SO_4)_2$	1.60	12	25.2	12
$RbBr$	113^{25}	0	191^{100}	0

Substance	Solubility, $100k_s$			
	20°C	N_{H_2O}	80°C	N_{H_2O}
$RbBrO_3$	2.93^{25}	0	5.08^{40}	0
Rb_2CO_3	223	2	...	...
$RbCl$	91.1	0	127.2	0
$RbClO_3$	5.4	0	62.8^{99}	0
$RbClO_4$	1.0	0	9.2	0
Rb_2CrO_4	73.6	0	95.6^{60}	0
$Rb_2Cr_2O_7$	4.96^{18}	0	27.3^{60}	0
$RbCr(SO_4)_2$	...	...	43.4^{25}	12
RbF	300^{18}	1.5	...	...
$RbFe(SO_4)_2$	...	...	31.9^{90}	12
$RbHCO_3$	116	1	...	...
RbI	169^{25}	0	219^{60}	0
$RbIO_3$	2.1^{28}	0	...	...
$RbIO_4$	0.65^{13}	0	...	...
$RbMnO_4$	0.5^0	0	4.7^{60}	0
RbN_3	107^{16}	0	...	...
$RbNO_3$	53.5	0	309	0
$RbOH$	179^{15}	2	282^{47}	1
Rb_2SO_4	48.2	0	75.0	0
Rb_2SeO_4	159^{12}	0	...	...
$RbV(SO_4)_2$	1.6^{10}	12	...	...
$[Ru(NO^+)Cl_5],K_2$	...	...	12.0^{25}	0
$RuO_4^{(24)}$	1.71^0	0	2.03^{20}	0
$SbCl_3^{(25)}$	920	0	1917^{50}	0
$SbF_3^{(25)}$	444.7	0	563.6^{30}	0
$[SbF_6],Na$	128.6	0	...	...
$Sb_2O_5^{(26)}$	0.3^*	n	...	...
$[SbS_4],Na_3$	130	9	...	...
$Sc_2(SO_4)_3$	64.6^{12}	5	32.2^{25}	5
$SeO_2^{(27)}$	264^{22}	–	471^{65}	–
$[SiF_6],Co$	...	...	76.8^{22}	6
$[SiF_6],Cu$	99.7^{17}	6	...	...
$[SiF_6],K_2$	0.11	0	0.38^{60}	0
$[SiF_6],Li_2$	59^{17}	2	...	...
$[SiF_6],Mg$	39.3^{18}	6	...	...
$[SiF_6],Mn$	90.4^{18}	6	...	...
$[SiF_6],(NH_4)_2$	22.7^{25}	0	48.2^{75}	0
$[SiF_6],Na_2$	0.67	0	1.96	0
$[SiF_6],Rb_2$	0.16	0	1.35^{100}	0
$[SiF_6],Sr$	2.6^{15}	2	...	...
$Sm(BrO_3)_3$	62.5	9	98.4^{40}	9

(Continued)

Substance	Solubility, $100k_s$			
	20°C	N_{H_2O}	80°C	N_{H_2O}
$SmCl_3$	93.4	6	99.9[50]	6
$Sm_2(SO_4)_3$	2.67	8	1.99[40]	8
$SnBr_2$[28]	85.2[0]	0	...	...
$SnCl_2$[29]	269.8[15]	2	...	...
$[SnCl_6],(NH_4)_2$	33.0[15]	0	...	...
$[SnF_6],K_2$	3.5[18]	1	31.5[100]	0
SnI_2[30]	0.98	0	2.95	0
$[Sn(OH)_6],K_2$[31]	85[10]	0	110.5[20]	0
$[Sn(OH)_6],Na_2$[31]	61.3[16]	0	50[100]	0
$SnSO_4$	18.8[19]	2	18.1[100]	2
$SrBr_2$	100	6	175	6
$Sr(BrO_3)_2$	33[16]	1	...	...
$Sr(CH_3COO)_2$	41.6	0.5	36.1	0.5
$SrCl_2$	53.1	6	93.1	2
$Sr(ClO_3)_2$	174[18]	8	...	...
$Sr(ClO_4)_2$	...	...	310[25]	n
$SrSrO_4$	0.091[25]	0	0.062[75]	0
SrF_2	0.011[0]	0	0.012[27]	0
SrI_2	179	6	277	6
$Sr(IO_3)_2$	0.03[15]	1	0.8[100]	0
$Sr(MnO_4)_2$	250[18]	3	...	...
$SrMoO_4$	0.01[17]	0	...	...
$Sr(NO_2)_2$	65.3	1	165[100]	1
$Sr(NO_3)_2$	70.4	4	98	0
$Sr(OH)_2$	0.81	8	8.3	8
$SrSO_4$	0.013	0	0.011[95]	0
SrS_2O_6	13.5	4	52[100]	0
SrS_4O_6	25[0]	6	64[30]	6
$Sr(SO_3S)$	1.7[13]	5	39[100]	2
$SrWO_4$	0.14[15]	0	...	...
$Tb_2(SO_4)_3$	3.56	8	2.55[40]	8
$[TcF_6],K_2$	...	...	1.5[25]	0
$ThBr_4$	65	4	...	...
$Th(NO_3)_4$	190.7	5	...	...
$Th(SO_4)_2$	1.38	9	0.81	4
$Th(SeO_4)_2$	0.498[0]	9	1.972[100]	9
$[TiF_6],K_2$	1.28	1	...	...
$TlAl(SO_4)_2$	6.40	12	35.36[60]	12
$TlBr$	0.048	0	0.25[68]	0
Tl_2CO_3	5.23[18]	0	27.2[100]	0

(Continued)

Substance	Solubility, $100k_s$			
	20°C	N_{H_2O}	80°C	N_{H_2O}
$Tl_2C_2O_4$	1.48^{15}	0	9.02^{100}	0
TlCl	0.32	0	1.60	0
$TlClO_3$	3.92	0	36.65	0
$TlClO_4$	19.72^{30}	0	81.49	0
TlF	245^{25}	0	285^{50}	0
TlN_3	0.30^{16}	0	...	...
TlNCS	0.315	0	0.73^{40}	0
$TlNO_2$	32.1^{25}	0	95.8^{98}	0
$TlNO_3$	9.55	0	111	0
TlOH	34.3^{18}	2	126.1^{90}	0
Tl_3PO_4	0.5^{15}	0	0.67^{100}	0
$Tl_4P_2O_7$	40^{10}	0	...	...
Tl_2S	0.02	0	...	...
Tl_2SO_4	4.87	0	14.61	0
$Tl_2S_2O_6$	41.8^{19}	0	...	...
Tl_2SeO_4	2.8	0	8.5	0
$TlVO_3$	0.87^{11}	0	0.21^{100}	0
UF_4	...	...	0.01^{25}	2.5
$UO_2(CH_3COO)_2$	7.07^{15}	2	...	...
$(UO_2)C_2O_4$	0.50	3	1.68^{75}	0
$(UO_2)F_2$	64.40	0	74.1^{100}	0
$UO_2(NO_3)_2$	119.3	6	203	6
$(UO_2)SO_4$	151.6^{16}	3	188.2^{22}	3
$U(SO_4)_2$	...	...	10.9^{24}	4
$V(NH_4)(SO_4)_2$	15.6	12	...	...
$V_2O_5^{(32)}$	0.07^{25}	n	0.07^{100}	n
$[W(CN)_8],K_4$	122^{18}	2	...	...
YBr_3	83.3^{30}	9	111.3^{75}	9
$Y(BrO_3)_3$	...	...	125^{25}	9
YCl_3	75.1	6	78.1	6
$Y(NO_3)_3$	...	...	96.7^{23}	5
$Y_2(SO_4)_3$	7.47^{16}	8	1.99^{95}	8
$Yb_2(SO_4)_3$	38.4	8	6.92	8
$ZnBr_2$	446.4	2	644.6	0
$Zn(CH_3COO)_2$	30.0	2	44.6^{100}	2
$ZnCl_2$	367	1.5	549	0
$Zn(ClO_3)_2$	200.3	4	273.2^{50}	4
$Zn(ClO_4)_2$	110.1	6	125.2^{50}	6
ZnF_2	0.9^{18}	4	...	...
$Zn(HCOO)_2$	5.2	2	21.2	2
ZnI_2	432^{18}	0	488	0

(*Continued*)

Substance	Solubility, $100k_s$			
	20°C	N_{H_2O}	80°C	N_{H_2O}
$Zn(NO_3)_2$	118.8	6	871^{70}	1
$ZnSO_4$	54.1	7	67.2	1
$ZnSeO_4$	63.6^{25}	1	55.3	0
$ZrCl_2O^{(33)}$	60	8	85^{60}	8
$ZrF_4^{(34)}$	1.5^{25}	–	1.39^{50}	–
$[ZrF_6],K_2$	0.78^2	0	25^{100}	0
$[Zr(H_2O)_4(SO_4)_2]$	80.3^{18}	0	99.1^{40}	0

(1) $As_2O_{3(sld)} + 3H_2O \rightleftarrows 2HAsO_2 + 2H_2O \rightleftarrows 2H_3AsO_3$

(2) 20°C: $As_2O_{5(sld)} + 4H_2O = 2(H_3AsO_4 \cdot H_2O)_{(sld)} \rightleftarrows 2H_3AsO_4 + H_2O$
80°C: $3As_2O_{5(sld)} + 9H_2O = 3(As_2O_5 \cdot 5H_2O)_{(sld)} + 4H_2O \rightleftarrows 6H_3AsO_4$

(3) $AuCl_{3(sld)} + 2H_2O = AuCl_3 \cdot 2H_2O_{(sld)} \rightleftarrows H[AuCl_3(OH)] + H_2O$

(4) 100°C: $M[BF_4]_{(sld)} + H_2O = M[B(OH)F_3]_{(sld)} + HF \rightleftarrows M[B(OH)F_3] + HF$
$(M^+ = K^+, NH_4^+, Na^+)$

(5) $B_2O_{3(sld)} + 5H_2O = 2B(OH)_{3(sld)} + 2H_2O \rightleftarrows 2[B(H_2O)(OH)_3]$

(6) 20°C: $Ba(HS)_{2(sld)} + 4H_2O = Ba(HS)_2 \cdot 4H_2O_{(sld)} \rightleftarrows Ba(HS)_2 + 4H_2O$
80°C: $Ba(HS)_{2(sld)} + 2H_2O \rightleftarrows Ba(OH)_2 + H_2S_{(gas)}$

(7) $BaO_{(sld)} + 9H_2O = Ba(OH)_2 \cdot 8H_2O_{(sld)} \rightleftarrows Ba(OH)_2 + 8H_2O$

(8) $2BaO_{2(sld)} + 16H_2O = 2\{BaO_2 \cdot 8H_2O\}_{(sld)} \rightleftarrows Ba(HO)_2 + Ba(OH)_2 + 14H_2O$

(9) $2BaS_{(sld)} + 14H_2O = Ba(HS)_2 \cdot 4H_2O_{(sld)} + Ba(OH)_2 \cdot 8H_2O_{(sld)} \rightleftarrows Ba(HS)_2 + Ba(OH)_2 + 12H_2O$

(10) $2Ca(BO_2)_{2(sld)} + 12H_2O = 2\{Ca(BO_2)_2 \cdot 6H_2O\}_{(sld)} \rightleftarrows (CaOH)_2B_4O_7 + 11H_2O$

(11) $[Co(NH_3)_3(NO_2)_3]_{(sld)} + 3H_2O \rightleftarrows [Co(H_2O)_3(NH_3)_3](NO_2)_3$

(12) $CrO_{3(sld)} + H_2O \rightleftarrows H_2CrO_4$

(13) 80°C: $2FeCl_{3(sld)} + (n+3)H_2O = Fe_2O_3 \cdot nH_2O_{(gel \rightleftarrows soln)} + 6HCl$

(14) $GeO_{2(sld)} + H_2O \rightleftarrows H_2GeO_3$

(15) 80°C: $Hg(CH_3COO)_{2(sld)} + H_2O = Hg(CH_3COO)OH_{(sld)} + CH_3COOH \rightleftarrows$
$(HgOH)CH_3COO + CH_3COOH$

(16) 80°C: $2Hg_2SO_{4(sld)} + 4H_2O = (Hg_2)_2SO_4(OH)_{2(sld)} + H_2SO_4 + 2H_2O \rightleftarrows$
$[Hg_2(H_2O)OH]_2SO_4 + H_2SO_4$

(17) $I_{2(sld)} + H_2O \rightleftarrows HI + HIO$

(18) $I_2O_{5(sld)} + H_2O = 2HIO_{3(sld)} \rightleftarrows 2HIO_3$

(19) 80°C: $M_2S_2O_{5(sld)} + H_2O = 2MHSO_{3(sld)} \rightleftarrows 2MHSO_3$ $(M = K, Na)$

(20) $7MoO_{3(sld)} + 3H_2O \rightleftarrows H_6Mo_7O_{24}$

(21) $OsO_{4(sld)} + 2H_2O \rightleftarrows [Os(H_2O)_2O_4]$

(22) $PtA_{4(sld)} + (n+2)H_2O = PtA_4 \cdot nH_2O_{(sld)} + 2H_2O \rightleftarrows [Pt(H_2O)_2A_4] + nH_2O$
$(A^- = Br^-; Cl^-, n = 10)$

(23) 80–100°C: $M_2[PtA_6]_{(sld)} + H_2O \rightleftarrows M[Pt(H_2O)A_5] + MA$
$(M^+ = Cs^+, K^+, NH_4^+, Rb^+; A^- = Br^-, Cl^-)$

(24) $RuO_{4(sld)} + 2H_2O \rightleftarrows [Ru(H_2O)_2O_4]$

(25) $SbA_{3(sld)} + nH_2O \rightleftarrows [Sb(H_2O)_nA_3]$ $(A = Cl, F)$

(26) $Sb_2O_{5(sld)} + nH_2O = Sb_2O_5 \cdot nH_2O_{(sld)} \rightleftarrows 2[Sb(H_2O)(OH)_5] + (n-7)H_2O$

(27) $SeO_{2(sld)} + H_2O = H_2SeO_{3(sld)} \rightleftarrows H_2SeO_3$

(28) $7SnBr_{2(sld)} + 10H_2O \rightleftarrows [Sn_3(OH)_4]Br_2 + [Sn_4(OH)_6]Br_2 + 10HBr$

(29) $7SnCl_{2(sld)} + 14H_2O = 7\{[Sn(H_2O)Cl_2] \cdot H_2O\}_{(sld)} \rightleftarrows [Sn_3(OH)_4]Cl_2 + [Sn_4(OH)_6]Cl_2 + 10HCl + 4H_2O$

(30) $20°C$: $SnI_{2(sld)} + H_2O \rightleftarrows [Sn(H_2O)I_2]$
$80°C$: $2SnI_{2(sld)} + H_2O \rightleftarrows (SnOH)I + H[SnI_3]$
(31) $M_2[Sn(OH)_6]_{(sld)} + H_2O \rightleftarrows M[Sn(H_2O)(OH)_5] + MOH$ (M = K, Na)
(32) $5V_2O_{5(sld)} + (n+3)H_2O = 5V_2O_5 \cdot nH_2O_{(sld)} + 3H_2O \rightleftarrows H_6V_{10}O_{28} + nH_2O$ (n = 1 – 3)
(33) $4ZrCl_2O_{(sld)} + 32H_2O = 4(ZrCl_2 \cdot 8H_2O)_{(sld)} \rightleftarrows [Zr_4(H_2O)_{16}(OH)_8]Cl_8 + 12H_2O$
(34) $2ZrF_{4(sld)} + 6H_2O = [Zr_2(H_2O)_6F_6]F_{2(sld)} \rightleftarrows H_2[Zr_2(H_2O)_2(OH)_4F_6] + 2HF$

5.3 Solubility of Gaseous and Liquid Substances in Water [5, 8, 16, 22, 23, 29, 36]

The data in this table are for the solubility of gases and liquids (at the end of the table) in water at different temperatures (°C) shown as superscripts at the values of solubility.

Solubility v_s for the majority of gases (at their partial pressure equal to the standard atmospheric pressure) is expressed in milliliters of the solute in 100 g water. For some gases (and also liquids), solubility is expressed by their solubility coefficient k_s multiplied by 100 (100 k_s), e.g., in grams of the dissolved substance per 100 g water.

Chemical reactions between the substances being dissolved and water (sometimes resulting in the formation of solid phases at the temperature of dissolution) are specified in the footnotes at the end of the table. Changes in the formula composition of substances are presented in the form of heterogeneous equilibria between the solute and the saturated solution.

Substance		v_s/100 g H_2O
Ar		5.2^0, 3.32^{20}, 2.54^{40}, 2.20^{50}
AsH$_3$		202^0
BF$_3$		106^0
CH$_4$		5.56^0, 3.31^{20}, 2.37^{40}, 1.95^{60}, 1.77^{80}, 1.70^{100}
C$_2$H$_2$		173^0, 131^{10}, 103^{20}, 84^{30}
C$_2$H$_4$		22.6^0, 19.1^5, 13.9^{15}, 12.2^{20}, 10.8^{25}, 9.8^{30}
C$_2$H$_6$		9.87^0, 4.72^{20}, 2.91^{40}, 2.18^{60}, 1.83^{80}, 1.72^{100}
C$_2$N$_2$		450^{20}
CO		3.54^0, 2.32^{20}, 1.77^{40}, 1.49^{60}, 1.43^{80}, 1.41^{100}
CO$_2$		171.3^0, 119.4^{10}, 87.8^{20}, 75.9^{25}, 53.0^{40}, 35.9^{60}
CSO		133^0, 54^{20}, 40.3^{30}
Cl$_2$(1)		461.0^0, 229.9^{20}, 201.9^{25}, 143.8^{40}, 102.3^{60}, 68.3^{80}
H$_2$		2.15^0, 1.82^{20}, 1.64^{40}, 1.6^{60-100}
HBr(2)	(100 k_s)	221.2^0, 198.2^{20}, 193.0^{25}, 171.4^{50}, 130^{100}
HCl(2)	(100 k_s)	82.3^0, 72.0^{20}, 67.3^{30}, 63.3^{40}, 59.6^{50}, 56.1^{60}
HI(2)	(100 k_s)	234^{10}
H$_2$S		467^0, 258.2^{20}, 166^{40}, 119^{60}, 91.7^{80}, 81^{100}
H$_2$Se		377^4, 331^{13}, 270^{25}
He		0.97^0, 0.99^{10-20}, 1.00^{30}, 1.02^{40}, 1.07^{50}, 1.21^{75}
Kr		11.0^0, 6.0^{25}, 4.67^{50}
N$_2$		2.35^0, 1.54^{20}, 1.18^{40}, 1.02^{60}, 0.96^{80}, 0.95^{100}
NH$_3$(3)	(100 k_s)	87.5^0, 52.6^{20}, 30.7^{40}, 18.4^{56}, 15.4^{80}, 7.4^{100}

(Continued)

Substance		$v_s/100$ g H_2O
NO		7.38^0, 4.71^{20}, 3.51^{40}, 2.95^{60}, 2.70^{80}, 2.63^{100}
N_2O		130.0^0, 104.8^5, 87.8^{10}, 73.8^{15}, 62.9^{20}, 54.4^{25}
Ne		1.23^0, 1.16^{25}, 0.987^4
O_2		4.89^0, 3.10^{20}, 2.31^{40}, 1.95^{60}, 1.78^{80}, 1.72^{100}
O_3		49.4^0, 45.4^{18}
PH_3		27^{20}
Rn		51.0^0, 22.4^{25}, 13.0^{50}
SF_6		1.47^0, 0.55^{25}
SO_2[(4)]	$(100\,k_s)$	22.8^0, 19.3^5, 13.5^{15}, 11.3^{20}, 4.5^{50}, 2.1^{90}
SO_2F_2	$(100\,k_s)$	10^9, 4.5^{17}
SbH_3		19.8^{20}
Xe	$(100\,k_s)$	24.1^0, 11.9^{25}, 8.4^{50}, 7.12^{80}
$Br_{2(lq)}$[(5)]	$(100\,k_s)$	2.30^0, 3.58^{20}, 3.45^{40}
$CS_{2(lq)}$	$(100\,k_s)$	0.179^{20}, 0.014^{50}
$[Ni(CO)_4]_{(lq)}$	$(100\,k_s)$	$0.018^{9.8}$

[(1)] $0°C$: $Cl_{2(gas)} + 6.75H_2O = Cl_2 \cdot 5.75H_2O_{(sld)} + H_2O \rightleftarrows H_3O^+ + Cl^- + HClO + 4.75H_2O$;
20–$80°C$: $Cl_{2(gas)} + 2H_2O \rightleftarrows H_3O^+ + Cl^- + HClO$
[(2)] $HA(gas) + H_2O \rightleftarrows H_3^+ + A^-$. [(3)] $NH_{3(gas)} + H_2O \rightleftarrows NH_3 \cdot H_2O$
[(4)] 0–$5°C$: $SO_{2(gas)} + nH_2O = SO_2 \cdot 7.67H_2O_{(sld)} + (n - 7.67)H_2O \rightleftarrows SO_2 \cdot nH_2O$;
15–$90°C$: $SO_{2(gas)} + nH_2O \rightleftarrows SO_2 \cdot nH_2O$
[(5)] $0°C$: $Br_{2(lq)} + 8.67H_2O = Br_2 \cdot 7.67H_2O_{(sld)} + H_2O \rightleftarrows H_3O^+ + Br^- + HBrO + 6.67H_2O$;
20–$40°C$: $Br_{2(lq)} + 2H_2O \rightleftarrows H_3O^+ + Br^- + HBrO$

5.4 Density of Water at Different Temperatures [4]

t, °C	ρ_{H_2O}, g/l	t, °C	ρ_{H_2O}, g/l	t, °C	ρ_{H_2O}, g/l
0	999.841	24	997.296	48	988.93
1	999.900	25	997.044	50	988.04
2	999.941	26	996.783	52	987.12
4	999.973	28	996.232	55	985.70
6	999.941	30	995.646	60	983.21
8	999.849	32	995.02	65	980.56
10	999.700	34	994.37	70	977.78
12	999.498	36	993.68	75	974.86
14	999.244	38	992.96	80	971.80
16	998.943	40	992.21	85	968.62
18	998.595	42	991.44	90	965.31
20	998.203	44	990.63	95	961.89
22	997.770	46	989.79	100	958.35

5.5 Density of Aqueous Solutions [10, 23, 36]

Density of aqueous solutions $\rho_{(soln)}$ different values of mass fraction w_B are listed for the following substances:

AgF	$FeSO_4$	K_2SO_4
$AgNO_3$	$Fe_2(SO_4)_3$	K_2SiO_3
$AlBr_3$	H_3AsO_4	K_2WO_4
$AlCl_3$	HBr	$La(NO_3)_3$
$AlNH_4(SO_4)_2$	HCN	$LiBr$
$Al(NO_3)_3$	$HCOOH$	$LiCl$
$Al_2(SO_4)_3$	HCl	LiI
$B(OH)_3$	$HClO_3$	$LiNO_3$
$BaBr_2$	$HClO_4$	$LiOH$
$Ba(CH_3COO)_2$	HF	Li_2SO_4
$BaCl_2$	HI	$MgBr_2$
BaI_2	HIO_3	$Mg(CH_3COO)_2$
$Ba(NO_3)_2$	HIO_4	$MgCl_2$
$BeCl_2$	HNO_3	$MgCrO_4$
$Be(NO_3)_2$	H_2O_2	MgI_2
$BeSO_4$	H_3PO_4	$Mg(NO_3)_2$
CH_3COOH	H_2SO_4	$MgSO_4$
$CaBr_2$	$H_2S_2O_6(O_2)$	$MnBr_2$
$Ca(CH_3COO)_2$	H_2SeO_4	$MnCl_2$
$CaCl_2$	$HgCl_2$	$Mn(NO_3)_2$
$Ca(HCOO)_2$	$InBr_3$	$MnSO_4$
CaI_2	$KAl(SO_4)_2$	NH_3
$Ca(NO_3)_2$	KBr	N_2H_4
$CdBr_2$	$KBrO_3$	NH_4Br
$CdCl_2$	$K(CH_3COO)$	$NH_4(CH_3COO)$
CdI_2	$K_2(C_4H_4O_6)$	$(NH_4)_2CO_3$
$Cd(NO_3)_2$	KCN	NH_4Cl
$CdSO_4$	K_2CO_3	$(N_2H_6)Cl_2$
$CoBr_2$	$K_2C_2O_4$	NH_4F
$CoCl_2$	KCl	$NH_4(HCOO)$
$Co(NO_3)_2$	$KClO_3$	NH_4I
$CoSO_4$	$KClO_4$	NH_4NCS
$CrBr_3$	K_2CrO_4	NH_4NO_3
$CrCl_3$	$K_2Cr_2O_7$	NH_2OH
$CrNH_4(SO_4)_2$	$KCr(SO_4)_2$	$(NH_3OH)Cl$
$Cr(NO_3)_3$	KF	$(NH_4)_2SO_4$
$Cr_2(SO_4)_3$	$KFe(SO_4)_2$	Na_3AsO_4
$CsBr$	$KHCO_3$	$Na_2B_4O_7$
$CsCl$	KH_2PO_4	$NaBr$
CsI	KHS	$NaBrO_3$
$CsNO_3$	$KHSO_4$	$Na(CH_3COO)$
Cs_2SO_4	KI	Na_2CO_3
$CuBr_2$	KIO_3	$Na_2C_2O_4$
$CuCl_2$	$KMnO_4$	$NaCl$
$Cu(NO_3)_2$	KN_3	$NaClO$
$CuSO_4$	$KNCS$	$NaClO_3$
$[Fe(CN)_6],K_3$	KNO_2	$NaClO_4$
$[Fe(CN)_6],K_4$	KNO_3	Na_2CrO_4
$FeCl_2$	$KNa(C_4H_4O_6)$	$Na_2Cr_2O_7$
$FeCl_3$	KOH	NaF
$FeNH_4(SO_4)_2$	K_2S	Na_2HAsO_4
$Fe(NO_3)_3$	K_2SO_3	$NaHCO_3$

Na(HCOO)	Na_2SO_3S	$[SiF_6],H_2$
NaH_2PO_4	Na_2SiO_3	$[Sn(OH)_6],Na_2$
Na_2HPO_4	Na_2WO_4	$SrBr_2$
$NaHSO_3$	$NiBr_2$	$SrCl_2$
$NaHSO_4$	$NiCl_2$	SrI_2
NaI	$Ni(NO_3)_2$	$Sr(NO_3)_2$
Na_2MoO_4	$NiSO_4$	$Th(NO_3)_4$
NaN_3	$Pb(CH_3COO)_2$	$TlNO_3$
NaNCS	$Pb(NO_3)_2$	Tl_2SO_4
$NaNO_2$	RbBr	$UO_2(CH_3COO)_2$
$NaNO_3$	RbCl	$UO_2(NO_3)_2$
NaOH	RbI	$ZnBr_2$
Na_3PO_4	$RbNO_3$	$ZnCl_2$
$Na_4P_2O_7$	RbOH	ZnI_2
Na_2SO_3	Rb_2SO_4	$Zn(NO_3)_2$
Na_2SO_4	SO_2	$ZnSO_4$

w_B	$\rho_{(soln)}$, g/l	w_B	$\rho_{(soln)}$, g/l	w_B	$\rho_{(soln)}$, g/l
		0.25	1254.5		
AgF (18°C)		0.30	1320.5	$AlNH_4(SO_4)_2$ (15°C)	
0.01	1008.8	0.35	1393.1	0.01	1007.9
0.02	1019.1	0.40	1474.3	0.02	1016.7
0.04	1039.9	0.50	1668.0	0.04	1034.8
0.06	1061.3	0.60	1910.0	0.06	1053.3
0.08	1083.8			0.08	1072.9
0.10	1107.1	$AlBr_3$ (20°C)		0.10	1099.1
0.12	1131.6	0.01	1005.7	0.12	1112.1
0.14	1157.2	0.02	1013.3	0.14	1132.0
0.16	1183.7	0.04	1028.9	0.16	1154.1
0.18	1211.0	0.06	1045.1	0.18	1175.7
0.20	1239.4	0.08	1061.9	0.20	1197.6
0.25	1314.6	0.10	1079.2	0.22	1219.7
0.30	1400.0	0.12	1096.9	0.24	1242.0
0.35	1480.0	0.14	1115.0	0.30	1309.8
0.40	1580.0	0.16	1133.6		
0.50	1850.0	0.18	1152.8	$Al(NO_3)_3$ (18°C)	
		0.20	1172.5	0.01	1006.5
$AgNO_3$ (20°C)		0.22	1192.8	0.02	1014.4
0.01	1007.0			0.04	1030.5
0.02	1015.4	$AlCl_3$ (18°C)		0.06	1046.9
0.04	1032.7	0.01	1005.7	0.08	1063.8
0.06	1050.6	0.02	1016.4	0.10	1081.1
0.08	1068.0	0.04	1034.4	0.12	1098.9
0.10	1088.2	0.06	1052.6	0.14	1117.1
0.12	1108.0	0.08	1071.1	0.16	1135.7
0.14	1128.4	0.10	1090.0	0.18	1154.9
0.16	1149.5	0.12	1109.3	0.20	1174.5
0.18	1171.5	0.14	1129.0	0.24	1215.3
0.20	1194.2	0.16	1149.1	0.28	1258.2

(Continued)

w_B	$\rho_{(soln)}$, g/l	w_B	$\rho_{(soln)}$, g/l	w_B	$\rho_{(soln)}$, g/l
0.24	1215.3	0.06	1043.3	Ba(NO$_3$)$_2$ (18°C)	
0.28	1258.2	0.08	1058.7		
0.30	1280.5	0.10	1074.5	0.01	1007.2
0.32	1303.6	0.12	1090.8	0.02	1015.1
Al$_2$(SO$_4$)$_3$ (19°C)		0.14	1107.5	0.04	1032.0
		0.16	1124.6	0.06	1049.4
0.01	1009.0	0.18	1142.1	0.08	1067.4
0.02	1019.0	0.20	1159.9	0.10	1086.0
0.04	1040.0	0.24	1197.0	BeCl$_2$ (18°C)	
0.06	1061.0	0.28	1235.6		
0.08	1083.0	0.30	1255.4	0.02	1011.8
0.10	1105.0	0.35	1306.9	0.04	1025.1
0.12	1129.0	0.40	1360.8	0.06	1038.6
0.14	1152.0	BaCl$_2$ (20°C)		0.08	1052.3
0.16	1176.0			0.10	1066.3
0.18	1201.0	0.02	1015.9	0.12	1080.6
0.20	1226.0	0.04	1034.1	0.14	1095.2
0.22	1252.0	0.06	1052.8	Be(NO$_3$)$_2$ (18°C)	
0.24	1278.0	0.08	1072.1		
0.26	1306.0	0.10	1092.1	0.02	1010.8
0.28	1333.0	0.12	1112.8	0.04	1023.9
B(OH)$_3$ (20°C)		0.14	1134.2	0.06	1036.1
		0.16	1156.4	0.08	1049.1
0.01	1002.2	0.18	1179.3	0.10	1062.4
0.02	1005.6	0.20	1203.1	0.12	1076.1
0.03	1009.1	0.22	1227.7	0.14	1090.2
0.04	1013.6	0.24	1253.1	0.16	1104.6
BaBr$_2$ (20°C)		0.26	1279.3	0.18	1119.3
		BaI$_2$ (20°C)		0.20	1134.4
0.02	1015.6			0.22	1149.0
0.04	1033.5	0.02	1015.4	0.24	1165.7
0.06	1051.9	0.04	1033.1	0.28	1198.5
0.08	1071.0	0.06	1051.3	BeSO$_4$ (20°C)	
0.10	1090.7	0.08	1070.1		
0.12	1111.1	0.10	1089.6	0.04	1034.2
0.14	1132.3	0.12	1109.9	0.06	1052.1
0.16	1154.3	0.14	1130.8	0.08	1071.3
0.18	1177.0	0.16	1152.5	0.10	1090.6
0.20	1200.6	0.18	1175.0	0.12	1109.8
0.25	1263.4	0.20	1198.4	CH$_3$COOH (20°C)	
0.30	1332.5	0.25	1261.0		
0.35	1408.7	0.30	1328.9	0.04	1004.1
0.40	1492.6	0.35	1404.0	0.08	1009.8
Ba(CH$_3$COO)$_2$ (18°C)		0.40	1490.0	0.12	1015.4
		0.50	1698.0	0.16	1020.8
0.01	1005.9	0.60	1970.0	0.20	1026.1
0.02	1013.3			0.24	1031.2
0.04	1028.2			0.28	1036.0

(*Continued*)

w_B	$\rho_{(soln)}$, g/l	w_B	$\rho_{(soln)}$, g/l	w_B	$\rho_{(soln)}$, g/l
0.32	1040.5	0.22	1126.3	0.06	1044.8
0.36	1044.8			0.08	1060.8
0.40	1048.8	CaCl$_2$ (20°C)		0.10	1077.1
0.44	1052.5	0.01	1007.0	0.12	1093.7
0.48	1055.9	0.02	1014.8	0.14	1110.6
0.52	1059.0	0.04	1031.6	0.16	1127.9
0.56	1061.8	0.06	1048.6	0.18	1145.5
0.60	1064.2	0.08	1065.9	0.20	1163.6
0.64	1066.3	0.10	1083.5	0.25	1211.0
0.68	1067.9	0.12	1101.5	0.30	1259.0
0.72	1069.1	0.14	1119.8	0.35	1311.0
0.76	1069.9	0.16	1138.6	0.40	1366.0
0.80	1069.9	0.18	1157.8	0.45	1423.0
0.84	1069.1	0.20	1177.5		
0.88	1067.4	0.25	1228.4	CdBr$_2$ (20°C)	
0.92	1064.3	0.28	1260.3	0.02	1015.8
0.96	1058.9	0.30	1281.6	0.04	1033.9
		0.35	1337.3	0.06	1052.4
CaBr$_2$ (20°C)		0.40	1395.7	0.08	1071.4
0.02	1015.2	Ca(HCOO)$_2$ (18°C)		0.10	1091.0
0.04	1032.6			0.12	1111.2
0.06	1050.4	0.01	1005.6	0.14	1132.2
0.08	1068.8	0.02	1012.6	0.16	1154.0
0.10	1087.7	0.04	1026.8	0.18	1176.6
0.12	1107.1	0.06	1041.3	0.20	1200.0
0.14	1127.2	0.08	1056.0	0.25	1260.5
0.16	1148.0	0.10	1070.8	0.30	1328.6
0.18	1169.6	0.12	1085.8	0.35	1404.9
0.20	1191.9			0.40	1490.2
0.25	1249.9	CaI$_2$ (20°C)			
0.30	1312.5	0.02	1015.0	CdCl$_2$ (20°C)	
0.35	1381.0	0.04	1032.3	0.02	1015.9
0.40	1457.0	0.06	1050.0	0.04	1033.9
0.45	1541.0	0.08	1068.3	0.06	1052.4
0.50	1635.0	0.10	1087.3	0.08	1071.5
		0.12	1106.9	0.10	1091.2
Ca(CH$_3$COO)$_2$ (18°C)		0.14	1127.3	0.12	1111.5
0.01	1004.3	0.16	1148.5	0.14	1132.4
0.02	1010.0	0.18	1170.3	0.16	1154.0
0.04	1021.5	0.20	1192.8	0.18	1176.2
0.06	1033.1	0.25	1253.0	0.20	1199.2
0.08	1044.0	0.30	1319.5	0.25	1260.4
0.10	1056.3	0.35	1392.8	0.30	1327.3
0.12	1067.9	0.40	1473.4	0.35	1401.0
0.14	1079.5			0.40	1483.3
0.16	1091.2	Ca(NO$_3$)$_2$ (18°C)		0.45	1574.8
0.18	1102.9	0.02	1013.7	0.50	1576.2
0.20	1114.6	0.04	1029.1		

(Continued)

w_B	$\rho_{(soln)}$, g/l	w_B	$\rho_{(soln)}$, g/l	w_B	$\rho_{(soln)}$, g/l
		0.40	1547.0	0.06	1058.8
CdI$_2$ (20°C)				0.08	1080.0
		CoBr$_2$ (18°C)			
0.02	1015.3			CrBr$_3$ (18°C)	
0.04	1032.8	0.01	1007.0		
0.06	1050.7	0.02	1016.0	0.01	1007.4
0.08	1069.0	0.04	1035.0	0.02	1016.2
0.10	1087.9	0.06	1054.0	0.04	1034.4
0.12	1107.5	0.08	1073.0	0.06	1053.2
0.14	1127.8	0.10	1094.0	0.08	1072.6
0.16	1148.9	0.12	1115.0	0.10	1092.6
0.18	1170.9	0.14	1137.0	0.12	1113.2
0.20	1193.7	0.16	1159.0	0.14	1134.5
0.25	1254.6	0.18	1182.0	0.16	1156.5
0.30	1321.9	0.20	1207.0	0.18	1179.4
0.35	1396.7	0.25	1272.0	0.20	1203.1
0.40	1480.1	0.30	1344.0	0.24	1253.2
				0.30	1335.5
Cd(NO$_3$)$_2$ (18°C)		CoCl$_2$ (18°C)			
				CrCl$_3$ (18°C)	
0.02	1015.4	0.01	1008.0		
0.04	1032.6	0.02	1017.0	0.01	1007.6
0.06	1050.2	0.04	1036.0	0.02	1016.6
0.08	1068.3	0.06	1055.0	0.04	1034.9
0.10	1086.9	0.08	1075.0	0.06	1053.5
0.12	1106.1	0.10	1095.0	0.08	1072.4
0.14	1126.1	0.12	1116.0	0.10	1091.7
0.16	1146.8	0.14	1137.0	0.12	1111.4
0.18	1168.2	0.16	1159.0	0.14	1131.6
0.20	1190.4	0.18	1182.0		
0.25	1248.8	0.20	1205.0	CrNH$_4$(SO$_4$)$_2$ (15°C)	
0.30	1312.4				
0.35	1382.2	Co(NO$_3$)$_2$ (18°C)		0.006	1008.1
0.40	1459.0			0.011	1017.2
0.50	1635.6	0.01	1007.0	0.022	1035.7
		0.02	1015.0	0.033	1054.5
CdSO$_4$ (18°C)		0.04	1032.0		
		0.06	1049.0	Cr(NO$_3$)$_3$ (15°C)	
0.02	1018.2	0.08	1067.0		
0.04	1038.3	0.10	1085.0	0.01	1007.0
0.06	1059.0	0.12	1104.0	0.02	1015.0
0.08	1080.3	0.14	1123.0	0.04	1032.0
0.10	1102.3	0.16	1143.0	0.06	1049.0
0.12	1125.0	0.18	1163.0	0.08	1066.0
0.14	1148.5	0.20	1184.0	0.10	1084.0
0.16	1172.9	0.25	1240.0	0.12	1103.0
0.18	1198.2	0.30	1300.0	0.14	1121.0
0.20	1224.2			0.16	1141.0
0.25	1294.0	CoSO$_4$ (25°C)		0.18	1161.0
0.30	1371.4			0.20	1181.0
0.35	1455.1	0.01	1007.2	0.22	1202.0
		0.02	1017.4	0.24	1223.0

(Continued)

w_B	$\rho_{(soln)}$, g/l	w_B	$\rho_{(soln)}$, g/l	w_B	$\rho_{(soln)}$, g/l
0.26	1246.0	0.12	1098.3	0.08	1067.6
0.28	1269.0	0.14	1116.8	0.10	1087.0
0.30	1293.0	0.16	1135.8	0.12	1107.1
		0.18	1155.5	0.14	1127.5
$Cr_2(SO_4)_3$ (15°C)		0.20	1175.8	0.16	1148.7
0.01	1009.1	0.22	1196.8	0.18	1169.6
0.02	1019.1	0.24	1218.5	0.20	1191.3
0.04	1039.5	0.30	1288.2	0.22	1213.7
0.06	1060.4	0.35	1352.2	0.24	1237.5
0.08	1081.7	0.40	1422.5	0.26	1264.3
0.10	1103.4	0.50	1585.8	$CuBr_2$ (0°C)	
0.12	1125.7	0.60	1788.6		
0.14	1148.6			0.002	1004.0
0.16	1172.2	CsI (20°C)		0.004	1007.0
0.18	1196.6			0.006	1011.0
0.20	1221.8	0.01	1006.1	0.008	1014.0
0.22	1247.9	0.02	1014.0	0.01	1018.0
0.24	1275.0	0.04	1030.8	0.02	1036.0
0.26	1303.2	0.06	1047.1	0.04	1072.0
0.28	1332.5	0.08	1064.4	0.06	1113.0
		0.10	1082.3		
CsBr (20°C)		0.12	1100.7	$CuCl_2$ (20°C)	
		0.14	1119.8		
0.01	1006.1	0.16	1139.5	0.01	1007.0
0.02	1014.1	0.18	1159.9	0.02	1017.0
0.04	1030.5	0.20	1181.1	0.04	1036.0
0.06	1047.3	0.22	1203.1	0.06	1056.0
0.08	1064.7	0.24	1225.8	0.08	1076.0
0.10	1082.7	0.30	1299.4	0.10	1096.0
0.12	1101.2	0.35	1367.8	0.12	1116.0
0.14	1120.3	0.40	1443.5	0.14	1138.0
0.16	1140.1	0.45	1528.0	0.16	1160.0
0.18	1160.5	0.50	1622.8	0.18	1182.0
0.20	1181.7			0.20	1205.0
0.22	1203.6	$CsNO_3$ (20°C)		0.22	1229.0
0.24	1226.3			0.24	1253.0
0.30	1299.7	0.01	1005.6	0.26	1278.0
0.35	1367.6	0.02	1013.2		
0.40	1442.8	0.04	1028.6	$Cu(NO_3)_2$ (20°C)	
0.45	1526.3	0.06	1044.4		
0.50	1619.7	0.08	1060.7	0.01	1007.0
		0.10	1077.5	0.02	1015.0
CsCl (20°C)		0.12	1094.6	0.04	1032.0
		0.14	1112.3	0.06	1050.0
0.01	1005.9			0.08	1069.0
0.02	1013.7	Cs_2SO_4 (20°C)		0.10	1088.0
0.04	1029.6			0.12	1107.0
0.06	1046.1	0.01	1006.1	0.14	1126.0
0.08	1062.9	0.02	1014.4	0.16	1147.0
0.10	1080.4	0.04	1031.6	0.18	1168.0
		0.06	1049.4		

(Continued)

w_B	$\rho_{(soln)}$, g/l	w_B	$\rho_{(soln)}$, g/l	w_B	$\rho_{(soln)}$, g/l
0.20	1189.0	0.16	1155.1	0.18	1155.1
0.25	1248.0	0.18	1177.1	0.20	1174.8
		0.20	1199.6	0.25	1228.1
CuSO$_4$ (20°C)		0.25	1259.6		
				FeSO$_4$ (18°C)	
0.01	1009.0	**FeCl$_3$ (20°C)**			
0.02	1019.0			0.01	1008.5
0.04	1040.0	0.01	1007.0	0.02	1018.0
0.06	1062.0	0.02	1015.0	0.04	1037.5
0.08	1084.0	0.04	1032.0	0.06	1057.5
0.10	1107.0	0.06	1049.0	0.08	1078.5
0.12	1131.0	0.08	1067.0	0.10	1100.0
0.14	1155.0	0.10	1085.0	0.12	1122.0
0.16	1180.0	0.12	1104.0	0.14	1144.5
0.18	1206.0	0.14	1123.0	0.16	1167.5
		0.16	1142.0	0.18	1190.5
[Fe(CN)$_6$],K$_3$ (20°C)		0.18	1162.0	0.20	1213.5
		0.20	1182.0		
0.01	1003.4	0.25	1234.0	**Fe$_2$(SO$_4$)$_3$ (17.$_5$°C)**	
0.02	1009.0	0.30	1291.0		
0.04	1020.1	0.35	1353.0	0.01	1007.0
0.06	1031.4	0.40	1417.0	0.02	1016.0
0.08	1042.7	0.45	1485.0	0.04	1033.0
0.10	1054.2	0.50	1551.0	0.06	1050.0
0.12	1065.6			0.08	1067.0
0.14	1078.9	**FeNH$_4$(SO$_4$)$_2$ (15°C)**		0.10	1084.0
0.16	1089.0			0.12	1103.0
0.18	1101.0	0.01	1007.0	0.14	1141.0
0.20	1113.0	0.02	1016.0	0.20	1181.0
		0.04	1032.0	0.25	1241.0
[Fe(CN)$_6$],K$_4$ (20°C)		0.06	1050.0	0.30	1307.0
		0.08	1068.0	0.35	1376.0
0.01	1005.1	0.10	1086.0	0.40	1449.0
0.02	1011.9	0.12	1104.0	0.50	1613.0
0.04	1025.6	0.14	1122.0		
0.06	1039.5	0.16	1141.0	**H$_3$AsO$_4$ (15°C)**	
0.08	1053.6	0.18	1161.0		
0.10	1067.8	0.20	1181.0	0.01	1005.7
0.12	1082.3	0.40	1380.0	0.02	1012.4
0.14	1097.1			0.04	1026.0
0.16	1112.0	**Fe(NO$_3$)$_3$ (18°C)**		0.06	1039.8
				0.08	1053.8
FeCl$_2$ (18°C)		0.01	1006.5	0.10	1068.1
		0.02	1014.4	0.12	1082.6
0.01	1007.5	0.04	1030.4	0.14	1097.5
0.02	1016.5	0.06	1046.8	0.16	1112.8
0.04	1034.8	0.08	1063.6	0.18	1128.5
0.06	1053.5	0.10	1081.0	0.20	1144.7
0.08	1072.6	0.12	1098.9	0.22	1161.4
0.10	1092.3	0.14	1117.2	0.26	1196.1
0.12	1112.6	0.16	1135.9	0.30	1233.1
0.14	1133.6				

(*Continued*)

w_B	$\rho_{(soln)}$, g / l	w_B	$\rho_{(soln)}$, g / l	w_B	$\rho_{(soln)}$, g / l
0.35	1282.9	0.14	1034.6	0.02	1010.3
0.40	1337.0	0.18	1044.2	0.04	1022.3
0.45	1395.9	0.22	1053.8	0.06	1034.4
0.50	1460.2	0.26	1063.4	0.08	1046.8
0.55	1530.4	0.30	1073.0	0.10	1059.4
0.60	1607.0	0.34	1082.4	0.12	1072.3
0.65	1690.4	0.38	1092.0	0.14	1085.6
0.70	1781.1	0.42	1101.6	0.16	1099.1
		0.46	1110.9	0.18	1113.0
HBr (20°C)		0.50	1120.8	0.20	1127.3
		0.54	1129.6	0.22	1141.9
0.01	1005.3	0.58	1138.2	0.24	1156.8
0.02	1012.4	0.62	1147.3		
0.04	1026.9	0.66	1156.6	**HClO$_4$ (20°C)**	
0.06	1041.7	0.70	1165.6		
0.08	1056.8	0.74	1175.3	0.001	1005.0
0.10	1072.3	0.78	1181.9	0.036	1020.0
0.12	1088.3	0.82	1189.7	0.069	1040.0
0.14	1104.8	0.86	1197.7	0.101	1060.0
0.16	1121.9	0.90	1204.5	0.131	1080.0
0.18	1139.6	0.94	1211.8	0.160	1100.0
0.20	1157.9	0.98	1218.4	0.189	1120.0
0.24	1196.1			0.216	1140.0
0.28	1236.7	**HCl (20°C)**		0.243	1160.0
0.30	1258.0			0.268	1180.0
0.35	1315.0	0.004	1000.0	0.293	1200.0
0.40	1377.2	0.024	1010.0	0.316	1220.0
0.45	1444.6	0.044	1020.0	0.339	1240.0
0.50	1517.3	0.064	1030.0	0.360	1260.0
0.55	1595.3	0.085	1040.0	0.381	1280.0
0.60	1678.7	0.105	1050.0	0.401	1300.0
0.65	1767.5	0.125	1060.0	0.420	1320.0
		0.145	1070.0	0.439	1340.0
HCN (20°C)		0.165	1080.0	0.451	1360.0
		0.184	1090.0	0.475	1380.0
0.80	759.0	0.204	1100.0	0.492	1400.0
0.82	752.0	0.223	1110.0	0.509	1420.0
0.84	745.0	0.243	1120.0	0.525	1440.0
0.86	738.0	0.262	1130.0	0.540	1460.0
0.88	731.0	0.282	1140.0	0.556	1480.0
0.90	724.0	0.301	1150.0	0.571	1500.0
0.92	717.0	0.321	1160.0	0.585	1520.0
0.94	711.0	0.342	1170.0	0.600	1540.0
0.96	704.0	0.362	1180.0	0.615	1560.0
0.98	697.0	0.383	1190.0	0.630	1580.0
		0.400	1198.0	0.645	1600.0
HCOOH (20°C)				0.660	1620.0
		HClO$_3$ (18°C)		0.675	1640.0
0.04	1009.4			0.690	1660.0
0.08	1019.7	0.01	1004.4		
0.10	1024.7				

w_B	$\rho_{(soln)}$, g/l	w_B	$\rho_{(soln)}$, g/l	w_B	$\rho_{(soln)}$, g/l
0.702	1675.0	0.04	1033.4	0.390	1240.0
		0.06	1051.7	0.421	1260.0
HF (20°C)		0.08	1070.6	0.453	1280.0
0.02	1005.0	0.10	1090.0	0.484	1300.0
0.04	1012.0	0.12	1110.0	0.517	1320.0
0.06	1021.0	0.14	1130.6	0.551	1340.0
0.08	1028.0	0.16	1151.9	0.588	1360.0
0.10	1036.0	0.18	1174.0	0.627	1380.0
0.12	1043.0	0.20	1196.9	0.670	1400.0
0.14	1050.0	0.22	1220.6	0.716	1420.0
0.16	1057.0	0.24	1245.0	0.767	1440.0
0.18	1064.0	0.26	1270.0	0.824	1460.0
0.20	1070.0	0.28	1295.6	0.891	1480.0
0.24	1084.0	0.30	1321.8	0.967	1500.0
0.28	1096.0	0.35	1390.0	0.972	1502.0
0.32	1107.0	0.40	1464.0	0.977	1504.0
0.36	1118.0			0.983	1506.0
0.40	1123.0	HIO_4 (17°C)		0.988	1508.0
0.42	1134.0			0.993	1510.0
0.44	1139.0	0.01	1007.6	0.998	1512.0
0.50	1155.0	0.02	1016.5		
		0.04	1034.9	H_2O_2 (18°C)	
HI (20°C)		0.06	1053.9		
		0.08	1073.7	0.01	1002.2
0.01	1005.4	0.10	1094.4	0.02	1005.8
0.02	1012.7	0.12	1116.1	0.04	1013.1
0.04	1027.7	0.14	1138.8	0.06	1020.4
0.06	1043.1	0.16	1162.3	0.08	1027.7
0.08	1058.9	0.18	1186.5	0.10	1035.1
0.10	1074.1	0.20	1211.6	0.12	1042.5
0.12	1091.8	0.24	1264.7	0.14	1049.9
0.14	1109.1	0.26	1293.1	0.16	1057.4
0.16	1127.0	0.28	1323.0	0.18	1064.9
0.18	1145.6	0.30	1354.5	0.20	1072.5
0.20	1164.9	0.32	1387.5	0.22	1080.2
0.24	1205.9			0.24	1088.0
0.28	1250.3	HNO_3 (20°C)		0.26	1095.9
0.30	1273.7			0.28	1104.0
0.35	1335.7	0.033	1000.0	0.30	1112.2
0.40	1402.9	0.040	1020.0	0.35	1132.7
0.45	1475.5	0.075	1040.0	0.40	1153.6
0.50	1560.0	0.110	1060.0	0.45	1179.9
0.55	1655.0	0.143	1080.0	0.50	1196.6
0.60	1770.0	0.176	1100.0	0.55	1218.8
0.65	1901.0	0.208	1120.0	0.60	1241.6
		0.240	1140.0	0.65	1265.2
HIO_3 (18°C)		0.270	1160.0	0.70	1289.7
		0.300	1180.0	0.75	1314.9
0.01	1007.1	0.329	1200.0	0.80	1340.6
0.02	1015.7	0.359	1220.0		

w_B	$\rho_{(soln)}$, g/l	w_B	$\rho_{(soln)}$, g/l	w_B	$\rho_{(soln)}$, g/l
0.85	1366.7	0.525	1420.0		
0.90	1393.1	0.545	1440.0	\multicolumn{2}{c}{H_2SeO_4 (20°C)}	
0.95	1419.7	0.564	1460.0	0.01	1005.9
\multicolumn{2}{c}{H_3PO_4 (20°C)}	0.583	1480.0	0.02	1013.6
		0.602	1500.0	0.04	1029.1
0.01	1003.8	0.620	1520.0	0.06	1044.7
0.02	1009.2	0.638	1540.0	0.08	1060.5
0.04	1020.0	0.656	1560.0	0.10	1076.6
0.06	1030.9	0.674	1580.0	0.12	1093.1
0.08	1042.0	0.691	1600.0	0.14	1110.1
0.10	1053.2	0.708	1620.0	0.16	1127.6
0.12	1064.7	0.725	1640.0	0.18	1145.5
0.14	1076.4	0.742	1660.0	0.20	1163.9
0.16	1088.4	0.759	1680.0	0.24	1206.2
0.18	1100.8	0.776	1700.0	0.28	1243.8
0.20	1113.4	0.794	1720.0	0.32	1287.4
0.24	1139.5	0.812	1740.0	0.36	1333.3
0.28	1166.5	0.831	1760.0	0.40	1381.9
0.30	1180.5	0.852	1780.0	0.44	1433.6
0.35	1216.0	0.877	1800.0	0.48	1489.2
0.40	1254.0	0.911	1820.0	0.52	1549.0
0.45	1293.0	0.916	1822.0	0.56	1614.0
0.50	1335.0	0.920	1824.0	0.60	1685.0
0.55	1379.0	0.925	1826.0	0.64	1761.0
0.60	1426.0	0.930	1828.0	0.68	1844.0
\multicolumn{2}{c}{H_2SO_4 (20°C)}	0.936	1830.0	0.72	1932.0
		0.943	1832.0	0.76	2025.0
0.003	1000.0	0.951	1834.0	0.80	2122.0
0.032	1020.0	0.957	1835.0	\multicolumn{2}{c}{$HgCl_2$ (20°C)}	
0.062	1040.0	\multicolumn{2}{c}{$H_2S_2O_6(O_2)$ (14°C)}		
0.091	1060.0			0.01	1006.5
0.120	1080.0	0.01	1005.0	0.02	1015.0
0.147	1100.0	0.02	1011.0	0.03	1023.6
0.174	1120.0	0.04	1022.0	0.04	1032.3
0.201	1140.0	0.06	1034.0	0.05	1041.1
0.227	1160.0	0.08	1046.0	\multicolumn{2}{c}{$InBr_3$ (18°C)}	
0.252	1180.0	0.10	1059.0		
0.277	1200.0	0.12	1072.0	0.01	1007.0
0.302	1220.0	0.14	1085.0	0.02	1015.5
0.326	1240.0	0.16	1099.0	0.04	1032.9
0.350	1260.0	0.18	1113.0	0.06	1050.9
0.374	1280.0	0.20	1127.0	0.08	1069.5
0.397	1300.0	0.22	1142.0	0.10	1088.8
0.420	1320.0	0.24	1157.0	0.12	1108.8
0.442	1340.0	0.26	1173.0	0.14	1129.5
0.463	1360.0	0.28	1189.0	0.16	1150.8
0.485	1380.0	0.30	1205.0	0.18	1172.8
0.505	1400.0	0.35	1245.0	0.20	1195.6

(Continued)

w_B	$\rho_{(soln)}$, g/l	w_B	$\rho_{(soln)}$, g/l	w_B	$\rho_{(soln)}$, g/l
0.30	1323.0	0.28	1146.6	0.24	1232.0
		0.35	1186.8	0.28	1275.6
$KAl(SO_4)_2$ (19°C)		0.40	1216.2	0.35	1354.8
0.01	1007.9	0.50	1276.1	0.40	1414.1
0.02	1017.4	0.60	1337.2	0.45	1475.9
0.03	1027.0			0.50	1540.4
0.04	1036.9	$K_2(C_4H_4O_6)$ (20°C)		0.53	1567.3
0.05	1046.5	0.01	1004.8		
0.06	1056.5	0.02	1011.4	$K_2C_2O_4$ (18°C)	
		0.04	1024.8	0.01	1006.1
KBr (20°C)		0.06	1038.3	0.02	1013.6
0.01	1005.4	0.08	1051.9	0.04	1028.8
0.02	1012.7	0.10	1065.7	0.06	1044.1
0.04	1027.5	0.12	1079.8	0.08	1059.6
0.06	1042.6	0.16	1108.7	0.10	1075.3
0.08	1058.1	0.18	1123.6	0.12	1091.2
0.10	1074.0	0.20	1138.7	0.14	1107.2
0.12	1090.3	0.22	1154.0		
0.14	1107.0	0.26	1185.5	KCl (20°C)	
0.16	1124.2	0.30	1218.1	0.01	1004.6
0.18	1141.6	0.35	1260.6	0.02	1011.0
0.20	1160.1	0.40	1305.1	0.04	1023.9
0.24	1198.0	0.50	1400.1	0.06	1036.9
0.28	1238.3			0.08	1050.0
0.30	1259.3	KCN (15°C)		0.10	1063.3
0.35	1314.7	0.01	1004.1	0.12	1076.8
0.40	1374.6	0.02	1009.2	0.14	1090.5
		0.04	1019.4	0.16	1104.3
$KBrO_3$ (20°C)		0.06	1029.7	0.18	1118.5
0.01	1005.6	0.08	1040.1	0.20	1132.3
0.02	1013.1	0.10	1050.6	0.22	1147.4
0.03	1020.6	0.12	1066.2	0.24	1162.3
0.04	1028.2	0.14	1071.8		
0.05	1035.9	0.16	1082.5	$KClO_3$ (18°C)	
		0.18	1093.1	0.01	1004.9
$K(CH_3COO)$ (18°C)				0.02	1011.3
		K_2CO_3 (20°C)		0.03	1017.8
0.01	1003.8	0.01	1007.2	0.04	1024.5
0.02	1008.9	0.02	1016.3	0.05	1031.2
0.04	1019.1	0.04	1034.5	0.06	1038.0
0.06	1029.3	0.06	1052.9		
0.08	1039.5	0.08	1071.5	$KClO_4$ (15°C)	
0.10	1049.7	0.10	1090.4	0.002	1000.4
0.12	1059.9	0.12	1109.4	0.004	1001.6
0.14	1070.3	0.14	1129.1	0.006	1002.9
0.16	1080.8	0.16	1149.0	0.008	1004.1
0.18	1091.4	0.18	1169.2	0.010	1005.4
0.20	1102.2	0.20	1189.8	0.012	1006.7
0.24	1124.4				

(*Continued*)

w_B	$\rho_{(soln)}$, g/l	w_B	$\rho_{(soln)}$, g/l	w_B	$\rho_{(soln)}$, g/l
0.014	1007.9	0.06	1051.2	0.10	1058.3
0.016	1009.2	0.08	1069.3	0.12	1070.4
0.018	1010.5	0.10	1087.7	0.14	1082.6
		0.12	1106.4	0.16	1094.9
K$_2$CrO$_4$ (18°C)		0.14	1125.4	0.18	1107.2
0.01	1006.6	0.16	1144.8	0.20	1119.6
0.02	1014.7	0.18	1164.6	0.24	1144.7
0.04	1031.1	0.20	1184.7	0.28	1170.1
0.06	1047.7	0.22	1205.2	0.35	1215.0
0.08	1064.7	0.26	1247.1	0.40	1247.9
0.10	1082.1			0.50	1314.4
0.12	1099.9	**KFe(SO$_4$)$_2$ (15°C)**			
0.14	1118.1			**KHSO$_4$ (18°C)**	
0.16	1136.6	0.01	1008.0		
0.18	1155.5	0.02	1017.0	0.01	1006.0
0.20	1174.8	0.04	1034.0	0.02	1014.0
0.24	1214.7	0.06	1052.0	0.04	1028.0
0.28	1256.6	0.08	1071.0	0.06	1042.0
0.30	1278.4	0.10	1090.0	0.08	1057.0
0.32	1301.0	0.12	1110.0	0.10	1072.0
0.36	1347.8	0.14	1130.0	0.12	1086.0
0.40	1396.3	0.16	1151.0	0.14	1102.0
		0.18	1173.0	0.16	1117.0
K$_2$Cr$_2$O$_7$ (20°C)		0.20	1196.0	0.18	1133.0
				0.20	1151.0
0.01	1005.2	**KHCO$_3$ (15°C)**		0.22	1167.0
0.02	1012.2			0.24	1184.0
0.03	1019.3	0.01	1005.8	0.26	1201.0
0.04	1026.4	0.02	1012.5	0.27	1211.0
0.05	1033.6	0.04	1026.0		
0.06	1040.8	0.06	1039.6	**KI (20°C)**	
0.07	1048.1	0.08	1053.4		
0.08	1055.4	0.10	1067.4	0.01	1005.5
0.09	1062.8			0.02	1013.0
0.10	1070.3	**KH$_2$PO$_4$ (10°C)**		0.04	1028.1
0.11	1077.9			0.06	1043.7
0.12	1085.5	0.01	1007.0	0.08	1059.7
		0.02	1014.2	0.10	1076.1
KCr(SO$_4$)$_2$ (15°C)		0.04	1028.4	0.12	1093.0
		0.06	1042.5	0.14	1110.4
0.01	1008.6	0.08	1056.7	0.16	1128.4
0.02	1018.2	0.10	1071.1	0.20	1166.0
0.04	1037.6	0.12	1086.3	0.24	1206.0
0.06	1057.3	0.14	1102.8	0.28	1248.7
0.08	1077.3			0.35	1330.8
		KHS (18°C)		0.40	1395.9
KF (18°C)				0.50	1545.8
		0.01	1004.5	0.60	1731.0
0.01	1007.2	0.02	1010.5		
0.02	1015.9	0.04	1022.4		
0.04	1033.4	0.06	1034.3		
		0.08	1046.3		

w_B	$\rho_{(soln)}$, g/l	w_B	$\rho_{(soln)}$, g/l	w_B	$\rho_{(soln)}$, g/l
		0.02	1011.0	0.28	1205.9
KIO$_3$ (18°C)		0.04	1024.0	0.32	1239.4
0.01	1007.7	0.06	1037.0	0.36	1274.2
0.02	1015.1	0.08	1049.0	KOH (20°C)	
0.03	1024.5	0.10	1062.0		
0.04	1033.3	0.12	1075.0	0.002	1000.0
0.05	1042.4	0.14	1088.0	0.024	1020.0
0.06	1051.5	0.16	1102.0	0.046	1040.0
KMnO$_4$ (15°C)		0.18	1116.0	0.067	1060.0
		0.22	1144.0	0.089	1080.0
0.01	1006.0	0.26	1172.0	0.110	1100.0
0.02	1013.0	0.30	1203.0	0.131	1120.0
0.03	1020.0	0.35	1242.0	0.152	1140.0
0.04	1027.1	0.40	1284.0	0.173	1160.0
0.05	1034.2	0.50	1378.0	0.194	1180.0
0.06	1041.4	0.60	1484.0	0.214	1200.0
KN$_3$ (20°C)		0.70	1598.0	0.234	1220.0
		KNO$_3$ (20°C)		0.254	1240.0
0.01	1003.0			0.273	1260.0
0.02	1009.0	0.01	1004.5	0.293	1280.0
0.04	1020.0	0.02	1010.8	0.312	1300.0
0.06	1031.0	0.04	1023.4	0.330	1320.0
0.08	1043.0	0.06	1036.3	0.349	1340.0
0.10	1055.0	0.08	1049.4	0.367	1360.0
0.12	1066.0	0.10	1062.7	0.386	1380.0
0.14	1072.0	0.12	1076.2	0.404	1400.0
0.16	1091.0	0.14	1089.9	0.422	1420.0
0.20	1116.0	0.16	1103.0	0.439	1440.0
0.24	1141.0	0.18	1118.1	0.457	1460.0
0.30	1180.0	0.20	1132.6	0.474	1480.0
KNCS (18°C)		0.22	1147.3	0.491	1500.0
		0.24	1162.3	0.508	1520.0
0.01	1003.5	KNa(C$_4$H$_4$O$_6$) (20°C)		0.521	1535.0
0.02	1008.5			K$_2$S (18°C)	
0.06	1028.8	0.01	1004.9		
0.10	1049.5	0.02	1011.6	0.01	1009.0
0.14	1070.8	0.04	1025.2	0.02	1017.0
0.18	1092.7	0.06	1039.0	0.04	1033.0
0.22	1115.2	0.08	1053.0	0.06	1049.0
0.26	1138.2	0.10	1067.3	0.08	1066.0
0.30	1161.8	0.12	1081.8	0.10	1083.0
0.40	1220.0	0.14	1096.5	0.12	1100.0
0.50	1284.9	0.16	1111.4	0.14	1118.0
0.60	1355.4	0.18	1126.5	0.16	1136.0
0.70	1430.7	0.20	1141.9	0.18	1154.0
KNO$_2$(17.5°C)		0.22	1157.6	0.20	1173.0
		0.24	1173.5	0.24	1211.0
0.01	1005.0	0.26	1189.6	0.28	1250.0

(*Continued*)

w_B	$\rho_{(soln)}$, g/l	w_B	$\rho_{(soln)}$, g/l	w_B	$\rho_{(soln)}$, g/l
0.35	1320.0				
0.40	1372.0				
0.45	1432.0				

K_2SO_3 (15°C) — column 1

w_B	$\rho_{(soln)}$, g/l
0.01	1007.0
0.02	1016.0
0.04	1032.0
0.06	1049.0
0.08	1067.0
0.10	1085.0
0.12	1103.0
0.14	1121.0
0.16	1140.0
0.18	1160.0
0.20	1179.0
0.22	1199.0
0.24	1220.0
0.26	1240.0

K_2SO_4 (20°C)

w_B	$\rho_{(soln)}$, g/l
0.01	1006.0
0.02	1014.0
0.03	1022.0
0.04	1031.0
0.06	1039.0
0.06	1047.0
0.07	1056.0
0.08	1064.0
0.09	1073.0
0.10	1081.0

K_2SiO_3 (20°C)

w_B	$\rho_{(soln)}$, g/l
0.01	1007.0
0.02	1016.0
0.04	1035.0
0.06	1054.0
0.08	1073.0
0.10	1092.0
0.12	1112.0
0.14	1133.0
0.16	1153.0
0.18	1175.0
0.20	1196.0
0.24	1241.0
0.28	1288.0

K_2WO_4 (15°C)

w_B	$\rho_{(soln)}$, g/l
0.01	1007.7
0.02	1016.4
0.04	1034.1
0.06	1052.3
0.08	1071.1
0.10	1090.5
0.12	1110.5
0.14	1131.2
0.16	1152.7
0.18	1175.0

$La(NO_3)_3$ (18°C)

w_B	$\rho_{(soln)}$, g/l
0.01	1007.6
0.02	1016.7
0.04	1035.3
0.06	1054.5
0.08	1074.2
0.10	1094.5
0.12	1115.3
0.14	1136.8
0.16	1158.9
0.18	1181.7
0.20	1205.2
0.22	1229.5
0.24	1254.7
0.26	1280.9
0.28	1308.0
0.30	1336.0

$LiBr$ (20°C)

w_B	$\rho_{(soln)}$, g/l
0.01	1005.5
0.02	1012.8
0.04	1027.7
0.06	1042.9
0.08	1058.5
0.10	1074.6
0.12	1091.0
0.14	1107.9
0.16	1125.3
0.18	1143.2
0.22	1180.6
0.26	1220.5
0.30	1262.9
0.35	1320.4
0.40	1383.6
0.45	1453.5

$LiCl$ (20°C)

w_B	$\rho_{(soln)}$, g/l
0.01	1004.1
0.02	1009.9
0.04	1021.5
0.06	1033.0
0.08	1044.4
0.10	1055.9
0.12	1067.5
0.14	1079.2
0.16	1091.0
0.18	1102.9
0.20	1115.0
0.24	1139.9
0.28	1165.8
0.32	1194.7
0.36	1224.0
0.40	1254.0
0.42	1269.0

LiI (20°C)

w_B	$\rho_{(soln)}$, g/l
0.01	1005.6
0.02	1013.1
0.04	1028.4
0.06	1044.2
0.08	1060.4
0.10	1077.1
0.12	1094.3
0.14	1112.0
0.16	1130.3
0.18	1149.2
0.22	1189.0
0.26	1231.5
0.30	1277.2
0.35	1339.3
0.40	1407.3
0.50	1569.2
0.60	1774.8

$LiNO_3$ (20°C)

w_B	$\rho_{(soln)}$, g/l
0.01	1004.1
0.02	1010.0
0.04	1022.0
0.06	1031.1
0.08	1046.5
0.10	1059.0
0.12	1071.8
0.14	1084.8

(Continued)

w_B	$\rho_{(soln)}$, g/l	w_B	$\rho_{(soln)}$, g/l	w_B	$\rho_{(soln)}$, g/l
0.16	1098.1		Mg(CH$_3$COO)$_2$ (18°C)	0.06	1044.1
0.18	1111.6			0.08	1060.0
0.20	1125.4	0.01	1004.0	0.10	1076.2
0.22	1139.5	0.02	1010.0	0.12	1092.8
0.26	1168.6	0.04	1022.0	0.14	1109.8
0.30	1198.8	0.06	1033.0	0.16	1127.2
0.35	1239.2	0.08	1045.0	0.18	1144.6
0.40	1283.7	0.10	1056.0	0.20	1163.0
				0.22	1181.5
	LiOH (20°C)		MgCl$_2$ (20°C)	0.24	1200.4
0.01	1010.2	0.02	1015.0		MgSO$_4$ (20°C)
0.02	1021.7	0.08	1065.4		
0.04	1043.7	0.14	1119.8	0.02	1018.6
0.06	1065.0	0.20	1175.7	0.04	1039.2
0.08	1086.2	0.26	1235.3	0.06	1060.2
0.10	1107.4	0.32	1297.9	0.08	1081.6
				0.10	1103.4
	Li$_2$SO$_4$ (20°C)		MgCrO$_4$ (18°C)	0.12	1125.6
0.01	1006.8	0.02	1018.0	0.14	1148.4
0.02	1015.5	0.04	1038.0	0.16	1171.7
0.04	1032.9	0.06	1059.0	0.18	1195.5
0.06	1050.5	0.08	1080.0	0.20	1219.8
0.08	1068.4	0.10	1102.0	0.22	1244.7
0.10	1086.3	0.14	1147.0	0.24	1270.1
0.12	1104.4	0.16	1170.0	0.26	1296.1
0.14	1122.8	0.20	1219.0		MnBr$_2$ (18°C)
0.16	1141.1		MgI$_2$ (20°C)		
0.18	1159.9			0.01	1007.1
0.20	1178.9	0.02	1014.9	0.02	1015.1
0.22	1198.4	0.04	1032.1	0.04	1033.2
0.24	1218.2	0.06	1049.8	0.06	1051.1
	MgBr$_2$ (20°C)	0.08	1078.0	0.08	1069.5
		0.10	1086.9	0.10	1088.6
0.02	1015.1	0.12	1106.5	0.12	1108.3
0.04	1032.4	0.14	1126.8	0.14	1128.7
0.06	1050.1	0.15	1137.3	0.16	1149.8
0.08	1068.3	0.18	1169.5	0.18	1171.6
0.10	1087.1	0.20	1192.0	0.20	1194.2
0.12	1106.5	0.25	1251.9	0.22	1217.6
0.14	1126.5	0.30	1318.0	0.24	1241.9
0.16	1147.1	0.35	1391.4	0.30	1320.6
0.18	1168.3	0.40	1473.7	0.32	1348.9
0.20	1190.3	0.50	1660.0		MnCl$_2$ (18°C)
0.25	1248.2	0.55	1776.0		
0.30	1311.0		Mg(NO$_3$)$_2$ (20°C)	0.01	1006.9
0.35	1379.0			0.02	1015.3
0.40	1452.0	0.02	1013.2	0.04	1032.4
0.45	1532.0	0.04	1028.5	0.06	1049.8

(Continued)

w_B	$\rho_{(soln)}$, g/l	w_B	$\rho_{(soln)}$, g/l	w_B	$\rho_{(soln)}$, g/l
0.08	1067.6	0.28	1327.7	1.00	1011.0
0.10	1085.9	0.30	1356.5		
0.12	1104.6			NH₄Br (18°C)	
0.14	1123.8	NH₃ (20°C)			
0.16	1143.5			0.01	1004.3
0.18	1163.8	0.005	0.996	0.02	1010.0
0.20	1184.6	0.019	0.990	0.04	1021.5
0.22	1206.1	0.038	0.982	0.06	1033.2
0.24	1228.3	0.058	0.974	0.08	1045.1
0.26	1251.1	0.078	0.966	0.10	1057.2
0.28	1274.6	0.099	0.958	0.14	1082.2
0.30	1298.8	0.120	0.950	0.18	1108.1
		0.143	0.942	0.22	1135.2
Mn(NO₃)₂ (18°C)		0.167	0.934	0.26	1163.5
		0.191	0.926	0.30	1193.3
0.01	1006.3	0.215	0.918	0.34	1224.7
0.02	1014.0	0.240	0.910		
0.04	1029.8	0.267	0.902	NH₄(CH₃COO) (18°C)	
0.06	1045.9	0.293	0.894		
0.08	1062.4	0.321	0.886	0.01	1000.8
0.10	1079.4	0.344	0.880	0.02	1003.0
0.12	1096.9			0.04	1007.4
0.14	1114.9	N₂H₄ (15°C)		0.06	1011.7
0.16	1133.3			0.08	1015.9
0.18	1152.2	0.01	1000.2	0.10	1020.0
0.20	1171.7	0.02	1001.3	0.12	1024.0
0.24	1212.5	0.04	1003.4	0.14	1027.9
0.28	1255.7	0.06	1005.6	0.16	1031.8
0.30	1278.1	0.08	1007.7	0.18	1035.6
0.35	1336.7	0.10	1009.9	0.20	1039.3
0.40	1399.3	0.12	1012.1	0.22	1042.9
0.50	1537.8	0.14	1014.3	0.24	1046.5
0.55	1614.6	0.16	1016.4	0.26	1050.0
		0.18	1018.6	0.28	1053.5
MnSO₄ (15°C)		0.22	1022.8	0.30	1056.9
		0.26	1026.7		
0.01	1008.9	0.30	1030.5	(NH₄)₂CO₃ (15°C)	
0.02	1018.8	0.35	1035.0		
0.04	1038.9	0.40	1038.0	0.01	1002.6
0.06	1059.5	0.45	1042.0	0.02	1006.1
0.08	1080.7	0.50	1044.0	0.04	1013.0
0.10	1102.5	0.55	1046.0	0.06	1019.0
0.12	1124.8	0.60	1047.0	0.08	1026.7
0.14	1147.8	0.65	1047.0	0.10	1033.5
0.16	1171.4	0.70	1046.0	0.12	1040.3
0.18	1195.6	0.75	1043.0	0.14	1047.1
0.20	1220.5	0.80	1040.0	0.16	1053.9
0.22	1246.1	0.85	1036.0	0.18	1060.7
0.24	1272.5	0.90	1030.0	0.20	1067.5
0.26	1299.7	0.95	1022.0	0.24	1080.8
				0.28	1094.0

(Continued)

w_B	$\rho_{(soln)}$, g/l	w_B	$\rho_{(soln)}$, g/l	w_B	$\rho_{(soln)}$, g/l
0.30	1100.6	0.12	1031.4	0.02	1006.4
0.35	1115.7	0.14	1036.6	0.04	1014.7
0.40	1129.4	0.16	1041.8	0.06	1023.0
0.45	1141.7	0.18	1046.9	0.08	1031.3
		0.22	1056.8	0.10	1039.7
NH$_4$Cl (20°C)		0.26	1066.5	0.12	1048.2
0.01	1001.3	0.30	1076.0	0.14	1056.7
0.02	1004.5	0.35	1087.4	0.16	1065.3
0.04	1010.7	0.40	1098.4	0.18	1074.0
0.06	1016.8	0.45	1108.9	0.20	1082.8
0.08	1022.7	0.50	1118.9	0.24	1100.5
0.10	1028.6			0.28	1118.6
0.12	1034.4	**NH$_4$I (18°C)**		0.30	1127.5
0.14	1040.1			0.35	1151.2
0.16	1045.7	0.01	1005.0	0.40	1175.4
0.18	1051.2	0.02	1011.4	0.50	1225.8
0.20	1056.7	0.04	1024.4	0.55	1252.0
0.22	1062.1	0.06	1037.7	0.60	1278.5
0.26	1072.6	0.10	1065.2		
		0.14	1094.2	**NH$_2$OH (20°C)**	
(N$_2$H$_6$)Cl$_2$ (20°C)		0.18	1124.8		
		0.22	1157.0	0.01	1000.2
0.01	1002.6	0.26	1190.8	0.02	1002.3
0.02	1007.0	0.30	1226.5	0.04	1006.5
0.04	1015.8	0.35	1274.5	0.06	1010.5
0.06	1024.6	0.40	1326.4	0.08	1014.9
0.08	1033.4	0.50	1442.3	0.10	1019.2
0.10	1042.2			0.12	1023.5
0.12	1050.9	**NH$_4$NCS (18°C)**		0.14	1027.8
0.14	1059.6			0.16	1032.2
0.16	1068.3	0.01	1000.9	0.18	1036.6
0.18	1077.0	0.02	1003.2	0.22	1045.4
		0.04	1007.8	0.26	1054.4
NH$_4$F (18°C)		0.06	1012.4	0.28	1059.1
		0.08	1017.0	0.30	1063.7
0.01	1003.4	0.10	1021.6	0.35	1075.5
0.02	1008.5	0.12	1026.3	0.40	1087.5
0.04	1017.8	0.14	1030.9	0.45	1099.7
0.06	1026.5	0.16	1035.6	0.50	1112.2
0.08	1034.6	0.18	1040.2	0.55	1124.9
0.12	1048.7	0.22	1049.5		
0.14	1054.7	0.26	1058.9	**(NH$_3$OH)Cl (17°C)**	
		0.30	1064.5		
NH$_4$(HCOO) (15°C)		0.38	1081.8	0.01	1004.6
		0.46	1100.7	0.02	1008.4
0.01	1001.9	0.50	1110.8	0.04	1016.7
0.02	1004.6	0.58	1132.2	0.06	1025.3
0.04	1010.1			0.08	1034.0
0.06	1015.5	**NH$_4$NO$_3$ (20°C)**		0.10	1043.7
0.08	1020.9			0.15	1065.5
0.10	1026.2	0.01	1002.3		

(*Continued*)

w_B	$\rho_{(soln)}$, g/l	w_B	$\rho_{(soln)}$, g/l	w_B	$\rho_{(soln)}$, g/l
0.20	1088.8	0.16	1135.2	0.135	1140.0
0.25	1112.6	0.20	1174.5	0.152	1160.0
		0.24	1216.3	0.169	1180.0
$(NH_4)_2SO_4$ (20°C)		0.28	1260.8	0.177	1190.0
0.01	1004.1	0.30	1284.1		
0.02	1010.1	0.35	1346.2	$Na_2C_2O_4$ (20°C)	
0.04	1022.0	0.40	1413.8	0.01	1006.4
0.06	1033.8			0.02	1014.7
0.08	1045.6	$NaBrO_3$ (20°C)		0.03	1022.9
0.10	1057.4			0.04	1031.2
0.12	1069.1	0.01	1006.4		
0.14	1080.8	0.02	1014.7	NaCl (20°C)	
0.16	1092.4	0.04	1030.5		
0.18	1103.9	0.06	1047.1	0.01	1005.3
0.20	1115.4	0.08	1064.1	0.02	1012.5
0.24	1138.3	0.10	1081.6	0.04	1026.8
0.28	1160.9	0.12	1099.6	0.06	1041.3
0.30	1172.1	0.14	1118.3	0.08	1055.9
0.35	1200.0	0.16	1137.3	0.10	1070.7
0.40	1227.7	0.18	1156.9	0.12	1085.7
0.50	1282.5	0.20	1177.1	0.14	1100.9
		0.22	1197.9	0.16	1116.2
Na_3AsO_4 (17°C)		0.24	1219.3	0.18	1131.9
				0.20	1147.8
0.01	1009.7	$Na(CH_3COO)$ (18°C)		0.22	1164.0
0.02	1020.7			0.24	1180.4
0.04	1043.1	0.01	1003.3	0.26	1197.2
0.06	1065.9	0.02	1008.4		
0.08	1089.2	0.04	1018.6	NaClO (18°C)	
0.10	1113.3	0.06	1028.9		
0.12	1137.3	0.08	1039.2	0.01	1005.3
		0.10	1049.5	0.02	1012.1
$Na_2B_4O_7$ (15°C)		0.12	1059.8	0.04	1025.8
		0.14	1070.2	0.06	1039.7
0.005	1004.2	0.16	1080.7	0.08	1053.8
0.010	1008.4	0.18	1091.3	0.10	1068.1
0.015	1013.1	0.20	1102.1	0.14	1097.7
0.020	1017.9	0.22	1113.0	0.18	1128.8
0.025	1022.6	0.24	1124.0	0.22	1161.4
0.030	1027.4	0.26	1135.1	0.26	1195.3
0.035	1032.1	0.28	1146.2	0.30	1230.7
				0.34	1268.0
NaBr (20°C)		Na_2CO_3 (20°C)		0.38	1308.5
				0.40	1328.5
0.01	1006.0	0.002	1000.0		
0.02	1013.9	0.021	1020.0	$NaClO_3$ (18°C)	
0.04	1029.8	0.040	1040.0		
0.06	1046.2	0.060	1060.0	0.01	1005.3
0.08	1063.1	0.079	1080.0	0.02	1012.1
0.10	1080.3	0.098	1100.0	0.04	1025.8
0.12	1098.1	0.116	1120.0	0.06	1039.7

(Continued)

w_B	$\rho_{(soln)}$, g/l	w_B	$\rho_{(soln)}$, g/l	w_B	$\rho_{(soln)}$, g/l
0.08	1053.8	0.08	1056.0	0.10	1063.0
0.10	1068.1	0.10	1070.0	0.12	1076.2
0.14	1097.7	0.12	1084.0	0.14	1089.5
0.18	1128.8	0.14	1098.0	0.16	1102.9
0.22	1161.4	0.16	1112.0	0.18	1116.4
0.26	1195.3	0.18	1126.0	0.20	1130.0
0.30	1230.7	0.22	1153.0	0.22	1143.9
0.34	1268.0	0.26	1179.0	0.24	1158.0
0.38	1308.5	0.30	1207.0		
0.40	1328.5	0.35	1244.0	NaH_2PO_4 (25°C)		
		0.40	1279.0		
$NaClO_4$ (18°C)		0.45	1312.0	0.01	1004.5
		0.50	1342.0	0.02	1012.0
0.01	1005.1			0.04	1027.0
0.02	1011.6	NaF (18°C)		0.06	1042.2
0.04	1024.7			0.08	1057.5
0.06	1038.1	0.01	1009.2	0.10	1073.0
0.08	1051.7	0.02	1019.8		
0.10	1065.6	0.03	1030.4	Na_2HPO_4 (18°C)		
0.12	1079.8	0.04	1040.9		
0.14	1094.3	0.05	1051.5	0.01	1009.0
0.18	1124.1			0.02	1020.0
0.22	1155.4	Na_2HAsO_4 (14°C)		0.03	1031.0
0.26	1188.3			0.04	1043.0
0.30	1222.7	0.01	1009.0	0.05	1055.0
0.34	1259.1	0.02	1017.5	0.06	1067.0
0.38	1296.9	0.04	1035.5		
		0.06	1053.3	$NaHSO_3$ (15°C)		
Na_2CrO_4 (18°C)		0.08	1075.5		
		0.10	1096.4	0.02	1017.0
0.01	1007.4	0.12	1118.0	0.04	1044.0
0.02	1016.3	0.14	1140.6	0.06	1063.0
0.04	1034.4	0.16	1163.5	0.08	1084.0
0.06	1052.9			0.10	1104.0
0.08	1071.8	$NaHCO_3$ (18°C)		0.12	1124.0
0.10	1091.2			0.14	1144.0
0.12	1111.0	0.01	1005.9	0.16	1165.0
0.14	1131.2	0.02	1013.2	0.18	1185.0
0.16	1151.8	0.03	1020.6	0.20	1202.0
0.18	1172.8	0.04	1028.0	0.24	1235.0
0.20	1194.2	0.05	1035.4	0.28	1268.0
0.22	1216.0	0.06	1042.9	0.32	1300.0
0.24	1238.3	0.07	1050.5	0.36	1330.0
0.26	1261.1	0.08	1058.1	0.38	1345.0
$Na_2Cr_2O_7$ (15°C)		Na(HCOO) (18°C)		$NaHSO_4$ (20°C)		
		0.01	1004.9	0.01	1005.9
0.01	1006.0	0.02	1011.2	0.02	1013.7
0.02	1013.0	0.04	1023.9	0.04	1029.3
0.04	1027.0	0.06	1036.8	0.06	1045.1
0.06	1041.0	0.08	1049.8	0.08	1061.1

(*Continued*)

w_B	$\rho_{(soln)}$, g/l	w_B	$\rho_{(soln)}$, g/l	w_B	$\rho_{(soln)}$, g/l
0.10	1077.3	0.16	1099.0	0.12	1081.9
0.12	1093.7	0.18	1112.0	0.14	1096.7
0.14	1110.3	0.20	1126.0	0.16	1111.8
0.16	1127.1	0.22	1140.0	0.18	1127.2
0.18	1144.1	0.24	1155.0	0.20	1142.6
0.20	1161.4	0.30	1202.0	0.24	1175.2
0.22	1178.9			0.28	1208.5
		NaNCS (18°C)		0.30	1225.6
NaI (20°C)				0.35	1270.1
		0.01	1003.8	0.40	1317.5
0.01	1006.0	0.02	1009.0	0.45	1368.3
0.02	1013.8	0.04	1019.6		
0.04	1029.8	0.06	1030.3	NaOH (20°C)	
0.06	1046.3	0.08	1041.1		
0.08	1063.3	0.10	1052.0	0.002	1000.0
0.10	1080.8	0.14	1074.1	0.019	1020.0
0.12	1098.8	0.18	1096.6	0.037	1040.0
0.16	1136.6	0.22	1119.7	0.056	1060.0
0.20	1176.9	0.26	1143.3	0.074	1080.0
0.24	1220.1	0.30	1167.7	0.092	1100.0
0.28	1266.3	0.35	1196.0	0.110	1120.0
0.35	1355.6	0.40	1228.0	0.128	1140.0
0.40	1427.1	0.45	1265.0	0.146	1160.0
0.50	1594.2			0.164	1180.0
0.60	1803.8	NaNO$_2$ (20°C)		0.183	1200.0
		0.01	1005.0	0.201	1220.0
Na$_2$MoO$_4$ (15°C)		0.02	1011.0	0.219	1240.0
		0.04	1024.0	0.237	1260.0
0.01	1007.8	0.06	1038.0	0.256	1280.0
0.02	1016.5	0.08	1052.0	0.274	1300.0
0.04	1034.3	0.10	1065.0	0.293	1320.0
0.06	1052.6	0.12	1078.0	0.311	1340.0
0.08	1071.3	0.14	1092.0	0.331	1360.0
0.10	1090.5	0.16	1107.0	0.350	1380.0
0.12	1110.2	0.18	1122.0	0.370	1400.0
0.14	1130.4	0.20	1137.0	0.390	1420.0
0.16	1151.1	0.24	1168.0	0.410	1440.0
0.18	1172.4	0.28	1198.0	0.431	1460.0
0.20	1194.3	0.32	1230.0	0.452	1480.0
0.22	1216.8	0.36	1264.0	0.473	1500.0
		0.40	1299.0	0.494	1520.0
NaN$_3$ (20°C)				0.505	1530.0
		NaNO$_3$ (20°C)			
0.01	1004.0			Na$_3$PO$_4$ (15°C)	
0.02	1010.0	0.01	1004.9		
0.04	1022.0	0.02	1011.7	0.01	1008.7
0.06	1034.0	0.04	1025.4	0.02	1019.4
0.08	1047.0	0.06	1039.2	0.03	1029.9
0.10	1059.0	0.08	1053.2	0.04	1040.5
0.12	1072.0	0.10	1067.4	0.05	1051.5
0.14	1085.0				

(Continued)

w_B	$\rho_{(soln)}$, g/l	w_B	$\rho_{(soln)}$, g/l	w_B	$\rho_{(soln)}$, g/l
0.06	1062.4	0.12	1100.3	0.04	1035.9
0.07	1073.7	0.14	1118.2	0.06	1055.5
0.08	1085.0	0.16	1136.5	0.08	1075.8
0.09	1096.2	0.18	1155.1	0.10	1096.9
0.10	1108.3	0.20	1174.0	0.12	1118.8
		0.24	1212.8	0.14	1141.4
$Na_4P_2O_7$ (20°C)		0.28	1253.2	0.16	1164.8
0.01	1009.0	0.30	1273.9	0.18	1188.9
0.02	1019.0	0.35	1327.9	0.20	1214.0
0.03	1028.0	0.40	1382.7	0.25	1281.5
0.04	1037.0	Na_2SiO_3 (18°C)		$NiCl_2$ (18°C)	
Na_2SO_3 (19°C)		0.01	1009.4	0.01	1008.0
0.01	1007.8	0.02	1020.3	0.02	1018.0
0.02	1017.6	0.04	1042.5	0.04	1038.0
0.04	1036.0	0.06	1065.2	0.06	1058.0
0.06	1055.6	0.08	1088.4	0.08	1079.0
0.08	1075.1	0.10	1112.2	0.10	1100.0
0.10	1094.8	0.12	1136.5	0.12	1122.0
0.12	1114.6	0.14	1161.3	0.14	1144.0
0.14	1134.6	0.16	1186.6	0.16	1167.0
0.16	1154.9	0.18	1212.3	0.18	1191.0
0.18	1175.5	0.20	1238.5	0.20	1216.0
Na_2SO_4 (20°C)		0.22	1265.3	$Ni(NO_3)_2$ (18°C)	
		0.24	1292.6		
0.01	1007.9	0.26	1320.4	0.01	1007.0
0.02	1016.4			0.02	1016.0
0.03	1025.6	Na_2WO_4 (20°C)		0.04	1033.0
0.04	1034.8	0.01	1007.4	0.06	1051.0
0.05	1044.1	0.02	1016.6	0.08	1069.0
0.06	1053.5	0.04	1035.4	0.10	1088.0
0.07	1062.9	0.06	1054.5	0.12	1108.0
0.08	1072.4	0.08	1074.2	0.14	1128.0
0.09	1081.9	0.10	1094.4	0.16	1148.0
0.10	1091.5	0.12	1115.4	0.18	1169.0
0.11	1101.2	0.14	1137.2	0.20	1191.0
0.12	1110.9	0.16	1159.8	0.25	1249.0
0.13	1120.7	0.18	1183.3	0.30	1311.0
0.14	1130.6	0.20	1207.6	0.35	1378.0
0.16	1150.6	0.22	1232.8	$NiSO_4$ (18°C)	
Na_2SO_3S (20°C)		0.26	1286.2		
		0.30	1344.4	0.01	1009.0
0.01	1006.5	0.34	1408.4	0.02	1020.0
0.02	1014.8	0.38	1478.6	0.04	1042.0
0.04	1031.5	$NiBr_2$ (18°C)		0.06	1063.0
0.06	1048.3			0.08	1085.0
0.08	1065.4	0.01	1007.8	0.10	1109.0
0.10	1082.7	0.02	1017.0	0.12	1133.0

(Continued)

w_B	$\rho_{(soln)}$, g/l	w_B	$\rho_{(soln)}$, g/l	w_B	$\rho_{(soln)}$, g/l
0.14	1158.0		RbCl (20°C)	0.12	1089.2
0.16	1183.0			0.14	1105.7
0.18	1209.0	0.01	1005.6	0.16	1122.7
		0.02	1013.0	0.18	1140.1
Pb(CH₃COO)₂ (18°C)		0.04	1043.8	0.20	1158.0
		0.06	1059.7	0.22	1176.3
0.01	1006.1	0.08	1076.0		
0.02	1013.7	0.10	1092.8		RbOH (18°C)
0.04	1029.0	0.12	1110.0	0.01	1008.0
0.06	1044.6	0.14	1127.8	0.02	1017.4
0.08	1060.5	0.16	1145.9	0.04	1036.8
0.10	1076.8	0.18	1164.7	0.06	1056.8
0.12	1093.6	0.20	1183.9	0.08	1077.4
0.14	1110.9	0.22	1203.8	0.10	1098.7
0.16	1128.3	0.24	1266.9	0.12	1120.6
0.18	1147.3	0.30	1324.1	0.14	1143.2
0.20	1166.3	0.35	1385.9		
0.24	1206.3	0.40	1453.3		Rb₂SO₄ (20°C)
0.28	1248.9	0.50	1526.8	0.01	1006.6
0.30	1271.1			0.02	1015.0
0.35	1330.4		RbI (20°C)	0.04	1032.2
0.40	1399.4			0.06	1049.9
		0.01	1005.9	0.08	1068.0
Pb(NO₃)₂ (15°C)		0.02	1013.7	0.10	1086.4
		0.04	1029.6	0.12	1105.2
0.05	1044.9	0.06	1046.0	0.14	1124.6
0.10	1093.7	0.08	1062.9	0.16	1144.6
0.15	1146.7	0.10	1080.4	0.18	1165.2
0.25	1267.8	0.12	1098.4	0.20	1186.4
0.30	1335.8	0.14	1117.0	0.22	1208.3
		0.16	1136.2	0.26	1254.2
RbBr (20°C)		0.18	1156.0	0.30	1302.8
		0.20	1176.6		
0.01	1005.9	0.22	1197.8		SO₂ (15°C)
0.02	1013.7	0.24	1219.8	0.005	1002.8
0.04	1029.7	0.30	1290.6	0.010	1005.6
0.06	1046.0	0.35	1355.9	0.015	1008.5
0.08	1062.9	0.40	1428.1	0.020	1011.3
0.10	1080.3	0.45	1507.9	0.025	1014.1
0.12	1098.2	0.50	1596.9	0.030	1016.8
0.14	1116.6	0.60	1809.2	0.035	1019.4
0.16	1135.6			0.040	1022.1
0.18	1155.3		RbNO₃ (20°C)	0.045	1024.8
0.20	1175.5			0.050	1027.5
0.22	1196.5	0.01	1005.3	0.055	1030.2
0.24	1218.2	0.02	1012.5	0.060	1032.8
0.30	1287.8	0.04	1027.2	0.065	1035.3
0.35	1351.9	0.06	1042.2	0.070	1037.7
0.40	1422.3	0.08	1057.5		
0.45	1500.1	0.10	1073.1		
0.50	1586.4				

(*Continued*)

w_B	$\rho_{(soln)}$, g/l	w_B	$\rho_{(soln)}$, g/l	w_B	$\rho_{(soln)}$, g/l
				0.02	1016.9
[SiF$_6$],H$_2$ (17.5°C)		SrCl$_2$ (20°C)		0.04	1035.4
				0.06	1054.6
0.01	1007.0	0.02	1016.1	0.08	1074.7
0.02	1015.0	0.04	1034.4	0.10	1095.7
0.04	1031.0	0.06	1053.2	0.12	1117.6
0.06	1048.0	0.08	1072.6	0.14	1140.4
0.08	1065.0	0.10	1092.5	0.16	1164.0
0.10	1082.0	0.12	1113.0	0.18	1188.5
0.12	1100.0	0.14	1134.1		
0.14	1117.0	0.16	1155.8	TlNO$_3$ (25°C)	
0.16	1136.0	0.18	1178.1		
0.18	1154.0	0.20	1201.0	0.01	1005.6
0.20	1173.0	0.25	1260	0.02	1014.2
0.24	1212.0	0.30	1325	0.03	1023.0
0.28	1252.0	0.35	1396	0.04	1031.9
0.32	1293.0			0.05	1040.9
0.34	1314.0	SrI$_2$ (20°C)		0.06	1050.1
				0.07	1059.4
[Sn(OH)$_6$],Na$_2$ (20°C)		0.02	1015.4		
		0.04	1033.1	Tl$_2$SO$_4$ (20°C)	
0.013	1006.0	0.06	1051.3		
0.025	1015.0	0.08	1070.1	0.01	1007.6
0.050	1033.0	0.10	1089.6	0.02	1017.0
0.075	1051.0	0.12	1109.9	0.03	1026.5
0.100	1069.0	0.14	1130.8	0.04	1036.0
0.125	1088.0	0.16	1152.6	0.05	1045.6
0.151	1107.0	0.18	1175.3		
0.176	1126.0	0.20	1199.0	UO$_2$(CH$_3$COO)$_2$ (20°C)	
0.201	1146.0	0.30	1329.5		
0.226	1166.0	0.40	1490.4	0.01	1005.5
0.251	1187.0			0.02	1012.9
		Sr(NO$_3$)$_2$ (20°C)		0.03	1020.3
SrBr$_2$ (20°C)				0.04	1027.8
		0.02	1015.0		
0.02	1015.7	0.04	1031.0	UO$_2$(NO$_3$)$_2$ (17°C)	
0.04	1033.7	0.06	1048.0		
0.06	1052.2	0.08	1065.0	0.01	1008.0
0.08	1071.2	0.10	1083.0	0.02	1017.0
0.10	1090.7	0.12	1101.0	0.04	1036.0
0.12	1110.9	0.14	1119.0	0.06	1054.0
0.14	1131.7	0.16	1138.0	0.08	1072.0
0.16	1153.2	0.18	1158.0	0.10	1091.0
0.18	1175.7	0.20	1179.0	0.12	1110.0
0.20	1199.2	0.25	1233.0	0.14	1129.0
0.25	1262.0	0.30	1290.0	0.16	1149.0
0.30	1330.0	0.35	1352.0	0.18	1171.0
0.35	1405.0	0.40	1419.0	0.20	1194.0
0.40	1489.0			0.22	1218.0
0.50	1686.0	Th(NO$_3$)$_4$ (15°C)		0.24	1243.0
				0.30	1322.0
		0.01	1007.9	0.40	1466.0

(Continued)

w_B	$\rho_{(soln)}$, g/l	w_B	$\rho_{(soln)}$, g/l	w_B	$\rho_{(soln)}$, g/l
0.50	1649.0	0.18	1166.5	0.04	1032.2
		0.20	1186.6	0.06	1049.6
ZnBr$_2$ (20°C)		0.25	1238.0	0.08	1067.5
		0.30	1292.8	0.10	1085.9
0.02	1016.7	0.40	1417.3	0.12	1104.8
0.04	1035.4	0.50	1568.1	0.14	1124.4
0.06	1054.4	0.60	1749.0	0.16	1144.5
0.08	1073.8	0.70	1962.0	0.18	1165.2
0.10	1093.5			0.20	1186.5
0.12	1113.5	ZnI$_2$ (20°C)		0.25	1242.7
0.14	1138.8			0.30	1302.9
0.16	1154.4	0.02	1016.0	0.35	1367.8
0.18	1175.3	0.04	1034.0	0.40	1437.8
0.20	1196.5	0.06	1053.0	0.50	1594.4
0.25	1254.3	0.08	1072.0		
0.30	1317.0	0.10	1091.0	ZnSO$_4$ (20°C)	
0.35	1385.9	0.12	1111.0		
0.40	1462.0	0.14	1131.0	0.02	1019.0
0.50	1643.0	0.16	1152.0	0.04	1040.3
		0.18	1174.0	0.06	1062.0
ZnCl$_2$ (20°C)		0.20	1197.0	0.08	1084.2
		0.25	1258.0	0.10	1107.1
0.02	1016.7	0.30	1325.0	0.12	1130.8
0.04	1035.0	0.35	1398.0	0.14	1155.3
0.06	1053.2	0.40	1478.0	0.16	1180.6
0.08	1071.5	0.45	1566.0	0.20	1232.0
0.10	1089.9			0.25	1304.0
0.12	1108.5	Zn(NO$_3$)$_2$ (18°C)		0.30	1378.0
0.14	1127.5				
0.16	1146.8	0.02	1015.4		

6

Nomenclature

6.1 Chemical Elements [1, 28, 31–37]

The symbols and names of chemical elements are listed in the Periodic Table on the fly-leaf. Latin stems used for naming compounds of some elements:

Ag	argent(um)
Au	aur(um)
Cu	cupr(um)
Fe	ferr(um)
Ni	niccol(um)
Pb	plumb(um)
Sb	stib(ium)
Sn	stann(um)
W	wolfram(ium)

All chemical elements are conventionally divided into metals and non-metals. The following elements are *non-metals*:

He Ne Ar Kr Xe Rn F Cl Br I At O S
Se Te N P As C Si B H

all other elements being *metals*.

The elements, in which the valence electron fills the *s*-orbital, are designated as the *s*-elements. The names *p*-, *d*- and *f*-elements are derived in a similar manner.

The following group names of elements are currently employed:

actinoids	Ac Th Pa U Np Pu Am Cm Bk Cf Es Fm Md No Lr
noble gases	He Ne Ar Kr Xe Rn
halogens	F Cl Br I At
lanthanoids	La Ce Pr Nd Pm Sm Eu Gd Tb Dy Ho Er Tm Yb Lu
chalcogens	O S Se Te Po
alkali metals	Li Na K Rb Cs Fr
alkaline-earth metals	Ca Sr Ba Ra
iron-group metals	Fe Co Ni
platinum metals	Ru Rh Pd Os Ir Pt

6.2 Elementary Substances and Compounds [31, 32, 35, 37]

The nomenclature of inorganic compounds that relies on the chemical language to convey information on the composition of substances consists of two mutually augmenting parts – chemical formulas and names.

The *formula* presents the composition of substances and compounds by the symbols of chemical elements and also numerals, letters and other signs. The *name* is a presentation of the composition of substances by a word or a group of words.

Chemical formulas and names are drawn up according to the system of nomenclature rules.

6.2.1 The Rules of Constructing Formulas

Each substance is designated by a unique formula that should invariably be written in one and the same manner so as to represent fully and adequately the composition of a given substance (in some instances, the formula also shows the mutual arrangement of atoms in a molecule as discussed below). For example:

$$Al \quad P_4 \quad S_8 \quad HgCl_2 \quad Hg_2Cl_2 \quad NH_3 \quad (Pb_2^{II}Pb^{IV})O_4 \quad Na_3PO_4$$
$$K_3[Cr(OH)_6] \quad SO_2{\cdot}nH_2O$$

The formulas of simple (single-element) substances are represented by the symbol of a respective element, wherein the right lower index (an Arabic numeral) denotes the number of atoms in the molecule in question, the number 1 being omitted, e.g.

$$Na \quad Sb \quad Xe \quad Li_2 \quad O_2 \quad P_4$$

PRACTICAL ROW OF ELEMENTS

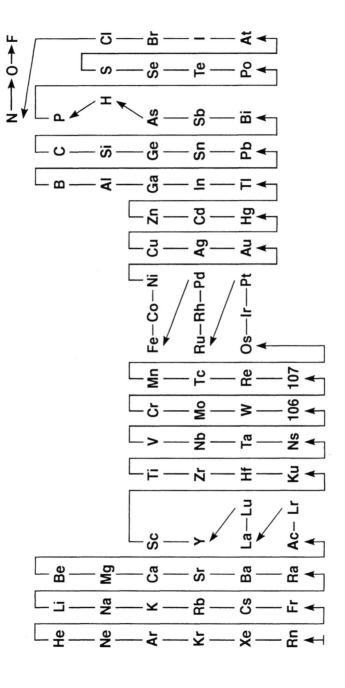

In the chemical formulas of compounds (multielement substances) the electropositive constituent (real or arbitrary cation) shall always be placed first, followed by the electronegative constituent (real or arbitrary anion), e.g.,

$$SF_6 \quad (Fe^{II}Fe_2^{III})O_4 \quad MgCl(OH) \quad [Cr(CO)_6] \quad AuCu_3$$

The chemical formulas of binary compounds designated in the general form as A_aB_b correspond to the arrangement of the symbols of the elements in the so-called Practical Row that essentially coincides with the arrangement of the elements in the Periodic Table of Elements with due regard to their electronegativity. Element A (the electropositive constituent of the formula) is invariably confined to the beginning of the Practical Row (on the left-hand side of the drawing), while Element B (the electronegative constituent of the formula) is any element located in the row further to the right-hand side thereof, as exemplified by the following formulas:

$$Li_2S \quad Al_2Cl_6 \quad MoO_3 \quad Cl_3N \quad P_4O_{10} \quad CuAl_2 \quad FeNi_3$$

In the formulas of the compounds containing two or more metallic or nonmetallic elements that form two or more independent constituents from the same group of elements, the symbols of elements should be arranged in a similar manner, viz.:

$$MgIBr \quad CrBr_2O_2 \quad Bi_2TeSe_2 \quad SOF_4 \quad Ni_3Bi_2S_2 \quad (CaTi)O_3 \quad Fe_6Co_2Ni$$

Traditionally, an exception from this rule is provided by the formulas of the nitrogen and phosphorus compounds with hydrogen and their derivatives:

$$NH_3 \quad ND_3 \quad N_2H_4 \quad N_2D_4 \quad NH_2Cl \quad NH_2OH \quad PH_3 \quad P_2H_4 \quad (PH)F_2$$

In the formulas of real multielement cations and anions, i.e. those existing in ionic crystals or solutions, the sequence of the ions in question is established by the rules of nomenclature for complex compounds (see Section 6.3.1), said rules being compulsory. Examples:

$$H_3O^+ \quad NO_2^+ \quad NH_4^+ \quad UO_2^{2+} \quad OH^- \quad CO_3^{2-} \quad CN^- \quad ClO_4^- \quad PHO_3^{2-}$$

The formulas of multicomponent substances containing such cations and/or anions comprise the formulas of these ions, wherein the sequence of components of the same kind (two dissimilar polyatomic cations or anions) shall be determined according to the practical row of elements on the basis of the first (left-hand) elements in the formulas of the components in question. Examples:

$$K(HF_2) \quad Fe(OH)_2 \quad Na_2(PHO_3) \quad H_2CO_3 \quad HMnO_4 \quad NH_4CN$$
$$UO_2(NO_3)_2 \quad (H_3O)ClO_4 \quad CaMg(CO_3)_2 \quad Fe(NH_4)_2(SO_4)_2$$
$$Ba_2(SeO_4)SO_4 \quad KAl_4(ASO_4)_3(OH)_4$$

Exceptions from this rule – a rare event in inorganic chemistry – pertain only to the formulas of chain compounds, wherein the sequence of elements conforms to the true structure of the molecules, particularly so in case the compounds are structural isomers. Thus, for the OCN^- anion the following compounds are known:

$$HOCN \quad HNCO \quad HCNO \quad Si(OCN)_4 \quad Si(NCO)_4 \quad AgOCN \quad AgNCO$$

Parentheses are used to indicate the number of polyatomic groups (if greater than unity), to separate cations from anions (once their composition is unobvious) and for isolating similar constituents, viz.:

$$(NH_4)_2S \quad Ca_3(PO_4)_2 \quad (CaTi)O_3 \quad Fe(S_2) \quad (NO)Cl \quad AlO(OH)$$

Parentheses should not be opened. Where double parentheses are required, use shall be made of a brace-parenthesis combination, such as $\{N(CH_3)_4\}_2SO_4$. Brackets are resorted to in order to isolate complexes.

The oxidation numbers of elements may be shown by the right upper index of the element symbol comprising Roman numerals (or 0) preceded by the + or − sign, but the + sign may be omitted, viz., $(Fe^{III}Cu^{I})S_2^{-II}$ and $K_4[Pd^0(CN)_4]$. When the oxidation numbers in compounds of the formula A_mM_a are ambiguous, the formula should be written as follows $M_m^v A_a^{-v}$. Indicating the oxidation numbers constitutes no obligatory nomenclature rule, but is often used to render the information provided by chemical formulas more succinct, e.g., in case the compounds contain the atoms of one and the same element in different oxidation states, such as $(Fe^{II}Fe_2^{III})O_4$ (using the empirical formula Fe_3O_4 a not recommended), and also in the case of complex anions, provided the charge of a counter ion in the compound is not obvious, e.g., in $Fe[Fe^{III}(CN)_6]$.

The ionic charge should be shown by Arabic numerals as the upper right index of the symbol, the numeral 1 being omitted, followed by the + or − sign, for example, Na^+, UO_2^{2+}, Cl^-, SO_4^{2-}. If the ionic charges are ambiguous, the notations of cations and anions will be M^{v+} and A^{v-}.

The formulas sometimes cited in the chemical literature either fail to conform to the nomenclature rules or else pertain to nonexistent substances and shall be regarded as obsolete. Their use in the modern nomenclature of inorganic substances is not recommended. The list of obsolete formulas is presented in Section 6.8.

6.2.2 The Rules of Composing Names

In compliance with the rules, the name of the compound represented by the formula should be read *from the left to the right*, viz.:

$HgCl_2$	mercury(II) chloride
Hg_2Cl_2	dimercury dichloride

Na_3PO_4 sodium orthophosphate
$K_3[Cr(OH)_6]$ potassium hexahydroxochromate(III)
$SO_2 \cdot nH_2O$ sulfur dioxide polyhydrate

The names of the majority of multicomponent substances consist of two words, viz., the name of the electropositive constituent (cation) which is left unmodified, and the name of the electronegative constituent (anion) which is changed to end in **-ide** in mononuclear anions and in **-ate** for polyatomic constituents (some traditional names fail to obey the latter rule), similarly to the names of free ions (see Section 6.2.3) but with the omission of the word **ion**. The names of electropositive constituents are analogous to the names of free cations (see Section 6.2.3) without the word **cation**. Exemplary two-word names are as follows:

SF_6 sulfur hexafluoride
SiO_2 silicon dioxide
Fe_3C triiron carbide
$Sn(OH)_2$ tin(II) hydroxide
$BaSO_4$ barium sulfate
NH_4VO_3 ammonium metavanadate
$Na_2S_2O_4$ disodium tetraoxodisulfate
$K_2Cr_3O_{10}$ potassium decaoxotrichromate(VI)

If two or more constituents of the same type are present, their names should be hyphenated, e.g.:

CSO   carbon sulfide-oxide
$S_2Cl_2O_5$ disulfur dichloride-pentaoxide
$(LiTmGe)O_4$ lithium-thulium-germanium tetraoxide
$K_2Mg(SO_4)_2$ dipotassium-magnesium sulfate

The systematic names of simple substances and numerous intermetallic compounds, the special names of common multicomponent substances, the names of complex compounds without the second sphere, and mineralogical names comprise one word only (see 6.2.3, 5.3.2, and 6.7), such as

S_8 octasulfur $AuCu_3$ goldtricopper
NH_3 ammonia $[Cr(CO)_6]$ hexacarbonylchromium

Numerical prefixes:

1	mono	5	penta	9	nona
2	di	6	hexa	10	deca
3	tri	7	hepta	11	undeca
4	tetra	8	octa	12	dodeca

and the indefinite numerals n (poly) are employed to indicate the number of electropositive and electronegative constituents in formulas. The terminal vowels being invariably retained in the names. Beyond 12, Greek prefixes are replaced by Arabic numerals (hyphenated), and the numerals are read as cardinal numbers. Examples:

B_4O_5	tetraboron pentaoxide
$Mn_2Cl(OH)_3$	dimanganese chloride-trihydroxide
S_nCl_2	polysulfur dichloride
$W_{20}O_{58}$	20-tungsten 58-oxide

The prefix mono is invariably omitted with the names of electropositive constituents (SCl_4 – sulfur tetrachloride instead of monosulfur tetrachloride), while other prefixes are not written in case the cation charge is obvious (Na_2S – sodium sulfide instead of disodium sulfide). In the names of electronegative constituents, the prefix mono is also omitted in the compounds, wherein the number of counterions is beyond unity (Br_3N – tribromine nitride instead of tribromine mononitride, and CCl_2O – carbon dichloride-oxide instead of monocarbon dichloride-oxide), and also in the case of compounds having at least two anions (one anion of each kind), for example, $PbClF$ – lead chloride-fluoride instead of lead monochloride-monofluoride. In all other instances the use of the prefix mono is compulsory, since otherwise the names of compounds become ambiguous (cf. CO – carbon monoxide, while the name carbon oxide is incomplete).

Multiplicative numerals:

1	monokis	4	tetrakis
2	bis	5	pentakis
3	tris	6	hexakis, etc.

shall be used instead of numerical prefixes when the latter, in combination with the name of a constituent, are likely to cause formula misinterpretation, or else when the name of the constituent includes numerical prefixes of a different significance; the names to which multiplicative prefixes refer should be placed in parentheses. Examples:

$KAl_3(SO_4)_2(OH)_6$	potassium-trialuminum bis(sulfate)-hexahydroxide
$[Os(C_6H_5)_2]$	bis(cyclopentadienyl)osmium

In the first example, the $(SO_4)_2$ constituent can not be named disulfate, in so far as this term is used to designate the ion $S_2O_7^{2-}$, while in the second example the name of the C_5H_5 – ligand itself contains the multiplicative prefix penta-.

The oxidation numbers of elements represented by Roman numerals shall be enclosed in parentheses immediately following the name of appropriate constituents (Stock's system). For example, the name of $(Pb_2^{II}Pb^{IV})O_4$ is dilead(II)-lead(IV) oxide. The oxidation numbers must be read as cardinal numerals. Oxidation numbers should be designated whenever the formula contains information thereon. Moreover, the Stock notation is recommended in place of numerical and multiplicative numerical prefixes in the names when the oxidation numbers may be calculated from the formula (generally for metallic electropositive constituents), so that the name becomes shorter. For example, Mn_2O_7 – manganese(VII) oxide instead of dimanganese heptaoxide, and $Cr_2(SO_4)_3$ – chromium(III) sulfate instead of dichromium tris(sulfate).

However, Stock's system is not versatile, since the oxidation numbers not always can be calculated from the formula, e.g., in the case of Fe_3C. Moreover this system of notations fails when in a formula the ratio of cation and anion numbers is intricate (2:2, 2:4, 4:10, etc.), e.g., for Hg_2Br_2 – dimercury dibromide that can not be designated as mercury(I) bromide, since the latter name is likewise valid for the HgBr formula.

The ion charge is indicated, according to the Ewens-Bassett system, in parentheses following the name. Examples: I_2^+ – diiodide(1+) cation and I_3^- – triiodide(1−) ion. The ion charges are read as cardinal numerals with an appropriate sign. The Ewens-Bassett notation is resorted to only in case other methods (the use of multiplicative prefixes or the citation of oxidation numbers) fail to provide information as regards the composition and structure of ions. This situation is usually observed with highly complicated ions, such as:

Bi_3^{3+}	tribismuth(3+) cation
$N_2H_5^+$	hydrazinium(1+) cation
$N_2H_6^{2+}$	hydrazinium(2+) cation
S_2^{2-}	disulfide(2−) ion
S_4^{2-}	tetrasulfide(2−) ion
$[Nb_6I_8]^{3+}$	octaiodohexaniobium(3+) cation

In the names of chain (catenary) compounds of identical composition, but having different sequences of atoms in the molecule, the sequence is indicated by the symbol of the atom, to which another constituent is bound. Example: HNCO – hydrogen cyanate-N in contrast to HOCN – hydrogen cyanate.

6.2.3 Kinds of Nomenclature Names

In the nomenclature of inorganic compounds, use is made of three kinds of names that may be composed for the single formula of a given compound.

Systematic names. These names fully and adequately represent the composition of a substance, are composed in conformity with the same set of rules as regards all substances, do not necessitate memorizing the chemical formulas and are, therefore, universal.

The systematic names of simple substances consist of the names of chemical elements and a numeral to show the number of atoms (in obvious instances, the prefix mono- shall be omitted):

Hg	mercury
Mg	magnesium
O	monooxygen
H	monohydrogen
H_2	dihydrogen
Na_2	disodium
S_6	hexasulfur
Se_n	polyselenium

The systematic names of single-element cations comprise the cation group name and the name of the respective chemical element with its oxidation number (Stock's method) for monoatomic cations or the number of atoms and the charge of the entire cation (the Ewens-Bassett method) for polyatomic cations; reference to the oxidation number which is constant for sp-elements shall be omitted, but should always be indicated for d-elements. Examples:

H^+	hydrogen cation
H_2^+	dihydrogen(1+) cation
Ba^{2+}	barium cation
Fe^{2+}	iron(II) cation
Fe^{3+}	iron(III) cation
S_8^{2+}	octasulfur(2+) cation

The systematic names of polyatomic cations are formed in compliance with the nomenclature rules for complex compounds (see Sec. 6.3.2). Examples:

UO_2^+	dioxouranium(V) cation
$FeOH^+$	hydroxoiron(II) cation
$HgNO_3^+$	nitratomercury(II) cation
VO_2^+	dioxovanadium(V) cation

The systematic names of monoatomic anions consist of the abbreviated English or Latin name of the element, with the termination **-ide** (see Sec. 6.1) and the group name **ion**, the number of atoms and the charge being specified in compliance with the Ewens-Bassett method (if obvious, the charge of monoatomic anions shall be omitted).

H^-	hydride ion
Cl^-	chloride ion
S^{2-}	sulfide ion
Te_6^{2-}	hexatelluride(2−) ion

The systematic names of polyatomic anions invariably end in the suffix **-ate** and should comply with the nomenclature names for complex compounds (see Sec. 6.3.2). Examples:

OF^-	fluorooxygenate(0) ion
MoO_3^{2-}	trioxomolybdate(IV) ion
RuO_4^-	tetraoxoruthenate(VII) ion
$S_2O_2^{2-}$	dioxodisulfate(2−) ion
PHO_3^{2-}	hydrogentrioxophosphate(III) ion
$PH_2O_2^-$	dihydrogendioxophosphate(I) ion
$I_3O_8^-$	octaoxotriiodate(V) ion
XeO_6^{4-}	hexaoxoxenonate(VIII) ion

The systematic names of compounds should be formed from the systematic names of electropositive constituents (cations) and electronegative constituents (anions). Examples:

CoF_2	cobalt(II) fluoride
CoF_3	cobalt(III) fluoride
$AlCl_3$	aluminum(III) chloride
Al_2Cl_6	dialuminum hexachloride
N_2O	dinitrogen oxide
P_4O_{10}	tetraphosphorus decaoxide
Cu_2O	copper(I) oxide
CuO	copper(II) oxide
Cr_2S_3	chromium(III) sulfide
$K_2(S_5)$	potassium pentasulfide(2−)
$(MgAl_2)O_4$	magnesium-dialuminum oxide
BaN_2	barium dinitride
Be_2C	diberyllium carbide
LiH	lithium hydride
Ca_2As_3	tricalcium diarsenide
FeP_2	iron diphosphide
Li_4RhH_5	tetralithium-rhodium pentahydride
K_3Sb	tripotassium stibide
MgB_2	magnesium diboride
Ag_4Ge	tetrasilver germanide
H_3NO_4	hydrogen tetraoxonitrate(V)
$LiCrO_2$	lithium dioxochromate(III)
K_2RuO_4	potassium tetraoxoruthenate(VI)
$K_2S_2O_2$	dipotassium dioxodisulfate
Na_2PoO_3	sodium trioxopolonate(IV)
$Na_8W_{12}O_{40}$	sodium 40-oxododecawolframate(VI)

To designate intermetallic compounds, provided the electronegativity values of their constituents are close to one another, use is made of systematic names written in one word, viz.:

Ag_5Al_3 pentasilvertrialuminum
$NiCu_3Al_6$ nickeltricopperhexaaluminum

The systematic names of elementary substances, cations, anions, and compounds are resorted to when no shorter nomenclature names are available (namely, traditional and special names cited below), for example:

O_3 trioxygen and ozone
NO_2^+ dioxonitrogen(V) cation and nitryl cation
N_3^- trinitride(1−) ion and azide ion
H_2O dihydrogen oxide and water
K_2CrO_4 potassium tetraoxochromate(VI) and potassium chromate

Traditional names. These names provide no precise information as regards the composition of compounds and call for memorizing the formulas, but are shorter than the systematic names [e.g. HNO_3 – nitric acid as against the systematic name – hydrogen trioxonitrate(V), and K_2SO_4 – potassium sulfate and the systematic name – potassium tetraoxosulfate(VI)], and therefore are recommended for use in the chemical nomenclature. However, traditional names may be used only for a limited range of common oxoacids and salts thereof, which are listed below. All other compounds of this class should be designated systematically.

Oxoacid		Anion	
$HAsO_2$	metaarsenous	AsO_2^-	metaarsenite
H_3AsO_3	orthoarsenous	AsO_3^{3-}	orthoarsenite
H_3AsO_4	arsenic	AsO_4^{3-}	arsenate
HBO_2	metaboric	BO_2^-	metaborate
–		BO_3^{3-}	orthoborate
–		$B_4O_7^{2-}$	tetraborate
–		BiO_3^-	bismuthate
$HBrO$	hypobromous	BrO^-	hypobromite
$HBrO_3$	bromic	BrO_3^-	bromate
$HBrO_4$	perbromic	BrO_4^-	perbromate
H_2CO_3	carbonic	CO_3^{2-}	carbonate
$HClO$	hypochlorous	ClO^-	hypochlorite
$HClO_2$	chlorous	ClO_2^-	chlorite
$HClO_3$	chloric	ClO_3^-	chlorate
$HClO_4$	perchloric	ClO_4^-	perchlorate
H_2CrO_4	chromic	CrO_4^{2-}	chromate
$H_2Cr_2O_7$	dichromic	$Cr_2O_7^{2-}$	dichromate
–		FeO_4^{2-}	ferrate

Oxoacid		Anion	
H_2GeO_3	germanic	GeO_3^{2-}	germanate
HIO	hypoiodous	IO^-	hypoiodite
HIO_3	iodic	IO_3^-	iodate
HIO_4	metaiodic	IO_4^-	metaperiodate
H_5IO_6	orthoperiodic	IO_6^{5-}	orthoperiodate
–		MnO_4^{2-}	manganate
$HMnO_4$	permanganic	MnO_4^-	permanganate
–		MoO_4^{2-}	molybdate
HNO_2	nitrous	NO_2^-	nitrite
HNO_3	nitric	NO_3^-	nitrate
$H_2N_2O_2$	hyponitrous	$N_2O_2^{2-}$	hyponitrite
HPO_3	metaphosphoric	PO_3^-	metaphophate
H_3PO_4	orthophosphoric	PO_4^{3-}	orthophosphate
$H_4P_2O_7$	diphosphoric	$P_2O_7^{4-}$	diphosphate
–		ReO_4^{2-}	rhenate
–		ReO_4^-	perrhenate
–		SO_3^{2-}	sulfite
H_2SO_4	sulfuric	SO_4^{2-}	sulfate
$H_2S_2O_7$	disulfuric	$S_2O_7^{2-}$	disulfate
$H_2S_nO_6$	polythionic	$S_nO_6^{2-}$	polythionate
H_2SeO_3	selenious	SeO_3^{2-}	selenite
H_2SeO_4	selenic	SeO_4^{2-}	selenate
$H_2Se_2O_7$	diselenic	$Se_2O_7^{2-}$	diselenate
H_2SiO_3	metasilicic	SiO_3^{2-}	metasilicate
H_4SiO_4	orthosilicic	SiO_4^{4-}	orthosilicate
–		$Si_2O_5^{6-}$	disilicate
$HTcO_4$	pertechnetic	TcO_4^-	pertechnetate
H_2TeO_3	tellurous	TeO_3^{2-}	tellurite
H_2TeO_4	metatelluruc	TeO_4^{2-}	metatellurate
H_6TeO_6	orthotelluric	TeO_6^{6-}	orthotellurate
–		VO_3^-	metavanadate
–		VO_4^{3-}	orthovanadate
–		WO_4^{2-}	wolframate

The traditional names of oxoacids are employed to form the names of substituted oxoacids, such as

$$H_2S_2O_6(O_2) \quad \text{peroxodisulfuric acid}$$
$$H_3AsS_4 \quad \text{tetrathioarsenic acid}$$
$$H_2SO_3S \quad \text{thiosulfuric acid}$$
$$HSeO_3F \quad \text{fluoroselenic acid}$$

Substituted sulfuric acids (except for peroxo- and thiosubstitued acids) are traditionally designated as sulfonic acids, viz.:

$$HSO_3F \quad \text{fluorosulfonic acid}$$
$$HSO_3(NH_2) \quad \text{aminosulfonic acid}$$

The traditional names of anions of oxoacid salts shall be employed for forming the names of salts of oxoacids and substituted oxoacids, the basic and acid salts inclusive. Examples:

Ag_2AsO_3	silver(I) orthoarsenite
$BaS_2O_6(O_2)$	barium peroxodisulfate
$Ca(HCO_3)_2$	calcium hydrogencarbonate
$Fe_2(SO_4)_3$	iron(III) sulfate
$Hg_2ClO_4(O)(OH)$	dimercury perchlorate-oxide-hydroxide
$Ti(NO_3)_2O$	titanium dinitrate-oxide
KSO_3F	potassium fluoroarsenate
$(NH_4)_3AsS_4$	ammonium tetrathioarsenate
NaH_2PO_4	sodium dihydrogenorthophosphate
Na_2SO_3S	sodium thiosulfate
$Zn_2(PO_4)OH$	dizinc orthophosphate-hydroxide

The name used here to designate the OH^- anion, viz., hydroxide, is the special name adopted for this anion (see below).

Special names. These names do not reflect the composition of compounds and, therefore, call for formula memorization, but are noted for their brevity as compared to the systematic names [cf. NH_4^+ is designated as ammonium cation, the systematic name being tetrahydrogennitrogen(-III); CN^- is cyanide ion, while the systematic name reads nitridocarbonate(II) ion; B_6H_{12} is hexaborane(12), as compared to the systematic name hexaboron dodecahydride], and are therefore recommended for use in the chemical nomenclature, albeit for a limited number of common ions and compounds. The list of authorized special names is given below. All other compounds should be designated by the systematic names.

Elementary substances

C (hex)	graphite	O_3	ozone
C (cub)	diamond	P_4	white phosphorus[1]

Cations

H_3O^+	oxonium
NF_4^+	tetrafluoroammonium[2,3]
NH_4^+	ammonium[3]
$N_2H_5^+$	hydrazinium(1+)
$N_2H_6^{2+}$	hydrazinium(2+)
NH_3OH^+	hydroxylaminium[4]
NO^+	nitrosyl
NO_2^+	nitryl

O_2^+	dioxygenyl
OH^+	hydroxyl
VO^{2+}	vanadyl
UO_2^{2+}	uranyl[5]

Anions

C_2^{2-}	acetylide
CN^-	cyanide
CN_2^{2-}	cyanamide
CNO^-	fulminate
HF_2^-	hydrogendifluoride
HO_2^{2-}	hydrogenperoxide
HS^-	hydrogensulfide[6]
N_3^-	azide
NCS^-	thiocyanate
NH^{2-}	imide
NH_2^-	amide
$N_2H_3^-$	hydrazide
NO^-	nitroside
O_2^{2-}	peroxide
O_2^-	hyperoxide or superoxide
O_3^-	ozonide
OCN^-	cyanate
OH^-	hydroxide

Compounds

$AlO(OH)$	aluminum metahydroxide[7]
B_2H_6	diborane(6)
B_4H_{10}	tetraborane(10)
B_5H_9	pentaborane(9)
B_5H_{11}	pentaborane(11)[8]
$C(NH_2)_2O$	carbamide
$CS(NH_2)_2$	thiocarbamide
HCN	hydrogen cyanide
HCN (aqua)	hydrocyanic acid[9]
HF	hydrogen fluorid[10]
HF (aqua)	hydrofluoric acid[9,10]
HN_3	hydrogen azide
HN_3 (aqua)	hydroazoic acid[9]
H_2O	water
H_2S	hydrogen sulfide[6,11]

H_2S (aqua)	hydrosulfuric acid
H_2S_n	polysulfane
ND_3	deuterioammonia
NH_3	ammonia
N_2H_4	hydrazine
NH_2Cl	chloroamine
NHF_2	difluoroamine
NH_2OH	hydroxylamine
PH_3	phosphine[12]
P_2H_4	diphosphane
SiH_4	monosilane[13]
Si_2H_6	disilane[13]
Si_3H_8	trisilane[13]
Si_4H_{10}	tetrasilane[13]

Notes:

[1] Other colored substances are designated similarly. [2] Other substituted NH_4^+ cations are designated similarly. [3] The compounds of the Group VA elements are named analogously. [4] The name hydroxylammonium is also permitted. [5] For other actinoids, the names are analogous. [6] For selenium and tellurium the names are analogous. [7] For other metals that form compounds of this composition, the names are analogous. [8] other boron hydrides are named analogously. [9] Aqueous solution. [10] For other halogens, the names are analogous. [11] Other names are also permitted: H_2S – monosulfane, H_2S – monoselane, H_2Te – monotellane. [12] For arsenic, antimony and bismuth, the names are analogous. [13] For germanium, tin and lead, the names are also analogous.

Apart from nomenclature names (systematic, traditional or special), in the scientific and technical literature, as well as in plant or laboratory practice recourse is made to other names (trivial, mineralogical or obsolete). The names in question are not embraced by the nomenclature of inorganic compounds. For recommendations as regards the use of these names, see Sec. 6.6–6.8.

6.3 Coordination Compounds [31–37]

A coordination compound comprises a *complex* (the first sphere) and the second sphere. The complex consists of a *central* atom (coordination center) and *ligands* (coordinating molecules or ions). Complexes having one central atom are designated as mononuclear complexes, while complexes with more than one central atom are named polynuclear complexes.

The complex may be cationic or anionic (the second sphere being a counter anion or a counter cation, respectively) or neutral, in which case the second sphere is absent. Coordination compounds may include two or more similarly charged coordinating ions, as well as contain both coordinating cations and anions.

6.3.1 The Rules of Constructing Formulas

In the formula of a mononuclear ionic or neutral complex comprising a single ligand kind L only, the symbol for the central atom M should be placed first followed by the formula of the ligand with the subscript n to indicate the number of ligands, and the formula for the whole complex entity should be placed in brackets: $[ML_n]$. In the formulas of multinuclear complexes, the number of centers of coordination m is likewise indicated: $[M_mL_n]$. In case the complex incorporates cationic L^+, anionic L^- and neutral L^0 ligands, the ligands should be placed after the center of coordination in the following sequence: $[M(L^+)(L^0)(L^-)]$.

The ligands having the same kind of charge but dissimilar chemical compositions shall be written in the sequence in which the first (left) elements in the formulas of the ligands appear in the practical row (see Sec. 6.2.1), but in case the first elements in the ligands are identical, the position of the second, third, etc. elements shall be considered (H_2O shall be written before NH_3, and C_6H_5N – before CO). Ligands having a simpler composition shall be disposed on the left (ahead) of more complicated ligands (N_2 ahead of NH_3, NH_3 – ahead of N_2H_4, and N_2H_4 – ahead of NH_2OH, etc.).

The use of letters to designate composite ligands (see below) shall not affect the order, in which the ligands in question are placed in the formulas of complexes; for example, the **en** ligand (ethylenedidiamine $C_2H_8N_2$) should precede the **diene** ligand ($C_4H_{13}N_3$).

Other rules of writing the formulas of coordination compounds coincide with the general nomenclature rules (see Sec. 6.2.1). Exemplary chemical formulas of coordination compounds are presented in Sec. 6.3.2.

The use of abbreviations (in small Latin letters) for organic ligands of a complicated composition is not an obligatory nomenclature rule and invariably calls for explanatory notes (in the same way as do the empirical formulas of ligands). Below are listed the most common abbreviations of ligand names:

acac	acetylacetonato $C_5H_7O_2^-$
bipy	2.2'-bipyridyl $(C_5H_4N)_2$
diene	diethylenetriamine $C_4H_{13}N_3$
Hdmg	dimethylglyoximato $C_4H_7O_2N_2^-$
en	ethylenediamine $C_2H_8N_2$
phen	1,10-phenanthroline $C_{12}H_8N_2$
py	pyridine C_5H_5N

In the nomenclature of organic compounds the most common groups are designated by letter symbols (Me – methyl CH_3, Et – ethyl C_2H_5, Ph – phenyl

C_6H_5, etc.), thereby rendering the formulas of ligands less cumbersome, e.g., PEt_3 – triethylphosphine.

6.3.2 The Rules of Composing Names

Central atoms. In complexes devoid of the second (outer) sphere and in complex cations, the centers of coordination have the names of respective chemical elements, such as

$[Pt(NH_3)_2Cl_2]$ diamminedichloroplatinum

$[Ni(NH_3)_6]^{2+}$ hexaamminenickel(II) cation

In complex anions, the names of central atoms are derived from English or Latin names of elements (for details, see Sec. 6.1). For example,

$[Au(CN)_2]^-$ dicyanoaurate(I) ion

$[TiCl_6]^{2-}$ hexachlorotitanate(IV) ion

Ligands. The names of *anionic ligands* end in **-o.** Generally, if the anion name ends in **-ide, -ite,** or **-ate,** the final **-e** is replaced by **-o.** Some anionic ligands do not follow this rule and are designated by special names. Examples:

CH_3COO^-	acetato	$NaSO_3^-$	sodiumsulfito
CN^-	cyano	O^{2-}	oxo
CO_3^-	carbonato	O_2^{2-}	peroxo
Cl^-	chloro	O_2^-	superoxo
H^-	hydrido	OH^-	hydroxo
HS^-	thiolo	S^{2-}	thio
NO^-	nitroso	SO_3^{2-}	sulfito
NO_2^-	nitro	SO_3S^{2-}	thiosulfato

Hydrocarbon anions used as ligands shall be named as follows: CH_3^- – methyl, $C_5H_5^-$ – cyclopentadienyl.

Neutral ligands are designated by the unmodified nomenclature names of the substances the ligands are composed of (N_2 – dinitrogen, N_2H_4 – hydrazine, C_2H_4 – ethylene, etc.), except for the following compounds that receive special names when used as neutral ligands:

H_2O	aqua	CO	carbonyl
NH_3	ammine	SO_2	dioxosulfur
NO	nitrosyl	PF_3	trifluorophosphorus

The following names are employed to designate *cationic ligands*:

N_2H_5 hydrazinium H^+ hydrogen
NO_2^+ nitrylium Ag^+ silver(I)
NO^+ nitrosylium $HgCl^+$ chloromercury(II)

Complexes. The names of complexes having no second sphere are also the names of coordination compounds and consist of a single word comprising first the number and names of ligands (for each ligand species separately) and then the central atom name (in the case of polynuclear complexes, the name of the central atom should be preceded by a numeral showing the number of said atoms). Examples:

$[Cr(C_6H_6)_2]$	dibenzenechromium
$[Fe(C_5H_5)_2]$	bis(cyclopentadienyl)iron
$[Mo_6Cl_{14}]$	14-chlorohexamolybdenum
$[Ni(CO)_4]$	tetracarbonylnickel
$[Rh(PEt_3)_3Cl]$	tris(triethylphosphine)chlororhodium
$[Zn(py)_2Cl_2]$	bis(pyridine)dichlorozinc

The names of complex cations comprise the number and names of ligands and the central atom name (in polynuclear complexes, the name of the central atom should be preceded by a numeral indicating the number of said atoms), followed by the oxidation number of the central atom (the Stock notation), or else the total cation charge (the Ewens-Bassett notation) should be indicated if the oxidation number is unknown. Examples:

$[Ag(NH_3)_2]^+$	diamminesilver(I) cation
$[Cr_2(NH_3)_9(OH)_2]^{4+}$	nonaamminedihydroxodichromium(III) cation
$[Bi_6(OH)_{12}]^{6+}$	dodecahydroxohexabismuth(III) cation
$[Mn(H_2O)_6]^{2+}$	hexaaquamanganese(II) cation
$[Nb_6Cl_{12}]^+$	dodecachlorohexaniobium(1+) cation
$[Pt(en)_2]^{2+}$	bis(ethylenediamine)platinum(II) cation
$[Ru(N_2)(NH_3)_5]^{2+}$	monokis(dinitrogen)pentaammineruthenium(II) cation

The names of complex anions comprise the number and names of ligands, the central atom name (preceded in multinuclear complexes by a numeral denoting the number of central atoms) modified to end in **-ate**, followed by the Stock oxidation number or by the Ewens-Bassett total anion charge. Examples:

$[AlH_6]^{3-}$	hexahydridoaluminate(III) ion
$[Nb_6Cl_{18}]^{2-}$	18-chloroniobate(2−) ion
$[Ni(CN)_4]^{2-}$	tetracyanoniccolate(II) ion
$[Sb_3F_{16}]^-$	16-fluorostibate(V) ion
$[SbS_4]^{5-}$	tetrathiostibate(III) ion
$[Zn(OH)_4]^{2-}$	tetrahydroxozincate(II) ion

In the names of multinuclear complex anions of heteropolycompounds of the composition $[MM_y^{VI}O_z]^{x-}$ (where $M - B^{III}$, Ce^{IV}, Ge^{IV}, I^{VII}, Mn^{IV}, P^V, Si^{IV}, Te^{VI} or Ti^{IV}, and $M^{VI} - Mo$ or W), the central atoms M^{VI} are designated molybdo (Mo) and wolframo (W). Examples:

$[GeMo_{12}O_{40}]^{4-}$	40-oxododecamolybdogermanate(IV) ion
$[PW_{12}O_{40}]^{3-}$	40-oxododecawolframophosphate(V) ion
$[TeMo_6O_{24}]^{6-}$	24-oxohexamolybdotellurate(VI) ion

Coordination compounds. The rules of citing the names of these compounds with the second sphere from their formulas and also of using the names of constituent anions and cations to form the names of said compounds are in full agreement with the general nomenclature rules (see Sec. 6.2.2). The rules of naming the complex compounds without the second sphere (neutral complexes) are presented above. The names of second sphere noncomplex ions are given in Sec. 6.2.3. The following examples illustrate the names of coordination compounds with the second sphere.

The compounds with complex cations:

$[Co(NH_3)_6]_2(SO_4)_3$	hexaamminecobalt(III) sulfate
$[Fe(H_2O)_5(NO)]SO_4$	pentaaquanitrosyliron(II) sulfate
$[Fe_2(H_2O)_8(OH)_2]Br_4$	octaaquadihydroxodiiron(III) bromide
$[Ga(H_2O)_6](NO_3)_3$	hexaaquagallium(III) nitrate
$[Pt(N_2H_5)_2Cl_2]Cl_2$	dihydraziniumdichloroplatinum(II) chloride
$[Pt^{II}(NH_3)_4][Pt^{IV}(NH_3)_4Cl_2]Cl_4$	tetraammineplatinum(II)-tetraamminedichloroplatinum(IV) chloride
$[SbAg_6](NO_3)_3$	hexasilver(I)antimony(III) nitrate
$[Sb_6F_{13}]F_5$	13-fluorohexaantimony(III) fluoride
$[W_6Br_{14}]Br_2$	14-bromohexatungsten dibromide

The compounds with complex anions:

$Na_3[Ag(SO_3S)_2]$	sodium bis(thiosulfato)argentate(I)
$Na_2[B_4O_5(OH)_4]$	sodium pentaoxotetrahydroxotetraborate(III)
$Cs[(I(I)_2]$	cesium diiodoiodate(I)
$Cs_4[Mg_3F_{10}]$	cesium decafluorotrimagnesate(II)
$K[Nd(SO_4)_2]$	potassium bis(sulfato)neodymate(III)
$Na_2[Pt_6(CO)_{12}]$	disodium dodecacarbonylhexaplatinate
$H_2[PtCl_6]$	hydrogen hexachloroplatinate(IV)
$O_2[Pt^VF_6]$	dioxygenyl hexafluoroplatinate(V)
$Rb_4[Sb^{III}Cl_6][Sb^VCl_6]$	rubidium hexachlorostibate(III)-hexachlorostibate(V)
$Na_3[V(NCS)_6]$	sodium hexakis(thiocyanato)vanadate(III)
$Li_3H[SiW_{12}O_{40}]$	trilithium-hydrogen 40-oxododecawolframosilicate

The compounds with complex cations and anions:

$[Cu(en)_2][Pt^{II}Cl_4]$	bis(ethylenediamine)copper(II) tetrachloroplatinate(II)
$[Pt(py)_4]_2[Fe^{II}(CN)_6]$	tetrakis(pyridine)platinum(II) hexacyanoferrate(II)
$[XeF_5][Co^{III}F_4]$	pentafluoroxenon(VI) tetrafluorocobaltate(III)
$[Ni(CH_3CN)_6][Ni^{III}I_4]$	hexaacetonitrilenickel(II) tetraiodoniccolate(II)
$[Co(NH_3)_6]_2[Re^{VI}(CN)_8]_3$	hexaamminecobalt(III) octacyanorhenate(VI)

6.4 Adducts (Addition Products) [31, 37]

Formulas. The adducts of the molecule-molecule or ion-molecule (hydrates, solvates, clathrates, etc.) types are designated by the formulas of their components with a dot in between; the number of molecules (or ions) of the second component should be inserted before its formula. Examples:

$$Kr\cdot 5.75H_2O \quad SO_2\cdot nH_2O \quad Ca^{2+}\cdot 8NH_3 \quad Na^+\cdot nH_2O \quad BiCl_3\cdot nNO_2$$
$$Na_2B_4O_7\cdot 10H_2O \quad Mn_2O_7\cdot 2H_2O \quad PCl_3\cdot 2.9Br_2$$

Names. For the $A\cdot nH_2O$ hydrates wherein A is the unit formula of the compound, the names consist of the name of said compound, followed by the word "hydrate" preceded by the numeral denoting the number of water molecules. Examples:

$Kr\cdot 5.75H_2O$	krypton 5.75-hydrate
$Mg^{2+}\cdot nH_2O$	magnesium cation polyhydrate
$H_3O^+\cdot 3H_2O$	oxonium cation trihydrate
$NH_3\cdot H_2O$	ammonia hydrate
$Cu_2S\cdot 5H_2O$	copper(I) sulfide pentahydrate
$MoO_3\cdot 2H_2O$	molybdenum(VI) oxide dihydrate
$HMnO_4\cdot 2H_2O$	permanganic acid dihydrate
$Ca(NO_3)_2\cdot 4H_2O$	calcium nitrate tetrahydrate
$CdSO_4\cdot 2.67H_2O$	cadmium(II) sulfate 2.67-hydrate

Composing the names of other adducts having any composition comprises listing from the formula (from the left to the right) the names of adduct components with a dash in between and terminating the name with the component ratio (in parentheses). Examples:

$Na_2CO_3\cdot 1.5H_2O_2$	sodium carbonate – hydrogen peroxide (1/1.5)
$I_3N\cdot nNH_3$	triiodine nitride – ammonia (1/n)
$BiCl_3\cdot nNO_2$	bismuth(III) chloride – nitrogen dioxide (1/n)
$Mn_2(SO_4)_3\cdot H_2SO_4\cdot 6H_2O$	manganese(III) sulfate – sulfuric acid – water (1/1/6)

6.5 Berthollides (Non-Stoichiometric Compounds) [31, 37]

Berthollides (non-stoichiometric compounds) comprise crystalline phases of a variable composition that depart significantly from stoichiometry (from the so-called ideal composition of berthollides relying on the ideal crystal lattice with an integral-valued stoichiometric ratio, or of daltonide). A general notation for the berthollides is to put the sign ≈ (circa) before the ideal composition formula, for example, ≈FeS, ≈Nb_2N, and ≈TiO.

A still more complete notation of non-stoichiometry is attained by introducing the subscript x:

$$Cu_{x-2}O \quad Fe_{1-x}S \quad Fe_xZn_{1-x}S \quad PaO_{2\pm x} \quad PbS_{1+x} \quad TiO_{1+x}$$

Where the value of x is known, the formulas of berthollides may be written as follows:

$$Fe_{0.877}S \quad \text{for } Fe_{1-x}S \quad \text{when } x = 0.123$$
$$Fe_{0.102}Zn_{0.898}S \quad \text{for } Fe_xZn_{1-x}S \quad \text{when } x = 0.102$$

The known ranges of x variations shall be indicated after the formula in parentheses, e.g., TiO_{1+x} ($-0.23 \leq x \leq 0.30$).

In the formulas of isomorphic substitution phases, an atom or groups of atoms should be separated by a comma and enclosed jointly in parentheses. For example, (Ni, Cu) denotes the complete homogeneity range from nickel to copper, and $(Li_2, Mg)Cl_2$ denotes the homogeneous region from lithium chloride to magnesium chloride.

6.6 Trivial Names [4, 7, 8, 10, 28, 31, 35]

Trivial names are the historically traditional names of chemical compounds and mixtures, solutions and processing products thereof employed in technical literature, laboratory and plant practice, and household. In the nomenclature of inorganic compounds, the employment of trivial names is not recommended.

6.6.1 Individual Compounds and Mixtures

Name	Composition
Acetate of lime, grey	$Ca(CH_3COO)_2 \cdot 2H_2O$
Acid	
Caro's	$H_2SO_3(O_2)$
glacial (acetic)	$CH_3COOH_{(sld)}$
Agate	product of mineral *opal* ageing
Alabaster	a) fine-grained mineral *gypsum*

(Continued)

Name	Composition
	b) erroneous name of burnt gypsum
Alanate, lithium	Li[AlH$_4$]
Alcohol, wood	CH$_3$OH
Alexandrite	mineral *chrysoberyl* with Cr^{III} impurity
Algaroth	Sb(Cl)O
Alum	mineral *potash alum*
ammonia	AlNH$_4$(SO$_4$)$_2$·12H$_2$O
burnt	KAl(SO$_4$)$_2$
ferric potassium	mineral *yavapaiite*
sodium	mineral *soda alum*
Alumina	mineral *corundum* (impure)
Alumogel	disperse Al$_2$O$_3$·0.25H$_2$O
Alumohydride, lithium	Li[AlH$_4$]
Alumstone	mineral *alunite*
Alundum	*corundum*-based refractory and chemically resistant material
Amethyst	mineral *quartz* with impurities
Ammophos	(NH$_4$)H$_2$PO$_4$ and (NH$_4$)$_2$HPO$_4$ mixture
Anhydride	
carbonic	CO$_2$
chromic	CrO$_3$
phosphoric	P$_4$O$_{10}$
sulfuric	SO$_3$
sulfurous	SO$_2$
Anhydrone	Mg(ClO$_4$)$_2$
Antichlor	Na$_2$SO$_3$S·5H$_2$O
Antimonite	mineral *stibnite*
Arsenic, white	As$_2$O$_3$ (amor)
Ash, soda	Na$_2$CO$_3$
Azophoska	mixture of (NH$_4$)$_2$HPO$_4$ and (NH$_4$)H$_2$PO$_4$ with KNO$_3$
Baryta, caustic	Ba(OH)$_2$
Bauxite	rock Al$_2$O$_3$·nH$_2$O
Benzene, inorganic	B$_3$H$_6$N$_3$
Bicarbonate	NaHCO$_3$
Bichromate	K$_2$Cr$_2$O$_7$
Black	
mercury	mineral *metacinnabar*
palladium	disperse Pd
platinum	disperse Pt
Blanc fixe	precipitated BaSO$_4$
Blende	
cadmium	mineral *greenockite*
manganese	mineral *alabandine*
nickel	mineral *millerite*
zinc	mineral *sphalerite*
Blue	
Paris (Prussian)	KFe[Fe^{II}(CN)$_6$]
Thenard's	(CoAl$_2$)O$_4$
Turnbull's	KFe[Fe^{II}(CN)$_6$]

(Continued)

Name	Composition
Bluestone	mineral *chalcanthite*
Borax	mineral *tincal*
burnt	$Na_2B_4O_7$
jewelry	mineral *tincalconite*
Borazol	$B_3H_6N_3$
Borazon	β-BN
Calamine	
common	mineral *hemimorphite*
noble	mineral *smithsonite*
Carbide, calcium	CaC_2
Carbonado	mineral *diamond* with impurities
Carborundum	SiC
Carnelian	product of mineral *opal* ageing
Caustic	NaOH
lunar	melt of $AgNO_3$ and KNO_3 mixture
Cement, magnesia (Sorel's)	calcined magnesia and $MgCl_2$ mixture (2:1 by volume)
Cementite	Fe_3C
Chalcedony	product of mineral *opal* ageing
Chalk	cake of impure precipitated $CaCO_3$
Chloride	
gold	$H[AuCl_4] \cdot 4H_2O$
luteo	$[M(NH_3)_6]Cl_3$ (M = Co, Cr, Ir)
praseo	*trans*-$[Co(NH_3)_4Cl_2]Cl \cdot H_2O$
purpureo	$[M(NH_3)_5Cl]Cl_2$ (M = Co, Cr)
Reise base I	$[Pt(NH_3)_4]Cl_2 \cdot H_2O$
Reise base II	*trans*-$[Pt(NH_3)_2Cl_2]$
violeo	*cis*-$[Co(NH_3)_4Cl_2]Cl \cdot H_2O$
Chromocene	$[Cr(C_5H_5)_2]$
Cinder, pyrite	Fe_2O_3
Clay, China (white)	mineral *kaolinite* and sand mixture
Coagulant	
aluminum	$Al_2(SO_4)_3 \cdot 18H_2O$
ferric	$Fe_2(SO_4)_3 \cdot 9H_2O$
Cobaltocene	$[Co(C_5H_5)_2]$
Crocus	Fe_2O_3 powder
Crown	
green	Cr_2O_3
lead	$PbCrO_4$
Crystal, rock	mineral *quartz*
Cyanide, soda	NaCN
Cymantrene	$[Mn(CO)_3(C_5H_5)]$
Diamond, black	mineral *diamond* with impurities
Diatomite	SiO_2 (amor)
Earth	
diatomaceous (infusorial)	SiO_2 (amor)
porcelain	mineral *kaolinite* and sand mixture
zirconium	mineral *baddeleyite*
Earths, rare	mixture of oxides of rare-earth elements

(Continued)

Name	Composition
Elements, rare-earth	Y and lanthanoids (sometimes, Sc)
Emery	alumina and mineral *magnetite* mixture
Ferricyanide	$K_3[Fe(CN)_6]$
Ferrocene	$[Fe(C_5H_5)_2]$
Ferrocyanide	$K_4[Fe(CN)_6] \cdot 3H_2O$
Filler, silica	disperse SiO_2 (amor)
Fixing agent	$Na_2SO_3S \cdot 5H_2O$
Flint	impure mineral *quartz*
Flowers of sulfur	α-S_8 powder
Fluospar	mineral *fluorite*
Gas	
carbonic acid	CO_2
Dawson	CO, H_2, CO_2, and N_2 mixture
detonating	H_2 and O_2 mixture (2:1 by volume)
laughing	N_2O
marsh (mine)	CH_4
producer	CO, N_2, and CO_2 mixture
suffocating	CO
water	CO and H_2 mixture
Gases, inert	He, Ne, Ar, Kr, Xe, and Rn
Gel, silica	$SiO_2 \cdot nH_2O$ ($n \leq 6$)
Glance	
antimony	mineral *stibnite*
bismuth	mineral *bismuthinite*
cobalt	mineral *cobaltite*
copper	mineral *chalcosine*
iron	mineral *hematite*
lead	mineral *galena*
manganese	mineral *alabandine*
molybdenum	mineral *molybdenite*
silver	mineral *argentite*
Glass, water	M_2SiO_3 (M = Na, K) and SiO_2 mixture
Gold	
mock	yellow plates of SnS_2 for gold imitation
mosaic	gold foil (for surface coating)
Graphite, white	α-BN
Green	
Guignet's	$Cr_2O_3 \cdot nH_2O$
Kassel (manganese)	$BaMnO_4$
Veronese	$Cu_3(AsO_4) \cdot 4H_2O$
Gypsum	
burnt	$CaSO_4 \cdot 0.5H_2O$
hydraulic	burnt gypsum and quicklime mixture
Heliotrope	product of mineral *opal* ageing
Hyacinth	impure mineral *zircon*
Hydrosulfite	$Na_2S_2O_4 \cdot 2H_2O$
Hyposulfite	$Na_2SO_3S \cdot 5H_2O$
Ice, dry	$CO_{2(sld)}$

(Continued)

Name	Composition
Iodide, nitrogen	$I_3N \cdot nNH_3$
Ironstone	
bog (brown)	rock $Fe_2O_3 \cdot nH_2O$
chrome	mineral *chromite*
magnetic	mineral *magnetite*
red	mineral *hematite*
titanous	mineral *ilmenite*
Jasper	product of mineral *opal* ageing
Kaolin	mineral *kaolinite* and sand mixture
Kieselguhr	SiO_2 (amor)
Lead, red	mineral *minium*
Lime	
air-hardening	CaO
bleaching	$Ca(ClO)_2$, $CaCl_2$ and $Ca(OH)_2$ mixture
burnt	CaO
caustic	$Ca(OH)_2$
chlorinated	$Ca(ClO)_2$, $CaCl_2$ and $Ca(OH)_2$ mixture
common	CaO
qrey	$Ca(CH_3COO)_2 \cdot 2H_2O$
magnesian	unslaked lime and burnt magnesia mixture
slaked	$Ca(OH)_2$
soda	mixture of slaked or unslaked lime with caustic soda
unslaked	CaO
Vienna	unslaked lime and burnt magnesia mixture
Limestone	mineral *calcite* with impurities
Lithopone	precipitated $BaSO_4$ and ZnS mixture
Liver of sulfur	Na_2S_n
Magnesia (calcined)	MgO
potash	mineral *picromerite*
Magnesia alba	$MgCO_3 \cdot 5H_2O$ and $Mg(OH)_2$ mixture
Mangan-blende	mineral *alabandine*
Mangan-spinel	mineral *galaxite*
Marble	mineral *calcite*
Meal, phosphorite	mineral *whitlockite* with impurities
Mercury, fulminating	$Hg(CNO)_2 \cdot 0.5H_2O$
Metabisulfite	$K_2S_2O_5$
Mixture, Eschka	calcined magnesia and calcined soda mixture
Monohydrate	H_2SO_4
Nickel, arsenical (copper)	mineral *nickeline*
Nickelocene	$[Ni(C_5H_5)_2]$
Nitroprusside, sodium	$Na_2[Fe(NO^+)(CN)_5]$
Ocher	
bismuth	mineral *bismite*
molybdic	mineral *molybdite*
tungstic	mineral *tungstite*
Oil of antimony	$SbCl_{3(lq)}$
Oil of tin	$SnCl_{4(lq)}$
Onyx	product of mineral *opal* ageing

(Continued)

Name	Composition
Ore	
acicular iron	mineral *goethite*
black iron	mineral *magnetite*
blue iron	mineral *vivianite*
grey antimony	mineral *stibnite*
grey manganese	mineral *pyrolusite*
horn quicksilver	mineral *calomel*
phosphorite	mineral *whitlockite* with impurities
red lead	mineral *crocoite*
Scheele's lead	mineral *stolzite*
siliceous zinc	mineral *hemimorphite*
vitriol lead	mineral *anglesite*
white lead	mineral *cerussite*
yellow lead	mineral *wulfenite*
Osmocene	$[Os(C_5H_5)_2]$
Oxide, red	mineral *cuprite*
Phosgene	CCl_2O
Pitchblende	mineral uraninite
Plaster	
lime	mixture of slaked lime, sand, and water
gypsum wall (Paris)	$CaSO_4 \cdot 0.5H_2O$
Potash	K_2CO_3
caustic	KOH
Powder	
Algaroth	$Sb(Cl)O$
bleaching	$Ca(ClO)_2$, $CaCl_2$, and $Ca(OH)_2$ mixture
Precipitate	precipitated $CaHPO_4 \cdot 2H_2O$
Prussiate of potash	
red	$K_3[Fe(CN)_6]$
yellow	$K_4[Fe(CN)_6] \cdot 3H_2O$
Pyrites	
arsenical	mineral *arsenopyrite*
cobalt	mineral *linnaeite*
copper	mineral *chalcopyrite*
hairy	mineral *millerite*
iron	mineral *pyrite*
magnetic	mineral *pyrrhotine*
manganese	mineral *hauerite*
poisonous	mineral *arsenopyrite*
red	mineral *nickeline*
sulfur	mineral *pyrite*
yellow	mineral *millerite*
Pyrolusite, lustrous	mineral *hausmannite*
Pyrophoren	Ni_3C
Quicklime	CaO
Rectificate	C_2H_5OH
Ruby	mineral *corundum* with Cr^{III} impurity
Ruthenocene	$[Ru(C_5H_5)_2]$

(Continued)

Name	Composition
Salammoniac, platinum	$(NH_4)_2[PtCl_6]$
Salt	
Bertollet's	$KClO_3$
bitter	mineral *epsomite*
Cleve's	*cis*-$[Pt(NH_3)_2Cl_4]$
common	NaCl
Epsom	mineral *epsomite*
Fisher's	$K_3[Co(NO_2)_6]$
fixing	$Na_2SO_3 \cdot 5H_2O$
Gérard's	*trans*-$[Pt(NH_3)_2Cl_4]$
Glauber's	mineral *mirabilite*
Gmelin's	$K_3[Fe(CN)_6]$
gold(en)	$Na[AuCl_4] \cdot 2H_2O$
Johnson's	$K[I(I)_2] \cdot H_2O$
Magnus	$[Pt(NH_3)_4][Pt^{II}Cl_4]$
Mohr's	mineral *mohrite*
Peyronet's	*cis*-$[Pt(NH_3)_2Cl_2]$
pink	$(NH_4)_2[SnCl_6]$
preparing	$Na_2[Sn(OH)_6]$
Reinecke's	$NH_4[Cr(NH_3)_2(NCS)_4] \cdot H_2O$
rock	mineral *halite*
Schlippe's	$Na_3[SbS_4] \cdot 9H_2O$
Seignette's	$KNa(C_4H_4O_6) \cdot 4H_2O$
table	NaCl
tin	$[Sn(H_2O)Cl_2] \cdot H_2O$
Zeise's	$K[Pt(C_2H_4)Cl_3] \cdot H_2O$
Saltpeter	mineral *niter*
ammonia	mineral *nitrammite*
baryta	mineral *nitrobarite*
Chile	mineral *nitratine*
Indian	mineral *niter*
lime	mineral *nitrocalcite*
magnesia	mineral *nitromagnesite*
Norwegian	mineral *nitrocalcite*
potash	mineral *niter*
soda	mineral *nitratine*
Salts, Toutton's	$M_2^I M^{II}(SO_4)_2 \cdot 6H_2O$, where M^I = K-Cs, Tl, NH_4; M^{II} = Mg, V-Zn, Cd
Sand	arenaceous mineral *quartz* with impurities
quartz	arenaceous mineral *quartz*
Sapphire	mineral *corundum* with Ti^{IV} and Fe^{II} impurities
Sard	product of mineral *opal* ageing
Schoenite	mineral *picromerite*
Silica	impure mineral *quartz*
Silver	
fulminating	Ag_3N
horn	mineral *chlorargyrite*
Soda	$Na_2CO_3 \cdot 10H_2O$

(Continued)

Name	Composition
baking	$NaHCO_3$
calcined	Na_2CO_3
caustic	$NaOH$
washing	Na_2CO_3
Spar	
bitter	mineral *dolomite*
brown	mineral *dolomite*
heavy	mineral *barite*
Iceland	mineral *calcite*
iron	mineral *siderite*
lead molybdic	mineral *wulfenite*
lime	mineral *calcite*
scarlet	mineral *rhodochrosite*
Scheele's	mineral *scheelite*
talc	mineral *magnesite*
yittrium	mineral *xenotime*
zinc	mineral *smithsonite*
Spinels (noble)	$(MgAl_2)O_4$
iron	mineral *hercynite*
lead	mineral *minium*
manganese	mineral *galaxite*
titanium-iron	mineral *ulvöspinel*
zinc	mineral *gahnite*
Spirit of wine	C_2H_5OH
Sublimate, corrosive	$HgCl_2$
Sugar of lead	$Pb(CH_3COO)_2 \cdot 3H_2O$
Sulfate	Na_2SO_4
Sulfite	Na_2SO_3
Superphosphate	
double	$Ca(H_2PO_4)_2 \cdot H_2O$
normal	$Ca(H_2PO_4)_2 \cdot H_2O$ and $CaSO_4$ mixture
Talc, bitter	mineral *magnesite*
Tanning agent	
alum	$KAl(SO_4)_2 \cdot 12H_2O$
chrome	$KCr(SO_4)_2 \cdot 12H_2O$
ferric	$Fe_2(SO_4)_3 \cdot 9H_2O$
Tartar	$K(HC_4H_4O_6)$
Texton	$NaClO_2 \cdot 3H_2O$
Thermit	powdered Al and $(Fe^{II}Fe_2^{III})O_4$ mixture
Thiophosgene	$CSCl_2$
Thiourea	$CS(NH_2)_2$
Tinsel	yellow plates of SnS_2 for gold imitation
Tinstone	mineral *cassiterite*
Titanocene	$[Ti(C_5H_5)_2]$
Topaz	mineral *cyanite* with H_2O and F^- impurities
Trisoda-phosphate	$Na_3PO_4 \cdot 12H_2O$
Tungstein	mineral *scheelite*
Urea	$C(NH_2)_2O$

(Continued)

Name	Composition
Vanadocene	$[V(C_5H_5)_2]$
Vinegar, wood	mixture of CH_3OH and CH_3COOH
Vitriol	
blue	mineral *chalcanthite*
emerald green	mineral *morenosite*
green	mineral *melanterite*
lead	mineral *anglesite*
pink-red	mineral *bieberite*
sky-blue	$CrSO_4 \cdot 7H_2O$
violet	$VSO_4 \cdot 7H_2O$
white	mineral *goslarite*
Water	
heavy	D_2O
ultraheavy	T_2O
White	
baryta	precipitated $BaSO_4$
lead	$PbCO_3$ and $Pb(OH)_2$ mixture
pearl (Spanish)	$BiNO_3(OH)_2$
titanium	γ-TiO_2 powder
zink	ZnO powder
zirconium	ZrO_2 powder
Zirconia	ZrO_2

6.6.2 Solutions

Name	Composition
Acid	
fuming	solution of SO_3 in concentrated H_2SO_4
hydrochloric	concentrated aqueous solution of HCl
hydrocyanic	aqueous solution of HCN
hydrofluoric	concentrated aqueous solution of HF
hydrosulfuric	aqueous solution of H_2S
Aqua regia	concentrated HNO_3 and hydrochloric acid mixture (1:3 by volume)
Liquor, Wackenroder's	aqueous solution of $H_2S_nO_6$ ($n = 3 \div 6$)
Lye	
potash	aqueous solution of KOH
soda	aqueous solution of NaOH
Milk, lime	$Ca(OH)_2$ suspension in lime water
Mixture	
Bordeaux	solution of $CuSO_4$ in lime milk
chrome	1) concentrated H_2SO_4 and saturated aqueous solution of $K_2Cr_2O_7$ (1:1 by volume) 2) CrO_3 and concentrated H_2SO_4 mixture
magnesian	solution of $MgCl_2$ and NH_4Cl in ammonia water
Oil of vitriol	commercial concentrated H_2SO_4
Oleum	solution of SO_3 in anhydrous H_2SO_4 (contains $H_2S_2O_7$)

(Continued)

Name	Composition
Perhydrol	30% aqueous solution of H_2O_2
Reagent	
Nessler	akaline aqueous solution of $K_2[HgI_4]$
Tschugaev	solution of dimethylglyoxime $C_4H_8O_2N_2$
	in saturatred aqueous solution of NH_3
Water	
ammonia	aqueous solution of NH_3
baryta	saturated aqueous solution of $Ba(OH)_2$
bromine	aqueous solution of Br_2 (contains HBrO and HBr)
chlorine	aqueous solution of Cl_2 (contains HClO and HCl)
gypsum	saturated aqueous solution of $CaSO_4$
iodine	aqueous solution of KI saturated with I_2
	(contains $K[I(I)_2]$)
Javel	aqueous solution of KOH saturated with Cl_2
	(contains KClO and KCl)
Labarraque	aqueous solution of NaOH saturated with Cl_2
	(contains NaClO and NaCl)
lime	saturated aqueous solution of $Ca(OH)_2$

6.6.3 Metal Alloys

Quantitative composition of alloys is cited in mass per cent

Name	Composition
Acid-resistant alloy	Sn ≤ 20, Tl ≤ 10; remainder Pb
Alumel	Al 1.8–2.5, Mn 1.8–2.2, Si 0.85–0.2, occasionally Fe 0.5;
	remainder Ni
Amalgam	Hg with metals of Groups IA, IIA, IB and IIB
Arndt's alloy	Cu 60 Mg 40
Babbitt (lead-base)	Pb 80–82, Sb 16–18, Cu 2
Babbitt (tin-base)	Sn 82–84, Sb 10–12, Cu 6
Brass	Zn 40 and also Al, Fe, Mn, Ni, Pb, etc.; remainder Cu
Bronze	Al ≤ 6 (aluminum), Be ≤ 6 (beryllium), Mn 5 (manganese),
	Sn ≤ 11 (tin), and also Cr, P, Pb, Si, Zn, etc.;
	remainder Cu
Cast-iron	C > 2 (generally 3–4.5), Mn 15–20 (spiegel iron),
	and also Si, S, P, etc.; remainder Fe
Chromel	Cr 9–10, Co 1, and also Fe; remainder Ni
Coinage alloy	Au 90, Ag 10 (gold); Ag 90, Cu 10 (silver); Cu 95, Sn 4, Zn 1
	(copper)
Constantan	Ni 40, Mn 1; remainder Cu
Copel	Cu 55, Ni 44
Devard's alloy	Al 45, Zn 5; remainder Cu
Duraluminum	Cu 2.5–5.5, Mg 0.5–2.0, Mn 0.5–1.2, Si 0.2–1.0;
	rermainder Al
Electron	Al ≤ 10, Zn ≤ 4, Mn ≤ 1.7, Be 0.02–0,05; remainder Mg
Ferroboron	Al 3–5, Si 3–4, B 12, C 0.1, S 0.02; remainder Fe

(*Continued*)

Name	Composition
Ferrochrome	Cr 60–85; remainder Fe
Ferromanganese	Mn ≥ 70, C 7–8, Si ≤ 2, P ≥ 0.35, S 0.03; remainder Fe
Ferromolybdenum	Mo ≥ 55, Si 1–2, P 0.2, C 0.1, S 0.1; remainder Fe
Ferroniobium	Nb + Ta 23–75, Si 10–11, Ti 7, Al ≤ 7, S ≤ 0.5, P 0.15; remainder Fe
Ferrosilicium	Si 12–90; remainder Fe
Ferrotantalum	cf. Ferroniobium
Ferrotitanium	Ti 18–25 or 40–45, Al 5–8, Si 3–6, Cu 3, C ca. 0.1, S ca. 0.1, P ca. 0.1; remainder Fe
Ferrotungsten	W 65–80, Si 0.4–1.0, Mn 0.2–0.6, C 0.5; remainder Fe
Ferrovanadium	V 35–80, Sn 3.5, Al ≥ 2, C 1, P 0.1–0.2, S 0.1–0.2; remainder Fe
Ferrozirconium	Zr 40, Si 10, Al 8–10; remainder Fe
German silver (Ni-Cu alloy)	Ni 18–30 and also Fe, Mn; remainder Cu
German silver (Ni-Zn-Cu alloy)	Ni 15–35, Zn 13–45; remainder Cu
Invar	Ni 36, Mn 0.5, C 0.5; remainder Al
Low-melting alloy	Sn 12, Zn 6 (tin, m.p. 17°C) or In 24 (indium, m.p. 16°C); remainder Ga
Magnalium	Mg 5–30 and also Cu, Mn, Ti; remainder Al
Magnetic alloy	Ni 22–34, Al 11–15; remainder Fe
Manganin	Mn 13, Ni 4; remainder Cu
Mond's alloy	Cu 26, Mn 4; remainder Ni
Monel metal	Cu 27–29, Fe 2, Mn 1.5; remainder Ni
Nichrome	Cr 15, Mn ≤ 4, and also Al, Fe, Si; remainder Ni
Nickeline	Ni 25–35, Mn 1, and also Fe, Zn; remainder Cu
Permalloy	Ni 40–50 or 70–83; remainder Fe
Platinite	Ni 46; remainder Fe
Platinum-rhodium alloy	Pt 90, Rh 10
Pobedit	Co 6–15, C 5–7: remainder W (contains WC)
Potal	Foil from Cu-base alloys (brass, tombac, etc.)
Rose's alloy	Pb 32, Sn 18; remainder Bi
Silumin	Si ≤ 23 and also Cu, Mg, Mn, Na; remainder Al
Solder	Sn 30–70; remainder Pb
Steel	C ≤ 2 (ordinary), W 17.5–18.5 (high-speed), Cr 13–27 (high-temperature), Cr ≤ 12 (tool, contains metal-like carbides), Mn 12–15 (manganese), Cr 18, Ni 9 (chromium-nickel) and also Al, Si, V; remainder Fe
Tombac	Zn 3–12; remainder Cu
Type-metal alloy	Sb 10–20, Sn 5–30; remainder Pb
Wood's alloy	Pb 11.2–40.3, Cd 8.85–9, Sn 8.85–9.75; remainder Bi

6.7 Mineralogical Names [24, 25, 27]

Mineralogical names are employed for designating natural substances (minerals) to distinguish them from man-made substances of the same composition (reagents) and also for designating allotropic forms of carbon (diamond and graphite) as well as polymorphic modifications of some substances.

Name	Formula	Name	Formula
Acanthite	Ag_2S (mon)	Bixbyite	Mn_2O_3
Afwillite	$Ca_3Si_2O_7 \cdot 3H_2O$	Bloedite	$Na_2Mg(SO_4)_2 \cdot 4H_2O$
Akdalaite	$Al_2O_3 \cdot 0.25H_2O$	Bobierrite	$Mg_3(PO_4)_2 \cdot 8H_2O$
Alabandine	MnS	Boehmite	$\gamma\text{-}AlO(OH)$
Alaite	$V_2O_5 \cdot H_2O$	Boussingaultite	$Mg(NH_4)_2(SO_4)_2 \cdot 6H_2O$
Alamosite	$PbSiO_3$	Bracewellite	$CrO(OH)$ (rhomb)
Albrittonnite	$CoCl_2 \cdot 6H_2O$	Bredigite	Ca_2SiO_4 (rhomb)
Alstonite	$BaCa(CO_3)_2$ (tricl)	Breithauptite	$NiSb$
Altaite	$PbTe$	Bromargyrite	$AgBr$
Aluminite	$Al_2SO_4(OH)_4 \cdot 7H_2O$	Bromellite	BeO
Alunite	$KAl_3(SO_4)_2(OH)_6$	Brookite	$\alpha\text{-}TiO_2$
Alunogen	$Al_2(SO_4)_3 \cdot 18H_2O$	Brucite	$Mg(OH)_2$
Amakinite	$Fe(OH)_2$	Bruggenite	$Ca(IO_3)_2 \cdot H_2O$
Anatase	$\beta\text{-}TiO_2$	Brushite	$CaHPO_4 \cdot 2H_2O$
Andalusite	$\gamma\text{-}Al_2(SiO_4)O$	Bunsenite	$Ni_{1-x}O$
Anglesite	$PbSO_4$	Cadmoselite	$CdSe$
Anhydrite	$CaSO_4$	Cadwaladerite	$AlCl(OH)_2 \cdot 4H_2O$
Annabergite	$Ni_3(AsO_4)_2 \cdot 8H_2O$	Calaverite	$AuTe_2$ (mon)
Antarcticite	$CaCl_2 \cdot 6H_2O$	Calciborite	$Ca(BO_2)_2$
Aragonite	$CaCO_3$ (rhomb)	Calcite	$CaCO_3$ (trig)
Arcanite	K_2SO_4	Calomel	Hg_2Cl_2
Argentite	Ag_2S (rhomb)	Carlinite	Tl_2S
Arsenolamprite	As_4	Carlsbergite	CrN
Arsenolite	$\beta\text{-}As_2O_3$	Carnallite	$KMgCl_3 \cdot 6H_2O$
Arsenopyrite	$Fe(As)S$	Carobbiite	KF
Avicennite	Tl_2O_3	Cassiterite	SnO_2
Avogadrite	$K[BF_4]$	Cattierite	$Co(S_2)$
Baddeleyite	ZrO_2	Celestine	$SrSO_4$
Bararite	$(NH_4)_2[SiF_6]$ (trig)	Cerianite	CeO_2
Barite	$BaSO_4$	Cerussite	$PbCO_3$
Barytocalcite	$BaCa(CO_3)_2$ (mon)	Cervantite	$(Sb^{III}Sb^V)O_4$
Bassanite	$CaSO_4 \cdot 0.5H_2O$	Chalcanthite	$CuSO_4 \cdot 5H_2O$
Bayerite	$Al(OH)_3$ (trig)	Chalcocyanite	$CuSO_4$
Behoite	$Be(OH)_2$	Chalcopyrite	$(FeCu)S_2$
Bellidoite	$Cu(IO_3)_2 \cdot 0.67H_2O$	Chalcosine	Cu_2S
Bellitodolite	Cu_2Se (tetr)	Chloraluminite	$AlCl_3 \cdot 6H_2O$
Berlinite	$AlPO_4$	Chlorargyrite	$AgCl$
Berndtite	SnS_2 (hex or trig)	Chloromagnesite	$MgCl_2$
Berzelianite	Cu_2Se (cub)	Chromatite	$CaCrO_4$
Bieberite	$CoSO_4 \cdot 7H_2O$	Chromite	$(Cr_2Fe)O_4$
Bishofite	$MgCl_2 \cdot 6H_2O$	Chrysoberyl	$(BeAl_2)O_4$
Bismite	Bi_2O_3 (mon)	Cinnabar	$\alpha\text{-}HgS$
Bismoclite	$Bi(Cl)O$	Claudetite	$\alpha\text{-}As_2O_3$
Bismuthinite	Bi_2S_3	Clausthalite	$PbSe$
Bismutite	$Bi_2CO_3(OH)_4$	Cobaltite	$Co(As)S$

Name	Formula	Name	Formula
Coccinite	Hg_2I_2	Fluorite	CaF_2
Coecite	SiO_2 (mon)	Forsterite	Mg_2SiO_4
Coffinite	$USiO_4$	Frankdicksonite	BaF_2
Cohenite	Fe_3C	Franklinite	$(Fe_2Zn)O_4$
Coloradoite	$HgTe$	Freboldite	$CoSe$
Cooperite	PtS	Gahnite	$(ZnAl_2)O_4$
Coquimbite	$Fe_2(SO_4)_3 \cdot 9H_2O$	Galaxite	$(MnAl_2)O_4$
Corundum	Al_2O_3	Galena	PbS_{1+x}
Cotunnite	$PbCl_2$	Gaylussite	$Na_2Ca(CO_3)_2 \cdot 5H_2O$
Covellite	CuS	Gersdorffite	$Ni(As)S$
Crednerite	$(Mn_2Cu)O_4$	Gibbsite	$Al(OH)_3$ (mon)
Cristobalite	SiO_2 (cub; tetr)	Glauberite	$Na_2Ca(SO_4)_2$
Crocoite	$PbCrO_4$	Glushinskite	$MgC_2O_4 \cdot 2H_2O$
Cryolite	$Na_3[AlF_6]$	Goethite	$FeO(OH)$ (rhomb)
Cryptohalite	$(NH_4)_2[SiF_6]$ (cub)	Goslarite	$ZnSO_4 \cdot 7H_2O$
Cuprite	Cu_2O	Graphite	α-C
Cyanite	α-$Al_2(SiO_4)O$	Greenockite	CdS (hex)
Diamond	β-C	Grimaldiite	$CrO(OH)$ (trig)
Diaspore	α-$AlO(OH)$	Groutite	$MnO(OH)$ (rhomb)
Dickite	$Al_2Si_2O_7 \cdot 2H_2O$ (mon)	Guanajuatite	Bi_2Se_3
Digenite	$Cu_{1.8}S$	Gypsum	$CaSO_4 \cdot 2H_2O$
Diopside	$CaMg(SiO_3)_2$	Halite	$NaCl$
Djurleite	$Cu_{1.94}S$	Halloysite	$Al_2Si_2O_7 \cdot 6H_2O$
Dolomite	$CaMg(CO_3)_2$	Hauerite	$Mn(S_2)$
Doloresite	$VO_2 \cdot 0.67H_2O$	Hausmannite	$(Mn^{II}Mn_2^{III})O_4$
Domeykite	Cu_3As	Hawleyite	CdS (cub)
Dorfmanite	$Na_2HPO_4 \cdot 2H_2O$	Heazlewoodite	Ni_3S_2
Downeyite	SeO_2	Hematite	Fe_2O_3
Duttonite	$VO(OH)_2$	Hemimorphite	$Zn_2SiO_4 \cdot H_2O$
Dzhalindite	$In(OH)_3$	Hercynite	$(FeAl_2)O_4$
Eitelite	$Na_2Mg(CO_3)_2$	Herzenbergite	SnS
Enargite	Cu_3AsS_4 (rhomb)	Hessite	Ag_2Te
Epsomite	$MgSO_4 \cdot 7H_2O$	Hetaerolithe	$(Mn_2Zn)O_4$
Eriochalcite	$CuCl_2 \cdot 2H_2O$	Heterogenite	$CoO(OH)$
Erlichmanite	$Os(S_2)$	Heterosite	$FePO_4$
Erythrite	$Co_3(AsO_4)_2 \cdot 8H_2O$	Hexahydroborite	$Ca(BO_2)_2 \cdot 6H_2O$
Eskolaite	Cr_2O_3	Hieratite	$K_2[SiF_6]$
Farringtonite	$Mg_3(PO_4)_2$	Hillebrandite	$Ca_2SiO_4 \cdot H_2O$
Fayalite	Fe_2SiO_4	Hongquiite	TiO_{1+x}
Feitknechtite	$MnO(OH)$ (hex)	Hopeite	$Zn_3(PO_4)_2 \cdot 4H_2O$
Ferberite	$FeWO_4$	Huebnerite	$MnWO_4$
Feroxyhyte	$FeO(OH)$ (hex)	Humboldtine	$FeC_2O_4 \cdot 2H_2O$
Ferrosilicite	$FeSi$	Huttonite	$ThSiO_4$ (mon)
Ferrosilite	$FeSiO_3$	Hydrohalite	$NaCl \cdot 2H_2O$
Ferrotantalite	$(Ta_2Fe)O_6$	Hydromolysite	$FeCl_3 \cdot 6H_2O$
Ferruccite	$Na[BF_4]$	Hydrophilite	$CaCl_2$ ($\cdot nH_2O$?)
Fluocerite	CeF_3	Hydrotungstite	$WO_3 \cdot 2H_2O$ (mon)
Fluorapatite	$Ca_5(PO_4)_3F$	Ice	H_2O (hex)

(*Continued*)

Name	Formula	Name	Formula
Ilmenite	$(TiFe)O_3$	Malachite	$Cu_2CO_3(OH)_2$
Iodargyrite	AgI	Malladrite	$Na_2[SiF_6]$
Iranite	$PbCrO_4 \cdot H_2O$	Mallardite	$MnSO_4 \cdot 7H_2O$
Jacobsite	$(MnFe_2)O_4$	Manganite	$MnO(OH)$ (mon)
Jaipurite	CoS_{1+x}	Manganosite	MnO
Jokokuite	$MnSO_4 \cdot 5H_2O$	Manganotantalite	$(Ta_2Mn)O_6$
Joliotite	$(UO_2)CO_3 \cdot (1.5 \div 2)H_2O$	Mansfieldite	$AlAsO_4 \cdot 2H_2O$
Jordisite	MoS_2 (amor)	Marcasite	$\alpha\text{-}Fe(S_2)$
Jurbanite	$Al(SO_4)OH \cdot 5H_2O$ (mon)	Marshite	CuI
Kainite	$KMg(SO_4)Cl \cdot 3H_2O$	Mascagnite	$(NH_4)_2SO_4$
Kalicinite	$KHCO_3$	Massicot	$\beta\text{-}PbO$
Kaneite	$MnAs$	Matlockite	$Pb(Cl)F$
Kaolinite	$Al_2Si_2O_7 \cdot 2H_2O$ (tricl)	Matteuccite	$NaHSO_4 \cdot H_2O$
Karelianite	V_2O_3	McConnellite	$CuCrO_2$
Keatite	SiO_2 (tetr)	Melanophlogite	SiO_2 (cub)
Kernite	$Na_2B_4O_7 \cdot 4H_2O$	Melanterite	$FeSO_4 \cdot 7H_2O$
Kerstenite	$PbSeO_4$	Melonite	$Ni(Te_2)$
Kieserite	$MgSO_4 \cdot H_2O$	Mercallite	$KHSO_4$
Kilchoanite	$Ca_3Si_2O_7$ (rhomb)	Meta-alunogen	$Al_2(SO_4)_3 \cdot 13.5H_2O$
Klockmannite	$CuSe$	Metaborite	HBO_2
Kolbeckite	$ScPO_4 \cdot 2H_2O$	Metacinnabar	$\beta\text{-}HgS$
Kotoite	$Mg_3(BO_3)_2$	Metastibnite	Sb_2S_3 (amor)
Köttigite	$Zn_3(AsO_4)_2 \cdot 8H_2O$	Meymacite	$WO_3 \cdot 2H_2O$ (amor)
Krennerite	$AuTe_2$ (rhomb)	Miargyrite	$(AgSb)S_2$
Kullerudite	$Ni(Se_2)$	Millerite	NiS
Lansfordite	$MgCO_3 \cdot 5H_2O$	Minasragrite	$[V(H_2O)_4(SO_4)O] \cdot H_2O$
Larnite	Ca_2SiO_4 (mon)	Minguzzite	$K_3[Fe(C_2O_4)_3] \cdot 3H_2O$
Laurionite	$Pb(Cl)OH$	Minium	$(Pb_2^{II}Pb^{IV})O_4$
Laurite	$Ru(S_2)$	Mirabilite	$Na_2SO_4 \cdot 10H_2O$
Lautarite	$Ca(IO_3)_2$	Mohrite	$Fe(NH_4)_2(SO_4)_2 \cdot 6H_2O$
Lawrencite	$FeCl_2$	Moissanite	SiC
Lechatelierite	SiO_2 (amor)	Molybdenite	MoS_2 (hex or trig)
Lime	CaO	Molybdite	MoO_3
Linnaeite	$(Co^{II}Co_2^{III})S_4$	Molysite	$FeCl_3$
Litharge	$\alpha\text{-}PbO$	Monetite	$CaHPO_4$
Lithiophosphate	Li_3PO_4	Monohydrocalcite	$CaCO_3 \cdot H_2O$
Löllingite	$FeAs_2$	Monteponite	CdO
Lopezite	$K_2Cr_2O_7$	Monticellite	$CaMgSiO_4$
Lueschite	$(NaNb)O_3$	Montroseite	$VO(OH)$
Luzonite	Cu_3AsS_4 (tetr)	Montroydite	HgO
Macedonite	$(TiPb)O_3$	Morenosite	$NiSO_4 \cdot 7H_2O$
Machatschkiite	$Ca_3(AsO_4)_2 \cdot 9H_2O$	Nahcolite	$NaHCO_3$
Maghemite	$Fe_{2.67}O_4$	Nantokite	$CuCl$
Magnesiochromite	$(MgCr_2)O_4$	Nasturan	$(U_2^{IV}U^{VI})O_8$
Magnesioferrite	$(MgFe_2)O_4$	Natron	$Na_2CO_3 \cdot 10H_2O$
Magnesite	$MgCO_3$	Natrosilite	$Na_2Si_2O_5$
Magnetite	$(Fe^{II}Fe_2^{III})O_4$	Naumannite	Ag_2Se
Mäkinenite	$NiSe$	Navajoite	$V_2O_5 \cdot 3H_2O$

(Continued)

Name	Formula	Name	Formula
Newberyite	$MgHPO_4 \cdot 3H_2O$	Raspite	$PbWO_4$ (mon)
Nickelbischofite	$NiCl_2 \cdot 6H_2O$	Realgar	$\alpha\text{-}As_4S_4$
Nickeline	$NiAs$	Rhodochrosite	$MnCO_3$
Niningerite	MgS	Rhodonite	$MnSiO_3$
Niter	KNO_3	Rokühnite	$FeCl_2 \cdot 2H_2O$
Nitrammite	NH_4NO_3	Romarchite	SnO
Nitratine	$NaNO_3$	Rooseveltite	$BiAsO_4$
Nitrobarite	$Ba(NO_3)_2$	Rosickyite	$\gamma\text{-}S$
Nitrocalcite	$Ca(NO_3)_2 \cdot 4H_2O$	Rösslerite	$MgHAsO_4 \cdot 7H_2O$
Nitromagnesite	$Mg(NO_3)_2 \cdot 6H_2O$	Rostite	$Al(SO_4)OH \cdot 5H_2O$ (rhomb)
Northupite	$Na_2Mg(CO_3)_2 \cdot H_2O$?	Rutherfordine	$(UO_2)CO_3$
Nyerereite	$Na_2Ca(CO_3)_2$	Rutile	$\gamma\text{-}TiO_2$
Oldhamite	CaS	Safflorite	$CoAs_2$
Olympite	Na_3PO_4	Salammoniac	NH_4Cl
Opal	$SiO_2 \cdot nH_2O$	Sassolite	$B(OH)_3$
Orangite	$ThSiO_4 \cdot nH_2O$	Sborgite	$NaB_5O_8 \cdot 5H_2O$
Orpiment	As_2S_3	Scacchite	$MnCl_2$
Osbornite	TiN	Scheelite	$CaWO_4$
Otavite	$CdCO_3$	Schoepite	$UO_2(OH)_2 \cdot H_2O$
Oxammite	$(NH_4)_2C_2O_4 \cdot H_2O$	Schreibersite	Fe_3P
Palladinite	PdO ?	Schultenite	$PbHAsO_4$
Paramontroseite	VO_2	Scorodite	$FeAsO_4 \cdot 2H_2O$
Pararealgar	$\beta\text{-}As_4S_4$	Sedercholmite	$Ni_{0.95}Se$
Patronite	VS_{2+x}	Sellaite	MgF_2
Periclase	MgO	Senarmontite	$\beta\text{-}Sb_2O_3$
Perovskite	$(CaTi)O_3$	Shcherbinaite	V_2O_5
Petzite	$(AuAg_3)Te_2$	Siderazot	$Fe_{2+x}N$
Phenakite	Be_2SiO_4	Siderite	$FeCO_3$
Phosgenite	$Pb_2(CO_3)Cl_2$	Siderotil	$FeSO_4 \cdot 5H_2O$
Phosphammite	$(NH_4)_2HPO_4$	Silhydrite	$SiO_2 \cdot 0.33H_2O$
Phosphosiderite	$FePO_4 \cdot 2H_2O$ (mon)	Sillenite	Bi_2O_3 (cub)
Picromerite	$K_2Mg(SO_4)_2 \cdot 6H_2O$	Sillimanite	$\beta\text{-}Al_2(SiO_4)O$
Plattnerite	PbO_2	Sinjarite	$CaCl_2 \cdot 2H_2O$
Portlandite	$Ca(OH)_2$	Skutterudite	$CoAs_3$
Potash alum	$KAl(SO_4)_2 \cdot 12H_2O$	Smaltite	$CoAs_{3-x}$
Powellite	$CaMoO_4$	Smithite	$AgAsS_2$ (mon)
Proustite	Ag_3AsS_3 (trig)	Smithsonite	$ZnCO_3$
Pyrargyrite	$Ag_3[SbS_3]$	Soda alum	$NaAl(SO_4)_2 \cdot 12H_2O$
Pyrite	$\beta\text{-}Fe(S_2)$	Söhngeite	$Ga(OH)_3$
Pyrochroite	$Mn(OH)_2$	Sperrylite	$PtAs_2$
Pyrolusite	MnO_{2-x} (tetr)	Sphalerite	$\alpha\text{-}ZnS$
Pyromorphite	$Pb_5(PO_4)_3Cl$	Spherocobaltite	$CoCO_3$
Pyrophanite	$(TiMn)O_3$	Spinel (noble)	$(MgAl_2)O_4$
Pyrrhotine	$Fe_{1-x}S$	Stephanite	$Ag_5[SbS_4]$
Quartz	SiO_2 (hex; trig)	Stibnite	Sb_2S_3 (rhomb)
Rammelsbergite	$NiAs_2$	Stilleite	$ZnSe$
Ramsdellite	MnO_{2-x} (rhomb)	Stishovite	SiO_2 (tetr)
Rankinite	$Ca_3Si_2O_7$ (mon)	Stolzite	$PbWO_4$ (tetr)

Name	Formula	Name	Formula
Strengite	$FePO_4 \cdot 2H_2O$ (rhomb)	Ullmannite	$(NiSb)S$
Strontianite	$SrCO_3$	Ulvöspinel	$(TiFe_2)O_4$
Struvite	$MgNH_4PO_4 \cdot 6H_2O$	Uraninite	UO_{2+x}
Suolunite	$CaSiO_3 \cdot H_2O$	Vaesite	$Ni(S_2)$
Sylvanite	$(AuAg)Te_4$	Valentinite	$\alpha\text{-}Sb_2O_3$
Sylvite	KCl	Vanadinite	$Pb_5(VO_4)_3Cl$
Syngenite	$K_2Ca(SO_4)_2 \cdot H_2O$	Variscite	$AlPO_4 \cdot 2H_2O$
Szmikite	$MnSO_4 \cdot H_2O$	Vaterite	$CaCO_3$ (hex)
Tarapacaite	K_2CrO_4	Villiaumite	NaF
Tellurantimony	Sb_2Te_3	Vivianite	$Fe_3(PO_4)_2 \cdot 8H_2O$
Tellurite	TeO_2	Wattevillite	$Na_2Ca(SO_4)_2 \cdot 4H_2O$
Tellurobismutite	Bi_2Te_3	Wavellite	$Al_3(PO_4)_2(OH)_3 \cdot 5H_2O$
Tenorite	CuO	Westerveldite	$FeAs$
Tephroite	Mn_2SiO_4	Whewellite	$CaC_2O_4 \cdot H_2O$
Teschemacherite	NH_4HCO_3	Whitlockite	$Ca_3(PO_4)_2$
Thenardite	Na_2SO_4	Willemite	Zn_2SiO_4
Thermonatrite	$Na_2CO_3 \cdot H_2O$	Witherite	$BaCO_3$
Thorianite	ThO_2	Wollastonite	$CaSiO_3$
Thorite	$ThSiO_4$ (tetr)	Wulfenite	$PbMoO_4$
Tiemannite	$HgSe_{1\pm x}$	Wurtzite	$\beta\text{-}ZnS$
Tincal	$Na_2B_4O_7 \cdot 10H_2O$	Wüstite	$Fe_{1-x}O$
Tincalconite	$Na_2B_4O_7 \cdot 5H_2O$	Xanthiosite	$Ni_3(AsO_4)_2$
Trechmannite	$AgAsS_2$ (trig)	Xanthoconite	Ag_3AsS_3 (mon)
Trevorite	$(Fe_2Ni)O_4$	Xenotime	YPO_4
Tridymite	SiO_2 (hex; rhomb)	Yavapaiite	$KFe(SO_4)_2 \cdot 12H_2O$
Troilite	FeS	Zavaritskite	$Bi(O)F$
Trona	$Na_3CO_3(HCO_3) \cdot 2H_2O$	Zincite	ZnO
Tugarinovite	MoO_2	Zinkosite	$ZnSO_4$
Tungstenite	WS_2	Zircon	$ZrSiO_4$
Tungstite	$WO_3 \cdot H_2O$	Zircosulfate	$[Zr(H_2O)_4(SO_4)_2]$

6.8 Obsolete Formulas and Names [4, 6–8, 28, 29, 31]

Obsolete formulas and names of individual substances and ions thereof, as well as obsolete names of groups of elements and obsolete names of compounds and ions thereof continue to be encountered in chemical literature. Moreover, the names of compounds that heretofore have not been synthesized and also ions, whose presence in crystals or aqueous solutions has not been ascertained, are likewise met with in the literature and should be regarded as obsolete (the ions formed in high-temperature melts are not considered here).

These formulas and names are not recommended for use in the modern nomenclature of inorganic compounds and should be replaced by the names (systematic, traditional and special) adopted in this nomenclature.

6.8.1 Individual Substances

This section lists obsolete formulas and/or names of individual substances and their modern recommended equivalents.

The substances are arranged in the alphabetical order of the symbols of elements that determine the formulas in obsolete nomenclature. Thus, the list of obsolete names commences with the compounds of silver (Ag), followed by the compounds of aluminum (Al), arsenic (As), gold (Au), etc. For example, the obsolete formula and name of $HAlO_2$ should be looked for among aluminum compounds. A dash in the column "Modern nomenclature" means that the compound in question has not been synthesized (or is non-existent).

Obsolete nomenclature		Modern nomenclature	
AgO	argentic oxide	$(Ag^IAg^{III})O_2$	silver(I)-silver(III) oxide
AlO_2^-	metaaluminate	AlO_2^-	dioxoaluminate(III)
AlO_3^{3-}	orthoaluminate	–	
$HAlO_2$	metaaluminic acid	$AlO(OH)$	aluminum metahydroxide
H_3AlO_3	orthoaluminic acid	$Al(OH)_3$	aluminum hydroxide
AsH_3	arsenic hydride	AsH_3	arsine
$AuCl_3 \cdot HCl$	gold chloride	$H[AuCl_4]$	hydrogen tetrachloroaurate(III)
$Au(OH)_3$	auric acid	$Au(OH)_3$	gold(III) hydroxide
BO^+	boryl cation	–	
$H_2B_4O_6$	subtraboric acid	$H_2B_4O_6$	dihydrogen hexaoxotetraborate
H_3BO_3	orthoboric acid	$B(OH)_3$	boron trihydroxide
$H_4B_2O_4$	hypoboric acid	$B_2(OH)_4$	diboron tetrahydroxide
BeO_2^{2-}	beryllate	BeO_2^{2-}	dioxoberyllate(II)
H_2BeO_2	beryllic acid	$Be(OH)_2$	beryllium hydroxide
BiH_3	bismuth hydride	BiH_3	bismuthine
BiO^+	bismuthyl cation	–	
BrO_2^+	bromyl cation	–	
BrO_3^+	perbromyl cation	–	
HBr	hydrobromic acid	HBr (aqua)	hydrobromic acid
CO	carbon oxide	CO	carbon monoxide
CO^{2+}	carbonyl cation	–	
COS	carbon sulfoxide	CSO	carbon sulfide-oxide
CS_4^{2-}	perthiocarbonate	$CS_2(S_2)^{2-}$	dithio{disulfido(2–)}carbonate(IV)
$HCNO$	fulminating acid	$HCNO$	hydrogen fulminate
HCO_3^-	bicarbonate	HCO_3^-	hydrogencarbonate
$HNCO$	isocyanic acid	$HNCO$	hydrogen cyanate-N
$HSCN$	rhodanic acid	$HNCS$	hydrogen thiocyanate
NC^-	isocyanide	CN^-	cyanide
NCO^-	isocyanate	OCN^-	cyanate
SCN^-	rhodanide	NCS^-	thiocyanate
$Ce(OH)_4$	ceric hydroxide	$Ce(OH)_4$	cerium(IV) hydroxide
$ClNO_3$	chlorine nitrate	$(NO_2)ClO$	nitryl hypochlorite
ClO_2^+	chloryl cation	–	
ClO_3^+	perchloryl cation	–	

Obsolete nomenclature		Modern nomenclature	
HCl	hydrochloric acid	HCl (aqua)	hydrochloric acid
$Co(OH)_2$	cobaltous hydrate	$Co(OH)_2$	cobalt(II) hydroxide
$Co(OH)_3$	cobaltic hydrate	$Co(OH)_3$	cobalt(III) hydroxide
CrO_2^-	chromite	CrO_2^-	dioxochromate(III)
CrO_2^+	chromyl cation	–	
CrO_4^{3-}	hypochromate	CrO_4^{3-}	tetraoxochromate(V)
CrO_5^{4-}	orthochromate	CrO_5^{4-}	pentaoxochromate(VI)
$Cr_2O_7^{2-}$	bichromate	$Cr_2O_7^{2-}$	dichromate
$H_2Cr_2O_7$	bichromic acid	$H_2Cr_2O_7$*	dichromic acid
CuI_2	copric iodide	–	
CuO_2^{2-}	cuprate	CuO_2^{2-}	dioxocuprate(II)
FNO_3	fluorine nitrate	$(NO_2)OF$	nitryl fluorooxygenate(0)
HF_2^-	bifluoride	HF_2^-	hydrodifluoride
HFO	hypofluoric acid	HOF	hydrogen fluorooxygenate(0)
FeO_2^-	ferrite	FeO_2^-	dioxoferrate(III)
Fe_3O_4	ferrous-ferric oxide	$(Fe^{II}Fe_2^{III})O_4$	iron(II)-diiron(III) oxide
$Fe(OH)_2$	ferrous hydrate	$Fe(OH)_2$	iron(II) hydroxide
$Fe(OH)_3$	ferric hydrate	$Fe(OH)_3$*	iron(III) hydroxide
$HFeO_2$	metaferric acid	FeO(OH)	iron metahydroxide
GeH_4	germanium hydride	GeH_4	monogermane
H_4GeO_4	orthogermanic acid	$Ge(OH)_4$*	germanium(IV) hydroxide
H^+	hydrogen	H^+	hydrogen cation
H_3O^+	hydronium cation; hydroxonium cation	H_3O^+	oxonium cation
HfO^{2+}	hafnyl cation	–	
$Hf(OH)_4$	hafnium hydroxide	$Hf(OH)_4$	hafnium(IV) hydroxide
HgO	mercuric oxide	HgO	mercury(II) oxide
Hg_2O	mercurous oxide	–	
$Hg(OH)_2$	mercuric hydrate	$Hg(OH)_2$*	mercury(II) hydroxide
$Hg_2(OH)_2$	mercurous hydrate	$Hg_2(OH)_2$*	dimercury dihydroxide
Hg_2S	mecurous sulfide	–	
IO_2^+	iodyl cation	–	
IO_3^+	periodyl cation	–	
IO_5^{3-}	mesoperiodate	IO_5^{3-}	pentaoxoiodate(VII)
HI	hydroiodic acid	HI (aqua)	hydroiodic acid
$[InH_4]^-$	indanate	$[InH_4]^-$	tetrahydridoindate(III)
MnO	manganous oxide	MnO	manganese(II) oxide
MnO_2^-	hypomanganite	MnO_2^-	dioxomanganate(III)
MnO_2	manganese dioxide	MnO_2	manganese(IV) oxide
MnO_3^{2-}	metamanganite	–	
MnO_3^+	manganyl cation	–	
MnO_4^{4-}	orthomanganite	–	
MnO_4^{3-}	hypomanganate	MnO_4^{3-}	tetraoxomanganate(V)
Mn_2O_3	manganic oxide	Mn_2O_3	manganese(III) oxide
$Mn(OH)_2$	manganous hydroxide	$Mn(OH)_2$	manganes(II) hydroxide
$Mn(OH)_3$	manganic hydroxide	$Mn(OH)_3$*	manganese(III) hydroxide
H_2MnO_3	metamanganous acid	–	
H_2MnO_4	manganic acid	–	

Obsolete nomenclature		Modern nomenclature	
H_3MnO_4	hypomanganic acid	–	
H_4MnO_4	orthomanganous acid	$Mn(OH)_4^{\bullet}$	manganese(IV) hydroxide
MoO_2^{2+}	molybdenyl cation	–	
$N_2H_4 \cdot HCl$	hydrazine hydrochloride	N_2H_5Cl	hydrazinium(1+) chloride
$N_2H_4 \cdot 2HCl$	hydrazine hydrochloride	$N_2H_6Cl_2$	hydrazinium(2+) chloride
$N_2H_5^+$	hydrazonium cation	$N_2H_5^+$	hydrazinium(1+) cation
$N_2H_6^{2+}$	hydrazinium cation	$N_2H_6^{2+}$	hydrazinium(2+) cation
NH_4OH	ammonium hydroxide	$NH_4OH^{\bullet}$	ammonium hydroxide
NO	nitrogen oxide	NO	nitrogen monooxide
NO^+	nitrozonium cation	NO^+	nitrosyl cation
NO_2^+	nitronium cation; nitroxyl cation; nitronyl cation	NO_2^+	nitryl cation
N_2O	nitrous oxide	N_2O	dinitrogen oxide
$N_2O_3^{2-}$	hyponitrate	$N_2O_3^{2-}$	trioxodinitrate(2−)
N_2O_3	nitrous anhydride	N_2O_3	dinitrogen trioxide
$N_2O_4^{4-}$	hydronitrite	$N_2O_4^{4-}$	tetraoxodinatrate(II)
N_2O_4	hyponitric oxide; nitrogen tetroxide	N_2O_4	dinitrogen tetraoxide
N_2O_5	nitric anhydride	N_2O_5	dinitrogen pentaoxide
HN_3	hydrazoic acid	NH_3	hydrogen azide
$Ni(OH)_2$	nickelous hydroxide	$Ni(OH)_2$	nickel(II) hydroxide
O_3F_2	ozone fluoride	–	
O_4F_2	oxozone fluoride	O_4F_2	tetraoxygen difluoride
OH^-	hydroxyl	OH^-	hydroxide
HO_2	hydroperoxyl	HO_2	hydrogen superoxide
$[Os(N)O_3],H$	osmiamic acid	$H[Os(N)O_3]$	hydrogen nitridotrioxoosmate(VII)
PH_3	hydrogen phosphide	PH_3	phosphine
P_2H_4	disphosphine	P_2H_4	diphosphane
PO^{3+}	phosphoryl cation	$P(OH)_4^+$	tetrahydroxophosphorus(V) cation
$P_2O_6^{4-}$	hypophosphate	$P_2O_6^{4-}$	hexaoxodiphosphate(4−)
$P_2O_7^{4-}$	pyrophosphate	$P_2O_7^{4-}$	diphosphate
HPO_3^{2-}	phosphite	PHO_3^{2-}	hydrogentrioxophosphate(III); phosphonate
$HP_2O_5^{3-}$	diphosphite $P_2HO_5^{3-}$	$P_2HO_5^{3-}$	hydrogenpentaoxophosphate(III)
$HP_2O_6^{3-}$	isohypophosphate; phosphite-phosphate	$P_2HO_6^{3-}$	hydrogenhexaoxophosphate(3−)
$H_2PO_2^-$	hypophosphite	$PH_2O_2^-$	dihydrogendioxophosphate(I); phosphinate
$H_2PO_3^-$	biphosphite	$H(PHO_3)^-$	hydrogen trioxohydrogenphosphate(III); hydrogenphosphonate
$H_2P_2O_5^{2-}$	pyrophosphite	$P_2H_2O_5^{2-}$	dihydrogenpentaoxodiphosphate(III)
H_3PO_2	hypophosphorous acid	$H(PH_2O_2)$	hydrogen dihydrogendioxophosphate(I); phosphinic acid
H_3PO_3	phosphorous acid	$H_2(PHO_3)$	hydrogen hydrogentrioxophosphate(III), phosphonic acid

(Continued)

Obsolete nomenclature		Modern nomenclature	
$H_4P_2O_5$	pyrophosphorous acid	$H_2(P_2H_2O_5)$	hydrogen dihydrogenpentaoxodiphosphate(III)
$H_4P_2O_6$	phosphorous-phosphoric acid	$H_3(P_2HO_6)$	trihydrogen hydrogenhexaoxodiphosphate
$H_4P_2O_6$	hypophosphoric acid	$H_4P_2O_6$	tetrahydrogen hexaoxodiphosphate
$H_4P_2O_7$	pyrophosphoric acid	$H_4P_2O_7$	diphosphoric acid
PbO_2^{2-}	plumbite	PbO_2^{2-}	dioxoplumbate(II)
PbO_3^{2-}	metaplumbate	PbO_3^{2-}	trioxoplumbate(IV)
PbO_4^{4-}	orthoplumbate	PbO_4^{4-}	tetraoxoplumbate(IV)
Pb_3O_4	plumbous-plumbic oxide	$(Pb_2^{II}Pb^{IV})O_4$	dilead(II)-lead(IV) oxide
$Pb_2(PbO_4)$	lead orthoplumbate	$(Pb_2^{II}Pb^{IV})O_4$	dilead(II)-lead(IV) oxide
H_2PbO_3	metaplumbic acid	$-$	
H_4PbO_4	orthoplumbic acid	$Pb(OH)_4^•$	lead(IV) hydroxide
ReO_3^-	metahyporhenate	ReO_3^-	trioxorhenate(V)
ReO_3^{2-}	metarhenite	ReO_3^{2-}	trioxorhenate(IV)
ReO_4^{4-}	orthorhenite	ReO_4^{4-}	tetraoxorhenate(IV)
ReO_4^{3-}	orthohypothenate	ReO_4^{3-}	tetraoxorhenate(V)
ReO_5^{6-}	mesorhenite	ReO_5^{6-}	pentaoxorhenate(IV)
ReO_5^{3-}	mesoperrhenate	ReO_5^{3-}	pentaoxorhenate(VII)
$Re_2O_7^{4-}$	pyrohyporhenate	$Re_2O_7^{4-}$	heptaoxodirhenate(V)
$HReO_4$	rhenic acid	$H_4Re_2O_9$	hydrogen nonaoxodirhenate(VII)
RuO_4^-	perruthenate	RuO_4^-	tetraoxoruthenate(VII)
SO^{2+}	thionyl cation	$-$	
SO_2^{2-}	sulfoxylate	SO_2^{2-}	dioxosulfate(II)
SO_2^{2+}	sulfuryl cation	$-$	
SO_5^{2-}	permonosulfate	$SO_3(O_5)^{2-}$	peroxosulfate
$S_2O_3^{2-}$	hyposulfite	$S_2O_3^{2-}$	thiosulfate
$S_2O_4^{2-}$	dithionite; hydrosulfite	$S_2O_4^{2-}$	tetraoxodisulfate(2−)
$S_2O_5^{2-}$	pyrosulfite; metabisulfite	$S_2O_5^{2-}$	pentaoxoxisulfate(IV)
$S_2O_7^{2-}$	pyrosulfate	$S_2O_7^{2-}$	disulfate
$S_2O_8^{2-}$	persulfate	$S_2O_6(O_2)^{2-}$	peroxodisulfate
$S_4O_{14}^{2-}$	pertetrasulfate	$S_4O_{12}(O_2)^{2-}$	dodecaoxoperoxosulfate(IV)
HS^-	bisulfide	HS^-	hydrosulfide
H_2S_2	hydrogen disulfide	H_2S_2	disulfane
HSO_3^-	bisulfite	HSO_3^-	hydrosulfite
HSO_4^-	bisulfate	HSO_4^-	hydrosulfate
H_2SO_2	sulfoxylic acid	H_2SO_2	hydrogen dioxosulfate(II)
H_2SO_3	sulfurous acid	$SO_2 \cdot nH_2O$	sulfur dioxide polyhydrate
H_2SO_5	permonosulfuric acid	$H_2SO_3(O_2)$	peroxosulfuric acid
$H_2S_2O_2$	thiosulfurous acid	$H_2S_2O_2$	dihydrogen dioxodisulfate
$H_2S_2O_3$	hyposulfurous acid	H_2SO_3S	thiosulfuric acid
$H_2S_2O_4$	dithionous acid	$H_2S_2O_4^•$	dihydrogen tetraoxodisulfate
$H_2S_2O_7$	pyrosulfuric acid	$H_2S_2O_7$	disulfuric acid
$H_2S_2O_8$	persulfuric acid	$H_2S_2O_6(O_2)$	peroxodisulfuric acid
$H_2S_nO_6$	sulfandisulfonic acids	$H_2S_nO_6^•$	polythionic acids
HSO_2F	fluorosulfinic acid	HSO_2F	hydrogen fluorosulfite

(Continued)

Obsolete nomenclature		Modern nomenclature	
$HSO_3(NH_2)$	sulfonamic acid	$HSO_3(NH_2)$	aminosulfonic acid
$NOHSO_4$	nitrosylsulfuric acid	$NOHSO_4$	nitrosyl hydrogensulfate
SbH_3	antimony hydride	SbH_3	stibine
SbO^+	antimonyl cation	—	
SbO_2^-	antimonite	SbO_2^-	dioxostibate(III)
SbO_3^-	metaantimonate	SbO_3^-	trioxostibate(V)
SbO_4^{3-}	orthoantimonate	SO_4^{3-}	tetraoxostibate(V)
$Sb(OH)_3$	antimony hydroxide	$Sb(OH)_3^*$	antimony(III) hydroxide
SbS_4^{3-}	thioantimonate	$[SbS_4]^{3-}$	tetrathiostibate(V)
$HSbO_2$	metastibous acid	$SbO(OH)$	antimony metahydroxide
H_3SbO_4	stibic acid	$Sb(OH)_5^*$	antimony(V) hydroxide
SeO^{2+}	selenyl cation	—	
$Se_2O_7^{2-}$	pyroselenate	$Se_2O_7^{2-}$	diselenate
$H_2Se_2O_7$	pyroselenic acid	$H_2Se_2O_7$	diselenic acid
SiH_4	silicon hydride	SiH_4	monosilane
SiO^{2+}	silicyl cation	—	
H_2SiO_3	silicic acid	$H_4SiO_4^*$	orthosilic acid
SnO_2^{2-}	stannite	SnO_2^{2-}	dioxostannate(II)
SnO_3^{2-}	metastannate	SnO_3^{2-}	trioxostannate(IV)
SnO_4^{4-}	orthostannate	SnO_4^{4-}	tetraoxostannate(IV)
SnS_3^{2-}	thiostannate	$[SnS_3]^{2-}$	trithiostannate(IV)
H_2SnO_3	metastannic acid	—	
H_4SnO_4	orthostannic acid	$Sn(OH)_4^*$	tin(IV) hydroxide
TeO^{2+}	telluryl cation	—	
H_2TeO_4	allotelluric acid	H_2TeO_4	metatelluric acid
TiO^{2+}	titanyl cation	—	
TiO_3^{2-}	metatitanate	TiO_3^{2-}	trioxotitanate(IV)
TiO_4^{4-}	orthotitanate	TiO_4^{4-}	tetraoxotitanate(IV)
H_2TiO_3	thallic acid	$TiO(OH)_2$	titanium oxide-dihydroxide
$Tl(OH)_3$	thallium hydroxide	$Tl(OH)_3^*$	thallium(III) hydroxide
U_3O_8	uranium(5, 6) oxide	$(U_2^VU^{VI})O_8$	diuranium(V)-uranium(VI) oxide
H_2UO_4	uranic acid	$UO_2(OH)_2$	uranyl hydroxide
VO_3^{2-}	metavanadite	VO_3^{2-}	trioxovanadate(IV)
VO_4^{4-}	orthovanadite	VO_4^{4-}	tetraoxovanadate(IV)
$V_4O_9^{2-}$	hvpovanadate; vanadite	$V_4O_9^{2-}$	nonaoxotetravanadate(IV)
VS_4^{3-}	thiovanadate	VS_4^{3-}	tetrathiovanadate(V);
			tetrathioorthovanadate
HVO_3	metavanadic acid	$V_2O_5 \cdot H_2O$	vanadium(V) oxide hydrate
H_3VO_4	orthovanadic acid	$V_2O_5 \cdot 3H_2O$	vanadium(V) oxide trihydrate
$H_4V_2O_7$	pyrovanadic acid	$V_2O_5 \cdot 2H_2O$	vanadium(V) oxide dihydrate
WO_2^{2+}	wolframyl cation	—	
XeO_6^{4-}	perxenate	XeO_6^{4-}	hexaoxoxenonate(VIII)
ZnO_2^{2-}	zincate	ZnO_2^{2-}	dioxozincate(II)
H_2ZnO_2	zincic acid	$Zn(OH)_2$	zinc hydroxide
ZrO^{2+}	zirconyl cation	—	
$Zr(OH)_4$	zirconium hydroxide	$Zr(OH)_4^*$	zirconium(IV) hydroxide

* The compound exists only in a dilute aqueous solution in the form of molecules and/or ions.

6.8.2 Collective of Substances

Obsolete names for the groups of elements, compounds and ions are listed below together with their recommended nomenclature counterparts.

Actinide	actinoid
Antimonide	stibide
Bichromate	dichromate
Boron hydride	borane
Halide	halogenide
Haloid	halogen
Hydrochloride	hydride-chloride or adduct with HCl
Hydrohalide	hydride-halogenide
Hypophosphite	phosphinate
Lanthanide	lanthanoid
Mixed oxide	double oxide
Mixed sulfide	double sulfide
Oxychloride	chloride-oxide
Phosphate, primary	dihydrogenorthophosphate
Phosphate, secondary	hydrogenorthophosphate
Phosphate, tertiary	orthophosphate
Pyrophosphate	diphosphate
Pyrosulfate	disulfate
Silicon hydride	silane
Subhalide	halogenide
Subnitride	nitride
Suboxide	oxide
Subsulfide	sulfide
Thioanhydride	sulfide-oxide
Thiohalide	sulfide-halogenide

6.9 The Names of Inorganic Substances in French, German, and Russian

Below there are lists of the names of chemical elements, the systematic, special, and traditional names of inorganic substances, the names of groups of elements, the special names of polyelement cations, the systematic traditional, and special names of anions, and special names of ligands, some mineralogical and trivial names of inorganic substances, and also numerical, multiplying, and other prefixes.

Abbreviations:

m, м	masculine	*n, c*	neuter
f, ж	feminine	*мн*	plural

English name	French, German, and Russian names
Acanthite	acanthite *f*, Akanthit *m*, акантит *м*
Acetate	acétate *m*, Azetat *n*, ацетат *м*
Acetylide	acétylure *m*, Azetylid *n*, ацетиленид *м*
Acid	acide *m*, Säure *f*, кислота *ж*
aminosulfonic ~	~ aminosulfonique, Aminosulfonsäure *f*, аминосульфоновая
arsenic ~	~ arsénique, Arsensäure *f*, мышьяковая ~
arsenious ~	~ arsénieux, arsenige ~, мышьяковистая ~
boric ~	~ borique, Borsäure *f*, борная ~
bromic ~	~ bromique, Bromsäure *f*, бромноватая ~
carbonic ~	~ carbonique, Kohlensäure *f*, угольная ~
chloric ~	~ chlorique, Chlorsäure *f*, хлорноватая ~
chlorous ~	~ chloreux, chlorige ~, хлористая ~
chromic ~	~ chromique, Chromsäure *f*, хромовая ~
dichromic ~	~ dichromique, Dichromsäure *f*, дихромовая ~
diphosphoric ~	~ diphosphorique, Diphosphorsäure *f*, дифосфорная ~
diselenic ~	~ disélénique, Diselensäure *f*, диселеновая ~
disulfuric ~	~ sulfurique, Dischwefelsäure *f*, дисерная ~
fluorselenic ~	~ fluorsélénique, Fluorselensäure *f*, фторселеновая ~
fluorsulfonic ~	~ fluorsulfonique, Fluorsulfonsäure *f*, фторсульфоновая ~
formic ~	~ formique, Ameisensäure *f*, муравьиная ~
germanic ~	~ germanique, Germaniumsäure *f*, германиевая ~
hydroazoic ~	~ azidhydrique, Stickstoffwasserstoffsäure *f*, азидоводородная ~
hydrobromic ~	~ bromhydrique, Bromwasserstoffsäure *f*, бромоводородная ~
hydrochloric ~	~ chlorhydrique, Chlorwasserstoffsäure *f*, хлороводородная ~
hydrocyanic ~	~ cyanhydrique, Zyanwasserstoffsäure *f*, циановодородная ~
hydrofluoric ~	~ fluorhydrique, Fluorwasserstoffsäure *f*, фтороводородная ~
hydroiodic ~	~ iodhydrique, Jodwasserstoffsäure *f*, иодоводородная ~
hydroselenic ~	~ sélenhydrique, Selenwasserstoffsäure *f*, селеноводородная ~
hydrosulfuric ~	~ sulfhydrique, Schwefelwasserstoffsäure *f*, сероводородная ~
hydrotelluric ~	~ tellurhydrique, Tellurwasserstoffsäure *f*, теллуроводородная ~
hypobromous ~	~ hypobromeux, hypobromige ~, бромноватистая ~
hypochlorous ~	~ hypochloreux, hypochlorige ~, хлорноватистая ~
hypoiodous ~	hypo-iodeux, hypojodige, иодноватистая ~
hyponitrous ~	~ hyponitreux, hyposalpetrige ~, азотноватистая ~
iodic ~	~ iodique, Jodsäure *f*, иодноватая ~
metaarsenious ~	~ méta-arsénieux, metaarsenige ~, метамышьяковистая ~
metaboric ~	~ métaborique, Metaborsäure *f*, метаборная ~
metaperiodic ~	~ métaperiodique, Metaperjodsäure *f*, метаиодная ~
metaphosphoric ~	~ métaphosphorique, Metaphosphorsäure *f*, метафосфорная ~
metatelluric ~	~ métatellurique, Metatellursäure *f*, метателлуровая ~
nitric ~	~ nitrique, Salpetersäure *f*, азотная ~
nitrous ~	~ nitreux, salpetrige ~, азотистая ~
orthoarsenious ~	~ ortho-arsénieux, orthoarsenige ~, ортомышьяковистая ~
orthoperiodic ~	~ orthoperiodique, Orthoperjodsäure *f*, ортоиодная ~
orthophosphoric ~	~ orthophosphorique, Orthophosphorsäure *f*, ортофосфорная ~
orthosilicic ~	~ orthosilicique, Orthokieselsäure *f*, ортокремниевая ~
orthotelluric ~	~ orthotellurique, Orthotellursäure *f*, ортотеллуровая ~
oxalic ~	~ oxalique, Oxalsäure *f*, щавелевая ~
perbromic ~	~ perbromique, Perbromsäure *f*, бромная ~

(*Continued*)

English name	French, German, and Russian names
perchloric ~	~ perchlorique, Perchlorsäure *f*, хлорная ~
periodic ~	~ periodique, Perjodsäure *f*, иодная ~
permanganic ~	~ permanganique, Permagansäure *f*, марганцовая ~
peroxodisulfuric ~	~ peroxodisulfurique, Peroxodischwefelsäure *f*, пероксодисерная ~
peroxosulfuric ~	~ peroxosulfurique, Peroxoschwefelsäure *f*, пероксосерная ~
pertechnetic ~	~ pertechnétique, Pertechnetiumsäure *f*, технециевая ~
phosphoric ~	~ phosphorique, Phosphorsäure *f*, фосфорная ~
polythionic ~	~ polythionique, Polythionsäure *f*, политионовая ~
selenic ~	~ sélénique, Selensäure *f*, селеновая ~
selenious ~	~ sélénieux, selenige ~, селенистая ~
silicic ~	~ silicique, Kieselsäure *f*, кремниевая ~
sulfonic ~	~ sulfonique, Sulfonsäure *f*, сульфоновая ~
sulfuric ~	~ sulfurique, Schwefelsäure *f*, серная ~
telluric ~	~ tellurique, Tellursäure *f*, теллуровая ~
tellurious ~	~ tellureux, tellurige ~, теллуристая ~
thionic ~	~ thionique, Thionsäure *f*, тионовая ~
thiosulfuric ~	~ thiosulfurique, Thioschwefelsäure *f*, тиосерная ~
Actinium	actinium *m*, Aktinium *n*, актиний *м*
Actinoid	actinoïde *m*, Aktinoid *n*, актиноид *м*
Agate	agate *f*, Achat *m*, агат *м*
Alebaster	albâtre *m*, Alabaster *m*, алебастр *м*
Alexandrite	alexandrite *f*, Alexandrit *m*, александрит *м*
Alum	alun *m*, Alaun *m*, квасцы *мн*
Alumina	alumine *f*, Tonerde *f*, глинозём *м*
Aluminate	aluminate *m*, Aluminat *n*, алюминат *м*
Aluminum	aluminium *m*, Aluminium *n*, алюминий *м*
Alundum	alundum *m*, Alundum *n*, алунд *м*
Amalgam	amalgame *m*, Amalgam *n*, амальгама *ж*
Americium	américium *m*, Amerizium *n*, америций *м*
Amethyst	améthyste *f*, Amethyst *m*, аметист *м*
Amide	amidure *m*, Amid *n*, амид *м*
Ammine-	ammine-, ammin-, аммин-
Ammonia	ammoniac *m*, Ammoniak *n*, аммиак *м*
Ammonium	ammonium *m*, Ammonium *n*, аммоний *м*
Anatase	anatase *f*, Anatas *m*, анатаз *м*
Anhydrite	anhydrite *m*, Anhydrit *m*, ангидрит *м*
Anion	anion *m*, Anion *n*, анион *м*
Antimony	antimoine *m*, Antimon *n*, сурьма *ж*
Aqua-	aqua-, aqua-, аква-
Aragonite	aragonite *f*, Aragonit *m*, арагонит *м*
Argentate	argentate *m*, Argentat *n*, аргентат *м*
Argentite	argentite *f*, Argentit *m*, аргентит *м*
Argon	argon *m*, Argon *n*, аргон *м*
Arsenate	arséniate *m*, Arsenat *n*, арсенат *м*
Arsenic	arsenic *m*, Arsen *n*, мышьяк *м*
Arsenide	arséniure *m*, Arsenid *n*, арсенид *м*
Arsenite	arsénite *m*, Arsenit *n*, арсенит *м*
Arsenopyrite	mispickel *m*, Arsenopyrit *m*, арсенопирит *м*

English name	French, German, and Russian names
Arsine	arsine *f*, Arsin *n*, арсин *м*
Astatine	astate *m*, Astat *n*, астат *м*
Aurate	aurate *m*, Aurat *n*, аурат *м*
Azide	azoture *m*, Azid *n*, азид *м*
Babbitt	babbit *m*, Babbitt *n*, баббит *м*
Baddeleyite	baddéléyite *f*, Baddeleyit *m*, бадделеит *м*
Barite	baryte *f*, Baryt *m*, барит *м*
Barium	baryum *m*, Barium *n*, барий *м*
Bauxite	bauxite *f*, B(e)auxit *m*, боксит *м*
Berkelium	berkélium *m*, Berkelium *n*, берклий *м*
Berlin blue	bleu *m* de Berlin, Berliner Blau *n*, лазурь *ж*, берлинская
Beryllate	béryllate *m*, Beryllat *n*, бериллат *м*
Beryllium	béryllium *m*, Beryllium *n*, бериллий *м*
Bis-	bis-, bis-, бис-
Bismuth	bismuth *m*, Wismut *n*, висмут *м*
Bismuthate	bismuthate *m*, Bismutat *n*, висмутат *м*
Bismuthinite	bismuthinite *f*, Bismuthin(it) *m*, висмутин *м*
Blende	blende *f*, Blende *f*, обманка *ж*
Boehmite	bohémite *f*, Böhmit *m*, бёмит *м*
Borane	borane *m*, Boran *n*, боран *м*
Borate	borate *m*, Borat *n*, борат *м*
Borax	borax *m*, Borax *m*, бура *ж*
Borazine	borazine *f*, Borazin *n*, боразин *м*
Borazol	borazole *m*, Borazol *n*, боразол *м*
Borazon	borazone *m*, Borazon *n*, боразон *м*
Boride	borure *m*, Borid *n*, борид *м*
Boron	bore *m*, Bor *n*, бор *м*
Brass	laiton *m*, Messing *n*, латунь *ж*
Bromate	bromate *m*, Bromat *n*, бромат *м*
Bromide	bromure *m*, Bromid *n*, бромид *м*
Bromine	brome *m*, Brom *n*, бром *м*
Bromite	bromite *m*, Bromit *n*, бромит *м*
Bronze	bronze *m*, Bronze *f*, бронза *ж*
Brookite	brookite *f*, Brookit *m*, брукит *м*
Cadmate	cadmate *m*, Kadmat *n*, кадмат *м*
Cadmium	cadmium *m*, Kadmium *n*, кадмий *м*
Calcite	calcite *f*, Kalzit *m*, кальцит *м*
Calcium	calcium *m*, Kalzium *n*, кальций *м*
Californium	californium *m*, Kalifornium *n*, калифорний *м*
Calomel	calomel *m*, Kalomel *n*, каломель *ж*
Carbamide	carbamide *m*, Karbamid *n*, карбамид *м*
Carbide	carbure *m*, Karbid *n*, карбид *м*
Carbon	carbone *m*, Kohlenstoff *m*, углерод *м*
Carbonate	carbonate *m*, Karbonat *n*, карбонат *м*
Carbonyl-	carbonyl-, karbonyl-, карбонил-
Carborundum	carborundum *m*, Karborundum *n*, карборунд *м*
Carnallite	carnallite *f*, Karnallit *m*, карналлит *м*
Cassiterite	cassitérite *f*, Kassiterit *m*, касситерит *м*

(Continued)

English name	French, German, and Russian names
Cation	cation *m*, Kation *n*, катион *м*
Celestine	célestine *f*, Zölestin *m*, целестин *м*
Cementite	cémentite *f*, Zementit, цементит *м*
Cerium	cérium *m*, Zer *n*, церий *м*
Cesium	césium *m*, Zäsium *n*, цезий *м*
Chalcedony	calcédoine *f*, Chalzedon *m*, халцедон *м*
Chalcogen	chalcogène *m*, Chalkogen *n*, халькоген *м*
Chalcopyrite	chalcopyrite *f*, Chalkopyrit *m*, халькопирит *м*
Chalk	craie *f*, Kreide *f*, мел *м*
Chloramine	chloramine *f*, Chloramin *n*, хлорамин *м*
Chlorate	chlorate *m*, Chlorat *n*, хлорат *м*
Chloride	chlorure *m*, Chlorid *n*, хлорид *м*
Chlorine	chlore *m*, Chlor *n*, хлор *м*
Chlorite	chlorite *m*, Chlorit *n*, хлорит *м*
Chromate	chromate *m*, Chromat *n*, хромат *м*
Chromium	chrome *m*, Chrom *n*, хром *м*
Chrysoberyl	chrysobéryl *m*, Chrysoberyll *m*, хризоберилл *м*
Cinnabar	cinabre *m*, Zinnober *m*, киноварь *ж*
Cis-	cis-, cis-, цис-
Coagulant	coagulant *m*, Koagulant *m*, коагулянт *м*
Cobalt	cobalt *m*, Kobalt *n*, кобальт *м*
Cobaltate	cobaltate *m*, Kobaltat *n*, кобальтат *м*
Constantan	constantan *m*, Konstantan *n*, константан *м*
Copper	cuivre *m*, Kupfer *n*, медь *ж*
Corundum	corindon *m*, Korund *m*, корунд *м*
Cristobalite	cristobalite *f*, Kristobalit *m*, кристобалит *м*
Crocoite	crocoïte *f*, Krokoit *m*, крокоит *м*
Cryolite	cryolithe *f*, Kryolith *m*, криолит *м*
Cuprate	cuprate *m*, Kuprat *n*, купрат *м*
Curium	curium *m*, Curium *n*, кюрий *м*
Cyanamide	cyanamide *m*, Zyanamid *n*, цианамид *м*
Cyanate	cyanate *m*, Zyanat *n*, цианат *м*
Cyanide	cyanure *m*, Zyanid *n*, цианид *м*
Deca-	déca-, deka-, дека-
Deuterium	deutérium *m*, Deuterium *n*, дейтерий *м*
Di-	di-, di-, ди-
Diamond	diamant *m*, Diamant *m*, алмаз *м*
Diaspore	diaspore *m*, Diaspor *m*, диаспор *м*
Diatomite	diatomite *f*, Diatomit *m*, диатомит *м*
Dicyanogen	dicyanogène *m*, Dizyan *n*, дициан *м*
Difluoramine	difluoramine *f*, Difluoramin *n*, дифторамин *м*
Dioxygenyl	dioxygénile *m*, Dioxygenyl *n*, диоксигенил *м*
Dodeca-	dodéca-, dodeka-, додека-
Dysprosium	dysprosium *m*, Dysprosium *n*, диспрозий *м*
Einsteinium	einsteinium *m*, Einsteinium *n*, эйнштейний *м*
Element	élément *m*, Element *n*, элемент *м*
alkaline-earth ~	~ alcalino-terreux, Erdalkalielement *n*, щёлочноземельный ~
alkaline ~	~ alcalin, Alkalielement *n*, щелочной ~

English name	French, German, and Russian names
Emery	émeri *m*, Schmirgel *m*, наждак *м*
Epsomite	epsomite *f*, Epsomit *m*, эпсомит *м*
Erbium	erbium *m*, Erbium *n*, эрбий *м*
Europium	europium *m*, Europium *n*, Европий *м*
Fermium	fermium *m*, Fermium *n*, фермий *м*
Ferrate	ferrate *m*, Ferrat *n*, феррат *м*
Ferrocene	ferrocène *m*, Ferrozen *n*, ферроцен *м*
Fluoride	fluorure *m*, Fluorid *n*, фторид *м*
Fluorine	fluor *m*, Fluor *n*, фтор *м*
Fluorite	fluorite *f*, Fluorit *m*, флюорит *м*
Fluoronium	fluoronium *m*, Fluoronium *n*, фтороний *м*
Formiate	formiate *m*, Formiat *n*, формиат *м*
Francium	francium *m*, Franzium *n*, франций *м*
Fulminate	fulminate *m*, Fulminat *n*, фульминат *м*
Gadolinium	gadolinium *m*, Gadolinium *n*, гадолиний *м*
Galenite	galénite *f*, Galenit *m*, галенит *м*
Gallate	gallate *m*, Gallat *n*, галлат *м*
Gallium	gallium *m*, Gallium *n*, галлий *м*
Gas	gaz *m*, Gas *n*, газ *м*
carbonic acid	~ carbonique, Kohlensäuregas *n*, углекислый ~
laughing ~	~ hilarant, Lachgas *n*, веселящий ~
noble ~	~ noble, Edelgas *n*, благородный ~
sulfurous ~	~ sulfureux, Schwefligsäuregas *n*, сернистый ~
Gaylussite	gay-lussite *f*, Gaylussit *m*, гейлюссит *м*
Germanate	germanate *m*, Germanat *n*, германат *м*
Germane	germane *m*, German *n*, герман *м*
Germanium	germanium *m*, Germanium *n*, германий *м*
Gold	or *m*, Gold *n*, золото *c*
Graphite	graphite *m*, Graphit *m*, графит *м*
Greenockite	greenockite *f*, Greenockit *m*, гринокит *м*
Gypsum	gypse *m*, Gips *m*, гипс *м*
Hafnate	hafnate *m*, Hafnat *n*, гафнат *м*
Hafnium	hafnium *m*, Hafnium *n*, гафний *м*
Halite	halite *f*, Halit *m*, галит *м*
Halogen	halogène *m*, Halogen *n*, галоген *м*
Hausmannite	hausmannite *f*, Hausmannit *m*, гаусманнит *м*
Heliotrope	héliotrope *m*, Heliotrop *m*, гелиотроп *м*
Helium	hélium *m*, Helium *n*, гелий *м*
Hematite	hématite *f*, Hämatit *m*, гематит *м*
Hepta-	hepta-, hepta-, гепта-
Hexa-	hexa-, hexa-, гекса-
Hexakis-	hexakis-, hexakis-, гексакис-
Holmium	holmium *m*, Holmium *n*, гольмий *м*
Hydrate	hydrate *m*, Hydrat *n*, гидрат *м*
Hydrazine	hydrazine *f*, Hydrazin *n*, гидразин *м*
Hydrazinium	hydrazinium *m*, Hydrazinium *n*, гидразиний *м*
Hydride	hydrure *m*, Hydrid *n*, гидрид *м*
Hydrogen	hydrogène *m*, Wasserstoff *m*, водород *м*

(Continued)

English name	French, German, and Russian names
Hydrogen-	hydrogéno-, hydrogen-, гидро-
Hydrogendifluoride	hydrogénodifluorure *m*, Hydrogendifluorid *n*, гидродифторид *м*
Hydrogenperoxide	hydrogénoperoxyde *m*, Hydrogenperoxid *n*, гидропероксид *м*
Hydrogensulfide	hydrogénosulfure *m*, Hydrogensulfid *n*, гидросульфид *м*
Hydroxide	hydroxyde *m*, Hydroxid *n*, гидроксид *м*
Hydroxo-	hydroxo-, hydroxo-, гидроксо-
Hydroxyl	hydroxyle *m*, Hydroxyl *n*, гидроксил *м*
Hydroxylamine	hydroxylamine *f*, Hydroxylamin *n*, гидроксиламин *м*
Hydroxylaminium	hydroxylaminium *m*, Hydroxylaminium *n*, гидроксиламиний *м*
Hyperoxide	hyperoxyde *m*, Hyperoxid *n*, надпероксид *м*
Hypobromite	hypobromite *m*, Hypobromit *n*, гипобромит *м*
Hypochlorite	hypochlorite *m*, Hypochlorit *n*, гипохлорит *м*
Hypoiodite	hypo-iodite *m*, Hypojodit *n*, гипоиодит *м*
Hyponitrite	hyponitrite *m*, Hyponitrit *n*, гипонитрит *м*
Hyposulfite	hyposulfite *m*, Hyposulfit *n*, гипосульфит *м*
Imide	imidure *m*, Imid *n*, имид *м*
Indate	indate *m*, Indat *n*, индат *м*
Indium	indium *m*, Indium *n*, индий *м*
Iodate	iodate *m*, Iodat[1] *n*, иодат *м*
Iodide	iodure *m*, Iodid[2] *n*, иодид *м*
Iodine	iode *m*, Jod *n*, иод *м*
Iodite	iodite *m*, Jodit *n*, иодит *м*
Ion	ion *m*, Ion *n*, ион *м*
Iridate	iridate *m*, Iridat *n*, иридат *м*
Iridium	iridium *m*, Iridium *n*, иридий *м*
Iron	fer *m*, Eisen *n*, железо *c*
Jasper	jaspe *m*, Jaspis *m*, яшма *ж*
Kaolin	kaolin *m*, Kaolin *m*, каолин *м*
Kieselguhr	kieselguhr *m*, Kieselguhr *f*, кизельгур *м*
Krypton	krypton *m*, Krypton *n*, криптон *м*
Kurchatovium	kurtchatovium *m*, Kurtschatowium *n*, курчатовий *м*
Lanthanoid	lanthanoïde *m*, Lanthanoid *n*, лантаноид *м*
Lanthanum	lanthane *m*, Lanthan *n*, лантан *м*
Lead	plomb *m*, Blei *n*, свинец *м*
Lime	chaux *f*, Kalk *m*, известь *ж*
bleaching powder	poudre *f* de blanchiment, Bleichkalk *m*, белильная
chlorinated ~	chlorure *m* de chaux, Chlorkalk *m*, хлорная ~
slaked ~	~ éteinte, gelöschter ~, гашёная ~
soda ~	~ sodée, Natronkalk *m*, натронная ~
unslaked	~ vive, ungelöschter ~, негашёная ~
Limestone	calcaire *m*, Kalkstein *m*, известняк *м*
Litharge	litharge *f*, glätte *f*, глёт *м*
Lithium	lithium *m*, Lithium *n*, литий *м*
Lithopone	lithopone *m*, Lithopone *f*, литопон *м*
Lutetium	lutécium *m*, Lutetium *n*, лютеций *м*
Magnesia	magnésie *f*, Magnesia *f*, магнезия *ж*
Magnesite	magnésite *f*, Magnesit *m*, магнезит *м*
Magnesium	magnésium *m*, Magnesium *n*, магний *м*

(Continued)

English name	French, German, and Russian names
Magnetite	magnétite *f*, Magnetit *m*, магнетит *м*
Malachite	malachite *f*, Malachit *m*, малахит *м*
Manganate	manganate *m*, Manganat *n*, манганат *м*
Manganese	manganèse *m*, Mangan *n*, марганец *м*
Marble	marbre *m*, Marmor *m*, мрамор *м*
Massicot	massicot *m*, Massicot *m*, массикот *м*
Mendelevium	mendélévium *m*, Mendelevium *n*, менделевий *м*
Mercapto-	mercapto-, merkapto-, меркапто-
Mercurate	mercurate *m*, Merkurat *n*, меркурат *м*
Mercury	mercure *m*, Quecksilber *n*, ртуть *ж*
Metaarsenite	méta-arsénite *m*, Metaarsenit *n*, метаарсенит *м*
Metaborate	métaborate *m*, Metaborat *n*, метаборат *м*
Metacinnabar	métacinabre *m*, Metazinnabarit *m*, метациннабарит *м*
Metahydroxide	métahydroxyde *m*, Metahydroxid *n*, метагидроксид *м*
Metaperiodate	métaperiodate *m*, Metaperjodat *n*, метапериодат *м*
Metaphosphate	métaphosphate *m*, Metaphosphat *n*, метафосфат *м*
Metasilicate	métasilicate *m*, Metasilikat *n*, метасиликат *м*
Metatellurate	métatellurate *m*, Metatellurat *n*, метателлурат *м*
Metavanadate	métavanadate *m*, Metavanadat *n*, метаванадат *м*
Minium	minium *m*, Mennige *f*, сурик *м*
Mirabilite	mirabilite *f*, Mirabilit *m*, мирабилит *м*
Molybdate	molybdate *m*, Molybdat *n*, молибдат *м*
Molybdenum	molybdène *m*, Molybdän *n*, молибден *м*
Mono-	mono-, mono-, моно-
Monokis-	monokis-, monokis-, монокис-
Natron	natron *m*, Natron *n*, натрон *м*
Neodymium	néodyme *m*, Neodym *n*, неодим *м*
Neon	néon *m*, Neon *n*, неон *м*
Neptunium	neptunium *m*, Neptunium *n*, нептуний *м*
Niccolate[3]	niccolate[3] *m*, Niccolat[3] *n*, никколат *м*
Nickel	nickel *m*, Nickel *n*, никель *м*
Niobate	niobate *m*, Niobat *n*, ниобат *м*
Niobium	niobium *m*, Niob *n*, ниобий *м*
Nitrate	nitrate *m*, Nitrat *n*, нитрат *м*
Nitride	nitrure *m*, Nitrid *n*, нитрид *м*
Nitrite	nitrite *m*, Nitrit *n*, нитрит *м*
Nitro-	nitro-, nitro-, нитро-
Nitrogen	azote *m*, Stickstoff *m*, азот *м*
Nitrosyl	nitrosyle *m*, Nitrosyl *n*, нитрозил *м*
Nitryl	nitryle *m*, Nitryl *n*, нитроил *м*
Nona-	nona-, nona-, нона-
Ocher	ocre *f*, Ocker *m*, охра *ж*
Octa-	octa-, okta-, окта-
Oleum	oléum *m*, Oleum *n*, олеум *м*
Onyx	onyx *m*, Onyx *m*, оникс *м*
Opal	opale *f*, Opal *m*, опал *м*
Ore	minerai *m*, Erz *n*, руда *ж*
Orpiment	orpiment *m*, Auripigment *n*, аурипигмент *м*

(*Continued*)

English name	French, German, and Russian names
Orthoarsenite	ortho-arsénite *m*, Orthoarsenit *n*, ортоарсенит *м*
Orthoborate	orthoborate *m*, Orthoborat *n*, ортоборат *м*
Orthoperiodate	orthoperiodate *m*, Orthoperjodat *n*, ортопериодат *м*
Orthophosphate	orthophosphate *m*, Orthophosphat *n*, ортофосфат *м*
Orthosilicate	orthosilicate *m*, Orthosilikat *n*, ортосиликат *м*
Orthotellurate	orthotellurate *m*, Orthotellurat *n*, ортотеллурат *м*
Orthovanadate	orthovanadate *m*, Orthovanadat *n*, ортованадат *м*
Osmate	osmate *m*, Osmat *n*, осмат *м*
Osmium	osmium *m*, Osmium *n*, осмий *м*
Oxalate	oxalate *m*, Oxalat *n*, оксалат *м*
Oxide	oxyde *m*, Oxid *n*, оксид *м*
Oxo-	оxо-, охо-, оксо-
Oxonium	oxonium *m*, Oxonium *n*, оксоний *м*
Oxygen	oxygène *m*, Sauerstoff *m*, кислород *м*
Oxygenate	exygénate *m*, Oxygenat *n*, оксигенат *м*
Ozone	ozone *m*, Ozon *n*, озон *м*
Ozonide	ozonure *m*, Ozonid *n*, озонид *м*
Palladate	palladate *m*, Palladat *n*, палладат *м*
Palladium	palladium *m*, Palladium *n*, палладий *м*
Penta-	penta-, penta-, пента-
Pentakis-	pentakis-, pentakis-, пентакис-
Perbromate	perbromate *m*, Perbromat *n*, пербромат *м*
Perchlorate	perchlorate *m*, Perchlorat *n*, перхлорат *м*
Periodate	periodate *m*, Perjodat *n*, периодат *м*
Permalloy	permalloy *m*, Permalloy *n*, пермаллой *м*
Permanganate	permanganate *m*, Permanganat *n*, перманганат *м*
Perovskite	perowskite *f*, Perowskit *m*, перовскит *м*
Peroxide	peroxyde *m*, Peroxid *n*, пероксид *м*
Peroxo-	peroxo-, peroxo-, пероксо-
Perrhenate	perrhénate *m*, Perrhenat *n*, перренат *м*
Pertechnetate	pertechnétate *m*, Pertechnetat *n*, пертехнетат *м*
Phenakite	phénacite *f*, Phenakit *m*, фенакит *м*
Phosgene	phosgène *m*, Phosgen *n*, фосген *м*
Phosphane	phosphane *m*, Phosphan *n*, фосфан *м*
Phosphate	phosphate *m*, Phosphat *n*, фосфат *м*
Phosphide	phosphure *m*, Phosphid *n*, фосфид *м*
Phosphine	phosphine *f*, Phosphin *n*, фосфин *м*
Phosphonium	phosphonium *m*, Phosphonium *n*, фосфоний *м*
Phosphorus	phosphore *m*, Phosphor *m*, фосфор *м*
Platinate	platinate *m*, Platinat *n*, платинат *м*
Platinum	platine *m*, Platin *n*, платина *ж*
Plumbate	plumbate[4] *m*, Plumbat *n*, плюмбат *м*
Plutonium	plutonium *m*, Plutonium *n*, плутоний *м*
Polonium	polonium *m*, Polonium *n*, полоний *м*
Poly-	poly-, poly-, поли-
Potash	potasse *f*, Pottasche *f*, поташ *м*
Potassium	potassium *m*, Kalium *n*, калий *м*
Praseodymium	praséodyme *m*, Praseodym *n*, празеодим *м*

(Continued)

English name	French, German, and Russian names
Promethium	prométhium *m*, Promethium *n*, прометий *м*
Protactinium	protactinium *m*, Protaktinium *n*, протактиний *м*
Protium	protium *m*, Protium *n*, протий *м*
Pyrite	pyrite *f*, Pyrit *m*, пирит *м*
Pyrolusite	pyrolusite *f*, Pyrolusit *m*, пиролюзит *м*
Pyrrhotine	pyrrhotine *f*, Pyrrhotin *m*, пирротин *м*
Quartz	quartz *m*, Quarz *m*, кварц *м*
Radium	radium *m*, Radium *n*, радий *м*
Radon	radon *m*, Radon *n*, радон *м*
Realgar	réalgar *m*, Realgar *m*, реальгар *м*
Rhenate	rhénate *m*, Rhenat *n*, ренат *м*
Rhenium	rhénium *m*, Rhenium *n*, рений *м*
Rhodium	rhodium *m*, Rhodium *n*, родий *м*
Rubidium	rubidium *m*, Rubidium *n*, рубидий *м*
Ruby	rubis *m*, Rubin *m*, рубин *м*
Ruthenate	ruthénate *m*, Ruthenat *n*, рутенат *м*
Ruthenium	ruthénium *m*, Ruthenium *n*, рутений *м*
Rutile	rutile *m*, Rutil *m*, рутил *м*
Salammoniac	salmiac *m*, Salmiak *m*, нашатырь *м*
Salpeter	salpêtre *m*, Salpeter *m*, селитра *ж*
Salt	sel *m*, Salz *n*, соль *ж*
Berthollet's ~	~ de Berthollet, Bertholletsches ~, бертоллетова ~
common ~	~ (de cuisine), Kochsalz *n*, поваренная ~
Glauber's ~	~ de Glauber, Glaubersalz *n*, глауберова ~
Mohr's ~	~ de Mohr, Mohrsches ~, ~ Мора
Samarium	samarium *m*, Samarium *n*, самарий *м*
Sand	sable *m*, Sand *m*, песок *м*
quartz ~	~ quartzeux, Quarzsand *m*, кварцевый ~
Sapphire	saphire *m*, Sa(*p*)phir *m*, сапфир *м*
Scandate	scandate *m*, Skandat *n*, скандат *м*
Scandium	scandium *m*, Skandium *n*, скандий *м*
Scheelite	scheelite *f*, Scheelit *m*, шеелит *м*
Schoenite	schénite *f*, Schönit *m*, шёнит *м*
Selane	sélane *m*, Selan *n*, селан *м*
Selenate	séléniate *m*, Selenat *n*, селенат *м*
Selenide	séléniure *m*, Selenid *n*, селенид *м*
Selenite	sélénite *m*, Selenit *n*, селенит *м*
Selenium	sélénium *m*, Selen *n*, селен *м*
Silane	silane *m*, Silan *n*, силан *м*
Silica	silice *f*, Kieselerde *f*, кремнезём *м*
Silicate	silicate *m*, Silikat *n*, силикат *м*
Silicide	siliciure *m*, Silizid *n*, силицид *м*
Silicon	silicium *m*, Silizium *n*, кремний *м*
Silver	argent *m*, Silber *n*, серебро *c*
Soda	soude *f*, Soda *f*, сода *ж*
~ ash	~ Solvay, kalzinierte ~, кальцинированная ~
caustic ~	~ caustique, kaustische ~, каустическая ~
drinking ~	sel *m* de Vichy, Speisesoda *f*, питьевая ~

(Continued)

English name	French, German, and Russian names
Sodium	sodium *m*, Natrium *n*, натрий *м*
Spar	spath *m*, Spat *m*, шпат *м*
Iceland ~	~ *d'*Islande, Islandspat *m*, исландский ~
fluor ~	~ fluor, Flußspat *m*, плавиковый ~
Sphalerite	sphalérite *f*, Sphalerit *m*, сфалерит *м*
Spinel	spinelle *f*, Spinell *m*, шпинель *ж*
Stannate	stannate *m*, Stannat *n*, станнат *м*
Steel	acier *m*, Stahl *m*, сталь *ж*
Stibate[5]	stibiate[5] *m*, Stibat[5] *n*, стибат *м*
Stibide[6]	stibiure[6] *m*, Stibid[6] *n*, стибид *м*
Strontium	strontium *m*, Strontium *n*, стронций *м*
Sublimate	sublimé *m*, Sublimat *n*, сулема *ж*
Sulfane	sulfane *m*, Sulfan *n*, сульфан *м*
Sulfate	sulfate *m*, Sulfat *n*, сульфат *м*
Sulfide	sulfure *m*, Sulfid *n*, сульфид *м*
Sulfite	sulfite *m*, Sulfit *n*, сульфит *м*
Sulfonate	sulfonate *m*, Sulfonat *n*, сульфонат *м*
Sulfonium	sulfonium *m*, Sulfonium *n*, сульфоний *м*
Sulfur	soufre *m*, Schwefel *m*, сера *ж*
Superphosphate	superphosphate *m*, Superphosphat *n*, суперфосфат *м*
Sylvite	sylvine *f*, Sylvin *m*, сильвин *м*
Tantalate	tantalate *m*, Tantalat *n*, танталат *м*
Tantalum	tantale *m*, Tantal *n*, тантал *м*
Tartrate	tartrate *m*, Tartrat *n*, тартрат *м*
Technetate	technétate *m*, Technetat *n*, технетат *м*
Technetium	technétium *m*, Technetium *n*, технеций *м*
Tellane	tellane *m*, Tellan *n*, теллан *м*
Tellurate	tellurate *m*, Tellurat *n*, теллурат *м*
Telluride	tellurure *m*, Tellurid *n*, теллурид *м*
Tellurite	tellurite *m*, Tellurit *n*, теллурит *м*
Tellurium	tellure *m*, Tellur *n*, теллур *м*
Terbium	terbium *m*, Terbium *n*, тербий *м*
Tetra-	tétra-, tetra-, тетра-
Tetrakis-	tétrakis-, tetrakis-, тетракис-
Thallium	thallium *m*, Thallium *n*, таллий *м*
Thioarsenate	thio-arséniate *m*, Thioarsenat *n*, тиоарсенат *м*
Thiocarbamide	thiocarbamide *m*, Thiocarbamid *n*, тиокарбамид *м*
Thiocyanate	thiocyanate *m*, Thiozyanat *n*, тиоцианат *м*
Thionate	thionate *m*, Thionat *n*, тионат *м*
Thiosulfate	thiosulfate *m*, Thiosulfat *n*, тиосульфат *м*
Thiourea	thio-urée *f*, Thioharnstoff *m*, тиомочевина *ж*
Thorium	thorium *m*, Thorium *n*, торий *м*
Thulium	thulium *m*, Thulium *n*, тулий *м*
Tin	étain *m*, Zinn *n*, олово *c*
Tincal	tincal *m*, Tinkal *m*, тинкал *м*
Titanate	titanate *m*, Titanat *n*, титанат *м*
Titanium	titane *m*, Titan *n*, титан *м*
Tombac	tombac *m*, Tombak *m*, томпак *м*

(Continued)

English name	French, German, and Russian names
Topaz	topaze *f*, Topas *m*, топаз *м*
Trans-	trans-, trans-, транс-
Tri-	tri-, tri-, три-
Tridymite	tridymite *f*, Tridymit *m*, тридимит *м*
Tris-	tris-, tris-, трис-
Tritium	tritium *m*, Tritium *n*, тритий *м*
Trona	trona *m*, Trona *m*, трона *ж*
Tungsten	tungstène *m*, Wolfram *n*, вольфрам *м*
Undeca-	undéca-, undeka-, ундека-
Uranate	uranate *m*, Uranat *n*, уранат *м*
Uraninite	uraninite *f*, Uraninit *m*, ураннинит *м*
Uranium	uranium *m*, Uran *n*, уран *м*
Uranyl	uranyle *m*, Uranyl *n*, уранил *м*
Urea	urée *f*, Harnstoff *m*, мочевина *ж*
Vanadate	vanadate *m*, Vanadat *n*, ванадат *м*
Vanadium	vanadium *m*, Vanadin *n*, ванадий *м*
Vanadyl	vanadyle *m*, Vanadyl *n*, ванадил *м*
Vitriol	vitriol *m*, Vitriol *m*, купорос *м*
Water	eau *f*, Wasser *n*, вода *ж*
chlorine ~	~ de chlore, Chlorwasser *n*, хлорная ~
lime ~	~ de chaux, Kalkwasser *n*, известковая ~
White	blanc *m*, Weiss *n*, белила *мн*
Wolframate[7]	wolframate[7] *m*, Wolframat *n*, вольфрамат *м*
Wulfenite	wulfénite *f*, Wulfenit *m*, вульфенит *м*
Wurtzite	wurtzite *f*, Wurtzit *m*, вюрцит *м*
Xenon	xénon *m*, Xenon *n*, ксенон *м*
Xenotime	xénotime *m*, Xenotim *m*, ксенотим *м*
Ytterbium	ytterbium *m*, Ytterbium *n*, иттербий *м*
Yttrium	yttrium *m*, Yttrium *n*, иттрий *м*
Zinc	zinc *m*, Zink *n*, цинк *м*
Zircon	zircon *m*, Zircon *m*, циркон *м*
Zirconium	zirconium *m*, Zirkonium *n*, цирконий *ж*

Old fashioned names: [1] Jodat, [2] Jodid, [3] nickelate, nickélate, Nickelat, [4] plombate, [5] antimonate, antimoniate, Antimonat, [6] antimonide, antiromiure, Antimonid; [7] tungstate.

References

1. *Pure and Appl. Chem.*, 1984, **56**, No. 5, 654.
2. Termicheskiye konstanty veshchestv. Spravochnik (*Thermal Constants of Substances. Reference book*), Glushko, V. P. , ed., Izd. VINITI, Moscow, 1965–1982.
3. Brauer, G., Rukovodstvo po neorganicheskomu sintezu (*Manual of Inorganic Synthesis*), Mir Publishing House, Moscow, 1985–1986, Vols. 1–6 (translation from German).
4. Spravochnik khimika (Chemist's Handbook), Nikol'skiy, B. P., ed., 3rd ed., Khimiya Publishing House, Leningrad, 1971, Vols. I, II.
5. *Handbook of Chemistry and Physics*, Cleveland, 1982.
6. *Gmelins Handbuch der anorganischen Chemie*, Berlin, 1924.
7. Nekrasov, B. V., Osnovy neorganisheskoy khimii (*Fundamentals of Inorganic Chemistry*), Khimiya Publishing House, Moscow, 1973, Vols. 1, 2.
8. Remy, H., Kurs neorganicheskoy khimii (*Inorganic Chemistry Treatise*), Mir Publishing House, Moscow, 1972–1974, Vols. I, II (translation from German).
9. Turova, N. Ya., Spravochniye tablitsy po neorganicheskoy khimii (*Reference Tables on Inorganic Chemistry*), Khimiya Publishing House, Leningrad, 1977.
10. Efimov, L. P., et al., Svoystva neorganicheskikh soedineniy. Spravochnik (*Properties of Inorganic Compounds. Reference Book*), Khimiya Publishing House, Leningrad, 1983.
11. Gurvich, L. V., Karachentsev, G. V., Kondratyev, V. I., et al., Energii razryva khimicheskikh svyazey. Potentsialy ionizatsii i srodstvo k elektronu (*Bond Dissociation Energies, Ionization Potentials and Electron Affinity*), Nauka Publishing House, Moscow, 1974.
12. Osipov, O. A., et al., Spravochnik po dipol'nym momentam (*Reference Book on Dipole Moments*), Vysshaya Shkola Publishing House, Moscow, 1971.
13. Krasnov, K. S., Filipenko, N. V., Bobkova, V. A., et al., Molekulyarniye postoyanniye neorganicheskikh soyedineniy (*Molecular Constants of Inorganic Compounds*), Khimiya Publishing House, Leningrad, 1979.
14. Termodinamicheskiye svoystva individual'nykh veshchestv. Spravochnoye izdaniye (*Thermodynamic Properties of Individual Substances. Reference book*), Nauka Publishing House, Moscow, 1978–1983, Vols. 1–4.
15. Naumov, G. V., et al., Spravochnik termodinamicheskikh velichin (dlya geologov), (*Reference Book on Thermodynamic Quantities (For Use by Geologists)*), Atomizdat Publishing House, Moscow, 1971.
16. Reed, R., et al., Svoystva gazov i zhidkostey (*Properties of Gases and Liquids*), Khimiya Publishing House, Leningrad, 1982 (translation from English).
17. Karapetyants, M. Kh., Vvedeniye v teoriyu khimicheskikh protsessov (*Introduction to the Theory of Chemical Processes*), Vysshaya Shkola Publishing House, Moscow, 1981.
18. Dobosz, D., Elektrokhimicheskiye konstanty (*Electrochemical Constants*), Mir Publishing House, Moscow, 1982 (translation from Hungarian).
19. Albert, A., and Sergent, E., Konstanty ionizatsii kislot i osnovaniy (*Ionization Constants of Acids and Bases*), Khimiya Publishing House, Leningrad-Moscow, 1967 (translation from English).
20. Nazarenko, V. A., et al., Gidroliz ionov metallov v razbavlennykh rastvorakh (*Hydrolysis of Metal Ions in Dilute Solutions*), Atomizdat Publishing House, Moscow, 1979.
21. Yatsimirskiy, K. B., and Vasil'yev, V. P., Konstanty nestoykosti kompleksnykh soedineniy (*Instability Constants of Complex Compounds*), AN SSSR Publishing House, Moscow, 1959.
22. Lur'ye, Yu. Yu., Spavochnik po analiticheskoy khimii (*Reference Book on Analytical Chemistry*), Khimiya Publishing House, Moscow, 1989.
23. Spravochnik khimika (*Chemist's Handbook*), Nikol'skiy, B. P., ed., 2nd ed., Khimiya Publishing House, Leningrad, 1964, Vol. III.
24. Mineraly. Spravochnik (*Minerals. Reference book*), Chukhrov, F. V., ed., Nauka Publishing House, Leningrad, 1960–1981, Vols. I-III.

25. Mineralogicheskiye tablitsy. Spravochnik (*Mineralogical Tables. Reference book*), Semenov, E. I., ed., Nedra Publishing House, 1981.

26. Fundamental'niye fizicheskiye konstanty (*Fundamental Physical Constants*), Izd. Standartov, Moscow, 1979.

27. Mineralogocheskaya entsiklopediya (*Mineralogical Encyclopedia*), Frye, K., ed., Sovetskaya Entsiklopediya Publishing House, Moscow, 1983 (translation from German).

28. Kimicheskiy entsiklopedicheskiy slovar' (*Chemical Encyclopedical Dictionary*), Knunyants, I. L., ed., Sovetskaya Entsiklopaedia Publishing House, Moscow, 1983.

29. Kolditz, L., Anorganikum (*Anorganicum*), Mir Publishing House, Moscow, 1985, Vols. 1, 2 (translation from German).

30. Johnson, D., Termodinamicheskiye aspekty neorganicheskoy khimii (*Thermodynamic Aspects of Inorganic Chemistry*), Mir Publishing House, Moscow, 1985 (translation from English).

31. Lidin, R. A., Molochko, V. A., Andreyeva, L. L., and Tsvetkov, A. A., Osnovy nomenclatury neorganicheskikh soedineniy (*Fundamentals of Nomenclature of Inorganic Compounds*), Khimiya Publishing House, Moscow, 1983.

32. Kahn, R., and Dermer, O., Vvedeniye v khimicheskuyu terminologiyu (*Introduction to Chemical Terminology*), Khimiya Publishing House, Moscow, 1983 (translation from English).

33. Huheey, J. E., Neorganicheskaya khimiya. Stroyeniye veshchestva i reaktsionnaya sposobnost' (*Inorganic Chemistry. Principles of Structure and Reactivity*), Khimiya Publishing House, Moscow, 1977 (translation from English).

34. Lidin, R. A., Molochko, V. A., and Andreyeva, L. L., Zadachi po neorganicheskoy khimii (*Book of Problems in Inorganic Chemistry*), Vysshaya Shkola Publishing House, Moscow, 1990.

35. Khimiya. Spravochnoye izdaniye (*Chemistry. Reference book*), Schröter, W. , ed., Khimiya Publishing House, Moscow, 1989 (translation from German).

36. Khimicheskaya entsiklopediya (*Chemical Encyclopedia*), Knunyants, I. L., ed., Sovetskaya Entsiklopediya Publishing House, Moscow, 1988–1995, Vols. I–IV.

37. Nomenklaturniye pravila IUPAC po khimii (*IUPAC Rules for Nomenclature of Chemistry*), Izd. VINITI, Moscow, 1979, Vol. 1, Book 1; 1983, Vol. 3, Book 1; 1988, Vol. 6 (translation from English).